MATHEMATICAL PAPERS.

London: C. J. CLAY AND SONS,
CAMBRIDGE UNIVERSITY PRESS WAREHOUSE,
AVE MARIA LANE.

Cambridge: DEIGHTON, BELL AND CO.
Leipzig: F. A. BROCKHAUS.
New York: MACMILLAN AND CO.

THE COLLECTED

MATHEMATICAL PAPERS

OF

ARTHUR CAYLEY, Sc.D., F.R.S.,

SADLERIAN PROFESSOR OF PURE MATHEMATICS IN THE UNIVERSITY OF CAMBRIDGE.

VOL. VI.

CAMBRIDGE:
AT THE UNIVERSITY PRESS.
1893

CAMBRIDGE:
PRINTED BY C. J. CLAY, M.A. AND SONS,
AT THE UNIVERSITY PRESS.

ADVERTISEMENT.

THE present volume contains 33 papers numbered 384 to 416 published for the most part in the years 1865 to 1872; the last paper 416, of the year 1882, is inserted in the present volume on account of its immediate connexion with the papers 411 and 415 on Reciprocal Surfaces.

The Table for the six volumes is

Vol.	I.	Numbers	1	to	100.
,,	II.	,,	101	,,	158.
,,	III.	,,	159	,,	222.
,,	IV.	,,	223	,,	299.
,,	V.	,,	300	,,	383.
,,	VI.	,,	384	,,	416.

CONTENTS.

CLASSIFICATION.

384.

ON THE TRANSFORMATION OF PLANE CURVES.

[From the *Proceedings of the London Mathematical Society*, vol. I. (1865—1866), No. III. pp. 1—11. Read Oct. 16, 1865.]

1. THE expression a "double point," or, as I shall for shortness call it, a "dp," is to be throughout understood to include a cusp: thus, if a curve has δ nodes (or double points in the restricted sense of the expression) and κ cusps, it is here regarded as having $\delta+\kappa$ dps.

2. It was remarked by Cramer, in his "Théorie des Lignes Courbes" (1750), that a curve of the order n has at most $\frac{1}{2}(n-1)(n-2)$, $=\frac{1}{2}(n^2-3n)+1$, dps.

3. For several years past it has further been known that a curve such that the coordinates $(x : y : z)$ of any point thereof are as rational and integral functions of the order n of a variable parameter θ, is a curve of the order n having this maximum number $\frac{1}{2}(n-1)(n-2)$ of dps.

4. The converse theorem is also true, viz.: in a curve of the order n, with $\frac{1}{2}(n-1)(n-2)$ dps, the coordinates $(x : y : z)$ of any point are as rational and integral functions of the order n of a variable parameter θ—or, somewhat less precisely, the coordinates are expressible rationally in terms of a parameter θ.

5. The foregoing theorem, as a particular case of Riemann's general theorem, to be presently referred to, dates from the year 1857; but it was first explicitly stated only last year (1864) by Clebsch, in the Paper, "Ueber diejenigen ebenen Curven deren Coordinaten rationale Functionen eines Parameters sind," *Crelle*, t. LXIV. (1864), pp. 43—63.

6. The demonstration is, in fact, very simple; it depends merely on the remark that we may, through the $\frac{1}{2}(n-1)(n-2)$ dps, and through $2n-3$ other points on the given curve of the order n, together $\frac{1}{2}(n^2+n)-2$, $=\frac{1}{2}(n-1)(n+2)-1$, points, draw

a series of curves of the order $n-1$, given by an equation $U+\theta V=0$, containing an arbitrary parameter θ; any such curve intersects the given curve in the dps, each counting as two points, in the $2n-3$ points, and in *one* other point; hence, as there is only one variable point of intersection, the coordinates of this point, viz., the coordinates of an arbitrary point on the given curve, are expressible rationally in terms of the parameter θ. The demonstration may also be effected in a similar manner by means of curves of the order $n-2$.

7. Before going further, it will be convenient to introduce the term "Deficiency," viz., a curve of the order n with $\frac{1}{2}(n-1)(n-2)-D$ dps, is said to have a deficiency $=D$: the foregoing theorem is that for curves with a deficiency $=0$, the coordinates are expressible rationally in terms of a parameter θ. Since in such a curve the different points succeed each other in a certain definite order, viz., in the order obtained by giving to the parameter its different real values from $-\infty$ to ∞, the curve may be termed a *unicursal* curve.

8. Riemann's general theorem, as applied to plane curves, is stated, but not in its complete form, by Schwarz, in the Paper, "De superficiebus in planum explicabilibus primorum septem ordinum," *Crelle*, t. LXIV. (1864), pp. 1—17: to complete the enunciation it is necessary to refer to page 137 of Riemann's own Paper, "Theorie der Abelschen Functionen," *Crelle*, t. LIV. (1857), pp. 115—155, viz., the enunciation will be:

9. For a curve of any order with a given deficiency D, the coordinates may be expressed as follows:

$D=0$, rationally in terms of a parameter θ, or what comes to the same thing, rationally in terms of the parameters (ξ, η), connected by an equation of the form $(1, \xi)(1, \eta)=0$.

$D>0$, rationally in terms of the parameters (ξ, η) connected by an equation of a certain form, viz.:

$D=1$, the equation is $(1, \xi)^2(1, \eta)^2=0$, or (what comes to the same thing) η is the square root of a quartic function of ξ.

$D=2$, the equation is $(1, \xi)^3(1, \eta)^2=0$, or (what comes to the same thing) η is the square root of a sextic function of ξ.

$D>2$, viz.:

D odd, $=2\mu-3$, the equation is $(1, \xi)^\mu(1, \eta)^\mu=0$, and is besides such, that treating (ξ, η) as Cartesian coordinates, the curve thereby represented has $(\mu-2)^2$ dps.

D even, $=2\mu-2$, the equation is $(1, \xi)^\mu(1, \eta)^\mu=0$, and is besides such, that treating (ξ, η) as Cartesian coordinates, the curve thereby represented has $(\mu-1)(\mu-3)$ dps.

10. To see more clearly the meaning of this, write $\frac{\xi}{\eta}, \frac{\eta}{\zeta}$, in place of (ξ, η), so that the coordinates $(x : y : z)$ are expressible rationally and homogeneously in terms

of (ξ, η, ζ), connected by an equation of the form $(\zeta, \xi)^\mu (\zeta, \eta)^\mu = 0$. Such an equation, treating therein (ξ, η, ζ) as coordinates, belongs to a curve of the order 2μ, with a μ-tuple point at $(\xi=0, \zeta=0)$, a μ-tuple point at $(\eta=0, \zeta=0)$, and which has besides $(\mu-2)^2$ or $(\mu-1)(\mu-3)$ dps, according as $D=2\mu-3$, or $2\mu-2$. The coordinates $(x:y:z)$ of a point of the given curve are expressible rationally in terms of the coordinates $(\xi:\eta:\zeta)$ of a point on the new curve; and we may say that the original curve is by means of the equations which give $(x:y:z)$ in terms of $(\xi:\eta:\zeta)$ *transformed* into the new curve.

11. A curve of the order 2μ may have $\tfrac{1}{2}(2\mu-1)(2\mu-2)$, $=2\mu^2-3\mu+1$ dps; hence in the new curve, observing that the μ-tuple points each count for $\tfrac{1}{2}(\mu^2-\mu)$ dps, we have

In the case $D = 2\mu-3$,

$$\begin{aligned} \text{Deficiency} = \ & 2\mu^2-3\mu+1 \\ & -\mu^2+\mu \\ & -\mu^2+4\mu-4 \\ \hline = \ & 2\mu-3, = D \end{aligned}$$

In the case $D = 2\mu-2$,

$$\begin{aligned} \text{Deficiency} = \ & 2\mu^2-3\mu+1 \\ & -\mu^2+\mu \\ & -\mu^2+4\mu-3 \\ \hline = \ & 2\mu-2, = D \end{aligned}$$

Moreover for $D=0$, the transformed curve is a conic, with 0 dps, and therefore with deficiency $=0$; in the case $D=1$, it is a quartic with 2 dps, and therefore deficiency $=2$; in the case $D=2$ it is a quintic with a triple point $=3$, and a double point $=1$, together 4 dps, and therefore deficiency $=2$. Hence in every case the new curve has the same deficiency as the original curve.

12. The theorem thus is that the given curve of the order n, with deficiency D, may be rationally transformed into a curve of an order depending only on the deficiency, and having the same deficiency with the given curve, viz.: $D=0$, the new curve is of the order $2\,(=D+2)$; $D=1$, it is of the order $4\,(=D+3)$; $D=2$, it is of the order $5\,(=D+3)$; and $D>2$, it is for D odd, of the order $D+3$; and for D even, of the order $D+2$. It will presently appear that these are not the lowest values which it is possible to give to the order of the new curve. Riemann's object was, not that the order of the transformed curve might be as low as possible, but that the equation in (ξ, η) might be in each of these parameters separately of the lowest possible order; and this he effected by giving to the transformed curve the two μ-tuple points.

13. It is to be noticed that the theorem that for any rational transformation of one curve into another the two curves have the same deficiency is in effect given (as a consequence of Riemann's theory) by Clebsch in the Paper, "Ueber die Singularitäten algebraischer Curven," *Crelle*, t. LXIV., pp. 98—100. I have, by the assistance of a formula communicated to me by Dr Salmon, obtained a direct analytical demonstration of this theorem.

14. I remark that (x, y, z) being connected by an equation, if $(x:y:z)$ are given rationally in terms of $(\xi:\eta:\zeta)$, then it follows that $(\xi:\eta:\zeta)$ are also

expressible rationally in terms of $(x : y : z)$: it is convenient to consider the transformation from this point of view, and I now proceed to the independent development of the theory, as follows:

15. We have a given curve $U = (x, y, z)^n = 0$, with deficiency D, which is by the transformation $\xi : \eta : \zeta = P : Q : R$ (where P, Q, R are given functions $(x, y, z)^k$ each of the same order k) transformed into the curve $\Upsilon = (\xi, \eta, \zeta)^\nu = 0$. The transformed curve has, as we know, the same deficiency D as the original curve.

16. To find the order of the transformed curve, we must find the number of its intersections with an arbitrary line $a\xi + b\eta + c\zeta = 0$. Writing in this equation $\xi : \eta : \zeta = P : Q : R$, we obtain the equation $aP + bQ + cR = 0$, and combining therewith the equation $U = 0$, the two equations, being of the orders k and n respectively, give kn systems of values of $(x : y : z)$, and to each of these, in virtue of the equations $\xi : \eta : \zeta = P : Q : R$, there corresponds a single set of values of $(\xi : \eta : \zeta)$, and therefore a single point of intersection; hence the number of intersections, that is, the order of the transformed curve, is $= kn$.

17. If, however, the curves $P = 0$, $Q = 0$, $R = 0$, meet in an ordinary point of the curve $U = 0$, then it is easy to see there is a reduction $= 1$ in the foregoing value; and so if they meet in a dp of the curve $U = 0$, then there is a reduction $= 2$. More generally if the curves $P = 0$, $Q = 0$, $R = 0$ each pass through the same α dps and β ordinary points of the curve $U = 0$, then there is a reduction $= 2\alpha + \beta$. In fact the curve $aP + bQ + cR = 0$, meets the curve $U = 0$, in kn points; but among these are included the α dps, each counting as 2 intersections, and the β points; the number of the remaining intersections is $= kn - 2\alpha - \beta$, and the order of the transformed curve is equal to this number.

I assume that we have $k < n$:

18. A curve of the order k may be made to pass through $\frac{1}{2}k(k+3)$ points; it is moreover known that if any three curves, $P = 0$, $Q = 0$, $R = 0$, of the order k each pass through the same $\frac{1}{2}k(k+3) - 1$ points, then the three curves have all their intersections common, the equations being, in fact, connected by an identical relation of the form $\alpha P + \beta Q + \gamma R = 0$. To make the order of the transformed curve as low as possible, we must make the curves $P = 0$, $Q = 0$, $R = 0$, meet on the curve $U = 0$ in as many points as possible, and it appears from the remark just made, that the greatest possible number is $= \frac{1}{2}k(k+3) - 2$; in particular, for $k = n-1$, $n-2$, $n-3$, the number of points on the curve $U = 0$ will be at most equal to $\frac{1}{2}(n^2+n) - 3$, $\frac{1}{2}(n^2-n) - 3$, $\frac{1}{2}(n^2-3n) - 2$, respectively.

19. Hence, considering the curve $U = 0$ with deficiency D, or with $\frac{1}{2}(n^2-3n) - D + 1$ dps, *first* if $k = n-1$, we may assume that the transforming curves $P = 0$, $Q = 0$, $R = 0$ of the order $n-1$, each pass

through the	$\frac{1}{2}(n^2-3n) - D + 1$	dps,
and through	$2n + D - 4$	other points,
together	$\frac{1}{2}(n^2+n) \qquad - 3$	points of the curve $U = 0$.

This being so, each of the three curves will meet the curve $U=0$

in the dps, counting as	$n^2-3n-2D+2$	points,
in the	$2n+\ D-4$	points,
and in	$D+2$	other points,
together	n^2-n	points;

whence the order of the transformed curve is $=D+2$.

20. In precisely the same manner, *secondly*, if $k=n-2$, then we may assume that the transforming curves $P=0$, $Q=0$, $R=0$, of the order $n-2$, each pass

through the	$\tfrac{1}{2}(n^2-3n)-D+1$	dps,
and through	$n+D-4$	other points,
together	$\tfrac{1}{2}(n^2-n)\quad -3$	points of the curve $U=0$;

and this being so, each of the three curves will meet the curve $U=0$

in the dps counting as	$n^2-3n-2D+2$	points,
in the	$n+\ D-4$	points,
and in	$D+2$	other points,
together	n^2-2n	points;

whence the order of the transformed curve is also in this case $=D+2$.

21. I was under the impression that the order of the transformed curve could not be reduced below $D+2$, but it was remarked to me by Dr Clebsch, that in the case $D>2$, the order might be reduced to $D+1$. In fact, considering, *thirdly*, the case $k=n-3$, we see that the transforming curves $P=0$, $Q=0$, $R=0$ of the order $n-3$ may be made to pass

through the	$\tfrac{1}{2}(n^2-3n)-D+1$	dps,
and through	$D-3$	other points,
together	$\tfrac{1}{2}(n^2-3n)\quad -2$	points of the curve $U=0$;

and this being so, each of the three curves meets the curve $U=0$,

in the dps counting as	$(n^2-3n)-2D+2$	points,
in the	$D-3$	points,
and in	$D+1$	other points,
together	n^2-3n	points;

whence the order of the transformed curve is in this case $=D+1$.

22. The general theorem thus is that a curve of the order n with deficiency D, can be, by a transformation of the order $n-1$ or $n-2$, transformed into a curve of

the order $D+2$; and if $D>2$, then the given curve can be by a transformation of the order $n-3$ transformed to a curve of the order $D+1$: the transformed curve having in each case the same deficiency D as the original curve.

23. In particular, if $D=1$, a curve of the order n with deficiency 1, or with $\frac{1}{2}(n^2-3n)$ dps, can be transformed into a cubic curve with the same deficiency, that is with 0 dps; or the given curve can be transformed into a cubic. This case is discussed by Clebsch in the Memoir "Ueber diejenigen Curven deren Coordinaten elliptische Functionen eines Parameters sind," *Crelle*, t. LXIV., pp. 210—271. And he has there given in relation to it a theorem which I establish as follows:

24. Using the transformation of the order $n-1$, if besides the $2n+D-4\ (=2n-3)$ points on the given curve $U=0$, we consider another point O on the curve, then we may, through the $\frac{1}{2}(n^2-3n)$ dps, the $2n-3$ points and the point O, draw a series of curves of the order $n-1$, viz., if P_0, Q_0, R_0, are what the functions P, Q, R, become on substituting therein for (x, y, z), the coordinates (x_0, y_0, z_0) of the given point O, then the equation of any such curve will be $aP+bQ+cR=0$, with the relation $aP_0+bQ_0+cR_0$ between the parameters a, b, c; or (what is the same thing) eliminating c, the equation will be $a(PR_0-P_0R)+b(QR_0-Q_0R)=0$, which contains the single arbitrary parameter $a:b$. In the cubic which is the transformation of the given curve we have a point O' corresponding to O and if (ξ_0, η_0, ζ_0) be the coordinates of this point, then corresponding to the series of curves of the order $n-1$, we have a series of lines through the point O' of the cubic, viz., the lines $a\xi+b\eta+c\zeta=0$ with the relation $a\xi_0+b\eta_0+c\zeta_0=0$ between the parameters; or, what is the same thing, we have the series of lines $a(\xi\zeta_0-\zeta\xi_0)+b(\eta\zeta_0-\zeta\eta_0)=0$, containing the same single parameter $a:b$. By determining this parameter, the curves of the order $n-1$, will be the curves of this order through the dps, the $2n-3$ points, and the point O, which touch the given curve $U=0$; and the lines will be the tangents to the cubic from the point O'; as the number of tangents to a cubic from a point on the cubic is $=4$, it is clear that the values of the parameter $a:b$ will be determined by a certain quartic equation; and there will of course be 4 tangent curves of the order $n-1$ corresponding to the 4 tangents to the cubic. Now by Dr Salmon's anharmonic property of the tangents of a cubic, if on the cubic we vary the position of the point O', the absolute invariant $I^3\div J^2$ of the quartic in $(a:b)$ remains unaltered; that is the absolute invariant $I^3\div J^2$ of the quartic which determines the 4 tangent curves of the order $n-1$ is independent of the position of the point O on the given curve $U=0$, and since the tangent curves in question have the same relation to each of the $2n-3$ points and to the point O, it follows that the invariant is also independent of the position of each of the $2n-3$ points; that is, we have the following theorem, viz.:

25. Considering a curve of the order n with deficiency $=1$; we may, through the $\frac{1}{2}(n^2-3n)$ dps, and through any $2n-2$ points on the curve, draw so as to touch the curve, four curves of the order $n-1$; viz., these are given by an equation $aP'+bQ'=0$, where the ratio $a:b$ is determined by a certain quartic equation $(*\between a, b)^4=0$; then *theorem*, the absolute invariant $I^3\div J^2$ of the quartic function, is independent of the

positions of the $2n-2$ points on the curve $U=0$, and it is consequently a function of only the coefficients of the curve $U=0$, being, as is obvious, an absolute invariant of the curve $U=0$.

26. And, moreover, if the curve $U=0$ is by a transformation of the order $n-1$, by means of $2n-3$ points on the curve as above, transformed into a cubic, then the absolute invariant $I^3 \div J^2$ of the quartic equation which determines the tangents to the cubic from any point O' on the cubic (or, what is the same thing, the absolute invariant $S^3 \div T^2$ of the cubic, taken with a proper numerical multiplier) is independent of the positions of the $2n-3$ points on the curve $U=0$, being in fact equal to the above-mentioned absolute invariant of the curve $U=0$. The like results apply to the transformation of the order $n-2$.

27. Suppose now that we have $D>2$, and consider a curve of the order n with the deficiency D, that is with $\frac{1}{2}(n^2-3n)-D+1$ dps, transformed by a transformation of the order $n-3$ into a curve of the order $D+1$ with deficiency D; then, assuming the truth of the subsidiary theorem to be presently mentioned, it may be shown by very similar reasoning to that above employed, that the absolute invariants of the transformed curve of the order $D+1$ (the number of which is $=4D-6$), will be independent of the positions of the $D-3$ points used in the transformation, and will be equal to absolute invariants[1] of the given curve $U=0$.

28. The subsidiary theorem is as follows: consider a curve of the order $D+1$, with deficiency D, that is, with $\frac{1}{2}D(D-1)-D=\frac{1}{2}(D^2-3D)$ dps; the number of tangents to the curve from any point O' on the curve is $=(D+1)D-(D^2-3D)-2$, $=4D-2$, (this assumes however, that the dps are proper dps, not *cusps*,) the pencil of tangents has $4D-5$ absolute invariants, and of these all but one, that is, $4D-6$, absolute invariants of the pencil are independent of the position of the point O' on the curve, and are respectively equal to absolute invariants of the curve.

29. To establish it, I observe that a curve of the order $D+1$ with deficiency D, or with $\frac{1}{2}(D^2-3D)$ dps, contains $\frac{1}{2}(D+1)(D+4)-\frac{1}{2}(D^2-3D)$, $=4D+2$ arbitrary constants, and it may therefore be made to satisfy $4D+2$ conditions. Now imagine a given pencil of $4D-2$ lines, and let a curve of the form in question be determined so as to pass through the centre of the pencil, and touch each of the $4D-2$ lines; the curve thus satisfies $4D-1$ conditions, and its equation will contain $4D+2-(4D-1)$, $=3$ arbitrary constants. But if we have any particular curve satisfying the $4D-1$ conditions, then by transforming the whole figure homologously, taking the centre of the pencil as pole and any arbitrary line as axis of homology, so as to leave the pencil of lines unaltered (analytically if at the centre of the pencil $x=0$, $y=0$, then by writing $\alpha x+\beta y+\gamma z$ in place of z) the transformed curve still satisfies the $4D-1$ conditions, and we have by the homologous transformation introduced into its equation 3 arbitrary constants, that is, we have obtained the most general curve which satisfies the conditions in question. The absolute invariants of the general curve are independent of the

[1] It is right to notice that the absolute invariants spoken of here, and in what follows, are not in general rational ones.

3 arbitrary constants introduced by the homologous transformation; and they are consequently functions of only the coefficients of the given pencil of $4D-2$ lines; this being so, it is obvious that they will be respectively equal to absolute invariants of the pencil of $4D-2$ lines. The number of the absolute invariants of the general curve of the order $D+1$ is $=\frac{1}{2}(D+1)(D+4)+1-9$, but there is a reduction $=1$, for each of the dps, hence in the present case the number is $\frac{1}{2}(D+1)(D+4)-\frac{1}{2}(D^2-3D)-8$, $=4D-6$; and there are thus $4D-6$ absolute invariants of the curve, each of them equal to an absolute invariant of the pencil; that is, of the $4D-5$ absolute invariants of the pencil, there are $4D-6$, each of them equal to an absolute invariant of the curve, and consequently independent of the position of the point O' on the curve; which is the theorem which was to be proved. I believe the reasoning is quite correct, but there are some points in it which require further examination, it is therefore given subject to any correction which may hereafter appear to be necessary.

30. The general subject may be illustrated by considerations belonging to solid geometry. If we imagine the original curve and the transformed curve as situate in different planes, then joining each point of the original curve with the corresponding point on the transformed curve, we have a series of lines forming a scroll (skew surface): if the two curves are of the orders n, n' respectively, then the complete section by the plane of the original curve is made up of this curve of the order n, and of n' generating lines; and similarly the complete section by the plane of the transformed curve is made up of this curve of the order n', and of n generating lines. Conversely, given a scroll of the order $n+n'$, any two sections of this scroll, being in general curves of the same order $n+n'$, are rational transformations the one of the other; but for the *general* scroll of the order $n+n'$, it is not possible to find sections breaking up as above.

385.

ON THE CORRESPONDENCE OF TWO POINTS ON A CURVE.

[From the *Proceedings of the London Mathematical Society*, vol. I. (1865—1866), No. VII. pp. 1—7. Read April 16, 1866.]

1. In a unicursal curve the coordinates (x, y, z) of any point of the curve are proportional to rational and integral functions of a variable parameter θ. Hence, if two points of the curve correspond in suchwise that to a given position of the first point there correspond α' positions of the second point, and to a given position of the second point α positions of the first point, the number of points which correspond each to itself is $=\alpha+\alpha'$. For let the two points be determined by their parameters θ, θ' respectively—then to a given value of θ there correspond α' values of θ', and to a given value of θ' there correspond α values of θ; hence the relation between (θ, θ') is of the form $(\theta, 1)^{\alpha}(\theta', 1)^{\alpha'}=0$; and writing therein $\theta'=\theta$, then for the points which correspond each to itself, we have an equation $(\theta, 1)^{\alpha+\alpha'}=0$ of the order $\alpha+\alpha'$; that is, the number of these points is $=\alpha+\alpha'$.

2. Hence for a unicursal curve we have a theorem similar to that of M. Chasles' for a line, viz., the theorem may be thus stated:

If two points of a unicursal curve have an (α, α') correspondence, the number of united points is $=\alpha+\alpha'$.

3. But a unicursal curve is nothing else than a curve with a deficiency $D=0$, and we thence infer

Theorem. If two points of a curve with deficiency D have an (α, α') correspondence, the number of united points is $=\alpha+\alpha'+2kD$; in which theorem $2k$ is a coefficient to be determined.

4. Suppose that the corresponding points are P, P' and imagine that when P is given, the corresponding points P' are the intersections of the given curve by a

curve Θ (the equation of the curve Θ will of course contain the coordinates of P as parameters, for otherwise the position of P' would not depend upon that of P). I find that if the curve Θ has with the given curve k intersections at the point P, then in the system of (P, P'), the number of united points is

$$\text{a} = \alpha + \alpha' + 2kD,$$

whence in particular, if the curve Θ does not pass through the point P, then the number of united points is $= \alpha + \alpha'$, as in a unicursal curve.

4*. The foregoing theorem is easily proved in the particular case where the k intersections at the point P take place in consequence of the curve θ having a k-tuple point at P. Taking $U = (x, y, z)^m = 0$ as the equation of the given curve (which for greater simplicity is assumed to be a curve without singularities), then if we suppose (x, y, z) to be the coordinates of the point P, and (x', y', z') to be the coordinates of the point P', write $U = (x, y, z)^m$, $U' = (x', y', z')^m$, U' being what U becomes on writing therein (x', y', z') in place of (x, y, z); and

$$\Theta = (x, y, z)^a (x', y', z')^{a'} (yz' - y'z,\ zx' - z'x,\ xy' - x'y)^k,$$

viz., Θ is a function of the order k in $yz' - yz'$, $zx' - z'x$, $x'y - x'y$, the coefficients of the several powers and products of these quantities being functions of the order a in (x, y, z) and of the order a' in (x', y', z'), which functions are such that they do not all of them vanish, identically, or in virtue of the equation $U = 0$, on writing therein $(x', y', z') = (x, y, z)$. Taking for a moment (x, y, z) as current coordinates, suppose that the equation of the given curve is $U = 0$; then if (x, y, z) are the coordinates of the point P, we have $U = 0$, and similarly if (x', y', z') are the coordinates of the point P' we have $U' = 0$. The equation $\Theta = 0$, considering therein (x, y, z) as the coordinates of the given point P (and so as parameters satisfying the equation $U = 0$) and (x', y', z') as current coordinates, will be a curve having a k-tuple point at P, we have thus the case above supposed; and P being given, the corresponding points P' are given as the intersections of the curves $U' = 0$, $\Theta = 0$, which are respectively of the orders m and $a' + k$; the total number of intersections is thus $= m(a' + k)$, but inasmuch as the curve $\Theta = 0$ has a k-tuple point at P, k of these intersections coincide with the point P, and the number of the remaining intersections, that is the number of positions of the point P', is $= ma' + (m-1)k$. Similarly when P' is given, the number of positions of the point P is $= ma + (m-1)k$: and we have therefore

$$\alpha + \alpha' = m(a + a') + 2(m-1)k.$$

To find the united points, it is to be observed, that upon writing $(x', y', z') = (x, y, z)$, the function Θ becomes identically $= 0$; but if we suppose, in the first instance, that P', P, are consecutive points on the curve $U = 0$, then we have

$$yz' - y'z : zx' - z'x : xy' - x'y = \delta_x U;\ \delta_y U;\ \delta_z U;$$

and the equation $\Theta = 0$ assumes the form

$$\Theta = (x, y, z)^a (x, y, z)^{a'} (\delta_x U,\ \delta_y U,\ \delta_z U)^k = 0,$$

which, (x, y, z) being current coordinates, is the equation of a curve of the order $a+a'+(m-1)k$; the united points are the intersection of this curve with the given curve $U=0$, and the number of the united points is thus

$$\text{a} = m(\alpha+\alpha') + m(m-1)k.$$

Hence attending to the above-mentioned value of $\alpha+\alpha'$, we have

$$\text{a} = \alpha+\alpha' + (m-1)(m-2)k.$$

But in the case of a curve $U=0$, without singularities, we have $2D=(m-1)(m-2)$, and we have thus the required formula

$$\text{a} = \alpha+\alpha'+2kD.$$

The investigation in the case where the k intersections at P arise wholly or in part from a contact of the curve Θ, or any branch or branches thereof, with the given curve U, is more difficult, and I abstain from entering upon it.

I apply the theorem to some examples:

5. Investigation of the class of a curve of the order m with δ dps. Take as corresponding points on the curve two points, such that the line joining them passes through a fixed point O: the united points will be the points of contact of the tangents through O; that is, the number of the united points is equal to the class of the curve. The curve Θ is here the line OP, which has with the given curve a single intersection at P; that is, we have $k=1$. The points P' corresponding to a given position of P are the remaining $m-1$ intersections of OP with the curve; that is, we have $\alpha'=m-1$; and in like manner $\alpha=m-1$. Hence the class is $=2(m-1)+2D$, viz., this is $=(2m-2)+(m^2-3m+2-2\delta)$, which is $=m^2-m-2\delta$, as it should be.

6. It is in the foregoing example assumed that the dps are none of them cusps; if the curve has $\delta+\kappa$ dps, κ of which are cusps (or what is the same thing, δ nodes and κ cusps); then the number of united points is equal $2(m-1)+2D$, $=m^2-m-2\delta-2\kappa$; but in this case each of the cusps is reckoned as a united point, and we have, therefore, class $+\kappa=m^2-m-2\delta-2\kappa$, that is, class $=m^2-m-2\delta-3\kappa$. This will serve as an instance of the special considerations which are required in the case of a curve with cusps, but in what follows, I shall assume that the dps are none of them cusps and thus attend to the case of a curve of the order m, with δ dps, and therefore of the class $n=m^2-m-2\delta$, and of the deficiency $D=\frac{1}{2}(m-1)(m-2)-\delta$, $=\frac{1}{2}(n-2m+2)$.

7. Investigation of the number of inflexions. Taking the point P' to be a tangential of P (that is, an intersection of the curve by the tangent at P), the united points are the inflexions, and the number of the united points is equal to the number of inflexions. The curve Θ is here the tangent at P, having with the given curve two intersections at P; that is $k=2$. P' is any one of the $m-2$ tangentials of P, hence $\alpha'=m-2$; and P is the point of contact of any one of the $n-2$

tangents from P' to the curve, that is, $\alpha = n - 2$. Hence the number of inflexions is $= (m-2) + (n-2) + 4D$, $= m + n - 4 + 2(n - 2m + 2)$, $= 3(n-m)$, which is right.

8. For the purpose of the next example it is necessary to present the fundamental equation $\mathrm{a} = \alpha + \alpha' + 2kD$ under a more general form. The curve Θ may intersect the given curve in a system of points P' each p times, a system of points Q' each q times, &c., in such manner that the points (P, P'), the points (P, Q'), &c., are pairs of points corresponding to each other according to distinct laws; and we shall then have the numbers $(\mathrm{a}, \alpha, \alpha')$, $(\mathrm{b}, \beta, \beta')$, &c., belonging to these pairs respectively; viz. (P, P') are points having an (α, α') correspondence, and the number of united points is $= \mathrm{a}$; similarly (P, Q') are points having a (β, β') correspondence, and the number of united points is $= \mathrm{b}$; and so on. The theorem then is

$$p(\mathrm{a} - \alpha - \alpha') + q(\mathrm{b} - \beta - \beta') + \ldots = 2kD.$$

9. Investigation of the number of double tangents:—Take P', an intersection with the curve of a tangent drawn from P to the curve (or what is the same thing, P, P' cotangentials of any point of the curve); the united points are here the points of contact of the several double tangents of the curve; or if τ be the number of double tangents, then the number of united points is $= 2\tau$. The curve Θ is the system of the $n-2$ tangents from P to the curve; each tangent has with the curve 1 intersection at P, that is, $k = n-2$; each tangent, besides, meets the curve in the point of contact Q' twice, and in $(m-3)$ points P'. Hence, if $(\mathrm{a}, \alpha, \alpha')$ refer to the points (P, Q'), and $(2\tau, \beta, \beta')$ to the points (P, P'), we have

$$2(\mathrm{a} - \alpha - \alpha') + 2\tau - \beta - \beta' = 2(n-2)D.$$

Moreover, from the last example the value of $\mathrm{a} - \alpha - \alpha'$ is $= 4D$, and the formula thus becomes

$$2\tau - \beta - \beta' = 2(n-6)D;$$

but from above it appears that we have $\beta = \beta' = (n-2)(m-3)$, whence

$$\begin{aligned} 2\tau &= 2(n-2)(m-3) + 2(n-6)D, \\ &= 2(n-2)(m-3) + (n-6)(n-2m+2), \\ &= n^2 - 10n + 8m, \end{aligned}$$

which is right; in fact, observing that ι (the number of inflexions) is $= 3n - 3m$, the formula is equivalent to $2\tau + 3\iota = n^2 - n - m$, that is, $m = n^2 - n - 2\tau - 3\iota$.

In the foregoing examples the curve Θ is a line or system of lines; but I give an example in which Θ is a system of conics, and in which, as will appear, we have to consider the two characteristics (μ, ν) of the system.

10. Investigation of the number of conics which can be drawn, satisfying any four conditions, and touching the given curve; or say of the number of the conics $(4Z)(1)$. Take P', an intersection of the given curve by a conic $(4Z)$ passing through the point P, then the number of the united points is equal to that of the conics $(4Z)(1)$. The curve Θ is here the system of the conics $(4Z)$ which pass through P;

hence, if (μ, ν) be the characteristics of the system of conics $(4Z)$, the number of the conics through P is $=\mu$; each of these has with the given curve 1 intersection at P, and consequently $k=\mu$. Moreover, each of the conics besides meets the curve in $(2m-1)$ points, and consequently $\alpha=\alpha'=\mu(2m-1)$. Hence the formula gives the number of united points

$$=2\mu(2m-1)+\mu(n-2m+2),$$
$$=\mu(n+2m);$$

or, as this may be written,

$$=\mu n+\nu m+m(2\mu-\nu).$$

But the system of conics $(4Z)$ contains $(2\mu-\nu)$ point-pairs (*coniques infiniment aplaties*), each of which, regarded as a pair of coincident lines, meets the given curve in m pairs of coincident points; that is, the point-pair is to be considered as a conic *touching* the given curve in m points; and there is on this account a reduction $=m(2\mu-\nu)$ in the number of the united points; whence, finally, the number of the conics $(4Z)$ (1) is $=\mu n+\nu m$. It is hardly necessary to remark that it is assumed that the conditions $(4Z)$ are conditions having no special relation to the given curve.

11. As a final example, suppose that the point P on a given curve of the order m, and the point Q on a given curve of the order m', have an (α, α') correspondence, and let it be required to find the class of the curve enveloped by the line PQ. Take an arbitrary point O, join OQ, and let this meet the curve m in P', then (P, P') are points on the curve m, having a $(m'\alpha, m\alpha')$ correspondence; in fact, to a given position of P there correspond α' positions of Q, and to each of these m positions of P', that is, to each position of P there correspond $m\alpha'$ positions of P'; and similarly to each position of P' there correspond $m'\alpha$ positions of P. The curve Θ is the system of the lines drawn from each of the α' positions of Q to the point O, hence the curve Θ does not pass through P, and we have $k=0$. Hence the number of the united points (P, P'), that is, the number of the lines PQ which pass through the point O, is $=m\alpha'+m'\alpha$, or this is the class of the curve enveloped by PQ.

12. It may be remarked, that if the two curves are curves in space (plane or of double curvature), then the like reasoning shows that the number of the lines PQ which meet a given line O is $=m\alpha'+m'\alpha$, that is, the order of the scroll generated by the line PQ is $=m\alpha'+m'\alpha$.

386.

ON THE LOGARITHMS OF IMAGINARY QUANTITIES.

[From the *Proceedings of the London Mathematical Society*, vol. II. (1866—1869), pp. 50—54. Read Dec. 12, 1867.]

THE theory of the logarithms of imaginary quantities admits of a remarkably simple representation.

Let P denote at pleasure the imaginary quantity $x+iy$, or else the point the coordinates of which are (x, y); viz., P regarded as a quantity will denote $x+iy$; but we may also speak of the point P.

Writing thus

$$P = x + iy,$$

and similarly

$$P' = x' + iy',$$

we have of course

$$\frac{P}{P'} = \frac{x+iy}{x'+iy'},$$

an imaginary quantity $X+iY$; and the point $\frac{P}{P'}$ will be the point the coordinates of which are (X, Y).

We have

$$P = re^{i\theta},$$

viz., r is $=\sqrt{x^2+y^2}$, the radical being positive, and θ is an arc such that

$$\cos\theta = \frac{x}{\sqrt{x^2+y^2}},\quad \sin\theta = \frac{y}{\sqrt{x^2+y^2}},$$

and moreover θ may be taken to be an arc between the limits $-\pi$, $+\pi$. The arc so defined may be denoted by $\tan^{-1}\frac{y}{x}$, so that we have $\theta = \tan^{-1}\frac{y}{x}$.

It is to be observed that θ has always a determinate unique value, except in the single case $y=0$, x negative, where we have indeterminately $\theta=\pm\pi$.

It is further to be remarked that, taking A for the origin of coordinates, we have $\theta=$ angle xAP, considered as positive or as negative according as P lies above or below the axis of x.

Starting from the equation

$$P=re^{i\theta},$$

we have similarly

$$P'=r'e^{i\theta'},$$

and

$$\frac{P}{P'}=\frac{r}{r'}e^{i\phi},$$

where ϕ is derived from $\frac{P}{P'}$ in the same way as θ from P, or θ' from P'.

Consequently

$$e^{i(\theta-\theta'-\phi)}=1,$$

and therefore $\theta-\theta'-\phi$ a multiple of 2π, say

$$\theta-\theta'-\phi=2m\pi,$$

and in this equation the value of m is determined by the limiting conditions above imposed on the values of θ, θ', ϕ. To see how this is, suppose in the first instance that the finite line or chord $P'P$, considered as drawn from P' to P, cuts the negative part of the axis of x upwards; P is then above, P' below, the axis of x; that is, θ, $-\theta'$ are each positive; and drawing the figure, it at once appears that the sum $\theta+(-\theta')$, that is $\theta-\theta'$, is a positive quantity greater than π. And in this case the angle ϕ will be equal to $2\pi-(\theta-\theta')$ taken negatively, that is, $\phi=-\{2\pi-(\theta-\theta')\}$, or $\theta-\theta'-\phi=2\pi$. But, in like manner, if $P'P$ cut the negative part of the axis of x downwards, P will be below, P' above, the axis of x; $-\theta$ and θ' are here each positive, and the figure shows that the sum $-\theta+\theta'$ is greater than π; and in this case the angle ϕ is $=2\pi-(-\theta+\theta')$; that is, we have $\theta-\theta'-\phi=-2\pi$. In every other case, (that is, if the chord $P'P$ either does not meet the axis of x, or if it meets the positive part of the axis of x,) $\theta-\theta'$ and ϕ are each in absolute magnitude less than π, and we have $\theta-\theta'-\phi=0$. So that we see that, according as the chord $P'P$, considered as drawn from P' to P, meets the negative part of the axis of x upwards or downwards, or as it does not meet the negative part of the axis of x, the value of $\theta-\theta'-\phi$ is $=2\pi$, $=-2\pi$, or $=0$.

Taking now $\log r$ to represent the real logarithm of the positive real quantity r, we may, as a definition of the logarithm of the imaginary quantity $P\,(=x+iy)$, write

$$\log P=\log r+i\theta.$$

The value so defined is of course one out of the infinite number of values of the logarithm, and it may for distinction be termed the "selected" value. In all that follows, the symbol "log" is to be understood to denote the selected value. We have

$$\log P=\log r+i\theta,$$

and similarly

$$\log P' = \log r' + i\theta',$$

and

$$\log \frac{P}{P'} = \log \frac{r}{r'} + i\phi.$$

Hence

$$\log P - \log P' = \log \frac{P}{P'} + i(\theta - \theta' - \phi),$$

so that, by what precedes, $\log P - \log P'$, if the chord $P'P$, considered as drawn from P' to P, cuts the negative part of the axis of x upwards, is $= \log \frac{P}{P'} + 2i\pi$; if the chord cuts the negative part of the axis of x downwards, it is $= \log \frac{P}{P'} - 2i\pi$, and in every other case it is $= \log \frac{P}{P'}$.

It is to be remarked that $\log P$, as above defined, is a continuous function of $P (= x + iy)$, with the single exception that, if the point P move from above to below or from below to above the negative part of the axis of x, the imaginary part of the logarithm changes from $+i\pi$ to $-i\pi$, or from $-i\pi$ to $+i\pi$, in the two cases respectively. And we are thus led to another mode of looking at the question.

Consider the integral

$$\int_{P'}^{P} \frac{dz}{z}.$$

The value of the integral may depend on the series of values assumed by the variable z as it passes from the limit $z = P'$ to the limit $z = P$, or say it may depend on the path of the variable z; in order to give the notation a precise signification, we must therefore fix the path of the variable z; and I do this by taking the path to be the right line $P'P$. Write now $z = P'.u$, we have $\frac{dz}{z} = \frac{du}{u}$; $z = P'$ gives $u = 1$; $z = P$ gives $u = \frac{P}{P'}$; and it is easy to see that, the path of z being along the right line P' to P, that of u is along the right line 1 to $\frac{P}{P'}$ $\Big($that is, from the point the coordinates whereof are $x = 1$, $y = 0$, to the point $\frac{P}{P'}\Big)$.

We have thus

$$\int_{P'}^{P} \frac{dz}{z} = \int_{1}^{P \div P'} \frac{du}{u},$$

the path in each case being a right line as above. The indefinite integral $\int \frac{du}{u} = \log u$; and as u passes from 1 to $\frac{P}{P'}$, there is no discontinuity in the value of $\log u$; the value of the right-hand side is thus $=\log\frac{P}{P'}$. As regards the left-hand side, the indefinite integral is in like manner $=\log z$; but here, if the chord $P'P$ cuts the negative part of the axis of x, there is a discontinuity in the value of $\log z$, viz., if the chord $P'P$, considered as drawn from P' to P, cuts the negative part of the axis of x upwards, there is an abrupt change in the value of $\log z$ from $-i\pi$ to $+i\pi$; and similarly, if the chord cut the negative part of the axis of x downwards, there is an abrupt change from $+i\pi$ to $-i\pi$; in the former case, by taking the definite integral to be $\log P - \log P'$, we take its value too large by $2i\pi$, in the latter case we take it too small by $2i\pi$; that is, the true value of the definite integral is in the former case $=\log P - \log P' - 2i\pi$, in the latter case it is $=\log P - \log P' + 2i\pi$. But if the chord PP' does not cut the negative part of the axis of x, then there is not any discontinuity, and the true value of the definite integral is $=\log P - \log P'$. We have thus in the three cases respectively

$$\begin{aligned}\log P - \log P' &= \log\frac{P}{P'} + 2i\pi,\\ &= \log\frac{P}{P'} - 2i\pi,\\ &= \log\frac{P}{P'},\end{aligned}$$

which agrees with the previous results.

It may be remarked, that it is merely in consequence of the particular definition adopted that there is in the value of $\log P$ a discontinuity at the passage over the negative part of the axis of x; with a different definition of the logarithm, there would be a discontinuity at the passage over some other line from the origin; but a discontinuity somewhere there must be. For if, as above, the chord $P'P$ meet the negative part of the axis of x, then forming a closed quadrilateral by joining by right lines the points 1 to P, P to P', P' to $\frac{P}{P'}$, and $\frac{P}{P'}$ to 1; the only side meeting the negative part of the axis of x is the side $P'P$; the integral $\int\frac{dz}{z}$, taken through the closed circuit in question, or say the integral

$$\left(\int_P^1 + \int_{P'}^P + \int_{P\div P'}^{P'} + \int_1^{P\div P'}\right)\frac{dz}{z}$$

has, by what precedes, a value in consequence of the discontinuity in passing from P' to P; viz., this is $=-2i\pi$ or $=2i\pi$, according as the chord $P'P$, considered as drawn from P' to P, cuts the negative part of the axis of x upwards or downwards; but this value $-2i\pi$ or $+2i\pi$ must be altogether independent of the definition of the logarithm; whereas if, by any alteration in the definition, the discontinuity could be avoided, the value of the integral, instead of being as above, would be $=0$. The foregoing value $-2i\pi$ or $+2i\pi$ is in fact that of the integral taken along (in the one or the other direction) any closed curve surrounding the point $z=0$ for which the function $\frac{1}{z}$ under the integral sign becomes infinite: but in obtaining the value as above, no use is made of the principles relating to the integration of functions which thus become infinite.

The equation

$$\log P = \log r + i\theta$$

gives

$$P^m = e^{m \log P} = r^m e^{im\theta},$$

or say

$$(x+iy)^m = r^m e^{im\theta},$$

where, m being any real quantity whatever, r^m denotes the positive real value of r^m. We have thus a definition of the value of $(x+iy)^m$, and the value so defined may be called the selected value. And similarly, for an imaginary exponent $m=p+qi$, we have

$$\begin{aligned}(x+iy)^{p+qi} &= e^{(p+qi)(\log r + i\theta)} \\ &= e^{p\log r - q\theta + i(p\theta + q\log r)} \\ &= r^p e^{-q\theta} \,.\, e^{i(p\theta+q\log r)},\end{aligned}$$

which is the selected value of $(x+iy)^{p+qi}$.

It may be remarked, in illustration of the advantage (or rather the necessity) of having a selected value, that in an integral $\int Zdz$, taken between given limits along a given path, it is necessary that we know, for the real or imaginary value of z corresponding to each point of the path, the value of the function Z; and consequently, if Z is a function involving $\log z$ or z^m, the indeterminateness which presents itself in these symbols (considered as belonging to a single value of z) is, so to speak, indefinitely multiplied, and $\int Zdz$ is really an unmeaning combination of symbols, unless by selecting, as above or otherwise, a unique value of $\log z$ or z^m, we render the function to be integrated a determinate function of the variable.

387.

NOTICES OF COMMUNICATIONS TO THE LONDON MATHEMATICAL SOCIETY.

[From the *Proceedings of the London Mathematical Society*, vol. II. (1866—1869), pp. 6—7, 25—26, 29, 61—63, 103—104, 123—125.]

December 13, 1866. pp. 6—7.

PROF. CAYLEY exhibited and explained some geometrical drawings. Thinking that the information might be convenient for persons wishing to make similar drawings, he noticed that the paper used was a tinted drawing paper, made in continuous lengths up to 24 yards, and of the breadth of about 56 inches(1); the half-breadth being therefore sufficient for ordinary figures, and the paper being of a good quality and taking colour very readily. Among the drawings was one of the conics through four points forming a convex quadrangle. The plane is here divided into regions by the lines joining each of the six pairs of points, and by the two parabolas through the four points; and the regions being distinguished by different colours, the general form of the conics of the system is very clearly seen. (Prof. Cayley remarked that it would be interesting to make the figures of other systems of conics satisfying four conditions; and in particular for the remaining elementary systems of conics, where the conics pass through a number 3, 2, 1 or 0 of points and touch a number 1, 2, 3 or 4 of lines: the construction of some of these figures is, however, practically a great deal more difficult.) Other figures related to Cartesians and Bicircular Quartics. One of these was a figure of a system of triconfocal Cartesians; and derived from this by inversion in regard to a circle, there was a figure of a system of quadriconfocal bicircular quartics: in the assumed position of the inverting circle, each quartic consists (like the Cartesian which gives rise to it) of an exterior and an interior continuous curve, and the general aspect of the figure is that of a distortion of the original figure of the Cartesians. Another figure was that of the bicircular quartic, for which the

1 Sold at Messrs Lechertier-Barbe's, Regent Street, at 6*d*. per yard, or 9*s*. the piece.

algebraical sum of the distances of a point thereof from three given foci is = 0 (this was selected for facility of construction, by the intersections of circles and confocal conics). The quartic consists of two equal and symmetrically situated pear-shaped curves, exterior to each other, and including the one of them two of the three given foci, the other of them the third given focus, and a fourth focus lying in a circle with the given foci: by inversion in regard to a circle having its centre at a focus the two pear-shaped curves became respectively the exterior and the interior ovals of a Cartesian. There was also a figure of the two circular cubics, having for foci four given points on a circle; and a figure (coloured in regions) in preparation for the construction of the analogous sextic curve derived from four given points not in a circle.

March 28, 1867. pp. 25—26.

Professor Cayley mentioned a theorem included in Prof. Sylvester's theory of derivation of the points of a cubic curve. Writing down the series of numbers 1, 2, 4, 5, 7, 8, 10, 11, 13, 14, 16, 17, &c., viz., all the numbers not divisible by 3, then (repetitions of the same number being permissible) taking any two numbers of the series, we have in the series a third number, which is the sum or else the difference of the two numbers (for example, 2, 2 give their sum 4, but 2, 7 give their difference 5), and we have thus a series of triads, in each of which one number is the sum of the other two. The theorem is, that it is possible on a cubic curve to construct a series of points, such that denoting them by the above numbers respectively, then for any triad of numbers as aforesaid the points denoted by the three numbers respectively lie *in lineâ*. And the theorem gives its own construction: in fact the series of triads is 112, 224, 145, 257, 178, 248, &c. Take 1, an arbitrary point on the cubic, then (by the theorem) the triad 112 shows that 2 is the tangential of 1; 224 shows that 4 is the tangential of 2; 145 that 5 is the third point of 1 and 4; 257 that 7 is the third point of 2 and 5. So far we have no theorem; we have merely, starting from the point 1, constructed by an arbitrary process the points 2, 4, 5, and 7. But going a step further; 178 and 448 show, the first of them, that 8 is the third point of 1 and 7, the second of them, that 8 is the tangential of 4. We have here the theorem that the third point of 1 and 7 is also the tangential of 4. Similarly, 10, 11, 13 are each of them (like 8) determined by two constructions; 14, 16, 17, 19, each of them by three constructions, and so on; the number of constructions increasing by unity for each group of four numbers. And the theorem is, that these constructions, 2, 3, or more, as the case may be, give always one and the same point. Prof. Cayley mentioned that on a large figure of a cubic curve he had, in accordance with the theorem, constructed the series of points 1, 2, 4, 5, 7, 8, 10, 11, 13, 14.

April 15, 1867. p. 29.

Prof. Cayley communicated a theorem relating to the locus of the ninth of the points of intersection of two cubics, seven of these points being fixed, while the eighth moves on a straight line.

March 26, 1868. pp. 61—63.

Prof. Cayley made some remarks on a mode of generation of a sibi-reciprocal surface, that is, a surface the reciprocal of which is of the same order and has the same singularities as the original surface.

If a surface be considered as the envelope of a *plane* varying according to given conditions, this is a mode of generation which is essentially not sibi-reciprocal; the reciprocal surface is given as the locus of a *point* varying according to the reciprocal conditions. But if a surface be considered as the envelope of a *quadric surface* varying according to given conditions, then the reciprocal surface is given as the envelope of a *quadric surface* varying according to the reciprocal conditions; and if the conditions be sibi-reciprocal, it follows that the surface is a sibi-reciprocal surface. For instance, considering the surface which is the envelope of a quadric surface touching each of 8 given lines; the reciprocal surface is here the envelope of a quadric surface touching each of 8 given lines; that is, the surface is sibi-reciprocal. So again, when a quadric surface is subjected to the condition that 4 given points shall be in regard thereto a conjugate system, this is equivalent to the condition that 4 given planes shall be in regard thereto a conjugate system—or the condition is sibi-reciprocal; analytically the quadric surface $ax^2+by^2+cz^2+dw^2=0$ is a quadric surface subjected to a sibi-reciprocal system of six conditions. Impose on the quadric surface two more sibi-reciprocal conditions,—for instance, that it shall pass through a given point and touch a given plane,—the envelope of the quadric will be a sibi-reciprocal surface. It was noticed that in this case the envelope was a surface of the order (= class) 12, and having (besides other singularities) the singularities of a conical point with a tangent cone of the class 3, and of a curve of plane contact of the order 3. In the foregoing instances the number of conditions imposed upon the quadric surface is 8; but it may be 7, or even a smaller number. An instance was given of the case of 7 conditions, viz.,—the quadric surface is taken to be $ax^2+by^2+cz^2+dw^2=0$ (6 conditions) with a relation of the form

$$Abc+Bca+Cab+Fad+Gbd+Hcd=0$$

between the coefficients (1 condition); this last condition is at once seen to be sibi-reciprocal; and the envelope is consequently a sibi-reciprocal surface—viz., it is a surface of the order (= class) 4, with 16 conical points and 16 conics of plane contact. It is the surface called by Prof. Cayley the "tetrahedroid," (see his paper "Sur la surface des ondes," *Liouv.* tom. XI. (1846), pp. 291—296 [47]), being in fact a homographic transformation of Fresnel's Wave Surface.

{Prof. Cayley adds an observation which has since occurred to him. If the quadric surface $ax^2+by^2+cz^2+dw^2=0$, be subjected to touch a given line, this imposes on the coefficients a, b, c, d, a relation of the above form, viz., the relation is

$$A^2bc+B^2ca+C^2ab+F^2ad+G^2bd+H^2cd=0;$$

where A, B, C, F, G, H are the "six coordinates" of the given line, and satisfy therefore the relation $AF+BG+CH=0$. It is easy to see that there are 8 lines for which the squared coordinates have the same values A^2, B^2, C^2, F^2, G^2, H^2; these 8 lines are symmetrically situate in regard to the tetrahedron of coordinates, and moreover they lie in a hyperboloid. The quadric surface, instead of being defined as above, may, it is clear, be defined by the equivalent conditions of touching each of the 8 given lines: that is, we have the envelope of a quadric surface touching each of 8 given lines; these lines not being arbitrary lines, but being a system of a very special form. By what precedes, the envelope is a quartic surface. It appears, however, that in virtue of the relation $AF + BG + CH = 0$, this is no longer a proper quartic surface, but that it resolves itself into the above-mentioned hyperboloid taken twice. That is, restoring the original A, B, &c., in place of A^2, B^2, &c., the envelope of the quadric $ax^2 + by^2 + cz^2 + dw^2 = 0$, where a, b, c, d vary, subject to the condition $Abc + Bca + Cab + Fad + Gbd + Hcd = 0$, (which is in general a tetrahedroid), is when A, B, C, F, G, H are the squared coordinates of a line (or, what is the same thing, when $\sqrt{AF} + \sqrt{BG} + \sqrt{CH} = 0$) a hyperboloid taken twice, viz., this is the hyperboloid passing through the given line and through the symmetrically situate seven other lines.}

November 12, 1868. pp. 103, 104.

Professor Cayley gave an account to the Meeting of a Memoir by Herr Listing, "Census räumlicher Complexe oder Verallgemeinerung des Euler'schen Satzes von den Polyedern," published in the *Göttingen Transactions* for 1862. The fundamental theorem is a relation $a - (b - \kappa) + (c - \kappa' + \pi) - (d - \kappa'' + \pi' - \omega) = 0$ existing in any figure whatever between a the number of points, b the number of lines, c the number of areas, d the number of spaces, and certain supplementary quantities κ, κ', κ'', π, π', ω. In an extensive class of figures these last are each $= 0$, and the relation is $a - b + c - d = 0$; thus, in a closed box, $a = 8$, $b = 12$, $c = 6$, $d = 2$ (viz., there is the finite space inside, and the infinite space outside, the box): if the box be opened, $a = 10$, $b = 15$, $c = 6$, $d = 1$; if the lid be taken away, $a = 8$, $b = 12$, $c = 5$, $d = 1$; in each case, $a - b + c - d = 0$. If the bottom be also taken away, $a = 8$, $b = 12$, $c = 4$, $d = 1$; but here one of the supplementary quantities comes in, $\kappa'' = 1$, and the theorem is $a - b + c - (d - \kappa'') = 0$. The chief difficulty and interest of the Memoir consist in the determination of the supplementary quantities κ, κ', κ'', π, π', ω.

December 10, 1868. pp. 123—125. Appended to Paper by Mr T. Cotterill "On a Correspondence of Points etc."

Observations by Professor Cayley and Mr W. K. Clifford on the connexion of the transformation with Cremona's general theory, and the analytical formulæ.

According to Cremona's general theory,—taking (x_1, y_1, z_1) and (x_2, y_2, z_2) as current coordinates in the two planes respectively,—if we take in the first plane, any three points 1, 2, 3, and any other three points 4′, 5′, 6′, then if $X_1 = 0$, $Y_1 = 0$, $Z_1 = 0$ are quartic curves, each having the double points 1, 2, 3, and the simple points 4′, 5′, 6′, we have a transformation $x_2 : y_2 : z_2 = X_1 : Y_1 : Z_1$ leading to a converse system

$$x_1 : y_1 : z_1 = X_2 : Y_2 : Z_2$$

of the like form; viz., there will be in the second plane three points 4, 5, 6, and three other points 1′, 2′, 3′, such that $X_2 = 0$, $Y_2 = 0$, $Z_2 = 0$, are quartics having the double points 4, 5, 6, and the simple points 1′, 2′, 3′.

Analytically, Cremona's transformation is obtained by assuming the reciprocals of x_2, y_2, z_2 to be proportional to linear functions of the reciprocals of x_1, y_1, z_1—(of course, this being so, the reciprocals of x_1, y_1, z_1 will be proportional to linear functions of the reciprocals of x_2, y_2, z_2). Solving this under the theory as above explained, write

$$\left.\begin{array}{l}\dfrac{1}{x_2}\\ :\dfrac{1}{y_2}\\ :\dfrac{1}{z_2}\end{array}\right\} = \left\{\begin{array}{l}\dfrac{a}{x_1}+\dfrac{b}{y_1}+\dfrac{c}{z_1}\\ :\dfrac{d}{x_1}+\dfrac{e}{y_1}+\dfrac{f}{z_1}\\ :\dfrac{g}{x_1}+\dfrac{h}{y_1}+\dfrac{i}{z_1}\end{array}\right\} = \left\{\begin{array}{l}P_1\\ :Q_1\\ :R_1\end{array}\right.$$

if

$$P_1 = ay_1z_1 + bz_1x_1 + cx_1y_1,$$
$$Q_1 = dy_1z_1 + ez_1x_1 + fx_1y_1,$$
$$R_1 = gy_1z_1 + hz_1x_1 + ix_1y_1.$$

Hence

$$x_2 : y_2 : z_2 = Q_1R_1 : R_1P_1 : P_1Q_1.$$

$Q_1R_1 = 0$, &c., are quartics, or generally $\alpha Q_1R_1 + \beta R_1P_1 + \gamma P_1Q_1 = 0$ is a quartic, having three double points $(y_1 = 0,\ z_1 = 0)$, $(z_1 = 0,\ x_1 = 0)$, $(x_1 = 0,\ y_1 = 0)$, and having besides the three points which are the remaining points of intersection of the conics $(Q_1 = 0,\ R_1 = 0)$, $(R_1 = 0,\ P_1 = 0)$, $(P_1 = 0,\ Q_1 = 0)$ respectively; viz., these last are the points

$$\frac{1}{x_1} : \frac{1}{y_1} : \frac{1}{z_1} = ei - hf : fg - id : dh - ge, \text{ \&c. \&c.}$$

The double and simple points are fixed points (that is, independent of α, β, γ), and the formulæ come under Cremona's theory. It is, however, necessary to show that if the points 4′, 5′, 6′ are in a line, the points 1′, 2′, 3′ are also in a line. This may be done as follows:

Let there be three planes A, B, C, and let the points of the first two correspond by ordinary triangular inversion in respect of the triangle α_1 on the plane A, and β_1 on the plane B. Let also the planes B, C correspond by ordinary triangular inversion in respect of the triangle β_2 on the plane B, and γ_2 on the plane C. Then the correspondence between A and C is the one considered, the points 1 2 3 forming the triangle α_1 and the points 4 5 6 forming the triangle γ_2. The points 4′5′6′ and 1′2′3′ in the planes A, C respectively correspond to the triangles β_1, β_2; and the conditions that 4′, 5′, 6′ shall be in a line and that 1′, 2′, 3′ shall be in a line, are the same condition, namely, that the triangles β_1, β_2 shall be inscribed in the same conic. Analogous properties must apparently belong to Cremona's other transformations, and the investigation of them will form an interesting part of the theory.

It is important, also, to notice the relation of the transformation to Hesse's "Uebertragungsprincip," *Crelle*, tom. LXVI. p. 15, which establishes a correspondence between the points of a plane and the point-pairs on a line. If $Ax^2 + 2Bxy + Cy^2 = 0$ is the equation of a point-pair, the coordinates in the plane are taken by Hesse *directly*, but in the present Paper *inversely* proportional to A, B, C.

388.

NOTE ON THE COMPOSITION OF INFINITESIMAL ROTATIONS.

[From the *Quarterly Journal of Pure and Applied Mathematics*, vol. VIII. (1867), pp. 7—10.]

THE following is a solution of a question proposed by me in the last Smith's Prize Examination:

"Show that infinitesimal rotations impressed upon a solid body may be compounded together according to the rules for the composition of forces."

DEFINITION. The "six coordinates" of a line passing through the point (x_0, y_0, z_0), and inclined at angles (α, β, γ), to the axes, are

$$\begin{aligned} a &= \cos\alpha, & f &= y_0 \cos\gamma - z_0 \cos\beta, \\ b &= \cos\beta, & g &= z_0 \cos\alpha - x_0 \cos\gamma, \\ c &= \cos\gamma, & h &= x_0 \cos\beta - y_0 \cos\alpha. \end{aligned}$$

I use, throughout, the term rotation to denote an infinitesimal rotation; this being so,

LEMMA 1. A rotation ω round the line (a, b, c, f, g, h), produces in the point (x, y, z), rigidly connected with the line, the displacements

$$\begin{aligned} \delta x &= \omega\,(\quad . \quad cy - bz + f), \\ \delta y &= \omega\,(-cx \quad . \quad + az + g), \\ \delta z &= \omega\,(\quad bx - ay \quad . \quad + h). \end{aligned}$$

LEMMA 2. Considering in a solid body the point (x, y, z), situate in the line (a, b, c, f, g, h), then for any infinitesimal motion of the solid body, the displacement of the point in the direction of the line is

$$= al + bm + cn + fp + gq + hr,$$

where l, m, n, p, q, r are constants depending on the infinitesimal motion of the solid body.

Hence, *first*, for a system of rotations

ω_1 about the line $(a_1, b_1, c_1, f_1, g_1, h_1)$,
ω_2 " " $(a_2, b_2, c_2, f_2, g_2, h_2)$,
&c.

the displacements of the point (x, y, z), are

$$\begin{aligned}
\delta x &= \quad . \quad y\Sigma c\omega - z\Sigma b\omega + \Sigma f\omega, \\
\delta y &= -x\Sigma c\omega \quad . \quad + z\Sigma a\omega + \Sigma g\omega, \\
\delta z &= x\Sigma b\omega + y\Sigma a\omega \quad . \quad + \Sigma h\omega;
\end{aligned}$$

and when the rotations are in equilibrium, the displacements $(\delta x, \delta y, \delta z)$ of any point (x, y, z) whatever must each of them vanish; that is, we must have

$$\Sigma\omega a = 0,\quad \Sigma\omega b = 0,\quad \Sigma\omega c = 0,\quad \Sigma\omega f = 0,\quad \Sigma\omega g = 0,\quad \Sigma\omega h = 0,$$

which are therefore the conditions for the equilibrium of the rotations ω_1, ω_2, &c.

Secondly, for a system of forces

P_1 along the line $(a_1, b_1, c_1, f_1, g_1, h_1)$,
P_2 " " $(a_2, b_2, c_2, f_2, g_2, h_2)$,
&c.

the condition of equilibrium as given by the principle of virtual velocities is

$$\Sigma P\,(al + bm + cn + fp + gq + hr) = 0;$$

or, what is the same thing, we must have

$$\Sigma Pa = 0,\quad \Sigma Pb = 0,\quad \Sigma Pc = 0,\quad \Sigma Pf = 0,\quad \Sigma Pg = 0,\quad \Sigma Ph = 0,$$

which are therefore the conditions for the equilibrium of the forces P_1, P_2, &c.

Comparing the two results we see that the conditions for the equilibrium of the rotations ω_1, ω_2, &c. are the same as those for the equilibrium of the forces P_1, P_2, &c.; and since, for rotations and forces respectively, we pass at once from the theory of equilibrium to that of composition; the rules of composition are the same in each case.

Demonstration of Lemma 1.

Assuming for a moment that the axis of rotation passes through the origin, then for the point P, coordinates (x, y, z), the square of the perpendicular distance from the axis is

$$\begin{aligned}
= &\ (\quad . \quad - y\cos\gamma + z\cos\beta)^2 \\
+ &\ (\ x\cos\gamma \quad . \quad - z\cos\alpha)^2 \\
+ &\ (-x\cos\beta + y\cos\alpha \quad . \quad)^2,
\end{aligned}$$

and the expressions which enter into this formula denote as follows; viz. if through the point P, at right angles to the plane through P and the axis of rotation, we draw a line PQ, = perpendicular distance of P from the axis of rotation, then the coordinates of Q referred to P as origin are

$$\begin{array}{ccc} . & -y\cos\gamma & +z\cos\beta, \\ x\cos\gamma & . & -z\cos\alpha, \\ -x\cos\beta+y\cos\alpha & & . \quad , \end{array}$$

respectively. Hence the foregoing quantities each multiplied by ω are the displacements of the point P in the directions of the axes, produced by the rotation ω. Suppose that the axis of rotation (instead of passing through the origin) passes through the point (x_0, y_0, z_0); the only difference is that we must in the formulæ write $(x-x_0, y-y_0, z-z_0)$ in place of (x, y, z): and attending to the significations of the six coordinates (a, b, c, f, g, h) it thus appears that the displacements produced by the rotation are equal to ω into the expressions

$$\begin{array}{cccc} . & -cy & +bz & +f, \\ cx & . & -az & +g, \\ -bx & +ay & . & +h, \end{array}$$

respectively.

Demonstration of Lemma 2.

For any infinitesimal motion whatever of a solid body, the displacements of the point (x, y, z) in the directions of the axes are

$$\begin{array}{l} \delta x = l \quad . \quad -ry+qz, \\ \delta y = m+rx \quad . \quad -pz, \\ \delta z = n-qx+py \quad . \ , \end{array}$$

and hence the displacement in the direction of the line (α, β, γ), is

$$\delta x\cos\alpha+\delta y\cos\beta+\delta z\cos\gamma,$$

which, attending to the signification of the six coordinates (a, b, c, f, g, h), is

$$= al+bm+cn+fp+gq+hr,$$

which is the required expression.

It is proper to remark that the last-mentioned expressions of $(\delta x, \delta y, \delta z)$ are in fact the displacements produced by a translation and a rotation. If we *assume* that every infinitesimal motion of a solid body can be resolved into a translation and a rotation, then, since a translation can be produced by two rotations, every infinitesimal motion of a solid body can be resolved into rotations alone, and the foregoing expressions for the displacements produced by a rotation, combining any number of them and writing $(\Sigma\omega a, \Sigma\omega b, \Sigma\omega c, \Sigma\omega f, \Sigma\omega g, \Sigma\omega h) = (-p, -q, -r, l, m, n)$ respectively, lead to the expressions for the displacements $\delta x, \delta y, \delta z$ produced by the infinitesimal motion of the solid body.

389.

ON A LOCUS DERIVED FROM TWO CONICS.

[From the *Quarterly Journal of Pure and Applied Mathematics*, vol. VIII. (1867), pp. 77—84.]

REQUIRED the locus of a point which is such that the pencil formed by the tangents through it to two given conics has a given anharmonic ratio.

Suppose, for a moment, that the equation of the tangents to the first conic is $(x-ay)(x-by)=0$, and that of the tangents to the second conic is $(x-cy)(x-dy)=0$, and write

$$A=(a-b)(c-d),$$
$$B=(a-c)(d-b),$$
$$C=(a-d)(b-c),$$

so that

$$A+B+C=0,$$

write also

$$k_1=\frac{B}{A},\quad k_2=\frac{C}{A},$$

then the anharmonic ratio of the pencil will have a given value k if

$$(k-k_1)(k-k_2)=0;$$

that is, if

$$k^2+k+\frac{BC}{A^2}=0,$$

or, what is the same thing, if

$$A^2(2k+1)^2+4BC-A^2=0;$$

that is, if

$$A^2(2k+1)^2-(B-C)^2=0,$$

where

$$A^2=(a-b)^2(c-d)^2,$$
$$B-C=(a+b)(c+d)-2(ab+cd),$$

are each of them symmetrical in regard to a, b, and in regard to c, d, respectively.

Let the equations of the two conics be

$$U=(a,\ b,\ c,\ f,\ g,\ h \between x,\ y,\ z)^2=0,$$
$$U'=(a',\ b',\ c',\ f',\ g',\ h' \between x,\ y,\ z)^2=0,$$

and let (α, β, γ) be the coordinates of the variable point. Putting as usual

$$(A,\ B,\ C,\ F,\ G,\ H)=(bc-f^2,\ ca-g^2,\ ab-h^2,\ gh-af,\ hf-bg,\ fg-ch),$$
$$K=abc-af^2-bg^2-ch^2+2fgh,$$

the equation of the tangents to the first conic is

$$(A,\ B,\ C,\ F,\ G,\ H \between X,\ Y,\ Z)^2=0,$$

where

$$X=\gamma y-\beta z,\quad Y=\alpha z-\gamma x,\quad Z=\beta x-\alpha y,$$

and therefore

$$\alpha X+\beta Y+\gamma Z=0.$$

Hence substituting for Z the value $-\frac{1}{\gamma}(\alpha X+\beta Y)$, we find, for the equation of the tangents, an equation of the form $\mathrm{a}X^2+2\mathrm{h}XY+\mathrm{b}Y^2=0$, which has, in effect, been taken to be $(X-aY)(X-bY)=0$; that is, we have

$$1 : a+b : ab=\mathrm{a} : -2\mathrm{h} : \mathrm{b};$$

and, in like manner, if the accented letters refer to the second conic

$$1 : c+d : cd=\mathrm{a}' : -2\mathrm{h}' : \mathrm{b}'.$$

Substituting for a, h, b their values, and for a′, h′, b′ the corresponding values, we find

$$\begin{aligned} 1 : a+b : ab &\\ = \ & A\gamma^2-2G\gamma\alpha+C\alpha^2\\ : \ & -2(H\gamma^2-F\alpha\gamma-G\beta\gamma+C\alpha\beta)\\ : \ & B\gamma^2-2F\beta\gamma+C\beta^2.\end{aligned}$$

$$\begin{aligned} 1 : c+d : cd &\\ = \ & A'\gamma^2-2G'\gamma\alpha+C'\alpha^2\\ : \ & -2(H'\gamma^2-F'\alpha\gamma-G'\beta\gamma+C'\alpha\beta)\\ : \ & B'\gamma^2-2F'\beta\gamma+C'\beta^2.\end{aligned}$$

We then have

$$\begin{aligned}(a-b)^2=(a+b)^2-4ab,&\\ = \ & 4(H\gamma^2-F\alpha\gamma-G\beta\gamma+C\alpha\beta)^2\\ & -4(A\gamma^2-2G\gamma\alpha+C\alpha^2)(B\gamma^2-2F\beta\gamma+C\beta^2),\\ = \ & -4\gamma^2(BC-F^2,\ \ldots \between \alpha,\ \beta,\ \gamma)^2,\\ = \ & -4\gamma^2K(a,\ \ldots \quad \between \alpha,\ \beta,\ \gamma)^2,\end{aligned}$$

and similarly

$$(c-d)^2 = -4\gamma^2 K' (a', \ldots \between \alpha, \beta, \gamma)^2.$$

We have, moreover,

$$\begin{aligned}(a+b)(c+d) &- 2(ab+cd)\\ &= 4(H\gamma^2 - F\alpha\gamma - G\beta\gamma + C\alpha\beta)(H'\gamma^2 - F'\alpha\gamma - G'\beta\gamma + C'\alpha\beta)\\ &\quad - 2(B\gamma^2 - 2F\beta\gamma + C\beta^2)(A'\gamma^2 - 2G'\gamma\alpha + C'\alpha^2)\\ &\quad - 2(B'\gamma^2 - 2F'\beta\gamma + C'\beta^2)(A\gamma^2 - 2G\gamma\alpha + C\alpha^2),\\ &= -2\gamma^2(BC' + B'C - 2FF', \ldots \between \alpha, \beta, \gamma)^2,\end{aligned}$$

and substituting the foregoing values, we find

$$4(2k+1)^2 KK'(a, \ldots \between \alpha, \beta, \gamma)^2 (a', \ldots \between \alpha, \beta, \gamma)^2 - \{(BC' + B'C - 2FF', \ldots \between \alpha, \beta, \gamma)^2\}^2 = 0,$$

or putting for shortness

$$\Theta = (BC' + B'C - 2FF', ., ., GH' + G'H - AF' - A'F, ., . \between \alpha, \beta, \gamma)^2,$$

the equation of the locus is

$$4(2k+1)^2 KK' . UU' - \Theta^2 = 0,$$

where (α, β, γ) are current coordinates. The locus is thus a quartic curve having quadruple contact with each of the conics $U=0$, $U'=0$; viz. it touches them at their points of intersection with the conic $\Theta = 0$, *which is the locus of the point such that the four tangents form a harmonic pencil.*

The equation may be written somewhat more elegantly under the form

$$4(2k+1)^2 . KU . K'U' - \Theta^2 = 0;$$

viz. in this equation we have

$$\begin{aligned} KU &= (BC - F^2, \ldots \between \alpha, \beta, \gamma)^2,\\ K'U' &= (B'C' - F'^2, \ldots \between \alpha, \beta, \gamma)^2,\\ \Theta &= (BC' + B'C - 2FF', \ldots \between \alpha, \beta, \gamma)^2.\end{aligned}$$

In the last form the equation is expressed in terms of the coefficients $(A, \ldots)$, $(A', \ldots)$ of the *line* equations of the conics, viz. these may be taken to be

$$(A, \ldots \between \xi, \eta, \zeta)^2 = 0, \quad (A', \ldots \between \xi, \eta, \zeta)^2 = 0.$$

In particular, if each of the conics break up into a pair of points, viz. (l, m, n) and (p, q, r) for the first conic, (l', m', n') and (p', q', r') for the second conic, then the line equations are

$$2(l\xi + m\eta + n\zeta)(p\xi + q\eta + r\zeta) = 0,$$

$$2(l'\xi + m'\eta + n'\zeta)(p'\xi + q'\eta + r'\zeta) = 0,$$

so that

$$A\ = 2lp, \;..\; F\ = mr\ + nq, \;..$$
$$A' = 2l'p', \;..\; F' = m'r' + n'q', ..$$

$$(BC\ - F^2, \;...) = -(mr\ - nq\ ,\ np\ - lr\ ,\ lq\ - mp\ )^2,$$
$$(B'C' - F'^2, \;...) = -(m'r' - n'q',\ n'p' - l'r',\ l'q' - m'p')^2,$$
$$BC' + B'C - 2FF' = 2\,\{(mn' - m'n)(qr' - q'r) - (mr' - nq')(m'r - n'q),\ ...\},$$

and substituting these values the equation is

$$(2k+1)^2 \begin{vmatrix} \alpha, & \beta, & \gamma \\ l, & m, & n \\ p, & q, & r \end{vmatrix}^2 \begin{vmatrix} \alpha, & \beta, & \gamma \\ l', & m', & n' \\ p', & q', & r' \end{vmatrix}^2 - \left\{ \begin{vmatrix} \alpha, & \beta, & \gamma \\ l, & m, & n \\ l', & m', & n' \end{vmatrix} \begin{vmatrix} \alpha, & \beta, & \gamma \\ p, & q, & r \\ p', & q', & r' \end{vmatrix} - \begin{vmatrix} \alpha, & \beta, & \gamma \\ l, & m, & n \\ p', & q', & r' \end{vmatrix} \begin{vmatrix} \alpha, & \beta, & \gamma \\ l', & m', & n' \\ p, & q, & r \end{vmatrix} \right\}^2 = 0,$$

which, if A, B, C denote

$$\begin{vmatrix} \alpha, & \beta, & \gamma \\ l, & m, & n \\ p, & q, & r \end{vmatrix} \begin{vmatrix} \alpha, & \beta, & \gamma \\ l', & m', & n' \\ p', & q', & r' \end{vmatrix}, \quad \begin{vmatrix} \alpha, & \beta, & \gamma \\ l, & m, & n \\ l', & m', & n' \end{vmatrix} \begin{vmatrix} \alpha, & \beta, & \gamma \\ p', & q', & r' \\ p, & q, & r \end{vmatrix}, \quad \begin{vmatrix} \alpha, & \beta, & \gamma \\ l, & m, & n \\ p', & q', & r' \end{vmatrix} \begin{vmatrix} \alpha, & \beta, & \gamma \\ l', & m', & n' \\ p, & q, & r \end{vmatrix}$$

respectively, $(A+B+C=0)$ is, in fact, the equation

$$(2k+1)^2 A^2 - (B-C)^2 = 0,$$

or, what is the same thing,

$$\left(k - \frac{B}{A}\right)\left(k - \frac{C}{A}\right) = 0,$$

that is

$$k = \frac{B}{A} \text{ or } k = \frac{C}{A},$$

either of which expresses the anharmonic property of the points of a conic in the form given by the theorem *ad quatuor lineas*.

Reverting to the case of two conics, then if these be referred to a set of conjugate axes, the equations will be

$$a\,x^2 + b\,y^2 + c\,z^2 = 0,$$
$$a'x^2 + b'y^2 + c'z^2 = 0,$$

we have $K = abc$, $K' = a'b'c'$,

$$\Theta = (bc' + b'c)\,aa'x^2 + (ca' + c'a)\,bb'y^2 + (ab' + a'b)\,cc'z^2,$$

and the equation of the quartic curve is

$$4\,(2k+1)^2\,abca'b'c'\,(ax^2 + by^2 + cz^2)\,(a'x^2 + b'y^2 + c'z^2)$$
$$- \{(bc' + b'c)\,aa'x^2 + (ca' + c'a)\,bb'y^2 + (ab' + a'b)\,cc'z^2\}^2 = 0.$$

I suppose in particular that the two conics are

$$x^2+my^2-1=0,$$
$$mx^2+\ \ y^2-1=0,$$

the equation of the quartic is

$$4(2k+1)^2 m^2(x^2+my^2-1)(mx^2+y^2-1)-\{(m^2+m)(x^2+y^2)-m^2-1\}^2=0;$$

or putting $\lambda=\dfrac{(m+1)^2}{4(2k+1)^2}$, this is

$$\lambda\left(x^2+y^2-\frac{m^2+1}{m^2+m}\right)^2-(x^2+my^2-1)(mx^2+y^2-1)=0.$$

To fix the ideas, suppose that m is positive and >1, so that each of the conics is an ellipse, the major semi-axis being $=1$, and the minor semi-axis being $=\dfrac{1}{\sqrt{(m)}}$. For any real value of k the coefficient λ is positive, and it may accordingly be assumed that λ is positive.

We have $\dfrac{m^2+1}{m(m+1)}>\dfrac{1}{m}<1$, or the radius of the circle is intermediate between the semi-axes of the ellipses, hence the points of contact on each ellipse are real points.

Writing for shortness

$$\alpha=\frac{m^2+1}{m^2+m},$$

the equation is

$$(x^2+my^2-1)(mx^2+y^2-1)-\lambda(x^2+y^2-\alpha)^2=0.$$

For the points on the axis of x, we have

$$(x^2-1)(mx^2-1)-\lambda(x^2-\alpha)^2=0,$$

that is

$$(m-\lambda)x^4+\{-(1+m)+2\lambda\alpha\}x^2+(1-\lambda\alpha^2)=0,$$

and thence

$$(m-\lambda)x^2=\tfrac{1}{2}(1+m)-\lambda\alpha\pm\tfrac{1}{2}\sqrt{\{(m-1)^2+4\lambda(1-\alpha)(1-m\alpha)\}},$$

or, substituting for α its value, this is

$$(m-\lambda)x^2=\tfrac{1}{2}(m+1)-\frac{\lambda\left(m+\dfrac{1}{m}\right)}{m+1}\pm\frac{\tfrac{1}{2}(m-1)}{m+1}\sqrt{\{(m+1)^2-4\lambda\}}.$$

Remarking that the values $\dfrac{(m+1)^2}{\left(m+\dfrac{1}{m}\right)^2}$, m, $\tfrac{1}{4}(m+1)^2$ are in the order of increasing magnitude, and considering successive values of λ; first the value $\lambda=\frac{1}{\alpha^2}, =\frac{(m+1)^2}{\left(m+\frac{1}{m}\right)^2}$, we have

$$(m-\lambda)\,x^2=\tfrac{1}{2}(m+1)-\frac{m+1}{m+\frac{1}{m}}\pm\frac{\tfrac{1}{2}(m-1)\left(m-\frac{1}{m}\right)}{\left(m+\frac{1}{m}\right)}$$

$$=\frac{(m+1)\tfrac{1}{2}\left(m+\frac{1}{m}-2\right)\pm\tfrac{1}{2}(m-1)\left(m-\frac{1}{m}\right)}{\left(m+\frac{1}{m}\right)};$$

or observing that

$$(m+1)\left(m+\frac{1}{m}-2\right)=(m+1)\frac{1}{m}(m-1)^2=\frac{1}{m}(m-1)(m^2-1)=(m-1)\left(m-\frac{1}{m}\right),$$

this is

$$(m-\lambda)\,x^2=0,\ \text{or}\ \frac{(m-1)\left(m-\frac{1}{m}\right)}{m+\frac{1}{m}},$$

or, what is the same thing,

$$\frac{(m-1)(m^3+2m^2-1)}{m\left(m+\frac{1}{m}\right)^2}\,x^2=0,\ \text{or}\ \frac{(m-1)\left(m-\frac{1}{m}\right)}{m+\frac{1}{m}},\qquad x^2=0,\ \text{or}\ \frac{\left(m^2-\frac{1}{m^2}\right)m}{m^3+2m^2-1}.$$

The next critical value is $\lambda=m$. The curve here is

$$(x^2+my^2-1)(mx^2+y^2-1)-m(x^2+y^2-\alpha)^2=0,$$

that is

$$\begin{aligned}&m(x^4+y^4)+(1+m^2)\,x^2y^2-(m+1)(x^2+y^2)+1\\ &-m(x^4+y^4)-\ \ 2m\ \ x^2y^2+\ \ 2m\alpha\ \ (x^2+y^2)-m\alpha^2=0,\end{aligned}$$

that is

$$(m-1)^2\,x^2y^2+(2m\alpha-m-1)(x^2+y^2)+1-m\alpha^2=0,$$

or, substituting for α its value,

$$2m\alpha-m-1=\frac{2m^2+2}{m+1}-(m+1)=\frac{(m-1)^2}{m+1},$$

$$1-m\alpha^2=1-\frac{(m^2+1)^2}{m(m+1)^2}=-\frac{(m-1)^2(m^2+m+1)}{m(m+1)^2};$$

the equation is

$$x^2y^2+\frac{1}{m+1}(x^2+y^2)-\frac{m^2+m+1}{m(m+1)^2}=0,$$

or, as this may also be written,

$$\left(x^2+\frac{1}{m+1}\right)\left(y^2+\frac{1}{m+1}\right)-\frac{1}{m}=0,$$

which has a pair of imaginary asymptotes parallel to the axis of x, and a like pair parallel to the axis of y, or what is the same thing, the curve has two isolated points at infinity, one on each axis.

The next critical value is $\lambda = \frac{1}{4}(m+1)^2$; the curve here reduces itself to the four lines

$$\left\{(x+y)^2 - \frac{m+1}{m}\right\}\left\{(x-y)^2 - \frac{m+1}{m}\right\} = 0;$$

and it is to be observed that when λ exceeds this value, or say $\lambda > \frac{1}{4}(m+1)^2$, the curve has no real point on either axis; but when $\lambda = \infty$, the curve reduces itself to $(x^2+y^2-\alpha)^2 = 0$, i.e. to the circle $x^2+y^2-\alpha=0$ twice repeated, having in this special case real points on the two axes.

It is now easy to trace the curve for the different values of λ. The curve lies in every case within the unshaded regions of the figure (except in the limiting cases after-mentioned); and it also touches the two ellipses and the four lines at the eight points k, at which points it also cuts the circle; but it does not cut or touch the

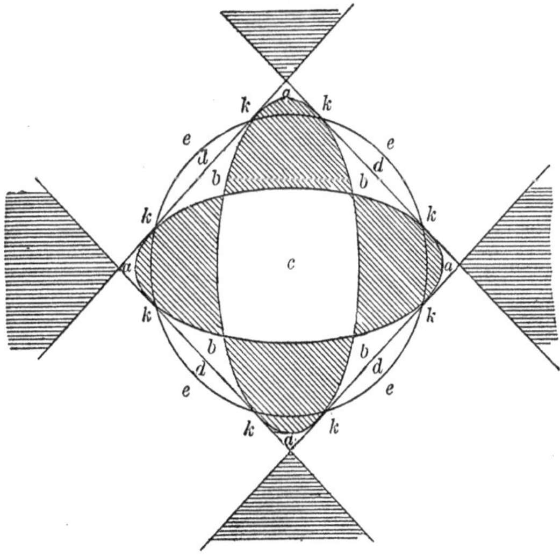

four lines, the two ellipses, or the circle, except at the points k. Considering λ as varying by successive steps from 0 to ∞;

$\lambda = 0$, the curve is the two ellipses.

$\lambda < \dfrac{(m+1)^2}{\left(m+\dfrac{1}{m}\right)^2}$, the curve consists of two ovals, an exterior sinuous oval lying in the four regions a and the four regions b; and an interior oval lying in the region c.

$\lambda = \dfrac{(m+1)^2}{\left(m+\dfrac{1}{m}\right)^2}$, there is still a sinuous oval as above, but the interior oval has dwindled to a conjugate point at the centre.

$\lambda > \dfrac{(m+1)^2}{\left(m+\dfrac{1}{m}\right)^2} < m$; $\lambda = m$; $\lambda > m < \dfrac{(m+1)^2}{4}$; there is no interior oval, but only a sinuous oval as above; which, as λ increases, approaches continually nearer to the four sides of the square. For the critical value $\lambda = m$, there is no change in the general form, but the curve has for this value of λ, two conjugate points, one on each axis at infinity.

$\lambda = \frac{1}{4}(m+1)^2$, the curve becomes the four lines.

$\lambda > \frac{1}{4}(m+1)^2$, the curve lies wholly in the four regions a and the four regions e, consisting thereof of four detached sinuous ovals. As λ deviates less from the value $\frac{1}{4}(m+1)^2$, each oval approaches more nearly to the infinite trilateral formed by the side and infinite line-portions which bound the regions d, e to which the oval belongs. And as λ departs from the limit $\frac{1}{4}(m+1)^2$, and approaches to ∞, each sinuous oval approaches more nearly to the circular arc which separates the two regions d, e, which contains the sinuous oval.

Finally, $\lambda = 0$, the curve is the circle twice repeated.

390.

THEOREM RELATING TO THE FOUR CONICS WHICH TOUCH THE SAME TWO LINES AND PASS THROUGH THE SAME FOUR POINTS.

[From the *Quarterly Journal of Pure and Applied Mathematics*, vol. VIII. (1867), pp. 162—167.]

THE sides of the triangle formed by the given points meet one of the given lines in three points, say P, Q, R; and on this same line we have four points of contact, say A_1, A_2, A_3, A_4; any two pairs, say A_1, A_2; A_3, A_4, form with a properly selected pair, say Q, R, out of the above-mentioned three points, an involution; and we have thus the three involutions

$$(A_1,\ A_2;\ \ A_3,\ A_4;\ \ Q,\ R),$$
$$(A_1,\ A_3;\ \ A_4,\ A_2;\ \ R,\ P),$$
$$(A_1,\ A_4;\ \ A_2,\ A_3;\ \ P,\ Q).$$

To prove this, let $x=0$, $y=0$ be the equations of the given lines, and take for the equations of the sides of the triangle formed by the given points

$$b\,x+a\,y-a\,b\ =0,$$
$$b'\,x+a'\,y-a'\,b'\ =0,$$
$$b''x+a''y-a''b''=0:$$

the equation of any one of the four conics may be written

$$\frac{Lab}{bx+ay-ab}+\frac{L'a'b'}{b'x+a'y-a'b'}+\frac{L''a''b''}{b''x+a''y-a''b''}=0,$$

and if this touches the axis of x, say at the point $x=\alpha$, then we must have

$$\frac{La}{x-a}+\frac{L'a'}{x-a'}+\frac{L''a''}{x-a''}=\frac{-K(x-\alpha)^2}{(x-a)(x-a')(x-a'')};$$

or, assuming as we may do, $K=-(a'-a'')(a''-a)(a-a')$, this gives

$$L\,a=(a-\alpha)^2(a'-a''),$$
$$L'a'=(a'-\alpha)^2(a''-a),$$
$$L''a''=(a''-\alpha)^2(a-a').$$

But in the same manner, if the conic touch the axis of y, say at the point $y=\beta$, we have

$$L\,b=(b-\beta)^2(b'-b''),$$
$$L'b'=(b'-\beta)^2(b''-b),$$
$$L''b''=(b''-\beta)^2(b-b');$$

and thence

$$b(a-\alpha)^2(a'-a''):b'(a'-\alpha)^2(a''-a):b''(a''-\alpha)^2(a-a')$$
$$=a(b-\beta)^2(b'-b''):a'(b'-\beta)^2(b''-b):a''(b''-\beta)^2(b-b').$$

Putting

$$P=a\,b\,(a'-a'')(b'-b''),$$
$$P'=a'b'(a''-a)(b''-b),$$
$$P''=a''b''(a-a')(b-b'),$$

we have

$$(a-\alpha)^2\frac{P}{a^2}:(a'-\alpha)^2\frac{P'}{a'^2}:(a''-\alpha)^2\frac{P''}{a''^2}=(b-\beta)^2(b'-b'')^2:(b'-\beta)^2(b''-b)^2:(b''-\beta)^2(b-b')^2;$$

and thence

$$(a-\alpha)\frac{\sqrt{(P)}}{a}:(a'-\alpha)\frac{\sqrt{(P')}}{a'}:(a''-\alpha)\frac{\sqrt{(P'')}}{a''}$$
$$=(b-\beta)(b'-b''):(b'-\beta)(b''-b):(b''-\beta)(b-b'),$$

which gives

$$(a-\alpha)\frac{\sqrt{(P)}}{a}+(a'-\alpha)\frac{\sqrt{(P')}}{a'}+(a''-\alpha)\frac{\sqrt{(P'')}}{a''}=0,$$

and we have in like manner

$$(b-\beta)\frac{\sqrt{(P)}}{b}+(b'-\beta)\frac{\sqrt{(P')}}{b'}+(b''-\beta)\frac{\sqrt{(P'')}}{b''}=0,$$

but the first of these equations is alone required for the present purpose. Putting for shortness

$$P=a^2X,\qquad P'=a'^2X',\qquad P''=a''^2X'',$$

the equation is

$$(a-\alpha)\sqrt{(X)}+(a'-\alpha)\sqrt{(X')}+(a''-\alpha)\sqrt{(X'')},$$

and by attributing the signs + and − to the radicals, we have, corresponding to the four conics, the equations

$$\begin{aligned} (a-\alpha_1)\sqrt{(X)}+(a'-\alpha_1)\sqrt{(X')}+(a''-\alpha_1)\sqrt{(X'')}&=0,\\ -(a-\alpha_2)\sqrt{(X)}+(a'-\alpha_2)\sqrt{(X')}+(a''-\alpha_2)\sqrt{(X'')}&=0,\\ (a-\alpha_3)\sqrt{(X)}-(a'-\alpha_3)\sqrt{(X')}+(a''-\alpha_3)\sqrt{(X'')}&=0,\\ (a-\alpha_4)\sqrt{(X)}+(a'-\alpha_4)\sqrt{(X')}-(a''-\alpha_4)\sqrt{(X'')}&=0,\end{aligned}$$

where $\alpha_1, \alpha_2, \alpha_3, \alpha_4$ are the values of α for the four conics respectively.

Eliminating a'' we obtain the system of three equations

$$\begin{aligned} (2a-\alpha_1-\alpha_2)\sqrt{(X)}+(\alpha_2-\alpha_1)\sqrt{(X')}+(\alpha_2-\alpha_1)\sqrt{(X'')}&=0,\\ (\alpha_3-\alpha_1)\sqrt{(X)}+(2a'-\alpha_1-\alpha_3)\sqrt{(X')}+(\alpha_3-\alpha_1)\sqrt{(X'')}&=0,\\ (\alpha_1+\alpha_2-\alpha_3-\alpha_4)\sqrt{(X)}+(\alpha_1+\alpha_3-\alpha_2-\alpha_4)\sqrt{(X')}+(\alpha_1+\alpha_4-\alpha_2-\alpha_3)\sqrt{(X'')}&=0,\end{aligned}$$

and then eliminating the radicals we have

$$\begin{vmatrix} 2a-\alpha_1-\alpha_2, & \alpha_2-\alpha_1, & \alpha_2-\alpha_1 \\ \alpha_3-\alpha_1, & 2a'-\alpha_1-\alpha_3, & \alpha_3-\alpha_1 \\ \alpha_1+\alpha_2-\alpha_3-\alpha_4, & \alpha_1+\alpha_3-\alpha_2-\alpha_4, & \alpha_1+\alpha_4-\alpha_2-\alpha_3 \end{vmatrix}=0,$$

which is in fact

$$-4\begin{vmatrix} 1, & a+a', & aa' \\ 1, & \alpha_1+\alpha_4, & \alpha_1\alpha_4 \\ 1, & \alpha_2+\alpha_3, & \alpha_2\alpha_3 \end{vmatrix}=0,$$

as may be verified by actual expansion; the transformation of the determinant is a peculiar one.

The foregoing result was originally obtained as follows, viz. writing for a moment

$$\begin{aligned} a\sqrt{(X)}+a'\sqrt{(X')}+a''\sqrt{(X'')}&=\Theta,\\ \sqrt{(X)}+\sqrt{(X')}+\sqrt{(X'')}&=\Phi,\end{aligned}$$

the four equations are

$$\begin{aligned} \Theta-\alpha_1\Phi&=0,\\ \Theta-\alpha_2\Phi&=2(a-\alpha_2)\sqrt{(X)},\\ \Theta-\alpha_3\Phi&=2(a'-\alpha_3)\sqrt{(X')},\\ \Theta-\alpha_4\Phi&=2(a''-\alpha_4)\sqrt{(X'')};\end{aligned}$$

these give

$$\begin{aligned} (\alpha_1-\alpha_2)\Phi&=2(a-\alpha_2)\sqrt{(X)},\\ (\alpha_1-\alpha_3)\Phi&=2(a'-\alpha_3)\sqrt{(X')},\\ (\alpha_1-\alpha_4)\Phi&=2(a''-\alpha_4)\sqrt{(X'')}.\end{aligned}$$

From the last equation we have

$$(\alpha_1-\alpha_4)\Phi = 2\{\Theta - a\sqrt{(X)} - a'\sqrt{(X')}\} - 2\alpha_4\{\Phi - \sqrt{(X)} - \sqrt{(X')}\}$$
$$= 2(\alpha_1-\alpha_4)\Phi - 2(a-\alpha_4)\sqrt{(X)} - 2(a'-\alpha_4)\sqrt{(X')};$$

that is

$$(\alpha_1-\alpha_4)\Phi - 2(a-\alpha_4)\sqrt{(X)} - 2(a'-\alpha_4)\sqrt{(X')} = 0;$$

or substituting for $\sqrt{(X)}$, $\sqrt{(X')}$ their values in terms of Φ, we find

$$\alpha_1-\alpha_4 - \frac{(a-\alpha_4)(\alpha_1-\alpha_2)}{a-\alpha_2} - \frac{(a'-\alpha_4)(\alpha_1-\alpha_3)}{a'-\alpha_3} = 0,$$

which may be written

$$\alpha_1-\alpha_4-(\alpha_1-\alpha_2)\left(1+\frac{\alpha_2-\alpha_4}{a-\alpha_2}\right)-(\alpha_1-\alpha_3)\left(1+\frac{\alpha_3-\alpha_4}{a'-\alpha_3}\right)=0,$$

that is

$$\alpha_2+\alpha_3-\alpha_1-\alpha_4+\frac{(\alpha_2-\alpha_1)(\alpha_2-\alpha_4)}{a-\alpha_2}+\frac{(\alpha_3-\alpha_1)(\alpha_3-\alpha_4)}{a'-\alpha_3}=0;$$

or again

$$(\alpha_2-\alpha_1)\left(1+\frac{\alpha_2-\alpha_4}{a-\alpha_2}\right)+(\alpha_3-\alpha_4)\left(1+\frac{\alpha_3-\alpha_1}{a'-\alpha_3}\right)=0,$$

that is

$$(\alpha_2-\alpha_1)\frac{a-\alpha_4}{a-\alpha_2}+(\alpha_3-\alpha_4)\frac{a'-\alpha_1}{a'-\alpha_3}=0;$$

or finally

$$(\alpha_2-\alpha_1)(a-\alpha_4)(a'-\alpha_3)+(\alpha_3-\alpha_4)(a-\alpha_2)(a'-\alpha_1)=0,$$

which is a known form of the relation

$$\begin{vmatrix} 1, & a+a', & aa' \\ 1, & \alpha_1+\alpha_4, & \alpha_1\alpha_4 \\ 1, & \alpha_2+\alpha_3, & \alpha_2\alpha_3 \end{vmatrix} = 0,$$

which gives the involution of the quantities a, a'; α_1, α_4; α_2, α.

We have in like manner

$$\begin{vmatrix} 1, & a'+a'', & a'a'' \\ 1, & \alpha_1+\alpha_2, & \alpha_1\alpha_2 \\ 1, & \alpha_3+\alpha_4, & \alpha_3\alpha_4 \end{vmatrix} = 0,$$

and

$$\begin{vmatrix} 1, & a''+a, & a''a \\ 1, & \alpha_1+\alpha_3, & \alpha_1\alpha_3 \\ 1, & \alpha_2+\alpha_4, & \alpha_2\alpha_4 \end{vmatrix} = 0,$$

which give the involutions of the systems a', a''; α_1, α_2; α_3, α_4 and a'', a; α_1, α_3; α_2, α_4 respectively.

It may be remarked that the equation of the conic passing through the three points and touching the axis of x in the point $x = \alpha$ is

$$\frac{(a-\alpha)^2(a'-a'')\,b}{bx+ay-ab}+\frac{(a'-\alpha)^2(a''-a)\,b'}{b'x+a'y-a'b'}+\frac{(a''-\alpha)^2(a-a')\,b''}{b''x+a''y-a''b''}=0,$$

and when this meets the axis of y we have

$$\frac{\frac{b}{a}(a-\alpha)^2(a'-a'')}{y-b}+\frac{\frac{b'}{a'}(a'-\alpha)^2(a''-a)}{y-b'}+\frac{\frac{b''}{a''}(a''-\alpha)^2(a-a')}{y-b''}=0.$$

Hence, if this touches the axis of y in the point $y = \beta$, the left-hand side must be

$$=\frac{\left[\frac{b}{a}(a-\alpha)^2(a'-a'')+\frac{b'}{a'}(a'-\alpha)^2(a''-a)+\frac{b''}{a''}(a''-\alpha)^2(a-a')\right](y-\beta)^2}{(y-b)(y-b')(y-b'')},$$

and equating the coefficients of $\frac{1}{y^2}$, we have

$$\frac{b^2}{a}(a-\alpha)^2(a'-a'')+\frac{b'^2}{a'}(a'-\alpha)^2(a''-a)+\frac{b''^2}{a''}(a''-\alpha)^2(a-a')$$

$$=\left[\frac{b}{a}(a-\alpha)^2(a'-a'')+\frac{b'}{a'}(a'-\alpha)^2(a''-a)+\frac{b''}{a''}(a''-\alpha)^2(a-a')\right](b+b'+b''-2\beta),$$

or what is the same thing,

$$\frac{b(b'+b'')}{a}(a-\alpha)^2(a'-a'')+\frac{b'(b''+b)}{a'}(a'-\alpha)^2(a''-a)+\frac{b''(b+b')}{a''}(a''-\alpha)^2(a-a')$$

$$=2\beta\left[\frac{b}{a}(a-\alpha)^2(a'-a'')+\frac{b'}{a'}(a'-\alpha)^2(a''-a)+\frac{b''}{a''}(a''-\alpha)^2(a-a')\right],$$

which gives β in terms of α, that is $\beta_1, \beta_2, \beta_3, \beta_4$ in terms of $\alpha_1, \alpha_2, \alpha_3, \alpha_4$ respectively.

Cambridge, 30 *November*, 1863.

391.

SOLUTION OF A PROBLEM OF ELIMINATION.

[From the *Quarterly Journal of Pure and Applied Mathematics*, vol. VIII. (1867), pp. 183—185.]

It is required to eliminate x, y from the equations

$$\left\|\begin{array}{ccccc} x^4, & x^3y, & x^2y^2, & xy^3, & y^4 \\ a, & b, & c, & d, & e \\ a', & b', & c', & d', & e' \\ a'', & b'', & c'', & d'', & e'' \end{array}\right\| = 0.$$

This system may be written

$$\begin{aligned} x^4 &= \Sigma\lambda a, \\ x^3y &= \Sigma\lambda b, \\ x^2y^2 &= \Sigma\lambda c, \\ x\,y^3 &= \Sigma\lambda d, \\ y^4 &= \Sigma\lambda e; \end{aligned}$$

if for shortness

$$\Sigma\lambda a = \lambda a + \lambda' a' + \lambda'' a'', \text{ \&c.}$$

Or putting

$$\frac{x}{y} = -k,$$

we have

$$\begin{aligned} \Sigma\lambda\,(a + kb) &= 0, \\ \Sigma\lambda\,(b + kc) &= 0, \\ \Sigma\lambda\,(c + kd) &= 0, \\ \Sigma\lambda\,(d + ke) &= 0; \end{aligned}$$

or, what is the same thing,

$$\lambda(a+kb)+\lambda'(a'+kb')+\lambda''(a''+kb'')=0,$$
$$\lambda(b+kc)+\lambda'(b'+kc')+\lambda''(b''+kc'')=0,$$
$$\lambda(c+kd)+\lambda'(c'+kd')+\lambda''(c''+kd'')=0,$$
$$\lambda(d+ke)+\lambda'(d'+ke')+\lambda''(d''+ke'')=0;$$

and representing the columns

$$\begin{matrix} a & b, & a' & b', & a'' & b'', \\ b & c, & b' & c', & b'' & c'', \\ c & d, & c' & d', & c'' & d'', \\ d & e, & d' & e', & d'' & e'', \end{matrix}$$

by

$$1,\ 2,\quad 3,\ 4,\quad 5,\ 6,$$

each equation is of the type

$$\lambda(1+k2)+\lambda'(3+k4)+\lambda''(5+k6)=0.$$

Multiplying the several equations by the minors of 135, each with its proper sign, and adding, the terms independent of k disappear, the equation divides by k, and we find

$$\lambda\, 2135+\lambda'\, 4135+\lambda''\, 6135=0;$$

operating in a similar manner with the minors of 246, the terms in k disappear, and we find

$$\lambda\, 1246+\lambda'\, 3246+\lambda''\, 5246=0;$$

again, operating with the minors of $(146+236+245+k246)$, we find

$$\begin{aligned} &\lambda\ \{1236+1245+k(2146+1246)\} \\ +&\lambda'\ \{3146+3245+k(4236+3246)\} \\ +&\lambda''\ \{5146+5236+k(6245+5246)\}=0, \end{aligned}$$

where the terms in k disappear, and this is

$$\lambda(1236+1245)+\lambda'(3146+3245)+\lambda''(5146+5236)=0.$$

We have thus three linear equations, which written in a slightly different form are

$$\begin{aligned} &\lambda\ 1235 &&+\lambda'\ 3451 &&+\lambda''\ 5613 &&=0, \\ &\lambda(1236+1245) &&+\lambda'(3452+3461) &&+\lambda''(5614+5623) &&=0, \\ &\lambda\ 1246 &&+\lambda'\ 3462 &&+\lambda''\ 5624 &&=0, \end{aligned}$$

and thence eliminating λ, λ', λ'', we have

$$\begin{vmatrix} 1235, & 1236+1245, & 1246 \\ 3451, & 3452+3461, & 3462 \\ 5613, & 5614+5623, & 5624 \end{vmatrix}=0,$$

which is the required result. It may be remarked that the second and third column are obtained from the first by operating on it with Δ, $\frac{1}{2}\Delta^2$, if $\Delta = 2\delta_1 + 4\delta_3 + 6\delta_5$. Or say the result is

$$(1,\ \Delta,\ \tfrac{1}{2}\Delta^2)\left|\begin{matrix} 1235 \\ 3451 \\ 5613 \end{matrix}\right| = 0.$$

In like manner for the system

$$\left\|\begin{matrix} x^5, & x^4y, & x^3y^2, & x^2y^3, & xy^4, & y^5 \\ a, & b, & c, & d, & e, & f \\ a', & b', & c', & d', & e', & f' \\ a'', & b'', & c'', & d'', & e'', & f'' \\ a''', & b''', & c''', & d''', & e''', & f''' \end{matrix}\right\| = 0,$$

if the columns are

$$\begin{matrix} a\ b, & a'\ b', & a''\ b'', & a'''\ b''', \\ b\ c, & b'\ c', & b''\ c'', & b'''\ c''', \\ c\ d, & c'\ d', & c''\ d'', & c'''\ d''', \\ d\ e, & d'\ e', & d''\ e'', & d'''\ e''', \\ e\ f, & e'\ f', & e''\ f'', & e'''\ f''', \\ =1,\ 2, & 3,\ 4, & 5,\ 6, & 7,\ 8; \end{matrix}$$

then the result is

$$(1,\ \Delta,\ \tfrac{1}{2}\Delta^2,\ \tfrac{1}{6}\Delta^3)\left|\begin{matrix} 12357 \\ 34571 \\ 56713 \\ 78135 \end{matrix}\right| = 0,$$

where

$$\Delta = 2\delta_1 + 4\delta_3 + 6\delta_5 + 8\lambda_7.$$

392.

ON THE CONICS WHICH PASS THROUGH TWO GIVEN POINTS AND TOUCH TWO GIVEN LINES.

[From the *Quarterly Journal of Pure and Applied Mathematics*, vol. VIII. (1867), pp. 211—219.]

LET $x=0$, $y=0$ be the equations of the given lines; $z=0$ the equation of the line joining the given points. We may, to fix the ideas, imagine the implicit constants so determined that $x+y+z=0$ shall be the equation of the line infinity.

Take $x-my=0$, $x-ny=0$ as the equations of the lines which by their intersection with $z=0$ determine the given points. The equation of the conic is

$$\{\sqrt{(m)}+\sqrt{(n)}\}\sqrt{(xy)}=x+y\sqrt{(mn)}+\gamma z,$$

or, what is the same thing,

$$(x-my)(x-ny)+2\{x+y\sqrt{(mn)}\}\gamma z+\gamma^2 z^2=0,$$

so that there are two distinct series of conics according as $\sqrt{(mn)}$ is taken with the positive or the negative sign.

The equation of the chord of contact is

$$x+y\sqrt{(mn)}+\gamma z=0,$$

which meets $z=0$ in the point $\{x+y\sqrt{(mn)}=0,\ z=0\}$ that is in one of the centres of the involution formed by the lines $(x=0,\ y=0)$, $(x-my=0,\ x-ny=0)$. It is to be observed that the conic is only real when mn is positive, that is (the lines and points being each real) the two points must be situate in the same region or in opposite regions of the four regions formed by the two lines: there are however other real cases; e.g. if the lines $x=0$, $y=0$ are real, but the quantities m, n are conjugate imaginaries; included in this we have the circles which touch two real lines.

To fix the ideas I take m and n each positive and $mn > 1$; also I attend first to the series where $\sqrt{(mn)}$ is taken positively. At the points where the conic meets infinity, we have

$$\{\sqrt{(m)} + \sqrt{(n)}\} \sqrt{(xy)} = x + y\sqrt{(mn)} - \gamma(x+y),$$

which gives two coincident points, that is the conic is a parabola, if

$$(1-\gamma)\{\sqrt{(mn)} - \gamma\} = \tfrac{1}{4}\{\sqrt{(m)} + \sqrt{(n)}\}^2,$$

that is

$$\gamma^2 - \gamma\{1 + \sqrt{(mn)}\} = \tfrac{1}{4}\{\sqrt{(m)} - \sqrt{(n)}\}^2,$$

or

$$\gamma = \tfrac{1}{2}[1 + \sqrt{(mn)} \pm \sqrt{\{(1+m)(1+n)\}}],$$

where it is to be noticed that

$$\gamma = \tfrac{1}{2}[1 + \sqrt{(mn)} + \sqrt{\{(1+m)(1+n)\}}]$$

is a positive quantity greater than $\sqrt{(mn)}$, say $\gamma = p$,

$$\gamma = \tfrac{1}{2}[1 + \sqrt{(mn)} - \sqrt{\{(1+m)(1+n)\}}]$$

is a negative quantity, say $\gamma = -q$, q being positive.

The order of the lines is as shown in fig. 1, see plate facing p. 52.

$\gamma = -\infty$ to $\gamma = -q$, curve is ellipse; $\gamma = -q$, parabola P_2,
$\gamma = -q$ to p, curve is hyperbola; $\gamma = p$, parabola P_1,
$\gamma = p$ to $\gamma = \infty$, ellipse.

Resuming the equation

$$(x - my)(x - ny) + 2\{x + y\sqrt{(mn)}\}\gamma z + \gamma^2 z^2 = 0,$$

the coefficients are

$$(a, b, c, f, g, h) = \{1, mn, \gamma^2, \gamma\sqrt{(mn)}, \gamma, -\tfrac{1}{2}(m+n)\},$$

and thence the inverse coefficients are

$$(A, B, C, F, G, H) =$$
$$[0, 0, -\tfrac{1}{4}(m-n)^2, -\tfrac{1}{2}\gamma\{\sqrt{(m)} + \sqrt{(n)}\}^2, -\tfrac{1}{2}\gamma\sqrt{(mn)}\{\sqrt{(m)} + \sqrt{(n)}\}^2, \tfrac{1}{2}\gamma^2\{\sqrt{(m)} + \sqrt{(n)}\}^2],$$
$$K = -\tfrac{1}{4}\gamma^2\{\sqrt{(m)} + \sqrt{(n)}\}^4,$$

or, omitting a factor, the inverse coefficients are

$$(A, B, C, F, G, H) = \left[0, 0, \frac{1}{2\gamma}\{\sqrt{(m)} - \sqrt{(n)}\}^2, 1, \sqrt{(mn)}, -\gamma\right].$$

Considering the line

$$\lambda x + \mu y + \nu z = 0,$$

the coordinates of the pole of this line are

$$\begin{aligned} x : y : z = &\quad -\gamma\mu + \sqrt{(mn)}\,\nu \\ : &\quad -\gamma\lambda + \nu \\ : &\quad \sqrt{(mn)}\,\lambda + \mu + \frac{1}{2\gamma}\{\sqrt{(m)} - \sqrt{(n)}\}^2\nu, \end{aligned}$$

or (what is the same thing) introducing the arbitrary coefficient k, we have

$$\begin{aligned} &kx+\gamma\mu-\nu\sqrt{(mn)}=0,\\ &ky+\gamma\lambda-\nu \qquad\quad =0,\\ &kz-\lambda\sqrt{(mn)}-\mu-\frac{1}{2\gamma}\{\sqrt{(m)}-\sqrt{(n)}\}^2\nu=0\,; \end{aligned}$$

the first two equations give

$$k:\gamma:-1=\nu\{\mu-\lambda\sqrt{(mn)}\}:\nu\{y\sqrt{(mn)}-x\}:\lambda x-\mu y,$$

that is

$$k=\frac{-\nu\{\mu-\lambda\sqrt{(mn)}\}}{\lambda x-\mu y},\qquad \gamma=\frac{-\nu\{y\sqrt{(mn)}-x\}}{\lambda x-\mu y},$$

or, substituting this value of γ in the third equation,

$$\frac{\nu\{\mu-\lambda\sqrt{(mn)}\}\,z}{\lambda x-\mu y}+\{\mu+\lambda\sqrt{(mn)}\}+\frac{\lambda x-\mu y}{x-y\sqrt{(mn)}}\,\frac{\{\sqrt{(m)}-\sqrt{(n)}\}^2}{2}=0,$$

that is

$$(\lambda x-\mu y)^2.\tfrac{1}{2}\{\sqrt{(m)}-\sqrt{(n)}\}^2+\{x-y\sqrt{(mn)}\}(\lambda x-\mu y)\{\mu+\lambda\sqrt{(mn)}\}$$
$$+z\{x-y\sqrt{(mn)}\}\,\nu\{\mu-\lambda\sqrt{(mn)}\}=0,$$

which is the equation of the curve, the locus of the pole of the line $\lambda x+\mu y+\nu z=0$ in regard to the conic

$$(x-my)(x-ny)+2\{x+y\sqrt{(mn)}\}\gamma z+\gamma^2z^2=0.$$

In particular, if $\lambda=\mu=\nu=1$, then for the coordinates of the centre of the conic, we have

$$x:y:z=-\gamma+\sqrt{(mn)}:-\gamma+1:\sqrt{(mn)}+1+\frac{1}{2\gamma}\{\sqrt{(m)}-\sqrt{(n)}\}^2;$$

and for the locus of the centre,

$$(x-y)^2.\tfrac{1}{2}\{\sqrt{(m)}-\sqrt{(n)}\}^2+(x-y)\{x-y\sqrt{(mn)}\}\{1+\sqrt{(mn)}\}+z\{x-y\sqrt{(mn)}\}\{1-\sqrt{(mn)}\}=0,$$

so that the locus is a conic, and it is obvious that this conic is a hyperbola. Putting for greater simplicity

$$\begin{aligned} &x-y \qquad\quad =X,\\ &x-y\sqrt{(mn)}=Y,\\ &z \qquad\qquad\quad =Z, \end{aligned}$$

the equation of the curve of centres is

$$X^2.\tfrac{1}{2}\{\sqrt{(m)}-\sqrt{(n)}\}^2+XY\{1+\sqrt{(mn)}\}+YZ\{1-\sqrt{(mn)}\}=0,$$

or, writing this under the form

$$Y[X\{1+\sqrt{(mn)}\}+Z\{1-\sqrt{(mn)}\}]+\tfrac{1}{2}\{\sqrt{(m)}-\sqrt{(n)}\}^2X^2=0,$$

the equation is

$$YQ + X^2 = 0,$$

where

$$X = x - y,$$

$$Y = x - y\sqrt{(mn)},$$

$$Q = \frac{2}{\{\sqrt{(m)} - \sqrt{(n)}\}^2}\,[\{1 + \sqrt{(mn)}\}(x - y) + \{1 - \sqrt{(mn)}\}\,z]:$$

these values give

$$x - y = X,$$

$$x - y\sqrt{(mn)} = Y,$$

$$\{1 - \sqrt{(mn)}\}\,z = \{\sqrt{(m)} - \sqrt{(n)}\}^2\,Q + 2\,\{1 + \sqrt{(mn)}\}\,X,$$

or, what is the same thing,

$$\{1 - \sqrt{(mn)}\}\,x = -\sqrt{(mn)}\;X + Y,$$

$$\{1 - \sqrt{(mn)}\}\,y = -X + Y,$$

$$\{1 - \sqrt{(mn)}\}\,z = 2\,\{1 + \sqrt{(mn)}\}\,X + \{\sqrt{(m)} - \sqrt{(n)}\}^2 Q,$$

whence also

$$\{1 - \sqrt{(mn)}\}\,(x + y + z) = \{1 + \sqrt{(mn)}\}\,X + 2Y + \{\sqrt{(m)} - \sqrt{(n)}\}^2\,Q,$$

or the equation of the line infinity is

$$\{1 + \sqrt{(mn)}\}\,X + 2Y + \{\sqrt{(m)} - \sqrt{(n)}\}^2\,Q = 0,$$

a formula which may be applied to finding the asymptotes and thence the centre of the conic

$$YQ + X^2 = 0.$$

In fact we have identically

$$\{2kx + 2ky - (2k+1)\,z\}^2 - (1 + 4k)\,(2kx - z)^2 = 4k^2\,(x + y + z)^2 - 4k\,(1 + 4k)\,(kx^2 + yz),$$

that is

$$-\,4k\,(1 + 4k)\,(kx^2 + yz) = \{2kx + 2ky - (2k+1)\,z\}^2 - (1 + 4k)\,(2kx - z)^2 - 4k^2\,(x + y + z)^2,$$

which, if $x + y + z = 0$ is the equation of the line infinity, puts in evidence the asymptotes of the conic $kx^2 + yz = 0$. Hence writing αx, βy, γz in the place of x, y, z respectively, and $\dfrac{k\alpha^2}{\beta\gamma} = k'$, that is, $k = \dfrac{\beta\gamma}{\alpha^2}\,k'$, we have

$$-\,\frac{4\beta\gamma}{\alpha^2}\,k'\left(1 + \frac{4\beta\gamma}{\alpha^2}\,k'\right)\beta\gamma\,(k'x^2 + yz) = \left\{\frac{2\beta\gamma}{\alpha}\,k'x + \frac{2\beta^2\gamma}{\alpha^2}\,k'y - \left(2\,\frac{\beta\gamma}{\alpha^2}\,k' + 1\right)\gamma z\right\}^2$$

$$-\left(1 + \frac{4\beta\gamma}{\alpha^2}\,k'\right)\left(\frac{2\beta\gamma}{\alpha}\,k'x - \gamma z\right)^2 - 4\,\frac{\beta^2\gamma^2}{\alpha^4}\,k'^2\,(\alpha x + \beta y + \gamma z)^2,$$

that is

$$-4\beta^2\gamma^2k'(\alpha^2+4\beta\gamma k')(k'x^2+yz)=\{2\alpha\beta\gamma k'x+2\beta^2\gamma k'y-(2\beta\gamma k'+\alpha^2)\gamma z\}^2$$
$$-(\alpha^2+4\beta\gamma k')(2\beta\gamma k'x-\alpha\gamma z)^2-4\beta^2\gamma^2k'^2(\alpha x+\beta y+\gamma z)^2,$$

or, what is the same thing,

$$-4\beta^2k'(\alpha^2+4\beta\gamma k')(k'x^2+yz)=\{2\alpha\beta k'x+2\beta^2k'y-(2\beta\gamma k'+\alpha^2)z\}^2$$
$$-(\alpha^2+4\beta\gamma k')(2\beta k'x-\alpha z)^2-4\beta^2k'^2(\alpha x+\beta y+\gamma z)^2,$$

which, when $\alpha x+\beta y+\gamma z=0$ is the equation of the line infinity, puts in evidence the asymptotes of the conic $k'x^2+yz=0$.

Now writing X, Y, Q in the place of x, y, z; $k'=1$, and $\alpha=\{1+\sqrt{(mn)}\}$, $\beta=2$, $\gamma=\{\sqrt{(m)}-\sqrt{(n)}\}^2$, we have

$$-16\,[\{1+\sqrt{(mn)}\}^2+8\,\{\sqrt{(m)}-\sqrt{(n)}\}^2]\,(YQ+X^2)$$
$$=\quad[4\,\{1+\sqrt{(mn)}\}\,X+8Y-(4\,\{\sqrt{(m)}-\sqrt{(n)}\}^2+\{1+\sqrt{(mn)}\}^2)\,Q]^2$$
$$-[\{1+\sqrt{(mn)}\}^2+8\,\{\sqrt{(m)}-\sqrt{(n)}\}^2]\,[4X-\{1+\sqrt{(mn)}\}\,Q]^2$$
$$-16\,[\{1+\sqrt{(mn)}\}\,X+2Y+\{\sqrt{(m)}-\sqrt{(n)}\}^2\,Q]^2,$$

and the asymptotes are

$$4\,\{1+\sqrt{(mn)}\}\,X+8Y-[4\,\{\sqrt{(m)}-\sqrt{(n)}\}^2+\{1+\sqrt{(mn)}\}^2]\,Q$$
$$=\pm\sqrt{\{1+\sqrt{(mn)}\}^2+8\,\{\sqrt{(m)}-\sqrt{(n)}\}^2}\,[4X-\{1+\sqrt{(mn)}\}\,Q].$$

At the centre

$$4\,\{1+\sqrt{(mn)}\}\,X+8Y-[4\,\{\sqrt{(m)}-\sqrt{(n)}\}^2+\{1+\sqrt{(mn)}\}^2]\,Q=0,$$
$$4X-\{1+\sqrt{(mn)}\}\,Q=0,$$

but the first equation is

$$\{1+\sqrt{(mn)}\}\,[4X-Q\,\{1+\sqrt{(mn)}\}]+8Y-4\,\{\sqrt{(m)}-\sqrt{(n)}\}^2\,Q=0,$$

so that we have

$$4X=\{1+\sqrt{(mn)}\}\,Q,\quad 2Y=\{\sqrt{(m)}-\sqrt{(n)}\}^2\,Q,$$

the first of these is

$$2\,\{\sqrt{(m)}-\sqrt{(n)}\}^2\,(x-y)-\{1+\sqrt{(mn)}\}^2\,(x-y)-(1-mn)\,z=0,$$

and the two together give

$$2X\,\{\sqrt{(m)}-\sqrt{(n)}\}^2-\{1+\sqrt{(mn)}\}\,Y=0,$$

so that we have

$$2\,\{\sqrt{(m)}-\sqrt{(n)}\}^2\,(x-y)-\{1+\sqrt{(mn)}\}\,\{x-y\sqrt{(mn)}\}=0,$$
$$[\{1+\sqrt{(mn)}\}^2-2\,\{\sqrt{(m)}-\sqrt{(n)}\}^2]\,(x-y)+(1-mn)\,z=0,$$

to determine the coordinates of the centre.

The equation of the chord of contact is

$$x + y\sqrt{(mn)} + \gamma z = 0,$$

which for $\gamma = 1$ is parallel to $y = 0$ and for $\gamma = \sqrt{(mn)}$ is parallel to $x = 0$. But the coordinates of the centre are

$$x : y : z = -\gamma + \sqrt{(mn)} : -\gamma + 1 : \sqrt{(mn)} + 1 + \frac{1}{2\gamma}\{\sqrt{(m)} - \sqrt{(n)}\}^2,$$

which for $\gamma = 1$ give

$$y = 0,\ x : z = -1 + \sqrt{(mn)} : \sqrt{(mn)} + 1 + \tfrac{1}{2}\{\sqrt{(m)} - \sqrt{(n)}\}^2, = -2 + 2\sqrt{(mn)} : 2 + m + n,$$

and for $\gamma = \sqrt{(mn)}$ give

$$x = 0,$$

$$y : z = 1 - \sqrt{(mn)} : \sqrt{(mn)} + 1 + \frac{1}{2\sqrt{(mn)}}\{\sqrt{(m)} - \sqrt{(n)}\}^2, \ = 2 - 2\sqrt{(mn)} : 2\sqrt{(mn)} + \frac{m+n}{\sqrt{(mn)}}.$$

The line drawn from the fixed point on the chord of contact to the centre has for its equation

$$x + y\sqrt{(mn)} + \gamma' z = 0,$$

where, writing for x, y, z the coordinates of the centre, we have

$$-\gamma\{1 + \sqrt{(mn)}\} + 2\sqrt{(mn)} + \gamma'\left[\sqrt{(mn)} + 1 + \frac{1}{2\gamma}\{\sqrt{(m)} - \sqrt{(n)}\}^2\right] = 0;$$

that is

$$\gamma' = \frac{\gamma\{1 + \sqrt{(mn)}\} - 2\sqrt{(mn)}}{1 + \sqrt{(mn)} + \frac{1}{2\gamma}\{\sqrt{(m)} - \sqrt{(n)}\}^2},$$

or, what is the same thing,

$$\gamma' - \gamma = \frac{-\gamma\{\sqrt{(m)} + \sqrt{(n)}\}^2}{\{\sqrt{(m)} - \sqrt{(n)}\}^2 + 2\gamma\{1 + \sqrt{(mn)}\}},$$

and consequently $\gamma' = \gamma$ only for $\gamma = 0$.

It is now easy to trace the corresponding positions of the chord of contact through the fixed point $\{x + y\sqrt{(mn)} = 0,\ z = 0\}$, and of the centre on the hyperbola which is the curve of centres: see fig. 2 in the plate facing p. 52.

The lines OP_2, OL, $O\Theta$, OP_1, OX, OG, OH are positions of the chord of contact, and the points P_2, L, Θ, P_1, X, G, H, on the hyperbola which is the curve of centres are the corresponding positions of the centre.

Chord of Contact.	*Centre.*
OP_2.	P_2, at infinity on hyperbola.
$OL\ (z=0)$.	L, $(z=0,\ x-y=0)$.
$O\Theta$.	Θ, the line joining this with O being always behind $O\Theta$.
OP_1.	P_1, at infinity on hyperbola.
$OX\ \{x+y\sqrt{(mn)}=0\}$.	$X\ (x=0,\ y=0)$.
OG (parallel to $y=0$).	G (on line $y=0$).
OH (parallel to $x=0$) and so back to	H (on line $x=0$) and so on to
OP_2.	P_2.

I have treated separately the case $\sqrt{(mn)}=1$.

Consider the conics which touch the lines $y-x=0$, $y+x=0$ and pass through the points

$$\{x=1,\ y=\sqrt{(1-c^2)}\},\ \{x=1,\ y=-\sqrt{(1-c^2)}\}.$$

The equation is of the form

$$y^2-x^2+k(x-\alpha)^2=0,$$

and to determine k, we have

$$1-c^2-1+k(1-\alpha)^2=0, \text{ and therefore } k=\frac{c^2}{(1-\alpha)^2}.$$

The equation thus becomes

$$(1-\alpha)^2(y^2-x^2)+c^2(x-\alpha)^2=0,$$

that is

$$(1-\alpha)^2y^2+\{c^2-(1-\alpha)^2\}x^2-2c^2\alpha x+c^2\alpha^2=0,$$

or as this may be written

$$(1-\alpha)^2y^2+\{c^2-(1-\alpha)^2\}\left\{x-\frac{c^2\alpha}{c^2-(1-\alpha)^2}\right\}^2-\frac{c^2\alpha^2(1-\alpha)^2}{c^2-(1-\alpha)^2}=0.$$

Hence the nature of the conic depends on the sign of $c^2-(1-\alpha)^2$, viz. if this be positive, or α between the limits $1+c$, $1-c$, the curve is an ellipse,

$$x\text{-coordinate of centre }=\frac{c^2\alpha}{c^2-(1-\alpha)^2},$$

which is positive,

$$x\text{-semi-axis }=\frac{\pm c\alpha(1-\alpha)}{c^2-(1-\alpha)^2},$$

$$y\text{-semi-axis }=\frac{c\alpha}{\sqrt{\{c^2-(1-\alpha)^2\}}}.$$

The coordinate of centre for $\alpha=1+c$ is $=+\infty$ (the curve being in this case a parabola P_1) and for $\alpha=1-c$ it is also $=+\infty$ (the curve being in this case a parabola P_2). The coordinate has a minimum value corresponding to $\alpha=\sqrt{(1-c^2)}$, viz. this is $=\frac{1}{2}\{1+\sqrt{(1-c^2)}\}$.

Hence as (α) passes from $1+c$ to $\sqrt{(1-c^2)}$, the coordinate of the centre passes from ∞ to its minimum value $\frac{1}{2}\{1+\sqrt{(1-c^2)}\}$; in the passage we have $\alpha=1$ giving the coordinate $=1$, the conic being in this case a pair of coincident lines $(x-1)^2=0$. And as (α) passes from the foregoing value $\sqrt{(1-c^2)}$ to $1-c$, the coordinate of the centre passes from the minimum value $\frac{1}{2}\{1+\sqrt{(1-c^2)}\}$ to ∞.

The curve is a hyperbola if α lies without the limits $1+c$, $1-c$,

$$x\text{-coordinate of centre} = \frac{-c^2\alpha}{(1-\alpha)^2-c^2},$$

which has the sign of $-\alpha$,

$$x\text{-semi-axis} = \frac{\pm c\alpha(1-\alpha)}{(1-\alpha)^2-c^2},$$

$$y\text{-semi-axis} = \frac{\pm c\alpha}{\sqrt{\{(1-\alpha)^2-c^2\}}},$$

$$\text{semi-aperture of asymptotes} = \tan^{-1}\sqrt{\left\{1-\frac{c^2}{(1-\alpha)^2}\right\}},$$

which for $\alpha=1\pm c$ is $=0$ (parabola), but increases as $1-\alpha$ increases positively or negatively, becoming $=45^\circ$ for $\alpha=\pm\infty$ (the asymptotes being in this case the pair of lines $y^2-x^2=0$):

$\alpha=+\infty$, coordinate of centre is $=\ \ 0$,

$\alpha=1+c$, " " $=-\infty$,

so that α diminishing from ∞ to $1+c$, the coordinate of the centre moves constantly in the same direction from 0 to $-\infty$,

$\alpha=1-c$, coordinate of centre is $=-\infty$,

$\alpha=0$, " " $=\ \ 0$,

the hyperbola being in this case the pair of lines $y^2=(1-c^2)x^2$.

α negative, the coordinate of centre becomes positive, viz. as α passes from $\alpha=0$ to $\alpha=-\sqrt{(1-c^2)}$, the coordinate of centre passes from 0 to a maximum positive value $\frac{1}{2}\{1-\sqrt{(1-c^2)}\}$, and then as α passes from $-\sqrt{(1-c^2)}$ to $-\infty$, the coordinate of centre diminishes from $\frac{1}{2}\{1-\sqrt{(1-c^2)}\}$ to 0. It is to be remarked that α being negative, the lines $y^2-x^2=0$ are touched by the branch on the negative side of the origin, that is the branch not passing through the two points $x=1$, $y=\pm\sqrt{(1-c^2)}$.

393.

ON THE CONICS WHICH TOUCH THREE GIVEN LINES AND PASS THROUGH A GIVEN POINT.

[From the *Quarterly Journal of Pure and Applied Mathematics*, vol. VIII. (1867), pp. 220—222.]

CONSIDER the triangles which touch three given lines; the three lines form a triangle, and the lines joining the angles of the triangle with the points of contact of the opposite sides respectively meet in a point S: conversely given the three lines and the point S, then joining this point with the angles of the triangle the joining lines meet the opposite sides respectively in three points which are the points of contact with the three given lines respectively of a conic; such conic is determinate and unique. Suppose now that the conic passes through a given point; the point S is no longer arbitrary, but it must lie on a certain curve; and this curve being known, then taking upon it any point whatever for the point S, and constructing as before the conic which corresponds to such point, the conic in question will pass through the given point, and will thus be a conic touching the three given lines and passing through the given point. And the series of such conics corresponds of course to the series of points on the curve.

I proceed to find the curve which is the locus of the point S.

We may take $x=0$, $y=0$, $z=0$ for the equations of the given lines, and $x : y : z = 1 : 1 : 1$ for the coordinates of the given point. The equation of a conic touching the three given lines is

$$a\sqrt{(x)} + b\sqrt{(y)} + c\sqrt{(z)} = 0,$$

and the coordinates of the corresponding point S are as $\frac{1}{a^2} : \frac{1}{b^2} : \frac{1}{c^2}$, that is, taking (x, y, z) for the coordinates of the point in question, we have

$$a : b : c = \frac{1}{\sqrt{(x)}} : \frac{1}{\sqrt{(y)}} : \frac{1}{\sqrt{(z)}},$$

the condition in order that the conic may pass through the given point is $a+b+c=0$, and we thus find for the curve, which is the locus of the point S, the equation

$$\frac{1}{\sqrt{(x)}}+\frac{1}{\sqrt{(y)}}+\frac{1}{\sqrt{(z)}}=0,$$

or, what is the same thing,

$$\sqrt{(yz)}+\sqrt{(zx)}+\sqrt{(xy)}=0,$$

the rationalised form of which is

$$y^2z^2+z^2x^2+x^2y^2-2xyz(x+y+z)=0.$$

This is a quartic curve with three cusps, viz. each angle of the triangle is a cusp; and by considering for example the cusp ($y=0$, $z=0$) and writing the equation under the form

$$x^2(y-z)^2-2x(yz^2+y^2z)+y^2z^2=0,$$

we see that the tangent at the cusp in question is the line $y-z=0$; that is, the tangents at the three cusps are the lines joining these points respectively with the given point (1, 1, 1). Each cuspidal tangent meets the curve in the cusp counting as three points and in a fourth point of intersection, the coordinates whereof in the case of the tangent $y-z=0$, are at once found to be $x:y:z=1:4:4$, or say this is the point (1, 4, 4); the point on the tangent $z-x=0$ is of course (4, 1, 4), and that on the tangent $x-y=0$ is (4, 4, 1). To find the tangents at these points respectively, I remark that the general equation of the tangent is

$$(X\delta_x+Y\delta_y+Z\delta_z)\left\{\frac{1}{\sqrt{(x)}}+\frac{1}{\sqrt{(y)}}+\frac{1}{\sqrt{(z)}}\right\}=0,$$

that is

$$\frac{X}{x^{\frac{3}{2}}}+\frac{Y}{y^{\frac{3}{2}}}+\frac{Z}{z^{\frac{3}{2}}}=0,$$

or for the point (1, 4, 4) the equation of the tangent is $8X+Y+Z=0$, or say $8x+y+z=0$; that is, the tangent passes through the point $x=0$, $x+y+z=0$, being the point of intersection of the line $x=0$ with the line $x+y+z=0$, which is the harmonic of the given point (1, 1, 1) in regard to the triangle; the tangents at the points (1, 4, 4), (4, 1, 4), (4, 4, 1) respectively pass through the points of intersection of the harmonic line $x+y+z=0$ with the three given lines respectively.

In the case where the given point lies within the triangle, the curve the locus of S lies wholly within the triangle, and is of the form shown in fig. 3 in the plate opposite; it is clear that in this case the conics of the system are all of them ellipses; there are however three limiting forms, viz. the line joining the given point with any angle of the triangle, such line being regarded as a twofold line or pair of coincident lines, is a conic of the system. The discussion of the two cases in which the given point lies outside the triangle, viz. in the infinite space bounded by two sides produced, or in the infinite space bounded by a side and two sides produced, may be effected without much difficulty.

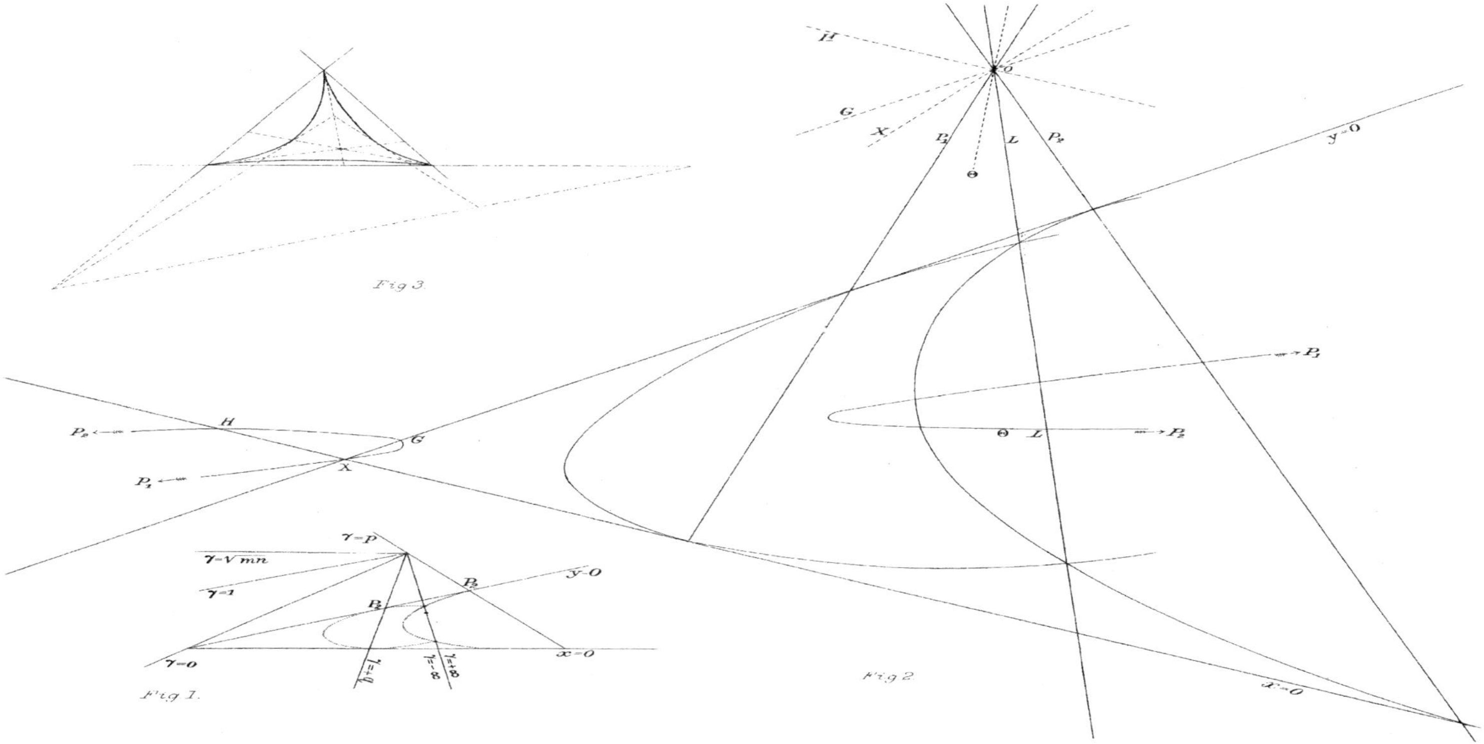

Fig 3.

Fig 1.

Fig 2.

394.

ON A LOCUS IN RELATION TO THE TRIANGLE.

[From the *Quarterly Journal of Pure and Applied Mathematics*, vol. VIII. (1867), pp. 264—277.]

IF from any point of a circle circumscribed about a triangle perpendiculars are let fall upon the sides, the feet of the perpendiculars lie in a line; or, what is the same thing, the locus of a point, such that the perpendiculars let fall therefrom upon the sides of a given triangle have their feet in a line, is the circle circumscribed about the triangle.

In this well known theorem we may of course replace the circular points at infinity by any two points whatever; or the Absolute being a point-pair, and the terms perpendicular and circle being understood accordingly, we have the more general theorem expressed in the same words.

But it is less easy to see what the corresponding theorem is, when instead of being a point-pair, the Absolute is a proper conic; and the discussion of the question affords some interesting results.

Take $(x=0,\ y=0,\ z=0)$ for the equations of the sides of the triangle, and let the equation of the Absolute be

$$(a,\ b,\ c,\ f,\ g,\ h\between x,\ y,\ z)^2=0,$$

then any two lines which are harmonics in regard to this conic (or, what is the same thing, which are such that the one of them passes through the pole of the other) are said to be perpendicular to each other, and the question is:

Find the locus of a point, such that the perpendiculars let fall therefrom on the sides of the triangle have their feet in a line.

Supposing, as usual, that the inverse coefficients are $(A,\ B,\ C,\ F,\ G,\ H)$, and that K is the discriminant, the coordinates of the poles of the three sides respectively are

(A, H, G), (H, B, F), (G, F, C). Hence considering a point P, the coordinates of which are (x, y, z), and taking (X, Y, Z) for current coordinates, the equation of the perpendicular from P on the side $X=0$ is

$$\begin{vmatrix} X, & Y, & Z \\ x\,, & y\,, & z \\ A, & H, & G \end{vmatrix} = 0,$$

and writing in this equation $X=0$, we find

$$(\quad 0 \quad, Ay - Hx,\ Az - Gx)$$

for the coordinates of the foot of the perpendicular. For the other perpendiculars respectively, the coordinates are

$$(Bx - Hy, \quad 0 \quad, Bz - Fy),$$

and

$$(Cx - Gz\,,\ Cy - Fz\,, \quad 0 \quad),$$

and hence the condition in order that the three feet may lie in a line is

$$\begin{vmatrix} 0 \quad, & Ay - Hx, & Az - Gx \\ Bx - Hy, & 0 \quad, & Bz - Fy \\ Cx - Gz\,, & Cy - Fz\,, & 0 \end{vmatrix} = 0\,;$$

or, what is the same thing,

$$(Ay - Hx)(Bz - Fy)(Cx - Gz) + (Az - Gx)(Bx - Hy)(Cy - Fz) = 0,$$

that is

$$\begin{aligned} &2\,(ABC - FGH)\,xyz \\ &+ A\,(\ FH - \ BG)\,yz^2 + A\,(FG - CH)\,y^2z \\ &+ B\,(\ FG - \ CH)\,zx^2 + B\,(GH - AF)\,z^2x \\ &+ C\,(\ GH - \ AF)\,xy^2 + C\,(HF - BG)\,x^2y = 0, \end{aligned}$$

which is the equation of the locus of P; the locus is therefore a cubic. Writing for a moment

$$(BC - F^2,\ CA - G^2,\ AB - H^2,\ GH - AF,\ HF - BG,\ FG - CH) = (A',\ B',\ C',\ F',\ G',\ H'),$$

and K' for the discriminant $ABC - AF^2 -$ &c., the equation is

$$2\,(ABC - FGH)\,xyz + Ayz\,(H'y + G'z) + Bzx\,(H'x + F'z) + Cxy\,(G'x + F'y) = 0,$$

or as this may also be written

$$\frac{2}{F'G'H'}(ABC - FGH)\,xyz + \frac{A}{F'}\,yz\left(\frac{y}{G'} + \frac{z}{H'}\right) + \frac{B}{G'}\,zx\left(\frac{x}{F'} + \frac{z}{H'}\right) + \frac{C}{H'}\,xy\left(\frac{x}{F'} + \frac{y}{G'}\right) = 0,$$

that is

$$\left[\frac{2}{F'G'H'}(ABC-FGH)-\frac{A}{F'^2}-\frac{B}{G'^2}-\frac{C}{H'^2}\right]xyz+\left(\frac{A}{F'}yz+\frac{B}{G'}zx+\frac{C}{H'}xy\right)\left(\frac{x}{F'}+\frac{y}{G'}+\frac{z}{H'}\right)=0,$$

and the cubic will therefore break up into a line and conic if only

$$\frac{2}{F'G'H'}(ABC-FGH)-\frac{A}{F'^2}-\frac{B}{G'^2}-\frac{C}{H'^2}=0,$$

and it is easy to see that conversely this is the necessary and sufficient condition in order that the cubic may so break up.

The condition is

$$\Omega = 2F'G'H'(ABC-FGH)-AG'^2H'^2-BH'^2F'^2-CF'^2G'^2=0,$$

we have

$$AA'+BB'+CC'=3ABC-AF^2-BG^2-CH^2, \ =K'+2(ABC-FGH),$$

and thence

$$\Omega = F'G'H'(AA'+BB'+CC'-K')-AG'^2H'^2-BH'^2F'^2-CF'^2G'^2,$$

that is

$$\begin{aligned}\Omega &= -AG'H'(G'H'-A'F')-BH'F'(H'F'-B'G')-CF'G'(F'G'-C'H')-K'F'G'H',\\ &= -AG'H'K'F-BH'F'K'G-CF'G'K'H-K'F'G'H',\\ &= -K'(AFG'H'+BGH'F'+CHF'G'+F'G'H'),\end{aligned}$$

so that the condition $\Omega=0$ is satisfied if $K'=0$, that is if the equation

$$(A,\ B,\ C,\ F,\ G,\ H\between\xi,\ \eta,\ \zeta)^2=0,$$

which is the line-equation of the Absolute breaks up into factors; that is, if the Absolute be a point-pair.

In the case in question we may write

$$(A,\ B,\ C,\ F,\ G,\ H\between\xi,\ \eta,\ \zeta)^2=2(\alpha\xi+\beta\eta+\gamma\zeta)(\alpha'\xi+\beta'\eta+\gamma'\zeta),$$

that is

$$(A,\ B,\ C,\ F,\ G,\ H)=(2\alpha\alpha',\ 2\beta\beta',\ 2\gamma\gamma',\ \beta\gamma'+\beta'\gamma,\ \gamma\alpha'+\gamma'\alpha,\ \alpha\beta'+\alpha'\beta),$$

whence also, putting for shortness,

$$(\beta\gamma'-\beta'\gamma,\ \gamma\alpha'-\gamma'\alpha,\ \alpha\beta'-\alpha'\beta)=\ (\lambda,\ \mu,\ \nu),$$

we have

$$(A',\ B',\ C',\ F',\ G',\ H')=-(\lambda^2,\ \mu^2,\ \nu^2,\ \mu\nu,\ \nu\lambda,\ \lambda\mu),$$

and also

$$K'=0,\ 2(ABC-FGH)=AA'+BB'+CC',\ =-2(\alpha\alpha'\lambda^2+\beta\beta'\mu^2+\gamma\gamma'\nu^2).$$

The original cubic equation is

$$(\alpha\alpha'\lambda^2+\beta\beta'\mu^2+\gamma\gamma'\nu^2)\,xyz+\alpha\alpha'\lambda yz\,(\mu y+\nu z)+\beta\beta'\mu zx\,(\lambda x+\nu z)+\gamma\gamma'\nu xy\,(\lambda x+\mu y)=0,$$

and this in fact is

$$(\alpha\alpha'\lambda yz+\beta\beta'\mu zx+\gamma\gamma'\nu xy)\,(\lambda x+\mu y+\nu z)=0.$$

The equation $\lambda x+\mu y+\nu z=0$ is that of the line through the two points which constitute the Absolute; the other factor gives

$$\alpha\alpha'\lambda yz+\beta\beta'\mu zx+\gamma\gamma'\nu xy=0,$$

which is the equation of a conic through the angles of the triangle ($x=0$, $y=0$, $z=0$), and which also passes through the two points of the Absolute; in fact, writing (α, β, γ) for (x, y, z) the equation becomes $\alpha\beta\gamma\,(\alpha'\lambda+\beta'\mu+\gamma'\nu)=0$, and so also writing ($\alpha'$, β', γ') for (x, y, z) it becomes $\alpha'\beta'\gamma'\,(\alpha\lambda+\beta\mu+\gamma\nu)=0$, which relations are identically satisfied by the values of (λ, μ, ν). Hence we see that the Absolute being a point-pair, the locus is the conic passing through the angles of the triangle, and the two points of the Absolute; that is, it is the *circle* passing through the angles of the triangle.

But assuming that K' is not $=0$, or that the Absolute is a proper conic, the equation $\Omega=0$ will be satisfied if

$$AFG'H'+BGH'F'+CHF'G'+F'G'H'=0,$$

we have F', G', $H'=Kf$, Kg, Kh respectively, or omitting the factor K^2, the equation becomes

$$AFgh+BGhf+CHfg+Kfgh=0,$$

which is

$$f^2g^2h^2-bcg^2h^2-cah^2f^2-abf^2g^2+2abcfgh=0,$$

or, as it may also be written,

$$abcf^2g^2h^2\left(\frac{1}{abc}-\frac{1}{af^2}-\frac{1}{bg^2}-\frac{1}{ch^2}+\frac{2}{fgh}\right)=0.$$

I remark that we have $ABC-FGH=K\,(abc-fgh)$; substituting also for F', G', H' the values Kf, Kg, Kh, the equation of the cubic curve is

$$2\,(abc-fgh)\,xyz+Ayz\,(hy+gz)+Bzx\,(hx+fy)+Cxy\,(gx+fy)=0,$$

and the transformed form is

$$\left[\frac{2}{fgh}\,(abc-fgh)-\frac{A}{f^2}-\frac{B}{g^2}-\frac{C}{h^2}\right]xyz+\left(\frac{A}{f}\,yz+\frac{B}{g}\,zx+\frac{C}{h}\,xy\right)\left(\frac{x}{f}+\frac{y}{g}+\frac{z}{h}\right)=0\,;$$

we have

$$\frac{2}{fgh}\,(abc-fgh)-\frac{A}{f^2}-\frac{B}{g^2}-\frac{C}{h^2}=abc\left(\frac{1}{abc}-\frac{1}{af^2}-\frac{1}{bg^2}-\frac{1}{ch^2}+\frac{2}{fgh}\right),$$

so that the foregoing condition

$$\frac{1}{abc}-\frac{1}{af^2}-\frac{1}{bg^2}-\frac{1}{ch^2}+\frac{2}{fgh}=0,$$

being satisfied, the cubic breaks up into the line $\frac{x}{f}+\frac{y}{g}+\frac{z}{h}=0$, and the conic

$$\frac{A}{f}yz+\frac{B}{g}zx+\frac{C}{h}xy=0.$$

It is to be remarked that in general a triangle and the reciprocal triangle are in perspective; that is, the lines joining corresponding angles meet in a point, and the points of intersections of opposite sides lie in a line; this is the case therefore with the triangle ($x=0$, $y=0$, $z=0$), and the reciprocal triangle

$$(ax+hy+gz=0,\ hx+by+fz=0,\ gx+fy+cz=0);$$

and it is easy to see that the line through the points of intersection of corresponding sides is in fact the above mentioned line $\frac{x}{f}+\frac{y}{g}+\frac{z}{h}=0$. It is to be noticed also that the coordinates of the point of intersection of the lines joining the corresponding angles are (F, G, H). The conic

$$\frac{A}{f}yz+\frac{B}{g}zx+\frac{C}{h}xy=0$$

is of course a conic passing through the angles of the triangle ($x=0$, $y=0$, $z=0$); it is *not*, what it might have been expected to be, a conic having double contact with the Absolute $(a, b, c, f, g, h \between x, y, z)^2$.

I return to the condition

$$\frac{1}{abc}-\frac{1}{af^2}-\frac{1}{bg^2}-\frac{1}{ch^2}+\frac{2}{fgh}=0,$$

this can be shown to be the condition in order that the sides of the triangle ($x=0$, $y=0$, $z=0$), and the sides of the reciprocal triangle ($ax+hy+gz=0$, $hx+by+fz=0$, $gx+fy+cz=0$) touch one and the same conic; in fact, using line coordinates, the coordinates of the first three sides are (1, 0, 0), (0, 1, 0), (0, 0, 1) respectively, and those of the second three sides are (a, h, g), (h, b, f), (g, f, c) respectively; the equation of a conic touching the first three lines is

$$\frac{L}{\xi}+\frac{M}{\eta}+\frac{N}{\zeta}=0,$$

and hence making the conic touch the second three sides, we have three linear equations from which eliminating L, M, N, we find

$$\begin{vmatrix} \frac{1}{a}, & \frac{1}{h}, & \frac{1}{g} \\ \frac{1}{h}, & \frac{1}{b}, & \frac{1}{f} \\ \frac{1}{g}, & \frac{1}{f}, & \frac{1}{c} \end{vmatrix}=0,$$

which is the equation in question.

We know that if the sides of two triangles touch one and the same conic, their angles must lie in and on the same conic. The coordinates of the angles are (1, 0, 0), (0, 1, 0), (0, 0, 1) and (A, H, G), (H, B, F), (G, F, C) respectively, and the angles will be situate in a conic if only

$$\begin{vmatrix} \frac{1}{A}, & \frac{1}{H}, & \frac{1}{G} \\ \frac{1}{H}, & \frac{1}{B}, & \frac{1}{F} \\ \frac{1}{G}, & \frac{1}{F}, & \frac{1}{C} \end{vmatrix} = 0,$$

an equation which must be equivalent to the last preceding one; this is easily verified. In fact, writing for shortness

$$\nabla = \begin{vmatrix} \frac{1}{a}, & \frac{1}{h}, & \frac{1}{g} \\ \frac{1}{h}, & \frac{1}{b}, & \frac{1}{f} \\ \frac{1}{g}, & \frac{1}{f}, & \frac{1}{c} \end{vmatrix}, \quad \square = \begin{vmatrix} \frac{1}{A}, & \frac{1}{H}, & \frac{1}{G} \\ \frac{1}{H}, & \frac{1}{B}, & \frac{1}{F} \\ \frac{1}{G}, & \frac{1}{F}, & \frac{1}{C} \end{vmatrix},$$

we have

$$-\square = \frac{1}{ABCF^2}(BC - F^2) + \frac{1}{CFGH^2}(FG - CH) + \frac{1}{BFG^2H}(HF - BG),$$

$$= \frac{K}{ABCF^2G^2H^2}(aG^2H^2 + hABFG + gCAHF),$$

and the second factor is

$$= aGH(AF + Kf) + AFhBG + AFgCH,$$

$$= AF(aGH + hBG + gCH) + KafGH.$$

But

$$aGH + hBG + gCH = G(aH + hB) + gCH = G - gF + gCH,$$

$$= G - gF + gCH,$$

$$= -g(FG - CH),$$

$$= -ghK,$$

so that the second factor is

$$= K(afGH - ghAF),$$

which is

$$= K(f^2g^2h^2 - bcg^2h^2 - cah^2f^2 - abf^2g^2 + 2abcfgh),$$

$$= Kabcf^2g^2h^2\left(\frac{1}{abc} - \frac{1}{af^2} - \frac{1}{bg^2} - \frac{1}{ch^2} + \frac{2}{fgh}\right),$$

$$= Kabcf^2g^2h^2\nabla,$$

so that we have identically

$$-ABCF^2G^2H^2\,\square = K^2abcf^2g^2h^2\,\nabla,$$

and the conditions $\nabla=0$, $\square=0$ are consequently equivalent.

The condition

$$\frac{1}{abc}-\frac{1}{af^2}-\frac{1}{bg^2}-\frac{1}{ch^2}+\frac{2}{fgh}=0,$$

is the condition in order that the function

$$\left(\frac{1}{a},\ \frac{1}{b},\ \frac{1}{c},\ \frac{1}{f},\ \frac{1}{g},\ \frac{1}{h}\between ax,\ by,\ cz\right)^2,$$

may break up into linear factors; the function in question is

$$\left(a,\ b,\ c,\ \frac{bc}{f},\ \frac{ca}{g},\ \frac{ab}{h}\between x,\ y,\ z\right)^2,$$

which is

$$=(a,\ b,\ c,\ f,\ g,\ h\between x,\ y,\ z)^2+2\left(\frac{A}{f}yz+\frac{B}{g}zx+\frac{C}{h}xy\right),$$

so that the condition is, that the conic

$$(a,\ b,\ c,\ f,\ g,\ h\between x,\ y,\ z)^2+2\left(\frac{A}{f}yz+\frac{B}{g}zx+\frac{C}{h}xy\right)=0,$$

(which is a certain conic passing through the intersections of the Absolute $(a,\ b,\ c,\ f,\ g,\ h\between x,\ y,\ z)^2=0$, and of the locus conic $\frac{A}{f}yz+\frac{B}{g}zx+\frac{C}{h}xy=0$) shall be a pair of lines. Writing the equation of the conic in question under the form

$$\left(a,\ b,\ c,\ \frac{bc}{f},\ \frac{ca}{g},\ \frac{ab}{h}\between x,\ y,\ z\right)^2=0,$$

the inverse coefficients A', B', C', F', G', H' of this conic, are

$$\left(-\frac{Abc}{f^2},\ -\frac{Bca}{g^2},\ -\frac{Cab}{h^2},\ -\frac{abc}{fgh}F,\ -\frac{abc}{fgh}G,\ -\frac{abc}{fgh}H\right),$$

so that we have $F' : G' : H' = F : G : H$. Hence, if in regard to this new conic we form the reciprocal of the triangle ($x=0$, $y=0$, $z=0$), and join the corresponding angles of the two triangles, the joining lines meet in a point which is the same point as is obtained by the like process from the triangle and its reciprocal in regard to the Absolute. But I do not further pursue this part of the theory.

It is to be noticed that the conic

$$\frac{A}{f}yz+\frac{B}{g}zx+\frac{C}{h}xy=0,$$

contains the angles of the reciprocal triangle, and is thus in fact the conic in which are situate the angles of the two triangles. For the coordinates of one of the angles of the reciprocal triangle are (A, H, G); we should therefore have

$$\frac{A}{f}HG+\frac{B}{g}GA+\frac{C}{h}AH=0,$$

which is

$$\frac{A}{fgh}(GHgh+BGhf+CHfg)=0,$$

or attending only to the second factor and writing

$$GH=Kf+AF,$$

the condition is

$$Kfgh+AFgh+BGhf+CHfg=0,$$

or substituting for K, A, B, C, F, G, H their values and reducing, this is

$$-abcf^2g^2h^2\left(\frac{1}{abc}-\frac{1}{af^2}-\frac{1}{bg^2}-\frac{1}{ch^2}+\frac{2}{fgh}\right)=0,$$

which is satisfied: hence the three angles of the reciprocal triangle lie on the conic in question.

Partially recapitulating the foregoing results, we see in the case where the Absolute is not a point-pair, that the locus of a point such that the perpendiculars from it on the sides of the triangle have their feet in a line, is in general a *cubic curve* passing through the angles of the triangle: if, however, the condition

$$\frac{1}{abc}-\frac{1}{af^2}-\frac{1}{bg^2}-\frac{1}{ch^2}+\frac{2}{fgh}=0$$

be satisfied, that is, if the triangle be such that the angles thereof and of the reciprocal triangle lie in a conic (or, what is the same thing, if the sides touch a conic) then the cubic locus breaks up into the line $\frac{x}{f}+\frac{y}{g}+\frac{z}{h}=0$, which is the line through the points of intersection of the corresponding sides of the two triangles, and into the conic

$$\frac{A}{f}yz+\frac{B}{g}zx+\frac{C}{h}xy=0,$$

which is the conic through the angles of the two triangles.

The question arises, given a conic (the Absolute) to construct a triangle such that its angles, and the angles of the reciprocal triangle in regard to the given conic, lie in a conic.

I suppose that two of the angles of the triangle are given, and I enquire into the locus of the remaining angle. To fix the ideas, let A, B, C be the angles of the triangle, A', B', C' those of the reciprocal triangle; and let the angles A and B be given. We have to find the locus of the point C: I observe however, that the lines AA', BB', CC' meet in a point O, and I conduct the investigation in such manner as to obtain simultaneously the loci of the two points C and O. The lines $C'B'$, $C'A'$ are the polars of A, B respectively, let their equations be $x=0$, and $y=0$, and let the equation of the line AB be $z=0$; this being so, the equation of the given conic will be of the form

$$(a, b, c, 0, 0, h \between x, y, z)^2=0.$$

I take (α, β, γ) for the coordinates of O and (x, y, z) for those of C; the coordinates of either of these points being of course deducible from those of the other.

Observing that the inverse coefficients are

$$(bc, ca, ab-h^2, 0, 0, -ch),$$

we find

$$\begin{aligned} &\text{coordinates of } A \text{ are } (\ \ b,\ -h,\ 0),\\ &\quad ,, \qquad\ \ B \quad ,, \ \ (-h,\ \ a,\ 0). \end{aligned}$$

The points A' and B' are then given as the intersections of AO with $C'A'\,(y=0)$ and of BO with $C'B'\,(x=0)$; we find

$$\begin{aligned} &\text{coordinates of } A' \text{ are } (h\alpha+b\beta,\ \ 0\ \ ,\ h\gamma),\\ &\quad ,, \qquad\ \ B' \quad ,, \ \ (\ \ 0\ \ ,\ a\alpha+h\beta,\ h\gamma). \end{aligned}$$

Moreover,

$$\begin{aligned} &\text{coordinates of } C' \text{ are } (0,\ 0,\ 1),\\ &\quad ,, \qquad\ \ C \quad ,, \ \ (x,\ y,\ z). \end{aligned}$$

The six points A, B, C, A', B', C' are to lie in a conic; the equations of the lines $C'A$, $C'B$, AB are $hX+bY=0$, $aX+hY=0$, $Z=0$, and hence the equation of a conic passing through the points C', A, B is

$$\frac{L}{aX+hY}+\frac{M}{hX+bY}+\frac{N}{Z}=0.$$

Hence, making the conic pass through the remaining points A', B', C, we find

$$\frac{L}{a(h\alpha+b\beta)}+\frac{M}{h(h\alpha+b\beta)}+\frac{N}{h\gamma}=0,$$

$$\frac{L}{h(a\alpha+h\beta)}+\frac{M}{b(a\alpha+h\beta)}+\frac{N}{h\gamma}=0,$$

$$\frac{L}{ax+hy}+\frac{M}{hx+by}+\frac{N}{z}=0,$$

and eliminating the L, M, N, we find

$$\begin{vmatrix} \frac{1}{a}, & \frac{1}{h}, & h\alpha + b\beta \\ \frac{1}{h}, & \frac{1}{b}, & a\alpha + h\beta \\ \frac{1}{ax+hy}, & \frac{1}{hx+by}, & \frac{hy}{z} \end{vmatrix} = 0,$$

or developing and reducing, this is

$$-\frac{(ab-h^2)}{hab}\frac{y}{z} + \frac{1}{h}\frac{a\alpha+h\beta}{ax+hy} + \frac{1}{h}\frac{h\alpha+b\beta}{hx+by}$$
$$-\frac{1}{a}\frac{a\alpha+h\beta}{hx+by} - \frac{1}{b}\frac{h\alpha+b\beta}{ax+hy} = 0.$$

We have still to find the relation between (α, β, γ) and (x, y, z); this is obtained by the consideration that the line $A'B'$, through the two points A', B' the coordinates of which are known in terms of (α, β, γ), is the polar of the point C, the coordinates of which are (x, y, z). The equation of $A'B'$ is thus obtained in the two forms

$$(a\alpha+h\beta)X + (h\alpha+b\beta)Y - \frac{(a\alpha+h\beta)(h\alpha+b\beta)}{hy}Z = 0,$$

and

$$(ax+hy)X + (hx+by)Z + \qquad cz \qquad Z = 0,$$

and comparing these, we have

$$x : y : z = \alpha : \beta : \frac{-(a\alpha+h\beta)(h\alpha+b\beta)}{chy},$$

or what is the same thing

$$\alpha : \beta : \gamma = x : y : \frac{-(ax+hy)(hx+by)}{chz},$$

(where it is to be observed that the equation $\alpha : \beta = x : y$ is the verification of the theorem that the lines AA', BB', CC' meet in a point O).

We may now from the above found relation eliminate either the (α, β, γ) or the (x, y, z); first eliminating the (α, β, γ), we find

$$-\frac{ab-h^2}{hab}\frac{Y}{Z} + \frac{2}{h} - \frac{1}{a}\frac{ax+hy}{hx+by} - \frac{1}{b}\frac{hx+by}{ax+by} = 0,$$

where

$$\frac{Y}{Z} = -\frac{(ax+hy)(hx+by)}{chz^2},$$

or, completing the elimination,

$$\frac{ab-h^2}{ch}\frac{(ax+hy)^2(hx+by)^2}{z^2} = (hb,\ -ab,\ ha \between ax+hy,\ hx+by)^2 = 0,$$

which is a quartic curve having a node at each of the points

$$(z=0,\ ax+hy=0),\ (z=0,\ hx+by=0),\ (ax+hy=0,\ hx+by=0),$$

that is, at each of the points B, A, C'. The right-hand side of the foregoing equation is

$$= -(ab-h^2)(ha,\ ab,\ hb\between x,\ y)^2,\ = -(ab-h^2)\,h\left(ax^2+by^2+\frac{2ab}{h}xy\right),$$

so that the equation may also be written

$$(ax+hy)^2(hx+by)^2+ch^2z^2\left(ax^2+by^2+\frac{2ab}{h}xy\right)=0.$$

Secondly, to eliminate the (x, y, z), we have

$$-\frac{ab-h^2}{hab}\frac{Y}{Z}+\frac{2}{h}-\frac{1}{a}\frac{a\alpha+h\beta}{h\alpha+b\beta}-\frac{1}{b}\frac{h\alpha+b\beta}{a\alpha+h\beta}=0,$$

where

$$\frac{Y}{Z}=-\frac{chy^2}{(a\alpha+h\beta)(h\alpha+b\beta)},$$

or, completing the elimination,

$$\begin{aligned}(ab-h^2)\,chy^2 &= \ (hb,\ -ab,\ ha\between a\alpha+h\beta,\ h\alpha+b\beta)^2\\ &= -(ab-h^2)\,h\left(a\alpha^2+b\beta^2+\frac{2ab}{h}\alpha\beta\right),\end{aligned}$$

that is

$$\left(a,\ b,\ c,\ 0,\ 0,\ \frac{ab}{h}\between \alpha,\ \beta,\ \gamma\right)^2=0.$$

Writing (x, y, z) in place of (α, β, γ), the locus of the point O is the conic

$$\left(a,\ b,\ c,\ 0,\ 0,\ \frac{ab}{h}\between x,\ y,\ z\right)^2=0,$$

which is a conic intersecting the Absolute

$$(a,\ b,\ c,\ 0,\ 0,\ h\between x,\ y,\ z)^2=0,$$

at its intersections with the lines $x=0$, $y=0$, that is the lines $C'B'$ and $C'A'$.

In regard to this new conic, the coordinates of the pole of $C'B'\,(x=0)$ are at once found to be $(-h,\ a,\ 0)$, that is, the pole of $C'B'$ is B; and similarly the coordinates of the pole of $C'A'\,(y=0)$ are $(b,\ -h,\ 0)$, that is, the pole of $C'A'$ is A. We may consequently construct the conic the locus of O, viz. given the Absolute and the points A and B, we have $C'A'$ the polar of B, meeting the Absolute in two points $(a_1,\ a_2)$, and $C'B'$ the polar of A meeting the Absolute in the points (b_1 and b_2); the lines $C'A'$ and $C'B'$ meet in C'. This being so, the required conic passes through the points a_1, a_2, b_1, b_2, the tangents at these points being Aa_1, Aa_2, Bb_1, Bb_2 respectively; eight conditions, five of which would be sufficient to determine the conic. It is to be remarked that the lines $C'B'$, $C'A'$ (which in regard to the Absolute are the polars of A, B respectively) are in regard to the required conic the polars of B, A respectively.

The conic the locus of O being known, the point O may be taken at any point of this conic, and then we have A' as the intersection of $C'A'$ with AO, B' as the intersection of $C'B'$ with BO, and finally, C as the pole of the line $A'B'$ in regard to the Absolute, the point so obtained being a point on the line $C'O$. To each position of O on the conic locus, there corresponds of course a position of C; the locus of C is, as has been shown, a quartic curve having a node at each of the points C', A, B.

The foregoing conclusions apply of course to spherical figures; we see therefore that on the sphere the locus of a point such that the perpendiculars let fall on the sides of a given spherical triangle have their feet in a line (great circle), is a spherical cubic. If, however, the spherical triangle is such that the angles thereof and the poles of the sides (or, what is the same thing, the angles of the polar triangle) lie on a spherical conic; then the cubic locus breaks up into a line (great circle), which is in fact the circle having for its pole the point of intersection of the perpendiculars from the angles of the triangle on the opposite sides respectively, and into the before-mentioned spherical conic. Assuming that the angles A and B are given, the above-mentioned construction, by means of the point O, is applicable to the determination of the locus of the remaining angle C, in order that the spherical triangle ABC may be such that the angles and the poles of the sides lie on the same spherical conic, but this requires some further developments. The lines $C'B'$, $C'A'$ which are the polars of the given angles A, B respectively, are the cyclic arcs of the conic the locus of O, or say for shortness the conic O; and moreover these same lines $C'B'$, $C'A'$ are in regard to the conic O, the polars of the angles B, A respectively. If instead of the conic O we consider the polar conic O', it follows that A, B are the foci, and $C'A'$, $C'B'$ the corresponding directrices of the conic O'. The distance of the directrix $C'A'$ from the centre of the conic, measuring such distance along the transverse axis is clearly $= 90° -$ distance of the focus A; it follows that the transverse semi-axis is $= 45°$, or what is the same thing, that the transverse axis is $= 90°$; that is, the conic O' is a conic described about the foci A, B with a transverse axis (or sum or difference of the focal distances) $= 90°$. Considering any tangent whatever of this conic, the pole of the tangent is a position of the point O, which is the point of intersection of the perpendiculars let fall from the angles of the spherical triangle on the opposite sides; hence, to complete the construction, we have only through A and B respectively to draw lines AC, BC perpendicular to the lines BO, CO respectively; the lines in question will meet in a point C, which is such that CO will be perpendicular to AB, and which point C is the required third angle of the spherical triangle ABC. In order to ascertain whether a given spherical triangle ABC has the property in question (viz. whether it is such that the angles thereof and of the polar triangle lie in a spherical conic), we have only to construct as before the conic O' with the foci A, B and transverse axis $= 90°$, and then ascertain whether the polar of the point O, the intersections of the perpendiculars from the angles of the triangle on the opposite sides respectively, is a tangent of the conic O'. It is moreover clear, that given a triangle ABC having the property in question, if with the foci A, B and transverse axis $= 90°$ we describe a conic, and if in like manner with the foci A, C and the same transverse axis, and with the foci B, C and the same transverse axis, we describe two other conics; then that the three conics will have a common tangent the pole whereof will be the point of intersection of the perpendiculars from the angles of the triangle ABC on the opposite sides respectively.

395.

INVESTIGATIONS IN CONNEXION WITH CASEY'S EQUATION.

[From the *Quarterly Journal of Pure and Applied Mathematics*, vol. VIII. (1867), pp. 334—341.]

In a paper read April 9, 1866, and recently published in the *Proceedings of the Royal Irish Academy*, Mr Casey has given in a very elegant form the equation of a pair of circles touching each of three given circles, viz. if $U=0$, $V=0$, $W=0$ be the equations of the three given circles respectively, and if considering the common tangents of $(V=0,\ W=0)$, of $(W=0,\ U=0)$, and of $(U=0,\ V=0)$ respectively, these common tangents being such that the centres of similitude through which they respectively pass lie in a line (viz. the tangents are all three direct, or one is direct and the other two are inverse), then if f, g, h are the lengths of the tangents in question, the equation

$$\surd(fU)+\surd(gV)+\surd(hW)=0,$$

belongs to a pair of circles, each of them touching the three given circles. (There are, it is clear, four combinations of tangents, and the theorem gives therefore the equations of four pairs of circles, that is of the eight circles which touch the three given circles.)

Generally, if $U=0$, $V=0$, $W=0$ are the equations of any three curves of the same order n, and if f, g, h are arbitrary coefficients, then the equation

$$\surd(fU)+\surd(gV)+\surd(hW)=0,$$

is that of a curve of the order $2n$, touching each of the curves $U=0$, $V=0$, $W=0$, n^2 times, viz. it touches

$$\begin{aligned} U&=0, \text{ at its } n^2 \text{ intersections with } & gV-hW&=0,\\ V&=0 \quad\quad ,, \quad\quad ,, & hW-fU&=0,\\ W&=0 \quad\quad ,, \quad\quad ,, & fU-gV&=0. \end{aligned}$$

If however the curves $U=0$, $V=0$, $W=0$ have a common intersection, then the curve in question has a node at this point, and besides touches each of the three curves in n^2-1 points; and similarly, if the curves $U=0$, $V=0$, $W=0$ have k common intersections, then the curve in question has a node at each of these points, and besides touches each of the three curves in n^2-k points.

In particular, if $U=0$, $V=0$, $W=0$ are conics having two common intersections, then the curve is a quartic having a node at each of the common intersections, and besides touching each of the given conics in two points; whence, if the coefficients f, g, h (that is, their ratios) are so determined that the quartic may have two more nodes, then the quartic, having in all four nodes, will break up into a pair of conics, each passing through the common intersections, and the pair touching each of the given conics in two points; that is, the component conics will each of them touch each of the given conics once. Taking the circular points at infinity for the common intersections, the conics will be circles, and we thus see that Casey's theorem is in effect a determination of the coefficients f, g, h, in such wise that the curve

$$\sqrt{(fU)}+\sqrt{(gV)}+\sqrt{(hW)}=0,$$

(which when $U=0$, $V=0$, $W=0$ are circles, is by what precedes a bicircular quartic) shall have two more nodes, and so break up into a pair of circles.

The question arises, given $U=0$, $V=0$, $W=0$, curves of the same order n, it is required to determine the ratios $f:g:h$ in such wise that the curve

$$\sqrt{(fU)}+\sqrt{(gV)}+\sqrt{(hW)}=0,$$

may have two nodes; or we may simply inquire as to the number of the sets of values of $(f:g:h)$, which give a binodal curve, $\sqrt{(fU)}+\sqrt{(gV)}+\sqrt{(hW)}=0$.

I had heard of Mr Casey's theorem from Dr Salmon, and communicated it together with the foregoing considerations to Prof. Cremona, who, in a letter dated Bologna, March 3, 1866, sent me an elegant solution of the question as to the number of the binodal curves. This solution is in effect as follows:

LEMMA. Given the curves $U=0$, $V=0$, $W=0$ of the same order n; consider the point (f, g, h), and corresponding thereto the curve $fU+gV+hW=0$. As long as the point (f, g, h) is arbitrary, the curve $fU+gV+hW=0$, will not have any node, and in order that this curve may have a node, it is necessary that the point (f, g, h) shall lie on a certain curve Σ; this being so, the node will lie on a curve J, the Jacobian of the curves U, V, W; and the curves J and Σ will correspond to each other, point to point; viz. taking for (f, g, h) any point whatever on the curve Σ, the curve $fU+gV+hW=0$ will be a curve having a node at some one point on the curve J; and conversely, in order that the curve $fU+gV+hW=0$ may be a curve having a node at a given point on the curve J, it is necessary that the point (f, g, h) shall be at some one point of the curve Σ. The curve Σ has however nodes and cusps; each node of Σ corresponds to two points of J, viz. the point (f, g, h) being at a node of Σ, the curve $fU+gV+hW=0$, is a binodal curve having a node at each of the corresponding points on J; and each cusp of Σ corresponds to two coincident points of J, viz. the point (f, g, h) being at a cusp of Σ, the curve $fU+gV+hW=0$ is a cuspidal curve having a cusp at the corresponding point of J. The number of the binodal curves $fU+gV+hW=0$ is thus equal to the number of the nodes of Σ, and the number of the cuspidal curves $fU+gV+hW=0$ is equal to the number of the cusps of Σ. The curve Σ is easily shown to be a curve of the order $3(n-1)^2$ and class $3n(n-1)$; and quà curve which corresponds point to point with J, it is a curve having the same deficiency as J, that is a deficiency $=\frac{1}{2}(3n-4)(3n-5)$; we have thence the Plückerian numbers of the curve Σ, viz.:

Order is	=	$3(n-1)^2$,
Class	=	$3n(n-1)$,
Cusps	=	$12(n-1)(n-2)$,
Nodes	=	$\frac{3}{2}(n-1)(n-2)(3n^2-3n-11)$,
Inflexions	=	$3(n-1)(4n-5)$,
Double tangents	=	$\frac{3}{2}(n-1)(n-2)(3n^2+3n-8)$.

Remarks. The consideration of the foregoing curve Σ is, I believe, first due to Prof. Cremona, it is a curve related to the three distinct curves $U=0$, $V=0$, $W=0$, in the same way precisely as Steiner's curve P_0 is related to the three curves $d_xU=0$, $d_yU=0$, $d_zU=0$. (Steiner, "Allgemeine Eigenschaften der algebraischen Curven," *Crelle*, t. XLVII. (1854), pp. 1—6; see also Clebsch, "Ueber einige von Steiner behandelte Curven," *Crelle*, t. LXIV. (1865), pp. 288—293), and the Plückerian numbers of P_0 (writing therein $n+1$ for n) are identical with those of Σ. The foregoing expressions $\frac{3}{2}(n-1)(n-2)(3n^2-3n-11)$ and $12(n-1)(n-2)$ for the numbers of the binodal and cuspidal curves $fU+gV+hW=0$, are given in my memoir "On the Theory of Involution," *Cambridge Philosophical Transactions*, t. XI. (1866), pp. 21—38, see p. 32, [348]; but the employment of the curve Σ very much simplifies the investigation.

Passing now to the proposed question, we have as before the curves $U=0$, $V=0$, $W=0$, of the same order n; and we may consider the point (f, g, h), and corresponding thereto the curve $\sqrt{(fU)}+\sqrt{(gV)}+\sqrt{(hW)}=0$, say for shortness the curve Ω, which is a curve of the order $2n$, having n^2 contacts with each of the given curves U, V, W. As long as the point (f, g, h) is arbitrary, the curve Ω has not any node; and in order that this curve may have a node, it is necessary that the point (f, g, h) shall lie on a certain curve Δ; this being so, the node will lie on the foregoing curve J, the Jacobian of the given curves U, V, W; and the curves J and Δ will correspond to each other, point to point, viz. taking for (f, g, h) any point whatever on the curve Δ, the curve Ω will have a node at some one point of J; and conversely, in order that the curve Ω may be a curve having a node at a given point of J, it is necessary that the point (f, g, h) shall be at some one point of the curve Δ. The curve Δ has however nodes and cusps; each node of Δ corresponds to two points of J, viz. for (f, g, h) at a node of Δ, the curve Ω is a binodal curve having a node at each of the corresponding points of J; each cusp of Δ corresponds to two coincident points of J, viz. for (f, g, h) at a cusp of Δ, the curve Ω is a cuspidal curve having a cusp at the corresponding point of J. The number of the binodal curves Ω is consequently equal to that of the nodes of Δ, and the number of the cuspidal curves Ω is equal to that of the cusps of Δ; we have consequently to find the Plückerian numbers of the curve Δ; and this Prof. Cremona accomplishes by bringing it into connexion with the foregoing curve Σ, and making the determination depend upon that of the number of the conics which satisfy certain conditions of contact in regard to the curve Σ.

Consider, as corresponding to any given point (f, g, h) whatever, the conic $\frac{f}{x}+\frac{g}{y}+\frac{h}{z}=0$ which passes through three fixed points, the angles of the triangle $x=0$, $y=0$, $z=0$. For points (f, g, h) which lie in an arbitrary line $Af+Bg+Ch=0$, the corresponding conics pass through the fourth fixed point $x:y:z=A:B:C$. Assume for the moment that to the points (f, g, h) which lie on the foregoing curve Δ, correspond conics which touch the foregoing curve Σ. Then 1°. to the points of intersection of the curve Δ with an arbitrary line, correspond the conics which pass through four arbitrary points and touch the curve Σ; or the order of the curve Δ is equal to the number of the conics which can be drawn through four arbitrary points to touch the curve Σ; viz. if m be the order, n the class of Σ, the number of these conics is $=2m+n$, or substituting for m, n the values $3(n-1)^2$ and $3n(n-1)$ respectively, the number of these conics, that is the order of Δ, is $=3(n-1)(3n-2)$. 2°. To the nodes of Δ correspond the conics which pass through three arbitrary points and have two contacts with Σ, viz. if m be the order, n the class, and κ the number of cusps of Σ, then the number of these conics is $=\frac{1}{2}(2m+n)^2-2m-5n-\frac{3}{2}\kappa$, or substituting for m, n their values as above, and for κ its value $=12(n-1)(n-2)$, the number of these conics, that is, the number of the nodes of Δ, is found to be

$$=\tfrac{3}{2}(n-1)(27n^3-63n^2+22n+16).$$

3°. To the cusps of Δ correspond the conics which pass through three arbitrary points, and have with Σ a contact of the second order; the number of these (m, n, κ as above) is $=3n+\kappa$, or substituting for n and κ their values as above, the number of these conics, that is the number of the cusps of Δ, is $=3(n-1)(7n-8)$. We have thence all the Plückerian numbers of the curve Δ, viz. these are

Order	$= 3(n-1)(3n-2)$,
Class	$= 6(n-1)^2$,
Nodes	$= \frac{3}{2}(n-1)(27n^3-63n^2+22n+16)$,
Cusps	$= 3(n-1)(7n-8)$,
Double tangents	$= \frac{3}{2}(n-1)(12n^3-36n^2+19n+16)$,
Inflexions	$= 12(n-1)(n-2)$,

and as a verification it is to be observed, that the deficiency of the curve Δ is equal to that of the curve J, viz. it has the value $\frac{1}{2}(3n-4)(3n-5)$. The foregoing numbers include the result that the number of the binodal curves

$$\sqrt{(fU)}+\sqrt{(gV)}+\sqrt{(hW)}=0,$$

is

$$=\tfrac{3}{2}(n-1)(27n^3-63n^2+22n+16).$$

The proof depended on the assumption, that to the points (f, g, h) which lie on the curve Δ, correspond the conics $\frac{f}{x}+\frac{g}{y}+\frac{h}{z}=0$ which touch the curve Σ; this M. Cremona proves in a very simple manner: the points of J correspond each to each with the points of Σ, or if we please they correspond each to each with the tangents of Σ. To the $6n(n-1)$ intersections of J with any curve Ω (viz. $\sqrt{(fU)}+\sqrt{(gV)}+\sqrt{(hW)}=0$) correspond the $6n(n-1)$ common tangents of Σ and the conic $\frac{f}{x}+\frac{g}{y}+\frac{h}{z}=0$; if Ω has a node, two of the $6n(n-1)$ intersections coincide, and the corresponding two tangents will also coincide, that is Ω having a node (or the point (f, g, h) being on the curve Δ), the conic touches the curve Σ. But it is not uninteresting to give an independent analytical proof. Write for shortness

$$\begin{aligned} dU &= A dx + B dy + C dz, \\ dV &= A' dx + B' dy + C' dz, \\ dW &= A'' dx + B'' dy + C'' dz, \end{aligned}$$

and let (x, y, z) be the coordinates of a point on J, (X, Y, Z) those of the corresponding point on Σ, (f, g, h) those of the corresponding point on Δ. Write also for shortness

$$BC'-B'C,\ CA'-C'A,\ AB'-A'B=P : Q : R,$$

then we have

$$\begin{aligned} AX + BY + CZ &= 0, \\ A'X + B'Y + C'Z &= 0, \\ A''X + B''Y + C''Z &= 0, \end{aligned}$$

$$\begin{aligned} A\sqrt{\left(\frac{f}{U}\right)} + B\sqrt{\left(\frac{g}{V}\right)} + C\sqrt{\left(\frac{h}{W}\right)} &= 0, \\ A' \quad ,, \quad + B' \quad ,, \quad + C' \quad ,, \quad &= 0, \\ A'' \quad ,, \quad + B'' \quad ,, \quad + C'' \quad ,, \quad &= 0, \end{aligned}$$

giving $\begin{vmatrix} A, & B, & C \\ A', & B', & C' \\ A'', & B'', & C'' \end{vmatrix}=0$, which is in fact the equation of the curve J; and moreover giving $X : Y : Z = P : Q : R$, to determine the point (X, Y, Z) on Σ; and

$$\sqrt{\left(\frac{f}{U}\right)} : \sqrt{\left(\frac{g}{V}\right)} : \sqrt{\left(\frac{h}{W}\right)} = P : Q : R,$$

or, what is the same thing, $f : g : h = P^2U : Q^2V : R^2W$, to determine the point (f, g, h) on Δ. Treating now (f, g, h) as constants, and (X, Y, Z) as current coordinates, the conic $\frac{f}{X}+\frac{g}{Y}+\frac{h}{Z}=0$, will touch the curve Σ at the point (P, Q, R), if only the equation of the conic is satisfied by these values and by the consecutive values $P+dP$, $Q+dQ$, $R+dR$; or what is the same thing, if we have

$$\frac{f}{P}+\frac{g}{Q}+\frac{h}{R}=0,$$

$$\frac{fdP}{P^2}+\frac{gdQ}{Q^2}+\frac{hdR}{R^2}=0,$$

that is

$$\frac{f}{P^2}:\frac{g}{Q^2}:\frac{h}{R^2}=QdR-RdQ : RdP-PdR : PdQ-QdP.$$

If the functions on the right-hand side are as $U : V : W$, then these equations give

$$f:g:h=P^2U : Q^2V : R^2W,$$

that is (f, g, h) will be a point on the curve Δ. It is therefore only necessary to show that in virtue of the equation $J=0$ of the curve J, and of the derived equation $dJ=0$, we have

$$QdR-RdQ : RdP-PdR : PdQ-QdP=U : V : W.$$

Take for instance the equation

$$V(QdR-RdQ)-U(RdP-PdR)=0,$$

that is

$$dR(UP+VQ+WR)-R(UdP+VdQ+WdR)=0,$$

and this, and the other two equations will be satisfied if only $UP+VQ+WR=0$, $UdP+VdQ+WdR=0$; we have, neglecting a numerical factor,

$$\begin{aligned} U &= Ax+A'y+A''z, \\ V &= Bx+B'y+B''z, \\ W &= Cx+C'y+C''z, \end{aligned}$$

whence, attending to the values of P, Q, R, we have

$$UP+VQ+WR=zJ=0;$$

hence also

$$UdP+VdQ+WdR+(PdU+QdV+RdW)=0,$$

so that

$$UdP+VdQ+WdR=0,$$

if only

$$PdU+QdV+RdW=0,$$

and substituting for P, Q, R, dU, dV, dW their values, the left-hand side is $=-Jdz$, which is $=0$; hence the equations in question are proved, and (f, g, h) is a point on the curve Δ.

It is to be noticed, that the two curves Σ, Δ are geometrically connected through the three arbitrary points as follows: viz. taking as axes the sides of the triangle formed by these three points, then starting from any point (f, g, h) of Δ, we take the inverse point $\left(\frac{1}{f}, \frac{1}{g}, \frac{1}{h}\right)$, the harmonic line thereof $fx+gy+hz=0$, and finally the inverse conic $\frac{f}{x}+\frac{g}{y}+\frac{h}{z}=0$, which by what precedes touches Σ in the point corresponding to the assumed point (f, g, h) of Δ: and conversely starting with an assumed point on Σ, we take the conic $\frac{f}{x}+\frac{g}{y}+\frac{h}{z}=0$ which passes through the angles of the triangle and touches Σ at the assumed point; the inverse line $fx+gy+hz=0$; the harmonic point $\left(\frac{1}{f}, \frac{1}{g}, \frac{1}{h}\right)$ of this line; and finally the inverse point (f, g, h), which will be on the curve Δ, the point corresponding to the assumed point on the curve Σ.

396.

ON A CERTAIN ENVELOPE DEPENDING ON A TRIANGLE INSCRIBED IN A CIRCLE.

[From the *Quarterly Journal of Pure and Applied Mathematics*, vol. IX. (1868), pp. 31—41 and 175—176.]

CONSIDERING a triangle and the circumscribed circle, and from any point of the circle drawing perpendiculars to the sides of the triangle; the feet of the three perpendiculars lie on a line; and (regarding the point as a variable point on the circle) the envelope of the line is a curve of the third class, having the line infinity for a double tangent, and being therefore a curve of the fourth order with three cusps, see Steiner's paper "Ueber eine besondere Curve dritter Klasse und vierten Grades," *Crelle*, t. LIII. (1857), pp. 231—237, which contains a series of very beautiful geometrical properties.

Mr Greer, in a paper in the last volume of the *Journal*, has expressed the equation of the line in a very elegant form, viz. if α, β, γ are the perpendicular distances of the point from the sides of the triangle; A, B, C the angles of the triangle; $(\lambda, \mu, \nu) = \left(\frac{\cos A}{\alpha}, \frac{\cos B}{\beta}, \frac{\cos C}{\gamma}\right)$; and (X, Y, Z) certain current coordinates, viz. these are the perpendicular distances from the sides, multiplied by $\sin A \tan A$, $\sin B \tan B$, $\sin C \tan C$ respectively; then the equation of the line is

$$X\lambda(\lambda-\mu)(\lambda-\nu) + Y\mu(\mu-\nu)(\mu-\lambda) + Z\nu(\nu-\lambda)(\nu-\mu) = 0,$$

where the parameters λ, μ, ν are connected by the equation $\lambda \tan A + \mu \tan B + \nu \tan C = 0$, or say by the equation

$$\lambda \mathrm{a} + \mu \mathrm{b} + \nu \mathrm{c} = 0.$$

We have a cubic equation in (λ, μ, ν) with coefficients which are linear functions of (X, Y, Z), and the required equation is that obtained by equating to zero the reciprocant of this cubic function, the facients of the reciprocant being the (a, b, c) of the linear relation; the reciprocant is of the degree 6 in (a, b, c) and of the degree 4 in the coefficients of the cubic function, that is in (X, Y, Z). But I remark that the equation in (λ, μ, ν), regarding these quantities as coordinates, is that of a cubic curve having a node at the point $\lambda = \mu = \nu$, or say the point (1, 1, 1); the corresponding value of $\lambda\mathrm{a} + \mu\mathrm{b} + \nu\mathrm{c}$ is $= \mathrm{a} + \mathrm{b} + \mathrm{c}$, and the reciprocant consequently contains the factor $(\mathrm{a} + \mathrm{b} + \mathrm{c})^2$, or dividing this out, the equation is only of the degree 4 in (a, b, c). The equation of the curve thus is

$$\frac{1}{(\mathrm{a}+\mathrm{b}+\mathrm{c})^2}\,\text{recip.}\,\{X\lambda(\lambda-\mu)(\lambda-\nu) + Y\mu(\mu-\nu)(\mu-\lambda) + Z\nu(\nu-\lambda)(\nu-\mu)\} = 0,$$

being of the degree 4 in (a, b, c), and also of the degree 4 in (X, Y, Z), that is, treating (X, Y, Z) as current coordinates, the envelope is as above stated a curve of the fourth order.

A symmetrical method for finding the reciprocant of a cubic function was given by Hesse, see my paper "On Homogeneous Functions of the Third Order with Three Variables," *Camb. and Dubl. Math. Jour.*, vol. I. (1846), pp. 97—104, [35]; the developed expression there given for the reciprocant is however erroneous; the correct value is given in my "Third Memoir on Quantics," *Phil. Trans.*, vol. CXLVI. (1856), see the Table 67, p. 644, [144] and we have only in the table to substitute for (ξ, η, ζ) the quantities (a, b, c), and for $(a, b, c, f, g, h, i, j, k, l)$ the coefficients of the cubic function of (λ, μ, ν), viz. multiplying by 6 in order to avoid fractions, these are

$$\begin{aligned}&(\;a,\;\;b,\;\;c,\;\;f,\;\;g,\;\;h,\;\;i,\;\;j,\;\;k,\;\;l\;)\\ &=(6X,\;6Y,\;6Z,\;-2Y,\;-2Z,\;-2X,\;-2Z,\;-2X,\;-2Y,\;X+Y+Z)\end{aligned}$$

respectively. The substitution might be performed as follows, viz. for the coefficient of a^6, we have

$$\left.\begin{array}{llll} b^2c^2 & +1\,.\,1296 & Y^2Z^2 & +1296 \\ bcfi & -6\,.\;\;144 & Y^2Z^2 & -\;\;864 \\ bi^3 & +4\,.-48 & YZ^3 & -\;\;192 \\ cf^3 & +4\,.-48 & Y^3Z & -\;\;192 \\ f^2i^2 & -3\,.\;\;\;16 & Y^2Z^2 & -\;\;\;48 \end{array}\right\} = -192YZ(Y-Z)^2,$$

and so for the other coefficients; but I have not gone through the labour of performing the calculation. Omitting the numerical factor -192, the coefficients of a^6, b^6, c^6 are of course

$$YZ(Y-Z)^2,\; ZX(Z-X)^2,\; XY(X-Y)^2;$$

and I find also that the coefficient of b^5c (the factor -192 being omitted) is

$$= ZX(3X^2 + 3Z^2 + 3YZ - 6ZX + 5XY),$$

whence those of the terms c^5a, &c. are also known.

I denote the result as follows:

$$(YZ(Y-Z)^2,\ ZX(Z-X)^2,\ XY(X-Y)^2,\ \ldots)(a,\ b,\ c)^6 = 0;$$

this equation divides as we have seen by $(a+b+c)^2$, and the quotient is

$$(YZ(Y-Z)^2,\ ZX(Z-X)^2,\ XY(X-Y)^2,\ \ldots)(a,\ b,\ c)^4 = 0;$$

and it may be remarked that the coefficient of b^3c in this quartic function of (a, b, c) is

$$= ZX(X^2 + Z^2 + 3YZ - 2ZX + 5XY.$$

The last mentioned equation, if the calculation were completed, would be analytically the best form for the equation of the envelope; but in view of what follows, I will change it by writing ax, by, cz in place of (X, Y, Z); x is therefore $=\frac{1}{\tan A}X$, that is, it is $=$ perpendicular distance $\times \sin A$; or, what is the same thing, the new coordinates (x, y, z) are proportional to the perpendicular distances from the sides, each distance divided by the perpendicular distance of the side from the opposite angle, the equation of the line infinity is thus $x+y+z=0$. I write also $(a, b, c) = \left(\frac{1}{a}, \frac{1}{b}, \frac{1}{c}\right)$, that is, we have $(a, b, c) = (\cot A, \cot B, \cot C)$. The system of equations is therefore

$$\frac{x}{a}\lambda(\lambda-\mu)(\lambda-\nu) + \frac{y}{b}\mu(\mu-\nu)(\mu-\lambda) + \frac{z}{c}\nu(\nu-\lambda)(\nu-\mu) = 0,$$

$$\frac{\lambda}{a} + \frac{\mu}{b} + \frac{\nu}{c} = 0,$$

giving for the envelope the equation

$$bcyz(cy-bz)^2 + cazx(az-cx)^2 + abxy(bx-ay)^2 + \&c. = 0;$$

and in this function, corresponding to the term

$$b^3cZX(X^2 + Z^2 + 3YZ - 2ZX + 5XY),$$

we have the term

$$azx(bc^2x^2 + a^2bz^2 + 3a^2cyz - 2abczx + 5ac^2xy).$$

It may be noticed that, arranging in powers of (x, y, z), the several portions of each coefficient are distinct literal functions; thus we see that the coefficient of z^3x is $= a^3c + a^3b +$ other combinations of (a, b, c): this is material in order to the comparison of the foregoing equation of the envelope in a different form which will be presently mentioned.

I proceed to find the tangential equation of the envelope. Representing the equation of the line by

$$\xi x+\eta y+\zeta z=0,$$

we have

$$\xi : \eta : \zeta = \frac{1}{a}\lambda(\lambda-\mu)(\lambda-\nu) : \frac{1}{b}\mu(\mu-\nu)(\mu-\lambda) : \frac{1}{c}\nu(\nu-\lambda)(\nu-\mu),$$

or, what is the same thing,

$$\xi : \eta : \zeta = \frac{1}{a}\quad\frac{\lambda}{\mu-\nu}\quad : \frac{1}{b}\quad\frac{\mu}{\nu-\lambda}\quad : \frac{1}{c}\quad\frac{\nu}{\lambda-\mu},$$

where as before

$$\frac{\lambda}{a}+\frac{\mu}{b}+\frac{\nu}{c}=0,$$

and eliminating λ, μ, ν, we find

$$a\xi(\eta-\zeta)^2+b\eta(\zeta-\xi)^2+c\zeta(\xi-\eta)^2=0.$$

In fact we find at once

$$\begin{aligned} a\xi(\eta-\zeta)^2 : b\eta(\zeta-\xi)^2 : c\zeta(\xi-\eta)^2 = {} & (\mu-\nu)\lambda\left\{\frac{1}{b}\mu(\lambda-\mu)-\frac{1}{c}\nu(\nu-\lambda)\right\} \\ & : (\nu-\lambda)\mu\left\{\frac{1}{c}\nu(\mu-\nu)-\frac{1}{a}\lambda(\lambda-\mu)\right\} \\ & : (\lambda-\mu)\nu\left\{\frac{1}{a}\lambda(\nu-\lambda)-\frac{1}{b}\mu(\mu-\nu)\right\}, \end{aligned}$$

and the sum of the three expressions on the right-hand side is

$$=-(\mu-\nu)(\nu-\lambda)(\lambda-\mu)\left(\frac{\lambda}{a}+\frac{\mu}{b}+\frac{\nu}{c}\right)=0,$$

which verifies the result just obtained.

The tangential equation of the envelope is thus

$$a\xi(\eta-\zeta)^2+b\eta(\zeta-\xi)^2+c\zeta(\xi-\eta)^2=0,$$

or the envelope is a curve of the third class having as a double tangent the line $\xi=\eta=\zeta$, that is the line infinity; in fact for these values the equation $\xi x+\eta y+\zeta z=0$ becomes $x+y+z=0$, which is the equation of the line infinity. The curve is *therefore* a curve of the fourth order, the equation of which is

$$\frac{1}{(x+y+z)^2}\text{ recip. }\{a\xi(\eta-\zeta)^2+b\eta(\zeta-\xi)^2+c\zeta(\xi-\eta)^2\}=0,$$

where the reciprocant in question may be calculated from the before mentioned table 67, viz. multiplying by 3 in order to avoid fractions, the coefficients of the table are

$$\begin{aligned}&(a,\ b,\ c,\ f,\ g,\ h,\ i,\ j,\ k,\quad l\qquad)\\ =&(0,\ 0,\ 0,\ c,\ a,\ b,\ b,\ c,\ a,\ -a-b-c)\end{aligned}$$

respectively, and for the facients $(\xi,\ \eta,\ \zeta)$ of the table we have to write $(x,\ y,\ z)$. The expression of the reciprocant is

$$= b^2c^2x^6 + c^2a^2y^6 + a^2b^2z^6 + \&c.,$$

and dividing by $(x+y+z)^2$ we have the equation of the envelope in the form

$$b^2c^2x^4 + c^2a^2y^4 + a^2b^2z^4 + \&c. = 0,$$

which must of course be identical with the former result

$$bcyz\,(cy-bz)^2 + cazx\,(az-cx)^2 + abxy\,(bx-ay)^2 + \&c. = 0.$$

Instead of discussing the curve of the third class

$$a\xi\,(\eta-\zeta)^2 + b\eta\,(\zeta-\xi)^2 + c\zeta\,(\xi-\eta)^2 = 0,$$

it will be convenient to write $(x,\ y,\ z)$ in place of $(\xi,\ \eta,\ \zeta)$, and discuss the curve of the third order, or cubic curve

$$U = ax\,(y-z)^2 + by\,(z-x)^2 + cz\,(x-y)^2 = 0,$$

which is of course a curve having a node at the point $(x=y=z)$, or say at the point $(1,\ 1,\ 1)$, and having therefore three inflexions lying in a line. The equation of the tangents at the node is found to be

$$a\,(y-z)^2 + b\,(z-x)^2 + c\,(x-y)^2 = 0,$$

that is, at the node the second derived functions of U are proportional to

$$(b+c,\ c+a,\ a+b,\ -a,\ -b,\ -c).$$

The equation of the Hessian may be found directly, or by means of the table, No. 61, in my memoir above referred to. It is as follows:

$$\begin{aligned}&(b+c)\,\{a\,(b+c)+2bc\}\,x^3\\ &+(c+a)\,\{b\,(c+a)+2ca\}\,y^3\\ &+(a+b)\,\{c\,(a+b)+2ab\}\,z^3\\ &-(3a^2+2bc+2ca+2ab)\,(cy^2z+byz^2)\\ &-(3b^2+2bc+2ca+2ab)\,(az^2x+czx^2)\\ &-(3c^2+2bc+2ca+2ab)\,(bx^2y+axy^2)\\ &+\{4\,(bc^2+b^2c+ca^2+c^2a+ab^2+a^2b)+6abc\}\,xyz=0.\end{aligned}$$

I find that in the function $b^2c^2x^6 + \&c.$ the term in z^5x is

$$= a\,(4a^2b + 4a^2c + 4ab^2 + 4abc - 2b^2c),$$

whence in the function $b^2c^2x^4 + \&c.$ the term in z^3x is

$$= z^3xa\,(4a^2b + 4a^2c + 2ab^2 + 4abc - 2b^2c),$$

a portion whereof is $= 4z^3x\,(a^3b + a^3c)$; and we thus obtain the numerical factor $= 4$, and thence the identity

$$b^2c^2x^4 + c^2a^2y^4 + a^2b^2z^4 + \&c. = 4bcyz\,(cy - bz)^2 + \ldots + \&c.$$

which equation I represent by

$$Ax^3 + By^3 + Cz^3 + 3\,(Fy^2z + Gz^2x + Hx^2y + Iyz^2 + Jzx^2 + Kxy^2) + 6\,Lxyz = 0,$$

or

$$(A,\ B,\ C,\ F,\ G,\ H,\ I,\ J,\ K,\ L \between x,\ y,\ z)^3 = 0,$$

viz. writing for shortness

$$M = bc + ca + ab,$$

the values of the coefficients are as follows:

$$\begin{aligned} A &= 3\,(b + c)\,(bc + M), & F &= -c\,(3a^2 + M), & I &= -b\,(3a^2 + M),\\ B &= 3\,(c + a)\,(ca + M), & G &= -a\,(3b^2 + M), & J &= -c\,(3b^2 + M),\\ C &= 3\,(a + b)\,(ab + M), & H &= -b\,(3c^2 + M), & K &= -a\,(3c^2 + M), \end{aligned}$$

$$L = 2\,(a + b + c)\,M - 3abc.$$

I remark that the cubic having a node at the point (1, 1, 1), the Hessian has at this point a node with the same tangents. The second derived functions for the Hessian are therefore at the node proportional to those of the cubic; it is easy to verify that we have in fact

$$\begin{aligned} A + H + J &= (b + c)\,M, & L + F + I &= -aM,\\ K + B + F &= (c + a)\,M, & J + L + G &= -bM,\\ G + I + C &= (a + b)\,M, & H + K + L &= -cM, \end{aligned}$$

these values give also

$$\begin{aligned} A + K + G + 2L + 2J + 2H &= 0,\\ H + B + I + 2F + 2L + 2K &= 0,\\ J + F + C + 2I + 2G + 2L &= 0, \end{aligned}$$

equations which merely express that the first derived functions vanish at the node. If, by these equations we express A, B, C in terms of the other coefficients, and substitute these values in the equation of the Hessian, this may be expressed in the form

$$\begin{aligned} &(y - z)^2\,\{\qquad L\,x + (2F + I + L)\,y + (F + 2I + L)\,z\}\\ &+ (z - x)^2\,\{(G + 2J + L)\,x + \qquad L\,y + (2G + J + L)\,z\}\\ &+ (x - y)^2\,\{(2H + K + L)\,x + (H + 2K + L)\,y + \qquad L\,z\} = 0, \end{aligned}$$

a form which puts in evidence the node (1, 1, 1).

I write

$$X = y - z, \quad Y = z - x, \quad Z = x - y,$$

so that we have identically

$$X + Y + Z = 0,$$

and that the equation of the tangents at the node is

$$aX^2 + bY^2 + cZ^2 = 0.$$

I write also for shortness

$$\begin{aligned} F' &= 2F + I + L, & I' &= F + 2I + L, \\ G' &= 2G + J + L, & J' &= G + 2J + L, \\ H' &= 2H + K + L, & K' &= H + 2K + L, \end{aligned}$$

the equation of the cubic is then

$$U = axX^2 + byY^2 + czZ^2, = 0,$$

and that of the Hessian is

$$\tilde{H}U = X^2(Lx + F'y + I'z) + Y^2(J'x + Ly + G'z) + Z^2(H'x + K'y + Lz).$$

Now observing that we have

$$\begin{aligned} L + 3aM &= L - 3(I + F + L) = -F' - I', \\ L + 3bM &= L - 3(G + J + L) = -G' - J', \\ L + 3cM &= L - 3(H + K + L) = -H' - K', \end{aligned}$$

we find

$$\begin{aligned} \tilde{H}U + 3MU = \;& X^2(I'Y - F'Z) \\ &+ Y^2(J'Z - G'X) \\ &+ Z^2(K'X - H'Y), \end{aligned}$$

which shows that the function $\tilde{H}U + 3MU$ is a cubic function of $y - z$, $z - x$, $x - y$, decomposable therefore into three linear factors; and the equation $\tilde{H}U + 3MU = 0$, is consequently that of the three lines drawn from the node to the three inflexions of the cubic (or the Hessian). We know also that the Hessian of the three lines is the pair of tangents at the node([1]), viz. that regarding any one of the variables X, Y, Z as a linear function of the third of them (in virtue of the equation $X + Y + Z = 0$), then that the cubic function of X, Y, Z has $aX^2 + bY^2 + cZ^2$ for its Hessian.

[1] Taking as the canonical form of a nodal cubic $U = x^3 + y^3 + 6lxyz = 0$, then we have $\tilde{H}U = x^3 + y^3 - 2lxyz = 0$; $x^3 + y^3 = 0$ is the equation of the lines from the node to the inflexions, and the Hessian of the binary cubic $x^3 + y^3$ is xy, where $xy = 0$ is the equation of the tangents at the node. We obtain as the only linear functions of U, $\tilde{H}U$ which are decomposable, $x^3 + y^3$ and xyz, the equation $xyz = 0$ gives $xy = 0$ which belongs to the tangents at the node or else $z = 0$, which is the equation of the line through the three inflexions: this line is so obtained a little further on in the text.

It is interesting to verify this; I write $Z=-X-Y$, the cubic function then assumes the form

$$(\alpha,\ \beta,\ \gamma,\ \delta \between X,\ Y)^3,$$

where $(\alpha, \beta, \gamma, \delta)$ have the values presently given.

The Hessian is

$$(2\alpha\gamma - 2\beta^2)\,X^2 + (\alpha\delta - \beta\gamma)\,.\,2XY + (2\beta\delta - 2\gamma^2)\,Y^2,$$

or writing $2XY = Z^2 - X^2 - Y^2$, this is

$$= (2\alpha\gamma - 2\beta^2 - \alpha\delta + \beta\gamma)\,X^2 + (2\beta\delta - 2\gamma^2 - \alpha\delta - \beta\gamma)\,Y^2 + (\alpha\delta - \beta\gamma)\,Z^2.$$

We find after some easy reductions,

$$\begin{aligned}
\tfrac{1}{3}\alpha &= \phantom{-H' + {}} K' + F', &&= -3\ \ (a+c)(ac+M),\\
\beta &= -H' + 2K' + I' + F', &&= 3a\,(a+c)(b\ \ +c\ \ ),\\
-\ \gamma &= -K' + 2H' + J' + G', &&= 3b\,(a+c)(b\ \ +c\ \ ),\\
-\tfrac{1}{3}\delta &= \phantom{-K' + {}} H' + J', &&= -3\ \ (b+c)(bc+M),
\end{aligned}$$

and hence

$$\alpha\delta - \beta\gamma = -81\,(a+c)(b+c)\,\{(ac+M)(bc+M) - ab\,(a+c)(b+c)\},$$

where the expression in { } is

$$\begin{aligned}
&= (ab + 2ac + bc)(ab + ac + 2bc) - ab\,(ab + ac + bc + c^2),\\
&= c\,\{ab\,(3a + 3b) + c\,(b + 2a)(a + 2b) - ab\,(a + b + c)\},\\
&= 2c\,(a + b)(bc + ca + ab),
\end{aligned}$$

and therefore

$$\alpha\delta - \beta\gamma = -162\,(b+c)(c+a)(a+b)\,Mc\,;$$

the other coefficients may be similarly calculated, and omitting the merely numerical factor, we have

$$\text{Hessian} = (b+c)(c+a)(a+b)\,M\,(aX^2 + bY^2 + cZ^2),$$

which is right.

I write next

$$\begin{aligned}
\tilde{H}U + 3MU - \vartheta U = {} & X^2\,(-a\vartheta x + I'Y\ - F'Z)\\
& + Y^2\,(-b\vartheta y + J'Z\ - G'X)\\
& + Z^2\,(-a\vartheta z + K'X - H'Y),
\end{aligned}$$

or writing $x = z - Y$, $y = z + X$, this is

$$\begin{aligned}\tilde{H}U + 3MU - \vartheta U = & - \vartheta z(aX^2 + bY^2 + cZ^2) \\ & + X^2\{(I' + a\vartheta)Y - F'Z\} \\ & + Y^2\{J'Z - (G' + b\vartheta)X\} \\ & + Z^2\{K'X - H'Y \qquad \},\end{aligned}$$

we may determine ϑ, so that the cubic function of X, Y, Z contains the factor $aX^2 + bY^2 + cZ^2$; writing $Z = -X - Y$, then

	Contains the factor	Quotient is
$X^3\ (\ K' + F' \qquad\qquad\qquad)$	$(a+c)\ X^2$	$-3(ac+M)X$
$+ X^2Y(-H' + 2K' + F' + I' + a\vartheta)$	$+\quad 2c\quad XY$	$+3(bc+M)Y.$
$+ XY^2(\ K' - 2H' - G' - J' - b\vartheta)$	$+(b+c)\ Y^2$	
$+ Y^3\ (-J' - H' \qquad\qquad)$		

We have seen that

$$K' + F' = -3(a+c)(ac+M),$$

$$J' + H' = -3(b+c)(bc+M),$$

whence the quotient is, as above stated,

$$= -3(ac+M)X + 3(bc+M)Y.$$

Comparing the coefficients of X^2Y, we have

$$\begin{aligned}a\vartheta &= -(-H' + 2K' + F' + I') + 3(a+c)(bc+M) - 6c(ac+M), \\ &= 9a(a+c)(b+c) + 3(a+c)(ab+ac+2bc) - 6c(ab+2ac+bc), \\ &= 12a(bc+ca+ab) = 12aM,\end{aligned}$$

that is $\vartheta = 12M$; and the same value would have been obtained by comparing the coefficients of XY^2. Hence $\tilde{H}U - \vartheta MU$ divides by $aX^2 + bY^2 + cZ^2$, the quotient being

$$-12Mz - 3(ac+M)X + 3(bc+M)Y,$$

which is

$$= -12Mz - 3(ac+M)(y-z) + 3(bc+M)(z-x),$$

or, finally it is

$$= -3\{(bc+M)x + (ca+M)y + (ab+M)z\},$$

and we thus have

$$\tilde{H}U - \vartheta MU = -3(aX^2 + bY^2 + cZ^2) \times \{(bc+M)x + (ca+M)y + (ab+M)z\},$$

so that the three inflexions are the intersections of the cubic curve by the line

$$(bc+M)x + (ca+M)y + (ab+M)z = 0.$$

It may be noticed, that if we write

$$\begin{aligned}x + y + z &= u, \\ bcx + cay + abz &= -Mu, \\ ax + by + cz &= v,\end{aligned}$$

then x, y, z will be as

$$\begin{aligned} &(b-c)\{(2M-bc)u-av\}\\ :\ &(c-a)\{(2M-ca)u-bv\}\\ :\ &(a-b)\{(2M-ab)u-cv\},\end{aligned}$$

and substituting these values in the equation

$$ax(y-z)^2+by(z-x)^2+cz(x-y)^2=0$$

of the cubic, we have a cubic equation for the ratio $(u:v)$; and thence the values (x, y, z) for the coordinates of the inflexions.

It may be added, that we have

$$\begin{aligned}12MU=-3(aX^2+bY^2+cZ^2)\{(bc+M)x+(ca+M)y+(ab+M)z\}\\ +\{X^2(I'Y-F'Z)+Y^2(J'Z-G'X)+Z^2(K'X-H'Y)=0,\end{aligned}$$

which is the equation of the cubic expressed in the canonical form.

Pp. 175—179. Effecting the process indicated p. 73, but writing for greater convenience (x, y, z) in place of (X, Y, Z), so that the substitution to be made is

$$\begin{aligned}&(a,\ b,\ c,\ f,\ g,\ h,\ i,\ j,\ k,\ l\ )\\ =&(6x,\ 6y,\ 6z,\ -2y,\ -2z,\ -2x,\ -2z,\ -2x,\ -2y,\ x+y+z),\end{aligned}$$

respectively (where I have corrected a misprint in the formula as originally given) I find the equation of the envelope to be

$$\begin{aligned}
&4yz\,(y-z)^2\,\mathrm{a}^4\\
+\ &4zx\,(z-x)^2\,\mathrm{b}^4\\
+\ &4xy\,(x-y)^2\,\mathrm{c}^4\\
+\ &4zx\,(z^2+x^2+3yz-2zx+5xy)\,\mathrm{b}^3\mathrm{c}\\
+\ &4xy\,(x^2+y^2+3zx-2xy+5yz)\,\mathrm{c}^3\mathrm{a}\\
+\ &4yz\,(y^2+z^2+3xy-2yz+5zx)\,\mathrm{a}^3\mathrm{b}\\
+\ &4xy\,(x^2+y^2+3yz-2xy+5zx)\,\mathrm{bc}^3\\
+\ &4yz\,(y^2+z^2+3zx-2yz+5xy)\,\mathrm{ca}^3\\
+\ &4zx\,(z^2+x^2+3xy-2zx+5yz)\,\mathrm{ab}^3\\
+\ &x\,(x^3-2x^2y-2x^2z+xy^2+38xyz+xz^2+12y^2z+12yz^2)\,\mathrm{b}^2\mathrm{c}^2\\
+\ &y\,(y^3-2y^2z-2y^2x+yz^2+38xyz+yx^2+12z^2x+12zx^2)\,\mathrm{c}^2\mathrm{a}^2\\
+\ &z\,(z^3-2z^2x-2z^2y+zx^2+38xyz+zy^2+12x^2y+12xy^2)\,\mathrm{a}^2\mathrm{b}^2\\
+\ &2yz\,(11x^2+y^2+z^2-2yz+24xy+24zx)\,\mathrm{a}^2\mathrm{bc}\\
+\ &2zx\,(11y^2+z^2+x^2-2zx+24yz+24xy)\,\mathrm{b}^2\mathrm{ca}\\
+\ &2xy\,(11z^2+x^2+y^2-2xy+24zx+24yz)\,\mathrm{c}^2\mathrm{ab}=0.
\end{aligned}$$

The function on the left-hand side is the quotient by $-48(\mathrm{a}+\mathrm{b}+\mathrm{c})^2$ of the sextic function, Table 67 of my third Memoir on Quantics, [144]; the foregoing quotient was calculated without using the coefficient of the term in $\mathrm{a}^2\mathrm{b}^2\mathrm{c}^2$ $(\xi\,\eta^2\zeta^2)$ of the table, but by way of verification, I calculated from the table the term in question, and found it to be

$$\begin{aligned} &(x+y+z)^4 \\ -\;\; &2(x+y+z)^2(yz+zx+xy) \\ +\; &296(x+y+z)\,xyz \\ -\;\; &8(y^2z^2+z^2x^2+x^2y^2), \end{aligned}$$

and this should consequently be equal to the coefficient of $\mathrm{a}^2\mathrm{b}^2\mathrm{c}^2$ in the product of $(\mathrm{a}+\mathrm{b}+\mathrm{c})^2$ into the foregoing quartic function of (a, b, c) that is, it should be

$$\begin{aligned} = \quad & x\,(x^3-2x^2y-2x^2z+xy^2+38xyz+xz^2+12y^2z+12yz^2) \\ + \quad & y\,(y^3-2y^2z-2y^2x+yz^2+38xyz+yx^2+12z^2x+12zx^2) \\ + \quad & z\,(z^3-2z^2x-2z^2y+zx^2+38xyz+zy^3+12x^2y+12xy^2) \\ & +4yz\,(11x^2+y^2+z^2-2yz+24xy+24zx) \\ & +4zx\,(11y^2+z^2+x^2-2zx+24yz+24xy) \\ & +4xy\,(11z^2+x^2+y^2-2xy+24zx+24yz), \end{aligned}$$

which is accordingly found to be the case.

397.

SPECIMEN TABLE $M \equiv a^{\alpha} b^{\beta}$ (MOD. N) FOR ANY PRIME OR COMPOSITE MODULUS.

[From the *Quarterly Journal of Pure and Applied Mathematics*, vol. IX. (1868), pp. 95—96 and plate.]

If N be a prime number, and a one of its primitive roots, then any number M prime to N, or what is the same thing, any number in the series $1, 2, \ldots N-1$, may be exhibited in the form $M \equiv a^{\alpha}$ (Mod. N); where α is said to be the index of M in regard to the particular root a. Jacobi's *Canon Arithmeticus* (Berlin, 1839), contains a series of tables, giving the indices of the numbers $1, 2, 3 \ldots N-1$ for every prime number N less than 1000, and giving conversely for each such prime number the numbers M which correspond to the indices $\alpha = 1, 2, \ldots (N-1)$ (*Tabulæ Numerorum ad Indices datos pertinentium et Indicum Numero dato correspondentium*). A similar theory applies, it is well known, to the composite numbers; the only difference is, that in order to exhibit for a given composite number N the different numbers less than N and prime to it, we require not a single root a, but two or more roots $a, b, \ldots$ and that in terms of these we have $M = a^{\alpha} b^{\beta} \ldots$ (Mod. N). For each root a there is an index A (or say the Indicator of the root), such that $a^{A} \equiv 1$ (Mod. N), A being the least index for which this equation is satisfied; and the indices $a, b, \ldots$ extend from 1 to $A, B, \ldots$ respectively; the number of different combinations or the product $AB \ldots$, being precisely equal to $\phi(N)$, the number of integers less than N and prime to it. The least common multiple of $A, B \ldots$, is termed the Maximum Indicator, and representing it by I, then for any number M not prime to N, we have $M^{I} \equiv 1$ (Mod. N), a theorem made use of by Cauchy for the solution of indeterminate equations of the first order. Thus $N = 20$, the roots may be taken to be 3, 11; the corresponding exponents are 4, 2 (viz. $3^4 \equiv 1$ (Mod. 20) $11^2 \equiv 1$ (Mod. 20)), and the product of these is 8, the number of integers less than 20 and prime to it; the series [*go to p.* 86]

NOS.	1	2	3	4	5	6	7	8	9	10	11	12	13	14	15	16	17	18	19	20	21	22	23	24	25	26	27	28	29	30
ROOTS		1	2	3	2	5	3	3, 5	2	3	2	5, 7	6	3	2, 11	3, 7	10	5	10	3, 11	2, 13	7	10	5, 7, 13	2	7	2	3, 13	10	7, 11
IND.		1	2	2	4	2	6	2, 2	6	4	10	2, 2	12	6	4, 2	4, 2	16	6	18	4, 2	6, 2	10	22	2, 2, 2	20	12	18	6, 2	28	4, 2
M. I.		1	2	2	4	2	6	2	6	4	10	2	12	6	4	4	16	6	18	4	6	10	22	2	20	12	18	6	28	4
ϕ	0	1	2	2	4	2	6	4	6	4	10	4	12	6	8	8	16	6	18	8	12	10	22	8	20	12	18	12	28	8
	1	0	0	0	0	0	0	0, 0	0	0	0	0, 0	0	0	0, 0	0, 0	0	0	0	0, 0	0, 0	0	0	0, 0, 0	0	0	0	0, 0	0	0, 0
		2	1		1		2		1		1		5		1, 0		10		17		1, 0		8		1		1		11	
			3	1	3		1	1, 0		1	8		8	1		1, 0	11		5	1, 0		4	20		7	8		1, 0	27	
				4	2		4		2		2		10		2, 0		4		16		2, 0		16		2		2		22	
					5	1	5	0, 1	5		4	1, 0	9	5		1, 1	7	1	2		1, 1	2	15	1, 0, 0		3	5	2, 1	18	
						6	3				9		1				5		4				6		8				10	
							7	1, 1	4	3	7	0, 1	7		1, 1	0, 1	9	2	12	3, 0		1	21	0, 1, 0	5	1	16		20	1, 0
								8	3		3		3		3, 0		14		15		3, 0		2		3		3		5	
									9	2	6		4	2		2, 0	6		10	2, 0		8	18		14	4		2, 0	26	
										10	5		2				1		1		2, 1		1				6		1	
											11	1, 1	11	4	0, 1	3, 0	13	5	6	0, 1	5, 0		3	1, 1, 0	16	5	13	1, 1	23	0, 1
												12	6				15		3				14		9				21	
													13	3	3, 1	3, 1	12	4	13	1, 1	0, 1	3	12	0, 0, 1	19		8	0, 1	2	3, 0
														14	2, 1		3		11				7		6		14		3	
															15	2, 1	2		7			6	13			11		3, 1	17	
																16	8		14		4, 0		10		4		4		16	
																	17	3	8	3, 1	5, 1	7	17	1, 0, 1	13	10	15	4, 1	7	1, 1
																		18	9				4		15				9	
																			19	2, 1	4, 1	9	5	0, 1, 1	8	7	12	5, 0	15	
																				20	3, 1		9				7		12	2. 0
																					21	5	19		12	9			19	
																						22	11		17		14		6	
																							23	1, 1, 1	11	2	11	5, 1	24	3, 1
																								24	10				4	
																									25	6	10	4, 0	8	
																										26	9		13	
																											27	3, 0	25	
																												28	14	
																													29	2, 1
																														30

31	32	33	34	35	36	37	38	39	40	41	42	43	44	45	46	47	48	49	50	
17	3, 15	2, 10	3	2, 6	5, 19	5	3	2, 14	3, 11, 21	6	5, 13	28	3, 21	2, 26	5	10	5, 7, 17	3	3	
30	8, 2	10, 2	16	12, 2	6, 2	36	18	12, 2	4, 2, 2	40	6, 2	42	10, 2	12, 2	22	46	4, 2, 2	42	20	
30	8	10	16	12	6	36	18	12	4	40	6	42	10	12	22	46	4	42	20	
30	16	20	16	24	12	36	18	24	16	40	12	42	20	24	22	46	16	42	20	
0	0, 0	0, 0	0	0, 0	0, 0	0	0	0, 0	0, 0, 0	0	0, 0	0	0, 0	0, 0	0	0	0, 0, 0	0	0	1
12		1, 0		1, 0		11		1, 0		26		39		1, 0		30		26		2
13	1, 0		1	11, 0		34	1		1, 0, 0	15		17	1, 0		16	18		1	1	3
24		2, 0		2, 0		22		2, 0		12		36		2, 0		14		10		4
20	7, 1	9, 1	5		1, 0	1	4	9, 0		22	1, 0	5	8, 0		1	17	1, 0, 0	29		5
25				0, 1		9				1		14				2		27		6
4	2, 1	2, 1	11		2, 1	28	6	11, 1	3, 0, 1	39		7	9, 1	1, 1	19	38	0, 1, 0		15	7
6		3, 0		3, 0		33		3, 0		38		33		3, 0		44		36		8
26	2, 0		2	10, 0		32	2		2, 0, 0	30		34	2, 0		10	36		2	2	9
2		0, 1				12		10, 0		8		2				1		13		10
29	7, 0		7	8, 0	5, 1	6	12	7, 0	0, 1, 0	3	5, 1	6		4, 1	9	27	3, 1, 0	40	8	11
7				1, 1		20				27		11				32		11		12
23	1, 1	6, 1	4	3, 1	4, 0	13	17		1, 1, 1	31	0, 1	40	2, 1	11, 1	14	3	3, 0, 1	33	17	13
16		3, 1				3		0, 1		25		4		2, 1		22				14
3	0, 1		6			35	5			37		22	9, 0		17	35		30		15
18		4, 0		4, 0		8		4, 0		24		30		4, 0		28		20		16
1	4, 0	9, 0		5, 1	3, 0	5	16	2, 1		33	5, 0	16	8, 1	9, 0	7	42	0, 0, 1	25	19	17
8				11, 0		7				16		31				20		28		18
22	5, 0	8, 1	14	10, 1	0, 1	25		5, 1	2, 1, 0	9	4, 1	29	1, 1	6, 0	15	29	1, 1, 1	35	14	19
14		1, 1				23		11, 0		34		41				31		39		20
17	3, 1		12			26	7		0, 0, 1	14		24	0, 1		13	10			16	21
11				9, 0		17		8, 0		29		3		5, 1		11		24		22
21	6, 1	5, 1	15	7, 0	1, 1	21	14	10, 1	1, 0, 1	36	1, 1	20	5, 0	11, 0		39	0, 1, 1	38	13	23
19				2, 1		31				13		8				16		37		24
10	6, 0	8, 0	10		2, 0	2	8	6, 0		4	2, 0	10	6, 0		2	34	2, 0, 0	16		25
5		7, 1		4, 1		24				17		37		0, 1		33		17		26
9	3, 0		3	9, 1		30	3		3, 0, 0	5		9	3, 0		4	8		3	3	27
28		4, 1				14		1, 1		11		1		3, 1		6				28
27	5, 1	7, 0	13	6, 0	5, 0	15	11	4, 1	2, 0, 1	7	3, 1	25	4, 1	10, 1	18	43	3, 0, 0	18	6	29
15						10				23		19				19		14		30
31	4, 1	6, 0	9	8, 1	4, 1	27	15	9, 1	0, 1, 1	28	2, 1	32	7, 0	8, 0	6	5	2, 1, 0	7	4	31
	32	5, 0		5, 0		19		5, 0		10		27		5, 0		12		4		32
		33	8	7, 1		4	13		1, 1, 0	18		23			3	45		41	9	33
			34	6, 1		16		3, 1		19		13		10, 0		26		9		34
				35	31	29	10	8, 1		21		12	7, 1		20	9	1, 1, 0			35
					36	18				2		28				4		12		36
						37	9	7, 1	3, 1, 1	32	4, 0	35	4, 0	9, 1	21	24	1, 0, 1	32	7	37
							38	6, 1		35		26		7, 0		13		19		38
								39	2, 1, 1	6		15	3, 1		8	21		34	18	39
									40	20		38				15		23		40
										41	3, 0	18	6, 1	8, 1	12	25	2, 0, 1	15	12	41
											42	21				40				42
												43	5, 1	7, 1	5	37	3, 1, 1	6	5	43
													44	6, 1		41		8		44
														45	11	7		31		45
															46			22		46
																47	2, 1, 1	5	11	47
																	48	21		48
																		49	10	49
																			50	50

[*from p.* 83] of these is in fact 1, 3, 7, 9, 11, 13, 17, 19, each of which is expressible in the required form, viz. $1 \equiv 3^0 . 11^0$, $3 \equiv 3^1 . 11^0$, $7 = 3^3 . 11^0$, &c. (Mod. 20): the maximum indicator is 4; viz. $1^4 \equiv 1$, $3^4 \equiv 1$, $7^4 \equiv 1$, &c. (Mod. 20).

The table pp. 84, 85 gives the Indices for the numbers less than N and prime to it, for all values of N from 1 to 50; the arrangement may be seen at a glance; of the five lines which form a heading, the first contains the numbers N; the second the root or roots belonging to each number N, the third the indicators of these roots, the fourth the maximum indicator, the fifth the number $\phi(N)$. The remaining lines contain the index or indices of each of the ϕN numbers M less than N and prime to it, the number corresponding to such index or indices, being placed outside in the same horizontal line. For example, 30 has the roots 7, 11, indices 4, 2 respectively; the Maximum Indicator is 4, and the number of integers less than 30 and prime to it is 8; taking any such number, say 17, the indices are 1, 1, that is, we have $17 = 7^1 . 11^1$ (Mod. 30).

The foregoing corresponds to the *Tabulæ Indicum Numero dato correspondentium* of Jacobi; on account of multiplicity of roots there does not appear to be any mode of forming a single table corresponding to the *Tabulæ Numerorum ad Indices datos pertinentium;* and there would be no adequate advantage in forming for each number N a separate table in some such form as

$N = 20.$

Roots		Nos.
3	11	
0	0	1
0	1	11
1	0	3
1	1	13
2	0	9
2	1	19
3	0	7
3	1	17

which I have written down in the form of a table of single entry; for although (whenever, as in the present case, the number of roots is only two) it might have been better exhibited as a table of double entry, when the number of roots is three or more it could not of course be exhibited as a table of corresponding multiple entry.

398.

ON A CERTAIN SEXTIC DEVELOPABLE, AND SEXTIC SURFACE CONNECTED THEREWITH.

[From the *Quarterly Journal of Pure and Applied Mathematics*, vol. IX. (1868), pp. 129—142 and 373—376.]

I PROPOSE to consider [first] the sextic developable derived from a quartic equation, viz. taking this to be $(a, b, c, d, e \between t, 1)^4 = 0$, where (a, b, c, d, e) are any linear functions of the coordinates (x, y, z, w), the equation of the developable in question is

$$(ae - 4bd + 3c^2)^3 - 27(ace - ad^2 - b^2e + 2bcd - c^3)^2 = 0.$$

I have already, in the paper "On a Special Sextic Developable," *Quarterly Journal of Mathematics*, vol. VII. (1866), pp. 105—113, [373], considered a particular case of this surface, viz. that in which c was $= 0$, the geometrical peculiarity of which is that the cuspidal edge is there an excubo-quartic curve (of a special form, having two stationary tangents), whereas in the general case here considered it is a sextic curve. There was analytically the convenience that the linear functions being only the four functions a, b, d, e, these could be themselves taken as coordinates, whereas in the present case we have the five linear functions a, b, c, d, e.

The developable

$$(ae - 4bd + 3c^2)^3 - 27(ace - ad^2 - b^2e + 2bcd - c^3)^2 = 0$$

is a sextic developable having for its cuspidal curve the sextic curve

$$ae - 4bd + 3c^2 = 0,$$

$$ace - ad^2 - b^2e + 2bcd - c^3 = 0,$$

(say $I = 0$, $J = 0$, as usual), and having besides a nodal curve the equations of which may be written

$$\begin{array}{cccccccccccc} 6(ac - b^2) & : & 3(ad - bc) & : & ae + 2bd - 3c^2 & : & 3(be - cd) & : & 6(ce - d^2) & : & 9J \\ = \quad a & : & b & : & c & : & d & : & e & : & I, \end{array}$$

viz. these equations are really equivalent to two equations, and they represent a curve of the fourth order which is an excubo-quartic. We may in fact find the equations of the nodal curve by assuming $(a, b, c, d, e \between t, 1)^4$ to be a perfect square, say to avoid fractions that it is $= 3(\alpha t^2 + 2\beta t + \gamma)^2$, then we have

$$a : b : c : d : e = 3\alpha^2 : 3\alpha\beta : \alpha\gamma + 2\beta^2 : 3\beta\gamma : 3\gamma^2,$$

which equations as involving the two arbitrary parameters $\alpha : \beta : \gamma$, give two equations between (a, b, c, d, e), and we may at once by means of them verify the above-mentioned equations of the nodal curve. It also hereby appears that the nodal curve is as stated an excubo-quartic curve; viz. we have between a, b, c, d, e a single linear relation, that is a quadric relation between α, β, γ, and this equation may be satisfied identically by taking for α, β, γ properly determined quadric functions of a variable parameter θ; whence a, b, c, d, e are proportional to quartic functions of the variable parameter θ, or the curve is an excubo-quartic.

The equations of the nodal curve may be presented under a somewhat different form; viz. the cubi-covariant of $(a, b, c, d, e \between t, 1)^4 = 0$ being

$$\left(\begin{array}{l} -\ a^2d + 3abc - 2b^3 \\ -\ a^2e - 2abd + 9ac^2 - 6b^2c \\ -\ 5abe + 15acd - 10b^2d \\ +\ 10ad^2 - 10b^2e \\ +\ 5ade + 10bd^2 - 15bce \\ +\ ae^2 + 2bde - 9c^2e + 6cd^2 \\ +\ be^2 - 3cde + 2d^3 \end{array}\right) \between (t, 1)^6 = 0,$$

say this function, multiplied by 6 to avoid fractions, is

$$(\mathrm{a}, \mathrm{b}, \mathrm{c}, \mathrm{d}, \mathrm{e}, \mathrm{f}, \mathrm{g} \between t, 1)^6,$$

that is

$$\begin{aligned} \mathrm{a} &= 6(-a^2d + 3abc - 2b^3), \\ \mathrm{b} &= 1(-a^2e - 2abd + 9ac^2 - 6b^2c), \\ \mathrm{c} &= 2(-abe + 3acd - 2b^2d), \\ \mathrm{d} &= 3(+ad^2 - b^2e), \\ \mathrm{e} &= 2(+ade + 2bd^2 - 3bce), \\ \mathrm{f} &= 1(+ae^2 + 2bde - 9c^2e + 6cd^2), \\ \mathrm{g} &= 6(+be^2 - 3cde + 2d^3), \end{aligned}$$

then the equations of the nodal curve may be written

$$\mathrm{a} = 0,\ \mathrm{b} = 0,\ \mathrm{c} = 0,\ \mathrm{d} = 0,\ \mathrm{e} = 0,\ \mathrm{f} = 0,\ \mathrm{g} = 0.$$

It may be mentioned that we have identically

$$\begin{aligned} &ae - 4bd + 3c^2 = 0,\\ &af - 3be + 2cd = 0,\\ &ag - 9ce + 8d^2 = 0,\\ &bg - 3cf + 2de = 0,\\ &bf - 4ce + 3d^2 = 0, \end{aligned}$$

and moreover

$$\begin{aligned} &ag - 6bf + 15ce - 10d^2\\ = &\quad -6(bf - 4ce + 3d^2), = +6(I^3 - 27J^2), \end{aligned}$$

so that the equation of the developable may be written in the form

$$ag - 6bf + 15ce - 10d^2 = 0,$$

or in the more simple form

$$bf - 4ce + 3d^2 = 0,$$

each of which puts in evidence the nodal curve on the surface.

The nodal and cuspidal curves meet in the points

$$\frac{a}{b} = \frac{b}{c} = \frac{c}{d} = \frac{d}{e},$$

being, as it is easy to show, a system of four points. The four points in question form a tetrahedron, the equations of the faces of which may be taken to be $x=0$, $y=0$, $z=0$, $w=0$; and the equation of the surface may be expressed in this system of quadriplanar coordinates.

We introduce these coordinates *ab initio*, by taking the quartic function of t to be

$$(a, b, c, d, e \between t, 1)^4 = x(t+\alpha)^4 + y(t+\beta)^4 + z(t+\gamma)^4 + w(t+\delta)^4,$$

that is, by writing

$$\begin{aligned} a &= x + y + z + w,\\ b &= \alpha x + \beta y + \gamma z + \delta w,\\ c &= \alpha^2 x + \beta^2 y + \gamma^2 z + \delta^2 w,\\ d &= \alpha^3 x + \beta^3 y + \gamma^3 z + \delta^3 w,\\ e &= \alpha^4 x + \beta^4 y + \gamma^4 z + \delta^4 w. \end{aligned}$$

Observe that (t_1, t_2, t_3, t_4) being any constant quantities, we thence have

$$\begin{aligned} &e - d\Sigma t_1 + c\Sigma t_1 t_2 - b\Sigma t_1 t_2 t_3 + a t_1 t_2 t_3 t_4\\ &\quad = x(\alpha - t_1)(\alpha - t_2)(\alpha - t_3)(\alpha - t_4)\\ &\quad + y(\beta - t_1)(\beta - t_2)(\beta - t_3)(\beta - t_4)\\ &\quad + z(\gamma - t_1)(\gamma - t_2)(\gamma - t_3)(\gamma - t_4)\\ &\quad + w(\delta - t_1)(\delta - t_2)(\delta - t_3)(\delta - t_4), \end{aligned}$$

and thence in particular

$$e - d\Sigma\alpha + c\Sigma\alpha\beta - b\Sigma\alpha\beta\gamma + a\,\alpha\beta\gamma\delta = 0,$$

viz. this is the linear relation which subsists identically between (a, b, c, d, e), the five linear functions of the coordinates (x, y, z, w).

Starting from the above values of (a, b, c, d, e), we find without difficulty

$$I = (\alpha-\beta)^4 xy + (\alpha-\gamma)^4 xz + (\alpha-\delta)^4 xw + (\beta-\gamma)^4 yz + (\beta-\delta)^4 yw + (\gamma-\delta)^4 zw,$$
$$J = (\alpha-\beta)^2(\beta-\gamma)^2(\gamma-\alpha)^2 xyz + (\alpha-\beta)^2(\beta-\delta)^2(\delta-\alpha)^2 xyw$$
$$+ (\alpha-\gamma)^2(\gamma-\delta)^2(\delta-\alpha)^2 xzw + (\beta-\gamma)^2(\gamma-\delta)^2(\delta-\beta)^2 yzw,$$

but we thus see the convenience of introducing constant multipliers into the expressions of the four coordinates respectively, viz. writing

$$x = (\beta\gamma\delta)^2 x',$$
$$y = (\gamma\delta\alpha)^2 y',$$
$$z = (\delta\alpha\beta)^2 z',$$
$$w = (\alpha\beta\gamma)^2 w',$$

where for shortness

$$(\beta\gamma\delta) = (\beta-\gamma)(\gamma-\delta)(\delta-\beta), \text{ \&c.},$$

or what is the same thing, taking the quartic to be

$$(a, b, c, d, e \between t, 1)^4 = x'(\beta\gamma\delta)^2(t+\alpha)^4 + y'(\gamma\delta\alpha)^2(t+\beta)^4 + z'(\delta\alpha\beta)^2(t+\gamma)^4 + w'(\alpha\beta\gamma)^2(t+\delta)^4,$$

we find

$$J = (\lambda'\mu'\nu')^2(x'y'z' + y'z'w' + z'x'w' + x'y'w'),$$
$$I = (\lambda'\mu'\nu')\{\lambda'(x'w' + y'z') + \mu'(y'w' + z'x') + \nu'(z'w' + x'y')\},$$

where for shortness

$$\lambda' = (\alpha-\delta)^2(\beta-\gamma)^2,$$
$$\mu' = (\beta-\delta)^2(\gamma-\alpha)^2,$$
$$\nu' = (\gamma-\delta)^2(\alpha-\beta),$$

or writing

$$\sqrt{(\lambda')} = (\alpha-\delta)(\beta-\gamma),$$
$$\sqrt{(\mu')} = (\beta-\delta)(\gamma-\alpha),$$
$$\sqrt{(\nu')} = (\gamma-\delta)(\alpha-\beta),$$

we have

$$\sqrt{(\lambda')} + \sqrt{(\mu')} + \sqrt{(\nu')} = 0,$$

and the equation of the developable is thus

$$\{\lambda'(x'w' + y'z') + \mu'(y'w' + z'x') + \nu'(x'w' + x'y')\}^3 - 27\lambda'\mu'\nu'(x'y'z' + y'z'w' + z'x'w' + x'y'w')^2 = 0.$$

Observe that $J = 0$ is a cubic surface passing through each edge of the tetrahedron, and having at each summit a conical point; $I = 0$ is a quadric surface passing through each summit of the tetrahedron, and at each of these points the tangent plane of the quadric surface touches the tangent cone of the cubic surface: to show this it is only necessary to observe that at the point ($x'=0$, $y'=0$, $z'=0$) the tangent cone is $y'z'+z'x'+x'y'=0$, and the tangent plane is $\lambda'x'+\mu'y'+\nu'z'=0$, and that these touch in virtue of the above-mentioned relation $\sqrt{(\lambda')}+\sqrt{(\mu')}+\sqrt{(\nu')}=0$. It follows that on the curve of intersection, or cuspidal edge of the developable, each of the summits is a cuspidal or stationary point, that is, the cuspidal curve has four stationary points; this agrees with the character of the curve as given, Salmon "On the Classification of Curves of Double Curvature," *Camb. and Dubl. Math. Jour.* vol. v. (1850), p. 39, viz. the character is there given

$$\mathrm{a}=6,\ m=6,\ n=4,\ r=6,\ g=3,\ h=6,\ \alpha=0,\ \beta=4,\ x=4,\ y=6,$$

($\beta=4$, that is, there are as stated 4 stationary points).

To find the equations of the nodal curve, instead of transforming the equations as given in terms of (a, b, c, d, e), it is better to deduce these from the equation of the surface; viz. if there is a nodal curve, we must have

$$\begin{aligned}&\delta_{x'}I : \delta_{y'}I : \delta_{z'}I : \delta_{w'}I : 18J\\ =\ &\delta_{x'}J : \delta_{y'}J : \delta_{z'}J : \delta_{w'}J : \quad I.\end{aligned}$$

Writing these under the form $\delta_{x'}I+\theta'\delta_{x'}J=0$, &c., where θ' is regarded as an arbitrary parameter [1], we have

$$\begin{aligned}\lambda'w'+\mu'z'+\nu'y'+\theta'(y'z'+y'w'+z'w')&=0,\\ \lambda'z'+\mu'w'+\nu'x'+\theta'(z'x'+z'w'+x'w')&=0,\\ \lambda'y'+\mu'x'+\nu'w'+\theta'(x'y'+x'w'+y'w')&=0,\\ \lambda'x'+\mu'y'+\nu'z'+\theta'(y'z'+z'x'+x'y')&=0,\end{aligned}$$

which equations (eliminating θ') must be equivalent to two equations only.

I remark that the first three equations may be regarded as a set of linear equations in 1, w', θ', $\theta'w'$; and determining from them the ratios of these quantities, we have, suppose,

$$1 : -w' : \theta' : -\theta'w' = A : B : C : D,$$

where

$$A=\begin{vmatrix}\lambda', & y'z', & y'+z'\\ \mu', & z'x', & z'+x'\\ \nu', & x'y', & x'+y'\end{vmatrix},\qquad B=\begin{vmatrix}y'z', & y'+z', & \mu'z'+\nu'y'\\ z'x', & z'+x', & \nu'x'+\lambda'z'\\ x'y', & x'+y', & \lambda'y'+\mu'x'\end{vmatrix},$$

$$C=\begin{vmatrix}y'+z', & \mu'z'+\nu'y', & \lambda'\\ z'+x', & \nu'x'+\lambda'z', & \mu'\\ x'+y', & \lambda'y'+\mu'x', & \nu'\end{vmatrix},\qquad D=\begin{vmatrix}\mu'z'+\nu'y', & \lambda', & y'z'\\ \nu'x'+\lambda'z', & \mu', & z'x'\\ \lambda'y'+\mu'x', & \nu', & x'y'\end{vmatrix}.$$

[1] The value of θ' is in fact $=-\dfrac{18J}{I}$, that is, instead of the four equations involving an arbitrary parameter θ', we have really four determinate equations.

We have thence $AD - BC = 0$ and (substituting in the fourth equation) $A(\lambda' x' + \mu' y' + \nu' z') + C(y'z' + z'x' + x'y') = 0$; each of these equations must contain the equation of the cone having $(x' = 0,\ y' = 0,\ z' = 0)$ for its vertex, and passing through the nodal curve. The two equations are of the orders 6 and 4 respectively; and as the curve is a quartic curve passing through the vertex in question, the equation of the cone is of the order 3. I have not effected the reduction of the sextic equation, but for the quartic equation, substituting for A, C their values, this is

$$\begin{aligned}
&-(\lambda' x' + \mu' y' + \nu' z')\,[\lambda' x'^2 (y' - z') + \mu' y'^2 (z' - x') + \nu' z'^2 (z' - x')]\\
&+ (y'z' + z'x' + x'y')\,[\lambda'^2 x' (y' - z') + \mu'^2 y' (z' - x') + \nu' z' \ (x' - y')]\\
&+ \{\mu'\nu' (y' - z') + \nu'\lambda' (z' - x') + \lambda'\mu' (x' - y')\}\,(x' + y' + z') = 0,
\end{aligned}$$

which is easily reduced to

$$\begin{aligned}
&\lambda'^2 x' (-x'^2 + y'z' + z'x' + x'y')(y' - z')\\
&+ \mu'^2 y' (-y'^2 + y'z' + z'x' + x'y')(z' - x')\\
&+ \nu'^2 z' \ (-z'^2 + y'z' + z'x' + x'y')(x' - y')\\
&+ \mu'\nu' \,[(x' + y' + z')(y'z' + z'x' + x'y') + x'y'z']\,(y' - z')\\
&+ \nu'\lambda' \,[(x' + y' + z')(y'z' + z'x' + x'y') + x'y'z']\,(z' - x')\\
&+ \lambda'\mu' \,[(x' + y' + z')(y'z' + z'x' + x'y') + x'y'z']\,(x' - y') = 0;
\end{aligned}$$

and I have found that this is transformable into

$$\begin{aligned}
2\,\{x'\sqrt{(\lambda')} + y'\sqrt{(\mu')} + z'\sqrt{(\nu')}\} \times [&y'z'\sqrt{(\lambda')}(\mu' y' - \nu' z')\, z'x'\sqrt{(\mu')}(\nu' z' - \lambda' x')\, x'y'\sqrt{(\nu')}(\lambda' x' - \mu' y')\\
&- x'y'z'\,\{\sqrt{(\mu')} - \sqrt{(\nu')}\}\,\{\sqrt{(\nu')} - \sqrt{(\lambda')}\}\,\{\sqrt{(\lambda')} - \sqrt{(\mu')}\}] = 0,
\end{aligned}$$

viz. the two functions are equivalent in virtue of the relation $\sqrt{(\lambda')} + \sqrt{(\mu')} + \sqrt{(\nu')} = 0$, or, what is the same thing, they only differ by a function $(x', y', z')^4$ into the evanescent factor $\lambda'^2 + \mu'^2 + \nu'^2 - 2\mu'\nu' - 2\nu'\lambda' - 2\lambda'\mu'$. The function in { } equated to zero is therefore the equation of the cubic cone.

I do not stop to give the steps of the investigation in the above form, as the investigation may be very much simplified as follows: by linear combinations of the four equations in x', y', z', w', θ', we deduce

$$\begin{aligned}
(\ \ \lambda' - \mu' - \nu')(x' + w' - y' - z') + 2\theta' (y'z' - x'w') &= 0,\\
(-\lambda' + \mu' - \nu')(y' + w' - z' - x') + 2\theta' (z'x' - y'w') &= 0,\\
(-\lambda' - \mu' + \nu')(z' + w' - x' - y') + 2\theta' (x'y' - z'w') &= 0,\\
(\ \ \lambda' + \mu' + \nu')(x' + y' + z' + w') + 2\theta' (y'z' + z'x' + x'y' + x'w' + y'w' + z'w') &= 0.
\end{aligned}$$

Hence writing

$$\begin{aligned}
\lambda &= \lambda' - \mu' - \nu', & x &= w' + x' - y' - z',\\
\mu &= -\lambda' + \mu' - \nu', & y &= w' - x' + y' - z',\\
\nu &= -\lambda' - \mu' + \nu', & z &= w' - x' - y' + z',\\
& & w &= w' + x' + y' + z',
\end{aligned}$$

we find

$$\begin{aligned}\mu\nu &= \lambda'^2-\mu'^2-\nu'^2+2\mu'\nu',\\ \nu\lambda &= -\lambda'^2+\mu'^2-\nu'^2+2\nu'\lambda',\\ \lambda\mu &= -\lambda'^2-\mu'^2+\nu' \ +2\lambda'\mu',\end{aligned}$$

and thence

$$\mu\nu+\nu\lambda+\lambda\mu=-(\lambda'^2+\mu'^2+\nu'^2-2\mu'\nu'-2\nu'\lambda'-2\lambda'\mu'),=0,$$

that is

$$\frac{1}{\lambda}+\frac{1}{\mu}+\frac{1}{\nu}=0,$$

the relation which connects the new constants λ, μ, ν. Moreover

$$\begin{aligned}yz-xw &= 4(y'z'-x'w'),\\ zx-yw &= 4(z'x'-y'w'),\\ xy-zw &= 4(x'y'-z'w'),\\ 3w^2-x^2-y^2-z^2 &= 8(y'z'+z'x'+x'y'+x'w'+y'w'+z'w'),\end{aligned}$$

and writing for greater convenience $\theta=-\dfrac{2}{\theta'}$, the equations are transformed into

$$\begin{aligned}\theta\lambda x &= xw-yz,\\ \theta\mu y &= yw-zx,\\ \theta\nu z &= zw-xy,\\ 2\theta(\lambda+\mu+\nu)w &= 3w^2-x^2-y^2-z^2,\end{aligned}$$

where

$$\frac{1}{\lambda}+\frac{1}{\mu}+\frac{1}{\nu}=0,$$

viz. these equations, eliminating θ, give the equations of the nodal curve.

From the first three equations eliminating θ, we deduce

$$\begin{aligned}yzw(\mu-\nu) &= x(\mu y^2-\nu z^2),\\ zxw(\nu-\lambda) &= y(\nu z^2-\lambda x^2),\\ xyw(\lambda-\mu) &= z(\lambda x^2-\mu y^2),\end{aligned}$$

or, as these equations may be written,

$$wxyz=\frac{x^2(\mu y^2-\nu z^2)}{\mu-\nu}=\frac{y^2(\nu z^2-\lambda x^2)}{\nu-\lambda}=\frac{z^2(\lambda x^2-\mu y^2)}{\lambda-\mu},$$

which equations, from the mode in which they are obtained, are it is clear equivalent to two equations only. Using the fourth equation, and eliminating θ by substituting therein for $\theta\lambda$, $\theta\mu$, $\theta\nu$ their values from the first three equations, we find

$$2w^2\left(3w-\frac{yz}{x}-\frac{zx}{y}-\frac{xy}{z}\right)=3w^2-x^2-y^2-z^2,$$

that is

$$3w^2 + x^2 + y^2 + z^2 = 2w\left(\frac{yz}{x} + \frac{zx}{y} + \frac{xy}{z}\right),$$

or, what is the same thing,

$$xyz(3w^2 + x^2 + y^2 + z^2) - 2w(y^2z^2 + z^2x^2 + x^2y^2) = 0,$$

we have to show that this is in fact included in the former system, for then the four equations with θ eliminated will it is clear give two equations only.

Observe that the former system may be written

$$w = \frac{x(\mu y^2 - \nu z^2)}{(\mu - \nu)yz},$$

$$(\mu - \nu)y^2z^2 + (\nu - \lambda)z^2x^2 + (\lambda - \mu)x^2y^2 = 0,$$

and that we have thus to show that substituting for w the value $w = \dfrac{x(\mu y^2 - \nu z^2)}{\mu - \nu}$ in the equation

$$xyz(3w^2 + x^2 + y^2 + z^2) - 2w(y^2z^2 + z^2x^2 + x^2y^2) = 0,$$

the result is

$$(\mu - \nu)y^2z^2 + (\nu - \lambda)z^2x^2 + (\lambda - \mu)x^2y^2 = 0.$$

The substitution in question gives

$$\frac{3x^2(\mu y^2 - \nu z^2)^2}{(\mu - \nu)^2 yz} + yz(x^2 + y^2 + z^2) - \frac{2(\mu y^2 - \nu z^2)}{(\mu - \nu)yz}(y^2z^2 + z^2x^2 + x^2y^2) = 0,$$

that is

$$3x^2(\mu y^2 - \nu z^2)^2 + (\mu - \nu)^2 y^2z^2(x^2 + y^2 + z^2) - 2(\mu - \nu)(\mu y^2 - \nu z^2)(y^2z^2 + z^2x^2 + x^2y^2) = 0,$$

which is in fact

$$-\mu^2y^2(y^2 - z^2)(z^2 - x^2) + 2\mu\nu x^2(y^2 - z^2)^2 - \nu^2z^2(y^2 - z^2)(x^2 - y^2) = 0,$$

that is, throwing out the factor $y^2 - z^2$, it is

$$-\mu^2y^2(z^2 - x^2) + 2\mu\nu x^2(y^2 - z^2) - \nu^2z^2(x^2 - y^2) = 0.$$

But in virtue of the equation $\dfrac{1}{\lambda} + \dfrac{1}{\mu} + \dfrac{1}{\nu} = 0$, we have

$$\frac{\mu\nu}{\lambda}\{(\mu - \nu)y^2z^2 + (\nu - \lambda)z^2x^2 + (\lambda - \mu)x^2y^2\}$$

$$= \frac{\mu\nu}{\lambda}[\lambda x^2(y^2 - z^2) + \mu y^2(z^2 - x^2) + \nu z^2(x^2 - y^2)],$$

$$= \mu\nu\, x^2(y^2 - z^2) - (\mu + \nu)[\mu y^2(z^2 - x^2) + \nu z^2(x^2 - y^2)],$$

$$= -\mu^2y^2(z^2 - x^2) + 2\mu\nu\, x^2(y^2 - z^2) - \nu^2z^2(x^2 - y^2),$$

and the required property thus holds good.

We thus see that the equations of the nodal curve are

$$wxyz = \frac{x^2(\mu y^2 - \nu z^2)}{\mu - \nu} = \frac{y^2(\nu z^2 - \lambda x^2)}{\nu - \lambda} = \frac{z^2(\lambda x^2 - \mu y^2)}{\lambda - \mu},$$

the nodal curve is thus the partial intersection of the two cubic scrolls (skew surfaces)

$$(\mu - \nu)\, wyz = x(\mu y^2 - \nu z^2), \quad (\nu - \lambda)\, wzx = y(\nu z^2 - \lambda x^2),$$

viz. taking A, B, C, D to be the summits of the tetrahedron the faces whereof are $x=0$, $y=0$, $z=0$, $w=0$, the first of these has AD for a nodal directrix, BC for a single directrix, BD, CD for generators; the second has BD for a nodal directrix, AC for a single directrix, AD, CD for generators; the surfaces intersect in the line AD twice, the line BD twice, and the line CD; the order of the residual curve, or nodal curve of the developable, is thus $9-(2+2+1)$, $=4$ as it should be.

I remark that the equation

$$(\mu - \nu)\, y^2z^2 + (\nu - \lambda)\, z^2x^2 + (\lambda - \mu)\, x^2y^2 = 0,$$

is the equation of the cone having its vertex at the point D, ($x=0$, $y=0$, $z=0$), and passing through the nodal curve; the lines DA, DB, DC are each of them a nodal line of the cone, or "line through two points" of the curve; for an excubo-quartic curve the number of lines through two points which pass through a given point not on the curve is in fact $=3$.

It remains to introduce the coordinates (x, y, z, w) into the equation of the developable. We have

$$\begin{aligned} 4x' &= w + x - y - z, \\ 4y' &= w - x + y - z, \\ 4z' &= w - x - y + z, \\ 4w' &= w + x + y + z, \end{aligned}$$

and thence

$$\begin{aligned} 16y'z' &= (w-x)^2 - (y-z)^2, \\ 16x'w' &= (w+x)^2 - (y+z)^2, \end{aligned}$$

giving

$$8(x'w' + y'z') = w^2 + x^2 - y^2 - z^2$$

and similarly

$$8(y'w' + z'x') = w^2 - x^2 + y^2 - z^2,$$

and

$$8(z'w' + x'y') = w^2 - x^2 - y^2 + z^2.$$

Moreover

$$\begin{aligned} 16(y'z' + z'x' + x'y') = &\ (w-x)^2 - (y-z)^2 \\ &+ (w-y)^2 - (z-x)^2 \\ &+ (w-z)^2 - (x-y)^2, \\ = &\ 3w^2 - 2w(x+y+z) \\ &- x^2 - y^2 - z^2 + 2yz + 2zx + 2xy. \end{aligned}$$

Consequently

$$64w'(y'z'+z'x'+x'y') = (w+x+y+z)$$
$$\times\{3w^2-2w(x+y+z)-x^2-y^2-z^2+2yz+2zx+2xy\},$$

$$64x'y'z' = (w+x-y-z)(w-x+y-z)(w-x-y-z),$$
$$= w^3$$
$$-w^2(x+y+z)$$
$$-w(x^2+y^2+z^2-2yz-2zx-2xy)$$
$$+ x^3+y^3+z^3-yz^2-y^2z-zx^2-z^2x-xy^2-x^2y+2xyz.$$

Putting for shortness

$$p=x+y+z, \quad \nabla = x^2+y^2+z^2-2yz-2zx-2xy,$$

the two expressions are

$3w^3$	w^3
$+\ w^2.+p$	$+w^2.-p$
$+\ w.-2p^2-\nabla$	$-w.\nabla$
$-p\nabla$	$+x^3+y^3+z^3-y^2z-yz^2-z^2x-zx^2-x^2y-xy^2+2xyz$

or observing that $-p\nabla$ is

$$=-x^3-y^3-z^3+y^2z+yz^2+z^2x+zx^2+x^2y+xy^2+6xyz,$$

we have

$$64(w'y'z'+w'z'x'+w'x'y'+x'y'z')$$
$$= 4w^3-2w(p^2+\nabla)+8xyz,$$
$$= 4w^3$$
$$-4w(x^2+y^2+z^2)$$
$$+8xyz,$$

that is

$$16(w'y'z'+w'z'x'+w'x'y'+x'y'z') = w^3$$
$$-w(x^2+y^2+z^2)$$
$$+2xyz.$$

Moreover

$$8\{\lambda'(x'w'+y'z')+\mu'(y'x'+z'x')+\nu'(z'w'+x'y')\}$$
$$= \lambda'(w^2+x^2-y^2-z^2)$$
$$+\mu'(w^2-x^2+y^2-z^2)$$
$$+\nu'(w^2-x^2-y^2+z^2),$$
$$=-(\lambda+\mu+\nu)w^2+\lambda x^2+\mu y^2+\nu z^2;$$

and we have

$$\lambda = 2(\beta-\delta)(\gamma-\delta)(\gamma-\alpha)(\alpha-\beta),$$
$$\mu = 2(\gamma-\delta)(\alpha-\delta)(\alpha-\beta)(\beta-\gamma),$$
$$\nu = 2(\alpha-\delta)(\beta-\delta)(\beta-\gamma)(\gamma-\alpha),$$

whence $\lambda\mu\nu = 8\lambda'\mu'\nu'$.

Hence finally λ, μ, ν denoting as just mentioned, and therefore satisfying $\frac{1}{\lambda}+\frac{1}{\mu}+\frac{1}{\nu}=0$, the equation of the developable is

$$\lambda\mu\nu\,\{w^3 - w(x^2+y^2+z^2) + 2xyz\}^2 + 108\,\{(\lambda+\mu+\nu)w^2 - \lambda x^2 - \mu y^2 - \nu z^2\}^3 = 0$$

(say this is $\lambda\mu\nu T^2 + 108S^3 = 0$), and this surface (which has obviously the cuspidal curve $S=0$, $T=0$) has also the nodal curve

$$wxyz = \frac{x^2(\mu y^2 - \nu z^2)}{\mu-\nu} = \frac{y^2(\nu z^2 - \lambda x^2)}{\nu-\lambda} = \frac{z^2(\lambda x^2 - \mu y^2)}{\lambda-\mu}.$$

I will show *à posteriori* that this is actually a nodal curve on the surface. Introducing an arbitrary parameter θ, the equations of the curve may be written *ut suprà*

$$\theta\lambda x = xw - yz,$$
$$\theta\mu y = yw - zx,$$
$$\theta\nu z = zw - xy,$$
$$2\theta(\lambda+\mu+\nu)w = 3w^2 - x^2 - y^2 - z^2,$$

and we have thence, as before,

$$2w\left(3w - \frac{yz}{x} - \frac{zx}{y} - \frac{xy}{z}\right) = 3w^2 - x^2 - y^2 - z^2.$$

Hence

$$2\theta = \frac{3w^2 - x^2 - y^2 - z^2}{(\lambda+\mu+\nu)w} = \frac{-2w(x^2+y^2+z^2) + 6xyz}{-\lambda x^2 - \mu y^2 - \nu z^2},$$
$$= \frac{3w^3 - w(x^2+y^2+z^2)}{(\lambda+\mu+\nu)w^2},$$
$$= \frac{3\{w^3 - w(x^2+y^2+z^2) + 2xyz\}}{(\lambda+\mu+\nu)w^2 - \lambda x^2 - \mu y^2 - \nu z^2},$$
$$= \frac{3T}{S}.$$

Hence writing

$$\theta(-2\lambda x) - (-2wx + 2yz) = 0,$$
$$\theta(-2\mu y) - (-2wy + 2zx) = 0,$$
$$\theta(-2\nu z) - (-2wz + 2xy) = 0,$$
$$\theta\,.\,2(\lambda+\mu+\nu)w - (3w^2 - x^2 - y^2 - z^2) = 0,$$

substituting for θ its value $=\dfrac{3T}{2S}$, and attending to the significations of S and T, we have

$$3T\delta_x S - 2S\delta_x T = 0,$$
$$3T\delta_y S - 2S\delta_y T = 0,$$
$$3T\delta_z S - 2S\delta_z T = 0,$$
$$3T\delta_w S - 2S\delta_w T = 0,$$

which are in fact the conditions to be satisfied in order that the point (x, y, z, w) may belong to a nodal curve of the surface $\lambda\mu\nu T^2 + 108 S^3 = 0$.

It is to be noticed that the coordinates of the before mentioned four points of intersection of the cuspidal and the nodal curves (being as already mentioned stationary points on the cuspidal curve) may be written $x, y, z, w = (1, 1, 1, 1), (1, -1, -1, 1), (-1, 1, -1, 1), (-1, -1, 1, 1)$.

We have thus far considered the developable, or torse, the equation of which is

$$\{\lambda'(x'w' + y'z') + \mu'(y'w' + z'x') + \nu'(z'w' + x'y')\}^3$$
$$- 27\lambda'\mu'\nu'(x'y'z' + x'y'w' + x'z'w' + y'z'w')^2 = 0,$$

where $\surd(\lambda') + \surd(\mu') + \surd(\nu') = 0$; or, what is the same thing, writing a, b, c, in place of $\surd(\lambda'), \surd(\mu'), \surd(\nu')$ respectively, the torse

$$\{a^2(x'w' + y'z') + b^2(y'w' + z'x') + c^2(z'w' + x'y')\}^3 - 27a^2b^2c^2(x'y'z' + x'y'w' + x'z'w' + y'z'w')^2 = 0,$$

where $a + b + c = 0$.

Inverting this by the equations $x', y', z', w' = \dfrac{1}{x}, \dfrac{1}{y}, \dfrac{1}{z}, \dfrac{1}{w}$, we obtain a sextic surface

$$\{a^2(xw + yz) + b^2(yw + zx) + c^2(zw + xy)\}^3 - 27a^2b^2c^2\, xyzw\,(x + y + z + w)^2 = 0,$$

where $a + b + c = 0$; which surface I propose [secondly] to consider in the present paper.

The surface has evidently the singular tangent planes $x = 0$, $y = 0$, $z = 0$, $w = 0$, each osculating the surface in a conic, that is, meeting it in the conic taken thrice, viz.,

$x = 0$, in a conic on the quadric cone $a^2yz + b^2yw + c^2zw = 0$,

$y = 0$, " " " $a^2xw + b^2zx + c^2zw = 0$,

$z = 0$, " " " $a^2xw + b^2yw + c^2xy = 0$,

$w = 0$, " " " $a^2yz + b^2zx + c^3xy = 0$;

and it has also a cuspidal conic, the intersection of the plane $x+y+z+w=0$ with the quadric surface

$$a^2(xw+yz)+b^2(yw+zx)+c^2(zw+yz)=0;$$

it may be observed that the four conics of osculation are also sections of this surface.

The surface has also a nodal curve, the equations of which might be obtained by inversion of those of the nodal curve of the sextic torse above referred to; but I prefer to obtain them independently, in a synthetical manner, as follows:

Take α, β, γ arbitrary, and write

$$\begin{aligned}
-A &= (b-c)\alpha + b\beta - c\gamma, & F &= b\gamma - c\beta,\\
-B &= (c-a)\beta + c\gamma - a\alpha, & G &= c\alpha - a\gamma,\\
-C &= (b-c)\gamma + a\alpha - b\beta, & H &= a\beta - b\alpha,
\end{aligned}$$
$$M=(b-c)\alpha+(c-a)\beta+(a-b)\gamma,$$
$$Q=a^2(b-c)\alpha+b^2(c-a)\beta+c^2(a-b)\gamma:$$

then it is to be shown, that not only the equation of the surface is satisfied, but that also each of the derived equations is satisfied, by the values

$$x:y:z:w=aAGHQ:bBHFQ:cCFGQ:abcFGHM;$$

each of the quantities A, B, C, M, Q is linearly expressible in terms of F, G, H, which are themselves connected by the equation $aF+bG+cH=0$; the foregoing values of x, y, z, w are consequently proportional to quartic functions of a single variable parameter, say $F\div G$; and there is thus an excubo-quartic nodal curve.

To establish the foregoing result, we have

$$\begin{aligned}
aA+bH+cG&=0,\\
aH+bB+cF&=0,\\
aG+bF+cC&=0,\\
aA+bB+cC&=0,\\
aF+bG+cH&=0,\\
F+G+H&=-M,\\
bcF+caG+abH&=Q,
\end{aligned}$$
$$\begin{aligned}
2bcF&=a^2A-b^2B-c^2C,\\
2caG&=-a^2A+b^2B-c^2C,\\
2abH&=-a^2A-b^2B+c^2C,
\end{aligned}$$
$$\begin{aligned}
aAGH+bHBF+cFG&=abcM(\alpha+\beta+\gamma)^2,\\
aGH+bHF+cFG&=-abc(\alpha+\beta+\gamma)^2,\\
a^3AGH+b^3HBF+c^3CFG&=-abc\{Q(\alpha+\beta+\gamma)^2+3FGH\},\\
aBCF+bCAG+cABH&=2Mabc(\alpha+\beta+\gamma)^2,
\end{aligned}$$
$$Q(\alpha+\beta+\gamma)^2+FGH=ABC,$$

which are all of them identical equations; but as to some of them the verification is rather complex.

Hence we have

$$x+y+z=Q(aAGH+bBHF+cCFG)$$
$$=abcMQ(\alpha+\beta+\gamma)^2,$$
$$w=abcMFGH,$$

and thence

$$x+y+z+w=abcM\{Q(\alpha+\beta+\gamma)^2+FGH\}$$
$$=abcMABC.$$

Moreover

$$xyzw=(abc)^2 ABC(FGH)^3 Q^3 M,$$

and

$$27a^2b^2c^2xyzw(x+y+z+w)^2=27(abc)^6(ABCFGHMQ)^3 \ (*).$$

Again

$$(a^2x+b^2y+c^2z)w=abcFGHMQ(a^3AGH+b^3BHF+c^3CFG)$$
$$=(abc)^2 FGHMQ\{Q(\alpha+\beta+\gamma)^2+3FGH\}$$
$$a^2yz+b^2zx+c^2xy=abcFGHQ^2(aBCF+bCAG+cABH)$$
$$=(abc)^2 FGHMQ\,.\,2Q(\alpha+\beta+\gamma)^2,$$

and thence

$$a^2(xw+yz)+b^2(yw+zx)+c^2(xy+zw)$$
$$=3(abc)^2 FGHMQ\{Q(\alpha+\beta+\gamma)^2+FGH\}$$
$$=3(abc)^2 ABCFGHMQ \qquad (*),$$

and the two equations marked (*) verify the equation of the surface.

To verify the derived equations, write for a moment $P=a^2(yz+xw)+b^2(zx+yw)+c^2(xy+zw)$, so that the equation of the surface is $P^3-27a^2b^2c^2xyzw(x+y+z+w)^2=0$, and the derived equation with respect to x is

$$\frac{3}{P}\frac{dP}{dx}=\frac{1}{x}+\frac{2}{x+y+z+w};$$

or substituting for P and $x+y+z+w$ their values, this is

$$\frac{dP}{dx}=\frac{(abc)^2ABCFGHMQ}{x}+2abcQFGH,$$

and similarly for y, z, and w. In particular, considering the derived equation in respect to w, this is

$$a^2x+b^2y+c^2z=abcABCQ+2abcQFGH$$
$$=abcQ(ABC+2FGH),$$

and we have as before

$$a^2x+b^2y+c^2z=Q(a^3AGH+b^3BHF+c^3CFG)$$
$$=abcQ\{\alpha+\beta+\gamma)^2+3FGH\}$$
$$=abcQ(ABC+2FGH),$$

which is thus verified; the verification of the derived equations for y, z, w can be effected, but not quite so easily.

The existence of the excubo-quartic nodal curve is thus established.

399.

ON THE CUBICAL DIVERGENT PARABOLAS.

[From the *Quarterly Journal of Pure and Applied Mathematics*, vol. IX. (1868), pp. 185—189.]

NEWTON reckons five forms, viz. these are the *simplex*, the *complex*, the *crunodal*, the *acnodal*, and the *cuspidal*, but as noticed by Murdoch, the simplex has three different forms, the *ampullate*, the *neutral*, and the *campaniform*. We have thus the 8 forms at once distinguishable by the eye.

Plücker has in all 13 species, the division into species being established or completed geometrically by reference to the asymptotic cuspidal curve (or asymptotic semi-cubical parabola), and analytically as follows, viz. writing the equation in the form

$$y^2 = x^3 - 3cx + 2d,$$

the different species are

simplex,	$y^2 = x^3 - 3cx + 2d$	$\left.\right\} c^2 < d^2,$	ampullate,
„	$y^2 = x^3 - 3cx - 2d$		campaniform,
„	$y^2 = x^3 + 2d,$		neutral,
„	$y^2 = x^3 - 2d,$		campaniform,
„	$y^2 = x^3 + 3cx + 2d,$		„
„	$y^2 = x^3 + 3cx,$		„
„	$y^2 = x^3 + 3cx - 2d,$		„
complex,	$y^2 = x^3 - 3cx + 2d$	$\left.\right\} c^3 > d^2,$	
„	$y^2 = x^3 - 3cx - 2d$		
„	$y^2 = x^3 - 3cx,$		
acnodal,	$y^2 = x^3 - 3cx - 2c\sqrt{(c)},$		
crunodal,	$y^2 = x^3 - 3cx + 2c\sqrt{(c)},$		
cuspidal,	$y^2 = x^3;$		

but of the simplex species, there are five which are to the eye campaniform, and the three complex species have with each other a close resemblance in form.

I remark as regards the simplex forms, that the tangents at the two inflexions meet in a point R on the axis, and that the ampullate, the neutral, and campaniform forms are distinguished from each other according to the position of R, viz. for the ampullate form, R lies within the curve, for the campaniform form R lies without the curve, and for the neutral form, R is at infinity. It is to be observed, as regards the complex forms, that here R always lies without the curve, between the infinite branch and the oval.

The further division of the simplex and complex forms so as to obtain the $7+3$ species of Plücker, may be effected by considering in conjunction with the point R a certain other point I on the axis; it is to be remarked that excluding the inflexion at infinity the cubical divergent parabola has in all eight inflexions, two real and six imaginary, viz. the inflexions lie by pairs on four ordinates, or if x be the abscissa corresponding to an inflexion, x is determined by a quartic equation; this equation has always two real and two imaginary roots, each of the imaginary roots gives a pair of imaginary inflexions; one of the real roots gives a positive value for y^2 and therefore two real inflexions, the tangents at these meet in the above-mentioned point R on the axis; the other real root gives a negative value for y^2 and therefore two imaginary inflexions, but the tangents at these meet in a real point on the axis, and this I call the point I. It is clear that for each of the four pairs of inflexions the tangents at the two inflexions meet at a point on the axis, so that if X be the abscissa of such point, then X is determined by a quartic equation; two of the roots of this equation are imaginary, the other two roots are real, and correspond to the points R and I respectively.

The equation of the curve being as above

$$y^2 = x^3 - 3cx + 2d,$$

then the coordinate x belonging to a pair of inflexions is found by the equation

$$x^4 - 6cx^2 + 8dx - 3c^2 = 0,$$

or what is the same thing,

$$(1,\ 0,\ -c,\ 2d,\ -3c^2 \between x,\ 1)^4 = 0,$$

(the invariant I is $=0$, and hence the discriminant, $=-27J^2$, is negative, or the roots are two real, two imaginary, as already mentioned): the corresponding value of X is easily found to be

$$X = \frac{x^3 + 3cx - 4d}{3(x^2 - c)},$$

and we thence obtain

$$3cX^4 - 4dX^3 - 6c^2X^2 + 12cdX - (c^3 + 4d^2) = 0,$$

or what is the same thing,

$$(3c,\ -d,\ -c^2,\ 3cd,\ -c^3 - 4d^2 \between X,\ 1)^4 = 0,$$

for the equation in X; the quadrinvariant I is $=0$, and hence the discriminant, $=-27J^2$, is negative; that is, the roots are two real, two imaginary, as already mentioned.

Considering the simplex forms, first, if $c=0$, then for the curve

$$y^2 = x^3 + 2d,$$

it appears that R lies at infinity, I within the curve; and for the curve

$$y^2 = x^3 - 2d,$$

that R lies without the curve, I at infinity.

It further appears that when $d=0$, or for the curve,

$$y^2 = x^3 + 3cx,$$

R, I lie equidistant from the vertex, R without, I within the curve.

Hence in the curve

$$y^2 = x^3 + 3cx + 2d,$$

since, when $d=0$, the points R, I are equidistant from the vertex, and for $c=0$, the point R is at infinity, it is easy to infer by continuity that the points R, I lie R without, I within the curve, I being nearer to the vertex.

And similarly in the curve

$$y^2 = x^3 + 3cx - 2d,$$

that the points R, I lie R without, I within the curve, R being nearer to the vertex.

Again, in the curve

$$y^2 = x^3 - 3cx + 2d,$$

since, in the curve $y^2 = x^3 + 3cx + 2d$, R is without, I within the curve, and as c becomes $=0$, R passes off to infinity, it appears that c having changed its sign, or for the curve now in question, R having passed through infinity, will be situate within the curve; that is, R, I lie each of them within the curve.

And similarly for the curve

$$y^2 = x^3 - 3cx - 2d,$$

it appears that R, I lie each without the curve.

Hence, finally, for the simplex forms, we have the 7 species of Plücker, viz.

$$y^2 = x^3 - 3cx + 2d, \ c^3 < d^2,$$

simplex ampullate, R, I within the curve;

$$y^2 = x^3 - 3cx - 2d, \ c^3 < d^2,$$

simplex campaniform, R, I without the curve;

$$y^2 = x^3 + 2d,$$

simplex neutral, I within the curve, R at infinity;

$$y^2 = x^3 - 2d,$$

simplex campaniform quasi-neutral, R without the curve, I at infinity;

$$y^2 = x^3 + 3cx + 2d,$$

simplex campaniform, R without and further from, I within and nearer to the curve;

$$y^2 = x^3 + 3cx,$$

simplex campaniform equidistant, viz. R and I are equidistant from the curve, R without and I within;

$$y^2 = x^3 + 3cx - 2d,$$

simplex campaniform, R without and nearer to, I within and further from the curve.

Passing to the complex forms, suppose for a moment that α is the diameter of the oval and β the distance of the oval from the vertex of the infinite branch; the equation of the curve then is $y^2 = x(x-\alpha)(x-\alpha-\beta)$, or changing the origin so as to make the term in x^2 to vanish, this is

$$y^2 = (x + \tfrac{2}{3}\alpha + \tfrac{1}{3}\beta)(x - \tfrac{1}{3}\alpha + \tfrac{1}{3}\beta)(x - \tfrac{1}{3}\alpha - \tfrac{2}{3}\beta),$$

or, what is the same thing,

$$y^2 = x^3 - \tfrac{2}{3}(\alpha^2 + \alpha\beta + \beta^2)x - \tfrac{1}{27}(\alpha - \beta)(2\alpha + \beta)(\alpha + 2\beta),$$

or comparing this with $y^2 = x^3 - 3cx + 2d$, d is $= +$, 0 or $-$, as $\alpha < \beta$, $\alpha = \beta$, $\alpha > \beta$, or say as the oval is smaller, mean, or larger; viz. the magnitude of the oval is estimated by the relation which the diameter thereof bears to the distance of the oval from the infinite branch. In the case $d = 0$, or for the curve $y^2 = x^3 - 3cx$ it appears (as for the corresponding simplex form $y^2 = x^3 + 3cx$) that the points R, I are equidistant from the point $x = 0$, which is in the present case the middle vertex, or vertex of the oval which vertex is nearest to the infinite branch. As the oval diminishes, so that the curve becomes ultimately acnodal, I remaining within the oval ultimately coincides with the acnode; and as the oval increases so that the curve becomes ultimately crunodal, R remaining between the oval and the infinite branch, ultimately coincides with the crunode; and it hence easily appears by continuity that for a smaller oval I is nearer to, R further from the middle vertex; while for a larger oval, I is further from, R nearer to the middle vertex. Hence for the complex forms the species are

$$y^2 = x^3 - 3cx + 2d,$$

smaller oval, I nearer to, R further from the middle vertex;

$$y^2 = x^3 - 3cx,$$

mean oval, R and I equidistant from the middle vertex;

$$y^2 = x^3 - 3cx - 2d,$$

larger oval, I further from, R nearer to the middle vertex: and the division into species is thus completed.

Cambridge, June 16, 1865.

400.

ON THE CUBIC CURVES INSCRIBED IN A GIVEN PENCIL OF SIX LINES.

[From the *Quarterly Journal of Pure and Applied Mathematics*, vol. IX. (1868), pp. 210—221.]

WE have to consider a pencil of six lines, that is, six lines meeting in a point, and a cubic curve touching each of the six lines. As a cubic curve may be made to satisfy nine conditions, the cubic curve will involve three arbitrary parameters; but if we have any particular curve touching the six lines, then transforming the whole figure homologously, the centre of the pencil being the pole and any line whatever the axis of homology, the pencil of lines remains unaltered, and the new curve touches the six lines of the pencil; the transformation introduces three arbitrary constants, and the general solution is thus given as such homologous transformation of a particular solution. To show the same thing analytically, take $(x=0,\ y=0,\ z=0)$ for the axes of coordinates, the lines $x=0$, $y=0$ being any two lines through the centre of the pencil, so that the equation of the pencil is $(*\between x,\ y)^6=0$, then if $\phi(x,\ y,\ z)=0$ is the equation of a cubic curve touching the six lines, the equation of the general curve touching the six lines will be $\phi(x,\ y,\ \alpha x+\beta y+\gamma z)=0$; or what is the same thing, considering the coordinate z as implicitly containing three arbitrary constants, viz. an arbitrary multiplier and the two arbitrary parameters of the line $z=0$, then the equation $\phi(x,\ y,\ z)=0$ may be taken to be that of the cubic touching the six lines.

Now the given binary sextic $(*\between x,\ y)^6$ may be expressed in the form P^2+Q^3, where P is a cubic function, Q a quadric function, of the coordinates $(x,\ y)$; or, what is the same thing, but introducing for homogeneity a constant c, we may write

$$(*\between x,\ y)^6=c\,[(a,\ h,\ k,\ b\between x,\ y)^3]^2+4\,[(j,\ l,\ f\between x,\ y)^2]^3;$$

in fact, comparing the two sides of this equation, we have each of the seven coefficients of the sextic equal to a function of the seven quantities $a\sqrt{(c)}$, $h\sqrt{(c)}$, $k\sqrt{(c)}$, $b\sqrt{(c)}$, j, l, f; so that conversely, these seven quantities are determinable (not however rationally) in terms of the coefficients of the given sextic. And when the sextic is expressed in the foregoing form, then it will presently be shown that we have

$$(a, h, k, b \between x, y)^3 + 3z(j, l, f \between x, y)^2 + cz^3 = 0,$$

or, what is the same thing,

$$(a, b, c, f, 0, h, 0, j, k \between x, y, z)^3 = 0,$$

as the equation of a cubic curve touching the six given lines; and by what precedes, it appears that this may be taken to be the equation of the general cubic curve which touches the six given lines. On account of the arbitrary constant c, it is sufficient to replace z by $\alpha x + \beta y + z$, or, what is the same thing, to consider $z=0$ as the equation of an arbitrary line, but without introducing therein an arbitrary multiplier.

To sustain the foregoing result, consider the cubic

$$(a, b, c, f, g, h, i, j, k, l \between x, y, z)^3 = 0,$$

then in general if $A = (,, \between x, y, z)^3$, $B = (,, \between x, y, z)^2 (\alpha, \beta, \gamma)$, $C = (,, \between x, y, z)(\alpha, \beta, \gamma)^2$, $D = (,, \between \alpha, \beta, \gamma)^3$, the equation of the pencil of tangents drawn from the point (α, β, γ) to the curve is

$$A^2D^2 - 6ABCD + 4AC^3 + 4B^3D - 3B^2C^2 = 0,$$

but writing for shortness

$$(,, \between x, y, z)^3 = (A', B', C', D' \between 1, z)^3,$$

so that

$$\begin{aligned} A' &= (a, h, k, b \between x, y)^3, \\ B' &= \quad (j, l, f \between x, y)^2, \\ C' &= \qquad (g, i \between x, y)\,, \\ D' &= \qquad\quad c \qquad\quad , \end{aligned}$$

then for the tangents from the point $(x=0,\ y=0)$, writing $(\alpha, \beta, \gamma) = (0, 0, 1)$, we have

$$\begin{aligned} A &= (A', B', C', D' \between 1, z)^3, \\ B &= \quad (B', C', D' \between 1, z)^2, \\ C &= \qquad (C', D' \between 1, z)\,, \\ D &= \qquad\quad D' \qquad\quad , \end{aligned}$$

and thence the equation of the pencil of tangents is

$$A'^2D'^2 - 6A'B'C'D' + 4A'C'^3 + 4B'^3D' - 3B'^2C'^2 = 0.$$

Hence for the curve

$$(a, b, c, f, 0, h, 0, j, k, l \between x, y)^3 = 0,$$

we have $g=0$, $i=0$, and therefore $C'=0$; the equation of the pencil of tangents is $A'^2D'^2+4B'^3D'=0$, or throwing out the constant factor D', and then replacing A', B', D' by their values, the equation of the pencil of tangents is

$$c\,[(a,\ h,\ k,\ b\between x,\ y)^3]^2+4\,[(j,\ l,\ f\between x,\ y)^2]^3=0,$$

which is the before-mentioned result.

The coefficients $a\sqrt{(c)}$, $h\sqrt{(c)}$, $k\sqrt{(c)}$, $b\sqrt{(c)}$, j, l, f, or (as we may call them) the coefficients of the cubic curve, are, it has been seen, functions of the coefficients of the given sextic $(*\between x,\ y)^6$; hence the invariants S and T of the cubic curve are also functions of the coefficients of the sextic, and it is easy to see that they are in fact invariants (not however rational invariants) of the sextic. To verify this, it is only necessary to show that the invariants S and T are functions of the invariants of the functions $\sqrt{(c)}\,.\,(a,\ h,\ k,\ b\between x,\ y)^3$ and $(j,\ l,\ f\between x,\ y)^2$; for if this be so, they will be invariants of the function

$$[c\,(a,\ h,\ k,\ b\between x,\ y)^3]^2+4\,[(j,\ l,\ f\between x,\ y)^2]^3,$$

that is of the sextic. We have in fact the general theorem, that if P, Q, R,... be any quantics in $(x,\ y,\ \ldots)$, and $\phi(P,\ Q,\ R,\ \ldots)$ a function of these quantics, homogeneous in regard to $(x,\ y,\ \ldots)$, then any function of the coefficients of ϕ, which is an invariant of the quantics P, Q, R,... is also an invariant of ϕ.

Considering for greater convenience the function

$$(a,\ h,\ k,\ b\between x,\ y)^3$$

in place of $\sqrt{(c)}\,.\,(a,\ h,\ k,\ b\between x,\ y)^3$, the invariants of the two functions $(a,\ h,\ k,\ b\between x,\ y)^3$ and $(j,\ l,\ f\between x,\ y)^2$ are as follows:

$$\begin{aligned}
\Box &= a^2b^2-6abhk+4ak^3+4bh^3-3h^2k^2,\\
\nabla &= fj-l^2,\\
\Theta &= j(bh-k^2)+l(hk-ab)+f(ak-h^2),\\
R &= +\ 1\ a^2f^3\\
&\quad +\ 6\ abflj\\
&\quad -\ 6\ ahf^2l\\
&\quad -\ 6\ akf^2j\\
&\quad +12\ akfl^2\\
&\quad +\ 1\ b^2j^3\\
&\quad -\ 6\ bhfj^2\\
&\quad +12\ bhjl^2\\
&\quad -\ 6\ bkj^2l\\
&\quad +\ 9\ h^2f^2j\\
&\quad -18\ hkfjl\\
&\quad +\ 9\ k^2fj^2\\
&\quad -\ 8\ abl^3,
\end{aligned}$$

viz. $\square$, ∇ are the discriminants of the two functions respectively, and Θ, R are simultaneous invariants of the two functions, R being in fact the resultant. The corresponding invariants of the functions $\sqrt{(c)} \,.\, (a, h, k, b \between x, y)^3$, and $(j, l, f \between x, y)^2$ are obviously $c^2\square$, ∇, $c\Theta$ and cR.

The values of S and T are obtained from the Tables 62 and 63 of my "Third Memoir on Quantics," *Phil. Trans.* vol. CXLVI. (1856), pp. 627—647, [144], by merely writing therein $g = i = 0$. It appears that they are in fact functions of $c^2\square$, ∇, $c\Theta$ and cR; viz. we have

$$S = \nabla^2 + c\Theta,$$

$$T = 8\nabla^3 + c(4R + 12\nabla\Theta) + c^2\square.$$

The invariants of the sextic $(* \between x, y)^6$, if for a moment the coefficients of this sextic are taken to be (a, b, c, d, e, f, g), that is, if the sextic be represented by $(a, b, c, d, e, f, g \between x, y)^6$ are the quadrinvariant $(= ag - 6bf + 15ce - 10d^2)$, Table No. 31 and Salmon's A., p. 203[1], the quartinvariant, No. 34, and Salmon's B., p. 203, the sextinvariant No. 35, and Salmon's C., p. 204, and the discriminant, which is a function of the tenth order $= a^5g^5 + \&c.$ recently calculated for the general form, Salmon, pp. 205—207, say these invariants are Q_2, Q_4, Q_6 and Q_{10}. These several invariants are functions of the above-mentioned expressions $c^2\square$, ∇, $c\Theta$ and cR; whence, conversely, these quantities are functions of the four invariants Q_2, Q_4, Q_6, Q_{10}; and the invariants S, T of the cubic curve, being functions of $c^2\square$, ∇, $c\Theta$ and cR, are also, as they should be, functions of the invariants Q_2, Q_4, Q_6 and Q_{10} of the sextic pencil $(* \between x, y)^6$.

To effect the calculation of Q_2, Q_4 and Q_6, I remark that inasmuch as by a linear transformation, the quadric $(j, l, f \between x, y)^2$ may be reduced to the form $2lxy$, and that the invariants of $(a, h, k, b \between x, y)^3$ and $2lxy$ are

$$\begin{aligned} \square &= a^2b^2 - 6abhk + 4ak^3 + 4bh^3 - 3h^2k^2, \\ \nabla &= -l^2, \\ \Theta &= -l(ab - hk), \\ R &= -8l^3ab, \end{aligned}$$

hence, writing $j = 0$, $f = 0$, and writing also $c = 1$, we may consider the sextic

$$[(a, h, k, b \between x, y)^3]^2 + 32l^3x^3y^3,$$

that is

$$(a^2,\ ah,\ \tfrac{1}{5}(2ak + 3h^2),\ \tfrac{1}{10}(ab + 9hk + 16l^3),\ \tfrac{1}{5}(2bh + 3k^2),\ bk,\ b^2 \between x, y)^6,$$

the invariants whereof are found to be functions of the last mentioned values of $\square$, ∇, Θ, R; to pass to the given sextic $(* \between x, y)^6$, put equal to

$$c\,[(a, h, k, b \between x, y)^3]^2 + 4\,[(j, l, f \between x, y)^2]^3,$$

we have only to consider $\square$, ∇, Θ, R as having their before-mentioned general values, and to restore the coefficient c by the principle of homogeneity.

[1] The pages refer to Salmon's Lessons Introductory to the Modern Higher Algebra (Second Edition, 1866). In the Fourth Edition, 1885, the values are given, pp. 260—265.

As regards the discriminant Q_{10}, this as already remarked, has been calculated for the general form, but for the present purpose it is easier, by dealing directly with the form $[(a, h, k, b \between x, y)^3]^2 + 32 l^3 x^3 y^3$, and then interpreting $\square$, ∇, Θ, R and restoring the coefficient c as above, to obtain the discriminant Q_{10} of the function

$$[c\,(a, h, k, b \between x, y)^3]^2 + 4\,[(j, l, f \between x, y)^2]^3$$

in the required form, as a function of $c^2\square$, ∇, $c\Theta$, cR.

I find after some laborious calculations

$$
\begin{aligned}
Q_2 = \quad 10 \text{ No. } 31 = \; & c^2 \left\{ \; 9\,\square \right. \\
+\, & c \left\{ \begin{array}{rr} & 40\,R \\ + & 288\,\nabla\Theta \end{array} \right. \\
+\, & \left\{ \; 256\,\nabla^3 \right.
\end{aligned}
$$

$$
\begin{aligned}
Q_4 = \quad 10000 \text{ No. } 34 = \; & c^4 \left\{ - \; 99\,\square^2 \right. \\
+\, & c^3 \left\{ \begin{array}{rr} - & 400\,R\square \\ + & 2304\,\nabla\Theta\square \\ + & 8640\,\Theta^3 \end{array} \right. \\
+\, & c^2 \left\{ \begin{array}{rr} & 12800\,R\nabla\Theta \\ + & 82944\,\nabla^2\Theta^2 \\ + & 4608\,\nabla^3\square \end{array} \right. \\
+\, & c \left\{ \begin{array}{rr} & 20480\,R\nabla^3 \\ + & 147456\,\nabla^4\Theta \end{array} \right. \\
+\, & \quad 65536\,\nabla^6
\end{aligned}
$$

$$
\begin{aligned}
Q_6 = 1000000 \text{ No. } 35 = \; & c^6 \left\{ + \; 7992\,\square^3 \right. \\
+\, & c^5 \left\{ \begin{array}{rr} + & 72000\,R\square^2 \\ + & 145152\,\nabla\Theta^2\square \\ - & 622080\,\Theta^3\square \end{array} \right. \\
+\, & c^4 \left\{ \begin{array}{rr} & 160000\,R^2\square \\ + & 691200\,R\nabla\Theta\square \\ + & 3456000\,R\Theta^3 \\ + & 3815424\,\nabla^2\Theta^2\square \\ + & 36080640\,\nabla\Theta^4 \\ + & 635904\,\nabla^3\square^2 \end{array} \right. \\
+\, & c^3 \left\{ \begin{array}{rr} + & 33177600\,R\nabla^2\Theta^2 \\ + & 4669440\,R\nabla^3\square \\ + & 217645056\,\nabla^3\Theta^3 \\ + & 23003136\,\nabla^4\Theta\square \end{array} \right.
\end{aligned}
$$

$$
\begin{aligned}
&+ c^2 \begin{cases} + 8192000\, R^2\nabla^3 \\ + 110100480\, R\nabla^4\Theta \\ + 509607936\, \nabla^5\Theta^2 \\ + 14155776\, \nabla^6\square \end{cases} \\
&+ c \begin{cases} 62914560\, R\nabla^6 \\ + 452984832\, \nabla^7\Theta \end{cases} \\
&+ \quad 134217728\, \nabla^9
\end{aligned}
$$

Q_{10} = multiple of discriminant

$$
\begin{aligned}
= \ & c^7 \{ - R^3\square^2 \\
&+ c^6 \begin{cases} - 8\, R^4\square \\ - 24\, R^3\nabla\Theta\square \\ + 64\, R^3\Theta^3 \end{cases} \\
&+ c^5 \begin{cases} - 16\, R^5 \\ - 96\, R^4\nabla\Theta \\ + 48\, R^4\nabla^2\Theta^2 \\ - 16\, R^4\nabla^3\square \end{cases} \\
&+ c^4 \{ - 64\, R^4\nabla^3,
\end{aligned}
$$

to which may be joined

$$
\begin{aligned}
Q_2^2 = \ & c^4 \{ \quad 81\, \square^2 \\
&+ c^3 \begin{cases} 720\, R\square \\ + 5184\, \nabla\Theta\square \end{cases} \\
&+ c^2 \begin{cases} 1600\, R^2 \\ + 23040\, R\nabla\Theta \\ + 82944\, \nabla^2\Theta^2 \\ + 4608\, \nabla^3\square \end{cases} \\
&+ c \begin{cases} 20480\, R\nabla^3 \\ + 147456\, \nabla^4\Theta \end{cases} \\
&+ \quad 65536\, \square^6
\end{aligned}
$$

$$
\begin{aligned}
Q_2^2 - Q_4 = \ & c^4 \{ \quad 180\, \square^2 \\
&+ c^3 \begin{cases} 1120\, R\square \\ + 2880\, \nabla\Theta\square \\ - 8640\, \Theta^3 \end{cases} \\
&+ c^2 \begin{cases} 1600\, R^2 \\ + 10240\, R\nabla\Theta \end{cases}
\end{aligned}
$$

and

$$
\begin{aligned}
\tfrac{1}{8}(Q_6 - 8\, Q_2^3) = \ & c^6 \{ \quad 270\, \square^3 \\
&+ c^5 \begin{cases} - 720\, R\square^2 \\ - 51840\, \nabla\Theta\square^2 \\ - 77760\, \Theta^3\square \end{cases}
\end{aligned}
$$

$$+c^4\left\{\begin{array}{l}-\quad 23200\, R^2\square\\ -\quad 535680\, R\Theta\nabla\square\\ +\quad 432000\, R\Theta^3\\ -1762560\, \nabla^2\Theta^2\square\\ +4510080\, \nabla\Theta^4\\ +\quad 17280\, \nabla^3\square^2\end{array}\right.$$

$$+c^3\left\{\begin{array}{l}-\quad 64000\, R^3\\ -1382400\, R^2\nabla\Theta\\ -5806080\, R\nabla^2\Theta^2\\ +\quad 30720\, R\nabla^3\square\\ +3317760\, \nabla^3\Theta^3\\ -1105920\, \nabla^4\Theta\square\end{array}\right.$$

$$+c^2\left\{\begin{array}{l}-\quad 204800\, R^2\nabla^3\\ -3932160\, R\nabla^4\Theta.\end{array}\right.$$

The foregoing values of S and T give

$$T^2-64S^3=\quad c^4\left\{\quad \square^2\right.$$

$$+c^3\left\{\begin{array}{l}\quad 8\, R\square\\ +24\, \nabla\Theta\square\\ -64\, \Theta^3\end{array}\right.$$

$$+c^2\left\{\begin{array}{l}\quad 16\, R^2\\ +96\, R\nabla\Theta\\ -48\, \nabla^2\Theta^4\\ +16\, \nabla^3\square\end{array}\right.$$

$$+c\quad 64\, R\nabla^3,$$

so that

$$Q_{10}=-c^3R^3(T^2-64S^3),$$

$$64S^3-T^2=\frac{Q_{10}}{c^3R^3},$$

$$S=\nabla^2+c\Theta,$$

and therefore

$$64-\frac{T^2}{S^3}=\frac{Q_{10}}{c^3R^3(\nabla^2+c\Theta)^3}=\frac{Q_{10}}{\{cR(\nabla^2+c\Theta)\}^3},$$

formulæ which are interesting in the theory.

We have

$$c\Theta = S - \nabla^2,$$
$$c^2\square = T - 12\nabla S + 4\nabla^3 - 4cR,$$

and if by means of these values we eliminate $c\Theta$ and $c^2\square$, we obtain Q_2, Q_4, Q_6 and Q_{10} as functions of S, T, ∇ and cR. Choosing instead of Q_4 and Q_6 the combinations $Q_2^2 - Q_4$ and $Q_6 - 8Q_2^3$, and forming also the expression for the combination $Q_2(Q_2^2 - Q_4)$, we have thus the system of formulæ

$$Q_2 = 9T + 180\nabla S + 4\nabla^3 + 4cR,$$

$$\begin{aligned}\tfrac{1}{20}(Q_2^2 - Q_4) = &+ 9T^2\\ &- 432S^3\\ &- 72T\nabla S\\ &- 72T\nabla^3\\ &- 16TcR\\ &+ 864\nabla^2S^2\\ &+ 144\nabla^4S\\ &+ 128\nabla ScR,\end{aligned}$$

$\tfrac{1}{80}(Q_6 - 8Q_2^3) =$	$\tfrac{1}{20}Q_2(Q_2^2 - Q_4) =$
$+ 27T^3$	$+ 81T^3$
$- 4212T^2\nabla S$	$+ 972T^2\nabla S$
$+ 6588T^2\nabla^3$	$- 612T^2\nabla^3$
$+ 252T^2cR$	$- 108T^2cR$
$- 7776TS^3$	$- 3888TS^3$
$- 16848T\nabla^2S^3$	$- 5184T\nabla^2S^3$
$+ 3456T\nabla^4S$	$- 11952T\nabla^4S$
$- 2592T\nabla ScR$	$- 2016T\nabla ScR$
$- 1296T\nabla^6$	$- 288T\nabla^6$
$- 1824T\nabla^3cR$	$- 352T\nabla^3cR$
$- 448Tc^2R^2$	$- 64Tc^2R^2$
$+ 544320\nabla S^4$	$- 77760\nabla S^4$
$- 461376\nabla^3S^3$	$+ 153792\nabla^3S^3$
$+ 74304cRS^3$	$- 1728cRS^3$
$+ 15552\nabla^5S^2$	$+ 29376\nabla^5S^2$
$- 10368\nabla^2S^2cR$	$+ 26996\nabla^2S^2cR$
$- 10728\nabla^7S$	$+ 576\nabla^7S$
$- 3264\nabla^4ScR$	$+ 1088\nabla^4ScR$
$- 1536\nabla Sc^2R^2,$	$+ 512\nabla Sc^2R^2,$

and, as mentioned above,

$$Q_{10} = c^3 R^3 (- T^2 + 64 S^3).$$

The just-mentioned value of Q_{10} should, I think, admit of being established *a priori*, and if this be so, then the substitution of the values of S and T in terms of $c^2\square$, ∇, $c\Theta$, cR, would be the easiest way of arriving at the before-mentioned expression of Q_{10} in terms of these same quantities. The calculation by which this expression was arrived at, is however not without interest, and it will be as well to indicate the mode in which it was effected.

Calculation of Q_{10}.

We have to find the discriminant of

$$c\,[(a,\ h,\ k,\ b \between x,\ y)^3]^2 + 32 l^3 x^3 y^3.$$

Consider for a moment the more general form $P^2 + 4Q^3$, then to find the discriminant, we have to eliminate between the equations

$$P\frac{dP}{dx} + 6Q^2\frac{dQ}{dx} = 0,$$

$$P\frac{dP}{dy} + 6Q^2\frac{dQ}{dy} = 0,$$

these are satisfied by the system $P = 0$, $Q^2 = 0$, and it follows that if R be the resultant of the equations $P = 0$, $Q = 0$, then the discriminant in question contains the factor R^2. For the other factor we may reduce the system to

$$P\frac{dP}{dx} + 6Q^2\frac{dQ}{dx} = 0,$$

$$\frac{dP}{dx}\frac{dQ}{dy} - \frac{dP}{dy}\frac{dQ}{dx} = 0.$$

Now writing $Q = 2lxy$, these equations become

$$P\frac{dP}{dx} + 48 l^3 x^2 y^3 = 0,$$

$$l\left(x\frac{dP}{dx} - y\frac{dP}{dy}\right) = 0,$$

the resultant of which is $= l^3$ into resultant of the system

$$P\frac{dP}{dx} + 48 l^3 x^2 y^3 = 0,$$

$$x\frac{dP}{dx} - y\frac{dP}{dy} \quad = 0,$$

but in virtue of the second equation, we have

$$P = \tfrac{1}{3}\left(x\frac{dP}{dx} + y\frac{dP}{dy}\right) = \tfrac{2}{3}y\frac{dP}{dy},$$

which reduces the first equation to

$$2y\frac{dP}{dx}\frac{dP}{dy} + 144l^3x^2y^3 = 0,$$

or omitting the factor $2y$, to

$$\frac{dP}{dx}\frac{dP}{dy} + 72l^3x^2y^3 = 0.$$

Hence, writing $P = \sqrt{(c)} \,.\, (a, h, k, b \between x, y)^3$, and therefore $\frac{dP}{dx} = 3\sqrt{(c)} \,.\, (a, h, k \between x, y)^2$, $\frac{dP}{dy} = 3\sqrt{(c)} \,.\, (h, k, b \between x, y)^2$; writing also $y = 1$, the two equations become

$$c\,(a, h, k \between x, 1)^2 (h, k, b \between x, 1)^2 + 8l^3x^2 = 0,$$

$$(a, h, k \between x, 1)^2 x - (h, k, b \between x, 1)^2 \quad = 0,$$

the second of which is more simply written

$$(a, h, -k, -b \between x, 1)^3 = 0.$$

Hence, restoring the factor l^3, and also to avoid fractions introducing the factor $8a^4$, the resultant of the two equations is

$$= 8l^3a^4\Pi\,\{8l^3x^2 + c\,(a, h, k \between x, 1)^2 (h, k, b \between x, 1)^2\},$$

where Π denotes the product of the factors corresponding to the three roots x_1, x_2, x_3 of the equation

$$(a, h, -k, -b \between x, 1)^3 = 0,$$

or what is the same thing,

$$ax^3 + hx^2 - kx - b = 0,$$

so that the symmetric functions are to be found from

$$\Sigma x_1 = -\frac{h}{a}, \quad \Sigma x_1x_2 = -\frac{k}{a}, \quad x_1x_2x_3 = \frac{b}{a}.$$

The required discriminant is the foregoing resultant multiplied by R, or say by c^2R^2, that is the discriminant Q_{10} is

$$= c^2R^2 \,.\, 8l^3a^4\Pi\,(8l^3x^2 + \Omega),$$

if for shortness we write

$$\Omega = (a, h, k \between x, 1)^2 \,.\, (a, k, b \between x, 1)^2,$$

and when the symmetric functions have been expressed in terms of the coefficients, the result is to be expressed as a function of $\square$, ∇, Θ, R by means of the values

$$\begin{aligned} c^2\square &= a^2b^2 + 4ak^3 + 4bh^3 - 6adhk - 3h^2k^2, \\ \nabla &= - l^2, \\ c\Theta &= - l(ab - hk), \\ cR &= - 8l^3ab. \end{aligned}$$

Thus, for instance, the first term of the result is

$$= c^2R^2 \,.\, 8l^3a^4 \,.\, 512l^3x_1{}^2x_2{}^2x_3{}^2,$$

which is

$$\begin{aligned} &= c^2R^2 \,.\, 4096l^{12}a^2b^2, \\ &= c^2R^2 \,.\, - 64c^2R^2\nabla^3, \\ &= - 64c^4R^4\nabla^3, \end{aligned}$$

which is a term in the before-mentioned expression for Q_{10}.

401.

A NOTATION OF THE POINTS AND LINES IN PASCAL'S THEOREM.

[From the *Quarterly Journal of Pure and Applied Mathematics*, vol. IX. (1868), pp. 268—274.]

TAKING six points 1, 2, 3, 4, 5, 6 on a conic; let $A, B, C, D, E, F, G, H, I, J, K, L, M, N, O$ denote each a combination of three lines, thus

$12.34.56 = A$	$12.35.64 = F$	$12.36.45 = K$
$13.45.62 = B$	$13.46.25 = G$	$13.42.56 = L$
$14.56.23 = C$	$14.52.36 = H$	$14.53.62 = M$
$15.62.34 = D$	$15.63.42 = I$	$15.64.23 = N$
$16.23.45 = E$	$16.24.53 = J$	$16.25.34 = O$

then any hexagon formed with the six points may be represented by a combination of some two of the letters A, B, &c., viz. the three alternate sides are the lines represented by one letter, and the other three alternate sides the lines represented by the other letter: for example, the hexagon 123456 is AE; and so for the other hexagons. Any duad AE thus representing a hexagon may be termed a hexagonal duad; the number of such duads is sixty. Each Pascalian line may be denoted by the symbol of the hexagon to which it belongs; thus, the line which belongs to the hexagon AE, is the line AE.

I form the following combinations:

$IMO.DHJ$	each involving all the duads 12, &c. except those of			123 . 456,
$DEG.BNO$	"	"	"	124 . 356,
$ELM.BCJ$	"	"	"	125 . 346,
$HLN.CGI$	"	"	"	126 . 345,
$EFI.JKN$	"	"	"	134 . 256,
$AEH.CKO$	"	"	"	135 . 246,
$AMN.CDF$	"	"	"	136 . 245,
$AGJ.ELO$	"	"	"	145 . 236,
$ABI.DKL$	"	"	"	146 . 235,
$GKM.BFH$	"	"	"	156 . 234,

and also the combinations:

AEGMI	involving all the duads 12, 13, &c.,	
ABHJN	,,	,,
BCFIO	,,	,,
CDGJK	,,	,,
DEFHL	,,	,,
KLMNO	,,	,,

which I call respectively the ten-partite and six-partite arrangements. It is to be remarked that (considering *IMO.DHJ* as standing for the six duads *IM*, *IO*, *MO* *DH*, *DJ*, *HJ*, and so for the others) the ten-partite arrangement contains all the sixty hexagonal duads: and in like manner, (considering *AEGMI* as standing for the ten duads *AE*, *AG*, *AM*, *AI*, *EG*, *EM*, *EI*, *GM*, *GI*, *MI*, and so for the others) the six-partite arrangement contains all the sixty hexagonal duads.

The 60 Pascalian lines intersect by 4's in the 45 Pascalian points p, by 3's in 20 points g and in 60 points h, and by 2's in 90 points m, 360 points r, 360 points t, 360 points z, and 9 points w.

The intersections of the Pascalian lines thus are

$45\,p$	counting as	270
$20\,g$	,, ,,	60
$60\,h$	,, ,,	180
$90\,m$	,, ,,	90
$360\,r$	,, ,,	360
$360\,t$	,, ,,	360
$360\,z$	,, ,,	360
$90\,w$	,, ,,	90
		$1770 = \tfrac{1}{2}60 \,.\, 59,$

and the intersections on each Pascalian line are

$3\,p$	counting as	9
$1\,g$	,, ,,	2
$3\,h$	,, ,,	6
$3\,m$	,, ,,	3
$12\,r$	,, ,,	12
$12\,t$	,, ,,	12
$12\,z$	,, ,,	12
$3\,w$	,, ,,	3
		59.

For the ten-partite arrangement, any double triad such as $ABI.DKL$ gives 15 intersections; $10 \times 15 = 150$; and any pair of double triads such as $ABI.DKL$ and $AEH.CKO$ gives 36 intersections; $45 \times 36 = 1620$; and these are

$$10 \times \begin{cases} 6\,g & \quad 60\,g \\ 9\,m & \quad 90\,m \end{cases}$$
$$\text{—— } 15 \qquad\qquad \text{—— } 150$$

$$45 \times \begin{cases} 6\,p & \quad 270\,p \\ 4\,h & \quad 180\,h \\ 8\,r & \quad 360\,r \\ 8\,t & \quad 360\,t \\ 8\,z & \quad 360\,z \\ 2\,w & \quad 90\,w \end{cases}$$
$$\text{—— } 36 \qquad\qquad \text{—— } 1620$$
$$1770.$$

For the six-partite arrangement any pentad such as $ABHJN$ gives 45 intersections; $6 \times 45 = 270$; and any two pentads such as $ABHJN$ and $AEGMI$ give 100 intersections; $15 \times 100 = 1500$; and these are

$$6 \times \begin{cases} 30\,h & \quad 180\,h \\ 15\,m & \quad 90\,m \end{cases}$$
$$\text{—— } 45 \qquad\qquad \text{—— } 270$$

$$15 \times \begin{cases} 4\,g & \quad 60\,g \\ 18\,p & \quad 270\,p \\ 24\,r & \quad 360\,r \\ 24\,t & \quad 360\,t \\ 24\,z & \quad 360\,z \\ 6\,w & \quad 90\,w \end{cases}$$
$$\text{—— } 100 \qquad\qquad \text{—— } 1500$$
$$1770.$$

I analyse the intersections of a Pascalian line, say AE, by the remaining 59 Pascalian lines as follows:

Observe that AE belongs to the triad AEH, the complementary triad whereof is CKO; it also belongs to the pentad $AEIMG$. We thus obtain, corresponding to AE, the arrangement

$$\begin{array}{cc|ccc} & & H & H & H \\ \hline H & A & B & N & J \\ H & E & F & L & D \\ \hline & & I & M & G \\ & & K & C & O \end{array}$$

viz. $HABNJ$, is the pentad which contains HA, the arrangement of the last three letters B, N, J thereof being arbitrary; $HEFLD$ is the pentad that contains HE, but the last three letters are so arranged that the columns HBF, HNL, HJD are each of them a triad, IMG is then the residue of the pentad $AEIMG$, and KCO is the complementary triad to AEH, but the arrangement of the letters IMG, and of the letters KCO, are each of them determinate; viz. these are such that we have $BFICO$, $NLMKO$, $JDGCK$, each of them a pentad.

And this being so we derive from the arrangement

2 g	AH, EH;
3 m	KC, KO, CO;
6 h	AI, AM, AG; EI, EM, EG;
12 z	IB, IF, MN, ML, GJ, GD; HB, HF, HN, HL, HJ, HD;
9 p	AB, AN, AJ; EF, EL, ED; BF, NL, JD;
12 r	CB, CF, CJ, CD; OB, OF, ON, OL; KN, KL, KJ, KD;
12 t	FL, FD, LD; BN, BJ, NJ; IC, IO; MK, MO; GK, GC;
3 w	IM, IG, MG;
59	

viz. the line AE in question meets AH, EH each of them in a point g; KC, KO, CO each in a point m; and so on. By constructing in the same way an arrangement for each of the lines AH, &c., we find the nature of the point of intersection of any two of the lines AB, AE, AH, &c.; and we may then present the results in a table (see Plate), which shows at a glance what is the point of intersection (whether a point g, m, h, z, p, r, t, or w) of any two of the Pascalian lines.

I further remark that representing the 45 Pascalian points as follows:

12 . 34 = a	13 . 24 = g	14 . 23 = m	15 . 23 = s	16 . 23 = y
12 . 35 = b	13 . 25 = h	14 . 25 = n	15 . 24 = t	16 . 24 = z
12 . 36 = c	13 . 26 = i	14 . 26 = o	15 . 26 = u	16 . 25 = α
12 . 45 = d	13 . 45 = j	14 . 35 = p	15 . 34 = v	16 . 34 = β
12 . 46 = e	13 . 46 = k	14 . 36 = q	15 . 36 = w	16 . 35 = γ
12 . 56 = f	13 . 56 = l	14 . 56 = r	15 . 46 = x	16 . 45 = δ

23 . 45 = ϵ	25 . 34 = λ	34 . 56 = ρ
23 . 46 = ζ	25 . 36 = μ	35 . 46 = σ
23 . 56 = η	25 . 46 = ν	36 . 45 = τ
24 . 35 = θ	26 . 34 = ξ	
24 . 36 = ι	26 . 35 = ω	
24 . 56 = κ	26 . 45 = π	

the sixty hexagons and their Pascalian lines then are

AE	123456	12.45 23.56 34.61	$d\eta\beta$
AH	125634	12.63 25.34 56.41	$c\lambda r$
EH	145236	14.23 45.36 52.61	$m\tau\alpha$
CK	123654	12.65 23.54 36.41	$f\epsilon q$
CO	143256	14.25 43.56 32.61	$n\rho y$
KO	125436	12.43 25.36 54.61	$a\mu\delta$
AM	126534	12.53 26.34 65.41	$b\xi r$
AG	125643	12.64 25.43 56.31	$e\lambda l$
AI	124365	12.36 24.65 43.51	$c\kappa v$
EG	132546	13.54 32.46 25.61	$j\zeta\alpha$
DF	126435	12.43 26.35 64.51	$a\omega x$
FL	124653	12.65 24.53 46.31	$f\theta k$
DL	134265	13.26 34.65 42.51	$i\rho t$
BN	132645	13.64 32.45 26.51	$k\epsilon u$
BJ	135426	13.42 35.26 54.61	$g\omega\delta$
JN	153246	15.24 53.46 32.61	$t\sigma y$
GK	125463	12.46 25.63 54.31	$e\mu j$
KM	126354	12.35 26.54 63.41	$b\pi q$
IO	152436	15.43 52.36 24.61	$v\mu z$
MO	143526	14.52 43.26 35.61	$n\xi\gamma$
EM	145326	14.32 45.26 53.61	$m\pi\gamma$
EI	154236	15.23 54.36 42.61	$s\tau z$
AN	123465	12.46 23.65 34.51	$e\eta v$
AJ	124356	12.35 24.56 43.61	$b\kappa\beta$
AB	126543	12.54 26.43 65.31	$d\zeta l$
DE	154326	15.32 54.26 43.61	$s\pi\beta$
EL	132456	13.45 32.56 24.61	$j\eta z$
EF	123546	12.54 23.46 35.61	$d\zeta\gamma$
CD	143265	14.26 43.65 32.51	$o\rho s$
CF	123564	12.56 23.64 35.41	$f\zeta p$

CG	132564	13.56	32.64	25.41	*lζn*
CI	142365	14.36	42.65	23.51	*qκs*
MN	146235	14.23	46.35	62.51	*mσu*
GJ	135246	13.24	35.46	52.61	*gσα*
BI	136245	13.24	36.45	62.51	*gτu*
DG	134625	13.62	34.25	46.51	*iλx*
LM	135624	13.62	35.24	56.41	*iθr*
FI	124635	12.63	24.35	46.51	*cθx*
BH	136254	13.25	36.54	62.41	*hτo*
FH	125364	12.36	25.64	53.41	*cνp*
FO	125346	12.34	25.46	53.61	*aνγ*
LO	134256	13.25	34.56	42.61	*hρz*
DK	126345	12.34	26.45	63.51	*aπw*
KL	124563	12.56	24.63	45.31	*fιj*
BO	134526	13.52	34.26	45.61	*hξδ*
NO	152346	15.34	52.46	23.61	*vνy*
BC	132654	13.65	32.54	26.41	*lεo*
CJ	142356	14.35	42.56	23.61	*pκy*
JK	124536	12.53	24.36	45.61	*bιδ*
KN	123645	12.64	23.45	36.51	*eεw*
DH	143625	14.62	43.25	36.51	*oλw*
HJ	142536	14.53	42.36	25.61	*pια*
HL	136524	13.52	36.24	65.41	*hιr*
HN	146325	14.32	46.25	63.51	*mνw*
BF	126453	12.46	26.53	64.31	*dωk*
DJ	153426	15.42	53.26	34.61	*tωβ*
LN	132465	13.46	32.65	24.51	*kηt*
GM	135264	13.26	35.64	52.41	*iσn*
IM	142635	14.63	42.35	26.51	*qθu*
GI	136425	13.42	36.25	64.51	*gμx*

Each Pascalian point belongs to four different hexagons; viz. a to the hexagons KD, KO, FD, FO; and so for the other points, thus:

a	$(K, F)(D, O)$	x	$(D, I)(F, G)$
b	$(A, K)(M, J)$	y	$(C, N)(J, O)$
c	$(A, F)(H, I)$	z	$(E, O)(I, L)$
d	$(A, F)(B, E)$	α	$(E, J)(G, H)$
e	$(A, K)(G, N)$	β	$(A, D)(E, J)$
f	$(C, L)(K, F)$	γ	$(E, O)(F, M)$
g	$(B, G)(I, J)$	δ	$(B, K)(J, O)$
h	$(B, L)(H, O)$	ϵ	$(B, K)(C, N)$
i	$(D, M)(G, L)$	ζ	$(C, E)(F, G)$
j	$(E, K)(G, L)$	η	$(A, L)(E, N)$
k	$(B, L)(F, N)$	θ	$(F, M)(I, L)$
l	$(A, C)(B, G)$	ι	$(H, K)(J, L)$
m	$(E, N)(H, M)$	κ	$(A, C)(I, J)$
n	$(C, M)(G, O)$	λ	$(A, D)(G, H)$
o	$(B, D)(C, H)$	μ	$(G, O)(I, K)$
p	$(C, H)(F, J)$	ν	$(F, N)(H, O)$
q	$(C, M)(I, K)$	ξ	$(A, O)(B, M)$
r	$(A, L)(H, M)$	ω	$(B, D)(F, J)$
s	$(C, E)(D, I)$	π	$(D, M)(E, K)$
t	$(J, L)(D, N)$	ρ	$(C, L)(D, O)$
u	$(B, M)(I, N)$	σ	$(G, N)(J, M)$
v	$(A, O)(N, I)$	τ	$(B, E)(H, I)$
w	$(D, N)(H, K)$		

I have constructed on a very large scale a figure of the sixty Pascalian lines, and the forty-five Pascalian points, marking them according to the foregoing notation; but the figure is from its complexity, and the inconvenient way in which the points are either crowded together or fly off to a great distance, almost unintelligible.

	AB	AE	AG	AH	AI	AJ	AM	AN	BC	BF	BH	BI	BJ	BN	BO	CD	CF	CG	CI	CJ	CK	CO	DE	DF	DG	DH	DJ	DK	DL	EF
AB		p	p	h	g	h	p	h	p	p	h	g	h	h	p	r	t	p	z	z	r	t	r	r	r	t	t	m	m	p
AE	p		h	g	h	p	h	p	r	p	z	z	t	t	r	r	r	t	t	r	m	m	p	t	z	z	p	r	t	p
AG	p	h		p	h	g	h	p	p	r	t	z	z	t	r	t	r	p	z	z	t	r	z	r	p	p	z	t	r	t
AH	h	g	p		p	h	p	h	t	z	h	z	w	w	t	r	r	r	r	t	m	m	z	t	p	p	z	r	t	z
AI	g	h	h	p		p	h	p	z	z	z	g	z	z	z	r	t	z	p	p	r	t	t	r	t	r	r	m	m	z
AJ	h	p	g	h	p		p	h	z	t	w	z	h	w	t	t	r	z	p	p	t	r	p	r	z	z	p	t	r	r
AM	p	h	h	p	h	p		g	r	r	t	z	t	z	p	m	m	t	t	r	r	r	t	m	t	r	r	r	r	t
AN	h	p	p	h	p	h	g		t	t	w	z	w	h	z	m	m	r	r	t	r	r	r	m	r	t	t	r	r	r
BC	p	r	p	t	z	z	r	t		h	p	h	g	p	h	p	h	p	h	g	p	h	r	z	t	p	z	t	r	t
BF	p	p	r	z	z	t	r	t	h		g	h	p	p	h	z	h	t	w	z	t	w	t	p	r	z	p	r	t	p
BH	h	z	t	h	z	w	t	w	p	g		p	h	h	p	p	z	r	t	z	r	t	t	z	r	p	z	r	t	z
BI	g	z	z	z	g	z	z	z	h	h	p		p	p	h	t	w	z	h	z	t	w	r	t	r	r	r	m	m	z
BJ	h	t	z	w	z	h	t	w	g	p	h	p		h	p	z	z	z	z	g	z	z	r	p	t	z	p	t	r	r
BN	h	t	t	w	z	w	z	h	p	p	h	p	h		g	r	t	r	t	z	p	z	m	r	m	t	t	r	r	r
BO	p	r	r	t	z	t	p	z	h	h	p	h	p	g		t	w	t	w	z	z	h	m	t	m	r	r	r	r	t
CD	r	r	t	r	r	t	m	m	p	z	p	t	z	r	t		g	h	p	h	h	p	p	g	h	p	h	h	p	z
CF	t	r	r	r	t	r	m	m	h	h	z	w	z	t	w	g		p	h	p	p	h	z	g	z	z	z	z	z	p
CG	p	t	p	r	z	z	t	r	p	t	r	z	z	r	t	h	p		g	h	h	p	z	z	h	t	w	w	t	p
CI	z	t	z	r	p	p	t	r	h	w	t	h	z	t	w	p	h	g		p	p	h	p	z	z	r	t	t	r	z
CJ	z	r	z	t	p	p	r	t	g	z	z	z	g	z	z	h	p	h	p		h	p	t	z	w	z	h	w	t	r
CK	r	m	t	m	r	t	r	r	p	t	r	t	z	p	z	h	p	h	p	h		g	t	z	w	t	w	h	z	r
CO	t	m	r	m	t	r	r	r	h	w	t	w	z	z	h	p	h	p	h	p	g		r	z	t	r	t	z	p	t
DE	r	p	z	z	t	p	t	r	r	t	t	r	r	m	m	p	z	z	p	t	t	r		h	g	h	p	p	h	h
DF	r	t	r	t	r	r	m	m	z	p	z	t	p	r	t	g	g	z	z	z	z	z	h		p	h	p	p	h	h
DG	r	z	p	p	t	z	t	r	t	r	r	r	t	m	m	h	z	h	z	w	w	t	g	p		p	h	h	p	z
DH	t	z	p	p	r	z	r	t	p	z	p	r	z	t	r	p	z	t	r	z	t	r	h	h	p		g	p	h	w
DJ	t	p	z	z	r	p	r	t	z	p	z	r	p	t	r	h	z	w	t	h	w	t	p	h	h	g		h	p	t
DK	m	r	t	r	m	t	r	r	t	r	r	m	t	r	r	h	z	w	t	w	h	z	p	p	h	p	h		g	t
DL	m	t	r	t	m	r	r	r	r	t	t	m	r	r	r	p	z	t	r	t	z	p	h	h	p	h	p	g		w
EF	p	p	t	z	z	r	t	r	t	p	z	z	r	r	t	z	p	p	z	r	r	t	h	h	z	w	t	t	w	
EG	t	h	h	z	w	z	w	t	r	r	r	t	r	m	m	z	p	p	z	t	t	r	g	z	g	z	z	z	z	p
EN	z	g	z	g	z	z	z	z	r	z	p	p	t	t	r	t	t	r	r	r	m	m	h	w	z	h	z	t	w	h
EI	z	h	w	z	h	t	w	t	t	z	p	p	r	r	t	p	z	z	p	r	r	t	p	z	z	t	r	r	t	g
EL	r	p	t	z	t	r	z	p	m	t	t	r	m	r	r	t	t	r	r	m	r	r	h	w	z	w	t	z	h	h
EM	t	h	w	z	w	t	h	z	m	r	r	t	m	r	r	r	r	t	t	m	r	r	p	t	z	t	r	p	z	p
FI	z	z	t	p	p	r	t	r	w	h	z	h	t	t	w	z	h	z	h	t	t	w	z	p	p	t	r	r	t	g
FH	z	z	r	p	p	t	r	t	z	g	g	z	z	z	z	z	p	r	t	p	r	t	w	h	t	h	z	t	w	h
FL	r	t	m	t	r	m	r	r	t	p	z	t	r	p	z	z	p	r	t	r	p	z	w	h	t	w	t	z	h	h
FO	t	r	m	r	t	m	r	r	w	h	z	w	t	z	h	z	h	t	w	t	z	h	t	p	r	t	r	p	z	p
GI	z	w	h	t	h	z	w	t	z	t	r	p	p	r	t	z	z	g	g	z	z	z	z	p	p	r	t	t	r	z
GJ	z	z	g	z	z	g	z	z	z	r	t	p	p	t	r	w	t	h	z	h	w	t	z	t	h	z	h	w	t	r
GK	r	t	p	r	t	z	z	p	t	m	m	r	t	r	r	w	t	h	z	w	h	z	z	t	h	t	w	h	z	r
GM	t	w	h	t	w	z	h	z	r	m	m	t	r	r	r	t	r	p	z	t	z	p	z	r	p	r	t	z	p	t
HJ	w	z	z	h	t	h	t	w	z	z	h	t	h	w	t	z	p	t	r	p	t	r	z	z	z	g	g	z	z	t
HL	t	z	r	p	r	t	p	z	r	z	p	r	t	z	p	t	t	m	m	r	r	r	w	w	t	h	z	z	h	w
HN	w	z	t	h	t	w	z	h	t	z	h	t	w	h	z	r	r	m	m	t	r	r	t	t	r	p	z	p	z	t
IM	z	w	w	t	h	t	h	z	t	t	r	p	r	p	z	r	t	z	p	r	p	z	t	r	t	m	t	r	r	z
IO	z	t	t	r	p	r	z	p	w	w	t	h	t	z	h	t	w	z	h	t	z	h	r	t	r	m	m	r	r	z
JH	t	r	z	t	r	p	p	z	z	r	t	r	p	z	p	w	t	w	t	h	h	z	t	t	w	z	h	h	z	m
JN	w	t	z	w	t	h	z	h	z	t	w	t	h	h	z	t	r	t	r	p	z	p	r	r	t	z	p	z	p	m
KL	m	r	r	r	m	r	t	t	r	r	r	m	r	t	t	z	p	t	r	t	p	z	z	z	z	z	z	g	g	t
KM	r	t	z	r	t	p	p	z	r	m	m	r	r	t	t	t	r	z	p	t	p	z	p	r	z	r	t	p	z	r
KN	t	r	p	t	r	z	z	p	p	r	t	r	z	p	z	t	r	t	r	z	p	z	r	r	t	p	z	p	z	m
KO	r	m	r	m	r	r	t	t	z	t	r	t	p	z	p	z	z	z	z	z	g	g	r	p	t	r	t	p	z	r
LM	r	z	t	p	t	r	p	z	m	r	r	r	m	t	t	r	r	r	r	m	t	t	z	t	p	t	r	z	p	z
LN	t	p	r	z	r	t	z	p	r	p	z	r	t	p	z	r	r	m	m	r	t	t	t	t	r	z	p	z	p	t
LO	r	r	m	r	r	m	t	t	t	z	p	t	r	z	p	p	z	r	t	r	z	p	t	z	r	t	r	z	p	z
MN	z	z	z	z	z	z	g	g	r	r	t	p	t	p	z	m	m	r	r	r	t	t	r	m	r	r	r	t	t	r
MO	p	t	t	r	z	r	p	z	t	t	r	z	r	z	p	r	t	p	z	r	z	p	r	r	r	m	m	t	t	p
NO	z	r	r	t	p	t	z	p	z	z	z	z	z	g	g	r	t	r	t	p	z	p	m	r	m	r	r	t	t	r

AB AE AG AH AI AJ AM AN BO BF BH BI BJ BN BO CD CF CG CI CJ CK CO DE DF DG DH DJ DK DL EF

Plate II.

EG	EH	EI	EL	EM	FI	FH	FL	FO	GI	GJ	GK	GM	HJ	HL	HN	IM	IO	JK	JN	KL	KM	KN	KO	LM	LN	LO	MN	MO	NO	
t	z	x	r	t	z	z	r	t	z	x	r	t	w	t	w	z	z	t	w	m	r	t	r	r	t	r	z	p	z	AB
h	g	h	p	h	x	x	t	r	w	z	t	w	z	z	z	w	t	r	t	r	t	r	m	z	p	r	z	t	r	AE
h	x	w	t	w	t	r	m	m	h	g	p	h	z	r	t	w	t	z	x	r	z	p	r	t	r	m	z	t	r	AG
z	g	z	z	x	p	p	t	r	t	z	r	t	h	p	h	t	r	t	w	r	r	t	m	p	z	r	z	r	t	AH
w	z	h	t	w	p	p	r	t	h	z	t	w	t	r	t	h	p	r	t	m	t	r	r	t	r	r	z	z	p	AI
z	z	t	r	t	r	t	m	m	z	g	z	z	h	t	w	t	r	p	h	r	p	z	r	r	t	m	z	r	t	AJ
w	z	w	z	h	t	r	r	r	w	z	z	h	t	p	z	h	z	p	z	t	p	z	t	p	z	t	g	p	z	AM
t	z	t	p	z	r	t	r	r	t	z	p	z	w	z	h	z	p	z	h	t	z	p	t	z	p	t	g	z	p	AN
r	r	t	m	m	w	z	t	w	z	z	t	r	z	r	t	t	w	z	z	r	r	p	z	m	r	t	r	t	z	BC
r	z	z	t	r	h	g	p	h	t	r	m	m	z	z	z	t	w	r	t	r	m	r	t	r	p	z	r	t	z	BF
r	p	p	t	r	z	g	z	z	r	t	m	m	h	p	h	r	t	t	w	r	m	t	r	r	z	p	t	r	z	BH
t	p	p	r	t	h	z	t	w	p	p	r	t	t	r	t	p	h	r	t	m	r	r	t	r	r	t	p	z	z	BI
r	t	r	m	m	t	z	r	t	p	p	t	r	h	t	w	r	t	p	h	r	r	z	p	m	t	r	t	r	z	BJ
m	t	r	r	r	t	z	p	z	r	t	r	r	w	z	h	p	z	z	h	t	t	p	z	t	p	z	p	z	g	BN
m	r	t	r	r	w	z	z	h	t	r	r	r	t	p	z	z	h	p	z	t	t	z	p	t	z	p	z	p	g	BO
z	t	p	t	r	z	z	z	z	z	w	w	t	z	t	r	r	t	w	t	z	t	t	z	r	r	p	m	r	r	CD
p	t	z	t	r	h	p	p	h	z	t	t	r	p	t	r	t	w	t	r	p	r	r	z	r	r	z	m	t	t	CF
p	r	z	r	t	z	r	r	t	g	h	h	p	t	m	m	z	z	w	t	t	z	t	z	r	m	r	r	p	r	CG
z	r	p	r	t	h	t	t	w	g	z	z	z	r	m	m	p	h	t	r	r	p	r	z	r	m	t	r	z	t	CI
t	r	r	m	m	t	p	r	t	z	h	w	t	p	r	t	r	t	h	p	t	t	z	z	m	r	r	r	r	p	CJ
t	m	r	r	r	t	r	p	z	z	w	h	z	t	r	r	p	z	h	z	p	p	p	g	t	t	z	t	z	z	CK
r	m	t	r	r	w	t	z	h	z	t	z	p	r	r	r	z	h	z	p	z	z	z	g	t	t	p	t	p	p	CO
g	h	p	h	p	z	w	w	t	z	z	z	z	z	w	t	t	r	t	r	z	p	r	r	z	t	t	r	r	m	DE
z	w	z	w	t	p	h	h	p	p	t	t	r	z	w	t	r	t	t	r	z	r	r	p	t	t	z	m	r	r	DF
g	z	z	z	z	p	t	t	r	p	h	h	p	z	t	r	t	r	w	t	z	z	t	t	p	r	r	r	r	m	DG
z	h	t	w	t	t	h	w	t	r	z	t	r	g	h	p	m	m	z	z	z	r	p	r	t	z	t	r	m	r	DH
z	z	r	t	r	r	z	t	r	t	h	w	t	g	z	z	t	m	h	p	z	t	z	t	r	p	r	r	m	r	DJ
z	t	r	z	p	r	t	z	p	t	w	h	z	z	z	p	r	r	h	z	g	p	p	p	z	z	z	t	t	t	DK
z	w	t	h	z	t	w	h	z	r	t	z	p	z	h	z	r	r	z	p	g	z	z	z	p	p	p	t	t	t	DL
p	h	g	h	p	g	h	h	p	z	r	r	t	t	w	t	z	z	m	m	t	r	m	r	z	t	z	r	p	r	EF
	p	h	p	h	z	t	t	r	h	p	p	h	p	t	r	w	t	t	r	p	z	r	r	z	r	r	t	t	m	EG
p		p	h	p	z	h	w	t	t	p	r	t	p	h	p	t	r	r	t	t	r	r	m	z	z	t	p	r	r	EH
h	p		p	h	g	z	z	z	h	t	t	w	r	t	r	h	p	m	m	r	t	m	r	z	r	p	t	z	r	EI
p	h	p		g	z	w	h	z	t	r	p	z	t	h	z	z	p	r	r	p	z	t	t	g	p	p	z	z	t	EL
h	p	h	g		z	t	z	p	w	t	z	h	r	z	p	h	z	r	r	z	p	t	t	g	z	z	p	p	t	EM
z	z	g	z	z		p	p	h	p	r	r	t	r	t	r	p	h	m	m	r	r	m	t	p	r	z	r	z	t	FI
t	h	z	w	t	p		h	p	r	r	m	m	p	h	p	r	t	r	t	t	m	r	r	t	z	z	r	r	p	FH
t	w	z	h	z	p	h		g	r	m	r	r	t	h	z	p	z	r	r	p	t	t	z	p	p	g	t	z	z	FL
r	t	z	z	p	h	p	g		t	m	r	r	r	z	p	z	h	r	r	z	t	t	p	z	z	g	t	p	p	FO
h	t	h	t	w	p	r	r	t		p	p	h	r	m	m	h	p	t	r	r	z	r	p	t	m	r	t	z	r	GI
p	p	t	r	t	r	r	m	m	p		h	p	p	r	t	m	r	h	p	t	z	z	t	r	r	m	p	r	r	GJ
p	r	t	p	z	r	m	r	r	p	h		g	t	r	r	z	p	h	z	p	g	p	p	z	t	t	z	z	t	GK
h	t	w	z	h	t	m	r	r	h	p	g		r	r	r	h	z	z	p	z	g	z	z	p	t	t	p	p	t	GM
p	p	r	t	r	r	p	t	r	r	p	t	r		p	h	m	m	p	h	p	r	z	r	r	z	r	t	m	t	HJ
t	h	t	h	z	t	h	h	z	m	r	r	r	p		g	r	r	p	z	p	t	z	t	p	g	p	z	t	z	HL
r	p	r	z	p	r	p	z	p	m	t	r	r	h	g		r	r	z	h	z	t	p	t	z	g	z	p	t	p	HN
w	t	h	z	h	p	r	p	z	h	m	z	h	m	r	r		g	r	r	t	p	t	z	p	t	z	p	g	z	IM
t	r	p	p	z	h	t	z	h	p	r	p	z	m	r	r	g		r	r	t	z	t	p	z	t	p	z	g	p	IO
t	r	m	r	r	m	r	r	r	t	h	h	z	p	p	z	r	r		g	p	p	g	p	t	z	t	z	t	z	JK
r	t	m	r	r	m	t	r	r	r	p	z	p	h	z	h	r	r	g		z	z	g	z	t	p	t	p	t	p	JN
p	t	r	p	z	r	t	p	z	r	t	p	z	p	p	z	t	t	p	z		h	h	h	h	h	h	w	w	w	KL
z	r	t	z	p	r	m	t	t	z	z	g	g	r	t	t	p	z	p	z	h		h	h	h	w	w	h	h	w	KM
r	r	m	t	t	m	r	t	t	r	z	p	z	z	z	p	t	t	g	g	h	h		h	w	h	w	h	w	h	KN
r	m	r	t	t	t	r	z	p	p	t	p	z	r	t	t	z	p	p	z	h	h	h		w	w	h	w	h	h	KO
z	z	z	g	g	p	t	p	z	t	r	z	p	r	p	z	p	z	t	t	h	h	w	w		h	h	h	h	w	LM
r	z	r	p	z	r	z	p	z	m	r	t	t	z	g	g	t	t	z	p	h	w	h	w	h		h	h	w	h	LN
r	t	p	p	z	z	z	g	g	r	m	t	t	r	p	z	z	p	t	t	h	w	w	h	h	h		w	h	h	LO
t	p	t	z	p	r	r	t	t	t	p	z	p	t	z	p	p	z	z	p	w	h	h	w	h	h	w		h	h	MN
t	r	z	z	p	z	r	z	p	z	r	z	p	m	t	t	g	g	t	t	w	h	w	h	h	w	h	h		h	MO
m	r	r	t	t	t	p	z	p	r	r	t	t	t	z	p	z	p	z	p	w	w	h	h	w	h	h	h	h		NO
EG	EN	EI	EL	EM	FI	FH	FL	FO	GI	GJ	GN	GM	NJ	HL	HN	IM	IO	JK	JN	KL	KM	KN	KO	LM	LN	LO	MN	MO	NO	

between pp. 122-123.

402.

ON A SINGULARITY OF SURFACES.

[From the *Quarterly Journal of Pure and Applied Mathematics*, vol. IX. (1868), pp. 332—338.]

A SURFACE having a nodal line has in general on this nodal line points where the two tangent planes coincide, or as I propose to term them "pinch-points." Thus, if the nodal line be the curve of complete intersection of any two surfaces $P=0$, $Q=0$, then the equation of the general surface having this curve for a nodal line is $(a, b, c \between P, Q)^2=0$ (where a, b, c are any functions of the coordinates), and the pinch-points are given as the intersections of the nodal line $P=0$, $Q=0$ with the surface $ac-b^2=0$. Consider the case where the nodal curve is a curve of partial intersection represented by the equations $\left\|\begin{matrix} P, & Q, & R \\ P', & Q', & R' \end{matrix}\right\|=0$, or say by the equations $p=0$, $q=0$, $r=0$ (viz. p, q, r denote the functions $QR'-Q'R$, $RP'-R'P$, $PQ'-P'Q$ respectively), and consequently we have identically

$$(P\,,\ Q\,,\ R\between p,\ q,\ r)=0,$$
$$(P',\ Q',\ R'\between p,\ q,\ r)=0,$$

or what is the same thing, (λ, μ) being arbitrary,

$$(\lambda P+\mu P',\ \lambda Q+\mu Q',\ \lambda R+\mu R'\between p,\ q,\ r)=0.$$

The general surface having the curve in question for its nodal line is represented by the equation

$$(a,\ b,\ c,\ f,\ g,\ h\between p,\ q,\ r)^2=0,$$

(where (a, b, c, f, g, h) are any functions of the coordinates), and it is easy to see that the condition for a pinch-point is the same as that which (considering p, q, r as coordinates and all the other quantities as constants), expresses that the line

$$(\lambda P+\mu P',\ \lambda Q+\mu Q',\ \lambda R+\mu R'\between p,\ q,\ r)=0,$$

touches the conic

$$(a, b, c, f, g, h)(p, q, r)^2 = 0,$$

viz. A, B, C, F, G, H being the inverse coefficients, $A = bc - f^2$, &c., this condition is

$$(A, B, C, F, G, H \between \lambda P + \mu P', \lambda Q + \mu Q', \lambda R + \mu R')^2 = 0,$$

or what is the same thing, the pinch-points are given as the common intersections of the nodal line $p = 0$, $q = 0$, $r = 0$ with each of the three surfaces

$$\begin{aligned}
&(A, B, C, F, G, H)(P, Q, R)^2 &= 0,\\
&(A, B, C, F, G, H)(P, Q, R)\ (P', Q', R') &= 0,\\
&(A, B, C, F, G, H)\qquad\qquad (P', Q', R')^2 &= 0,
\end{aligned}$$

these last three equations in fact, adding only a single relation to the relations expressed by the equations

$$p = 0,\ q = 0,\ r = 0.$$

If the functions P, Q, R, P', Q', R' are linear functions of the coordinates, then the curve $(p = 0,\ q = 0,\ r = 0)$ is a cubic curve in space, or skew cubic; and if moreover (a, b, c, f, g, h) are constants, then the equation

$$(a, b, c, f, g, h \between p, q, r)^2 = 0,$$

belongs to a quartic surface having the skew cubic for a nodal line: this surface is (it may be observed) a ruled surface, or scroll. With a view to ulterior investigations, I propose to study the theory of the pinch-points in regard to this particular surface; and to simplify as much as possible, I fix the coordinates as follows:

Considering the skew cubic as given, let any point O on the cubic be taken for the origin; let $x = 0$ be the equation of the osculating plane at O; $y = 0$ that of any other plane through the tangent line at O; $z = 0$, that of any other plane through O, not passing through the tangent line; and $w = 0$ that of a fourth plane; then the equation of the cubic will be

$$\left\|\begin{matrix} x, & y, & z \\ y, & z, & w \end{matrix}\right\| = 0,$$

or what is the same thing, the values of p, q, r are $yw - z^2$, $zy - xw$, and $xz - y^2$ respectively. And conversely, the cubic being thus represented, the point $(x = 0,\ y = 0,\ z = 0)$ may be considered as standing for any point whatever on the skew cubic; the osculating plane at this point being $x = 0$, and the tangent line being $x = 0$, $y = 0$. For the purpose of the present investigations, we may without loss of generality write $w = 1$; and for convenience I shall do this; the values of p, q, r thus become $y - z^2$, $yz - x$, $xz - y^2$, and the equation of the surface is

$$(a, b, c, f, g, h \between y - z^2, yz - x, xz - y^2)^2 = 0.$$

At a pinch-point, we have

$$(A,\ B,\ C,\ F,\ G,\ H)(x,\ y,\ z)^2 \qquad\qquad = 0,$$
$$(A,\ B,\ C,\ F,\ G,\ H)(x,\ y,\ z)\ (y,\ z,\ 1) = 0,$$
$$(A,\ B,\ C,\ F,\ G,\ H) \qquad\quad (y,\ z,\ 1)^2 = 0,$$

and hence the origin will be a pinch-point if $C=0$, that is, if $ab-h^2=0$. This however appears more readily by remarking, that the equation of the pair of tangent planes at the origin is

$$(a,\ b,\ c,\ f,\ g,\ h\between y,\ -x,\ 0)^2 = 0,$$

or what is the same thing,

$$(a,\ h,\ b\between y,\ -x)^2 = 0;$$

the two tangent planes therefore coincide, or there is a pinch-point, if only $ab-h^2=0$.

By what precedes, it appears that if we wish to study the form of the quartic surface, 1°, in the neighbourhood of an arbitrary point on the nodal line; 2°, in the neighbourhood of a pinch-point; it is sufficient in the first case to consider the general surface

$$(a,\ b,\ c,\ f,\ g,\ h\between y-z^2,\ yz-x,\ xz-y^2)^2 = 0,$$

in the neighbourhood of the origin; and in the second case, to study the special surface for which $ab-h^2=0$, or writing for convenience $a=1$, and therefore $b=h^2$, the surface

$$(1,\ h^2,\ c,\ f,\ g,\ h\between y-z^2,\ yz-x,\ xz-y^2)^2 = 0,$$

in the neighbourhood of the origin.

Consider first the surface

$$(a,\ b,\ c,\ f,\ g,\ h)\ (y-z^2,\ yz-x,\ xz-y^2)^2 = 0.$$

A plane through the origin is either a plane not passing through the tangent line $(x=0,\ y=0)$, and the equation $z=0$ will serve to represent any such plane; or if it pass through the tangent line, then it is either a non-special plane, which may be represented by the equation $y=0$; or it is a special plane: viz. either the osculating plane $x=0$ of the nodal line, or else one or the other of the two tangent planes $(a,\ h,\ b\between y,\ -x)^2=0$ of the surface. I consider therefore the sections of the surface by these planes $z=0$, $y=0$, $x=0$, $(a,\ h,\ b\between y,\ -x)^2=0$ respectively.

Section by the non-special plane $z=0$.

The equation is

$$(a,\ b,\ c,\ f,\ g,\ h\between y,\ -x,\ -y^2)^2 = 0,$$

which represents a curve having at the origin an ordinary node, the equations of the two tangents being $(a,\ h,\ b\between y,\ -x)^2=0$, viz. these are the intersections of the two tangent planes by the plane $z=0$.

Section by the non-special plane through the tangent line, viz. the plane $y=0$.

The equation is

$$(a,\ b,\ c,\ f,\ g,\ h\between -z^2,\ -x,\ xz)^2=0,$$

or what is the same thing,

$$bx^2-2fx^2z+2hxz^2+cx^2z^2-2gxz^3+az^4=0.$$

Writing as usual $x=Az^\mu+\&c.$ we have

$$\mu=2,\quad bA^2+2hA+a=0,$$

and since $ab-h^2$ is by hypothesis not $=0$, A has two unequal values; we have at the origin two branches $x=A_1z^2+B_1z^3+\&c.$, $x=A_2z^2+B_2z^3+\&c.$, having the common tangent $x=0$ (viz. this is the tangent $x=0$, $y=0$ of the nodal curve), and with a two-pointic intersection of the two branches, that is, the point at the origin is an ordinary tacnode.

Section by the osculating plane $x=0$.

The equation is

$$(a,\ b,\ c,\ f,\ g,\ h\between y-z^2,\ yz,\ -y^2)^2=0.$$

We may write $y=z^2+Az^\mu+\&c.$, we at once find $\mu=3$, and then

$$(a,\ b,\ c,\ f,\ g,\ h\between Az^3+\&c.,\ z^3+\&c.,\ -z^4+\&c.)^2=0,$$

that is

$$(a,\ h,\ b\between A,\ 1)^2=0.$$

A has two unequal values, and the branches through the origin are

$$y=z^2+A_1z^3+B_1z^4+\&c.,\quad y=z^2+A_2z^3+B_2z^4+\&c.\ \ldots,$$

viz. the branches have the common tangent line $y=0$ (the tangent $x=0$, $y=0$ of the nodal curve), but in the present case a three-pointic intersection.

Section by one of the tangent planes $(a,\ h,\ b\between y,\ -x)^2=0$.

Writing $y=-mx$, and therefore $(a,\ h,\ b\between m,\ -1)^2=0$, the equation is

$$(a,\ b,\ c,\ f,\ g,\ h\between -mx-z^2,\ -x-mzx,\ xz-m^2x^2)^2=0,$$

which represents of course the projection of the section on the plane $z=0$, $x=0$, but which (since there is no alteration in the singularities) may be considered as representing the section itself. Developing, the coefficient of x^2 is $am^2+2hm+b$, which is $=0$, and the equation becomes

$$\begin{array}{rlcll}
 & 2m^2(gm+f) & x^3 & + & cm^4 \quad x^4\\
+ & 2\{hm^2+(b-g)m-f\} & x^2z & + & 2m^2(fm-c) \quad x^3z\\
+ & 2(am+h) & xz^2 & + & (b+2g)m^2-2fm+c \quad x^2z^2\\
 & & & + & 2(hm-g) \quad xz^3\\
 & & & + & a \quad z^4=0,
\end{array}$$

so that the curve has at the origin a triple point, the tangent to one branch being the line $x=0$ (the tangent $x=0$, $y=0$ of the nodal curve).

Consider next the surface

$$(1,\ h^2,\ c,\ f,\ g,\ h \between y - z^2,\ yz - x,\ xz - y^2)^2 = 0,$$

being as already remarked, the general surface referred to a pinch-point as origin.

Section by the non-special plane $z = 0$.

The equation is

$$(1,\ h^2,\ c,\ f,\ g,\ h \between y,\ -x,\ -y^2)^2 = 0,$$

where, attending only to the terms of the lowest order, we find $(1,\ h,\ h^2 \between y,\ -x)^2 = 0$, that is $(y - hx)^2 = 0$, showing that the origin is a cusp.

Section by the non-special plane through the tangent line, viz. the plane $y = 0$.

The equation is

$$(1,\ h^2,\ c,\ f,\ g,\ h \between -z^2,\ -x,\ xz)^2 = 0,$$

or what is the same thing,

$$h^2x^2 + 2hxz^2 - 2fx^2z + cx^2z^2 - 2gxz^3 + z^4 = 0,$$

that is

$$(hx + z^2)^2 - 2fx^2z + cx^2z^2 - 2gxz^3 = 0,$$

writing $hx = -z^2 + Ax^\mu$, we find at once $\mu = \frac{5}{2}$, and then $A^2 = \dfrac{2f}{h^2} - \dfrac{2g}{h}$, so that the branches are $hx = -z^2 \pm Ax^{\frac{5}{2}}$; whence we have at the origin a cusp of the second order or node cusp.

Section by the osculating plane $x = 0$.

The equation is

$$(1,\ h^2,\ c,\ f,\ g,\ h \between y - z^2,\ yz,\ -y^2)^2 = 0;$$

writing $y = z^2 - hz^3 + Az_\mu$, we easily find $\mu = \frac{7}{2}$, and then

$$(1,\ h^2,\ c,\ f,\ g,\ h \between -hz^3 + Az^{\frac{7}{2}},\ z^3 - hz^4,\ -z^4)^2 = 0,$$

where the terms in z^6, and $z^{6+\frac{1}{2}}$ disappear of themselves, the terms in z^7 give $A^2 + 2gh = 0$, and the branches are

$$y = z^2 - hz^3 \pm Az^{\frac{7}{2}} \text{ \&c.},$$

viz. there is a cusp of a superior order.

Section by the tangent plane $y = hx$.

The equation is

$$(1,\ h^2,\ c,\ f,\ g,\ h \between hx - z^2,\ -x + hzx,\ xz - h^2x^2)^2 = 0,$$

representing the projection on the plane of zx. Developing, the equation is

$$\begin{array}{lll} 2(f-gh)\,x^2(h^2x-z) & +x^4ch^4 & \\ & +x^3z & -2h^2(hf+c) \\ & +x^2z^2 & (h^4+2g^2h^2+2fh+c) \\ & +xz^3 & -2(h^2+g) \\ & +z^4 & 1 \qquad\qquad =0, \end{array}$$

and there is at the origin a triple point (= cusp + 2 nodes) arising from the passage of an ordinary branch through a cusp; the tangent at the cusp being it will be noticed the line $x=0$, that is the tangent $x=0$, $y=0$ to the nodal curve at the pinch-point.

The results of the investigation may be presented in a tabular form as follows:

Plane of Section.	Nature of Section. Origin, an ordinary point.	Origin, a Pinch-point.
Non-special.	Node.	Cusp.
Ditto, through tangent line of nodal curve.	Tacnode = 2 nodes.	Node-cusp, = node + cusp.
Osculating plane of nodal curve.	$y=z^2+\left.\begin{matrix}A_1\\A_2\end{matrix}\right\\}z^3+$ &c.	$y=z^2+hz^3\pm Az^{\frac{7}{2}}$ &c.
Either of the two tangent planes.	Triple point, one branch touching the tangent of nodal line.	
The single tangent plane.		Triple point, = cusp + 2 nodes; the cuspidal branch touching the tangent of the nodal line.

I have not considered the special cases where one of the two tangent planes, or (as the case may be) the single tangent plane of the surface coincides with the osculating plane of the nodal curve.

403.

ON PASCAL'S THEOREM.

[From the *Quarterly Journal of Pure and Applied Mathematics,* vol. IX. (1868), pp. 348—353.]

I CONSIDER the following question: to find a point such that its polar plane in regard to a given system of three planes is the same as its polar plane in regard to another given system of three planes.

The equations of any six planes whatever may be taken to be $X=0$, $Y=0$, $Z=0$, $U=0$, $V=0$, $W=0$, where

$$X+Y+Z+U+V+W=0,$$
$$aX+bY+cZ+fU+gV+hW=0,$$

and so also any quantities X, Y, Z, U, V, W satisfying these relations may be regarded as the coordinates of a point in space; we pass to the ordinary system of quadriplanar coordinates by merely substituting for V, W their values as linear functions of X, Y, Z, U.

This being so, the equations of the given systems of three planes may be taken to be

$$XYZ=0, \qquad UVW=0,$$

and if we take for the coordinates of the required point (x, y, z, u, v, w), where

$$x+y+z+u+v+w=0,$$
$$ax+by+cz+fu+gv+hw=0,$$

then the equations of the two polar planes are

$$\frac{X}{x}+\frac{Y}{y}+\frac{Z}{z}=0, \qquad \frac{U}{u}+\frac{V}{v}+\frac{W}{w}=0,$$

respectively, and we have to find (x, y, z, u, v, w), such that these two equations may represent the same plane, or that the two equations may in virtue of the linear relations between (X, Y, Z, U, V, W) be the same equation.

The ordinary process by indeterminate multipliers gives

$$\frac{1}{x}+\lambda+\mu a=0,$$

$$\frac{1}{y}+\lambda+\mu b=0,$$

$$\frac{1}{z}+\lambda+\mu c=0,$$

$$\frac{k}{u}+\lambda+\mu f=0,$$

$$\frac{k}{v}+\lambda+\mu g=0,$$

$$\frac{k}{w}+\lambda+\mu h=0,$$

and we have the before-mentioned linear relations between (x, y, z, u, v, w); these last are satisfied by the values

$$(x, y, z, u, v, w)=\left(\frac{1}{a-\theta}, \frac{1}{b-\theta}, \frac{1}{c-\theta}, -\frac{1}{f-\theta}, -\frac{1}{g-\theta}, -\frac{1}{h-\theta}\right),$$

if only

$$\frac{1}{a-\theta}+\frac{1}{b-\theta}+\frac{1}{c-\theta}-\frac{1}{f-\theta}-\frac{1}{g-\theta}-\frac{1}{h-\theta}=0;$$

in fact, θ satisfying this equation, the relation

$$x+y+z+u+v+w=0$$

is obviously satisfied; and observing that we have

$$ax=\frac{a}{a-\theta}=1+\frac{\theta}{a-\theta}, \ldots, fu=\frac{-f}{f-\theta}=-1-\frac{\theta}{f-\theta}, \ldots,$$

we have

$$ax+by+cz+fu+gv+hw$$

$$=\Sigma\left(1+\frac{\theta}{a-\theta}\right)-\Sigma\left(1+\frac{\theta}{f-\theta}\right),$$

$$=\theta\left(\Sigma\frac{1}{a-\theta}-\Sigma\frac{1}{f-\theta}\right), =0,$$

so that the relation $ax+by+cz+fu+gv+hw=0$ is also satisfied. Substituting the foregoing values of (x, y, z, u, v, w) the six equations containing k, λ, μ, will be all of them satisfied if only

$$\mu=-1, \lambda=\theta, k=-1.$$

The coordinates of the required point thus are

$$\left(\frac{1}{a-\theta}, \frac{1}{b-\theta}, \frac{1}{c-\theta}, -\frac{1}{f-\theta}, -\frac{1}{g-\theta}, -\frac{1}{h-\theta}\right),$$

where

$$\frac{1}{a-\theta}+\frac{1}{b-\theta}+\frac{1}{c-\theta}-\frac{1}{f-\theta}-\frac{1}{g-\theta}-\frac{1}{h-\theta}=0;$$

and, the equation in θ being of the fourth order, there are thus four points, say the points O_1, O_2, O_3, O_4, which have each of them the property in question.

It will be convenient to designate the planes $X=0$, $Y=0$, $Z=0$, $U=0$, $V=0$, $W=0$ as the planes a, b, c, f, g, h respectively; the line of intersection of the planes $X=0$, $Y=0$ will then be the line ab, and the point of intersection of the planes $X=0$, $Y=0$, $Z=0$ the point abc; and so in other cases.

I say that from any one of the points O it is possible to draw

a line meeting the lines $af \,.\, bg \,.\, ch$ (1),
" " $ag \,.\, bh \,.\, cf$ (2),
" " $ah \,.\, bf \,.\, cg$ (3),
" " $af \,.\, bh \,.\, cg$ (4),
" " $ag \,.\, bf \,.\, ch$ (5),
" " $ah \,.\, bg \,.\, cf$ (6),

and consequently, that the four points O are the four common points of the six hyperboloids passing through these triads of lines respectively.

In fact, considering θ as determined by the foregoing quartic equation, and writing for shortness

$$\begin{aligned} (a-\theta)X &= A, & (f-\theta)U &= F, \\ (b-\theta)Y &= B, & (g-\theta)V &= G, \\ (c-\theta)Z &= C, & (h-\theta)W &= H, \end{aligned}$$

so that

$$A+B+C+F+G+H=0,$$

the equations $A+F=0$, $B+G=0$, $C+H=0$, are equivalent to two equations only, and it is at once seen, that these are in fact the equations of a line through the point O meeting the three lines af, bg, ch respectively.

The equation $A+F=0$, is in fact satisfied by the values $X : U = \frac{1}{a-\theta} : -\frac{1}{f-\theta}$, and by $X=0$, $U=0$; it is consequently the equation of the plane through O and the line af; similarly, $B+G=0$ is the equation of the plane through O and the line

bg; and $C+H=0$ is the equation of the plane through O and the line ch; and the three equations being equivalent to two equations only, the planes have a common line which is the line in question.

The equations of the six lines thus are:

$$\begin{array}{llll}(1) & A+F=0, & B+G=0, & C+H=0,\\ (2) & A+G=0, & B+H=0, & C+F=0,\\ (3) & A+H=0, & B+F=0, & C+G=0,\\ (4) & A+F=0, & B+H=0, & C+G=0,\\ (5) & A+G=0, & B+F=0, & C+H=0,\\ (6) & A+H=0, & B+G=0, & C+F=0.\end{array}$$

It is further to be noticed, that if in any one of these systems, for instance in the system $A+F=0$, $B+G=0$, $C+H=0$, we consider θ as an arbitrary quantity, then the equations are those of any line whatever cutting the lines af, bg, ch; and hence eliminating θ, we have the equation of the hyperboloid through the three lines af, bg, ch; the equations of the six hyperboloids are thus found to be

$$\begin{array}{lccccc}(1) & \dfrac{ax+fu}{x+u} & = & \dfrac{by+gv}{y+v} & = & \dfrac{cz+hw}{z+w},\\[2ex] (2) & \dfrac{ax+gv}{x+v} & = & \dfrac{by+hw}{y+w} & = & \dfrac{cz+fu}{z+u},\\[2ex] (3) & \dfrac{ax+hw}{x+w} & = & \dfrac{by+fu}{y+u} & = & \dfrac{cz+gv}{z+v},\\[2ex] (4) & \dfrac{ax+fu}{x+u} & = & \dfrac{by+hw}{y+w} & = & \dfrac{cz+gv}{z+v},\\[2ex] (5) & \dfrac{ax+gv}{x+v} & = & \dfrac{by+fu}{y+u} & = & \dfrac{cz+hw}{z+w},\\[2ex] (6) & \dfrac{ax+hw}{x+w} & = & \dfrac{by+gv}{y+v} & = & \dfrac{cz+fu}{z+u},\end{array}$$

respectively; the equations in the same line being of course equivalent to a single equation.

For each one of the six lines we have

$$(A,\ B,\ C)=(-F,\ -G,\ -H)$$

in some order or other, and it is thus seen that the six lines lie on a cone of the second order, the equation whereof is

$$A^2+B^2+C^2-F^2-G^2-H^2=0.$$

Consider now the six planes a, b, c, f, g, h, and taking in the first instance an arbitrary point of projection, and a plane of projection which is also arbitrary—the line of intersection ab of the planes a and b will be projected into a line ab, and the point of intersection of the planes a, b, c into a point abc; and so in other cases. We have thus a plane figure, consisting of the fifteen lines ab, ac, ... gh, and of the twenty points abc, abf, ... fgh; and which is such, that on each of the lines there lie four of the points, and through each of the points there pass three of the lines, viz. the points abc, abf, abg, abh lie on the line ab; and the lines bc, ca, ab meet in the point abc, and so in other cases. If now the point of projection instead of being arbitrary, be one of the above-mentioned four points O, then the projections of the lines af, bg, ch meet in a point, and the like for each of the six triads of lines; that is in the plane figure we have six points 1, 2, 3, 4, 5, 6, each of them the intersection of three lines as shown in the diagram,

$$1 = af \,.\, bg \,.\, ch,$$

$$2 = ag \,.\, bh \,.\, cf,$$

$$3 = ah \,.\, bf \,.\, cg,$$

$$4 = af \,.\, bh \,.\, cg,$$

$$5 = ag \,.\, bf \,.\, ch,$$

$$6 = ah \,.\, bg \,.\, cf,$$

and these six points lie in a conic. It is clear that the lines af, ag, ah; bf, bg, bh; cf, cg, ch are the lines 14, 25, 36; 35, 16, 24; 26, 34, 15 respectively.

Conversely, starting from the points 1, 2, 3, 4, 5, 6 on a conic, and denoting the lines 14, 25, 36; 35, 16, 24; 26, 34, 15 (being, it may be noticed, the sides and diagonals of the hexagon 162435) in the manner just referred to, then it is possible to complete the figure of the fifteen lines ab, ac, ... gh and of the twenty points abc, abf, ... fgh, such that each line contains upon it four points, and that through each point there pass three lines, in the manner already mentioned.

Of the fifteen lines, nine, viz. the lines af, ag, ah; bf, bg, bh; cf, cg, ch are, as has been seen, lines through two of the six points 1, 2, 3, 4, 5, 6; the remaining lines are bc, ca, ab; gh, hf, fg. These are Pascalian lines,

bc	of the hexagon	162435,
ca	„	152634,
ab	„	142536,
gh	„	152436,
hf	„	142635,
bg	„	162534,

which appears thus, viz.

$$\begin{aligned} \text{line } bc \text{ contains points } & bcf\ , \quad bcg\ , \quad bch\ , \\ & = bf\,.\,cf, \quad bg\,.\,cg, \quad bh\,.\,ch, \\ & = 35\,.\,26, \quad 16\,.\,34, \quad 24\,.\,15\,; \end{aligned}$$

that is, bc is the Pascalian line of the hexagon 162435; and the like for the rest of the six lines.

The twenty points abc, $abf, \ldots fgh$ are as follows, viz. omitting the two points abc, fgh, the remaining eighteen points are the Pascalian points (the intersections of pairs of lines each through two of the points 1, 2, 3, 4, 5, 6) which lie on the Pascalian lines bc, ca, ab, gh, hf, fg respectively; the point abc is the intersection of the Pascalian lines bc, ca, ab, and the point fgh is the intersection of the Pascalian lines gh, hf, fg, the points in question being two of the points P (Steiner's twenty points, each the intersection of three Pascalian lines).

We thus see that we have two triads of hexagons such that the Pascalian lines of each triad meet in a point, and that the two points so obtained, together with the eighteen points on the six Pascalian lines, form a system of twenty points lying four together on fifteen lines, and which points and lines are the projections of the points and lines of intersection of six planes; or, say simply that the figure is the projection of the figure of six planes.

It is to be added, that if the planes are a, b, c, f, g, h, then the point of projection is any one of the four points which have the same polar plane in regard to the system of the planes a, b, c, and in regard to the system of the planes f, g, h. The consideration of the solid figure affords a demonstration of the existence as well of the six Pascalian lines as of the two points each the intersection of three of these lines.

404.

REPRODUCTION OF EULER'S MEMOIR OF 1758 ON THE ROTATION OF A SOLID BODY.

[From the *Quarterly Journal of Pure and Applied Mathematics*, vol. IX. (1868), pp. 361—373.]

EULER'S Memoir "Du mouvement de rotation des corps solides autour d'un axe variable," *Mém. de Berlin*, 1758, pp. 154—193 (printed in 1765), seems to have been written subsequently to the memoir with a similar title in the Berlin *Memoirs* for 1760, and to the "Theoria Motus Corporum Solidorum &c.," *Rostock*, 1765, and there are contained in the first-mentioned memoir some very interesting results which appear to have escaped the notice of later writers on the subject; viz. Euler succeeds in integrating the equations of motion *without the assistance furnished by the consideration of the invariable plane*. In reproducing these results I make the following alterations in Euler's notation, viz. instead of x, y, z I write p, q, r; instead of Ma^2, Mb^2, Mc^2 (where M is the mass) I write A, B, C, these quantities denoting the principal moments, and in some equations where the omission or insertion of the factor M is really immaterial I write A, B, C in the place of a^2, b^2, c^2; moreover instead of Euler's A, B, C $\left(\text{which denote respectively } \frac{b^2-c^2}{a^2}, \frac{c^2-a^2}{b^2}, \frac{a^2-b^2}{c^2}\right)$ I write L, M, N; but in other respects Euler's notation is preserved. The equations of motion are

$$Adp + (C-B)\,qrdt = 0,$$

$$Bdq + (A-C)\,rpdt = 0,$$

$$Cdr + (B-A)\,pqdt = 0;$$

so that putting for shortness

$$L = \frac{B-C}{A}, \quad M = \frac{C-A}{B}, \quad N = \frac{A-B}{C},$$

and introducing the auxiliary quantity u such that $du = pqrdt$, we have

$$p^2 = \mathfrak{A} + 2Lu,$$
$$q^2 = \mathfrak{B} + 2Mu,$$
$$r^2 = \mathfrak{C} + 2Nu,$$

where $\mathfrak{A}$, $\mathfrak{B}$, $\mathfrak{C}$ are constants of integration, and thence

$$t = \int \frac{du}{\sqrt{\{(\mathfrak{A} + 2Lu)(\mathfrak{B} + 2Mu)\}(\mathfrak{C} + 2Nu)}},$$

where the integral may without loss of generality be taken from $u = 0$; u, and consequently p, q, r, are thus given functions of t; and it is moreover clear that $\mathfrak{A}$, $\mathfrak{B}$, $\mathfrak{C}$ are the initial values of p^2, q^2, r^2. We have also if ω be the angular velocity round the instantaneous axis

$$\omega^2 = \mathfrak{A} + \mathfrak{B} + \mathfrak{C} + 2(L + M + N)u.$$

Euler then assumes that the position in space of the principal axes is geometrically determined as follows, viz. (treating the axes as points on a sphere) it is assumed that the distances from a fixed point P of the sphere are respectively l, m, n, and that

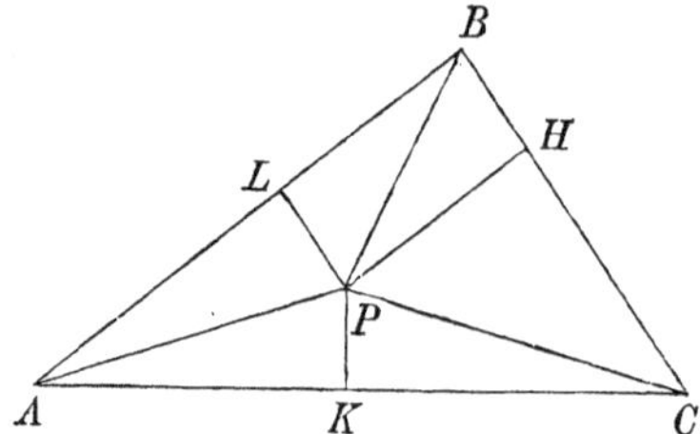

the inclinations of these distances to a fixed arc PQ are respectively λ, μ, ν. We have then the geometrical relations

$$\cos^2 l + \cos^2 m + \cos^2 n = 1;$$

$$\sin(\mu - \nu) = \frac{\cos l}{\sin m \sin n}, \quad \cos(\mu - \nu) = -\frac{\cos m \cos n}{\sin m \sin n},$$

$$\sin(\nu - \lambda) = \frac{\cos m}{\sin n \sin l}, \quad \cos(\nu - \lambda) = -\frac{\cos n \cos l}{\sin n \sin l},$$

$$\sin(\lambda - \mu) = \frac{\cos n}{\sin l \sin m}, \quad \cos(\lambda - \mu) = \frac{\cos l \cos m}{\sin l \sin m};$$

whence also

$$\sin \mu = \frac{-\cos\lambda \cos n - \sin\lambda \cos l \cos m}{\sin l \sin m},$$

$$\cos \mu = \frac{\sin\lambda \cos n - \cos\lambda \cos l \cos m}{\sin l \sin m},$$

$$\sin \nu = \frac{\cos\lambda \cos m + \sin\lambda \cos l \cos n}{\sin l \sin n},$$

$$\cos \nu = \frac{-\sin\lambda \cos m - \cos\lambda \cos l \cos n}{\sin l \sin n}.$$

The geometrical equations connecting the resolved angular velocities p, q, r with the differentials of l, m, n, λ, μ, ν are

$$\begin{aligned} dl\ \sin l\ &= dt\,(q\cos n - r\cos m), & d\lambda\sin^2 l &= -dt\,(q\cos m + r\cos n),\\ dm\sin m &= dt\,(r\cos l - p\cos n), & d\mu\sin^2 m &= -dt\,(r\cos n + p\cos l),\\ dn\ \sin n &= dt\,(p\cos m - q\cos l), & d\nu\sin^2 n &= -dt\,(p\cos l + q\cos m). \end{aligned}$$

Multiplying the equations of motion respectively by $\cos l$, $\cos m$, $\cos n$, and adding, we obtain an equation which is reducible to the form

$$d\,(Ap\cos l + Bq\cos m + Cr\cos n) = 0,$$

whence integrating

$$Ap\cos l + Bq\cos m + Cr\cos n = \mathfrak{D},$$

$\mathfrak{D}$ being a constant of integration. One other integral equation is necessary for the determination of the angles l, m, n. The expressions for dl, dm, dn give at once

$$p\,dl\sin l + q\,dm\sin m + r\,dn\sin n = 0.$$

Instead of the arcs l, m, n, Euler introduces a new variable v, such that

$$v = p\cos l + q\cos m + r\cos n;$$

by means of the last preceding equation, we find

$$dv = dp\cos l + dq\cos m + dr\cos n,$$

and then, substituting for dp, dq, dr, their values,

$$dv = \left(\frac{L\cos l}{p} + \frac{M\cos m}{q} + \frac{N\cos n}{r}\right) du,$$

from which the relation between v and u is to be determined. We have

$$\begin{aligned} \cos^2 l + \cos^2 m + \cos^2 n &= 1,\\ Ap\cos l + Bq\cos m + Cr\cos n &= \mathfrak{D},\\ p\cos l + q\cos m + r\cos n &= v, \end{aligned}$$

which give $\cos l$, $\cos m$, $\cos n$ in terms of u, v; the resulting formulæ contain the radical

$$\sqrt{\left\{\begin{aligned}&(L^2A^2q^2r^2 + M^2B^2r^2p^2 + N^2C^2p^2q^2) - \mathfrak{D}^2(x^2+y^2+z^2)\\ &\qquad + 2\mathfrak{D}v\,(Ap^2 + Bq^2 + Cr^2) - v^2(A^2p^2 + B^2q^2 + C^2r^2)\end{aligned}\right\}},$$

which for shortness is represented by $\sqrt{\{(\cdot\cdot)\}}$. We then have

$$\cos l = \frac{\mathfrak{D}p\,(NCq^2 - MBr^2) + BCpv\,(Mr^2 - Nq^2) + LAqr\sqrt{\{(\cdot\cdot)\}}}{L^2A^2q^2r^2 + M^2B^2r^2p^2 + N^2C^2p^2q^2},$$

$$\cos m = \frac{\mathfrak{D}q\,(LAr^2 - NCp^2) + CAqv\,(Np^2 - Lr^2) + MBrp\sqrt{\{(\cdot\cdot)\}}}{L^2A^2q^2r^2 + M^2B^2r^2p^2 + N^2C^2p^2q^2},$$

$$\cos n = \frac{\mathfrak{D}r\,(MBp^2 - LAq^2) + ABrv\,(Lq^2 - Mp^2) + NCpq\sqrt{\{(\cdot\cdot)\}}}{L^2A^2q^2r^2 + M^2B^2r^2p^2 + N^2C^2p^2q^2},$$

and substituting these values in the differential equation

$$\frac{dv}{du} = \frac{L\cos l}{p} + \frac{M\cos m}{q} + \frac{N\cos n}{r},$$

the equation to be integrated becomes

$$\frac{dv}{du}(L^2A^2q^2r^2 + M^2B^2r^2p^2 + N^2C^2p^2q^2) = LMN\mathfrak{D}(Ap^2 + Bq^2 + Cr^2) - LMNv(A^2p^2 + B^2q^2 + C^2r^2)$$

$$+ \frac{1}{pqr}(L^2Aq^2r^2 + M^2Br^2p^2 + N^2Cp^2q^2).$$

Now substituting for p, q, r their values, we have

$$L^2A^2q^2r^2 + M^2B^2r^2p^2 + N^2C^2p^2q^2 = L^2A^2\mathfrak{B}\mathfrak{C} + M^2B^2\mathfrak{C}\mathfrak{A} + N^2C^2\mathfrak{A}\mathfrak{B} - 2LMNu(\mathfrak{A}A^2 + \mathfrak{B}B^2 + \mathfrak{C}C^2),$$

$$L^2Aq^2r^2 + M^2Br^2p^2 + N^2Cp^2q^2 = L^2A\mathfrak{B}\mathfrak{C} + M^2B\mathfrak{C}\mathfrak{A} + N^2C\mathfrak{A}\mathfrak{B} - 2LMNu(\mathfrak{A}A + \mathfrak{B}B + \mathfrak{C}C),$$

$$p^2 + q^2 + r^2 = \mathfrak{A} + \mathfrak{B} + \mathfrak{C} + 2(L + M + N)u,$$

$$Ap^2 + Bq^2 + Cr^2 = \mathfrak{A}A + \mathfrak{B}B + \mathfrak{C}C,$$

$$A^2p^2 + B^2q^2 + C^2r^2 = \mathfrak{A}A^2 + \mathfrak{B}B^2 + \mathfrak{C}C^2:$$

and writing for shortness

$$\mathfrak{A} + \mathfrak{B} + \mathfrak{C} = E,$$

$$\mathfrak{A}A + \mathfrak{B}B + \mathfrak{C}C = F,$$

$$\mathfrak{A}A^2 + \mathfrak{B}B^2 + \mathfrak{C}C^2 = G,$$

$$L^2A\,\mathfrak{B}\mathfrak{C} + M^2B\mathfrak{C}\mathfrak{A} + N^2C\mathfrak{A}\mathfrak{B} = H,$$

$$L^2A^2\mathfrak{B}\mathfrak{C} + M^2B^2\mathfrak{C}\mathfrak{A} + N^2C^2\mathfrak{A}\mathfrak{B} = K,$$

where $K = EG - F^2$, substituting these values and observing that

$$L + M + N = -LMN,$$

the radical of the formula becomes

$$\sqrt{\{(\cdot\cdot)\}} = \sqrt{(K - 2LMNGu + 2\mathfrak{D}^2LMNu - \mathfrak{D}^2E + 2\mathfrak{D}Fv - Gv^2)},$$

and the differential equation becomes

$$\frac{dv}{du}(K - 2LMNGu) = LMN\mathfrak{D}F - LMNGv + \frac{1}{pqr}(H - 2LMNFu)\sqrt{\{(\cdot\cdot)\}},$$

which can be reduced to the form

$$\frac{Kdv - LMNF\mathfrak{D}du - 2LMNGudv + LMNGvdu}{\sqrt{\{K - \mathfrak{D}^2E + 2LMN(\mathfrak{D}^2 - G)u + 2\mathfrak{D}Fv - Gv^2\}}} = \frac{Hdu - 2LMNFudu}{\sqrt{\{(2Lu + \mathfrak{A})(2Mu + \mathfrak{B})(2Nu + \mathfrak{C})\}}}.$$

Euler remarks that as the right-hand side of the equation contains only the variable u, the solution will be effected if we can find a function of u, a multiplier of the left-hand side; he had elsewhere explained the method of finding such multipliers, and applying it to the equation in hand, the multiplier of the left-hand side, and therefore of the equation itself, is found to be $\frac{1}{K-2LMNGu}$, or what is the same thing $\frac{\sqrt{(G)}}{K-2LMNGu}$.

Multiplying by this quantity, the right-hand side may for shortness be represented by dU, so that

$$dU = \frac{(H-2LMNFu)\sqrt{(G)}\,du}{(K-2LMNGu)\sqrt{\{(2Lu+\mathfrak{A})(2Mu+\mathfrak{B})(2Nu+\mathfrak{C})\}}},$$

and U may be considered as a given function of u, or what is the same thing of t.

As regards the left-hand side, attending to the equation $K=EG-F^2$, the radical multiplied into $\sqrt{(G)}$ may be presented under the form

$$\sqrt{[\{(G-\mathfrak{D}^2)(K-2LMNGu)-(Gv-\mathfrak{D}F)^2\}]};$$

and consequently the left-hand side becomes

$$\frac{(K-2LMNGu)\,Gdv + LMNG\,(Gv-\mathfrak{D}F)\,du}{(K-LMNGu)\sqrt{\{(G-\mathfrak{D}^2)(K-2LMNGu)-(Gv-\mathfrak{D}F)^2\}}},$$

which putting for the moment $K-2LMNGu=p^2$, $Gv-\mathfrak{D}F=q$, $G-\mathfrak{D}^2=f^2$, becomes $\frac{pdq-qdp}{p\sqrt{(f^2p^2-q^2)}}$, the integral of which is $\sin^{-1}\frac{q}{fp}$; hence restoring the values of p, q, f, the integral is

$$\sin^{-1}\frac{Gv-\mathfrak{D}F}{\sqrt{(G-\mathfrak{D}^2)}\sqrt{(K-2LMNGu)}}.$$

Hence considering the constant of integration as included in U, or writing

$$U = \mathfrak{E} + \int \frac{(H-2LMNFu)\sqrt{(G)}\,du}{(K-2LMNGu)\sqrt{\{(2Lu+\mathfrak{A})(2Mu+\mathfrak{B})(2Nu+\mathfrak{C})\}}},$$

we have for the required integral of the differential equation

$$\sin^{-1}\frac{Gv-\mathfrak{D}F}{\sqrt{(G-\mathfrak{D}^2)}\sqrt{\{(K-2LMNGu)\}}} = U,$$

whence also

$$\frac{Gv-\mathfrak{D}F}{\sqrt{(G-\mathfrak{D}^2)}\sqrt{\{(K-2LMNGu)\}}} = \sin U,$$

and

$$\frac{\sqrt{[\{(G-\mathfrak{D}^2)(K-2LMNGu)-(Gv-\mathfrak{D}F)^2\}]}}{\sqrt{(G-\mathfrak{D}^2)}\sqrt{\{(K-2LMNGu)\}}} = \cos U,$$

so that the value of the original radical is

$$\sqrt{\{(\cdot\cdot)\}} = \frac{\sqrt{(G-\mathfrak{D}^2)}\sqrt{\{(K-2LMNGu)\}}}{\sqrt{(G)}}\cos U.$$

Substituting in the expressions for the cosines of the arcs l, m, n, these values of v and the radical; the formulæ after some reductions become

$$\cos l = \frac{\mathfrak{D}Ap}{G} + \frac{BCp\,(M\mathfrak{C}-N\mathfrak{B})\sqrt{(G-\mathfrak{D}^2)}}{G\sqrt{(K-2LMNGu)}}\sin U + \frac{LAqr\sqrt{(G-\mathfrak{D}^2)}}{\sqrt{(G)}\sqrt{(K-2LMNGu)}}\cos U,$$

$$\cos m = \frac{\mathfrak{D}Bq}{G} + \frac{CAq\,(N\mathfrak{A}-L\mathfrak{C})\sqrt{(G-\mathfrak{D}^2)}}{G\sqrt{(K-2LMNGu)}}\sin U + \frac{MBrp\sqrt{(G-\mathfrak{D}^2)}}{\sqrt{(G)}\sqrt{(K-2LMNGu)}}\cos U,$$

$$\cos n = \frac{\mathfrak{D}Cr}{G} + \frac{ABr\,(L\mathfrak{B}-M\mathfrak{A})\sqrt{(G-\mathfrak{D}^2)}}{G\sqrt{(K-2LMNGu)}}\sin U + \frac{NCpq\sqrt{(G-\mathfrak{D}^2)}}{\sqrt{(G)}\sqrt{(K-2LMNGu)}}\cos U,$$

where for shortness p, q, r are retained in place of their values $\sqrt{(2Lu+\mathfrak{A})}$, $\sqrt{(2Mu+\mathfrak{B})}$, $\sqrt{(2Nu+\mathfrak{C})}$.

The values of l, m, n being known, that of λ could be determined by the differential equation

$$d\lambda = -\frac{dt\,(q\cos m + z\cos n)}{\sin^2 l},$$

and then the values of μ, ν would be determined without any further integration; but it is better to consider, in the place of any one of the principal axes in particular, the instantaneous axis, which is a line inclined to these at angles α, β, γ, the cosines of which are $\frac{p}{\omega}$, $\frac{q}{\omega}$, $\frac{r}{\omega}$ (if as before $\omega^2 = p^2+q^2+r^2$). Considering the instantaneous axis as a point of the sphere, let j denote the distance OP from the fixed point P, and ϕ the inclination OPQ of this distance to the fixed arc PQ. We have

$$\cos j = \cos\alpha\cos l + \cos\beta\cos m + \cos\gamma\cos n,$$

$$\sin j\cos\phi = \cos\alpha\sin l\cos\lambda + \cos\beta\sin m\cos\mu + \cos\gamma\sin n\cos\nu,$$

$$\sin j\sin\phi = \cos\alpha\sin l\sin\lambda + \cos\beta\sin m\sin\mu + \cos\gamma\sin n\sin\nu,$$

$$\cos(\lambda-\phi) = \frac{\cos\alpha - \cos l\cos j}{\sin l\sin j}, \qquad \sin(\lambda-\phi) = \frac{\cos\gamma\cos m - \cos\beta\cos n}{\sin l\sin j},$$

$$\cos(\mu-\phi) = \frac{\cos\beta - \cos m\cos j}{\sin m\sin j}, \qquad \cos(\mu-\phi) = \frac{\cos\alpha\cos n - \cos\gamma\cos l}{\sin m\sin j},$$

$$\cos(\nu-\phi) = \frac{\cos\gamma - \cos n\cos j}{\sin n\sin j}, \qquad \cos(\nu-\phi) = \frac{\cos\beta\cos l - \cos\alpha\cos m}{\sin n\sin j},$$

so that λ, μ, ν are determined in terms of j and ϕ. These expressions give

$$d\phi = \frac{1}{\gamma^2 \sin^2 j}\{\cos l\,(qdr - rdq) + \cos m\,(rdp - pdr) + \cos n\,(pdq - qdp)\},$$

which is reducible to Euler's equation

$$d\phi = dt\,\frac{p\,(M\mathfrak{C} - N\mathfrak{B})\cos l + q\,(N\mathfrak{A} - L\mathfrak{C})\cos m + r\,(L\mathfrak{B} - M\mathfrak{A})\cos n}{E - 2LMNu - v^2},$$

and thence, substituting for $\cos l$, $\cos m$, $\cos n$ their values, and observing that

$$Ap^2(M\mathfrak{C} - N\mathfrak{B}) + Bq^2(N\mathfrak{A} - L\mathfrak{C}) + Cr^2(L\mathfrak{B} - M\mathfrak{A}) = -(H - 2LMNFu),$$

$$BCp^2(M\mathfrak{C} - N\mathfrak{B})^2 + CAq^2(N\mathfrak{A} - L\mathfrak{C})^2 + ABr^2(L\mathfrak{B} - M\mathfrak{A})^2 = F(H - 2LMNFu),$$

$$LA\;(M\mathfrak{C} - N\mathfrak{B}) + MB\;(N\mathfrak{A} - L\mathfrak{C}) + NC\;(L\mathfrak{B} - M\mathfrak{A}) = LMNF,$$

the equation becomes

$$d\phi\,(E - 2LMNu - v^2) \div dt = \frac{-\mathfrak{D}\,(H - 2LMFu)}{G} + \frac{F\,(H - 2LMNFu)\sqrt{(G - \mathfrak{D}^2)}}{G\sqrt{(K - 2LMNGu)}}\sin U$$
$$+ \frac{LMNFpqr\sqrt{(G - \mathfrak{D}^2)}}{\sqrt{(G)}\sqrt{(K - 2LMNGu)}}\cos U,$$

where it is to be remarked that

$$G^2(E - 2LMNu - v^2)$$
$$= (G - \mathfrak{D}^2)F^2 + G\,(K - 2LMNGu) - (G - \mathfrak{D}^2)(K - 2LMNGu)\sin^2 U$$
$$- 2\mathfrak{D}F\sqrt{(G - \mathfrak{D}^2)}\sqrt{\{(K - 2LMNGu)\}}\sin U.$$

Now

$$dU = \frac{dt\,(H - 2LMNFu)\sqrt{(G)}}{K - 2LMNGu}, \quad du = pqrdt,$$

the differential $d\phi$ can be expressed as a fraction, the numerator whereof is

$$-\mathfrak{D}dU\,(K - 2LMNGu)\sqrt{(G)} + FdU\sqrt{\{G\,(G - \mathfrak{D}^2)(K - 2LMNGu)\}}\sin U$$
$$+ \frac{LMNFGdu\sqrt{\{G\,(G - \mathfrak{D}^2)\}}}{\sqrt{(K - 2LMNGu)}}\cos U,$$

and the denominator

$$(G - \mathfrak{D}^2)F^2 + G\,(K - 2LMNGu) - 2\mathfrak{D}F\sqrt{(G - \mathfrak{D}^2)(K - 2LMNGu)}\sin U$$
$$-(G - \mathfrak{D}^2)(K - 2LMNGu)\sin^2 U.$$

To simplify, write

$$\sqrt{(K - 2LMNGu)} = s, \quad \sqrt{(G - \mathfrak{D}^2)} = h,$$

the numerator is

$$-\mathfrak{D}s^2\,dU\sqrt{(G)}+Fhs\,dU\sqrt{(G)}\sin U-Fhs\sqrt{(G)}\cos U,$$

and the denominator

$$h^2F^2+Gs^2-2\mathfrak{D}Fhs\sin U-h^2s^2\sin^2 U,$$

which, observing that $h^2=G-\mathfrak{D}^2$, is equal to

$$(Fh-\mathfrak{D}s\sin U)^2+Gs^2\cos^2 U,$$

and we have

$$d\phi=\frac{-\mathfrak{D}s^2dU+Fhs\sin U dU-Fhds\cos U}{(Fh-\mathfrak{D}s\sin U)^2+Gs^2\cos^2 U}\sqrt{(G)},$$

the integral of which is

$$\phi+\mathfrak{F}=\tan^{-1}\frac{Fh-\mathfrak{D}s\sin U}{s\cos U\sqrt{(G)}},$$

where $\mathfrak{F}$ is the constant of integration, or substituting for h, s their values, the equation is

$$\tan(\phi+\mathfrak{F})=\frac{F\sqrt{(G-\mathfrak{D}^2)}-\mathfrak{D}\sin U\sqrt{(K-2LMNGu)}}{\cos U\sqrt{\{G(K-2LMNGu)\}}}.$$

It may be added that

$$\omega\cos j=v=\frac{1}{G}\left[\mathfrak{D}F+\sqrt{\{(G-\mathfrak{D}^2)(K-2LMNGu)\}}\sin U\right],$$

and therefore

$$\cos j=\frac{\mathfrak{D}F+\sqrt{\{(G-\mathfrak{D}^2)(K-2LMNGu)\}}\sin U}{G\sqrt{(E-2LMNu)}}.$$

Euler remarks that the complexity of the solution owing to the circumstance that the fixed point P is left arbitrary; and that the formulæ may be simplified by taking this point so that $G-\mathfrak{D}^2=0$, and he gives the far more simple formulæ corresponding to this assumption; this is in fact taking the point P in the *direction of the normal to the invariable plane*, and the resulting formulæ are identical with the ordinary formulæ for the solution of the problem. The term *invariable plane* is not used by Euler, and seems to have first occurred in Lagrange's "Essai sur le problème de trois corps," *Prix de l'Acad. de Berlin*, t. IX., 1772.

To prove the before-mentioned equation for $d\phi$; starting from the equations

$$\cos j=\cos\alpha\cos l+\cos\beta\cos m+\cos\gamma\cos n=\frac{v}{\omega},$$

$$\sin j\cos\phi=\cos\alpha\sin l\cos\lambda+\cos\beta\sin m\cos\mu+\cos\gamma\sin n\cos\nu$$

$$\sin j\sin\phi=\cos\alpha\sin l\cos\lambda+\cos\beta\sin m\sin\mu+\sin\gamma\sin n\sin\nu,$$

we have

$$\cos j\,dj\cos\phi-\sin j\sin\phi\,d\phi$$

$$=-\sin\alpha\,d\alpha\sin l\cos\lambda-\&c.+\cos\alpha\cos\lambda\cos l\,dl+\&c.-\cos\alpha\sin l\sin\lambda\,d\lambda+\&c.,$$

the second term is

$$\frac{p}{\omega}\cos\lambda\cot\ l\,(q\cos n-r\cos m)$$

$$+\frac{q}{\omega}\cos\mu\cot m\,(r\cos\ l-p\cos n\,)$$

$$+\frac{r}{\omega}\cos\nu\ \cot\ n\,(p\cos m-q\cos l\,),$$

and the third term is

$$+\frac{p}{\omega}\sin\lambda\,\mathrm{cosec}\ \ l\,(q\cos m+r\cos n\,)$$

$$+\frac{q}{\omega}\sin\mu\ \mathrm{cosec}\,m\,(r\ \cos n\ +p\cos l)$$

$$+\frac{r}{\omega}\sin\nu\ \mathrm{cosec}\,n\,(p\cos\ \ l\ +q\cos m).$$

Hence the second and third terms together are

$$=\frac{pq}{\omega}\left(\cos\lambda\,\frac{\cos l\cos n}{\sin l}-\cos\mu\,\frac{\cos m\cos n}{\sin m}+\sin\lambda\,\frac{\cos m}{\sin l}+\sin\mu\,\frac{\cos l}{\sin m}\right)+\text{\&c.},$$

$$=\frac{pq}{\omega}\begin{Bmatrix}-\cos\lambda\sin n\cos(\nu-\lambda)+\sin\lambda\sin n\sin(\nu-\lambda)\\+\cos\mu\sin n\cos(\mu-\nu)+\sin\mu\sin n\sin(\mu-\nu)\end{Bmatrix}+\text{\&c.},$$

$$=\frac{pq}{\omega}\sin n\begin{Bmatrix}-\cos\lambda\cos(\nu-\lambda)+\sin\lambda\sin(\nu-\lambda)\\+\cos\mu\cos(\mu-\nu)+\sin\mu\sin(\mu-\nu)\end{Bmatrix}+\text{\&c.},$$

$$=\frac{pq}{\omega}\sin n\begin{Bmatrix}-\cos\{\lambda+(\nu-\lambda)\}\\+\cos\{\mu-(\mu-\nu)\}\end{Bmatrix}+\text{\&c.},$$

$$=\frac{pq}{\omega}\sin n\,(-\cos\nu+\cos\nu)+\text{\&c.},=0\,;$$

we have therefore

$$\cos j\,dj\cos\phi-\sin j\sin\phi\,d\phi$$

$$=-\sin\alpha\,d\alpha\sin l\cos\lambda-\sin\beta\,d\beta\sin m\cos\mu-\sin\gamma\,d\gamma\sin n\cos\nu,$$

$$=d\,\frac{p}{\omega}\,.\sin l\cos\lambda+d\,\frac{q}{\omega}\,.\sin m\cos\mu+d\,\frac{r}{\omega}\,.\sin n\sin\nu$$

$$=+\frac{1}{\omega}\ \ (\sin l\cos\lambda\,dp+\sin m\cos\mu\,dq+\sin n\cos\nu\,dr)$$

$$-\frac{d\omega}{\omega^2}(\sin l\cos\lambda\,p+\sin m\cos\mu\,q+\sin n\cos\nu\,r)$$

$$=-\cot j\cos\phi\,d\frac{v}{\omega}-\sin j\sin\phi\,d\phi.$$

Hence therefore

$$\sin j \sin\phi d\phi = -\cot j\cos\phi d\frac{v}{\omega}$$

$$-\frac{1}{\omega}(\sin l\cos\lambda\, dp + \sin m\cos\mu\, dq + \sin n\cos\nu\, dr)$$

$$+\frac{d\omega}{\omega^2}(\sin l\cos\lambda\, p + \sin m\cos\mu\, q + \sin n\cos\nu\, r)$$

$$= -\cot j\cos\phi\frac{1}{\omega}(\cos l\, dp + \cos m\, dq + \cos n\, dr)$$

$$+\cot j\cos\phi\frac{d\omega}{\omega^2}(p\cos l + q\cos m + r\cos n)$$

$$-\frac{1}{\omega}(\sin l\cos\lambda\, dp + \sin m\cos\mu\, dq + \sin n\cos\nu\, dr)$$

$$+\frac{d\omega}{\omega^2}(\sin l\cos\lambda\,.\,p + \sin m\cos\mu\,.\,q + \sin n\cos\nu\,.\,r)$$

$$= \frac{1}{\omega}\{(-\cot j\cos\phi\cos l - \sin l\cos\lambda)\, dp + \&c.\}$$

$$+\frac{d\omega}{\omega^2}\{(\quad\cot j\cos\phi\cos l + \sin l\cos\lambda)\, p\quad + \&c.\}.$$

But we have

$$\cos(\lambda-\phi) = \frac{\cos\alpha - \cos l\cos j}{\sin l\sin j},$$

$$= \frac{\cos\alpha}{\sin l\sin j} - \cot l\cot j,$$

$$\sin(\lambda-\phi) = \frac{\cos\gamma\cos m - \cos\beta\cos n}{\sin l\sin j},$$

and thence

$$\cos\lambda = \cos\{(\lambda-\phi)+\phi\} = \frac{\cos\phi(\cos\alpha - \cos l\cos j) - \sin\phi(\cos\gamma\cos m - \cos\beta\cos n)}{\sin l\sin j},$$

whence also

$$\cot j\cos\phi\cos l + \sin l\cos\lambda$$

$$= \frac{1}{\sin j}\{\cos\phi\cos l + \cos\phi(\cos\alpha - \cos l\cos j) - \sin\phi(\cos\gamma\cos m - \cos\beta\cos n)\},$$

$$= \frac{1}{\sin j}\{\cos\alpha\cos\phi - \sin\phi(\cos\gamma\cos m - \cos\beta\cos n)\},$$

$$= \frac{1}{\omega\sin j}\{p\cos\phi - \sin\phi(r\cos m - q\cos n)\}.$$

Hence the expression for $\sin j \sin\phi\, d\phi$ is

$$= -\frac{1}{\omega^2 \sin j}\,[\{p\cos\phi - \sin\phi\,(r\cos m - q\cos n)\}\,dp + \ldots]$$

$$+\frac{d\omega}{\omega^3 \sin j}\,[\{p\cos\phi - \sin\phi\,(r\cos m - q\cos n)\}\;\;p + \ldots]$$

$$= -\frac{1}{\omega^2 \sin j}\,[\omega\, d\omega\cos\phi - \sin\phi\,\{(r\cos m - q\cos n)\,dp + \ldots\}]$$

$$+\frac{d\gamma}{\omega^3 \sin j}\,\omega^2\cos\phi \qquad\qquad = \sin j\sin\phi\, d\phi,$$

or finally

$$\sin j\sin\phi\, d\phi = \frac{1}{\omega^2}\,\frac{\sin\phi}{\sin j}\,[(r\cos m - q\cos n)\,dp + \&c.],$$

that is

$$d\phi = \frac{1}{\omega^2\sin^2 j}\left\{\begin{array}{l}\;\;\,(r\cos m - q\cos n)\,dp\\ +(p\cos n - r\cos l\,)\,dq\\ +(q\cos l\, - p\cos m)\,dr\end{array}\right\},$$

which is the required expression for $d\phi$.

Recapitulating, A, B, C, p, q, r denote as usual,

$$L = \frac{B-C}{A},\quad M = \frac{C-A}{B},\quad N = \frac{A-B}{C},\quad du = pqrdt,$$

$$\begin{aligned} p &= \sqrt{(\mathfrak{A} + 2Lu)},\\ q &= \sqrt{(\mathfrak{B} + 2Mu)},\\ r &= \sqrt{(\mathfrak{C} + 2Nu)}; \end{aligned}$$

$$\begin{aligned} \mathfrak{A} + \mathfrak{B} + \mathfrak{C} &= E,\\ \mathfrak{A}A + \mathfrak{B}B + \mathfrak{C}C &= F,\\ \mathfrak{A}A^2 + \mathfrak{B}B^2 + \mathfrak{C}C^2 &= G; \end{aligned}$$

$$\begin{aligned} L^2A\mathfrak{B}\mathfrak{C} + M^2B\mathfrak{C}\mathfrak{A} + N^2C\mathfrak{A}\mathfrak{B} &= H,\\ L^2A^2\mathfrak{B}\mathfrak{C} + M^2B^2\mathfrak{C}\mathfrak{A} + N^2C^2\mathfrak{A}\mathfrak{B} &= K; \end{aligned}$$

so that

$$K = EG - F^2,$$

$$U = \mathfrak{C} + \int\frac{(H - 2LMNFu)\,du\,\sqrt{(G)}}{(K - 2LMNGu)\sqrt{\{(\mathfrak{A} + 2Lu)(\mathfrak{B} + 2Mu)(\mathfrak{C} + 2Nu)\}}},$$

$$\cos l = \frac{\mathfrak{D}Ap}{G} + \frac{BCp(B\mathfrak{C} - C\mathfrak{B})\sqrt{(G - \mathfrak{D}^2)}}{G\sqrt{(K - 2LMNGu)}} \sin U + \frac{LAqr\sqrt{(G - \mathfrak{D}^2)}}{\sqrt{\{G(K - 2LMNGu)\}}} \cos U,$$

$$\cos m = \frac{\mathfrak{D}Bq}{G} + \frac{CAq(C\mathfrak{A} - A\mathfrak{C})\sqrt{(G - \mathfrak{D}^2)}}{G\sqrt{(K - 2LMNGu)}} \sin U + \frac{MBrp\sqrt{(G - \mathfrak{D}^2)}}{\sqrt{\{G(K - 2LMNGu)\}}} \cos U,$$

$$\cos n = \frac{\mathfrak{D}Cr}{G} + \frac{ABr(A\mathfrak{B} - B\mathfrak{A})\sqrt{(G - \mathfrak{D}^2)}}{G\sqrt{(K - 2LMNGu)}} \sin U + \frac{NCpq\sqrt{(G - \mathfrak{D}^2)}}{\sqrt{\{G(K - 2LMNGu)\}}} \cos U,$$

$$\omega^2 = E - 2LMNu,$$

$$\cos j = \frac{\mathfrak{D}F + \sqrt{\{(G - \mathfrak{D}^2)(K - 2LMNGu)\}} \sin U}{G\sqrt{(E - 2LMNu)}},$$

$$v = p\cos l + q\cos m + r\cos n$$

$$= \frac{1}{G}[\mathfrak{D}F + \sqrt{\{(G - \mathfrak{D}^2)(K - 2LMNGu)\}} \sin U],$$

$$\tan(\phi + \mathfrak{F}) = \frac{F\sqrt{(G - \mathfrak{D}^2)} - \mathfrak{D}\sin U\sqrt{(K - 2LMNGu)}}{\cos U\sqrt{\{G(K - 2LMNGu)\}}}.$$

[The angles which determine the position of the body are thus expressed in terms of u, which is given as a function of t by the foregoing equation $du = pqrdt$, where p, q, r denote given functions of u.]

405.

AN EIGHTH MEMOIR ON QUANTICS.

[From the *Philosophical Transactions of the Royal Society of London*, vol. CLVII. (for the year 1867). Received January 8,—Read January 17, 1867.]

THE present Memoir relates mainly to the binary quintic, continuing the investigations in relation to this form contained in my Second, Third, and Fifth Memoirs on Quantics, [141], [144], [156]; the investigations which it contains in relation to a quantic of any order are given with a view to their application to the quintic. All the invariants of a binary quintic (viz. those of the degrees 4, 8, 12, and 18) are given in the Memoirs above referred to, and also the covariants up to the degree 5; it was interesting to proceed one step further, viz. to the covariants of the degree 6; in fact, while for the degree 5 we obtain 3 covariants and a single syzygy, for the degree 6 we obtain only 2 covariants, but as many as 7 syzygies; one of these is, however, the syzygy of the degree 5 multiplied into the quintic itself, so that, excluding this derived syzygy, there remain $(7-1=)$ 6 syzygies of the degree 6. The determination of the two covariants (Tables 83 and 84 *post*) and of the syzygies of the degree 6, occupies the commencement of the present Memoir. [These covariants 83, 84 are the covariants M and N of the paper 143, "Tables of the covariants M to W of the binary quintic", and they are accordingly not here reproduced.]

The remainder of the Memoir is in a great measure a reproduction (with various additions and developments) of researches contained in Professor Sylvester's Trilogy, and in a recent memoir by M. Hermite[1]. In particular, I establish in a more general form (defining for that purpose the functions which I call "Auxiliars") the theory which is the basis of Professor Sylvester's criteria for the reality of the roots of a quintic equation, or, say, the theory of the determination of the character of an equation of any order. By way of illustration, I first apply this to the quartic equation; and

[1] Sylvester "On the Real and Imaginary Roots of Algebraical Equations; a Trilogy," *Phil. Trans.* vol. CLIV. (1864), pp. 579—666. Hermite, "Sur l'Équation du 5e degré," *Comptes Rendus*, t. LXI. (1866), and in a separate form, Paris, 1866.

I then apply it to the quintic equation, following Professor Sylvester's track, but so as to dispense altogether with his amphigenous surface, and making the investigation to depend solely on the discussion of the bicorn curve, which is a principal section of this surface. I explain the new form which M. Hermite has given to the Tschirnhausen transformation, leading to a transformed equation the coefficients whereof are all invariants; and, in the case of the quintic, I identify with my Tables his cubicovariants $\phi_1(x, y)$ and $\phi_2(x, y)$. And in the two new Tables, 85 and 86, I give the leading coefficients of the other two cubicovariants $\phi_3(x, y)$ and $\phi_4(x, y)$, [these are now also identified with my Tables]. In the transformed equation the second term (or that in z^4) vanishes, and the coefficient $\mathfrak{A}$ of z^3 is obtained as a quadric function of four indeterminates. The discussion of this form led to criteria for the character of a quintic equation, expressed like those of Professor Sylvester in terms of invariants, but of a different and less simple form; two such sets of criteria are obtained, and the identification of these, and of a third set resulting from a separate investigation, with the criteria of Professor Sylvester, is a point made out in the present memoir. The theory is also given of the canonical form which is the mechanism by which M. Hermite's investigations were carried on. The Memoir contains other investigations and formulæ in relation to the binary quintic; and as part of the foregoing theory of the determination of the character of an equation, I was led to consider the question of the imaginary linear transformations which give rise to a real equation: this is discussed in the concluding articles of the memoir, and in an Annex I have given a somewhat singular analytical theorem arising thereout.

The paragraphs and Tables are numbered consecutively with those of my former Memoirs on Quantics. I notice that in the Second Memoir, p. 126, we should have No. 26 = (No. 19)^2 − 128 (No. 25), viz. the coefficient of the last term is 128 instead of 1152. [This correction is made in the present reprint, 141, where the equation is given in the form $Q' = G^2 - 128Q$.]

Article Nos. 251 to 254.—*The Binary Quintic, Covariants and Syzygies of the degree* 6.

251. The number of asyzygetic covariants of any degree is obtained as in my Second Memoir on Quantics, *Philosophical Transactions*, vol. CXLVI. (1856), pp. 101—126, [141], viz. by developing the function

$$\frac{1}{(1-z)(1-xz)(1-x^2z)(1-x^3z)(1-x^4z)(1-x^5z)},$$

as shown p. 114, and then subtracting from each coefficient that which immediately precedes it; or, what is the same thing, by developing the function

$$\frac{1-x}{(1-z)(1-xz)(1-x^2z)(1-x^3z)(1-x^4z)(1-x^5z)},$$

which would lead directly to the second of the two Tables which are there given; the Table is there calculated only up to z^5, but I have since continued it up to z^{18}, so as to show the number of the asyzygetic covariants of every order in the variables up to the degree 18 in the coefficients, being the degree of the skew invariant, the highest of the irreducible invariants of the quintic. The Table is, for greater convenience, arranged in a different form, as follows:

Table No. 81.

Table for the number of the Asyzygetic Covariants of any order, to the degree 18.

	0	1	2	3	4	5	6	7	8	9	10	11	12	13	14	15	16	17	18	
0	1		0		1		0		2		0		3		0		4		1	0
1		0		0		1		1		1		2		3		3		4		1
2			1		0		2		1		4		2		6		4		9	2
3		0		1		1		2		3		4		5		7		8		3
4			0		2		1		4		3		7		6		11		9	4
5		1		1		2		3		5		6		8		10		13		5
6			1		1		4		3		7		7		12		11		18	6
7				1		2		4		5		8		10		13		16		7
8			0		2		2		6		6		11		11		18		18	8
9				1		3		4		7		9		13		16		20		9
10			1		1		4		5		10		10		17		18		26	10
11				1		2		5		7		11		14		19		23		11
12					2		3		7		8		15		16		24		26	12
13				0		2		4		8		11		16		20		27		13
14					1		4		5		11		13		21		23		33	14
15				1		2		5		8		13		17		24		29		15
16					1		2		7		9		16		19		29		32	16
17						2		4		8		12		19		24		32		17
18					0		3		5		11		14		23		27		39	18
19						1		4		7		13		18		26		33		19
20					1		2		6		9		17		21		32		37	20
21						1		3		8		12		19		26		36		21
22							2		4		10		14		24		29		42	22
23						0		3		6		12		18		27		35		23
24							1		5		8		16		21		33		40	24
25						1		2		6		11		19		26		37		25
26							1		3		9		13		23		30		44	26
27								2		5		11		17		27		36		27
28							0		3		6		14		20		32		40	28
29								1		5		9		17		25		37		29
30							1		2		7		12		22		29		44	30
31								1		3		9		15		25		35		31
32									2		5		12		18		31		40	32
33								0		3		7		15		23		35		33
34									1		5		9		19		27		42	34
35								1		2		7		13		23		33		35
36									1		3		10		16		28		38	36
37										2		5		12		20		33		37
38									0		3		7		16		24		39	38
39										1		5		10		20		30		39
40									1		2		7		13		25		35	40
41										1		3		10		17		29		41
42											2		5		13		21		36	42
43										0		3		7		16		26		43
44											1		5		10		21		31	44
45										1		2		7		14		26		45
46											1		3		10		17		31	46
47												2		5		13		22		47
48											0		3		7		17		27	48
49												1		5		10		21		49
50											1		2		7		14		27	50
51												1		3		10		18		51
52													2		5		13		22	52
53												0		3		7		17		53
54													1		5		10		22	54
55												1		2		7		14		55
56													1		3		10		18	56
57														2		5		13		57
58													0		3		7		17	58
59														1		5		10		59
60													1		2		7		14	60
61														1		3		10		61
62															2		5		13	62
63														0		3		7		63
64															1		5		10	64
65														1		2		7		65
66															1		3		10	66
67																2		5		67
68															0		3		7	68
69																1		5		69
70															1		2		7	70
71																1		3		71
72																	2		5	72
73																0		3		73
74																	1		5	74
75																1		2		75
76																	1		3	76
77																		2		77
78																	0		3	78
79																		1		79
80																	1		2	80
81																		1		81
82																			2	82
83																		0		83
84																			1	84
85																		1		85
86																			1	86
87																				87
88																			0	88
89																				89
90																			1	90

[In regard to this table No. 81 it is hardly necessary to notice that for any column with an even heading the numbers of the column correspond to the even outside numbers, while for any column with an odd heading the numbers of the column correspond to the odd outside numbers. The table is in fact a table of the differences of the numbers of the *af*-table, 142; thus in this table writing down cols. 5 and 6 and in each of them forming the differences by subtracting from each number the number immediately below it, we have cols. 5 and 6 of the table No. 81, viz.:

1	1	2	3	5	7	9	11	14	16	18	19	20	col. 5, 13—12 of *af*-table.
1	0	1	1	2	2	2	2	3	2	2	1	1	col. 5 of table No. 81.

1	1	2	3	5	7	10	12	16	19	23	25	29	30	32	32	col. 6, 15 of *af*-table.
1	0	1	1	2	2	3	2	4	3	4	2	4	1	2	0	col. 6 of table No. 81.]

252. The interpretation up to the degree 6 is as follows:

[In the following Table No. 82 as originally printed, the heading of the fourth column was "Constitution. Nos. in ( ) refer to Tables in former Memoirs except (83) and (84) which are given *post*," and the covariants were referred to by their Nos. accordingly.]

Table No. 82.

Degree.	Order.	No.	Constitution. Notation is the alphabetic notation of 143, A the quintic itself, B, C quadricovariants, &c.	N=new covt. S=syzygy.
0	0	1	viz. the absolute constant unity.	
1	5	1	A	N.
2	10	1	A^2	
,,	6	1	C	N.
,,	2	1	B	N.
3	15	1	A^3	
,,	11	1	AC	
,,	9	1	F	N.
,,	7	1	AB	
,,	5	1	E	N.
,,	3	1	D	N.
4	20	1	A^4	
,,	16	1	A^2C	
,,	14	1	AF	
,,	12	2	C^2, A^2B	
,,	10	1	AE	
,,	8	2	AD, BC	
,,	6	1	I	N.
,,	4	2	H, B^2	N.
,,	0	1	G	N.
5	25	1	A^5	
,,	21	1	A^3C	
,,	19	1	A^2F	
,,	17	2	A^3B, AC^2	
,,	15	2	A^2E, CF	
,,	13	2	A^2D, ABC	
,,	11	2	$AI+BF-CE=0$	S.
,,	9	3	AB^2, AH, CD	
,,	7	2	BE, L	N.
,,	5	2	AG, BD	
,,	3	1	K	N.
,,	1	1	J	N.
6	30	1	A^6	
,,	26	1	A^4C	
,,	24	1	A^3F	
,,	22	2	A^3C^2, A^4B	
,,	20	2	ACF, A^3E	
,,	18	3	$E^2 + 4C^3 + A^3D - A^2BC = 0$	S.
,,	16	2	$A\{AI + BF - CE\} = 0$	S'.
,,	14	4	$-6ACD - 1EF - 4BC^2 + A^2H = 0$, A^2B^2	S.
,,	12	3	$AL + 3DF - 2CI = 0$, ABE	S.
,,	10	4	$4B^2C + 12ABD - A^2G + E^2 = 0$, CH	S.
,,	8	2	$AK + 2BI - 3DE = 0$	S.
,,	6	4	$AJ + 2BH - B^3 - CG - 9D^2 = 0$	S.
,,	4	1	N	N.
,,	2	1	M	N.

253. For the explanation of this I remark that the Table No. 81 shows that we have for the degree 0 and order 0 one covariant; this is the absolute constant unity; for the degree 1 and order 5, 1 covariant, this is the quintic itself, A; for degree 2 and order 10, 1 covariant; this is the square of the quintic, A^2; for same degree and order 6, 1 covariant, which had accordingly to be calculated, viz. this is the covariant C; and similarly whenever the Table No. 81 indicates the existence of a covariant of any degree and order, and there does not exist a product of the covariants previously calculated, having the proper degree and order, then in each such case (shown in the last preceding Table by the letter N) a new covariant had to be calculated. On coming to degree 5, order 11, it appears that the number of asyzygetic invariants is only $=2$, whereas there exist of the right degree and order the 3 combinations AI, BF, CE; there is here a syzygy, or linear relation, between the combinations in question; which syzygy had to be calculated, and was found to be as shown, $AI+BF-CE=0$, a result given in the Second Memoir, p. 126. Any such case is indicated by the letter S. At the place degree 6, order 16, we find a syzygy between the combinations A^2I, A^2BF, ACE; as each term contains the factor A, this is only the last-mentioned syzygy multiplied by A, not a new syzygy, and I have written S' instead of S. The places degree 6, orders 18, 14, 12, 10, 8, 6 indicate each of them a syzygy, which syzygies, as being of the degree 6, were not given in the Second Memoir, and they were first calculated for the present Memoir. It is to be noticed that in some cases the combinations which might have entered into the syzygy do not all of them do so; thus degree 6, order 14, the syzygy is between the four combinations ACD, EF, BC^2, A^2H, and does not contain the remaining combination A^2B. The places degree 6, orders 4, 2, indicate each of them a new covariant, and these, as being of the degree 6, were not given in the Second Memoir, but had to be calculated for the present Memoir.

254. I notice the following results:

$$\text{Quadrinvt. } 6H = 3G^2,$$

$$\text{Cubinvt. } 6H = -G^3+54GQ,$$

$$\text{Disct. } (\alpha B+\beta M) = (-G,\ Q,\ -3U \between \alpha,\ \beta)^2,$$

$$\text{Jac. } (B,\ H) = 6M,$$

$$\text{Hess. } 3D = N,$$

the last two of which indicate the formation of the covariants given in the new Tables $M=$ No. 83 and $N=$ No. 84: viz. if to avoid fractions we take 3 times the covariant D, being a cubic $(a, \ldots)^3(x, y)^3$, then the Hessian thereof is a covariant $(a, \ldots)^6(x, y)^2$, which is given in Table, M No. 83; and in like manner if we form the Jacobian of the Tables B and H which are respectively of the forms $(a,..)^2(x, y)^2$, and $(a,..)^6(x, y)^4$, this is a covariant $(a,..)^6(x, y)^4$, and dividing it by 6 to obtain the coefficients in their lowest terms, we have the new Table, N No. 84. I have in these, for greater distinctness, written the numerical coefficients *after* instead of *before*, the literal terms to which they belong.

The two new Tables are:

Table No. 83. $M = (*\between x, y)^2$. See 143.

Table No. 84. $N = (*\between x, y)^4$. See 143.

Article No. 255.—*Formulæ for the canonical form* $ax^5 + by^5 + cz^5 = 0$, *where* $x + y + z = 0$.

255. The quintic $(a, b, c, d, e, f\between x, y)^5$ may be expressed in the form

$$ru^5 + sv^5 + tw^5,$$

where u, v, w are linear functions of (x, y) such that $u + v + w = 0$. Or, what is the same thing, the quintic may be represented in the canonical form

$$ax^5 + by^5 + cz^5,$$

where $x + y + z = 0$; this is $= (a - c, \ -c, \ -c, \ -c, \ -c, \ b - c\between x, y)^5$, and the different covariants and invariants of the quintic may hence be expressed in terms of these coefficients (a, b, c).

For the invariants we have

$$\begin{aligned} G &= J = b^2c^2 + c^2a^2 + a^2b^2 - 2abc\,(a + b + c), \\ Q &= K = a^2b^2c^2\,(bc + ca + ab), \\ -U &= L = a^4b^4c^4, \\ W &= I = 4a^5b^5c^5\,(b - c)(c - a)(a - b). \end{aligned}$$

[Observe that throughout the present Memoir, the invariants, instead of being called $G, Q, -U, W$ are called I, J, K, L, viz. the I, J, K, L in all that follows denote the invariants, and not the covariants denoted by these letters in 142, 143. Moreover D is used to denote the invariant Q', which is in fact the discriminant of the quintic.]

Hence, writing for a moment

$$\begin{aligned} a + b + c &= p, \text{ and therefore } & J &= q^2 - 4pr, \\ bc + ca + ab &= q & K &= r^2q, \\ abc &= r & L &= r^4, \end{aligned}$$

we have

$$(a - b)^2(b - c)^2(c - a)^2 = p^2q^2 - 4q^3 - 4p^3r + 18pqr - 27r^2,$$

and thence

$$I^2 = 16r^{10}\,(p^2q^2 - 4q^3 - 4p^3r + 18pqr - 27r^2),$$

and

$$\begin{aligned} &J\,(K^2 - JL)^2 + 8K^3L - 72JKL^2 - 432L^3 \\ &\qquad = \ r^{10}\,\{(q^2 - 4pr)\,16p^2 + 8q^3 - (q^2 - 4pr)\,72q - 432r^2\}, \\ &\qquad = 8r^{10}\,\{(q^2 - 4pr)(2p^2 - 9q) + q^3 - 54r^2\}, \\ &\qquad = 16r^{10}\,\{p^2q^2 - 4q^3 - 4p^3r + 18pqr - 27r^2\}, \end{aligned}$$

that is,

$$I^2 = J(K^2 - JL)^2 + 8K^3L - 72JKL^2 - 432L^3,$$

which is the simplest mode of obtaining the expression for the square of the 18-thic or skew invariant I in terms of the invariants J, K, L of the degrees 4, 8, 12 respectively.

If instead of the invariant K of the degree 8 we consider the invariant D [$=Q'$ as before-mentioned] of the same degree, this is

$$Q' = D = \{b^2c^2 + c^2a^2 + a^2b^2 - 2abc(a+b+c)\}^2 - 128a^2b^2c^2(bc+ca+ab),$$
$$= q^4 - 8q^2pr - 128qr^2 + 16p^2r^2,$$
$$D = \text{Norm}\ \left((bc)^{\frac{1}{4}} + (ca)^{\frac{1}{4}} + (ab)^{\frac{1}{4}}\right);$$

and we have also the following covariants:

$$B = (-ac,\ ab - ac - bc,\ -bc \between x,\ y)^2,$$
$$= bcyz + cazx + abxy.$$

$$C = (-ac,\ -3ac,\ -3ac,\ ab - ac - bc,\ -3bc,\ -3bc,\ -bc \between x,\ y)^6$$
$$= bcy^3z^3 + caz^3x^3 + abx^3y^3.$$

$$D = (0,\ -abc,\ -abc,\ 0 \between x,\ y)^3 = abc\, xyz.$$

$$\begin{aligned} E = &\ (\ a^2b - \ \ ac^2 + \ \ bc^2 - a^2c - 2abc)\, x^5 \\ &+ (\qquad - \ 5ac^2 + \ 5bc^2 \qquad - 5abc)\, x^4y \\ &+ (\qquad - 10ac^2 + 10bc^2 \qquad - 2abc)\, x^3y^2 \\ &+ (\qquad - 10ac^2 + 10bc^2 \qquad + 2abc)\, x^2y^3 \\ &+ (\qquad - \ 5ac^2 + \ 5bc^2 \qquad + 5abc)\, xy^4 \\ &+ (-ab^2 - \ \ ac^2 + \ \ bc^2 + b^2c + 2abc)\, y^5 \end{aligned}$$

$$= (b-c)\,a^2x^5 + (c-a)\,b^2y^5 + (a-b)\,c^2z^5$$
$$- abc\,(y-z)(z-x)(x-y)(yz+zx+xy).$$

Article No. 256.—*Expression of the 18-thic Invariant in terms of the roots.*

256. It was remarked by Dr Salmon, that for a quintic $(a,\ b,\ c,\ d,\ e,\ f \between x,\ y)^5$ which is linearly transformable into the form $(a,\ 0,\ c,\ 0,\ e,\ 0 \between x,\ y)^5$, the invariant I is $=0$. Now putting for convenience $y=1$, and considering for a moment the equation

$$x(x-\beta)(x-\gamma)(x-\delta)(x-\epsilon) = 0,$$

then writing herein $\dfrac{x}{mx+n}$ for x, the transformed equation is

$$x(x-\beta')(x-\gamma')(x-\delta')(x-\epsilon') = 0,$$

where

$$\beta' = \frac{n\beta}{1-m\beta}, \quad \gamma' = \frac{n\gamma}{1-m\gamma}, \quad \&c.;$$

hence m may be so determined that $\beta'+\gamma'$ may be $=0$; viz. this will be the case if $\beta+\gamma=2m\beta\gamma$, or $m=\dfrac{\beta+\gamma}{2\beta\gamma}$. In order that $\delta'+\epsilon'$ may be $=0$, we must of course have $m=\dfrac{\delta+\epsilon}{2\delta\epsilon}$, and hence the condition that simultaneously $\beta'+\gamma'=0$ and $\delta'+\epsilon'=0$ is $\dfrac{\beta+\gamma}{2\beta\gamma}=\dfrac{\delta+\epsilon}{2\delta\epsilon}$; that is, $(\beta+\gamma)\,\delta\epsilon-\beta\gamma\,(\delta+\epsilon)=0$. Or putting $x-\alpha$ for x and $\beta-\alpha$, $\gamma-\alpha$, &c. for β, γ, &c., we have the equation

$$(x-\alpha)(x-\beta)(x-\gamma)(x-\delta)(x-\epsilon)=0,$$

which is by the transformation $x-\alpha$ into $\dfrac{x-\alpha}{m(x-\alpha)+n}$ changed into

$$(x-\alpha')(x-\beta')(x-\gamma')(x-\delta')(x-\epsilon')=0$$

(where $\alpha'=\alpha$), and the condition in order that in the new equation it may be possible to have simultaneously $\beta'+\gamma'-2\alpha'=0$, $\delta'+\epsilon'-2\alpha'=0$, is

$$(\beta+\gamma-2\alpha)(\delta-\alpha)(\epsilon-\alpha)-(\delta+\epsilon-2\alpha)(\beta-\alpha)(\gamma-\alpha)=0,$$

or, as this may be written,

$$\begin{vmatrix} 1, & 2\alpha\ \ , & \alpha^2 \\ 1, & \beta+\gamma, & \beta\gamma \\ 1, & \delta+\epsilon, & \delta\epsilon \end{vmatrix}=0.$$

Hence writing $x+\alpha'$ for x, the last-mentioned equation is the condition in order that the equation

$$(x-\alpha)(x-\beta)(x-\gamma)(x-\delta)(x-\epsilon)=0$$

may be transformable into

$$x(x-\beta')(x-\gamma')(x-\delta')(x-\epsilon')=0,$$

where $\beta'+\gamma'=0$, $\delta'+\epsilon'=0$, that is, into the form $x(x^2-\beta'^2)(x^2-\delta'^2)=0$. Or replacing y, if we have

$$(a, b, c, d, e, f\between x, y)^5=a(x-\alpha y)(x-\beta y)(x-\gamma y)(x-\delta y)(x-\epsilon y),$$

then the equation in question is the condition in order that this may be transformable into the form $(a', 0, c', 0, e', 0\between x, y)^5$; that is, in order that the 18-thic invariant I may vanish. Hence observing that there are 15 determinants of the form in question, and that any root, for instance α, enters as α^2 in 3 of them and in the simple power α in the remaining 12, we see that the product

$$a^{18}\,\Pi\begin{vmatrix} 1, & 2\alpha\ \ , & \alpha^2 \\ 1, & \beta+\gamma, & \beta\gamma \\ 1, & \delta+\epsilon, & \delta\epsilon \end{vmatrix}$$

contains each root in the power 18, and is consequently a rational and integral function of the coefficients of the degree 18, viz. save as to a numerical factor it is equal to the invariant I. And considering the equation $(a, \ldots \between x, y)^5 = 0$ as representing a range of points, the signification of the equation $I = 0$ is that, the pairs (β, γ) and (δ, ϵ) being properly selected, the fifth point α is a focus or sibiconjugate point of the involution formed by the pairs (β, γ) and (δ, ϵ).

Article Nos. 257 to 267.—*Theory of the determination of the Character of an Equation; Auxiliars; Facultative and Non-facultative space.*

257. The equation $(a, b, c \ldots \between x, y)^n = 0$ is a *real* equation if the ratios $a : b : c, \ldots$ of the coefficients are all real. In considering a given real equation, there is no loss of generality in considering the coefficients $(a, b, c \ldots)$ as being themselves real, or in taking the coefficient a to be $= 1$; and it is also for the most part convenient to write $y = 1$, and thus to consider the equation under the form $(1, b, c \ldots \between x, 1)^n = 0$. It will therefore (unless the contrary is expressed) be throughout assumed that the coefficients (including the coefficient a when it is not put $= 1$) are all of them real; and, in speaking of any functions of the coefficients, it is assumed that these are rational and integral real functions, and that any values attributed to these functions are also real.

258. The equation $(1, b, c \ldots \between x, 1)^n = 0$, with α real roots and 2β imaginary roots, is said to have the character $\alpha r + 2\beta i$; thus a quintic equation will have the character $5r$, $3r + 2i$, or $r + 4i$, according as its roots are all real, or as it has a single pair, or two pairs, of imaginary roots.

259. Consider any m functions $(A, B, \ldots K)$ of the coefficients, ($m =$ or $< n$). For given values of $(A, B, \ldots K)$, *non constat* that there is any corresponding equation (that is, the corresponding values of the coefficients $(b, c, \ldots)$ may be of necessity imaginary), but attending only to those values of $(A, B, \ldots K)$ which have a corresponding equation or corresponding equations, let it be assumed that the equations which correspond to a given set of values of $(A, B, \ldots K)$ have a determinate character (one and the same for all such equations): this assumption is of course a condition imposed on the form of the functions $(A, B, \ldots K)$; and any functions satisfying the condition are said to be "auxiliars." It may be remarked that the n coefficients $(b, c, \ldots)$ are themselves auxiliars; in fact for given values of the coefficients there is only a single equation, which equation has of course a determinate character. To fix the ideas we may consider the auxiliars $(A, B, \ldots K)$ as the coordinates of a point in m-dimensional space, or say in m-space.

260. Any given point in the m-space is either "facultative," that is, we have corresponding thereto an equation or equations (and if more than one equation then by what precedes these equations have all of them the same character), or else it is "non-facultative," that is, the point has no corresponding equation.

261. The entire system of facultative points forms a region or regions, and the entire system of non-facultative points a region or regions; and the m-space is thus divided into facultative and non-facultative regions. The surface which divides the

facultative and non-facultative regions may be spoken of simply as the bounding surface, whether the same be analytically a single surface, or consist of portions of more than one surface.

262. Consider the discriminant D, and to fix the ideas let the sign be determined in such wise that D is $+$ or $-$ according as the number of imaginary roots is $\equiv 0$ (mod. 4), or is $\equiv 2$ (mod. 4); then expressing the equation $D=0$ in terms of the auxiliars $(A, B, \ldots K)$, we have a surface, say the discriminatrix, dividing the m-space into regions for which D is $+$, and for which D is $-$, or, say, into positive and negative regions.

263. A given facultative or non-facultative region may be wholly positive or wholly negative, or it may be intersected by the discriminatrix and thus divided into positive and negative regions. Hence taking account of the division by the discriminatrix, but attending only to the facultative regions, we have positive facultative regions and negative facultative regions. Now using the simple term region to denote indifferently a positive facultative region or a negative facultative region, it appears from the very notion of a region as above explained that we may pass from any point in a given region to any other point in the same region without traversing either the bounding surface or the discriminatrix; and it follows that the equations which correspond to the several points of the same region have each of them one and the same character; that is, to a given region there correspond equations of a given character.

264. It is proper to remark that there may very well be two or more regions which have corresponding to them equations with the same character; any such regions may be associated together and considered as forming a kingdom; the number of kingdoms is of course equal to the number of characters, viz. it is $=\frac{1}{2}(n+2)$ or $\frac{1}{2}(n+1)$ according as n is even or odd; and this being so, the general conclusion from the preceding considerations is that the whole of facultative space will be divided into kingdoms, such that to a given kingdom there correspond equations having a given character; and conversely, that the equations with a given character correspond to a given kingdom. Hence (the characters for the several kingdoms being ascertained) knowing in what kingdom is situate a point $(A, B, \ldots K)$, we know also the character of the corresponding equations.

265. Any conditions which determine in what kingdom is situate the point $(A, B, \ldots K)$ which belongs to a given equation $(1, b, c \ldots \between x, 1)^n = 0$, determine therefore the character of the equation. It is very important to notice that the form of these conditions is to a certain extent indeterminate; for if to a given kingdom we attach any portion or portions of non-facultative space, then any condition or conditions which confine the point $(A, B, \ldots K)$ to the resulting aggregate portion of space, in effect confine it to the kingdom in question; for of the points within the aggregate portion of space it is only those within the kingdom which have corresponding to them an equation, and therefore, if the coefficients $(b, c, \ldots)$ of the given equation are such as to give to the auxiliars $(A, B, \ldots K)$ values which correspond to a point situate within the above-mentioned aggregate portion of space, such point will of necessity be within the kingdom.

266. In the case where the auxiliars are the coefficients $(b, c, \ldots)$, to any given values of the auxiliars there corresponds an equation, that is, all space is facultative space. And the division into regions or kingdoms is effected by means of the discriminatrix, or surface $D=0$, alone. Thus in the case of the quadric equation $(1, x, y \between \theta, 1)^2=0$ the m-space is the plane. We have $D=x^2-y$, and the discriminatrix is thus the parabola $x^2-y=0$. There are two kingdoms, each consisting of a single region, viz. the positive kingdom or region $(x^2-y=+)$ outside the parabola, and the negative kingdom or region $(x^2-y=-)$ inside the parabola, which have the characters $2r$ and $2i$, or correspond to the cases of two real roots and two imaginary roots, respectively. And the like as regards the cubic $(1, x, y, z \between \theta, 1)^3=0$; the m-space is here ordinary space, $D=-4x^3z+3x^2y^2+6xyz-4y^3-z^2$, and the division into kingdoms is effected by means of the surface $D=0$; but as in this case there are only the two characters $3r$ and $r+2i$, there can be only the two kingdoms $D=+$ and $D=-$ having these characters $3r$ and $r+2i$ respectively, and the determination of the character of the cubic equation is thus effected without its being necessary to proceed further, or inquire as to the form or number of the regions determined by the surface $D=0$: I believe that there are only two regions, so that in this case also each kingdom consists of a single region. But proceeding in the same manner, that is, with the coefficients themselves as auxiliars, to the case of a quartic equation, the m-space is here a 4-dimensional space, so that we cannot by an actual geometrical discussion show how the 4-space is by the discriminatrix or hypersurface $D=0$ divided into kingdoms having the characters $4r$, $2r+2i$, $4i$ respectively. The employment therefore of the coefficients themselves as auxiliars, although theoretically applicable to an equation of any order whatever, can in practice be applied only to the cases for which a geometrical illustration is in fact unnecessary.

267. I will consider in a different manner the case of the quartic, chiefly as an instance of the actual employment of a surface in the discussion of the character of an equation; for in the case of a quintic the auxiliars are in the sequel selected in such manner that the surface breaks up into a plane and cylinder, and the discussion is in fact almost independent of the surface, being conducted by means of the curve (Professor Sylvester's Bicorn) which is the intersection of the plane and cylinder.

Article Nos. 268—273.—*Application to the Quartic equation.*

268. Considering then the quartic equation $(a, b, c, d, e \between \theta, 1)^4=0$ (I retain for symmetry the coefficient a, but suppose it to be $=1$, or at all events positive), then if I, J signify as usual, and if for a moment

$$\vartheta = a^2d-3abc+2b^3,$$
$$X = 3aJ+2(b^2-ac)I,$$

we have identically

$$\tfrac{9}{4}(3a^2J^2+X^2)\vartheta^2 = 9(b^2-ac)^3X^2-a^2(b^2-ac)^3(I^3-27J^2)-a^2X^3$$

(see my paper, "A discussion of the Sturmian Constants for Cubic and Quartic Equations," *Quart. Math. Journ.*, vol. IV. (1861) pp. 7—12), [290]. And I write

$$x = b^2 - ac,$$

$$y = 3aJ + 2(b^2 - ac)I,$$

$$z = I^3 - 27J^2 (= D).$$

269. I borrow from Sturm's theorem the conclusion (but nothing else than this conclusion) that (x, y, z) possess the fundamental property of auxiliars (that is, that the quartic equations (if any) corresponding to a given system of values of (x, y, z) have one and the same character). The foregoing equation gives $9x^3y^2 - x^3z - y^3 =$ a square function, and therefore positive; that is, the facultative portion of space is that for which $9x^3y^2 - x^3z - y^3$ is $=+$. And the equation

$$x^3(9y^2 - z) - y^3 = 0$$

is that of the bounding surface, dividing the facultative and non-facultative portions of space.

270. To explain the form of the surface we may imagine the plane of xy to be that of the paper, and the positive direction of the axis of z to be in front of the paper. Taking z constant, or considering the sections by planes parallel to that of xy,

$z = 0$, gives $y^2(9x^3 - y) = 0$, viz. the section is the line $y = 0$, or axis of x twice, and the cubical parabola $y = x^3$.

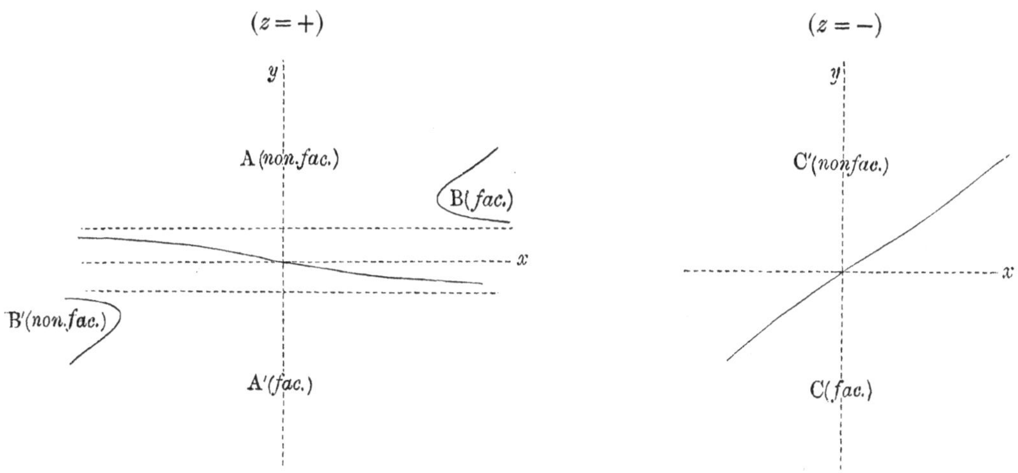

$z = +$, the curve $x^3 = \dfrac{y^3}{9y^2 - z}$ has two asymptotes $y = \pm\frac{1}{3}\sqrt{z}$, parallel to and equidistant from the axis of x, and consists of a branch included between the two parallel asymptotes, and two other portions branches outside the asymptotes, as shown in the figure $(z = +)$.

$z=-$, the curve $x^3=\dfrac{y^3}{9y^2-z}$ has no real asymptote, and consists of a single branch, resembling in its appearance the cubical parabola as shown in the figure ($z=-$).

Taking x as constant, or considering the sections by planes parallel to that of zy, the equation of the section is $z=9y^2-\dfrac{y^3}{x^3}$, which is a cubical parabola, meeting the plane of xy in a point on the cubical parabola $y=9x^3$, and also in a twofold point on the axis of x, that is, touching the plane of xy at the last-mentioned point.

271. The surface consists of a single sheet extending to infinity, the form of which is most easily understood by considering the sections by a system of spheres having the origin of coordinates for their common centre. These sections have all of them the same general form; and one of them is shown fig. 1 of the Plate at the end of the present Memoir, the projection being orthogonal on the plane of xy or plane of the paper, and the spherical curve being shown, the portion of it above the plane of the paper by a continuous line, that below it by a dotted line (the double point in the figure is thus of course only an apparent one): the same figure shows also the sections by planes parallel to that of xy previously shown in the figures ($z=+$) and ($z=-$).

272. Now considering the discriminatrix $D=0$, in this case the plane $z=0$, it appears that the bounding surface and this plane divide space into six regions, viz. above the plane of the paper we have the four regions, A non-facultative, B facultative, A' facultative, B' non-facultative, and below it the two regions, C facultative, C' non-facultative. There are thus in all three facultative regions A', B, C, and since A' and B correspond to $D=+$, these must have the characters $4r$ and $4i$, and it is easy by considering a particular case to show that B has the character $4r$, and A' the character $4i$; C corresponds to $D=-$, and can therefore only have the character $2r+2i$. Hence, for any given equation, (x, y, z) will lie in one of the regions (B, A', C), and if (x, y, z)

is in the region B, the character is $4r$,
„ A', „ $4i$,
„ C, „ $2r+2i$.

273. It is right to notice that the determination of the character is really made in what precedes; the determination of the analytical criteria of the different characters is a mere corollary; to obtain these it is only necessary to remark that

$z=+,\ x=+,\ y=+$ includes the whole of facultative region B,

that is, (x, y, z) being each positive, the character is $4r$;

$$\left.\begin{array}{r} z=+,\ x=+,\ y=- \\ x=-,\ y=- \\ x=-,\ y=- \end{array}\right\}\text{include each a part and together the whole of facultative region } A',$$

that is, z being $+$, but (x, y) not each positive, the character is $4i$;

$$\left.\begin{array}{l} z=-,\ x=+,\ y=+ \\ \quad ,,\quad x=+,\ y=- \\ \quad ,,\quad x=-,\ y=- \end{array}\right\} \text{include each a part and together the whole of facultative region } C,$$

$z=-,\ x=-,\ y=+$ does not include any facultative space,

that is, z being $-$, the character is $2r+2i$; and the combination of signs $z=-$, $x=-$, $y=+$ is one which does not exist.

The results thus agree with those furnished by Sturm's theorem; and in particular the impossibility of $z=-$, $x=-$, $y=+$ appears from Sturm's theorem, inasmuch as his combination would give a gain instead of a loss of changes of sign.

Article Nos. 274 to 285.—*Determination of the characters of the quintic equation.*

274. Passing now to the case of the quintic, I write

$$\begin{aligned} J &= G, \\ K &= Q, \\ D &= Q', \\ L &= -U, \\ I &= W; \end{aligned}$$

viz. J is the quartinvariant, K and D are octinvariants (D the discriminant), L is 12-thic invariant, and I is the 18-thic or skew invariant. Hence also J, D, $2^{11}L-J^3$ are invariants of the degrees 4, 8, 12 respectively; and forming the combinations

$$x=\frac{2^{11}L-J^3}{J^3},\ y=\frac{D}{J^2},\ z=J,$$

I assume that (x, y, z) are auxiliars, reserving for the concluding articles of the present memoir the considerations which sustain this assumption.

275. The separation into regions is effected as follows:—We have identically (see *ante*, No. 255)

$$16I^2=JK^4+8LK^3-2J^2LK^2-72JL^2K-432L^3+J^3L^2;$$

or putting for K its value $=\frac{1}{128}(J^2-D)$, this is

$$\begin{aligned} 2^{32}I^2 &= J(J^2-D)^4+\&\text{c.}, \\ &= (J^3-2^{11}L)^2(J^3-3^3.2^{10}L) \\ &\quad +DJ(-4J^6+61.2^{10}J^3L+144.2^{20}L^2) \\ &\quad +D^2J^2(\ \ 6J^3-2^{10}.29L) \\ &\quad +D^3\ \ (-4J^3-\ \ 2^{10}L) \\ &\quad +D^4J; \end{aligned}$$

or writing as above

$$x = \frac{2^{11} L - J^3}{J^3}, \quad y = \frac{D}{J^2},$$

whence also

$$1 + x = \frac{2^{11} L}{J^3},$$

this is

$$\begin{aligned} 2^{32} \frac{I^2}{J^9} = & - x^2 \{\tfrac{3}{2}(1+x) - 1\} \\ & + y \{ 36(1+x)^2 - \tfrac{61}{2}(1+x) - 4\} \\ & - y^2 \{ \tfrac{29}{2}(1+x) - 6\} \\ & + y^3 \{-\tfrac{1}{2}(1+x) - 4\} \\ & + y^4, \end{aligned}$$

or, what is the same thing,

$$\begin{aligned} 2 \cdot 2^{32} \frac{I^2}{J^9} = & - 3x^3 - x^2 \\ & + y(72x^2 + 205x + 125) \\ & + y^2(-29x - 17) \\ & + y^3(-x - 9) \\ & + y^4 \cdot 2 \\ = & \ \ \phi(x, y) \text{ suppose.} \end{aligned}$$

276. Hence also writing $z = J$, we have

$$z\phi(x, y) = 2 \cdot 2^{32} \frac{I^2}{J^8} = +,$$

or the equation of the bounding surface may be taken to be

$$z\phi(x, y) = 0,$$

that is, the bounding surface is composed of the plane $z = 0$, and the cylinder $\phi(x, y) = 0$. Taking the plane of the paper for the plane $z = 0$, the cylinder meets this plane in a curve $\phi(x, y) = 0$, which is Professor Sylvester's Bicorn: this curve divides the plane into certain regions, and if we attend to the solid figure and instead of the curve consider the cylinder, then to each region of the plane there correspond *in solido* two regions, one in front of, the other behind the plane region, and of these regions *in solido*, one is facultative, the other is non-facultative (viz. for given values of (x, y), whatever be the sign of $\phi(x, y)$, then for a certain sign of z, $z\phi(x, y)$ will be positive or the solid region will be facultative, and for the opposite sign of z, $z\phi(x, y)$ will be negative or the region will be non-facultative). It hence appears that we may attend only to the plane regions, and that (the proper sign being attributed to z, that is to J) each of these may be regarded as facultative. It is to be added that the discriminatrix is in the present case the plane $y = 0$, or, if we attend only to the plane figure, it is the line $y = 0$; so that in the plane figure the separation into regions is effected by means of the Bicorn and the line $y = 0$.

277. Reverting to the equation of the Bicorn, and considering first the form at infinity, the intersections of the curve by the line infinity are given by the equation $y^3(2y-x)=0$, viz. there is a threefold intersection $y^3=0$, and a simple intersection $2y-x=0$; the equation $y^3=0$ indicates that the intersection in question is a point of inflexion, the tangent at the inflexion (or stationary tangent) being of course the line infinity; the visible effect is, however, only that the direction of the branch is ultimately parallel to the axis of x. The equation $2y-x=0$ indicates an asymptote parallel to this line, and the equation of the asymptote is easily found to be $2y-x+5=0$.

278. The discussion of the equation would show that the curve has an ordinary cusp; and a cusp of the second kind, or node-cusp, equivalent to a cusp and node; the curve is therefore a unicursal curve, or the coordinates are expressible rationally in terms of a parameter ϕ; we in fact have

$$x=\frac{-(\phi+2)(\phi^3-\phi^2+2\phi-4)}{\phi^3(\phi+1)},\quad y=\frac{(\phi+2)^2(\phi-3)}{\phi^2(\phi+1)},$$

whence also

$$\frac{dy}{dx}=\tfrac{1}{2}\phi(\phi+2).$$

279. The curve may be traced from these equations (see Plate, fig. 2, where the bicorn is delineated along with a cubic curve afterwards referred to): as ϕ extends from an indefinitely small positive value ϵ through infinity to $-1-\epsilon$, we have the upper branch of the curve, viz.

$\phi=\epsilon$, gives $x=\infty$, $y=-\infty$, point at infinity, the tangent being horizontal,

$\phi=\infty$, gives $x=-1$, $y=+$, the node-cusp, tangent parallel to axis of y,

$\phi=-2$, gives $x=0$, $y=0$, the tangent at this point being the axis of x,

$\phi=-1-\epsilon$, gives $x=\infty$, $y=+$, point at infinity along the asymptote;

and as ϕ extends from $x=-1+\epsilon$ to $x=-\epsilon$, we have the lower branch, viz.

$\phi=-1+\epsilon$, gives $x=-\infty$, $y=-\infty$, point at infinity along the asymptote,

$\phi=-\frac{3}{4}$, $x=-76\frac{23}{27}$, $y=-41\frac{2}{3}$; the cusp, shown in the figure out of its proper position (observe that for $x=-76\frac{23}{27}$, we have for the asymptote $y=-40\frac{25}{27}$, so that the distance below the asymptote is $=\frac{20}{27}$; Professor Sylvester's value $y=-25$ for the ordinate of the cusp is an obvious error of calculation).

$\phi=-\epsilon$, $x=-\infty$, $y=-\infty$, point at infinity, the tangent being horizontal.

The class of the curve is $=4$.

280. The node-cusp counts as a node, a cusp, an inflexion, and a double tangent; the node-cusp absorbs therefore $(6+8+1=)$ 15 inflexions, and the other cusp 8 inflexions; there remains therefore $(24-15-8=)$ 1 inflexion, viz. this is the inflexion at infinity, having the line infinity for tangent; there is not, besides the tangent at the node-cusp, any other double tangent of the curve.

281. The form of the Bicorn, so far as it is material for the discussion, is also shown in the Plate, fig. 3, and it thereby appears that it divides the plane into three regions; viz. these are the regions PQR and S, for each of which $\phi(x, y)$ is $=-$, and the region TU, for which $\phi(x, y)$ is $=+$; that is, for PQR and S we must have $J=-$, and for TU we must have $J=+$. Hence in connexion with the bicorn, considering the line $y=0$, we have the six regions P, Q, R, S, T, U. It has just been seen that for P, Q, R, S we have $J=-$, and for T, U we have $J=+$; and the sign of J being given, the equations $x=\frac{2^{11}L-J^3}{J^3}$, $y=\frac{D}{J^2}$, then fix for the several regions the signs of $2^{11}L-J^3$ and D, as shown in the subjoined Table; by what precedes each of the six regions has a determinate character, which for R, S, and U (since here D is $=-$) is at once seen to be $3r+2i$, and which, as will presently appear, is ascertained to be $5r$ for P and $r+4i$ for Q and T.

282. We have thus the Table

$$\begin{array}{llll} P, & D=+, & J=-, & 2^{11}L-J^3=+ \,\}\, 5r, \\ Q, & D=+, & J=-, & 2^{11}L-J^3=- \\ T, & D=+, & J=+, & 2^{11}L-J^3=\pm \end{array}\Big\} r+4i,$$

$$\left.\begin{array}{llll} R, & D=-, & J=-, & 2^{11}L-J^3=\pm \\ S, & D=-, & J=-, & 2^{11}L-J^3=+ \\ U, & D=-, & J=+, & 2^{11}L-J^3=\pm \end{array}\right\} 3r+2i;$$

so that we have the kingdom $5r$ consisting of the single region P, the kingdom $r+4i$ consisting of the regions Q and T, and the kingdom $3r+2i$ consisting of the regions R, S, and U.

283. For a given equation if D is $=-$, the character is $3r+2i$; if $D=+$, $J=+$, the character is $r+4i$; if $D=+$, $J=-$, then, according as $2^{11}L-J^3$ is $=+$ or is $=-$, the character is $5r$ or $r+4i$. But in the last case the distinction between the characters $5r$ and $r+4i$ may be presented in a more general form, involving a parameter μ, arbitrary between certain limits. In fact drawing upwards from the origin, as in Plate, fig. 3, the lines $x-2y=0$ and $x+y=0$, and between them any line whatever $x+\mu y=0$, the point (x, y), assumed to lie in the region P or Q, will lie in the one or the other region according as it lies on the one side or the other side of the line in question, viz. in the region P if $x+\mu y$ is $=-$, in the region Q if $x+\mu y$ is $=+$. But we have

$$x+\mu y=\frac{2^{11}L-J^3+\mu JD}{J^3},$$

and J being by supposition negative, the sign of $2^{11}L-J^3+\mu JD$ is opposite to that of $x+\mu y$. The region is thus P or Q according to the sign of $2^{11}L-J^3+\mu JD$; and completing the enunciation, we have, finally, the following criteria for the number of real roots of a given quintic equation, viz.

If $D=-$, the character is $3r+2i$,

If $D=+$, $J=+$, then it is $r+4i$.

But if $D=+$, $J=-$, then μ being any number at pleasure between the limits $+1$ and -2, both inclusive, if

$$2^{11}L-J^3+\mu JD=+, \text{ the character is } 5r,$$
$$2^{11}L-J^3+\mu JD=-, \quad \text{,,} \quad \text{,,} \quad \text{,,} \quad r+4i.$$

284. The characters $5r$ of the region P and $r+4i$ of the regions Q and T may be ascertained by means of the equation $(a,\ 0,\ c,\ 0,\ e,\ 0 \between \theta,\ 1)^5=0$, that is

$$\theta\,(a\theta^4+10c\theta^2+5e)=0\,;$$

there is always the real root $\theta=0$, and the equation will thus have the character $5r$ or $r+4i$ according as the reduced equation $a\theta^4+10c\theta^2+5e=0$ has the character $4r$ or $4i$. It is clear that $(a,\ e)$ must have the same sign, for otherwise θ^2 would have two real values, one positive, the other negative, and the character would be $2r+2i$. And $(a,\ e)$ having the same sign, then the character will be $4r$, if θ^2 has two real positive values, that is, if $ae-5c^2$ is $=-$, and the sign of c be opposite to that of a and e, or, what is the same thing, if ce be $=-$; but if these two conditions are not satisfied, then the values of θ^2 will be imaginary, or else real and negative, and in either case the character will be $4r$.

285. Now, for the equation in question, putting in the Tables $b=d=f=0$, we find

$$\begin{aligned} D &= 256\,ae^3\,(ae-5c^2)^2, \\ J &= 16\,ce\,(ae+3c^2), \\ 2^{11}L-J^3 &= 2^{12}\,ce^3\,\{2\,(ae-c^2)^4-c^2\,(ae+3c^3)^3\} \\ &= 2^{12}\,ce^3\,(ae-5c^2)\,(2a^3e^3+a^2c^2e^2+8ac^4e+5c^6). \end{aligned}$$

We have by supposition $D=+$, that is, $ae=+$; hence J has the same sign as ce; whence if $J=+$, then also $ce=+$, and the character is $4i$; that is the character of the region T is $r+4i$. But if $J=-$, then also $ce=-$. But ae being $=+$, the sign of $2^{11}L-J^3$ is the same as that of $ce\,(ae-5c^2)$, and therefore the opposite of that of $ae-5c^2$: hence $D=+$, $J=-$, the quartic equation has the character $4r$ or $4i$ according as $2^{11}L-J^3$ is $=+$ or $=-$. Hence the region P has the character $5r$ and the region Q the character $r+4i$; and the demonstration is thus completed.

Article Nos. 286 to 293.—HERMITE'S *new form of* TSCHIRNHAUSEN'S *transformation, and application thereof to the quintic.*

286. M. Hermite demonstrates the general theorem, that if $f\,(x,\ y)$ be a given quantic of the n-th order, and $\phi\,(x,\ y)$ any covariant thereof of the order $n-2$, then considering the equation $f(x,\ 1)=0$, and writing

$$z=\frac{\phi\,(x,\ 1)}{f_x'\,(x,\ 1)}$$

(where $f_x'(x, 1)$ is the derived function of $f(x, 1)$ in regard to x), then eliminating x, we have an equation in z, the coefficients whereof are all of them invariants of $f(x, y)$.

287. In particular for the quintic $f(x, y) = (a, b, c, d, e, f \between x, y)^5$, if

$$\phi_1(x, y), \quad \phi_2(x, y), \quad \phi_3(x, y), \quad \phi_4(x, y)$$

are any four covariant cubics, writing

$$z = \frac{t\,\phi_1(x, 1) + u\,\phi_2(x, 1) + v\,\phi_3(x, 1) + w\,\phi_4(x, 1)}{f_x'(x, 1)}$$

(viz. the numerator is a covariant cubic involving the indeterminate coefficients t, u, v, w) then, in the transformed equation in z, the coefficients are all of them invariants of the given quintic. Conducting the investigation by means of a certain canonical form, which will be referred to in the sequel, he fixes the signification of his four covariant cubics, these being respectively covariant cubics of the degrees 3, 5, 7, and 9, defined as follows; viz. starting with the form

$$-3\begin{vmatrix} y^3, & -y^2x, & yx^2, & -x^3 \\ a, & b, & c, & d \\ b, & c, & d, & e \\ c, & d, & e, & f \end{vmatrix},$$

$$= -3D, \; = -3(A, B, C, D \between x, y)^3, \text{ or } (-3A, -B, -C, -3D \between x, y)^3, \text{ suppose,}$$

and considering also the quadric covariant

$$(\alpha, \beta, \gamma \between x, y)^2, \; = B,$$

then $\phi_1, \phi_2, \phi_3, \phi_4$ are derived from the form

$$(A, B, C, D \between \zeta x - \eta(\beta x + 2\gamma y), \; \zeta y + \eta(2\alpha x + \beta y))^3,$$

viz. we have

$$\begin{aligned}
\phi_1(x, y) &= -3(A, B, C, D \between x, y)^3, \\
\phi_2(x, y) &= +3(A, B, C, D \between x, y)^2(-\beta x - 2\gamma y, \; 2\alpha x + \beta y), \\
\{\phi_3(x, y)\} &= -3(A, B, C, D \between x, y)\,(-\beta x - 2\gamma y, \; 2\alpha x + \beta y)^2, \\
\{\phi_4(x, y)\} &= +3(A, B, C, D \between \quad -\beta x - 2\gamma y, \; 2\alpha x + \beta y)^3,
\end{aligned}$$

where $\{\phi_3(x, y)\}$ and $\{\phi_4(x, y)\}$ are the functions originally called by him $\phi_3(x, y)$ and $\phi_4(x, y)$: those ultimately so called by him are

$$\begin{aligned}
{}^{(1)}\,\phi_3(x, y) &= 4\{\phi_3(x, y)\} + J\phi_1(x, y), \qquad (J = G), \\
\phi_4(x, y) &= 4\{\phi_4(x, y)\} + 3J\phi_2(x, y) + 96\psi_1(x, y),
\end{aligned}$$

[1] M. Hermite, p. 17, has erroneously written $\phi_3(x, y) + 4A\phi_1(x, y)$, instead of $4\phi_3(x, y) + A\phi_1(x, y)$; the latter expression is that which he really makes use of, and the formula in the text is correct.

where $\psi_1(x, y)$ is the cubicovariant $(-27A^2D+9ABC-2B^3, \ldots \between x, y)^3$ of $\phi_1(x, y)$, $=(-3A, -B, -C, -3D \between x, y)^3$, *ut suprà*.

The covariant $\phi_2(x, y)$ has the property that if the given quintic $(a, \ldots \between x, y)^5$ contains a square factor $(lx+my)^2$, then $\phi_2(x, y)$ contains the factor $lx+my$: $\{\phi_3(x, y)\}$ and $\{\phi_4(x, y)\}$ are covariants not possessing the property in question, and they were for this reason replaced by $\phi_3(x, y)$ and $\phi_4(x, y)$ which possess it, viz. $\phi_3(x, y)$ contains the factor $lx+my$, and $\phi_4(x, y)$ contains $(lx+my)^3$, being thus a perfect cube when the given quintic contains a square factor.

288. The covariants $\phi_1(x, y)$ and $\phi_2(x, y)$ are included in my Tables, viz. we have

$$\phi_1(x, y) = -3D \text{ of } 142,\ 143,$$

$$\phi_2(x, y) = -\ K \quad \text{,,} \quad \text{,,}$$

(observe that in K the first coefficient vanishes if $a=0$, $b=0$, which is the property just referred to of $\phi_2(x, y)$); the other two covariants, as being of the degree 7 and 9, are not included in my Tables, but I have calculated the leading coefficients of these covariants respectively, viz.

Table No. 85 gives leading coefficient (or that of x^3) in $\phi_3(x, y)$, and

Table No. 86 gives leading coefficient (or that of x^3) in $\phi_4(x, y)$ [and by means thereof we have the values of the covariants in question].

The coefficients in question vanish for $a=0$, $b=0$, that is, $\phi_3(x, y)$ and $\phi_4(x, y)$ then each of them contain the factor y; if the remaining coefficients of $\phi_4(x, y)$ were calculated, it should then appear that for $a=0$, $b=0$, those of x^2y, xy^2 would also vanish, and thus that $\phi_4(x, y)$ would be a mere constant multiple of y^3.

Table No. 85 [= leading coefficient of $16BJ-15DG$].

a^3cef^2	$+\ 1$	$a^2b^2ef^2$	$-\ 1$	ab^3df^2	$+\ 64$	b^4cf^2	...
$a^3d^2f^2$	$+\ 15$	a^2bcdf^2	$-\ 94$	ab^3e^2f	$-\ 54$	b^4def	$-\ 144$
a^3de^2f	$-\ 32$	a^2bce^2f	$+\ 86$	$ab^2c^2f^2$	$-\ 48$	b^4e^3	$+\ 135$
a^3e^4	$+\ 16$	a^2bd^2ef	$+\ 106$	ab^2cdef	$+\ 184$	b^3c^2ef	$+\ 108$
		a^2bde^3	$-\ 96$	ab^2ce^3	$-\ 135$	b^3cd^2f	$+\ 288$
		$a^2c^3f^2$	$+\ 63$	ab^2d^3f	$-\ 272$	b^3cde^2	$-\ 450$
		a^2c^2def	$-\ 188$	$ab^2d^2e^2$	$+\ 243$	b^3d^3e	$+\ 80$
		$a^2c^2e^3$	$+\ 32$	abc^3ef	$-\ 66$	b^2c^3df	$-\ 360$
		a^2cd^3f	$+\ 60$	abc^2d^2f	$+\ 212$	$b^2c^3e^2$	$+\ 135$
		$a^2cd^2e^2$	$+\ 68$	abc^2de^2	$+\ 148$	$b^2c^2d^2e$	$+\ 360$
		a^2d^4e	$-\ 36$	$abcd^3e$	$-\ 412$	b^2cd^4	$-\ 160$
				abd^5	$+\ 144$	bc^5f	$+\ 108$
				ac^4df	$-\ 36$	bc^4de	$-\ 180$
				ac^4e^2	$-\ 48$	bc^3d^3	$+\ 80$
				ac^3d^2e	$+\ 124$		
				ac^2d^4	$-\ 48$		
	± 32		± 415		± 1119		± 1294

Table No. 86 [= leading coefficient of S'].

a^4cef^3	+ 9	a^3bef^3	− 9	$a^2b^3df^3$	+ 120	ab^4cf^3	− 576	b^6f^3	+ 192
$a^4d^2f^3$	+ 21	a^3bcdf^3	− 162	$a^2b^3e^2f^2$	− 21	ab^4def^2	+ 672	b^5cef^2	− 1440
a^4def^2	− 78	$a^3bce^2f^2$	+ 99	$a^2b^2c^2f^3$	+ 486	ab^4e^3f	− 359	$b^5d^4f^2$	− 192
a^4e^4f	+ 48	$a^3bd^2ef^2$	+ 309	$a^2b^2cdef^2$	− 2160	$ab^3c^2ef^2$	+ 3456	b^5de^2f	− 1080
		a^3bde^3f	+ 12	$a^2b^2ce^3f$	+ 1023	$ab^3cd^2f^2$	− 864	b^5e^4	+ 2025
		a^3be^5	− 240	$a^2b^2d^3f^2$	+ 120	ab^3cde^2f	+ 2094	$b^4c^2df^2$	+ 2592
		$a^3c^3f^3$	− 81	$a^2b^2d^2e^2f$	− 1053	ab^3ce^4	− 3915	$b^4c^2e^2f$	+ 3546
		$a^3c^2def^2$	+ 1026	$a^2b^2de^4$	+ 1314	ab^3d^3ef	+ 528	b^4cd^2ef	+ 5280
		$a^3c^2e^3f$	− 768	$a^2bc^2ef^2$	− 1863	$ab^3d^2e^3$	− 45	b^4cde^3	− 13500
		$a^3cd^3f^2$	− 738	$a^2bc^2d^2f^2$	+ 2538	$ab^2c^3df^2$	− 2592	b^4d^4f	− 4800
		$a^3cd^2e^2f$	− 564	$a^2bc^2de^2f$	+ 2340	$ab^2c^3e^2f$	− 9747	$b^4d^3e^2$	+ 7800
		a^3cde^4	+ 1056	$a^2bc^2e^4$	+ 672	$ab^2c^2d^2ef$	− 8496	$b^3c^4f^2$	− 648
		a^3d^4ef	+ 756	a^2bcd^3ef	+ 2820	$ab^2c^2de^3$	+ 26610	b^3c^3def	− 14040
		$a^3d^3e^3$	− 696	$a^2bcd^2e^3$	− 7812	ab^2cd^4f	+ 8544	$b^3c^3e^3$	+ 3075
				a^2bd^5f	− 3024	ab^2cd^4e	− 16650	$b^3c^2d^3f$	+ 9120
				$a^2bd^4e^2$	+ 4572	ab^2d^5e	+ 720	$b^3c^2d^2e^2$	+ 16350
				$a^2c^4df^2$	− 324	abc^5f^2	+ 972	b^3cd^4e	− 19200
				$a^2c^4e^2f$	+ 3888	abc^4def	+ 24048	b^3d^6	+ 4800
				$a^2c^3d^2ef$	− 8748	abc^4e^3	− 4464	b^2c^5ef	+ 4860
				$a^2c^3de^3$	− 4800	abc^3d^3f	− 15984	$b^2c^4d^2f$	− 3240
				$a^2c^2d^4f$	+ 4248	$abc^3d^2e^2$	− 30108	$b^2c^4de^2$	− 8100
				$a^2c^2d^3e^2$	+ 14520	abc^2d^4e	+ 35088	$b^2c^3d^3e$	+ 9000
				a^2cd^5e	− 11448	$abcd^6$	− 8640	bc^2d^5	− 2400
				a^2d^7	+ 2592	ac^6ef	− 7776		
						ac^5d^2f	+ 5184		
						ac^5de^2	+ 12960		
						ac^4d^3e	− 14400		
						ac^3d^5	+ 3840		
	$\pm$ 78		$\pm$ 3258		$\pm$ 41253		$\pm$ 124716		$\pm$ 68640

[The values thus are $\phi_3(x, y) = 16BJ - 15DG$; $\phi_4(x, y) = S'$.]

289. The equation in z is of the form

$$z^5 + \frac{\mathfrak{A}}{D} z^3 + \frac{\mathfrak{B}}{D} z^2 + \frac{\mathfrak{C}}{D} z + \frac{\mathfrak{D}}{D} = 0,$$

where D is the discriminant of the quintic and $\mathfrak{A}$, $\mathfrak{B}$, $\mathfrak{C}$, $\mathfrak{D}$ denote rational and integral functions of the coefficients (a, b, c, d, e, f). And the covariants $\phi_1(x, y)$, $\phi_2(x, y)$, $\phi_3(x, y)$, $\phi_4(x, y)$ having the values given to them above, the actual value of $\mathfrak{A}$ is obtained as a quadric function of the indeterminates (t, u, v, w), viz. this is

$$= [D_1 t^2 - 6BDtv - D(D_1 - 10AB) v^2] + D[-Bu^2 + 2D_1 uw + 9(BD - 10AD_1) w^2],$$

where $D_1 = 25AB + 16C$, these quantities, and the quantity $N (= D_1^2 - 10ABD_1 + 9B^2D)$ afterwards spoken of, being in the notation of the present Memoir as follows:

$$\begin{array}{lll} A = & J & (= \ G), \\ B = & -K & (= -Q), \\ C = & 9L + JK & (= -9U + GQ), \\ D = & D & (= \ Q'), \\ D_1 = & 9(16L - JK), & \\ N = & 1152(18L^2 - JKL - K^3). & \end{array}$$

290. If by establishing two linear relations between the coefficients (t, u, v, w) the equation $\mathfrak{A}=0$ can be satisfied (which in fact can be done by the solution of a quadric equation), then these quantities can be by means of the relations in question expressed as linear functions of any two of them, say of v and w; and then the next coefficient $\mathfrak{B}$ will be a cubic function $(v, w)^3$, and the equation $\mathfrak{B}=0$ will be satisfied by means of a cubic equation $(v, w)^3=0$, that is, the transformed equation in z can be by means of the solution of a quadric and a cubic equation reduced to the trinomial form

$$z^5+\frac{\mathfrak{C}}{D}z+\frac{\mathfrak{D}}{D}=0,$$

and M. Hermite shows that the equation $\mathfrak{A}=0$ can be satisfied as above very simply, and that in two different ways, viz.

291. 1°. $\mathfrak{A}=0$ if

$$D_1t^2-6BDtv-(\ \ D_1-10AB\ )\,v^2\ =0,$$
$$Bu^2-2D_1uw-(9BD\ -10AD_1)\,w^2=0,$$

that is, N denoting as above, if

$$t=\frac{3BD+\sqrt{ND}}{D_1}v,\qquad u=\frac{D_1+\sqrt{N}}{B}w.$$

292. 2°. Writing the expression for $\mathfrak{A}$ in the form

$$D_1(t^2-Dv^2+2Duw-10ADw^2)+BD(10Av^2-6tv-u^2+9Dw^2),$$

then $\mathfrak{A}=0$, if

$$t^2-Dv^2+2Duw-10ADw^2=0,$$
$$10Av^2-6tv-u^2+\ 9\ \ Dw^2=0.$$

These equations, writing therein

$$t=\frac{1}{\sqrt{2}}\sqrt{D}T,\qquad u=U+5AW,\qquad v=\frac{1}{\sqrt{2}}V,\qquad w=W,$$

become

$$T^2-V^2+4UW=0,$$
$$-5AV^2+3\sqrt{D}TV+U^2+10AUW+(25A^2-9D)\,W^2=0,$$

the first of which is satisfied by the values

$$T=\rho\,W-\frac{1}{\rho}U,\qquad V=\rho\,W+\frac{1}{\rho}U;$$

and then substituting for T and V, the second equation will be also satisfied if only

$$\rho^2=5A+3\sqrt{D}.$$

Article Nos. 293 to 295.—HERMITE'S *application of the foregoing results to the determination of the Character of the quintic equation.*

293. By considerations relating to the form

$$\frac{1}{D}\left\{[D_1 t^2 - 6BDtv - D(D_1 - 10AB)v^2] + D[-Bu^2 + 2D_1 uw + 9BD - 10AD_1 w^2]\right\},$$

M. Hermite obtains criteria for the character of the quintic equation $f(x, 1) = 0$.

294. If $D = -$, the character is $3r + 2i$, but if $D = +$, then expressing the foregoing form as a sum of four squares affected with positive or negative coefficients, the character will be $5r$ or $2 + 4i$, according as the coefficients are all positive, or are two positive and two negative. Whence, if N denote as above, then for

$$D = +,\ N = -,\quad D_1 = +,\ B = -,\ \text{character is } 5r,$$

and

$$\left.\begin{array}{l} D = +,\ N = -,\ BD_1 = + \\ \\ D = +,\ N = + \end{array}\right\}\quad \text{character is}\quad r + 4i;$$

and further, the combination $D = +,\ N = -,\ D_1 = -,\ B = +$ cannot arise (Hermite's first set of criteria).

295. Again, from the equivalent form

$$\frac{1}{D}\left\{D_1(t^2 - Dv^2 + 2Duw - 10Aw^2) + BD(10Av^2 - 6tv - u^2 + 9Dw^2)\right\},$$

which, if ω, ω' are the roots of the equation $9\theta^2 - 10A\theta + D = 0$, is

$$= \frac{1}{D}\left\{\frac{D_1\omega - BD}{\omega - \omega'}\left[(t - 3\omega' v)^2 - \omega'\left(u - \frac{D}{\omega'}w\right)^2\right] + \frac{D_1\omega' - BD}{\omega' - \omega}\left[(t - 3\omega v)^2 - \omega\left(u - \frac{D}{\omega}w\right)\right]^2\right\};$$

then by similar reasoning it is concluded that

$$D = +,\ 25A^2 - 9D = +,\ A = -,\ N = -,\ \text{character is } 5r,$$

$$\left.\begin{array}{l} D = +,\ 25A^2 - 9D = +,\ A = -,\ N = +, \\ D = +,\ 25A^2 - 9D = +,\ A = +, \\ D = +,\ 25A^2 - 9D = -, \end{array}\right\}\quad \text{,,} \qquad r + 4i.$$

(Hermite's second set of criteria.)

Article Nos. 296 to 303.—*Comparison with the Criteria* No. 283: *the Nodal Cubic.*

296. For the discussion of Hermite's results, it is to be observed that in the notation of the present Memoir we have

$$\begin{aligned} A &= J, \\ B &= -K = -\tfrac{1}{128}(J^2 - D), \\ D &= D, \\ D_1 &= 16L - JK = \tfrac{1}{128}(2^{11}L - J^3 + JD), \\ N &= 18L^2 - JKL - K^3 \\ &= \frac{1}{2^{21}}\{3^2 . 2^{22} L^2 - 14JL(J^2 - D) - (J^2 - D)^3\}, \end{aligned}$$

or, putting as above,

$$x = \frac{2^{11}L - J^3}{J^3},\ y = \frac{D}{J^2},\ \text{and therefore } 1 + x = \frac{2^{11}L}{J^3},\ 1 - y = \frac{J^2 - D}{J^2},$$

we have

$$\begin{aligned} A &= J, \\ B &= \tfrac{1}{128}J^2(y - 1), \\ D &= J^2 y, \\ D_1 &= \tfrac{1}{128}J^3(x + y), \\ N &= \frac{1}{2^{21}}J^6\{9(1 + x)^2 - 8(1 + x)(1 - y) - (1 - y)^3\}, \\ &= \frac{1}{2^{21}}J^6 . \{y^3 - 3y^2 + 8xy + 9x^2 + 11y + 10x\}. \end{aligned}$$

It thus becomes necessary to consider the curve

$$\psi(x,\ y) = y^3 - 3y^2 + 8xy + 9x^2 + 11y + 10x = 0,$$

the equation whereof may also be written

$$9x + 4y + 5 = (y - 1)\sqrt{25 - 9y}.$$

297. This is a cubic curve, viz. it is a divergent parabola having for axis the line $9x + 4y + 5 = 0$, and its ordinates parallel to the axis of x; and having moreover a node at the point $x = -1$, $y = +1$, that is, at the node-cusp of the bicorn; the curve is thus a nodal cubic; we may trace it directly from the equation, but it is to be noticed that *quâ* nodal cubic it is a unicursal curve; the coordinates x, y are therefore rationally expressible in terms of a parameter ψ; and it is easy to see that we in fact have

$$\begin{aligned} 81(x + 1) &= \psi^2(\psi - 8), \\ 9(y - 1) &= -\psi(\psi - 8), \end{aligned}$$

whence also

$$\frac{dy}{dx} = \frac{-18(\psi - 4)}{\psi(3\psi - 16)}.$$

298. We see that

$\psi=\infty$, gives $x=\infty$, $y=-\infty$, point at infinity, the direction of the curve parallel to axis of x.

$\psi=9$, „ $x=0$, $y=0$, the origin.

$\psi=8$, „ $x=-1$, $y=+1$, the node, tangent parallel to axis of y.

$\psi=\frac{16}{3}$, „ $x=\frac{4325}{2187}$, $y=\frac{209}{81}$, tangent parallel to the axis of y.

$\psi=4$, „ $x=-\frac{145}{81}$, $y=\frac{25}{9}$, tangent parallel to axis of x.

$\psi=0$, „ $x=-1$, $y=+1$, the node.

$\psi=-1$, „ $x=-\frac{10}{9}$, $y=0$.

$\psi=-16$, „ $x=-76\frac{23}{27}$, $y=-41\frac{2}{3}$, the cusp of the bicorn.

$\psi=-\infty$, „ $x=-\infty$, $y=-\infty$, point at infinity, direction of curve parallel to axis of x.

299. The Nodal Cubic is shown along with the Bicorn, Plate, fig. 2; it consists of one continuous line, passing from a point at infinity, through the cusp of the bicorn, on to the node-cusp, then forming a loop so as to return to the node-cusp, again meeting the bicorn at the origin, and finally passing off to a point at infinity, the initial and ultimate directions of the curve being parallel to the axis of x.

300. It may be remarked that, inasmuch as one of the branches of the cubic touches the bicorn at the node-cusp, the node-cusp counts as $(4+2=)$ 6 intersections; the intersections of the cubic with the bicorn are therefore the cusp, the node-cusp, and the origin, counting together as $(2+6+1=)$ 9 intersections, and besides these the point at infinity on the axis of x, counting as 3 intersections. This may be verified by substituting in the equation of the cubic the bicorn ϕ-values of x and y. But to include all the proper factors, we must first write the equation of the cubic in the homogeneous form

$$(9x+8y+5z)^2 z-(y-z)^2(25z-9y)=0,$$

and herein substitute the values

$$x : y : z=-(\phi+2)(\phi^3-\phi^2+2\phi-4) : (\phi+2)^2(\phi-3)\phi : (\phi+1)\phi^3;$$

the result is found to be

$$\phi^3\{(\phi+1)(4\phi^2+6\phi-9)^2-(2\phi+3)^2(4\phi^3+4\phi^2+18\phi+27)\}=0,$$

that is

$$-9\phi^3(\phi+2)(4\phi+3)^2=0;$$

and considering this as an equation of the order 12, the roots are $\phi=0$, 3 times, $\phi=-2$, 1 time; $\phi=-\frac{3}{4}$, 2 times, and $\phi=\infty$, 6 times.

301. The cubic curve divides the plane into 3 regions, which may be called respectively the loop, the antiloop, and the extra cubic; for a point within the loop or antiloop, $\psi(x, y)$ is $=-$, for a point in the extra cubic $\psi(x, y)$ is $=+$. If in conjunction with the cubic we consider the discriminatrix, or line $y=0$, then we have in all six regions, viz. y being $=+$, three which may be called the loop, the triangle,

and the upper region; and y being $=-$, three which may be called the right, left, and under regions respectively; the triangle and the other region form together the antiloop.

302. It is now easy to discuss Hermite's two sets of criteria; the first set becomes

$$y=+,\quad y-1=-,\quad J(x+y)=+,\quad \psi(x,y)=-,\quad \text{character } 5r,$$

$$\left.\begin{array}{ll} y=+, \quad J(y-1)(x+y)=+, & \psi(x,y)=- \\ y=+, & \psi(x,y)=+ \end{array}\right\}\ \text{character } r+4i,$$

$$y=+,\quad y-1=+,\quad J(x+y)=-,\quad \psi(x,y)=-,\quad \text{cannot exist.}$$

Referring to the Plate, fig. 4, which shows a portion of the cubic and the bicorn, then 1° the conditions $y=+, \psi(x, y)=-$ imply that the point (x, y) is within the loop or within the triangle of the cubic; the condition $y-1=-$ brings it to be within the triangle, and for any point within the triangle we have $x+y=-$, whence also the condition $J(x+y)=+$ becomes $J=-$; hence the conditions amount to $J=-$, (x, y) within the triangle; but by the general theory (x, y), being within the triangle, that is, in the region P or T, if $J=-$, will of necessity be within the region P; so that the conditions give $J=-$, (x, y) within the region P; the corresponding character being $5r$, which is right.

2°. $y=+$, $\psi(x, y)=-$, the point (x, y) must be within the loop, or within the triangle; if (x, y) is within the loop, then $y-1=+$, $x+y=1$, and the condition $J(y-1)(x+y)=+$ becomes $J=-$, that is, we have $J=-$ and (x, y) within the loop, that is, in the region T. And again, if (x, y) be within the triangle, then $y-1=-$, $x+y=+$, and the condition $J(y-1)(x+y)=+$ still gives $J=-$; but $J=-$, and (x, y) within the triangle, that is, in the region T or P, will of necessity be in the region T; so that in either case we have $J=-$, (x, y) in the region T, which agrees with the character $r+4i$.

3°. $y=+$, $\psi(x, y)=+$, (x, y) is in the upper region, that is, in the region Q or T; if (x, y) is in the region Q, then of necessity $J=-$, and if in the region T, then of necessity $J=+$, that is, we have

$$J=-,\ (x, y) \text{ in the region } Q, \text{ or}$$
$$J=+,\ (x, y) \text{ in the region } P,$$

which agrees with the character $r+4i$.

And it is to be observed that the portions of T under 2° and 3° respectively make up the whole of the region T, and that 3° relates to the whole of the region Q, so that the conditions allow the point (x, y) to be anywhere in Q or T, which is right.

4°. $y=+$, $\psi(x, y)=-$, (x, y) is in the loop or the triangle, and then $y-1=+$ implies that it is in the loop, whence $x+y=+$, and the condition $J(x+y)=-$ becomes $J=-$; we should therefore if the combination existed have $J=-$, (x, y) within the loop, that is, in the region T; but this is impossible.

303. Hermite's second set of criteria are

$$\left.\begin{array}{llll} y=+, & \tfrac{25}{9}-y=+, & J=-, & \psi(x,\ y)=-, \text{ character } 5r, \\ y=+, & \tfrac{25}{9}-y=+, & J=-, & \psi(x,\ y)=+ \\ y=+, & \tfrac{25}{9}-y=+, & J=+, & \\ y=+, & \tfrac{25}{9}-y=-, & & \end{array}\right\} \text{character } r+4i.$$

1°. If $y=+$, $\psi(x, y)=-$, then the point (x, y) must be situate within the loop or within the triangle; and recollecting that at the highest point of the loop we have $y=\tfrac{25}{9}$, the condition $\tfrac{25}{9}-y=+$ is satisfied for every such point, and may therefore be omitted. The conditions therefore are $J=-$, (x, y) within the loop, that is, in the region T, or within the triangle, that is, in the region P or the region T; but for any point of T the general theory gives $J=+$, and the conditions are therefore $J=-$, (x, y) within the region P; which agrees with the character $5r$.

2°. $y=+$, $\psi(x, y)=+$, that is, (x, y) is within the upper region, that is, in the region Q or T; and $\tfrac{25}{9}-y=+$, (x, y) will be within the portions of Q and T which lie beneath the line $y=\tfrac{25}{9}$; but $J=-$, and therefore (x, y) cannot lie in the region T; hence the conditions amount to $J=-$, (x, y) within that portion which lies beneath the line $y=\tfrac{25}{9}$ of the region Q.

3°. $y=+$, $\tfrac{25}{9}-y=+$, (x, y) lies beneath the line $y=\tfrac{25}{9}$, viz. in one of the regions P, Q or T; but $J=+$, (x, y) cannot lie in the region P or Q; hence the conditions give $J=+$, (x, y) within the portion which lies beneath the line $y=\tfrac{25}{9}$ of the region T.

4°. $y=+$, $\tfrac{25}{9}-y=-$, that is, (x, y) lies above the line $y=\tfrac{25}{9}$, and therefore in one of the regions T or Q; and by the general theory, according as (x, y) lies in T or in Q, we shall have $J=+$ or $J=-$, hence the conditions give

$J=-$, (x, y) within the portion which lies above the line $y=\tfrac{25}{9}$, of the region Q.

$J=+$, (x, y) within the portion which lies above the line $y=\tfrac{25}{9}$, of the region T.

2°, 3°, and 4°, each of them agree with the character $r+4i$, and together they imply $J=-$, (x, y) anywhere in the region Q, or else $J=+$, (x, y) anywhere in the region T; which is right.

Article Nos. 304 to 307.—Hermite's *third set of Criteria; comparison with* No. 283, *and remarks.*

304. In the concluding portion of his memoir, M. Hermite obtains a third set of criteria for the character of a quintic equation; this is found by means of the equation for the function

$$a^4(\theta_0-\theta_1)(\theta_1-\theta_2)(\theta_2-\theta_3)(\theta_3-\theta_4)(\theta_4-\theta_0)$$

of the roots $(\theta_0, \theta_1, \theta_2, \theta_3, \theta_4)$ of the given quintic equation $(a, b, c, d, e, f \between \theta, 1)^5=0$. The function in question has 12 pairs of equal and opposite values, or it is determined by an equation of the form $(u^2, 1)^{12}=0$, which equation is decomposable, not rationally but by the adjunction thereto of the square root of the discriminant, into two equations of the form $(u^2, 1)^6=0$; viz. one of these is

$$\begin{aligned}
&u^{12}\\
&+u^{10}(\mathrm{a}+3\sqrt{\Delta})\\
&+u^8\,[\tfrac{1}{4}(\mathrm{a}-\sqrt{\Delta})^2+\Delta]\\
&-u^6\ \mathrm{d}\\
&+u^4\,[\tfrac{1}{4}(\mathrm{a}+\sqrt{\Delta})^2+\Delta]\,\Delta\\
&+u^2\,(\mathrm{a}-3\sqrt{\Delta})\,\Delta^2\\
&+\qquad \Delta^3=0,
\end{aligned}$$

and the other is of course derived from it by reversing the sign of $\sqrt{\Delta}$. I have in the equation written (a, d) instead of Hermite's writing capitals A, D; the sign $-$ of the term in u^6 instead of $+$, as printed in his memoir, is a correction communicated to me by himself. The signification of the symbols is in the author's notation

$$\begin{aligned}
\mathrm{a}&=5^4A,\\
\mathrm{d}&=4\,.\,5^9\,(AD-\tfrac{80}{9}D_1),\\
\Delta&=5^5D,
\end{aligned}$$

whence, in the notation of the present memoir, the expressions of these symbols are

$$\begin{aligned}
\mathrm{a}&=\ \ 5^4J,\\
\mathrm{d}&=-\tfrac{1}{2}5^{10}\,(2^{11}L-J^3-\tfrac{3}{5}JD),\\
\Delta&=\ \ 5^5D.
\end{aligned}$$

305. From the equation in u, taking therein the radical $\sqrt{\Delta}$ as positive, M. Hermite obtains ($\mathrm{d}<0$ a mistake for $\mathrm{d}>0$) the following as the necessary and sufficient conditions for the reality of all the roots,

$$\Delta=+,\quad \mathrm{a}+3\sqrt{\Delta}=-,\quad \mathrm{d}=+,\quad \text{character } 5r$$

(Hermite's third set of criteria).

306. It is clear that $\mathrm{a}+3\sqrt{\Delta}=-$ is equivalent to ($\mathrm{a}=-$ and $\mathrm{a}^2-9\Delta=+$), and we have $\mathrm{a}^2-9\Delta=5^5(125J^2-9D)$, so that these conditions for the character $5r$ are

$$D=+,\quad J=-,\quad 125J^2-9D=+,\quad 2^{11}L-J^3-\tfrac{3}{5}JD=+.$$

Now, writing as above,

$$x=\frac{2^{11}L-J^3}{J^3},\quad y=\frac{D}{J^2},$$

these are $y=+$, $J=-$, $\frac{125}{9}-y=+$, $x-\frac{3}{5}y=-$; the conditions $y=+$, $J=-$ imply that (x, y) is in the region P or the region Q; and the condition $x-\frac{3}{5}y=-$ (observe the line $x - \frac{3}{5}y = 0$ lies between the lines $x + y = 0$, $x - 2y = 0$, and so does not cut either the region P or the region Q) restricts (x, y) to the region P; and for every point of P y is at most $= 1$, and the condition $\frac{125}{9} - y = +$ is of course satisfied. The condition, $125J^2 - 9D = +$, is thus wholly unnecessary, and omitting it, the conditions are

$$D = +, \quad J = -, \quad 2^{11}L - J^3 - \tfrac{3}{5}JD = 0, \quad \text{character } 5r,$$

which, $-\frac{3}{5}$ being an admissible value of μ, agrees with the result *ante*, No. 283.

307. It may be remarked in passing that if 12345 is a function of the roots $(x_1, x_2, x_3, x_4, x_5)$ of a quintic equation, which function is such that it remains unaltered by the cyclical permutation 12345 into 23451, and also by the reversal (12345 into 15432) of the order of the roots, so that the function has in fact the 12 values

$$\begin{array}{ll} \alpha_1 = 12345, & \beta_1 = 24135, \\ \alpha_2 = 13425, & \beta_2 = 32145, \\ \alpha_3 = 14235, & \beta_3 = 43125, \\ \alpha_4 = 21435, & \beta_4 = 13245, \\ \alpha_5 = 31245, & \beta_5 = 14325, \\ \alpha_6 = 41325, & \beta_6 = 12435, \end{array}$$

then $\phi(\alpha, \beta)$ being any unsymmetrical function of (α, β), the equation having for its roots the six values of $\phi(\alpha, \beta)$ (viz. $\phi(\alpha_1, \beta_1)$, $\phi(\alpha_2, \beta_2) \ldots \phi(\alpha_6, \beta_6)$) can be expressed rationally in terms of the coefficients of the given quintic equation and of the square root of the discriminant of this equation. In fact, v being arbitrary, write

$$L = \Pi_6 \{v - \phi(\alpha, \beta)\}, \quad M = \Pi_6 \{v - \phi(\beta, \alpha)\},$$

then the interchange of any two roots of the quintic produces merely an interchange of the quantities L, M; that is,

$$L + M \text{ and } (L - M) \div \zeta^{\frac{1}{2}}(x_1, x_2, x_3, x_4, x_5)$$

are each of them unaltered by the interchange of any two roots, and are consequently expressible as rational functions of the coefficients; or observing that $\zeta^{\frac{1}{2}}(x_1, x_2, x_3, x_4, x_5)$ is a multiple of $\sqrt{D}$, we have L a function of the form $P + Q\sqrt{D}$; the equation $L = 0$, the roots whereof are $v = \phi(\alpha_1, \beta_1) \ldots v = \phi(\alpha_6, \beta_6)$, is consequently an equation of the form $P + Q\sqrt{D} = 0$, viz. it is a sextic equation $(*\between v, 1)^6 = 0$, the coefficients of which are functions of the form in question. Hence in particular

$$u^2 = 12345 = (x_1 - x_2)^2 (x_2 - x_3)^2 (x_3 - x_4)^2 (x_4 - x_5)^2 (x_5 - x_1)^2$$

is determined as above by an equation $(*\between u^2, 1)^6 = 0$. Another instance of such an equation is given by my memoir "On a New Auxiliary Equation in the Theory of Equations of the Fifth Order," *Phil. Trans.* vol. CLI. (1861), pp. 263—276, [268].

Article Nos. 308 to 317.—HERMITE'S *Canonical form of the quintic.*

308. It was remarked that M. Hermite's investigations are conducted by means of a canonical form, viz. if $A\ (=J,\ =G)$ be the quartinvariant of the given quintic $(a,\ b,\ c,\ d,\ e,\ f \between x,\ y)^5$, then he in fact finds $(X,\ Y)$ linear functions of $(x,\ y)$ such that we have

$$(a,\ b,\ c,\ d,\ e,\ f \between x,\ y)^5 = (\lambda,\ \mu,\ \sqrt{k},\ \sqrt{k},\ \mu',\ \lambda' \between X,\ Y)^5$$

(viz. in the transformed form the two mean coefficients are equal; this is a convenient assumption made in order to render the transformation completely definite, rather than an absolutely necessary one); and where moreover the quadricovariant B of the transformed form is

$$= \sqrt{A}XY,$$

or, what is the same thing, the coefficients $(\lambda,\ \mu,\ \sqrt{k},\ \sqrt{k},\ \mu',\ \lambda')$ of the transformed form are connected by the relations

$$\left.\begin{aligned} \lambda\mu' - 4\mu\sqrt{k} + 3k &= 0, \\ \lambda'\mu - 4\mu'\sqrt{k} + 3k &= 0, \\ \lambda\lambda' - 3\mu\mu' \quad + 2k &= \sqrt{A}, \end{aligned}\right\}$$

the advantage is a great simplicity in the forms of the several covariants, which simplicity arises in a great measure from the existence of the very simple covariant operator $\frac{d}{dX} \cdot \frac{d}{dY}$ (viz. operating therewith on any covariant we obtain again a covariant).

309. Reversing the order of the several steps, the theory of M. Hermite's transformation may be established as follows:

Starting from the quintic

$$(a,\ b,\ c,\ d,\ e,\ f \between x,\ y)^5,$$

and considering the quadricovariant thereof

$$(\alpha,\ \beta,\ \gamma \between x,\ y)^2 \qquad B$$

$\big((\alpha,\ \beta,\ \gamma)$ are of the degree 2$\big)$, and also the linear covariant

$$Px + Qy \qquad J$$

$\big((P,\ Q)$ are of the degree 5$\big)$, we have

$$\beta^2 - 4\alpha\gamma = A, \qquad G$$

and moreover

$$(\alpha,\ \beta,\ \gamma \between Q,\ -P)^2 = -C,$$

viz. the expression on the left hand, which is of the degree 12, and which is obviously an invariant, is $= -C$, where C is (*ut suprà*)

$$C = 9L + JK = -9U + GM.$$

The Jacobian of the two forms, viz.

$$\begin{vmatrix} 2\alpha x+\beta y, & \beta x+2\gamma y \\ P \quad , & Q \end{vmatrix},$$

$$= x(2\alpha Q-\beta P)+y(\beta Q-2P\gamma),$$

is a linear covariant of the degree 7, say it is

$$= P'x+Q'y,$$

and it is to be observed that the determinant $PQ'-P'Q$ of the two linear forms is $= -2(\alpha,\ \beta,\ \gamma \between Q,\ -P)^2$, that is, it is $=2C$.

310. Hence writing

$$T=\frac{1}{2\sqrt{C}}(Px+Qy)=\frac{1}{2\sqrt[4]{A}}(\quad X+Y),$$

$$U=\frac{1}{2\sqrt{C}}(P'x+Q'y)=\frac{\sqrt[4]{A}}{2}\quad(-X+Y),$$

whence also

$$X=T\sqrt[4]{A}-\frac{U}{\sqrt[4]{A}},$$

$$Y=T\sqrt[4]{A}+\frac{U}{\sqrt[4]{A}},$$

the determinant of substitution from $(X,\ Y)$ to $(T,\ U)$ is $=2$, that from $(T,\ U)$ to $(x,\ y)$ is $\frac{1}{4C}2C,\ =\frac{1}{2}$, and consequently that from $(X,\ Y)$ to $(x,\ y)$ is $=1$.

We have

$$AT^2-U^2=\frac{1}{4C}\{(\beta^2-4\alpha\gamma)(Px+Qy)^2-(P'x+Q'y)^2\};$$

or putting for P', Q' their values, this is $=\frac{1}{4C}$ into $4(\alpha,\ \beta,\ \gamma\between Q,\ -P)^2(\alpha x^2+2\beta xy+\gamma y^2)$, that is, we have

$$AT^2-U^2=\alpha x^2+\beta xy+\gamma y^2;$$

and we have also

$$AT^2-U^2=\tfrac{1}{4}\sqrt{A}\,[(X+Y)^2-(X-Y)^2]=\sqrt{A}XY,$$

consequently

$$\alpha x^2+\beta xy+\gamma y^2=AT^2-U^2=\sqrt{A}XY.$$

311. We have

$$x=\frac{1}{\sqrt{C}}(\quad Q'T-QU),$$

$$y=\frac{1}{\sqrt{C}}(-P'T+PU),$$

so that, pausing a moment to consider the transformation from (x, y) to (T, U), we have

$$(a,\ b,\ c,\ d,\ e,\ f \between x,\ y)^5 = \frac{1}{\sqrt{C^5}}(a,\ b,\ c,\ d,\ e,\ f \between Q'T - QU,\quad -P'T + PU)^5$$

$$= \frac{1}{\sqrt{C^5}}(\mathrm{a},\ \mathrm{b},\ \mathrm{c},\ \mathrm{d},\ \mathrm{e},\ \mathrm{f} \between T,\ U)^5 \text{ suppose,}$$

where (a, b, c, d, e, f) are *invariants*, of the degrees 36, 34, 32, 30, 28, 26 respectively; it follows that b, d, f each of them contain as a factor the 18-thic invariant I, the remaining factors being of the orders 16, 12, 8 respectively.

312. That (a, b, c, d, e, f) are invariants is almost self-evident; it may however be demonstrated as follows. Writing

$$\{y\partial_x\} = \ a\partial_b + 2b\partial_c + 3c\partial_d + 4d\partial_e + 5e\partial_f,\ = \delta \text{ suppose,}$$
$$\{x\partial_y\} = 5b\partial_a + 4c\partial_b + 3d\partial_c + 2e\partial_d + f\partial_e,\ = \delta_1 \quad ,,$$

then $Px + Qy$, $P'x + Q'y$ being covariants, we have $\delta P = 0$, $\delta Q = P$, $\delta P' = 0$, $\delta Q' = P'$, whence, treating T, U as constants, $\delta(Q'T - QU) = P'T - PU$, $\delta(-P'T + PU) = 0$. Hence

$$\begin{aligned}
&\delta(a,\ b,\ c,\ d,\ e,\ f \between Q'T - QU,\ -P'T + PU)^5 \\
&= \ 5(a,\ b,\ c,\ d,\ e \between Q'T - QU,\ -P'T + PU)^4 . (-P'T + PU) \\
&\quad + 5(a,\ b,\ c,\ d,\ e \between \quad ,, \quad ,, \quad)^4 . (\ P'T - PU) \\
&\quad + 5(b,\ c,\ d,\ e,\ f \between \quad ,, \quad ,, \quad)^4 \ . \ 0,
\end{aligned}$$

the three lines arising from the operation with δ on the coefficients (a, b, c, d, e, f) and on the facients $Q'T - QU$ and $-P'T + PU$ respectively; the third line vanishes of itself, and the other two destroy each other, that is,

$$\delta\ (a,\ b,\ c,\ d,\ e,\ f \between Q'T - QU,\ -P'T + PU)^5 = 0, \text{ and similarly}$$
$$\delta_1(a,\ b,\ c,\ d,\ e,\ f \between Q'T - QU,\ -P'T + PU)^5 = 0,$$

or the function $(a, b, c, d, e, f \between Q'T - QU,\ -PT + PU)^5$, treating therein T and U as constants, is an invariant, that is, the coefficients of the several terms thereof are all invariants.

313. The expressions for the coefficients (a, b, c, d, e, f) are in the first instance obtained in the forms

$$\begin{aligned}
\mathrm{a} &= \ \ 2(L + 5MC + 10C^2), \\
\mathrm{b} &= -2(L' + 3M'C)\,A, \\
\mathrm{c} &= \ \ 2(L + MC - 2C^2)\,A^{-1}, \\
\mathrm{d} &= -2(L' - M'C), \\
\mathrm{e} &= \ \ 2(L - 3MC + 2C^2)\,A^{-2}, \\
\mathrm{f} &= -2(L' - 5M'C)\,A^{-1},
\end{aligned}$$

where, developing M. Hermite's expressions,

$72L=$	$24M=$	$24L'=$	$24M'=$
$A^7B\ \ +\ \ 1$	$A^4B\ \ -\ \ 1$	$ABI\ \ +\ \ 1$	$I+1$
$A^6C^2\ \ +\ \ 1$	$A^3C\ \ -\ \ 1$	$CI\ \ +\ \ 5$	
$A^5B^2\ \ +\ \ 6$	$A^2B^2\ \ -\ \ 3$		
$A^4BC\ \ -\ \ 24$	$ABC\ \ +\ 12$		
$A^3B^2\ \ +\ \ 9$	$C^2\ \ +\ 24$		
$A^3C^2\ \ -\ \ 39$			
$A^2B^2C\ \ +\ \ 9$			
$ABC^2\ \ +\ 108$			
$C^3\ \ +\ \ 72$			

and substituting these values, we find

36a =	36b =	36c =	36d =	36e =	36f =
$A^7B\ \ +\ \ 1$	$A^2BI\ \ -\ \ 3$	$A^6B\ \ +\ \ 1$	$ABI\ \ -\ \ 3$	$A^5B\ \ +\ \ 1$	$BI\ -\ 3$
$A^6C^2\ \ +\ \ 1$	$ACI\ \ -\ 24$	$A^5C\ \ +\ \ 1$	$CI\ \ -\ 12$	$A^4C\ \ +\ \ 1$	
$A^5B^2\ \ +\ \ 6$		$A^4B^2\ \ +\ \ 6$		$A^3B^2\ \ +\ \ 6$	
$A^4BC\ \ -\ \ 39$		$A^3BC\ \ -\ \ 27$		$A^2BC\ \ -\ 15$	
$A^3B^3\ \ +\ \ 9$		$A^2B^3\ \ +\ \ 9$		$AB^3\ \ +\ \ 9$	
$A^3C^2\ \ -\ \ 54$		$A^2C^2\ \ -\ \ 42$		$AC^2\ \ -\ 30$	
$A^2BC\ \ -\ \ 36$		$BC^2\ \ +\ 144$		$B^2C\ \ +\ 36$	
$ABC^2\ \ +\ 288$					
$C^3\ \ +\ 1152$					

I have not thought it worth while to make in these formulæ the substitutions $A=J$, $B=-K$, $C=9L+JK$, which would give the expressions for (a, b, c, d, e, f) in terms of J, K, L.

314. Substituting for (x, y) their values in terms of (X, Y), we have

$$(a,\ b,\ c,\ d,\ e,\ f\between x,\ y)^5$$

$$=(a,\ b,\ c,\ d,\ e,\ f\between \frac{1}{2\sqrt{C}}\left(\frac{Q'}{\sqrt[4]{A}}+Q\sqrt[4]{A}\right)X+\frac{1}{2\sqrt{C}}\left(\frac{Q'}{\sqrt[4]{A}}-Q\sqrt[4]{A}\right)Y,$$

$$\frac{1}{2\sqrt{C}}\left(\frac{-P'}{\sqrt[4]{A}}-P\sqrt[4]{A}\right)X+\frac{1}{2\sqrt{C}}\left(-\frac{P'}{\sqrt[4]{A}}+P\sqrt[4]{A}\right)Y$$

$$=(\lambda,\ \mu,\ \nu,\ \nu',\ \mu',\ \lambda'\between X,\ Y)^5 \text{ suppose,}$$

and by what precedes

$$\alpha x^2+\beta xy+\gamma y^2=\ \sqrt{A}XY;$$

this gives

$$\alpha\partial_y{}^2-\beta\partial_y\partial_x+\gamma\partial_x{}^2=-\sqrt{A}\partial_X\partial_Y,$$

and thence

$$(\alpha\partial_y{}^2-\beta\partial_y\partial_x+\gamma\partial_x{}^2)^2\,(a,\ b,\ c,\ d,\ e,\ f\between x,\ y)^5$$

$$=A\partial_X{}^2\partial_Y{}^2\,(\lambda,\ \mu,\ \nu',\ \mu',\ \nu,\ \lambda'\between X,\ Y)^5$$

$$=120A\,(\nu X+\nu' Y);$$

the left-hand side is a linear covariant of the degree 5, it is consequently a mere numerical multiple of $Px+Qy$; and it is easy to verify that it is $=120(Px+Qy)$. (In fact writing $b=d=e=0$, the expression is $(3c^2\partial_y{}^2 - af\partial_y\partial_x)^2(ax^5+10cx^3y^2+fy^5)$, and the only term which contains x is $a^2f^2 . \partial_y{}^2\partial_x{}^2 . 10cx^3y^2 = 120a^2cf^2 . x$; but for $b=d=e=0$, Table J gives $Px = a^2cf^2x$, and the coefficient 120 is thus verified.) But $Px+Qy$ is $=\dfrac{\sqrt{C}}{\sqrt[4]{A}}(X+Y)$, and we have thus $A\nu = A\nu' = \dfrac{\sqrt{C}}{\sqrt[4]{A}}$, whence not only $\nu=\nu'$, $=\sqrt{k}$ suppose, but we have further $k=\dfrac{C}{\sqrt[4]{A^5}}$, a result given by M. Hermite.

315. Substituting for $\nu=\nu'$ the value $\sqrt{k}$, we have

$$\begin{aligned}
&(a,\ b,\ c,\ d,\ e,\ f\between x,\ y)^5 \\
&= (a,\ b,\ c,\ d,\ e,\ f\between \frac{1}{2\sqrt{C}}\left(\frac{Q'}{\sqrt[4]{A}}+Q\sqrt[4]{A}\right)X + \frac{1}{2\sqrt{C}}\left(\frac{Q'}{\sqrt[4]{A}}-Q\sqrt[4]{A}\right)Y, \\
&\qquad \frac{1}{2\sqrt{C}}\left(\frac{-P'}{\sqrt[4]{A}}-P\sqrt[4]{A}\right)X + \frac{1}{2\sqrt{C}}\left(-\frac{P'}{\sqrt[4]{A}}+P\sqrt[4]{A}\right)Y)^5, \\
&= (\lambda,\ \mu,\ \sqrt{k},\ \sqrt{k},\ \mu',\ \lambda'\between X,\ Y)^5,
\end{aligned}$$

and we have then $\alpha x^2+\beta xy+\gamma y^2 = \sqrt{A}XY$, viz. the left-hand side being the quadricovariant of $(a,\ b,\ c,\ d,\ e,\ f\between x,\ y)^5$, the equation shows that the quadricovariant of the form $(\lambda,\ \mu,\ \sqrt{k},\ \sqrt{k},\ \mu',\ \lambda'\between X,\ Y)^5$ is $=\sqrt{A}XY$, and we thus arrive at the starting-point of Hermite's theory.

316. The coefficients $(\lambda,\ \mu,\ \sqrt{k},\ \sqrt{k},\ \mu',\ \lambda')$ of Hermite's form are by what precedes *invariants*; they are consequently expressible in terms of the invariants A, B, C (and I). M. Hermite writes

$$\lambda\lambda' = g,\quad \mu\mu' = h,$$

and he finds

$$\sqrt{A} = g-3h+2k,\quad \frac{B}{\sqrt{A^3}} = h-k,\quad \frac{C}{\sqrt{A^5}} = k,$$

or, what is the same thing,

$$g = \frac{A^3+3AB+C}{\sqrt{A^5}},\quad h = \frac{AB+C}{\sqrt{A^5}},\quad k = \frac{C}{\sqrt{A^5}},$$

which give g, h, k in terms of A, B, C, and then putting

$$\Delta = (9k^2+16hk-gh)^2 - 24hk^3,\ = \frac{I^2}{A^7}$$

(the equation $I^2 = A^7\Delta$ is in fact equivalent to the before-mentioned expression of I^2

in terms of the other invariants), the coefficients $(\lambda, \mu, \mu', \lambda')$ are expressed in terms of g, h, k, that is of A, B, C, viz. we have

$$\left\{\begin{aligned} 72\sqrt{k^5}\lambda\ &= h(g-16k)^2 - 9k(g+16k) + (g-16k)\sqrt{\Delta}, \\ 24\sqrt{k^3}\mu\ &= 9k^2 + 16hk - gh - \sqrt{\Delta}, \\ 24\sqrt{k^3}\mu' &= 9k^2 + 16hk - gh + \sqrt{\Delta}, \\ 72\sqrt{k^5}\lambda' &= h(g-16k)^2 - 9k(g+16k) - (g-16k)\sqrt{\Delta}; \end{aligned}\right.$$

these values of $(\lambda, \mu, \mu', \lambda')$ could of course be at once expressed in terms of (J, K, L), but I have not thought it necessary to make the transformation.

317. It has been already noticed that the linear covariant $(C, = Px + Qy)$, was

$$= \sqrt{A}\,(\sqrt{k},\ \sqrt{k}\,\between X,\ Y),$$

it is to be added that the septic covariant $(P'x + Q'y)$ is

$$= \sqrt{A^3}\,(\sqrt{k},\ -\sqrt{k}\,\between X,\ Y),$$

and that the canonical forms of the cubicovariants $\phi_1(x, y)$, &c. are as follows:

$$\begin{aligned} \phi_1(X, Y) &= \sqrt{A}\,(\mu,\ \ 3\sqrt{k},\ \ 3\sqrt{k},\ \ \mu'\between X,\ Y)^3, \\ \phi_2(X, Y) &= \ A\,(\mu,\ \ \sqrt{k},\ -\sqrt{k},\ -\mu'\between X,\ Y)^3, \\ \{\phi_3(X, Y)\} &= \sqrt{A^3}\,(\mu,\ -\sqrt{k},\ -\sqrt{k},\ \ \mu'\between X,\ Y)^3, \\ \{\phi_4(X, Y)\} &= \ A^2\,(\mu,\ -3\sqrt{k},\ \ 3\sqrt{k},\ -\mu'\between X,\ Y)^3, \end{aligned}$$

$$\psi_1(X, Y) = \sqrt{A^3}\left(\begin{array}{l} (2\sqrt{k^3} - 3\mu k + \mu'\mu^2), \\ 3(\sqrt{k^3} + \mu\mu'\sqrt{k} - 2\mu k), \\ -3(\sqrt{k^3} + \mu\mu'\sqrt{k} - 2\mu' k), \\ -(2\sqrt{k^3} - 3\mu' k + \mu\mu'^2), \end{array}\right)\between X,\ Y)^3,$$

$$\phi_3(X, Y) = \sqrt{A^3}\,(5\mu,\ -\sqrt{k},\ \sqrt{k},\ 5\mu\between X,\ Y)^3,$$

$$\phi_4(X, Y) = \sqrt{A^3}\left(\begin{array}{l} (7\sqrt{A}\mu + 96(2\sqrt{k^3} - 3\mu k + \mu'\mu^2)), \\ -3(3\sqrt{A}\sqrt{k} - 96(\sqrt{k^3} + \mu\mu'\sqrt{k} - 2\mu k)), \\ +3(3\sqrt{A}\sqrt{k} - 96(\sqrt{k^3} + \mu\mu'\sqrt{k} - 2\mu' k)), \\ -(7\sqrt{A}\mu' + 96(2\sqrt{k^3} - 3\mu' k + \mu\mu'^2)) \end{array}\right)\between X,\ Y)^3$$

or, as the last formula may also be written,

$$\phi_4(X, Y) = \sqrt{A^3}\left(\begin{array}{l} ((7g - 53h + 110k)\mu - 64\lambda\mu'\sqrt{k}), \\ -3((3g + 151h - 90k)\sqrt{k} - 64\lambda'\mu^2), \\ +3((3g + 151h - 90k)\sqrt{k} - 64\lambda\mu'^2), \\ -((7g - 53h + 110k)\mu' - 64\lambda'\mu\sqrt{k}) \end{array}\right)\between X,\ Y)^3.$$

It is in fact by means of these comparatively simple canonical expressions that M. Hermite was enabled to effect the calculation of the coefficient 𝔄.

Article Nos. 318 to 326.—*Theory of the imaginary linear transformations which lead to a real equation.*

318. An equation $(a, b, c, \ldots \between x, y)^n = 0$ is real if the ratios $a : b : c$, &c. of the coefficients are all real. In speaking of a given real equation there is no loss of generality in assuming that the coefficients $(a, b, c, \ldots)$ are all real; but if an equation presents itself in the form $(a, b, c, \ldots \between x, y)^n = 0$ with imaginary coefficients, it is to be borne in mind that the equation may still be real; viz. the coefficients may contain an imaginary common factor in such wise that throwing this out we obtain an equation with real coefficients.

In what follows I use the term *transformation* to signify a linear transformation, and speak of equations connected by a linear transformation as *derivable* from each other. An imaginary transformation will in general convert a real into an imaginary equation; and if the proposition were true universally,—viz. if it were true that the transformed equation was always imaginary—it would follow that a real equation derivable from a given real equation could then be derivable from it only by a real transformation, and that the two equations would have the same character. But any two equations having the same absolute invariants are derivable from each other, the two real equations would therefore be derivable from each other by a real transformation, and would thus have the same character; that is, all the equations (if any) belonging to a given system of values of the absolute invariants would have a determinate character, and the absolute invariants would form a system of auxiliars.

But it is not true that the imaginary transformation leads always to an imaginary equation; to take the simplest case of exception, if the given real equation contains only even powers or only odd powers of x, then the imaginary transformation $x : y$ into $ix : y$ gives a real equation. And we are thus led to inquire in what cases an imaginary transformation gives a real equation.

319. I consider the imaginary transformation $x : y$ into

$$(a+bi)\,x + (c+di)\,y \;:\; (e+fi)\,x + (g+hi)\,y,$$

or, what is the same thing, I write

$$x = (a+bi)\,X + (c+di)\,Y,$$
$$y = (e+fi)\,X + (g+hi)\,Y,$$

and I seek to find P, Q real quantities such that $Px + Qy$ may be transformed into a linear function $RX + SY$, wherein the ratio $R : S$ is real, or, what is the same thing, such that $RX + SY$ may be the product of an imaginary constant into a real linear function of (X, Y). This will be the case if

$$Px + Qy = (1+\theta i)\,\{P\,(aX + cY) + Q\,(eX + gY)\},$$

that is if

$$P(bX+dY)+Q(fX+hY)=\theta\{P(aX+cY)+Q(eX+fY)\},$$

which implies the relations

$$bP+fQ=\theta(aP+eQ),$$
$$dP+hQ=\theta(cP+gQ),$$

or, what is the same thing,

$$(b-a\theta)P+(f-e\theta)Q=0,$$
$$(d-c\theta)P+(h-g\theta)Q=0,$$

and if the resulting value of $P:Q$ be real, the last-mentioned equations give

$$(ag-ce)\theta^2-(ah+bg-cf-de)\theta+bh-df=0,$$

and θ being known, the ratio $P:Q$ is determined rationally in terms of θ.

320. The equation in θ will have its roots real, equal, or imaginary, according as

$$(ah+bg-cf-de)^2-4(ag-ce)(bh-df),$$

that is

$$\begin{aligned}&a^2h^2+b^2g^2+c^2f^2+d^2e^2\\&-2ahbg-2ahcf-2ahde-2bgcf-2bgde-2cfde\\&+4adfgh+4bceh\end{aligned}$$

is $=+$, $=0$, or $=-$; and I say that the transformation is subimaginary, neutral, and superimaginary in these three cases respectively. In the subimaginary case there are two functions $Px+Qy$ which satisfy the prescribed conditions; in the neutral case a single function; in the superimaginary case no such function. But in the last-mentioned case there are two conjugate imaginary functions, $Px+Qy$, which contain as factors thereof respectively two conjugate imaginary functions $UX+VY$.

321. Hence replacing the original x, y, X, Y by real linear functions thereof, the subimaginary transformation is reduced to the transformation $x:y$ into $kX:Y$, where k is imaginary; and the superimaginary transformation is reduced to $x+iy:x-iy$ into $k(X+iY):(X-iY)$, where k is imaginary. As regards the neutral transformation, it appears that this is equivalent to

$$x=(a+bi)X+(c+di)Y,$$
$$y=\qquad\qquad(g+hi)Y,$$

with the condition $0=(ah+bg)^2-4agbh$, $=(ah-bg)^2$, that is, we have $ah-bg=0$, or without any real loss of generality $g=a$, $h=b$, or the transformation is

$$x=(a+bi)X+(c+di)Y,$$
$$y=\qquad\qquad(a+bi)Y,$$

that is, $x:y=X+kY:Y$, where k is imaginary.

322. The original equation after any real transformation thereof, is still an equation of the form

$$(a, \ldots \between x, y)^n = 0;$$

and if we consider first the neutral transformation, the transformed equation is

$$(a, \ldots \between X + kY, Y)^n = 0;$$

this is not a real equation except in the case where k is real.

323. For the superimaginary transformation, starting in like manner from $(a, \ldots \between x, y)^n = 0$, this may be expressed in the form

$$(\alpha + \beta i, \gamma + \delta i, \ldots, \gamma - \delta i, \alpha - \beta i \between x + iy, x - iy)^n = 0,$$

viz. when in a real equation $(x, y)^n = 0$ we make the transformation $x : y$ into $x + iy : x - iy$, the coefficients of the transformed equation will form as above pairs of conjugate imaginaries. Proceeding in the last-mentioned equation to make the transformation $x + iy : x - iy$ into $k(X + iY) : X - iY$, I throw k into the form

$$\cos 2\phi + i \sin 2\phi, \quad = (\cos\phi + i \sin\phi) \div (\cos\phi - i \sin\phi)$$

(of course it is not here assumed that ϕ is real), or represent the transformation as that of $x + iy : x - iy$ into $(\cos\phi + i \sin\phi)(X + iY) : (\cos\phi - i\sin\phi)(X - iY)$; the transformed equation thus is

$$(\alpha + \beta i, \ldots \alpha - \beta i \between (\cos\phi + i \sin\phi)(X + iY), \quad (\cos\phi - i\sin\phi)(X - iY))^n = 0.$$

The left-hand side consists of terms such as $(X^2 + Y^2)^{n-2s}$ into

$$(\gamma + \delta i)(\cos s\phi + i \sin s\phi)(X + iY)^s + (\gamma - \delta i)(\cos s\phi - i \sin s\phi)(X - iY)^s,$$

viz. the expression last written down is

$$= (\gamma \cos s\phi - \delta \sin s\phi)\{(X + iY)^s + (X - iY)^s\}$$

$$- (\gamma \sin s\phi + \delta \cos s\phi)\left\{\frac{(X + iY)^s - (X - iY)^s}{s}\right\},$$

and observing that the expressions in { } are real, the transformed equation is only real if $(\gamma \cos s\phi - \delta \sin s\phi) \div (\gamma \sin s\phi + \delta \cos s\phi)$ be real, that is, in order that the transformed equation may be real, we must have $\tan s\phi =$ real; and observing that if $\tan s\phi$ be equal to any given real quantity whatever, then the values of $\tan\phi$ are all of them real, and that $\tan\phi$ real gives $\cos\phi$ and $\sin\phi$ each of them real, and therefore also ϕ real, it appears that the transformed equation is only real for the transformation

$$x + iy : x - iy = (\cos\phi + i \sin\phi)(X + iY) : (\cos\phi - i \sin\phi)(X - iY),$$

wherein ϕ is real; and this is nothing else than the *real* transformation $x : y$ into $X \cos\phi - Y \sin\phi : X \sin\phi + Y \cos\phi$. Hence neither in the case of the neutral transformation or in that of the superimaginary transformation can we have an imaginary transformation leading to a real equation.

324. There remains only the subimaginary transformation, viz. this has been reduced to $x : y$ into $kX : Y$, the transformed equation is

$$(a, \ldots \between kX,\ Y)^n = 0,$$

and this will be a real equation if some power k^p of k (p not greater than n) be real, and if the equation $(a, .. \between x,\ y)^n = 0$ contain only terms wherein the index of x (or that of y) is a multiple of p. Assuming that it is the index of y which is a multiple, the form of the equation is in fact $x^\alpha (x^p,\ y^p)^m = 0$, $(n = mp + \alpha)$, and the transformed equation is $X^\alpha (k^p X^p,\ Y^p)^m = 0$, which is a real equation.

325. It is to be observed that if p be odd, then writing $k^p = K$ (K real) and taking k' the real p-th root of K, then the very same transformed equation would be obtained by the real transformation $x : y$ into $k'X : Y$; so that the equation obtained by the imaginary transformation, being also obtainable by a real transformation, has the same character as the original equation.

326. Similarly if p be even, if K be real and positive, the equation $k^p = K$ has a real root k' which may be substituted for the imaginary k, and the transformed equation will have the same character as the original equation; but if K be negative, say $K = -1$ (as may be assumed without loss of generality), then there is no real transformation equivalent to the imaginary transformation, and the equation given by the imaginary transformation has not of necessity the same character as the original equation; and there are in fact cases in which the character is altered. Thus if $p = 2$, and the original equation be $x (x^2,\ y^2)^m = 0$, or $(x^2,\ y^2)^m = 0$, then making the transformation $x : y$ into $iX : Y$, the transformed equation will be $X (X^2,\ -Y^2)^m = 0$ or $(X^2,\ -Y^2)^m = 0$, giving imaginary roots $X^2 + aY^2 = 0$ corresponding to real roots $x^2 - ay^2 = 0$.

Article No. 327.—*Application to the auxiliars of a quintic.*

327. Applying what precedes to a quintic equation $(a, \ldots . \between x,\ y)^5 = 0$, this after any real transformation whatever will assume the form $(a', \ldots \between x',\ y')^5 = 0$; and the only cases in which we can have an imaginary transformation producing a real equation of an altered character is when this equation is $(a',\ 0,\ c',\ 0,\ e',\ 0 \between x',\ y')^5 = 0$ (c' not $= 0$), or when it is $(a',\ 0,\ 0,\ 0,\ e',\ 0 \between x',\ y')^5 = 0$, viz. when it is $x' (a'x'^4 + 10c'x'^2y'^2 + 5e'y'^4) = 0$, or $x'(a'x'^4 + 5e'y'^4) = 0$. In the latter case the transformation x', y' into $X \sqrt[4]{-1} : Y$ gives the real equation $X (a'X^4 - 5e'Y^4) = 0$. I observe however that for the form $(a',\ 0,\ 0,\ 0,\ e',\ 0 \between x,\ y)^4$, and consequently for the form $(a, \ldots \between x,\ y)^5$ from which it is derived we have $J = 0$; this case is therefore excluded from consideration. The remaining case is $(a',\ 0,\ c',\ 0,\ e',\ 0 \between x',\ y')^5 = 0$, which is by the imaginary transformation $x' : y'$ into $iX : Y$ converted into $(a',\ 0,\ -c',\ 0,\ e',\ 0 \between X,\ Y)^5 = 0$; for the first of the two forms we have $J = 16\, a'c'e'^2$, and for the second of the two forms $J = -16\, a'c'e'^2$, that is, the two values of J have opposite signs. Hence considering an equation $(a,\ b,\ c,\ d,\ e,\ f \between x,\ y)^5 = 0$ for which J is not $= 0$, whenever this is by an imaginary transformation converted into a real equation, the sign of J is reversed; and it follows that, given the values of the absolute invariants and the value of J (or what is sufficient, the sign of J), the

different real equations which correspond to these data must be derivable one from another by real transformations, and must consequently have a determinate character; that is, the Absolute Invariants, and J, constitute a system of auxiliars.

ANNEX.—*Analytical Theorem in relation to a Binary Quantic of any Order.*

The foregoing theory of the superimaginary transformation led me to a somewhat remarkable theorem. Take for example the function

$$(a,\ b,\ c\between x+k,\ 1-kx)^2,$$

or, as this may be written,

$$\begin{array}{c|ccc} & k^2 & k & 1 \\ \hline x^2 & c, & 2b, & a \\ x & 2b, & 2a-2c, & -2b \\ 1 & a, & -2b, & c, \end{array} \quad \text{or} \quad \left(\begin{array}{ccc} c, & 2b, & a \\ 2b, & 2a-2c, & -2b \\ a, & -2b, & c \end{array}\between k,\ 1)^2 (x,\ 1)^2,$$

then the determinant

$$\left|\begin{array}{ccc} c, & 2b, & a \\ 2b, & 2a-2c, & -2b \\ a, & -2b, & c \end{array}\right|$$

is a product of linear functions of the coefficients $(a,\ b,\ c)$; its value in fact is

$$=-2(a+c)(a+2bi+ci^2)(a-2bi+ci^2),\ =-2(a+c)[(a-c)^2+4b^2].$$

To prove this directly, I write

$$\begin{aligned} a' &= a-2bi+ci^2, \\ b' &= a \qquad\ -ci^2, \\ c' &= a+2bi+ci^2, \end{aligned}$$

and we then have

$$\left|\begin{array}{ccc} c, & 2b, & a \\ 2b, & 2a-2c, & -2b \\ a, & -2b, & c \end{array}\right| \left|\begin{array}{ccc} 1, & 2\ , & 1 \\ i\ , & 0\ , & -i \\ i^2, & -2i^2, & i^2 \end{array}\right|$$

$$=\begin{array}{l|l} & (1,\ i,\ i^2),\ (2,\ 0,\ -2i^2),\ (1,\ -i,\ i^2) \\ \hline (\ c, \quad 2b, \quad a) & \\ (2b, \quad 2a-2c, \quad -2b) & \\ (\ a, \quad -2b, \quad c) & \end{array},$$

$$=\left|\begin{array}{ccc} i^2a', & -2i^2b', & i^2c' \\ 2ia', & 0b', & -2ic' \\ a', & 2b', & c' \end{array}\right|, \quad =a'b'c'\left|\begin{array}{ccc} i^2, & -2i^2, & i^2 \\ 2i, & 0\ , & -2i \\ 1, & 2\ , & 1 \end{array}\right|,$$

whence observing that the determinants

$$\begin{vmatrix} 1, & 2, & 1 \\ i, & 0, & -i \\ i^2, & -2i^2, & i^2 \end{vmatrix}, \quad \begin{vmatrix} i^2, & -2i^2, & i^2 \\ 2i, & 0, & -2i \\ 1, & 2, & 1 \end{vmatrix}$$

are as $1 : -2$, we have the required relation,

$$\begin{vmatrix} c, & 2b, & a \\ 2b, & 2a-2c, & -2b \\ a, & -2b, & c \end{vmatrix} = -2a'b'c', \; = -2(a+c)\{(a-c)^2+4b^2\}.$$

It is to be remarked that the determinant

$$\begin{vmatrix} 1, & 2, & 1 \\ i, & 0, & -i \\ i^2, & -2i^2, & i^2 \end{vmatrix}, \text{ taken as the multiplier of } \begin{vmatrix} c, & 2b, & a \\ 2b, & 2a-2c, & -2b \\ a, & -2b, & c \end{vmatrix}$$

is obtained by writing therein $a=b=c, =1$; and multiplying the successive lines thereof by $1, \frac{1}{2}i, i^2$ ($1, \frac{1}{2}, 1$ are the reciprocals of the binomial coefficients 1, 2, 1), the proof is the same, and the multiplier is obtained in the like manner for a function of any order; thus for the cubic $(a, b, c, d \between k+x, 1-kx)^3$,

	k^3	k^2	k	1
$= x^3$	$-d,$	$3c,$	$-3b,$	a
x^2	$3c,$	$-6b+3d,$	$3a-6c,$	$3b$
x	$-3b,$	$3a-6c,$	$6b-3d,$	$3c$
1	$a,$	$3b,$	$3c,$	d

the multiplier is obtained from the determinant by writing therein $a=b=c=d=1$, and multiplying the successive lines by $1, \frac{1}{3}i, \frac{1}{3}i^2, i^3$, viz. the multiplier is

$$\begin{vmatrix} -1, & 3, & -3, & 1 \\ i, & -i, & -i, & i \\ -i^2, & -i^2, & i^2, & i^2 \\ i^3, & 3i^3, & 3i^3, & i^3 \end{vmatrix}$$

and the value of the determinant is found to be

$$9(a-3bi+3ci^2-di^3)(a-bi-ci^2+di^3)(a+bi-ci^2-di^3)(a+3bi+3ci^2+di^3),$$
$$= 9\left((a-3c)^2+(3b-d)^2\right)\left((a+c)^2+(b+d)^2\right).$$

But the theory may be presented under a better form; take for instance the cubic, viz. writing $\dfrac{x}{y}$ and $\dfrac{k}{l}$ for x and k respectively, we have $(a, b, c, d \between ky + lx,\ ly - kx)^3$

	k^3	k^2l	kl^2	l^3
$= x^3$	$-d,$	$3c,$	$-3b,$	a
x^2y	$3c,$	$-6b+3d,$	$3a-6c,$	$3b$
xy^2	$-3b,$	$3a-6c,$	$6b-3d,$	$3c$
y^3	$a,$	$3b,$	$3c,$	d

a bipartite cubic function $(* \between k, l)^3 (x, y)^3$; and the determinant formed out of the matrix is at once seen to be an invariant of this bipartite cubic function.

Assume now that we have identically

$$(a, b, c, d \between x, y)^3 = (a', b', c', d' \between \tfrac{1}{2}(x+iy),\ \tfrac{1}{2}(x-iy))^3,$$

viz. this equation written under the equivalent form

$$(a', b', c', d' \between X, Y)^3 = (a, b, c, d \between X+Y,\ i(X-Y))^3,$$

determines (a', b', c', d') as linear functions of (a, b, c, d), it in fact gives

$$\begin{aligned}
a' &= (a, b, c, d \between 1, -i)^3 &&= a - 3bi + 3ci^2 - di^3,\\
b' &= (a, b, c, d \between 1, -i)^2(1, i) &&= a - bi - ci^2 + di^3,\\
c' &= (a, b, c, d \between 1, -i)(1, i)^2 &&= a + bi - ci^2 - di^3,\\
d' &= (a, b, c, d \between 1, i)^3 &&= a + 3bi + 3ci^2 + di^3,
\end{aligned}$$

then observing that $ky + lx \pm i(ly - kx) = (x \pm iy)(\mp ik + l)$, we have

$$(a, b, c, d \between ky+lx,\ ly-kx)^3 = (a', b', c', d' \between \tfrac{1}{2}(x+iy)(-ik+l),\ \tfrac{1}{2}(x-iy)(ik+l))^3,$$

and if in the expression on the right-hand side we make the linear transformations

$$\begin{aligned}
x+iy &= x'\sqrt{2}, & -ik+l &= k'\sqrt{2},\\
x-iy &= -iy'\sqrt{2}, & ik+l &= -il'\sqrt{2},
\end{aligned}$$

which are respectively of the determinant $+1$, the transformed function is

$$= (a', b', c', d' \between k'x',\ -l'y')^3,$$

that is, we have

$$(a, b, c, d \between ky+lx,\ ly-kx)^3 = (a', b', c', d' \between k'x',\ -l'y')^3.$$

The last-mentioned function is

	k'^3	k'^2l'	$k'l'^2$	l'^3
x'^3	a'	.	.	.
x'^2y'	.	$-3b'$	.	.
$x'y'^2$	.	.	$+3c'$	.
y'^3	.	.	.	$-d'$

and (from the invariantive property of the determinant) the original determinant is equal to the determinant of this new form, viz. we have

$$\begin{vmatrix} -d, & 3c, & -3b, & a \\ 3c, & -6b+3d, & 3a-6c, & 3b \\ -3b, & 3a-6c, & 6b-3d, & 3c \\ a, & 3b, & 3c, & d \end{vmatrix} = 9a'b'c'd',$$

$$= 9\,[(a-3c)^2+(3b-d)^2]\,[(a+c)^2+(b+d)^2],$$

which is the required theorem. And the theorem is thus exhibited in its true connexion, as depending on the transformation

$$(a, \ldots \between x, y)^n = (a', \ldots \between \tfrac{1}{2}(x+iy),\ \tfrac{1}{2}(x-iy))^n.$$

Addition, 7th *October*, 1867.

Since the present Memoir was written, there has appeared the valuable paper by MM. Clebsch and Gordan "Sulla rappresentazione tipica delle forme binarie," *Annali di Matematica*, t. I. (1867) pp. 23—79, relating to the binary quintic and sextic. On reducing to the notation of the present memoir the formula 95 for the representation of the quintic in terms of the covariants α, β, which should give for (a, b, c, d, e, f) the values obtained *ante*, No. 312, I find a somewhat different system of values; viz. these are

36a =	36b =	36c =	36d =	36e =	35f =
A^7B + 1	$*A^4I$ − 1	A^6B + 1	$*A^3I$ − 1	A^5B + 1	$*A^2I$ − 1
A^6C + 1	A^2BI − 3	A^5C + 1	ABI − 3	A^4C + 1	ABI − 3
A^5B^2 + 6	$*ACI$ + 24	A^4B^2 + 6	$*CI$ + 12	A^3B^2 + 6	
A^4BC − 39		A^3BC − 27		A^2BC − 15	
A^3B^3 + 9		A^2B^3 + 9		AB^3 + 9	
A^3C^2 − 54		A^2C^2 − 42		AC^2 − 30	
A^2B^2C − 126		$*AB^2C$ − 90		$*B^2C$ − 54	
ABC^2 + 288		BC^2 + 144			
C^3 + 1152					

where I have distinguished with an asterisk the terms which have different coefficients in the two formulæ. I cannot at present explain this discrepancy.

Fig. 1.

Fig. 2.

The lower cusp of the Bicorn is drawn out of its true position which is much further off along the asymptote, the co-ordinates in fact are $x=-76\frac{23}{27}$, $y=-44\frac{2}{3}$ *(the co-ordinates of the upper or node-cusp being* $-1, 1$.

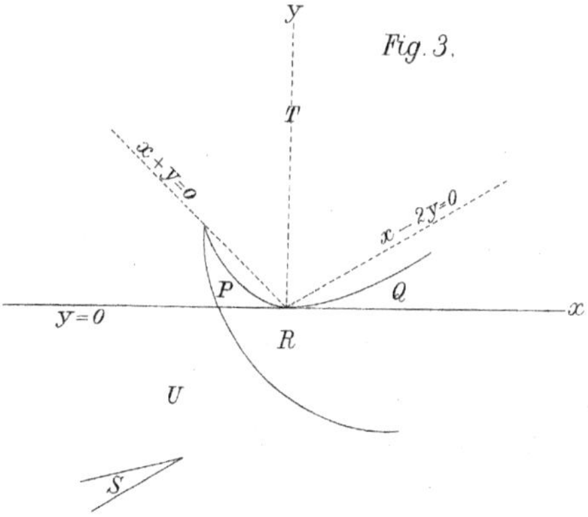

Fig. 3.

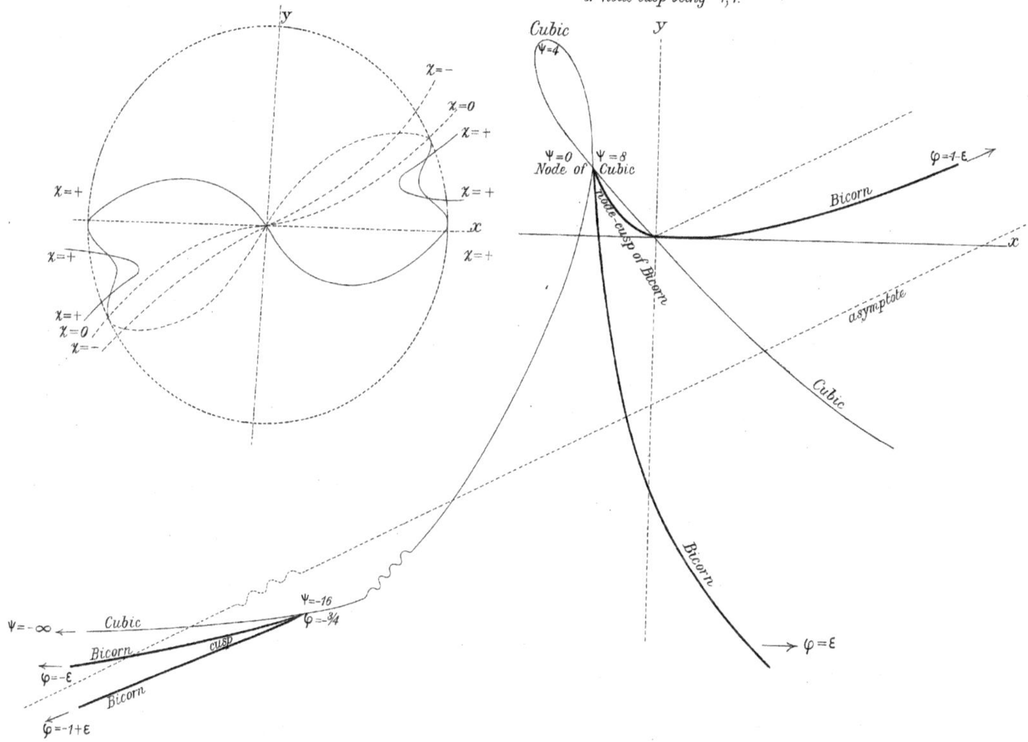

Fig. 4.

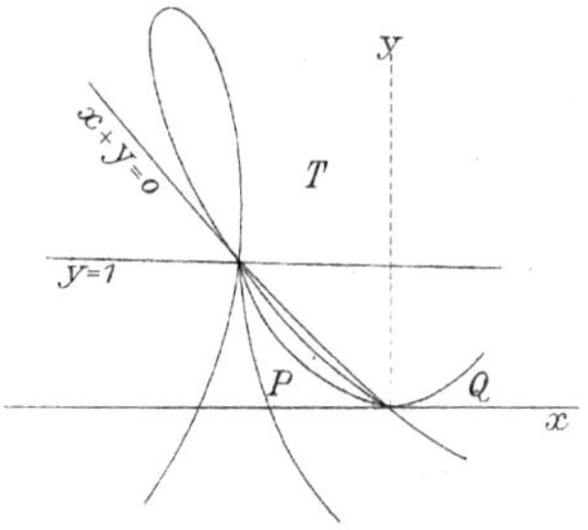

406.

ON THE CURVES WHICH SATISFY GIVEN CONDITIONS.

[From the *Philosophical Transactions of the Royal Society of London*, vol. CLVIII. (for the year 1868), pp. 75—143. Received April 18,—Read May 2, 1867.]

THE present Memoir relates to portions only of the subject of the curves which satisfy given conditions; but any other title would be too narrow: the question chiefly considered is that of finding the number of the curves which satisfy given conditions; the curves are either curves of a determinate order r (and in this case the conditions chiefly considered are conditions of contact with a given curve), or else the curves are conics; and here (although the conditions chiefly considered are conditions of contact with a given curve or curves) it is necessary to consider more than in the former case the theory of conditions of any kind whatever. As regards the theory of conics, the Memoir is based upon the researches of Chasles and Zeuthen, as regards that of the curves of the order r, upon the researches of De Jonquières: the notion of the quasi-geometrical representation of conditions by means of loci in hyper-space is employed by Salmon in his researches relating to the quadric surfaces which satisfy given conditions. The papers containing the researches referred to are included in the subjoined list. I reserve for a separate Second Memoir the application to the present question, of the Principle of Correspondence.

List of Memoirs and Works relating to the Curves which satisfy given conditions, with remarks

De Jonquières: "Théorèmes généraux concernant les courbes géométriques planes d'un ordre quelconque," *Liouv.* t. VI. (1861), pp. 113—134. In this valuable memoir is established the notion of a series of curves of the *index* N; viz. considering the curves of the order n which satisfy $\frac{1}{2}n(n+3)-1$ conditions, then if N denotes how many there are of these curves which pass through a given arbitrary point, the series is said to be of the index N.

In Lemma IV it is stated that all the curves C_n of a series of the index N can be analytically represented by an equation $F(y, x) = 0$, which is rational and integral of the degree N in regard to a variable parameter λ: this is not the case; see Annex No. 1.

Chasles: Various papers in the *Comptes Rendus*, t. LVIII. *et seq.* 1864—67. The first of them (Feb. 1864), entitled "Détermination du nombre des sections coniques qui doivent toucher cinq courbes données d'ordre quelconque, ou satisfaire à diverses autres conditions," establishes the notion of the two characteristics (μ, ν) of a system of conics which satisfy four conditions; viz. μ is the number of these conics which pass through a given arbitrary point, and ν the number of them which touch a given arbitrary line. The Principle of Correspondence for points on a line is established in the paper of June—July 1864. Many of the leading points of the theory are reproduced in the present Memoir. The series of papers includes one on the conics in space which satisfy seven conditions (Sept. 1865), and another on the surfaces of the second order which satisfy eight conditions (Feb. 1866).

Salmon: "On some Points in the Theory of Elimination," *Quart. Math. Journ.* t. VII. pp. 327—337 (Feb. 1866); "On the Number of Surfaces of the Second Degree which can be described to satisfy nine Conditions," Ibid. t. VIII. pp. 1—7 (June 1866),—which two papers are here referred to on account of the notion which they establish of the quasi-geometrical representation of conditions by means of loci in hyper-space.

Zeuthen: Nyt Bidrag... Contribution to the Theory of Systems of Conics which satisfy four conditions, 8°. pp. 1—97 (Copenhagen, Cohen, 1865), translated, with an addition, in the *Nouvelles Annales.*

The method employed depends on the determination of the line-pairs and point-pairs, and of the numerical coefficients by which these have to be multiplied, in the several systems of conics which satisfy four conditions of contact with a given curve or curves. It is reproduced in detail, with the enumeration called "Zeuthen's Capitals," in the present Memoir.

Cayley: "Sur les coniques déterminées par cinq conditions d'intersection avec une courbe donnée," *Comptes Rendus,* t. LXIII. pp. 9—12, July 1866. Results reproduced in the present Memoir.

De Jonquières: Two papers, *Comptes Rendus,* t. LXIII. Sept. 1866, reproduced and further developed in the "Mémoire sur les contacts multiples d'ordre quelconque des courbes du degré r qui satisfont à des conditions données de contact avec une courbe fixe du degré m; suivi de quelques réflexions sur la solution d'un grand nombre de questions concernant les propriétés projectives des courbes et des surfaces algébriques," *Crelle,* t. LXVI. (1866), pp. 289—322,—contain a general formula for the number of curves C^r having contacts of given orders $a, b, c, ..$ with a given curve U^m, which formula is referred to and considered in the present Memoir.

De Jonquières: Recherches sur les séries ou systèmes de courbes et de surfaces algébriques d'ordre quelconque; suivies d'une réponse &c. 4°. Paris, Gauthier Villars, 1866 (1).

1 The foregoing list is not complete, and the remarks are not intended to give even a sketch of the contents of the works comprised therein, but only to show their bearing on the present Memoir.

Article Nos. 1 to 23.—*On the quasi-geometrical representation of Conditions.*

1. A condition imposed upon a subject gives rise to a relation between the parameters of the subject; for instance, the subject may be, as in the present Memoir, a plane curve of a given order, and the parameters be any arbitrary parameters contained in the equation of the curve. The condition may be onefold, twofold,... or, generally, k-fold, and the corresponding relation is onefold, twofold, ... or k-fold accordingly. Two or more conditions, each of a given manifoldness, may be regarded as forming together a single condition of a higher manifoldness, and the corresponding relations as forming a single relation; and thus, though it is often convenient to consider two or more conditions or relations, this case is in fact included in that of a k-fold condition or relation. In dealing with such a condition or relation it is assumed that the number of parameters is at least $=k$; for otherwise there would not in general be any subject satisfying the condition: when the number of parameters is $=k$, the number of subjects satisfying the condition is in general determinate.

2. A subject which satisfies a given condition may for shortness be termed a solution of the condition; and in like manner any set of values of the parameters satisfying the corresponding relation may be termed a solution of the relation. Thus for a k-fold condition or relation, and the same number k of parameters, the number of solutions is in general determinate.

3. A condition may in some cases be satisfied in more than a single way, and if a certain way be regarded as the ordinary and proper one, then the others are *special* or *improper:* the two epithets may be used conjointly, or either of them separately, almost indifferently. For instance, the condition that a curve shall touch a given curve (have with it a two-pointic intersection) is satisfied if the curve have with the given curve a proper contact; or if it have on the given curve a node or a cusp (or, more specially, if it be or comprise as part of itself two coincident curves); or if it pass through a node or a cusp of the given curve: the first is regarded as the ordinary and proper way of satisfying the condition; the other two as special or improper ways; and the corresponding solutions are ordinary and proper solutions, or special or improper ones accordingly. This will be further explained in speaking of the locus which serves for the representation of a condition.

4. A set of any number, say ω, of parameters may be considered as the coordinates of a point in ω-dimensional space; and if the parameters are connected by a onefold, twofold, ... or k-fold relation, then the point is situate on a onefold, twofold, ... or k-fold locus accordingly; to the relation made up of two or more relations corresponds the locus which is the intersection or common locus of the loci corresponding to the several component relations respectively. A locus is at most ω-fold, viz. it is in this case a point-system. The relation made up of a k-fold relation, an l-fold relation, &c., is in general $(k+l+\&c.)$ fold, and the corresponding locus is $(k+l+\&c.)$ fold accordingly.

5. The order of a point-system is equal to the number of the points thereof, where, of course, coincident points have to be attended to, so that the distinct points of the system may have to be reckoned each its proper number of times. The locus

corresponding to any linear j-fold relation between the coordinates is said to be a j-fold *omal* locus; and if to any given k-fold relation we join an arbitrary $(\omega-k)$ fold linear relation, that is, intersect the k-fold locus by an arbitrary $(\omega-k)$ fold omal locus, so as to obtain a point-system, the order of the k-fold relation or locus is taken to be equal to the number of points of the point-system, that is, to the order of the point-system. And this being so, if a k-fold relation, an l-fold relation, &c. are *completely* independent, that is, if they are not satisfied by values which satisfy a less than $(k+l+\&c.)$ fold relation, or, what is the same thing, if the k-fold locus, the l-fold locus, &c., have no common less than $(k+l+\&c.)$ fold locus, then the relations make up together a $(k+l+\&c.)$ fold relation, and the loci intersect in a $(k+l+\&c.)$ fold locus, the orders whereof are respectively equal to the product of the orders of the given relations or loci. In particular if we have $k+l+\&c.=\omega$, then we have an ω-fold relation, and corresponding thereto a point-system, the orders whereof are respectively equal to the product of the orders of the given relations or loci.

6. A k-fold relation, an l-fold relation, &c., if they were together equivalent to a less than $(k+l+\&c.)$ fold relation, would not be independent; but the relations, assumed to be independent, may yet *contain* a less than $(k+l+\&c.)$ fold relation, that is, they may be satisfied by the values which satisfy a certain less than $(k+l+\&c.)$ fold relation (say the common relation), and exclusively of these, only by the values which satisfy a proper $(k+l+\&c.)$ fold relation, which is, so to speak, a residual equivalent of the given relations. This is more clearly seen in regard to the loci; the k-fold locus, the l-fold locus, &c. may have in common a less than $(k+l+\&c.)$ fold locus, and besides intersect in a residual $(k+l+\&c.)$ fold locus. (It is hardly necessary to remark that such a connexion between the relations is precisely what is excluded by the foregoing definition of *complete* independence.) In particular if $k+l+\&c.=\omega$, the several loci may intersect, say in an $(\omega-j)$ fold locus, and besides in a residual ω-fold locus, or point-system. The order (in any such case) of the residual relation or locus is equal to the product of the orders of the given relations or loci, *less* a reduction depending on the nature of the common relation or locus, the determination of the value of which reduction is often a complex and difficult problem.

7. Imagine a curve of given order, the equation of which contains ω arbitrary parameters: to fix the ideas, it may be assumed that these enter into the equation rationally, so that the values of the parameters being given, the curve is uniquely determined. Suppose, as above, that the parameters are taken to be the coordinates of a point in ω-dimensional space; so long as the curve is not subjected to any condition, the point in question, say the parametric point, is an arbitrary point in the ω-dimensional space; but if the curve be subjected to a onefold, twofold,... or k-fold condition, then we have a onefold, twofold,... or k-fold relation between the parameters, and the parametric point is situate on a onefold, twofold,... or k-fold locus accordingly: to each position of the parametric point on the locus there corresponds a curve satisfying the condition, that is, a solution of the condition. In the case where the condition is ω-fold, the locus is a point-system, and corresponding to each point of the point-system we have a solution of the condition; the number of solutions is equal to the number of points of the point-system.

8. Considering the general case where the condition, and therefore also the locus, is k-fold, it is to be observed that every solution whatever, and therefore each special solution (if any), corresponds to some point on the k-fold locus; we may therefore have on the k-fold locus what may be termed "special loci," viz. a special locus is a locus such that to each point thereof corresponds a special solution. A special locus may of course be a point-system, viz. there are in this case a determinate number of special solutions corresponding to the several points of this point-system. We may consider the other extreme case of a special k-fold locus, viz. the k-fold locus of the parametric point may break up into two distinct loci, the special k-fold locus, and another k-fold locus the several points whereof give the ordinary solutions: we can in this case get rid of the special solutions by attending exclusively to the last-mentioned k-fold locus and regarding it as the proper locus of the parametric point. But if the special locus be a more than k-fold locus, that is, if it be not a part of the k-fold locus itself, but (as supposed in the first instance) a locus on this locus, then the special solutions cannot be thus got rid of: we have the k-fold locus of the parametric point, a locus such that to every point thereof there corresponds a proper solution, save and except that to the points lying on the special locus there correspond special or improper solutions. It is to be noticed that the special locus may be, but that is not in every case, a singular locus on the k-fold locus.

9. Suppose that the conditions to be satisfied by the curve are a k-fold condition, an l-fold condition, &c. of a total manifoldness $=\omega$. If the conditions are *completely* independent (that is, if the corresponding relations, *ante*, No. 5, are completely independent), we have a k-fold locus, an l-fold locus, &c., having no common locus other than the point-system of intersection, and the number of curves which satisfy the given conditions, or (as this has been before expressed) the number of solutions, is equal to the number of points of the point-system, or to the order of the point-system, viz. it is equal to the product of the orders of the loci which correspond to the several conditions respectively; among these we may however have special solutions, corresponding to points situate on the special loci upon any of the given loci; but when this is the case the number of these special solutions can be separately calculated, and the number of proper solutions is equal to the number obtained as above, less the number of the special solutions.

10. If, however, the given conditions are not completely independent (that is, if the corresponding relations are not completely independent), then the k-fold locus, the l-fold locus, &c. intersect in a common $(\omega-j)$ fold locus, and besides in a residual point-system. The several points of the $(\omega-j)$ fold locus give special solutions—in fact the very notion of the conditions being *properly* satisfied by a curve implies that the curve shall satisfy a true $(k+l+\&c.)$ fold, that is, a true ω-fold condition; the proper solutions are therefore comprised among the solutions given by the residual point-system, and the number of them is as before equal to the order of the point-system, or number of the points thereof, *less* the number of points which give special solutions: the order of the point-system is, as has been seen, equal to the product of the orders of the k-fold locus, the l-fold locus, &c., *less* a reduction depending on the nature of

the common $(\omega - j)$ fold locus, and the difficulty is in general in the determination of the value of this reduction.

11. In all that precedes, the number of the parameters has been taken to be ω; but if the parameters are taken to be contained in the equation of the curve homogeneously, then the parameters before made use of are in fact the ratios of these homogeneous parameters; and using the term henceforward as referring to the homogeneous parameters, the numbers of the parameters will be $= \omega + 1$.

12. I assume also that the equation of the curve contains the parameters linearly: this being so, the condition that the curve shall pass through a given arbitrary point implies a linear relation between the parameters; and the condition that the curve shall pass through j given points, a j-fold linear relation between the parameters. It follows that the number of the curves which satisfy a given k-fold condition, and besides pass through $\omega - k$ given points, is equal to the order of the k-fold relation, or of the corresponding k-fold locus; and thus if we define the order of the k-fold condition to be the number of the curves in question, the condition, relation, and locus will be all of the same order, and in all that precedes we may (in place of the order of the relation or of the locus) speak of the order of the condition. Thus, subject to the modifications occasioned by common loci and special solutions as above explained, the order of the $(k + l + \&c.)$ fold condition made up of a k-fold condition, an l-fold condition, &c., is equal to the product of the orders of the component conditions; and in particular if $k + l + \&c. = \omega$, then the order of the ω-fold condition, or number of the solutions thereof, is equal to the product of the orders of the component conditions.

13. The conditions to be satisfied by the curve may be conditions of contact with a given curve or curves. In particular if the curve touch a given curve, the parametric point is then situate on a onefold locus. It is to be noticed in reference hereto that if the given curve have nodes or cusps, then we have special solutions, viz. if the sought for curve passes through a node or a cusp of the given curve; and each such node or cusp gives rise to a special onefold locus, presenting itself in the first instance as a factor of the onefold locus of the parametric point; this is, however, a case where the special locus is of the same manifoldness as the general locus (*ante*, No. 8), and is consequently separable; throwing off therefore all these special loci, we have a onefold locus which no longer comprises the points which correspond to curves passing through a node or a cusp of the given curve; the onefold locus, so divested of the special onefold factors, may be termed the "contact-locus" of the given curve. To each point of the contact-locus there corresponds a curve having with the given curve a two-pointic intersection, viz. this is either a proper contact, or it is a special contact, consisting in that the sought for curve has on the given curve a node or cusp, or (which is a higher speciality) in that the sought for curve is or contains as part of itself two or more coincident curves (*ante*, No. 3). To a point in general on the contact-locus there corresponds a curve having a proper contact with the given curve, save and except that to each point on any one of certain special loci on the contact-locus there corresponds a curve having some kind of special contact as above with the given curve. To fix the ideas, it may be mentioned that for the curves of

the order r which touch a given curve of the order m and class n, the order of the contact-locus is $= n + (2r - 2)m$.

14. If, then, the curve touch a given curve, the parametric point is situate on the contact-locus of that curve. If it touch a second given curve, the parametric point is in like manner situate on the contact-locus of the second given curve, that is, it is situate on the twofold locus which is the intersection of the two contact-loci; and the like in the case of any number of contacts each with a distinct given curve. But if the curve, instead of ordinary contacts with distinct given curves, has either a contact of the second, or third, or any higher order, or has two or more ordinary or other contacts with the same given curve, then if the total manifoldness be $= k$, the parametric point is situate on a k-fold locus, which is given as a singular locus of the proper kind on the onefold contact-locus; so that the theory of the contact-locus corresponding to the case of a single contact with a given curve, contains in itself the theory of any system whatever of ordinary or other contacts with the same given curve, viz. the last-mentioned general case depends on the discussion of the singular loci which lie on the contact-locus. And similarly, if the curve has any number of ordinary or other contacts with each of two or more given curves, we have here to consider the intersections of singular loci lying on the contact-loci which correspond to the several given curves respectively, or, what is the same thing, to the singular loci on the intersection of these contact-loci; that is, the theory depends on that of the contact-loci which belong to the given curves respectively.

15. Suppose that the curve which has to satisfy given conditions is a line; the equation is $ax + by + cz = 0$, and the parameters (a, b, c) are to be taken as the coordinates of a point in a plane. Any onefold condition imposed upon the line establishes a onefold relation between the coordinates (a, b, c), and the parametric point is situate on a curve; a second onefold condition imposed on the line establishes a second onefold relation between the coordinates (a, b, c), and the parametric point is thus situate on a second curve; it is therefore determined as a point of intersection of two ascertained curves. In particular if the condition imposed on the line is that it shall touch a given curve, the locus of the parametric point is a curve, the contact-locus; (this is in fact the ordinary theory of geometrical reciprocity, the locus in question being the reciprocal of the given curve;) and the case of the twofold condition of a contact of the second order, or of two contacts, with the given curve, depends on the singular points of the contact-locus, or reciprocal of the given curve; in fact according as the line has a contact of the second order, or has two contacts with the given curve (that is, as it is an inflexion-tangent, or a double tangent of the given curve), the parametric point is a cusp or a node on its locus, the reciprocal curve: this is of course a fundamental notion in the theory of reciprocity, and it is only noticed here in order to show the bearing of the remark (*ante*, No. 14) upon the case now in hand where the curve considered is a line.

16. If the curve which has to satisfy given conditions is a conic

$$(a, b, c, f, g, h \between x, y, z)^2 = 0,$$

we have here six parameters (a, b, c, f, g, h), which are taken as the coordinates of a point in 5-dimensional space. It may be remarked that in this 5-dimensional space we have the onefold cubic locus $abc - af^2 - bg^2 - ch^2 + 2fgh = 0$, which is such that to any position of the parametric point upon it there corresponds not a proper conic but a line-pair; this may be called the discriminant-locus. We have also the threefold locus the relation of which is expressed by the six equations

$$(bc - f^2 = 0,\quad ca - g^2 = 0,\quad ab - h^2 = 0,\quad gh - af = 0,\quad hf - bg = 0,\quad fg - ch = 0),$$

which is such that to any position of the parametric point thereon, there corresponds not a proper conic but a coincident line-pair. I call this the Bipoint-locus[1], and I notice that its order is $= 4$; in fact to find the order we must with the equations of the Bipoint combine two arbitrary linear relations,

$$(* \between a, b, c, f, g, h) = 0,$$
$$(*' \between a, b, c, f, g, h) = 0;$$

the equations of the locus are satisfied by

$$a : b : c : f : g : h = \alpha^2 : \beta^2 : \gamma^2 : \beta\gamma : \gamma\alpha : \alpha\beta$$

(where $\alpha : \beta : \gamma$ are arbitrary); and substituting these values in the linear relations, we have two quadric equations in (α, β, γ), giving four values of the set of ratios $(\alpha : \beta : \gamma)$; that is, the order is $= 4$, or the Bipoint is a threefold quadric locus.

17. The discriminant-locus does not in general present itself except in questions where it is a condition that the conic shall have a node (reduce itself to a line-pair); thus for the conics which have a node and touch a given curve (m, n), or, what is the same thing, for the line-pairs which touch a given curve (m, n), the parametric point is here situate on a twofold locus, the intersection of the discriminant-locus with the contact-locus. It may be noticed that this twofold locus is of the order $3(n + 2m)$, but that it breaks up into a twofold locus of the order $3n$, which gives the proper solutions; viz. the nodal conics which touch the given curve properly, that is, one of the two lines of the conic touches the curve; and into a twice repeated twofold locus of the order $3m$ which gives the special solutions, viz. in these the nodal conic has with the given curve a special contact, consisting in that the node or intersection of the two lines lies on the given curve. By way of illustration see Annex No. 2. But the consideration of the Bipoint-locus is more frequently necessary.

18. Suppose that the conic satisfies the condition of touching a given curve; the parametric point is then situate on a onefold contact-locus $(a, b, c, f, g, h)^q = 0$ (to fix the ideas, if the given curve is of the order m and class n, then the order q of the contact-locus is $= n + 2m$). The contact-locus of any given curve whatever passes through the Bipoint-locus; in fact to each point of the Bipoint-locus there corresponds a coincident line-pair, that is, a conic which (of course in a special sense) touches the given curve whatever it be; and not only so, but inasmuch as we have a special

[1] In framing the epithet Bipoint, the coincident line-pair is regarded as being really a point-pair: see *post*, No. 30.

contact at each of the points of intersection of the given curve with the coincident line-pair regarded as a single line, that is, in the case of a given curve of the m-th order, m special contacts, the Bipoint-locus is a multiple curve on the corresponding contact-locus.

19. If the conic has simply to touch a given curve of the order m_1 and class n_1, then the order of the condition (or number of the conics which satisfy the condition, and besides pass through four given points) is equal to the order of the contact-locus, that is, it is $= n_1 + 2m_1$. If the conic has also to touch a second given curve of the order m_2 and class n_2, then the order of the twofold condition (or number of the conics which satisfy the twofold condition, and besides pass through three given points) is equal to the order of the intersection or common locus of the two contact-loci; and these being of the orders $n_1 + 2m_1$ and $n_2 + 2m_2$ respectively, the order of the intersection and therefore that of the twofold condition is $= (n_1 + 2m_1)(n_2 + 2m_2)$. But in the next succeeding case it becomes necessary to take account of the singular locus.

20. If the conic has to touch three given curves of the order and class (m_1, n_1), (m_2, n_2), (m_3, n_3) respectively, we have here three contact-loci of the orders $n_1 + 2m_1$, $n_2 + 2m_2$, $n_3 + 2m_3$ respectively; these intersect in a threefold locus, but since each of the contact-loci passes through the threefold Bipoint-locus, this is part of the intersection of the three contact-loci; and not only so, but inasmuch as they pass through the Bipoint-locus m_1, m_2, m_3 times respectively, the Bipoint-locus must be counted $m_1 m_2 m_3$ times, and its order being $=4$, the intersection of the contact-locus is made up of the Bipoint reckoning as a threefold locus of the order $4m_1 m_2 m_3$, and of a residual threefold locus of the order

$$(n_1 + 2m_1)(n_2 + 2m_2)(n_3 + 2m_3) - 4m_1 m_2 m_3,$$

$$= n_1 n_2 n_3 + 2(n_1 n_2 m_3 + \&c.) + 4(n_1 m_2 m_3 + \&c.) + 4m_1 m_2 m_3;$$

and the order of the threefold condition (or number of the conics which touch the three given curves, and besides pass through two given points) is equal to the order of the residual threefold locus, and has therefore the value just mentioned.

21. In going on to the cases of the conics touching four or five given curves, the same principles are applicable; the contact-loci have the Bipoint (a certain number of times repeated) as a common threefold locus, and they besides intersect in a residual fourfold or (as the case is) fivefold locus, and the order of the condition is equal to the order of this residual locus; but the determination of the order of the residual locus presents the difficulties alluded to, *ante*, No. 10. I do not at present further examine these cases, nor the cases of the conics which have with a given curve or curves contacts of the second or any higher order, or more than a single contact with the same given curve.

22. The equation of the conic has been in all that precedes considered as containing the six parameters (a, b, c, f, g, h); but if the question as originally stated relates only to a class of conics the equation whereof contains linearly 2, 3, 4, or 5 parameters, or if, reducing the equation by means of any of the given conditions, it

can be brought to the form in question, then in the latter case we may employ the equation in such reduced form, attending only to the remaining conditions; and in either case we have the equation of a conic containing linearly 2, 3, 4, or 5 parameters, which parameters are taken as the coordinates of a point in 1-, 2-, 3-, or 4-dimensional space, and the discussion relates to loci in such dimensional space. This is in fact what is done in Annex No. 2 above referred to, where the conics considered being the conics which pass through three given points, the equation is taken to be $fyz + gzx + hxy = 0$, and we have only the three parameters (f, g, h); and also in Annex No. 3, where the conics pass through two given points, and are represented by an equation containing the four parameters (a, b, c, h): I give this Annex as a somewhat more elaborate example than any which is previously considered, of the application of the foregoing principles, and as an investigation which is interesting for its own sake. See also Annexes 4 and 5, which contain other examples of the theory. The remark as to the number of parameters is of course applicable to the case where the curve which satisfies the given conditions is a curve of any given order r; the number of the parameters is here at most $= \frac{1}{2}(r+1)(r+2)$, and the space therefore at most $\frac{1}{2}r(r+3)$ dimensional; but we may in particular cases have $\omega + 1$ parameters, the coordinates of a point in ω-dimensional space, where ω is any number less than $\frac{1}{2}r(r+3)$.

23. I do not at present consider the case of a curve of the order r, or further pursue these investigations; my object has been, not the development of the foregoing quasi-geometrical theory, so as to obtain thereby a series of results, but only to sketch out the general theory, and in particular to establish the notion of the order of condition, and to show that, as a rule (though as a rule subject to very frequent exceptions), the order of a compound condition is equal to the product of the orders of the component conditions. The last-mentioned theorem seems to me the true basis of the results contained in a subsequent part of this paper in connexion with the formulæ of De Jonquières, *post*, No. 74 *et seq.* But I now proceed to a different part of the general subject.

Article Nos. 24 to 72.—*Reproduction and Development of the Researches of* CHASLES *and* ZEUTHEN.

24. The leading points of Chasles's theory are as follows: he considers the conics which satisfy four conditions $(4X)$, and establishes the notion of the *characteristics* (μ, ν) of such a system, viz. $\mu, = (4X\,\cdot)$, denotes the number of conics in the system which pass through a given (arbitrary) point, and $\nu, = (4X/)$, the number of conics in the system which touch a given (arbitrary) line. We may say that μ is the parametric order, and ν the parametric class of the system.

25. The conics

$$(::),\quad (\therefore /),\quad (:/\!/),\quad (\cdot /\!/\!/),\quad (/\!/\!/\!/)$$

which pass through four given points, or which pass through three given points and touch a given line, &c., ... or touch four given lines, have respectively the characteristics

$$(1,\ 2),\quad (2,\ 4),\quad (4,\ 4),\quad (4,\ 2),\quad (2,\ 1).$$

26. A single condition (X) imposed upon a conic has two representative numbers, or simply representatives, (α, β); viz. if ($4Z$) be an *arbitrary* system of four conditions, and (μ, ν) the characteristics of ($4Z$), then the number of the conics which satisfy the five conditions (X, $4Z$) is $= \alpha\mu + \beta\nu$.

27. As an instance of the use of the characteristics, if X, X', X'', X''', X'''' be any five independent conditions, and (α, β), ... (α'''', β'''') the representatives of these conditions respectively, then the number of the conics which satisfy the five conditions (X, X', X'', X''', X'''') is

$$=(1,\ 2,\ 4,\ 4,\ 2,\ 1\between \alpha,\ \beta)(\alpha',\ \beta')(\alpha'',\ \beta'')(\alpha''',\ \beta''')(\alpha'''',\ \beta'''')$$

viz. this notation stands for $1\alpha\alpha'\alpha''\alpha'''\alpha'''' + 2\Sigma\alpha\alpha'\alpha''\alpha'''\beta'''' \ldots + 1\beta\beta'\beta''\beta'''\beta''''$.

28. In particular if X be the condition that a conic shall touch a given curve of the order m and class n, then the representatives of this condition are (n, m), whence the number of the conics which touch each of five given curves (m, n), ... (m'''', n'''') is

$$=(1,\ 2,\ 4,\ 4,\ 2,\ 1\between n,\ m)(n',\ m')(n'',\ m'')(n''',\ m''')(n'''',\ m'''').$$

29. A system of conics ($4X$) having the characteristics (μ, ν), contains

$2\nu-\mu$ line-pairs, that is, conics each of them a pair of lines; and

$2\mu-\nu$ point-pairs, that is, conics each of them a pair of points (*coniques infiniment aplaties*).

30. I stop to further explain these notions of the line-pair and the point-pair; and also the notion of the line-pair-point.

A conic is a curve of the second order and second class; *quà* curve of the second order it may degenerate into a pair of lines, or line-pair (but the class is then $=0$): *quà* curve of the second class it may degenerate into a pair of points, or point-pair (but the order is then $=0$). The two lines of a line-pair may be coincident, and we have then a coincident line-pair; such a line-pair (it must I think be postulated) ordinarily arises, not from a line-pair the two lines of which become coincident, but from a proper conic, flattening by the gradual diminution of its conjugate axis, while its transverse axis remains constant or approaches a limit different from zero; the conic thus tends (not to an indefinitely extended but) to a terminated line[1]; in other words, the tangents of the conic become more and more nearly lines through two fixed points, the terminations of the terminated line; and these terminating points, which continue to exist up to the instant when the conjugate axis takes its limiting value $=0$, are regarded as still existing at this instant, and the coincident line-pair as being in fact the point-pair formed by the two terminating points. Similarly the two points of a point-pair may be coincident, and we have then a coincident point-

[1] A line is regarded as extending from any point A thereof to B, and then in the same direction, from B through infinity to A; it thus consists of two portions separated by these points; and considering either portion as removed, the remaining portion is a terminated line.

pair; such a point-pair (it must in like manner be postulated) ordinarily arises, not from a point-pair the two points of which become coincident, but from a proper conic sharpening itself to coincide with its asymptotes, and so becoming ultimately a pair of lines through the coincident point-pair; and the coincident point-pair is regarded as being in fact the line-pair formed by some two lines through the coincident point-pair.

31. In accordance with the foregoing notions we may with propriety, and it will in the sequel be found convenient to speak of a point-pair as a line terminated by two points on this line, and similarly to speak of a line-pair as a point terminated (that is, the pencil of lines through the point is terminated) by two lines through the point.

32. If in a point-pair thus considered as a line terminated by two points the two points become coincident (the line continuing to exist as a definite line), or, what is the same thing, if in a line-pair thus considered as a point terminated by two lines, the two lines become coincident (the point continuing to exist as a definite point), we have a "line-pair-point;" viz. this is at once a coincident line-pair and a coincident point-pair; it may also be regarded as the limit of a conic the axes of which, and the ratio of the conjugate to the transverse axis, all ultimately vanish: it may be described as a line terminated each way at a point thereof, or as a point terminated each way at a line through it. The notion of a line-pair-point first presents itself in Zeuthen's researches, as will presently appear; but it may be noticed here that line-pair-points, and these the same line-pair-points, may present themselves among the $2\nu-\mu$ line-pairs, and among the $2\mu-\nu$ point-pairs of the system of conics $4X$.

33. Returning to the foregoing theory of characteristics, I remark that the fundamental notion may be taken to be, not the characteristics (μ, ν) of the conics which satisfy four conditions, but in every case the number of the conics which satisfy five conditions. Thus for the conics not subjected to any condition, we may consider the symbols

(⁙), (∷/), (∴//), (:///), (·////), (/////)

denoting the number of the conics which pass through five given points, or which pass through four given points and touch a given line, &c. ..., or which touch five given lines; these numbers are respectively

$$=1, \quad 2, \quad 4, \quad 4, \quad 2, \quad 1.$$

So for the conics which satisfy a given condition X, or two conditions $2X$, ..., or five conditions $5X$, we have respectively the numbers

X, (∷), (∴/), (://), (·///), (////)
$2X$, (∴), (: /), (·//), (///)
$3X$, (:), (· /), (//)
$4X$, (·), (/)
$5X$,

where the X, $2X$, &c. belong to the symbols which follow: read $(X ::)$, $(X .\because /)$, &c., or, as we may for shortness represent them,

$$\begin{array}{lllll} \mu''', & \nu''', & \rho''', & \sigma''', & \tau''' \\ \mu'', & \nu'', & \rho'', & \sigma'' & \\ \mu', & \nu', & \rho' & & \\ \mu, & \nu & & & \\ \mu_0 & & & & \end{array}$$

viz. the single condition X has the five characteristics $(\mu''', \ldots \tau'''), \ldots$; the four conditions $4X$, the characteristics (μ, ν) as in the original theory; and the five conditions $5X$ a single characteristic μ_0.

34. We thus see the origin of the notion of the representatives (α, β) of a single condition X; for considering the arbitrary four conditions $4Z$, the characteristics whereof are (μ, ν), and assuming that the single characteristic, or number of the conics $(X, 4Z)$, is $= \alpha\mu + \beta\nu$, and taking for $(4Z)$ successively the conditions

$$(::),\quad (\because /),\quad (: //),\quad (\cdot ///),\quad (////),$$

having respectively the characteristics

$$(1, 2),\quad (2, 4),\quad (4, 4),\quad (4, 2),\quad (2, 1),$$

we have

$$\begin{aligned} \mu''' &= 1\alpha + 2\beta, \\ \nu''' &= 2\alpha + 4\beta, \\ \rho''' &= 4\alpha + 4\beta, \\ \sigma''' &= 4\alpha + 2\beta, \\ \tau''' &= 2\alpha + 1\beta, \end{aligned}$$

that is, the characteristics $(\mu''', \nu''', \rho''', \sigma''', \tau''')$ of a single condition X are not independent, but are representable as above by means of two independent quantities (α, β); or, what is the same thing, we have

$$\nu''' = 2\mu''',\quad \sigma''' = 2\tau''',\quad \rho''' = \tfrac{2}{3}(\nu''' + \sigma'''),$$

which being satisfied, the representatives (α, β) are given by

$$\alpha = \tfrac{1}{3}(2\tau''' - \mu'''),\quad \beta = \tfrac{1}{3}(2\mu''' - \tau''').$$

35. I find that a like property exists as to the characteristics $(\mu'', \nu'', \rho'', \sigma'')$ of the two conditions $2X$, viz. these are not independent but are connected by a single linear relation,

$$\mu'' - \tfrac{3}{2}\nu'' + \tfrac{3}{2}\rho'' - \sigma'' = 0.$$

This may be proved in the case where the conditions $2X$ are two separate conditions (X, X'); viz. let the representatives of these be (α, β), (α', β') respectively, then

26—2

combining with them the three arbitrary conditions X'', X''', X'''' having respectively the representatives (α'', β''), (α''', β'''), (α'''', β''''), we have the general equation

$$(X, X', X'', X''', X'''') = (1, 2, 4, 4, 2, 1 \between \alpha, \beta)(\alpha', \beta')(\alpha'', \beta'')(\alpha''', \beta''')(\alpha'''', \beta'''');$$

taking herein

$$(X'', X''', X'''') = (.\cdot.), (:/), (\cdot//), (///)$$

successively, and observing that the representatives of $(\cdot)$ are $(1, 0)$ and those of $(/)$ are $(0, 1)$, we thus obtain for $(\mu'', \nu'', \rho'', \sigma'')$, characteristics of (X, X'), the values

$$\begin{aligned} \mu'' &= (1, 2, 4 \between \alpha, \beta)(\alpha', \beta'), \\ \nu'' &= (2, 4, 4 \between \alpha, \beta)(\alpha', \beta'), \\ \rho'' &= (4, 4, 2 \between \alpha, \beta)(\alpha', \beta'), \\ \sigma'' &= (4, 2, 1 \between \alpha, \beta)(\alpha', \beta'), \end{aligned}$$

(viz. $\mu'' = 1\alpha\alpha' + 2(\alpha\beta' + \alpha'\beta) + 4\beta\beta'$, &c.), and these values give identically

$$2\mu'' - 3\nu'' + 3\rho'' - 2\sigma'' = 0,$$

which is the foregoing equation. And I assume that the theorem extends to the case of two inseparable conditions $2X$, but in this case I do not even know where the proof is to be sought for.

The characteristics (μ', ν', ρ') of the three conditions $3X$ are in general independent.

36. It has been mentioned that if (α, β) are the representatives of the condition X, and (μ, ν) the characteristics of the conditions $4Z$, then

$$(X, 4Z) = \alpha\mu + \beta\nu;$$

this is the most convenient form of the theorem, but as (α, β) are known functions of the characteristics $(\mu''', \nu''', \rho''', \sigma''', \tau''')$ of the condition X, the equation is in effect an expression for $(X, 4Z)$ in terms of the characteristics of X and $4Z$ respectively.

There is, similarly, an expression for $(2X, 3Z)$ in terms of the characteristics $(\mu', \nu', \rho', \sigma')$ of $3Z$ (satisfying the relation $\mu' - \frac{3}{2}\nu' + \frac{3}{2}\rho' - \sigma' = 0$) and the characteristics (μ, ν, ρ) of $2X$, viz. we have

$$\begin{aligned} (2X, 3Z) = \quad & \mu\,(\qquad\qquad\qquad\quad - \tfrac{1}{4}\rho' + \tfrac{1}{2}\sigma') \\ & + \nu\,(-\tfrac{3}{8}\mu' + \tfrac{5}{16}\nu' + \tfrac{5}{16}\rho' - \tfrac{3}{8}\sigma') \\ & + \rho\,(\ \tfrac{1}{2}\mu' - \tfrac{1}{4}\nu' \qquad\qquad\qquad\quad). \end{aligned}$$

This may be easily proved in the case where the conditions $2X$ are two separable conditions X, X' having the representatives (α, β), (α', β') respectively, and the conditions $3Z$ three separable conditions Z'', Z''', Z'''' having the representatives (α'', β''), (α''', β'''), (α'''', β'''') respectively; we have, in fact,

$$\begin{aligned} \mu' &= (1, 2, 4 \between \alpha, \beta)(\alpha', \beta'), & \mu &= (1, 2, 4, 4 \between \alpha'', \beta'')(\alpha''', \beta''')(\alpha'''', \beta''''), \\ \nu' &= (2, 4, 4 \between \ \text{,,} \quad \text{,,}\ ), & \nu &= (2, 4, 4, 2 \between \ \text{,,} \quad \text{,,} \quad \text{,,}\ ), \\ \rho' &= (4, 4, 2 \between \ \text{,,} \quad \text{,,}\ ), & \rho &= (4, 4, 2, 1 \between \ \text{,,} \quad \text{,,} \quad \text{,,}\ ), \\ \sigma' &= (4, 2, 1 \between \ \text{,,} \quad \text{,,}\ ); \end{aligned}$$

and with these values the function

$$(X,\ X',\ Z'',\ Z''',\ Z''''),\ = (1,\ 2,\ 4,\ 4,\ 2,\ 1 \between \alpha,\ \beta)(\alpha',\ \beta')(\alpha'',\ \beta'')(\alpha''',\ \beta'''),\ (\alpha'''',\ \beta'''')$$

is found to be expressible as above in terms of $(\mu,\ \nu,\ \rho)$, $(\mu',\ \nu',\ \rho',\ \sigma')$; but I do not know how to conduct the proof for the inseparable conditions $2X$ and $3Z$.

37. It may be remarked by way of verification that writing successively

$$(3Z) = (\therefore),\ (:/),\ (\cdot //),\ (///),$$

that is,

$$(\mu,\ \nu,\ \rho) = (1,\ 2,\ 4),\ (2,\ 4,\ 4),\ (4,\ 4,\ 2),\ (4,\ 4,\ 1),$$

we have in the first case

$$\begin{aligned} (2X\ \therefore) = & \qquad\qquad\qquad\quad -\tfrac{1}{4}\rho' + \tfrac{1}{2}\sigma' \\ & -\tfrac{3}{4}\mu' + \tfrac{5}{8}\nu' + \tfrac{5}{8}\rho' - \tfrac{3}{4}\sigma' \\ & + 2\mu' - \nu' \\ = & \quad \mu' + \tfrac{1}{4}(\mu' - \tfrac{3}{2}\nu' + \tfrac{3}{2}\rho' - \sigma'),\ = \mu', \end{aligned}$$

and similarly in the other three cases,

$$(2X \cdot //) = \nu',\quad (2X \cdot //) = \rho',\quad (2X\ ///) = \sigma'.$$

38. Let $(\mu,\ \nu,\ \rho,\ \sigma)$ be the characteristics of $2Z$, $(\mu - \frac{3}{2}\nu + \frac{3}{2}\rho - \sigma = 0)$, and $(\mu',\ \nu',\ \rho',\ \sigma')$ the characteristics of $2X$, $(\mu' - \frac{3}{2}\nu' + \frac{3}{2}\rho' - \sigma' = 0)$. Then in the formula for $(2X,\ 3Z)$, writing successively for $3X$

$$(2X \cdot),\ \text{characteristics}\ (\mu,\ \nu,\ \rho),$$

and

$$(2X\ /),\qquad \text{,,} \qquad (\nu,\ \rho,\ \sigma),$$

we obtain expressions for the characteristics $(2X,\ 2Z \cdot)$ and $(2X,\ 2Z\ /)$ of $(2X,\ 2Z)$, viz. eliminating from the formulæ, first the $(\sigma,\ \sigma')$ and secondly the $(\mu,\ \mu')$, each of these may be expressed in two different forms as follows:

$$\begin{aligned} & (2X,\ 2Z \cdot) \\ = & \ \tfrac{1}{2}\mu\mu' \\ & + \tfrac{7}{8}\nu\nu' \\ & - \tfrac{3}{4}(\mu\nu' + \mu'\nu) \\ & + \tfrac{1}{2}(\mu\rho' + \mu'\rho) \\ & - \tfrac{1}{4}(\nu\rho' + \nu'\rho) \\ = & \ \tfrac{1}{2}\sigma\sigma' \\ & - \tfrac{3}{8}\rho\rho' \\ & - \tfrac{1}{4}\nu\nu' \\ & - \tfrac{1}{4}(\rho\sigma' + \rho'\sigma) \\ & + \tfrac{1}{2}(\nu\rho' + \nu'\rho), \end{aligned}$$

$$\begin{aligned} & (2X,\ 2Z\ /) \\ = & \ \tfrac{1}{2}\mu\mu' \\ & - \tfrac{3}{8}\nu\nu' \\ & - \tfrac{1}{4}\rho\rho' \\ & - \tfrac{1}{4}(\mu\nu' + \mu'\nu) \\ & + \tfrac{1}{2}(\nu\rho' + \nu'\rho) \\ = & \ \tfrac{1}{2}\sigma\sigma' \\ & + \tfrac{7}{8}\rho\rho' \\ & - \tfrac{3}{4}(\rho\sigma' + \rho'\sigma) \\ & + \tfrac{1}{2}(\nu\sigma' + \nu'\sigma) \\ & - \tfrac{1}{4}(\nu\rho' + \nu'\rho), \end{aligned}$$

the two expressions of the same quantity being of course equivalent in virtue of the relations between (μ, ν, ρ, σ) and (μ', ν', ρ', σ') respectively.

The characteristics of (X, Z), (X, $2Z$), (X, $3Z$) are at once deducible from the before-mentioned expression $\alpha\mu + \beta\nu$ of (X, $4Z$).

39. Zeuthen's investigations are based upon the before-mentioned theorem, that in a system of conics ($4X$), characteristics (μ, ν), there are $2\mu - \nu$ point-pairs and $2\nu - \mu$ line-pairs. If in the given system the number of point-pairs is $= \lambda$ and the number of line-pairs is $= \varpi$, then, conversely, the characteristics of the system are

$$\mu = \tfrac{1}{3}(2\lambda + \varpi), \quad \nu = \tfrac{1}{3}(\lambda + 2\varpi).$$

And by means of this formula he investigates the characteristics of the several systems of conics which satisfy four conditions ($4X$) of contact with a given curve or curves, viz. these are the conics

(1) (1) (1) (1), (1, 1) (1) (1), (1, 1) (1, 1), (1, 1, 1) (1), (1, 1, 1, 1),
(2) (1) (1) , (2) (1, 1) , (2, 1) (1) , (2, 1, 1),
(2) (2) , (2, 2) ,
(3) (1) , (3, 1) ,
(4) ,

where (1) denotes contact of the first order, (2) of the second order, (3) of the third order, (4) of the fourth order, with a given curve; (1) (1) denotes contacts of the first order with each of two given curves, (1, 1) two such contacts with the same given curve, and so on. A given curve is in every case taken to be of the order m and class n, with δ nodes, κ cusps, τ double tangents, and ι inflexions (m_1, n_1, δ_1, κ_1, τ_1, ι_1; m_2, n_2, &c., as the case may be). The symbols (1), &c. might be referred to the corresponding curves by a suffix; thus $(1)_m$ would denote that the contact is with a given curve of the order m (class n, &c.); but this is in general unnecessary.

40. In a system of conics satisfying four conditions of contact, as above, it is comparatively easy to see what are the point-pairs and line-pairs in these several systems respectively; but in order to find the values of λ and ϖ, each of these point-pairs and line-pairs has to be counted not once, but a proper number of times; and it is in the determination of these multiplicities that the difficulty of the problem consists. I do not enter into this question, but give merely the results.

41. For the statement of these I introduce what I call the notation of Zeuthen's Capitals. We have to consider several classes of point-pairs and the reciprocal classes of line-pairs. A point-pair may be described (*ante*, No. 31) as a terminated line, and a line-pair as a terminated point; and we have first the following point-pairs, viz.:

A, line terminated each way in the intersection of two curves or of a curve with itself (node).

B, tangent to a curve, terminated in a curve, and in the intersection of two curves or of a curve with itself.

C, common tangent of two curves, or double tangent of a curve, terminated each way in a curve.

D, inflexion tangent of a curve terminated each way in a curve:

and the corresponding line-pairs, viz.:

A', point terminated each way in the common tangent of two curves or the double tangent of a curve.

B', point of a curve terminated by the tangent of a curve, and by the common tangent of two curves or double tangent of a curve.

C', intersection of two curves, or of a curve with itself (node), terminated each way by the tangent to a curve.

D', cusp of a curve terminated each way by the tangent to a curve:

all which is further explained by what follows; thus in the case $(1)(1)(1)(1)$, $=(1)_{m_1}(1)_{m_2}(1)_{m_3}(1)_{m_4}$, the value of A is given as $\Sigma m_1 m_2 \,.\, m_3 m_4 \,(=3 m_1 m_2 m_3 m_4)$. Here A is the number of the point-pairs terminated one way in the intersection of any two m_1, m_2 of the four curves, and the other way in the intersection of the remaining two m_3, m_4 of the four curves. But in the case $(1,\ 1)(1)(1)$, $=(1,\ 1)_m(1)_{m_1}(1)_{m_2}$, the value of A is given as $=\delta m_1 m_2 + m m_1 \,.\, m m_2$. Here A denotes the number of the point-pairs, which are either $(\delta m_1 m_2)$ terminated one way at a node of m, and the other way at an intersection of m_1, m_2, or else $(m m_1 \,.\, m m_2)$ terminated one way at an intersection of m, m_1, and the other way at an intersection of m, m_2: and so in other cases.

42. This being so, we have

$(1)(1)(1)(1)$, $=(1)_{m_1}(1)_{m_2}(1)_{m_3}(1)_{m_4}$.

$A=\Sigma m_1 m_2 \,.\, m_3 m_4 \;(=3\; m_1 m_2 m_3 m_4)$,	1	$A'=\Sigma n_1 n_2 \,.\, n_3 n_4 \;(=3\; n_1 n_2 n_3 n_4)$,
$B=\Sigma m_1 m_2 \,.\, m_3 \,.\, n_4 \;(=3\Sigma m_1 m_2 m_3 n_4)$,	2	$B'=\Sigma n_1 n_2 \,.\, n_3 \,.\, m_4 \;(=3\Sigma n_1 n_2 n_3 m_4)$,
$C=\Sigma m_1 \,.\, m_2 \,.\, n_3 n_4 \;(=\Sigma m_1 m_2 n_3 n_4)$.	4	$C'=\Sigma n_1 \,.\, n_2 \,.\, m_3 m_4 \;(=\Sigma n_1 n_2 m_3 m_4)$.

$(1,\ 1)(1)(1)$, $=(1,\ 1)_m(1)_{m_1}(1)_{m_2}$.

$A=\delta m_1 m_2 + m m_1 \,.\, m m_2$,	1	$A'=\tau n_1 n_2 + n n_1 \,.\, n n_2$,
$B=\delta n_1 m_2 + \delta n_2 m_1$	2	$B'=\tau m_2 n_2 + \tau m_2 n_1$
$+ m m_1 (n-2) m_2 + m m_2 (n-2) m_1$		$+ n n_1 (m-2) n_2 + n n_2 (m-2) n_1$
$+ m m_1 n_2 (m-1) + m m_2 n_1 (m-1)$		$+ n n_1 m_2 (n-1) + n n_2 m_1 (n-1)$
$+ m_1 m_2 n (m-2)$,		$+ n_1 n_2 m (n-2)$,
$C=\tau m_1 m_2$	4	$C'=\delta n_1 n_2$
$+ n n_1 (m-2) m_2 + n n_2 (m-2) m_1$		$+ m m_1 (n-2) n_2 + m m_2 (n-2) n_1$
$+ n_1 n_2 \,.\, \tfrac{1}{2} m (m-1)$,		$+ m_1 m_2 \,.\, \tfrac{1}{2} n (n-1)$,
$D=\iota m_1 m_2$.	3	$D'=\kappa n_1 n_2$.

$(1,\ 1)(1,\ 1), = (1,\ 1)_m (1,\ 1)_{m_1}.$

$A = \delta\delta_1 \qquad + \tfrac{1}{2} mm_1 (mm_1 - 1),$	1	$A' = \tau\tau_1 \qquad + \tfrac{1}{2} nn_1 (nn_1 - 1),$
$B = \delta n_1 (m_1 - 2) \qquad + \delta_1 n (m - 2)$ $+ mm_1 (n-2)(m_1 - 1) + mm_1 (n_1 - 2)(m - 1),$	2	$B' = \tau m_1 (n - 2) \qquad + \tau_1 m (n - 2)$ $+ nn_1 \ (m-2)(n_1 - 1) + nn_1 (m_1 - 2)(n - 1),$
$C = \tau . \tfrac{1}{2} m_1 (m_1 - 1) \qquad + \tau_1 . \tfrac{1}{2} m (m - 1)$ $+ nn_1 (m - 2)(m_1 - 2),$	4	$C' = \delta . \tfrac{1}{2} n_1 (n_1 - 2) \qquad + \delta_1 . \tfrac{1}{2} n (n - 1)$ $+ mm_1 (n - 2)(n_1 - 2),$
$D = \iota . \tfrac{1}{2} m_1 (m_1 - 1) \qquad + \iota_1 . \tfrac{1}{2} m (m - 1).$	3	$D' = \kappa . \tfrac{1}{2} n_1 (n_1 - 1) \qquad + \kappa_1 . \tfrac{1}{2} n (n - 1).$

$(1,\ 1,\ 1)(1) = (1,\ 1,\ 1)_m (1)_{m_1}.$

$A = \delta mm_1,$	1	$A' = \tau nn_1,$
$B = \delta (n - 4) m_1 + \delta n_1 (m - 2),$ $+ mm_1 (n - 2)(m - 3),$	2	$B' = \tau (m - 4) n_1 + \tau m_1 (n - 2),$ $+ nn_1 (m - 2)(n - 3),$
$C = \tau (m - 4) m_1 + nn_1 . \tfrac{1}{2} (m - 2)(m - 3),$	4	$C' = \delta (n - 4) n_1 + mm_1 . \tfrac{1}{2} (n - 2)(n - 3),$
$D = \iota (m - 3) m_1.$	3	$D' = \kappa (n - 3) n_1.$

$(1,\ 1,\ 1,\ 1), = (1,\ 1,\ 1,\ 1)_m.$

$A = \tfrac{1}{2} \delta (\delta - 1),$	1	$A' = \tfrac{1}{2} \tau (\tau - 1),$
$B = \delta (n - 4)(m - 4),$	2	$B' = \tau (m - 4)(n - 4),$
$C = \tau . \tfrac{1}{2} (m - 4)(m - 5),$	3	$C' = \delta . \tfrac{1}{2} (n - 4)(n - 5),$
$D = \iota . \tfrac{1}{2} (m - 3)(m - 4).$	4	$D' = \kappa . \tfrac{1}{4} (n - 3)(n - 4).$

43. Secondly, we have the point-pairs:

E, tangent to curve from intersection of two curves or of a curve with itself (node), and terminated at the point of contact and the last-mentioned point.

F, tangent to a curve at intersection with another curve or with itself, and terminated there and at a curve.

G, common tangent of two curves or double tangent of a curve, terminated at one of the points of contact and at a curve.

D, *ut suprà.*

H, line joining cusp of a curve with intersection of two curves or of a curve with itself, and terminated at these points.

I, line from cusp of a curve touching a curve, and terminated at the cusp and at a curve.

J, Inflexion tangent of a curve, terminated there and at a curve:

and the corresponding line-pairs, viz.

E', point on a curve in common tangent of two curves or double tangent of a curve, and terminated by this tangent and by tangent to a curve.

F', point on a curve in common tangent of this and another curve or in double tangent of this curve, and terminated by this tangent and by tangent to a curve.

D', *ut suprà*.

H', intersection of inflexion tangent of a curve with common tangent of two curves or double tangent of a curve, and terminated by these lines.

I', intersection of inflexion tangent of a curve with a curve, and terminated by this tangent and by tangent of a curve:

and this being so,

$$(2)(1)(1), = (2)_m (1)_{m_1} (1)_{m_2}.$$

$E = n \cdot m_1 m_2,$	3	$E' = m \cdot n_1 n_2,$
$F = mm_1 \cdot m_2 + mm_2 \cdot m_1,$	3	$F' = nn_1 \cdot m_2 + nn_2 \cdot m_1,$
$G = nn_2 \cdot m_1 + nn_1 \cdot m_2,$	6	$G' = mm_2 \cdot n_1 + mn_1 \cdot n_2,$
$D = \iota m_1 m_2,$	2	$D' = \kappa n_1 n_2,$
$H = \kappa m_1 m_2,$	1	$H' = \iota n_1 n_2,$
$I = \kappa n_1 m_2 + \kappa n_2 m_1.$	2	$I' = \iota m_2 n_1 + \iota m_2 n_1.$

$$(2)(1,\ 1) = (2)_m (1,\ 1)_{m_1}.$$

$E = \delta_1 n,$	3	$E' = \tau_1 m,$
$F = m \cdot m_1 (m_1 - 1),$	3	$F' = n \cdot n_1 (n_1 - 1),$
$G = nn_1 (m_1 - 2),$	6	$G' = mm_1 (n_1 - 2),$
$D = \iota \cdot \frac{1}{2} m_1 (m_1 - 1),$	2	$D' = \kappa \cdot \frac{1}{2} n_1 (n_1 - 1),$
$H = \kappa \delta_1,$	1	$H' = \iota \tau_1,$
$I = \kappa n_1 (m_1 - 2).$	2	$I' = \iota m_1 (n_1 - 2).$

$$(2,\ 1)(1), = (2,\ 1)_m (1)_{m_1}.$$

$E = (n-2) \cdot mm_1,$	3	$E' = (m-2) \cdot nn_1,$
$F = mm_1 (m-2) + 2\delta m_1,$	3	$F' = nn_1 (n-2) + 2\tau n_1,$
$G = nn_1 (m-2) + 2\tau m_1,$	6	$G' = mm_1 (n-2) + 2\delta n_1,$
$D = \iota (m-3) m_1,$	2	$D' = \kappa (n-3) n_1,$
$H = \kappa mm_1,$	1	$H' = \iota nn_1,$
$I = \kappa (n-3) m_1 + \kappa n_1 (m-2),$	2	$I' = \iota (m-3) n_1 + \iota m_1 (n-2),$
$J = \iota m_1.$	5	$J' = \kappa n_1.$

$(2, 1, 1), =(2, 1, 1)_m$.

$E = \delta(n-4)$,	3	$E' = \tau(m-4)$,
$F = 2\delta(m-3)$,	3	$F' = 2\tau(n-3)$,
$G = 2\tau(m-4)$,	6	$G' = 2\delta(n-4)$,
$D = \iota . \tfrac{1}{2}(m-3)(m-4)$,	2	$D' = \kappa . \tfrac{1}{2}(n-3)(n-4)$,
$H = \delta\kappa$,	1	$H' = \iota\tau$,
$I = \kappa(n-3)(m-4)$,	2	$I' = \iota(m-3)(n-4)$,
$J = \iota(m-3)$.	5	$J' = \kappa(n-3)$.

44. Thirdly, we have the point-pairs:

K, common tangent of two curves or double tangent of a curve, terminated at points of contact.

L, line from cusp of a curve touching a curve, and terminated at cusp and point of contact.

M, line joining cusp of a curve with cusp of a curve, and terminated by the two cusps.

N, inflexion tangent terminated each way at inflexion, viz. this is a *line-pair-point.*

O, cuspidal tangent terminated each way at cusp, viz. this is a *line-pair-point:*

and the corresponding line-pairs:

K', intersection of two curves or of curve with itself (node), and terminated by the two tangents.

L', intersection of inflexion tangent of a curve with a curve, and terminated by the inflexion tangent and the tangent at the intersection.

M', intersection of inflexion tangent of a curve with inflexion tangent of a curve, and terminated by the two inflexion tangents.

N', $=O$, *line-pair-point* as above.

O', $=N$, *line-pair-point* as above:

which being so, we have

$(2)(2), =(2)_m(2)_{m_1}$.

$K = nn_1$,	9	$K' = mm_1$,
$L = \kappa n_1 + \kappa_1 n$,	3	$L' = \iota m_1 + \iota_1 m$,
$M = \kappa\kappa_1$.	1	$M' = \iota\iota_1$.

$(2,\ 2),\ = (2,\ 2)_m.$

$K = \tau,$	9	$K' = \delta,$
$L = \kappa(n-3),$	3	$L' = \iota(m-3),$
$M = \frac{1}{2}\kappa(\kappa-1),$	1	$M' = \frac{1}{2}\iota(\iota-1),$
$N = \iota,$	2	$N' = \kappa,$
$O = \kappa.$	1	$O' = \iota.$

45. Fourthly, we have the point-pairs:

P, tangent of a curve at its intersection with another curve or itself, terminated each way at the point of contact—*line-pair-point.*

Q, common tangent of two curves or double tangent of a curve, terminated each way at one of the points of contact—*line-pair-point.*

J, *ut suprà.*

R, cuspidal tangent terminated at cusp and at a curve:

and the corresponding line-pairs:

P', $= Q$, *line-pair-point.*

Q', $= P$, *line-pair-point.*

J', *ut suprà.*

R', inflexion of curve terminated by the inflexion tangent and by tangent to a curve:

which being so, we have

$(3)(1),\ = (3)_m(1)_{m_1}.$

$P = mm_1,$	2	$P' = nn_1,$
$Q = nn_1,$	2	$Q' = mm_1,$
$J = \iota m_1,$	5	$J' = \kappa n_1,$
$R = \kappa m_1.$	4	$D' = \iota n_1.$

$(3,\ 1),\ = (3,\ 1)_m.$

$P = 2\delta,$	2	$Q' = 2\tau,$
$Q = 2\tau,$	2	$P' = 2\delta,$
$J = \iota(m-3),$	5	$J' = \kappa(n-3),$
$R = \kappa(m-3).$	4	$R' = \iota(n-3).$

46. And lastly, we have the point-pairs N, O (*line-pair-points*) and the line-pairs N', O' (*line-pair-points*), *ut suprà*, and

$(4),\ = (4)_m.$

$N = \iota,$	4	$N' = \kappa,$
$O = \kappa.$	2	$O' = \iota.$

47. Where in all cases the central column of figures gives the numerical factors which multiply the corresponding capitals, thus we have

for (1)(1)(1)(1)

$$\lambda = 2\nu - \mu = A + 2B + 4C,$$

$$\varpi = 2\mu - \nu = A' + 2B' + 4C';$$

for (1, 1)(1)(1),

$$\lambda = 2\nu - \mu = A + 2B + 4C + 3D,$$

$$\varpi = 2\mu - \nu = A' + 2B' + 4C' + 3D',$$

and so on.

48. The elements $(m, n, \delta, \kappa, \tau, \iota)$ of a curve satisfy Plücker's six equations, and Zeuthen uses these equations, in a somewhat unsystematic way, to simplify the form of his results.

It is convenient in his formulæ to write $3m + \iota, = 3n + \kappa, = \alpha$, and to express everything in terms of (m, n, α), viz. we have for this purpose

$$2\delta = m^2 - m + 8n - 3\alpha,$$

$$2\tau = n^2 - 8m - n - 3\alpha.$$

But I make another alteration in the form of his results; he gives, for instance, the characteristics of (1, 1)(1)(1) as

$$\mu = \mu''' m_1m_2 + \mu'' (m_1n_2 + m_2n_1) + \mu' n_1n_2,$$

$$\nu = \nu''' m_1m_2 + \nu'' (m_1n_2 + m_2n_1) + \nu' n_1n_2,$$

where

$$\mu' = 2m(m + n - 3) + \tau, = (1, 1 .\!\cdot\!.),$$

$$\mu'' = \nu' = 2n(m + 2n - 5) + 2\tau, = (1, 1 :/),$$

$$\mu''' = \nu'' = 2n(2m + n - 5) + 2\delta, = (1, 1 \cdot //),$$

$$\nu''' = 2n(m + n - 3) + \delta, = (1, 1 ///),$$

viz. the four components have really the significations (1, 1 .·.) set opposite to them respectively; and accordingly, instead of giving the formulæ for the two characteristics of (1, 1)(1)(1), I give those for the four characteristics (1, 1 .·.), &c. of (1, 1), thus in every case obtaining formulæ which relate to a single curve only. Subject to the last-mentioned variation of form, I give Zeuthen's original expressions in Annex 6; but here in the text I express them as above in terms of (m, n, α), viz.

49. We have the formulæ

(1)

(::) = $n + 2m$,

(.·./) = $2n + 4m$,

(: //) = $4n + 4m$,

(· ///) = $4n + 2m$,

(////) = $2n + 2m$;

(1, 1)

$$(\therefore) = 2m^2 + 2mn + \tfrac{1}{2}n^2 - 2m - \tfrac{1}{2}n - \tfrac{3}{2}\alpha,$$
$$(\,:/) = 2m^2 + 4mn + \ \ n^2 - 2m - \ \ n - 3\alpha,$$
$$(\cdot\,//) = \ \ m^2 + 4mn + 2n^2 - \ \ m - 2n - 3\alpha,$$
$$(///) = \tfrac{1}{2}m^2 + 2mn + 2n^2 - \tfrac{1}{2}m - 2n - \tfrac{3}{2}\alpha.$$

(1, 1, 1)

$$(\,:) = \tfrac{2}{3}m^3 + 2m^2n + \ \ mn^2 + \tfrac{1}{6}n^3 - 2m^2 - 3mn - \tfrac{1}{2}n^2 - \tfrac{20}{3}m - \tfrac{29}{3}n + \alpha(-3m - \tfrac{3}{2}n + 13),$$
$$(\cdot/) = \tfrac{1}{3}m^3 + 2m^2n + 2mn^2 + \tfrac{1}{3}n^3 - \ \ m^2 - 4mn - \ \ n^2 - \tfrac{46}{3}m - \tfrac{46}{3}n + \alpha(-3m - 3n + 20),$$
$$(//) = \tfrac{1}{6}m^3 + \ \ m^2n + 2mn^2 + \tfrac{2}{3}n^3 - \tfrac{1}{2}m^2 - 3mn - 2n^2 - \tfrac{29}{3}m - \tfrac{20}{3}n + \alpha(-\tfrac{3}{2}m - 3n + 13);$$

(1, 1, 1, 1)

$$\begin{aligned}(\,\cdot\,) = {} & \tfrac{1}{12}m^4 + \tfrac{2}{3}m^3n + m^2n^2 + \tfrac{1}{3}mn^3 + \tfrac{1}{24}m^4 \\ & - \tfrac{1}{2}m^3 - 3m^2n - 2mn^2 - \tfrac{1}{4}n^3 - \tfrac{181}{12}m^2 - 21mn - \tfrac{229}{24}n^2 + \tfrac{191}{2}m + \tfrac{403}{4}n \\ & + \alpha(-\tfrac{3}{2}m^2 - 3mn - \tfrac{3}{4}n^2 + \tfrac{43}{2}m + \tfrac{55}{4}n - \tfrac{357}{4}) + \alpha\,.\,\tfrac{9}{8},\end{aligned}$$

$$\begin{aligned}(\,/\,) = {} & \tfrac{1}{24}m^4 + \tfrac{1}{3}m^3n + m^2n^2 + \tfrac{2}{3}mn^3 + \tfrac{1}{12}n^4 \\ & - \tfrac{1}{4}m^3 - 2m^2n - 3mn^2 - \tfrac{1}{2}n^3 - \tfrac{229}{24}m^2 - 21mn - \tfrac{181}{2}n^2 + \tfrac{403}{4}m + \tfrac{191}{2}n \\ & + \alpha(-\tfrac{3}{4}m^2 - 3mn - \tfrac{3}{2}n^2 + \tfrac{55}{4}m + \tfrac{43}{2}n - \tfrac{357}{4}) + \alpha^2\,.\,\tfrac{9}{8};\end{aligned}$$

(2)

$$(\therefore) = \ \ \alpha,$$
$$(\,:/) = 2\alpha,$$
$$(\cdot//) = 2\alpha,$$
$$(///) = \ \ \alpha;$$

(2, 1)

$$(\,:\,) = 12m + 12n + (2m + \ \ n - 14)\,\alpha,$$
$$(\,\cdot/) = 24m + 24n + (2m + 2n - 24)\,\alpha,$$
$$(\,//) = 12m + 12n + (\ \ m + 2n - 14)\,\alpha;$$

(2, 1, 1)

$$(\,\cdot\,) = 24m^2 + 36mn + 12n^2 - 168m - 168n + \alpha(\ \ m^2 + 2mn + \tfrac{1}{2}n^2 - 25m - \tfrac{29}{2}n + 138) - \tfrac{3}{2}\alpha^2,$$
$$(\,/\,) = 12m^2 + 36mn + 24n^2 - 168m - 168n + \alpha(\tfrac{1}{2}m^2 + 2mn + \ \ n^2 - \tfrac{29}{2}m - 25n + 138) - \tfrac{3}{2}\alpha^2;$$

(2, 2)

$$(\,\cdot\,) = 27m + 24n - 20\alpha + \tfrac{1}{2}\alpha^2,$$
$$(\,/\,) = 24m + 27n - 20\alpha + \tfrac{1}{2}\alpha^2;$$

(3)

$$(\,:\,) = -4m - 3n + 3\alpha,$$
$$(\,\cdot/) = -8m - 8n + 6\alpha,$$
$$(\,//) = -3m - 4n + 3\alpha;$$

(3, 1)

$$(\cdot\ ) = -8m^2 - 12mn - 3n^2 + 56m + 53n + \alpha\,(6m + 3n - 39),$$

$$(\,/\,) = -3m^2 - 12mn - 8n^2 + 53m + 56n + \alpha\,(3m + 6n - 39);$$

(4)

$$(\cdot\ ) = -10m - 8n + 6\alpha,$$

$$(\,/\,) = -8m - 10n + 6\alpha.$$

50. By means of the foregoing formulæ I obtain, as will presently be shown, the following formulæ for the number of the conics which satisfy five conditions, viz.:

$$(5) = -15m - 15n + 9\alpha;$$

$$(4,\ 1) = -8m^2 - 20mn - 8n^2 + 104m + 104n + \alpha\,(6m + 6n - 66);$$

$$(3,\ 2) = 120m + 120n + \alpha\,(-4m - 4n - 78) + 3\alpha^2;$$

$$\begin{aligned}(3,\ 1,\ 1) = &-\tfrac{3}{2}m^3 - 10m^2n - 10mn^2 - \tfrac{3}{2}n^3 + \tfrac{109}{2}m^2 + 116mn + \tfrac{109}{2}n^2 - 434m - 434n\\ &+ \alpha\,(\tfrac{3}{2}m^2 + 6mn + \tfrac{3}{2}n^2 - \tfrac{69}{2}m - \tfrac{69}{2}n + 291) - \tfrac{9}{2}\alpha^2;\end{aligned}$$

$$\begin{aligned}(2,\ 2,\ 1) = &\ 24m^2 + 54mn + 24n^2 - 468m - 468n\\ &+ \alpha\,(-8m - 8n + 327) + \alpha^2\,(\tfrac{1}{2}m + \tfrac{1}{2}n - 12);\end{aligned}$$

$$\begin{aligned}(2,\ 1,\ 1,\ 1) = &\ 6m^3 + 30m^2n + 30mn^2 + 6n^3 - 174m^2 - 348mn - 174n^2 + 1320m + 1320n\\ &+ \alpha\,(\tfrac{1}{6}m^3 + m^2n + mn^2 + \tfrac{1}{6}n^3 - \tfrac{15}{2}m^2 - 26mn - \tfrac{15}{2}n^2 + \tfrac{358}{3}m + \tfrac{358}{3}n - 960)\\ &+ \alpha^2\,(-\tfrac{3}{2}m - \tfrac{3}{2}n + 28);\end{aligned}$$

$$\begin{aligned}(1,\ 1,\ 1,\ 1,\ 1)\,(^1) = &\ \tfrac{1}{120}m^5 + \tfrac{1}{12}m^4n + \tfrac{1}{3}m^3n^2 + \tfrac{1}{3}m^2n^3 + \tfrac{1}{12}mn^4 + \tfrac{1}{120}n^5\\ &- \tfrac{1}{12}m^4 - \tfrac{5}{6}m^3n - 2m^2n^2 - \tfrac{5}{6}mn^3 - \tfrac{1}{12}n^4\\ &- \tfrac{113}{12}m^3 - \tfrac{209}{12}m^2n - \tfrac{209}{12}mn^2 - \tfrac{113}{25}n^3\\ &+ \tfrac{1267}{12}m^2 + \tfrac{593}{3}mn + \tfrac{1267}{12}n^2 - \tfrac{3159}{5}m - \tfrac{3159}{5}n\\ &+ \alpha(-\tfrac{1}{4}m^3 - \tfrac{3}{2}m^2n - \tfrac{3}{2}mn^2 - \tfrac{1}{4}n^3 + \tfrac{29}{4}m^2 + 23mn + \tfrac{29}{4}n^2 - \tfrac{337}{4}m - \tfrac{337}{4}n + 486)\\ &+ \alpha^2\,(\tfrac{9}{8}m + \tfrac{9}{8}n - 15).\end{aligned}$$

51. I observe that by means of the above-mentioned expressions of $(X,\ 4Z)$ and $(2X,\ 3Z)$, the foregoing results, other than those for (5), (4, 1), &c., may be presented in a somewhat different form, viz. we have

$$(4Z)\,(1) = n\,(\cdot\ ) + m\,(/),$$

where $(\cdot)$ denotes $(4Z\,\cdot)$, $(/)$ denotes $(4Z\,/)$, and so in other cases, the understood term being $3Z$ or $2Z$, as the case may be.

[1] In my paper in the *Comptes Rendus*, I gave erroneously the coefficients $-\tfrac{3259}{5}m - \tfrac{3259}{5}n\,.\,.+\alpha\,(\ldots+\tfrac{195}{2})$.

$$
\begin{aligned}
(3Z)(2) &= (\cdot /)\,\tfrac{1}{2}\alpha\,; \\
(3Z)(1,\ 1) &= (\ :\ )(\tfrac{1}{2}n^2 - \tfrac{1}{2}n) \\
&\quad + (\cdot /)(mn - \tfrac{3}{4}\alpha) \\
&\quad + (\ //)(\tfrac{1}{2}m^2 - \tfrac{1}{2}m)\,; \\
(2Z)(3) &= (\therefore)(\ \tfrac{3}{2}m + \ n - \tfrac{3}{4}\alpha) \\
&\quad + (:/)(-\tfrac{7}{4}m - \tfrac{3}{2}n + \tfrac{9}{8}\alpha) \\
&\quad + (\cdot //)(-\tfrac{3}{2}m - \tfrac{7}{4}n + \tfrac{9}{8}\alpha) \\
&\quad + (///)(\ \ m + \tfrac{3}{2}n - \tfrac{3}{4}\alpha)\,; \\
(2Z)(2,\ 1) &= (\therefore)\{-3m - 3n + \alpha(-\tfrac{1}{4}m + \tfrac{1}{4}n + 2)\} \\
&\quad + (:/)\{\ \tfrac{9}{2}m + \tfrac{9}{2}n + \alpha(\ \tfrac{3}{8}m + \tfrac{1}{8}n - 4)\} \\
&\quad + (\cdot //)\{\ \tfrac{9}{2}m + \tfrac{9}{2}n + \alpha(\ \tfrac{1}{8}m + \tfrac{3}{8}n - 4)\} \\
&\quad + (\ ///)\{-3m - 3n + \alpha(\ \tfrac{1}{4}m - \tfrac{1}{4}n + 2)\}\,;
\end{aligned}
$$

$$(2Z)(1,\ 1,\ 1) =$$

$$
\begin{aligned}
&(\therefore)\{-\tfrac{1}{24}m^3 - \tfrac{1}{4}m^2n + \tfrac{1}{4}mn^2 + \tfrac{5}{24}n^3 + \tfrac{1}{8}m^2 + 0mn - \tfrac{5}{8}n^2 + \tfrac{11}{12}m + \tfrac{29}{12}n + \alpha(\ \tfrac{3}{8}m - \tfrac{3}{8}n - 1)\} \\
&+ (:/)\{\ \tfrac{1}{16}m^3 + \tfrac{3}{8}m^2n + \tfrac{1}{8}mn^2 - \tfrac{1}{16}n^3 - \tfrac{3}{16}m^2 - \tfrac{1}{2}mn + \tfrac{3}{16}n^2 - \tfrac{19}{8}m - \tfrac{25}{8}n + \alpha(-\tfrac{9}{16}m - \tfrac{3}{16}n + 3)\} \\
&+ (\cdot //)\{\ \tfrac{1}{16}m^3 + \tfrac{1}{8}m^2n + \tfrac{3}{8}mn^2 + \tfrac{1}{16}n^3 + \tfrac{3}{16}m^2 - \tfrac{1}{2}mn - \tfrac{3}{16}n^2 - \tfrac{25}{8}m - \tfrac{19}{8}n + \alpha(-\tfrac{3}{16}m - \tfrac{9}{16}n + 3)\} \\
&+ (\ ///)\{\ \tfrac{5}{24}m^3 + \tfrac{1}{4}m^2n - \tfrac{1}{4}mn^2 - \tfrac{1}{24}n^3 - \tfrac{5}{8}m^2 + 0mn + \tfrac{1}{8}n^2 + \tfrac{29}{12}m + \tfrac{11}{12}n + \alpha(-\tfrac{3}{8}m + \tfrac{3}{8}n - 1)\}\,;
\end{aligned}
$$

in all which formulæ it is to be recollected that we have

$$(\therefore) - \tfrac{3}{2}(:/) + \tfrac{3}{2}(\cdot //) - (///) = 0,$$

to which may be joined

$$(Z)(4X) = a\,(4X\cdot) + b\,(4X/),$$

where a, b are the representatives of the condition (Z), and where $(4X)$ is to be considered as standing successively for (4), (3, 1), (2, 2), (2, 1, 1), and (1, 1, 1, 1), the values of $(4X\cdot)$ and $(4X/)$ being in each case given by the foregoing Table.

52. The formulæ are very convenient for the calculation of the numbers of the conics which satisfy five conditions of contact with two given curves; thus if, for example, $(3Z)$, $=(3)_{m_1}$, denotes the condition of a contact of the third order with a given curve (m_1), then writing for symmetry $(2)_m$ in place of (2), we have

$$
\begin{aligned}
(3)_{m_1}(2)_m &= \tfrac{1}{2}\alpha\,(3\cdot /)_{m_1} \\
&= \alpha\,(-4m_1 - 4n_1 + 3\alpha_1).
\end{aligned}
$$

53. To obtain the foregoing expressions of (5), (4, 1), (3, 2), (3, 1, 1), (2, 2, 1), (2, 1, 1, 1), and (1, 1, 1, 1, 1), I assume that the given curve breaks up into two curves (m, n, α) and (m', n', α'), or, as we may for shortness express it, into two curves m and m'.

We have then

$$(5)_{m+m'} = (5)_m + (5)_{m'},$$

viz. the conics which have contact of the 5th order with the aggregate curve $m+m'$ are made up of the conics which have this contact with the curve m and the conics which have this contact with the curve m'. Writing this under the form

$$(5)_{m+m'} - (5)_m - (5)_{m'} = 0,$$

and observing that $(5)_m$ is a function $\phi(m, n, \alpha)$, and that consequently this is a functional equation $\phi(m+m', n+n', \alpha+\alpha') - \phi(m, n, \alpha) - \phi(m', n', \alpha') = 0$, the solution is

$$\phi(m, n, \alpha) = am + bn + c\alpha,$$

where a, b, c are arbitrary constants; but as the solution should be symmetrical in regard to m, n, we have $a = b$, or the solution is $\phi(m, n, \alpha) = a(m+n) + c\alpha$.

54. Similarly we have

$$(4, 1)_{m+m'} - (4, 1)_m - (4, 1)_{m'} = (4)_m (1)_{m'} + (4)_{m'} (1)_m,$$

viz. the conics which have with the aggregate curve $m+m'$ the contacts (4, 1) are made up of the conics which have the two contacts 4 and 1 with the one curve or with the other curve, or the contact 4 with the one curve and the contact 1 with the other curve. The expression on the right-hand side is a known function of (m, n, α), (m', n', α'); hence the form of the functional equation is

$$\phi(m+m', n+n', \alpha+\alpha') - \phi(m, n, \alpha) - \phi(m', n', \alpha') = F(m, n, \alpha, m', n', \alpha');$$

and any particular solution of this equation being obtained, the general solution is found by adding to it the term $am + bn + c\alpha$. Assuming that the particular solution is symmetrical in regard to (m, n), then the term to be added is as before $= a(m+n) + c\alpha$. And similarly for (3, 2), (3, 1, 1), &c.; that is, in every case we have a solution containing two arbitrary constants a, c, which remain to be determined.

55. Now in every case except $(5)_m$ the number of intersections of the conic with the curve is > 6 (viz. for $(4, 1)_m$ and $(3, 2)_m$ the number is 7, for (3, 1, 1) and (2, 2, 1), it is 8, and for the remaining two cases it is 9 and 10 respectively); hence if the given curve m be a cubic, the number of conics satisfying the prescribed conditions is $= 0$; and since a cubic may be the general cubic or a nodal or a cuspidal cubic, we have the three cases $(m, n, \alpha) = (3, 6, 18)$, $(3, 4, 12)$, and $(3, 3, 10)$. We have thus in each case three conditions for the determination of the constants a, c; so that there is in each case a verification of the resulting formula.

56. In the omitted case $(5)_m$, when the curve m is a cubic, the theory of the conics $(5)_m$ is a known one, viz. the points of contact of these conics, or the "sextactic" points of the cubic, are the points of contact of the tangents from the points of inflexion; the number of the conics $(5)_m$ is thus $= (n-3)\iota$, viz. in the three cases

respectively it is $=27$, 3, and 0. Hence for determining the constants we have the three equations

$$9a+18c=27,$$
$$7a+12c=\ 3,$$
$$6a+10c=\ 0,$$

which are satisfied by $a=-15$, $c=9$, and the resulting formula is

$$(5)=-15m-15n+9\alpha.$$

In the particular case of a curve without nodes or cusps, this is $(5)=12n-15m$, $=m(12m-27)$, which agrees with the result obtained in my memoir "On the Sextactic Points of a Plane Curve," *Phil. Trans.* vol. CLV. (1865), pp. 545—578, [341].

57. The subsidiary results required for the remaining cases (4, 1), &c. are at once obtained from the foregoing formulæ for $(4Z)(1)$, $(3Z)(2)$, &c.; for example, we have

$$\begin{aligned}(4)_m(1)_{m'} &= n'(-10m-\ 8n+6\alpha)\\ &\quad+m'(-\ 8m-10n+6\alpha),\end{aligned}$$

with like expressions for $(3,\ 1)_m(1)_{m'}$, &c.,

$$\begin{aligned}(3)_m(2)_{m'} &= \tfrac{1}{2}\alpha'\ (-8m-8n+6\alpha),\\ (3)_m(1,\ 1)_{m'} &= (\tfrac{1}{2}n'^2-\tfrac{1}{2}n')(-4m-3n+3\alpha)\\ &\quad+(m'n'-\tfrac{3}{4}\alpha')(-8m-8n+6\alpha)\\ &\quad+(\tfrac{1}{2}m'^2-\tfrac{1}{2}m')(-3m-4n+3\alpha);\end{aligned}$$

with like expressions for $(2,\ 1)_m(2)_{m'}$, $(2,\ 1)_m(1,\ 1)_{m'}$, &c. &c.

58. Calculation of (4, 1). We have

$$\begin{aligned}(4,\ 1)_{m+m'}-(4,\ 1)_m-(4,\ 1)_{m'} &= (4)_m(1)_{m'}+(4)_{m'}(1)_m,\\ &=-16mm'-20(mn'+m'n)-16nn'+6(\alpha n'+\alpha'n)+6(\alpha m'+\alpha'm),\end{aligned}$$

the integral of which is

$$(4,\ 1)_m=-8m^2-20mn-8n^2+a(m+n)+\alpha(6m+6n+c).$$

The particular cases $(m, n, \alpha)=(3, 6, 18)$, $(3, 4, 12)$, $(3, 3, 10)$ give respectively

$$0=252+9a+18c,$$
$$0=\ 64+7a+12c,$$
$$0=\ 36+6a+10c,$$

satisfied by $a=104$, $c=-66$.

59. Calculation of (3, 2). We have

$$(3,\ 2)_{m+m'} - (3,\ 2)_m - (3,\ 2)_{m'} = (3)_m\,(2)_{m'} + (3)_{m'}\,(2)_m,$$
$$= -4\,(m\alpha' + m'\alpha) - 4\,(n\alpha' + n'\alpha) + 6\alpha\alpha':$$

the integral is

$$(3,\ 2)_m = a\,(m+n) + \alpha\,(-4m - 4n + c) + 3\alpha^2,$$

and, as before,

$$324 + 9a + 18c = 0,$$
$$96 + 7a + 12c = 0,$$
$$60 + 6a + 10c = 0,$$

satisfied by $a = \mathbf{120}$, $c = -78$.

60. For the calculation of (3, 1, 1) we have similarly

$$(3,\ 1,\ 1)_{m+m'} - (3,\ 1,\ 1)_m - (3,\ 1,\ 1)_{m'} = (3)_m\,(1,\ 1)_{m'} + (3)_{m'}\,(1,\ 1)_m$$
$$+ (3,\ 1)_m\,(1)_{m'} + (3,\ 1)_{m'}\,(1)_m.$$

The function on the right-hand side was of course calculated from the values of $(3)_m(1,\ 1)_m$, &c.; but there is no use in this (and the more complicated cases which follow) in actually writing down the values of the function in question; it can in each case be calculated *backwards* from the foregoing expressions of (3, 1, 1), &c., and the values so obtained be verified by actual substitution. But assuming it to be known, the solution of the functional equation gives of course the foregoing expression for (3, 1, 1), except that the terms in $m+n$ and α are therein $a\,(m+n) + c\alpha$; and I shall in this and the subsequent cases give only the three equations which determine the constants. In the present case these are

$$-332 + 9a + 18c = 0,$$
$$-454 + 7a + 12c = 0,$$
$$-306 + 6a + 10c = 0,$$

satisfied by $a = -\mathbf{434}$, $c = 291$.

61. The remaining cases are (2, 2, 1), (2, 1, 1, 1) and (1, 1, 1, 1, 1). We have

$$(2,\ 2,\ 1)_{m+m'} - (2,\ 2,\ 1)_{m'} = (2,\ 2)_m\,(1)_{m'} + (2,\ 2)_{m'}\,(1)_m$$
$$+ (2,\ 1)_m\,(2)_{m'} + (2,\ 1)_{m'}\,(2)_m,$$

and

$$-1674 + 9a + 18c = 0,$$
$$-\ \ 648 + 7a + 12c = 0,$$
$$-\ \ 462 + 6a + 10c = 0,$$

satisfied by $a = -468$, $c = 327$.

Again,

$$(2,\ 1,\ 1,\ 1)_{m+m'} - (2,\ 1,\ 1,\ 1)_m - (2,\ 1,\ 1,\ 1)_{m'} = (2,\ 1,\ 1)_m\,(1)_{m'} + (2,\ 1,\ 1)_{m'}\,(1)_m$$
$$+ (2,\ 1)_m\,(1,\ 1)_{m'} + (2,\ 1)_{m'}\,(1)_m$$
$$+ (2)_m\,(1,\ 1,\ 1)_{m'} + (2)_{m'}\,(1,\ 1,\ 1)_m,$$

and

$$5400 + 9a + 18c = 0,$$
$$2280 + 7a + 12c = 0,$$
$$1680 + 6a + 10c = 0,$$

satisfied by $a = 1320$, $c = -960$; and finally,

$$(1, 1, 1, 1, 1)_{m+m'} - (1, 1, 1, 1, 1)_m - (1, 1, 1, 1, 1)_{m'} = (1, 1, 1, 1)_m (1)_{m'} + (1, 1, 1, 1)_{m'} (1)_m + (1, 1, 1)_m (1, 1)_{m'} + (1, 1, 1)_{m'} (1, 1)_m,$$

and

$$-30618 + 90a + 180c = 0,$$
$$-14094 + 70a + 120c = 0,$$
$$-10692 + 60a + 120c = 0,$$

satisfied by $10a = 6318$, $10c = 4860$, that is, $a = -\frac{3159}{5}$, $c = 486$.

62. The contacts of a conic with a given curve which have been thus far considered are contacts at unascertained points of the curve; but a conic may have with the given curve *at a given point thereof* a contact of the first order, the condition will be denoted by $(\bar{2})$; or a contact of the second order, the condition will be denoted by $(\bar{3})$, and so on. It is to be observed that the conditions $(\bar{2})$, $(\bar{3})$, &c. are sibireciprocal, the contact at a given point of the curve is the same thing as contact with a given tangent of the curve; but if we write $(\bar{1})$ to denote the condition of passing through a given point of the curve, this is *not* the same thing as the condition of touching a given tangent of the curve; and this last condition, if it were necessary to deal with it, might be denoted by $(\underline{1})$. But I attend only to the condition $(\bar{1})$. The expressions for the number of conics which satisfy such conditions as $(\bar{1})$, $(\bar{2})$, &c. are obtainable in several ways.

63. (1°) When the total number of conditions is 4, the question may be solved by Zeuthen's method, viz. by determining the line-pairs and point-pairs of the system $4Z$, with the proper numerical coefficients, and thence deducing the values of the characteristics $(4Z\cdot)$ and $(4Z/)$. A few cases are in fact thus solved in Zeuthen's work.

64. (2°) By the foregoing functional method. It is to be observed that there is a difference in the form of the functional equation, and that the general solution is always given in the form, Particular Solution + Constant, so that there is only a single constant to be determined by special considerations. To take the simplest example, let it be required to find the number of the conics $(3Z)(\bar{1}, 1)$: writing for shortness in place hereof $(\bar{1}, 1)$, or (in order to mark the curve (m) to which the symbol has reference) $(\bar{1}, 1)_m$, let the curve (m) be the aggregate of the curves (m) and (m'). Regarding the point $\bar{1}$ as a given point on the curve (m), that is, an arbitrary point in regard to the curve (m'), we have thus the equation

$$(\bar{1}, 1)_{m+m'} - (\bar{1}, 1)_m = (\cdot 1)_{m'},$$

where the right-hand side is known; and so in general the form of the functional equation is always $\phi(m+m') - \phi(m) =$ *given value*, that is,

$$\phi(m+m', n+n', \alpha+\alpha') - \phi(m, n, \alpha) = \text{given function of } (m, n, \alpha, m', n', \alpha');$$

whence, as stated, the general solution is Particular Solution + Constant. In the case in hand, taking successively $(3Z) = (\therefore)$, $(:/)$, $(\cdot//)$, and $(///)$, we have in the first of these cases

$$(\bar{1},\ 1)_{m+m'} - (\bar{1},\ 1)_m = n' + 2m',$$

whence $(\bar{1},\ 1)_m = n + 2m + \text{const.} = (\bar{1},\ 1)(\therefore)$; and the value of the constant being in any way ascertained to be $= -2$, we have $(\bar{1},\ 1)(\therefore) = n + 2m - 2$; and the like for the other three cases.

65. (3°) The expressions for the number of conics which satisfy such conditions as $(\bar{1})$, $(\bar{2})$, &c. are deducible with more or less facility from the corresponding expressions wherein $(\bar{1})$, $(\bar{2})$, &c. are replaced by $(\cdot)$, $(:)$, &c.; thus from [1] $(::1) = n + 2m$ we deduce

$$(\therefore \bar{1},\ 1) = (::/) - 2(\therefore \bar{2}) = n + 2m - 2,$$

viz. if one of the four arbitrary points of $(::/)$ becomes a point on the curve, then the condition $(::/)$ is satisfied specially by the conic $(\therefore \bar{2})$ which passes through the remaining three points and touches the curve at the point in question; 2 of the conics $(::/)$ coincide with the conic in question. We have thus a reduction $2(\therefore \bar{2})$, $= 2$, and the number of the conics $(\therefore \bar{1},\ 1)$ is $= n + 2m - 2$. Similarly, we have the system

$$\begin{aligned}
(\therefore \bar{1},\ 1 \quad\quad\ ) &= n + 2m - 2,\\
(:\ \bar{1},\ \bar{1},\ 1 \quad) &= n + 2m - 4,\\
(\cdot\ \bar{1},\ \bar{1},\ \bar{1},\ 1) &= n + 2m - 6,\\
(\bar{1},\ \bar{1},\ \bar{1},\ \bar{1},\ 1) &= n + 2m - 8.
\end{aligned}$$

Again, two or even three of the given points on the curve may come together without any reduction being thereby caused, that is, we have

$$\begin{aligned}
(:\ \bar{2},\ 1 \quad\ ) &\qquad\qquad\qquad = n + 2m - 4,\\
(\cdot\ \bar{2},\ \bar{1},\ 1 \quad) &= (\cdot\ \bar{3},\ 1 \quad) = n + 2m - 6,\\
(\ \bar{2},\ \bar{1},\ \bar{1},\ 1) &= (\ \bar{3},\ \bar{1},\ 1) = n + 2m - 8;
\end{aligned}$$

but if the four points on the curve coincide in pairs, or, what is the same thing, if in $(\bar{2},\ \bar{1},\ \bar{1},\ 1)$ the points $\bar{1}$ and $\bar{1}$ come to coincide, then there is a special reduction, and we have

$$(\bar{2},\ \bar{2},\ 1) = n + 2m - 8\,[-(m-2)] = m + n - 6,$$

viz. here $(m-2)$ of the conics come to coincide with the two points considered as a point-pair or infinitely thin conic. If the points $\bar{2}$ and $\bar{2}$ come to coincide, that is, if the four given points on the curve all coincide, there is no further reduction, but we have

$$(\bar{4},\ 1) = m + n - 6.$$

[1] I write indifferently $(1)(::)$, $(1::)$ or $(::1)$; and so in other cases.

66. The expressions involving a single $(\bar{1})$ may in every case be reduced by the foregoing method to depend upon other expressions; thus we have

$$\begin{array}{llll}
(3Z\ )(\bar{1},\ 1) & =(\cdot 1) & -2(\bar{2}) & , \\
(2Z\ )(\bar{1},\ 2) & =(\cdot 2) & -3(\bar{3}) & , \\
\quad ,,\quad (\bar{1},\ 1,\ 1) & =(\cdot 1,\ 1) & -2(\bar{2},\ 1) & , \\
(\ Z\ )(\bar{1},\ 1,\ 2) & =(\cdot 1,\ 2) & -2(\bar{2},\ 2) & -3(1,\ \bar{3}), \\
\quad ,,\quad (\bar{1},\ 1,\ 1,\ 1) & =(\cdot 1,\ 1,\ 1) & -2(\bar{2},\ 1,\ 1), & \\
\quad ,,\quad (\bar{1},\ 3) & =(\cdot 3) & -4(\bar{4}) & , \\
(\bar{1},\ 4) & =(\cdot 4) & -5(\bar{5}) & , \\
\&c., & & &
\end{array}$$

where, comparing for example the equations for $(Z)(\bar{1},\ 1,\ 2)$ and $(2Z)(\bar{1},\ 1,\ 1)$, it will be observed that in the first case the contacts 1, 2 of the symbol $(\bar{1},\ 1,\ 2)$ successively coalesce with the point $\bar{1}$, giving respectively $2(\bar{2},\ 2)$ and $3(1,\ \bar{3})$, the exterior factor being in each case the barred number, whereas the second case, where the contacts 1, 1 of the symbol $(\bar{1},\ 1,\ 1)$ are of the *same* order, we do not consider each of these symbols separately (thus obtaining $2(\bar{2},\ 1)+2(1,\ \bar{2}),\ =4(\bar{2},\ 1)$), but the identical symbol is taken only once, giving $2(\bar{2},\ 1)$. Thus we have also

$$(\bar{1},\ 1,\ 1,\ 1,\ 1)=(\cdot 1,\ 1,\ 1,\ 1)-2(\bar{2},\ 1,\ 1,\ 1).$$

67. The value of a symbol involving $(\bar{2})$, say the symbol $(3Z)(\bar{2})$, is connected with that of $\tfrac{1}{2}(3Z\cdot /)$; but as an instance of the correction which is sometimes required I notice the equation

$$(\bar{2},\ 1,\ 1,\ 1)=\tfrac{1}{2}(1,\ 1,\ 1\cdot /)-\{\tfrac{1}{2}(m-2)(m-3)+\tfrac{1}{2}(n-2)(n-3)+3(\bar{3},\ 1,\ 1)+2(\bar{4},\ 1)\},$$

which I have verified by other considerations.

68. We obtain the series of results:

$(\bar{1})$

$(\ ::\ )=1,$

$(\ .\!\cdot\!.\ /)=2,$

$(\ :\ //)=4,$

$(\cdot ///)=4,$

$(////)=2;$

$(\bar{1},\ 1)$

$(\ .\!\cdot\!.\ )=\ n+2m-2,$

$(\ :\ /\ )=2n+4m-4,$

$(\cdot //\ )=4n+4m-4,$

$(\ ///\ )=4n+2m-2;$

$(\bar{1}, 2)$

$(\ : \ ) = \alpha - 3,$

$(\cdot /) = 2\alpha - 6,$

$(//) = 2\alpha - 3;$

$(\bar{1}, 1, 1)$

$(\ : \ ) = 2m^2 + 2mn + \tfrac{1}{2}n^2 - 6m - \tfrac{5}{2}n + 8 - \tfrac{3}{2}\alpha,$

$(\cdot /) = 2m^2 + 4mn + n^2 - 6m - 5n + 12 - 3\alpha,$

$(//) = m^2 + 4mn + 2n^2 - 3m - 6n + 8 - 3\alpha;$

$(\bar{1}, 3)$

$(\ \cdot \ ) = - 4m - 3n - 4 + 3\alpha,$

$(\ / \ ) = - 8m - 8n - 4 + 6\alpha;$

$(\bar{1}, 1, 2)$

$(\ \cdot \ ) = 6m + 9n + 30 + \alpha(2m + n - 16),$

$(\ / \ ) = 21m + 18n + 30 + \alpha(2m + 2n - 26);$

$(\bar{1}, 1, 1, 1)$

$(\ \cdot \ ) = \tfrac{2}{3}m^3 + 2m^2n + mn^2 + \tfrac{1}{6}n^3 - 4m^2 - 7mn - \tfrac{3}{2}n^2 + \tfrac{22}{3}m - \tfrac{2}{3}n - 36 + \alpha(-3m - \tfrac{3}{2}n + 16),$

$(\ / \ ) = \tfrac{1}{3}m^3 + 2m^2n + 2mn^2 + \tfrac{1}{3}n^3 - 2m^2 - 8mn - 3n^2 - \tfrac{19}{3}m - \tfrac{4}{3}n - 36 + \alpha(-3m - 3n + 23);$

$(\bar{2})$

$(\therefore) = 1,$

$(: /) = 2,$

$(\cdot //) = 2,$

$(///) = 1;$

$(\bar{2}, 1)$

$(\ : \ ) = 2m + n - 4,$

$(\cdot /) = 2m + 2n - 6,$

$(//) = m + 2n - 4;$

$(\bar{2}, 2)$

$(\ \cdot \ ) = \alpha - 6,$

$(\ / \ ) = \alpha - 6;$

$(\bar{2}, 1, 1)$

$(\ \cdot \ ) = m^2 + 2mn + \tfrac{1}{2}n^2 - 7m - \tfrac{9}{2}n + 18 - \tfrac{3}{2}\alpha,$

$(\ / \ ) = \tfrac{1}{2}m^2 + 2mn + n^2 - \tfrac{9}{2}m - 7n + 18 - \tfrac{3}{2}\alpha;$

$(\bar{3})$

$(\ : \ ) = 1,$

$(\cdot /) = 2,$

$(//) = 1;$

$(\bar{3},\ 1)$

$$(\,\cdot\,) \qquad = n + 2m - 6,$$

$$(\,/\,) \qquad = 2n + m - 6\,;$$

$(\bar{4})$

$$(\,\cdot\,) \qquad = 1,$$

$$(\,/\,) \qquad = 1\,;$$

which are the several cases for the conics which satisfy not more than four conditions, and

69. For the conics satisfying 5 conditions, we have

$$(\bar{5}) \qquad = 1,$$

$$(\bar{4},\ 1) \qquad = m + n - 6,$$

$$(\bar{3},\ 2) \qquad = -9 + \alpha,$$

$$(\bar{3},\ 1,\ 1) \qquad = \tfrac{1}{2}m^2 + 2mn + \tfrac{1}{2}n^2 - \tfrac{13}{2}m - \tfrac{13}{2}n + 27 - \tfrac{3}{2}\alpha,$$

$$(\bar{2},\ 3) \qquad = -4m - 4n - 6 + 3\alpha,$$

$$(\bar{2},\ 2,\ 1) \qquad = 6m + 6n + 54 + \alpha(m + n - 15),$$

$$(\bar{2},\ 1,\ 1,\ 1) \qquad = \tfrac{1}{6}m^3 + m^2n + mn^2 + \tfrac{1}{6}n^3 - \tfrac{5}{2}m^2 - 8mn - \tfrac{5}{2}n^2 + \tfrac{37}{3}m + \tfrac{37}{3}n - 75$$
$$+ \alpha\left(-\tfrac{3}{2}m - \tfrac{3}{2}n + \tfrac{29}{2}\right),$$

$$(\bar{1},\ 4) \qquad = -10m - 8n - 5 + 6\alpha,$$

$$(\bar{1},\ 1,\ 3) \qquad = -8m^2 - 12mn - 3n^2 + 60m + 57n + 36 + \alpha(6m + 3n - 45),$$

$$(\bar{1},\ 2,\ 2) \qquad = 27m + 24n + 27 - 23\alpha + \tfrac{1}{2}\alpha^2,$$

$$(\bar{1},\ 1,\ 1,\ 2) \qquad = \tfrac{45}{2}m^2 + 30mn + \tfrac{21}{2}n^2 - \tfrac{321}{2}m - \tfrac{321}{2}n - 189$$
$$+ \alpha\left(m^2 + 2mn + \tfrac{1}{2}n^2 - 27m - \tfrac{33}{2}n + \tfrac{345}{2}\right) - \tfrac{3}{2}\alpha^2,$$

$$(\bar{1},\ 1,\ 1,\ 1,\ 1) = \tfrac{1}{12}m^4 + \tfrac{2}{3}m^3n + m^2n^2 + \tfrac{1}{3}mn^3 + \tfrac{1}{24}n^4 - \tfrac{5}{6}m^3 - 5m^2n - 4mn^2 - \tfrac{7}{2}n^3$$
$$- \tfrac{121}{12}m^2 - 5mn - \tfrac{209}{4}n^2 + \tfrac{425}{6}m + \tfrac{913}{4}n + 150$$
$$+ \alpha\left(-\tfrac{3}{2}m^2 - 3mn - \tfrac{3}{4}n^2 + \tfrac{49}{2}m + \tfrac{67}{4}n - \tfrac{473}{4}\right) + \tfrac{9}{8}\alpha^2.$$

70. The given point on the curve to which the symbols $\bar{1}$, $\bar{2}$, &c. refer may be a singular point, and in particular it is proper to consider the case where the point is a cusp. I use in this case an appropriate notation; a conic which simply passes through a cusp, in fact meets the curve at the cusp in two points; and I denote the condition of passing through the cusp by $\overline{1\kappa 1}$; similarly, a conic which touches the curve at the cusp, in fact there meets it in three points, and I denote the condition by $\overline{2\kappa 1}$; $\overline{1\kappa 1}$, $\overline{2\kappa 1}$ are thus special forms of $\bar{1}$, $\bar{2}$, and the annexed $\bar{1}$ indicates the additional point of intersection arising *ipso facto* from the point $\bar{1}$ or $\bar{2}$ being a cusp. Similarly, we should have the symbols $\overline{3\kappa 1}$, $\overline{4\kappa 1}$, $\overline{5\kappa 1}$; but it is to be observed that at a cusp of the curve there is no *proper* conic having a higher contact than

$\overline{2\kappa 1}$; thus if the symbol contains $\overline{3\kappa 1}$, or *à fortiori*, if it contain $\overline{4\kappa 1}$ or $\overline{5\kappa 1}$, the number of the conics is in every case $=0$; it is thus only the cases $\overline{1\kappa 1}$ and $\overline{2\kappa 1}$ which need to be considered.

71. The several modes of investigation which apply to the case of contact at a given ordinary point of the curve are applicable to the case of contact at a cusp: we may if we please employ the functional method; we have here a functional equation of the foregoing form, $\phi(m+m')-\phi m=$ given value (that is, $\phi(m+m',\ n+n',\ \alpha+\alpha')-\phi(m,\ n,\ \alpha)=$ given function of $(m,\ n,\ \alpha,\ m',\ n',\ \alpha')$), and the general solution is as before $=$ Particular Solution $+$ Constant; so that there is in each case a single arbitrary constant to be determined by special considerations. The determination of the constant is in some instances conveniently effected by means of the case of the cuspidal cubic: see Annexes Nos. 4 and 5.

The formation of the functional equation itself is similar to that in the corresponding case where the given point on the curve is an ordinary point. For example, we have

$$
\begin{aligned}
(2Z)\,(1,\ \bar{1},\ 1)_{m+m'}-(1,\ \bar{1},\ 1)_m = \ &(1,\ \bar{1})_m\,(1)_{m'} &&= n'\,(\bar{1},\ 1\,\cdot)_m+m'\,(\bar{1},\ 1\,/)_m\\
&+(\bar{1})_m\,(1,\ 1)_{m'} &&\quad+\tfrac{1}{2}\,(n'^2-n')\,(\bar{1}\,:)_m\\
& &&\quad+\ (m'n'-\tfrac{3}{4}\alpha')\,(\bar{1}\,\cdot\,/)_m\\
& &&\quad+\tfrac{1}{2}\,(m'^2\ -m')\,(\bar{1}\,//)_m,
\end{aligned}
$$

and we may herein simply change $\bar{1}$ into $\overline{1\kappa 1}$. Writing successively $2Z=(\,:\,)$, $(\cdot\,/)$ and $(//)$, we find

$$
\begin{aligned}
(\overline{1\kappa 1},\ 1,\ 1\,:)_{m+m'}-(\ :\ )_m&=n'(\ n+2m-3)+m'(2n+2m-6)+(\tfrac{1}{2}n'^2-\tfrac{1}{2}n')1+(m'n'-\tfrac{3}{4}\alpha')2+(\tfrac{1}{2}m'^2-\tfrac{1}{2}m')4,\\
(\qquad\quad \cdot\,/)_{m+m'}-(\cdot/\ )_m&=n'(2n+4m-6)+m'(4n+4m-6)+(\tfrac{1}{2}n'^2-\tfrac{1}{2}n')2+(m'n'-\tfrac{3}{4}\alpha')4+(\tfrac{1}{2}m'^2-\tfrac{1}{2}m')4,\\
(\qquad\quad //)_{m+m'}-(\ //)_m&=n'(4n+4m-6)+m'(4n+2m-3)+(\tfrac{1}{2}n'^2-\tfrac{1}{2}n')4+(m'n'-\tfrac{3}{4}\alpha')4+(\tfrac{1}{2}m'^2-\tfrac{1}{2}m')4,
\end{aligned}
$$

which only differ from the corresponding expressions with $\bar{1}$ in that they contain

$$n+2m-3,\ 2n+4m-6,\ 4n+4m-6,\ 4m+2n-3$$

in place of

$$n+2m-2,\ 2n+4m-4,\ 4n+4m-4,\ 4m+2n-2$$

respectively, and they lead to the expressions for $(\overline{1\kappa 1},\ 1,\ 1\,:)$, &c., the arbitrary constant being in each case properly determined.

72. We have

$(\overline{1\kappa 1})$

$(\ ::\ )=1,$

$(\,\therefore/\,)=2,$

$(\,://\,)=4,$

$(\cdot\,///)=4,$

$(\,////)=2;$

$(\overline{1\kappa 1},\ 1)$

$$(\ \therefore\ ) = n + 2m - 3,$$
$$(\ :/\ ) = 2n + 4m - 6,$$
$$(\ \cdot//\ ) = 4n + 4m - 6,$$
$$(\ ///\ ) = 4n + 2m - 3;$$

$(\overline{1\kappa 1},\ 2)$

$$(\ :\ ) = \alpha - 4,$$
$$(\ \cdot/\ ) = 2\alpha - 8,$$
$$(\ //\ ) = 2\alpha - 4;$$

$(\overline{1\kappa 1},\ 1,\ 1)$

$$(\ :\ ) = 2m^2 + 2mn + \tfrac{1}{2}n^2 - 8m - \tfrac{7}{2}n + 13 - \tfrac{3}{2}\alpha,$$
$$(\ \cdot/\ ) = 2m^2 + 4mn + n^2 - 8m - 7n + 18 - 3\alpha,$$
$$(\ //\ ) = m^2 + 4mn + 2n^2 - 4m - 8n + 12 - 3\alpha;$$

$(\overline{1\kappa 1},\ 3)$

$$(\ \cdot\ ) = -4m - 3n - 5 + 3\alpha,$$
$$(\ /\ ) = -8m - 8n - 6 + 6\alpha;$$

$(\overline{1\kappa 1},\ 1,\ 2)$

$$(\ \cdot\ ) = 4m + 8n + 44 + \alpha\,(2m + n - 17),$$
$$(\ /\ ) = 20m + 16n + 42 + \alpha\,(2m + 2n - 27);$$

$(\overline{1\kappa 1},\ 1,\ 1,\ 1)$

$$(\ \cdot\ ) = \tfrac{2}{3}m^3 + 2m^2n + mn^2 + \tfrac{1}{6}n^3 - 5m^2 - 9mn - 2n^2 + \tfrac{43}{3}m + \tfrac{29}{6}n - 57 + \alpha(-3m - \tfrac{3}{2}n + \tfrac{35}{2}),$$
$$(\ /\ ) = \tfrac{1}{3}m^3 + 2m^2n + 2mn^2 + \tfrac{1}{3}n^3 - \tfrac{5}{2}m^2 - 10mn - 4n^2 - \tfrac{11}{6}m + \tfrac{17}{3}n - 54 + \alpha(-3m - 3n + \tfrac{49}{2});$$

$(\overline{2\kappa 1})$

$$(\ \therefore\ ) = 1,$$
$$(\ :/\ ) = 2,$$
$$(\ \cdot//\ ) = 2,$$
$$(\ ///\ ) = 1;$$

$(\overline{2\kappa 1},\ 1)$

$$(\ :\ ) = 2m + n - 5,$$
$$(\ \cdot/\ ) = 2m + 2n - 6,$$
$$(\ //\ ) = m + 2n - 4;$$

$(\overline{2\kappa 1},\ 2)$

$$(\ \cdot\ ) = \alpha - 7,$$
$$(\ /\ ) = \alpha - 6;$$

$(\overline{2\kappa 1}, 1, 1)$

$$
\begin{aligned}
(\ \cdot\ ) &= m^2 + 2mn + \tfrac{1}{2}n^2 - 7m - \tfrac{11}{2}n + 21 - \tfrac{3}{2}\alpha, \\
(\ /\ ) &= \tfrac{1}{2}m^2 + 2mn + n^2 - \tfrac{9}{2}m - 7n + 18 - \tfrac{3}{2}\alpha.
\end{aligned}
$$

73. The remainder of this table, being the part where the symbols $(\cdot)$ and $(/)$ do not occur, I present under a somewhat different form as follows:

$$
\begin{array}{lll}
(\overline{5\kappa 1}) & = 0, & \\
(\overline{4\kappa 1}, 1) & = 0, & \\
(\overline{3\kappa 1}, 2) & = 0, & \\
(\overline{3\kappa 1}, 1, 1) & = 0, & \\
(\overline{2}, 3) & - (\overline{2\kappa 1}, 3) & = 0, \\
(\overline{2}, 2, 1) & - (\overline{2\kappa 1}, 2, 1) & = n - 3, \\
(\overline{2}, 1, 1, 1) & - (\overline{2\kappa 1}, 1, 1, 1) & = \tfrac{1}{2}(n-3)(n-4), \\
(\overline{1}, 4) & - (\overline{1\kappa 1}, 4) & = 1, \\
(\overline{1}, 1, 3) & - (\overline{1\kappa 1}, 1, 3) & = (\overline{2\kappa 1}, 3) + (n-3), \\
(\overline{1}, 2, 2) & - (\overline{1\kappa 1}, 2, 2) & = 3(n-3) + \kappa - 1, \\
(\overline{1}, 1, 1, 2) & - (\overline{1\kappa 1}, 1, 1, 2) & = (\overline{2\kappa 1}, 1, 2) + \tfrac{1}{2}(n-3)(n-4) + \delta + 2\overline{n-3}\,\overline{m-4}, \\
(\overline{1}, 1, 1, 1, 1) & - (\overline{1\kappa 1}, 1, 1, 1, 1) & = (\overline{2\kappa 1}, 1, 1, 1).
\end{array}
$$

These results relating to a cusp, are useful for the investigations contained in the Second Memoir.

It will be noticed that the symbols which contain $\overline{2\kappa 1}$ are not, like those which contain $\overline{2}$, symmetrical in regard to (m, n): the interchange of (m, n) would of course imply the change of a cusp into an inflexion, and would therefore give rise to a new symbol such as $\overline{2\iota 1}$; but I have not thought it necessary to consider the formulæ which contain this new symbol.

Investigations in extension of those of De Jonquières *in relation to the contacts of a Curve of the order* r *with a given curve.* Article Nos. 74 to 93.

74. De Jonquières has given a formula for the number of curves C^r of the order r which have with a given curve U^m of the mth order t contacts of the orders a, b, c, &c. respectively, which besides pass through p points distributed at pleasure on the curve U^m (this includes the case of contacts of any orders at given points of the curve U^m), and which moreover satisfy any other $\tfrac{1}{2}r(r+3) - (a+b+c+\&c.) - p$

conditions; viz. the number of the curves C^r is $= \mu(a+1)(b+1)(c+1)\ldots$ into

$$\begin{cases} \quad [rm-(a+b+c\,..)-p \quad]^t \\ +[rm-(a+b+c\,..)-p-1]^{t-1}(a \;+b \;+c \;..)[D]^1 \\ +[rm-(a+b+c\,..)-p-2]^{t-2}(ab+ac+bc\,..)[D]^2 \\ \quad\vdots \\ +[rm-(a+b+c\,..)-p-t]^{\circ} \;\; (abc\ldots \qquad)[D]^t, \end{cases}$$

where the curve U^m is a curve *without cusps*, and having therefore a deficiency $D=\frac{1}{2}(m-1)(m-2)-\delta$; the numbers $a, b, c,..$ are assumed to be all of them unequal, but if we have α of them each $=a$, β of them each $=b$, &c., then the foregoing expression is to be divided by $[\alpha]^\alpha[\beta]^\beta\ldots$; and μ denotes the number of the curves C^r which satisfy the system of conditions obtained from the given system by replacing the conditions of the t contacts of the orders a, b, c, &c. respectively by the condition of passing through $a+b+c\ldots$ arbitrary points. In order that the formula may give the number of the *proper curves* C^r which satisfy the prescribed conditions, it is sufficient that the $\frac{1}{2}r(r+3)-(a+b+c\,..)-p$ conditions shall include the conditions of passing through at least a certain number T of arbitrary points: this restriction applies to all the formulæ of the present section.

75. I will for convenience consider this formula under a somewhat less general form, viz. I will put $p=0$, and moreover assume that the $\frac{1}{2}r(r+3)-(a+b+c\,..)$ conditions are the conditions of passing through this number of arbitrary points; whence $\mu=1$.

We have thus a curve C^r having with the given curve U^m t contacts of the orders $a, b, c\,..$ respectively, and besides passing through $\frac{1}{2}r(r+3)-(a+b+c\,..)$ arbitrary points; and the number of such curves is by the formula $=(a+1)(b+1)(c+1),\ldots$ into

$$\begin{cases} \quad [rm-(a+b+c\,..) \quad]^t \\ +[rm-(a+b+c\,..)-1]^{t-1}(a \;+b \;+c \;..)[D]^1 \\ +[rm-(a+b+c\,..)-2]^{t-2}(ab+ac+bc\,..)[D]^2 \\ \quad\vdots \\ +[rm-(a+b+c\,..)-t\,]^{\circ} \;\; (abc\ldots \qquad)[D]^t, \end{cases}$$

where, as before, in the case of any equalities between the numbers $a, b, c, \ldots$, the expression is to be divided by $[\alpha]^\alpha[\beta]^\beta\ldots$.

76. I have succeeded in extending the formula to the case of a curve with cusps: instead of writing down the general formula, I will take successively the cases of a single contact a, two contacts a, b, three contacts a, b, c, &c.; and then denoting the numbers of the curves C^r by (a), (a, b), (a, b, c), &c. in these cases respectively, I say that we have

$$(a) = \quad (a+1)\begin{Bmatrix} rm-a \\ +aD \end{Bmatrix}$$
$$-a \;\;.\;\;.\;\; \kappa;$$

$$(a,\ b) = (a+1)(b+1)\left\{\begin{array}{l} [rm-a-b]^2 \\ +\,[rm-a-b-1]^1\,(a+b)\,[D]^1 \\ +\qquad\qquad ab\quad [D]^2 \end{array}\right\}$$

$$-\left\{\begin{array}{l} a\,(b+1)\left\{\begin{array}{l}[rm-a-b-1]^1 \\ +\qquad\qquad bD\end{array}\right\} \\ +\,b\,(a+1)\left\{\begin{array}{l}[rm-a-b-1]^1 \\ +\qquad\qquad aD\end{array}\right\}\end{array}\right\}[\kappa]^1$$

$$+\,ab \quad . \quad . \quad . \quad . \quad . \quad . \quad . \quad . \quad . \quad . \quad [\kappa]^2,$$

$$(a,\ b,\ c) = (a+1)(b+1)(c+1)\left\{\begin{array}{l} [rm-a-b-c\quad]^3 \\ +\,[rm-a-b-c-1]^2\,(a\ +\ b+\ c)\,[D]^1 \\ +\,[rm-a-b-c-2]^1\,(ab+ac+bc)\,[D]^2 \\ +\qquad\qquad abc\qquad [D]^3 \end{array}\right\}$$

$$-\,[\Sigma c\,(a+1)(b+1)\left\{\begin{array}{l} [rm-a-b-c-1]^2 \\ +\,[rm-a-b-c-2]^1\,(a+b)\ [D]^1 \\ +\qquad\qquad ab\quad [D]^2 \end{array}\right\}]\,[\kappa]^1$$

$$+\,[\Sigma bc\,(a+1)\left\{\begin{array}{l} [rm-a-b-c-2]^1 \\ +\qquad\qquad aD \end{array}\right\}]\,[\kappa]^2$$

$$-\,abc \quad . \quad . \quad . \quad . \quad . \quad . \quad . \quad . \quad . \quad . \quad [\kappa]^3.$$

77. The foregoing examples are sufficient to exhibit the law; but as I shall have to consider the cases of four and five contacts, I will also write down the formula for $(a,\ b,\ c,\ d)$, putting therein for shortness

$$a+b+c+d=\alpha,\ ab+..+cd=\beta,\ abc..+bcd=\gamma,\ abcd=\delta,$$
$$a+b+c=\alpha',\ ab+ac+bc=\beta',\ abc=\gamma',\ a+b=\alpha'',\ ab=\beta'',\ a=\alpha''';$$

and also the formula for $(a,\ b,\ c,\ d,\ e)$, putting therein in like manner

$$(\alpha,\ \beta,\ \gamma,\ \delta,\ \epsilon),\ (\alpha',\ \beta',\ \gamma',\ \delta'),\ (\alpha'',\ \beta'',\ \gamma''),\ (\alpha''',\ \beta'''),\ (\alpha'''')$$

for the combinations of $(a,\ b,\ c,\ d,\ e)$, $(a,\ b,\ c,\ d)$, $(a,\ b,\ c)$, $(a,\ b)$ and (a) respectively. We have

$$(a,\ b,\ c,\ d) = (a+1)(b+1)(c+1)(d+1)\left\{\begin{array}{l} [rm-\alpha\quad]^4 \\ +\,[rm-\alpha-1]^3\,\alpha\,[D]^1 \\ +\,[rm-\alpha-2]^2\,\beta\,[D]^2 \\ +\,[rm-\alpha-3]^1\,\gamma\,[D]^3 \\ +\qquad\qquad \delta\,[D]^4 \end{array}\right\}$$

$$[\Sigma d\,(a+1)(b+1)(c+1)\left\{\begin{array}{l}\;\;[rm-\alpha-1]^3 \\ +[rm-\alpha-2]^2\,\alpha'\,[D]^1 \\ +[rm-\alpha-3]^1\,\beta'\,[D]^2 \\ +\qquad\qquad\quad\gamma'\,[D]^3\end{array}\right\}][\kappa]^1$$

$$+[\Sigma cd\,.\,(a+1)(b+1)\left\{\begin{array}{l}\;\;[rm-\alpha-2]^2 \\ +[rm-\alpha-3]^1\,\alpha''\,[D]^1 \\ +\qquad\qquad\quad\beta''\,[D]^2\end{array}\right\}][\kappa]^2$$

$$-[\Sigma bcd\,(a+1)\left\{\begin{array}{l}\;\;[rm-\alpha-3]^1 \\ +\qquad\qquad\quad\alpha'''\,[D]^1\end{array}\right\}][\kappa]^3$$

$$+\,abcd\;\;.\;\;.\;\;.\;\;.\;\;.\;\;.\;\;.\;\;.\;\;.\;\;.\;\;[\kappa]^4,$$

$$(a,\,b,\,c,\,d,\,e)=(a+1)(b+1)(c+1)(d+1)(e+1)\left\{\begin{array}{l}\;\;[rm-\alpha\quad]^5 \\ +[rm-\alpha-1]^4\,\alpha\;\;[D]^1 \\ +[rm-\alpha-2]^3\,\beta\;\;[D]^2 \\ +[rm-\alpha-3]^2\,\gamma\;\;[D]^3 \\ +[rm-\alpha-4]^1\,\delta\;\;[D]^4 \\ +\qquad\qquad\quad\epsilon\;\;[D]^5\end{array}\right\}$$

$$-[\Sigma e\,(a+1)(b+1)(c+1)(d+1)\left\{\begin{array}{l}\;\;[rm-\alpha-1]^4 \\ +[rm-\alpha-2]^3\,\alpha'\,[D]^1 \\ +[rm-\alpha-3]^2\,\beta'\,[D]^2 \\ +[rm-\alpha-4]^1\,\gamma'\,[D]^3 \\ +\qquad\qquad\quad\delta'\,[D]^4\end{array}\right\}][\kappa]^1$$

$$+[\Sigma de\,(a+1)(b+1)(c+1)\left\{\begin{array}{l}\;\;[rm-\alpha-2]^3 \\ +[rm-\alpha-3]^2\,\alpha''\,[D]^1 \\ +[rm-\alpha-4]^1\,\beta''\,[D]^2 \\ +\qquad\qquad\quad\gamma''\,[D]^3\end{array}\right\}][\kappa]^2$$

$$-[\Sigma cde\,(a+1)(b+1)\left\{\begin{array}{l}\;\;[rm-\alpha-3]^2 \\ +[rm-\alpha-4]^1\,\alpha'''\,[D]^1 \\ +\qquad\qquad\quad\beta'''\,[D]^2\end{array}\right\}][\kappa]^3$$

$$+[\Sigma bcde\,(a+1)\left\{\begin{array}{l}\;\;[rm-\alpha-4]^1 \\ +\qquad\qquad\quad\alpha''''\,[D]^1\end{array}\right\}][\kappa]^4$$

$$-\,abcde\;.\;\;.\;\;.\;\;.\;\;.\;\;.\;\;.\;\;.\;\;.\;\;.\;\;.\;\;.\;\;[\kappa]^5.$$

78. In all these formulæ there is, as before, a numerical divisor in the case of any equalities among the numbers a, b, c, &c. And D denotes, as before, the deficiency, viz. its value now is $D=\frac{1}{2}(m-1)(m-2)-\delta-\kappa$; or observing that the class n is $=m^2-m-2\delta-3\kappa$, we have $D=\frac{1}{2}n-m+1+\frac{1}{2}\kappa$, or say $D=1-m+\frac{1}{2}n+\frac{1}{2}\kappa$, $=1+\Delta$ if $\Delta=-m+\frac{1}{2}n+\frac{1}{2}\kappa$.

79. It is to be observed with reference to the applicability of these formulæ within certain limits only, that the formulæ are the *only* formulæ which are *generally* true; thus taking the simplest case, that of a single contact a, the only algebraical expression for the number of the curves C^r which have with a given curve U^m a contact of the order a, and besides pass through the requisite number $\frac{1}{2}r(r+3)-a$ of arbitrary points, is that given by the formula, viz.

$$(a)=(a+1)(rm-a+aD)-a\kappa.$$

Considering the curve U^m and the order r of the curve C^r as given, if a has successively the values 1, 2,... up to a limiting value of a, the formula gives the number of the proper curves C^r which have with the given curve U^m a contact of the required order a: beyond this limiting value the formula no longer gives the number of the proper curves C^r which satisfy the required condition, and it thus ceases to be applicable; but there is no algebraic function of a which would give the number of the proper curves C^r as well beyond as up to the foregoing limiting value of a.

80. The formulæ are applicable provided only the conditions include the conditions of passing through a sufficient number of arbitrary points; viz. when the number of arbitrary points is sufficiently great, it is not possible to satisfy the conditions specially by means of improper curves C^r, being or comprising a pair of coincident curves. Thus to take a simple example, suppose it is required to find the number of the conics which touch a given curve t times and besides pass through $5-t$ given points: if the number of the given points be 4 or 3 there is no coincident line-pair through the given points, and therefore no coincident line-pair satisfying the given conditions; if the number of the given points is $=2$, then the line joining these points gives a coincident line-pair having at each of its m intersections with the given curve a special contact therewith, that is, having in $\frac{1}{6}m(m-1)(m-2)$ ways three special contacts with the given curve; if the number of the given points is 1 or 0, then in the first case any line whatever through the given point, and in the second case any line whatever, regarded as a coincident line-pair, has m special contacts with the given curve; and so in general there is a certain value for the number of given points, for which value the conditions of contact may be satisfied by a determinate number of improper curves C^r, and for values inferior to it the conditions may be satisfied by infinite series of improper curves C^r. It is by such considerations as these that De Jonquières has determined the minimum value T of the number of arbitrary points to which the conditions should relate in order that the formulæ may be applicable: I refer for his investigation and results to paragraphs XVII and XVIII of his memoir. I remark that in the case where the number of improper solutions is finite, the formula can be corrected so as to give the number of proper solutions by simply

subtracting the number of the improper solutions: but this is not so when the improper solutions are infinite in number; the mode of obtaining the approximate formula is here to be sought in the considerations contained in the first part of the present Memoir; see in particular *ante*, Nos. 8, 9 and 10.

81. The expressions for (a), (a, b), &c. may be considered as functions of rm, $1+\Delta$, and κ, and they vanish upon writing therein $rm=0$, $\Delta=0$, $\kappa=0$; they are consequently of the form $(rm, \Delta, \kappa)^1+(rm, \Delta, \kappa)^2+$ &c., and I represent by $[a]$, $[a, b]$, &c. the several terms $(rm, \Delta, \kappa)^1$, which are the portions of (a), (a, b), &c. respectively, linear in rm, Δ, and κ. The terms in question are obtained with great facility; thus, to fix the ideas, considering the expressions for (a, b, c, d),

1°. To obtain the term in rm, we may at once write $D=1$, $\kappa=0$, the expression is thus reduced to

$$(a+1)(b+1)(c+1)(d+1)\{[rm-\alpha]^4+[rm-\alpha-1]^3\alpha\},$$

and the factor in $\{\ \}$ being $=rm\,[rm-\alpha-1]^3$, the coefficient of rm is

$$(a+1)(b+1)(c+1)(d+1)[-\alpha-1]^3,$$

which is

$$=-(a+1)(b+1)(c+1)(d+1).(\alpha+1)(\alpha+2)(\alpha+3).$$

2°. To obtain the term in Δ, writing $rm=0$, $\kappa=0$, and observing that

$$[D]^1=\Delta+1,\quad [D]^2=(\Delta+1)\Delta,\quad [D]^3=(\Delta+1)\Delta(\Delta-1),\quad [D]^4=(\Delta+1)\Delta(\Delta-1)(\Delta-2),$$

&c. give the terms Δ, Δ, $-\Delta$, $+2\Delta$, -6Δ, &c. respectively, the term in Δ is

$$(a+1)(b+1)(c+1)(d+1)\left\{\begin{array}{lr} [-\alpha-1]^3\alpha\,. & 1 \\ +[-\alpha-2]^2\beta\,. & 1 \\ +[-\alpha-3]^1\gamma\,. & -1 \\ +\qquad\qquad\delta\,. & 2 \end{array}\right\}\Delta$$

$$=(a+1)(b+1)(c+1)(d+1)\left\{\begin{array}{lr} -\ \alpha & (\alpha+1)(\alpha+2)(\alpha+3) \\ +\ \beta & (\alpha+2)(\alpha+3) \\ +\ \gamma & (\alpha+3) \\ +2\delta & \end{array}\right\}\Delta.$$

3°. For the term in κ, writing $rm=0$, $D=1$, and observing that $[\kappa]^1$, $[\kappa]^2$, $[\kappa]^3$, $[\kappa]^4$ give respectively the terms κ, $-\kappa$, 2κ, -6κ, this is

$$=\left[\begin{array}{llr} -\Sigma d & (a+1)(b+1)(c+1)\{[-\alpha-1]^3+[-\alpha-2]^2\alpha'\}. & 1 \\ +\Sigma cd & (a+1)(b-1)\qquad\{[-\alpha-2]^2+[-\alpha-3]^1\alpha''\}. & -1 \\ -\Sigma bcd & (a-1)\qquad\{[-\alpha-3]^1+\qquad\alpha'''\}. & 2 \\ +\ abcd & & .-6 \end{array}\right]\kappa,$$

where the terms in { } are

$$-(\alpha+1-\alpha')(\alpha+2)(\alpha+3),\quad (\alpha+2-\alpha'')(\alpha+3) \text{ and } -(\alpha+3-\alpha'''),$$

that is,

$$-(d+1)(\alpha+2)(\alpha+3),\quad (c+d+2)(\alpha+3) \text{ and } -(b+c+d+3)$$

respectively; whence the whole expression is

$$=\begin{pmatrix} \Sigma d & (a+1)(b+1)(c+1)(d+1).(\alpha+2)(\alpha+3) \\ -\ \Sigma cd & (c+d+2)(a+1)(b+1).\qquad (\alpha+3) \\ +\ 2\Sigma bcd & (b+c+d+3)(a+1) \\ -\ 6\ abcd & \end{pmatrix}\kappa,$$

the expression multiplying $(\alpha+2)(\alpha+3)$ is

$$(a+1)(b+1)(c+1)(d+1)\,\Sigma d, \ =(a+1)(b+1)(c+1)(d+1)\,\alpha;$$

and we have moreover

$$(a+1)(b+1)(c+1)(d+1)=(1+\alpha+\beta+\gamma+\delta);$$

the other lines are of course expressible in terms of $(\alpha, \beta, \gamma, \delta)$, but as the law of their formation would then be hidden, I abstain from completing the reduction.

82. The series of formulæ is

$$\begin{aligned}[a] \ &= \ (a+1)\,rm \\ &\quad +(a+1)\,a\Delta \\ &\quad - \qquad a\kappa, \end{aligned}$$

$$\begin{aligned}[a,\ b] &= -(a+1)(b+1)(\alpha+1)\quad ..\,rm \\ &\quad -(a+1)(b+1)\begin{Bmatrix} \alpha(\alpha+1) \\ -\qquad \beta \end{Bmatrix}\Delta \\ &\quad + \qquad \begin{Bmatrix} \Sigma b(a+1)(b+1) \\ -\ ab \end{Bmatrix}\kappa, \end{aligned}$$

where $\alpha=a+b$, $\beta=ab$; and coeff. of κ expressed in terms of α, β is $=\alpha(1+\alpha+\beta)-\beta$.

$$\begin{aligned}[a,\ b,\ c] &= \ (a+1)(b+1)(c+1)(\alpha+1)(\alpha+2)\quad ..\,rm \\ &\quad +(a+1)(b+1)(c+1)\begin{Bmatrix} \alpha(\alpha+1)(\alpha+2) \\ -\qquad \beta(\alpha+2) \\ -\qquad \gamma \end{Bmatrix}\Delta \\ &\quad +\begin{Bmatrix} -\Sigma\ c(a+1)(b+1)(c+1)(\alpha+2) \\ +\Sigma bc(b+c+2)(a+1) \\ -2abc \end{Bmatrix}\kappa, \end{aligned}$$

where $\alpha=a+b+c$, $\beta=ab+ac+bc$, $\gamma=abc$; and the coefficient of κ expressed in terms of α, β, γ is $=-\alpha^3-\alpha^2\beta-\alpha^2\gamma-3\alpha^2-\alpha\beta-2\alpha+2\beta+\gamma$.

$$[a,\ b,\ c,\ d] = -(a+1)(b+1)(c+1)(d+1)(\alpha+1)(\alpha+2)(\alpha+3) \qquad ..rm$$
$$-(a+1)(b+1)(c+1)(d+1)\left\{\begin{array}{lr} & \alpha(\alpha+1)(\alpha+2)(\alpha+3) \\ -\ \beta & (\alpha+2)(\alpha+3) \\ -\ \gamma & (\alpha+3) \\ -2\delta & \end{array}\right\}\Delta$$
$$+\left\{\begin{array}{llll} +\ \Sigma & d(a+1)(b+1)(c+1)(d+1) & & (\alpha+2)(\alpha+3) \\ -\ \Sigma & cd(c+d+2) & (a+1)(b+1) & (\alpha+3) \\ +2\Sigma & bcd(b+c+d+3)(a+1) & & \\ -6 & abcd & & \end{array}\right\}\kappa,$$

where $\alpha = a+b+c+d,\ ..\ \delta = abcd.$

$$[a,\ b,\ c,\ d,\ e] = (a+1)(b+1)(c+1)(d+1)(e+1)(\alpha+1)(\alpha+2)(\alpha+3)(\alpha+4) \qquad ..rm$$
$$+(a+1)(b+1)(c+1)(d+1)(e+1)\left\{\begin{array}{lr} & \alpha(\alpha+1)(\alpha+2)(\alpha+3)(\alpha+4) \\ -\ \beta & (\alpha+2)(\alpha+3)(\alpha+4) \\ -\ \gamma & (\alpha+3)(\alpha+4) \\ -2\delta & (\alpha+4) \\ -6\epsilon & \end{array}\right\}\Delta$$
$$+\left\{\begin{array}{llr} -\ \Sigma e & (a+1)(b+1)(c+1)(d+1)(e+1) & (\alpha+2)(\alpha+3)(\alpha+4) \\ +\ \Sigma de & (d+e+2)(a+1)(b+1)(c+1) & (\alpha+3)(\alpha+4) \\ -2\Sigma cde & (c+d+e+3)(a+1)(b+1) & (\alpha+4) \\ +6\Sigma bcde & (b+c+d+e+4)(a+1) & \\ -24abcde & & \end{array}\right\}\kappa,$$

where $\alpha = a+b+c+d+e,\ \beta = \&c., \ldots\ \epsilon = abcde.$

83. The complete functions (a), $(a,\ b)$, $(a,\ b,\ c)$, &c. may be expressed by means of the linear terms $[a]$, $[a,\ b]$, $[a,\ b,\ c]$, &c. as follows, viz. we have

$$\begin{array}{lll} (a) & = & [a], \\ (a,\ b) & = & [a]\,[b] \\ & + & [a,\ b], \\ (a,\ b,\ c) & = & [a]\,[b]\,[c] \\ & + & [a]\,[b,\ c] + [b]\,[a,\ c] + [c]\,[a,\ b] \\ & + & [a,\ b,\ c], \\ (a,\ b,\ c,\ d) & = & [a]\,[b]\,[c]\,[d] \\ & + & \Sigma[a]\,[b]\,[c,\ d] \\ & + & \Sigma[a,\ b]\,[c,\ d] \\ & + & \Sigma[a]\,[b,\ c,\ d] \\ & + & [a,\ b,\ c,\ d], \end{array}$$

and so on: this is easily verified for (a, b), and without much difficulty for (a, b, c), but in the succeeding cases the actual verification would be very laborious.

84. The theoretical foundation is as follows. Writing for greater distinctness $(a)_m$ in place of (a), we have $(a)_m$ to denote the number of the curves C^r which have with a given curve U^m a contact of the order a, and which besides pass through $\frac{1}{2}r(r+3)-a$ points. Let the curve U^m be the aggregate of two curves of the orders m, m' respectively, or say let the curve U^m be the two curves m, m', then we have

$$(a)_{m+m'} = (a)_m + (a)_{m'},$$

a functional equation, the solution of which is

$$(a)_m = [a]_m,$$

where $[a]_m$ is a linear function of n, m, κ, or, what is the same thing, of m, Δ, κ. I assume for the moment that when the coefficients are determined $[a]_m$ would be found to have the value $=[a]$.

Similarly, if $(a, b)_m$ denote the number of the curves C^r which have with the given curve U^m contacts of the orders a and b respectively, and which besides pass through $\frac{1}{2}r(r+3)-a-b$ points, then if the given curve break up into the curves m, m', then we have

$$(a,\ b)_{m+m'} - (a,\ b)_m - (a,\ b)_{m'} = \{(a)_m (b)_{m'}\} + \{(a)_{m'} (b)_m\},$$

where $\{(a)_m (b)_{m'}\}$ is the number of the curves C^r which have with m a contact of the order a and with m' a contact of the order b, and which pass through the $\frac{1}{2}r(r+3)-a-b$ points; and the like for $\{(a)_{m'} (b)_m\}$. Then, not universally, but for values of a and b which are not too great, the order of the aggregate condition is equal to the product of the orders of the component conditions (*ante*, No. 12), that is, we have

$$\{(a)_m (b)_{m'}\} = (a)_m \cdot (b)_{m'} = [a]_m [b]_{m'},$$
$$\{(a)_{m'} (b)_m\} = (a)_{m'} \cdot (b)_m = [a]_{m'} [b]_m,$$

and thence the functional equation

$$(a,\ b)_{m+m'} - (a,\ b)_m - (a,\ b)_{m'} = [a]_m [b]_{m'} + [a]_{m'} [b]_m.$$

But $[a]_m$, &c. being linear functions of m, Δ, κ, we have

$$[a]_{m+m'} = [a]_m + [a]_{m'},\quad [b]_{m+m'} = [b]_m + [b]_{m'},$$

and thence a particular solution of the equation is at once seen to be $[a]_m [b]_m$; the general solution is therefore

$$(a,\ b)_m = [a]_m [b]_m + [a,\ b]_m,$$

where $[a, b]_m$ is an arbitrary linear function of m, Δ, κ. Hence, assuming for the present that if determined its value would be found to be $=[a, b]$, we have the required formula $(a, b) = [a]\,[b] + [a, b]$.

The investigation of the expression for $(a, b, c)_m$ depends in like manner on the assumption that we have

$$\{(a)_m (b, c)_{m'}\} = (a)_m \cdot (b, c)_{m'} = [a]_m \{[b]_{m'} [c]_{m'} + [b, c]_{m'}\},$$

and so in the succeeding cases; and we thus, within the limits in which these assumptions are correct, obtain the series of formulæ for (a, b), (a, b, c)....

85. It is to be observed in the investigation of (a, b) that if $a = b$, the two terms $[a]_m [b]_{m'}$ and $[a]_{m'} [b]_m$ become equal, and the equal value must be taken not twice but only once, that is, the functional equation is

$$(a, a)_{m+m'} - (a, a)_m - (a, a)_{m'} = [a]_m [a]_{m'},$$

and the solution, writing $\frac{1}{2}[a, a]_m$ for the arbitrary linear function, is

$$(a, a)_m = \tfrac{1}{2}[a]_m [a]_m + \tfrac{1}{2}[a, a]_m,$$

in which solution it would appear, by the determination of the arbitrary function, that $[a, a]$ has the value obtained from $[a, b]$ by writing therein $b = a$. Writing the equation in the form

$$(a, a) = \tfrac{1}{2}[a][a] + \tfrac{1}{2}[a, a],$$

and comparing with the equation for (a, b), we see that $[a, b]$ is *not* to be considered as acquiring any divisor when b is put $= a$, but that the divisor is introduced as a divisor of the whole right-hand side of the equation in virtue of the remark as to the divisor of the functions (a, b), (a, b, c)... in the case of any equalities between the numbers $(a, b, c \ldots)$. This is generally the case, and the foregoing expressions for $[a, b]$, $[a, b, c]$, &c. are thus to be regarded as true without modification even in the case of any equalities among the numbers a, b, c....

86. To complete according to the foregoing method the determination of the expressions for (a), (a, b),.., we have to determine the linear functions $[a]$, $[a, b]$, &c., which are each of them of the form $fm + g\Delta + h\kappa$, where (f, g, h) are functions of r and of a, b, &c.; and I observe that the determination can be effected if we know the values of (a), (a, b), &c. in the cases of a unicursal curve without cusps and with a single cusp respectively. Thus assume that in these two cases respectively we have

$$(a) = (a+1)(rm - a),$$

$$(a) = (a+1)(rm - a) - a.$$

Writing first $\Delta = -1$, $\kappa = 0$, and secondly $\Delta = -1$, $\kappa = 1$, we have

$$(a+1)(rm - a) \quad = fm - g,$$

$$(a+1)(rm - a) - a = fm - g + h,$$

whence

$$f = (a+1)r, \quad g = (a+1)a, \quad h = -a,$$

giving the foregoing value

$$[a] = (a+1)\,rm + (a+1)\,a\Delta - a\kappa.$$

Similarly, for two contacts assume that we have in the two cases respectively

$$(a,\ b) = (a+1)(b+1)\,[rm-a-b]^2,$$
$$(a,\ b) = (a+1)(b+1)\,[rm-a-b]^2 - \{a\,(b+1) + b\,(a-1)\}\,[rm-a-b-1]^1.$$

Starting here from the formula $[a,\ b] = (a,\ b) - [a]\,[b] = fm + g\Delta + h\kappa$, and writing successively $\Delta = -1$, $\kappa = 0$, and $\Delta = -1$, $\kappa = 1$, we have

$$(a+1)(b+1)\,[rm-a-b]^2 - \{(a+1)(rm-a)\}\,\{(b+1)(rm-b)\} \quad = fm - g,$$
$$(a+1)(b+1)\,[rm-a-b]^2 - \{a\,(b+1) + b\,(a+1)\}\,[rm-a-b-1]^1$$
$$- \{(a+1)(rm-a) - a\}\,\{(b+1)(rm-b) - b\} = fm - g + h\,;$$

the first of which, putting therein $a+b=\alpha$, $ab=\beta$, is at once reduced to

$$(a+1)(b+1)\,\{rm\,(-\alpha-1) + \alpha\,(\alpha+1) - \beta\} = fm - g,$$

whence $f = -(a+1)(b+1)(\alpha+1)\,r$, $g = -(a+1)(b+1)(\alpha\,(\alpha+1)-\beta)$. And taking the difference of the two equations, we have

$$- \{a\,(b+1) + b\,(a+1)\}\,(rm-a-b-1)$$
$$+ a\,(b+1)(rm-b) + b\,(a+1)(rm-a) - ab = h,$$

that is $h = (a+1)(b+1)(a+b) - ab$; whence $[a,\ b]$ has the value above assigned to it.

87. The actual calculation of $[a,\ b,\ c]$ would be laborious, and that of the subsequent terms still more so; but it is clear that the principle applies, and that the foregoing values, assuming them to be correct, would be obtained if only we know, for a unicursal curve without cusps, that

$$(a,\ b,\ c,\ldots) = (a+1)(b+1)(c+1)\ldots[rm-(a+b+c,\ldots)]^t$$

(t the number of contacts $a,\ b,\ c,\ldots$), and for a unicursal curve with a single cusp, that

$$(a,\ b,\ c,\ldots) = \ (a+1)(b+1)(c+1)\ldots[rm-(a+b+c\ \ldots)\qquad]^t$$
$$- \Sigma a\,(b+1)(c+1)\ldots[rm-(a+b+c\ \ldots)-1]^{t-1},$$

viz. that the diminution of $(a,\ b,\ c,\ldots)$ occasioned by the single cusp is

$$= [rm-(a+b+c,\ \ldots)-1]^{t-1}\,.\,\Sigma\,\{a\,(b+1)(c+1)\ldots\}.$$

88. Consider a unicursal curve U^m, and a curve C^r having therewith t contacts of the orders $a, b, c, \ldots$ respectively. The coordinates (x, y, z) of any point of the unicursal curve are given as functions of the order m of a variable parameter θ; and substituting these values in the equation of the curve C^r, we have an equation of the degree rm in θ, but containing the coefficients of C^r linearly; this equation gives of course the values of θ which correspond to the rm intersections of the two curves. Hence in order that the curve C^r may have the prescribed contacts with U^m, the equation of the degree rm in θ must have t systems of equal roots, viz. a system of a equal roots, another system of b equal roots, &c.: this implies between the coefficients of the equation an $(a+b+c\ldots)$ fold relation, which may be shown to be of the order $(a+1)(b+1)(c+1)\ldots[rm-(a+b+c\ldots)]^t$; and since the coefficients in question are linear in regard to the coefficients in the equation of the curve C^r, the order of the relation between the last-mentioned coefficients has the same value; that is, the number of the curves C^r which have the prescribed contacts with the unicursal curve U^m and besides pass through the requisite number of given points, is $=(a+1)(b+1)(c+1)\ldots[rm-(a+b+c\ldots)]^t$.

89. The reduction in the case of a cusp appears to be caused as follows:—Consider on the curve U^m a points indefinitely near to the cusp, and let the condition of the curve C^r having the contact of the a-th order be replaced by the condition of passing through the a points; that is, consider the curves C^r which have with the curve U^m $(t-1)$ contacts of the orders $b, c, \ldots$ respectively, which pass through the a points on the curve U^m in the neighbourhood of the cusp, and which also pass through the requisite number of arbitrary points. The number of these curves is $=(b+1)(c+1)\ldots[rm-a-(b+c+\ldots)]^{t-1}$ (the term $rm-a$ instead of rm, on account of the given a points on the curve: compare herewith De Jonquières' formula containing $rm-p$). Each of these curves, in that it passes through a points in the neighbourhood of the cusp, will *ipso facto* pass through $\overline{a+1}$ points (viz. a curve which simply passes through the cusp of a cuspidal curve meets the cuspidal curve there in two points, a curve which touches the cuspidal tangent meets the curve in three points, &c.), and be consequently, in an improper sense, a curve having a contact of the a-th order with the given curve U^m. I assume that it counts as such curve a times, and this being so, we have, on account of the curves in question, a reduction $=a(b+1)(c+1)\ldots[rm-(a+b+c\ldots)]^{t-1}$. We have in like manner for the curves passing through b points in the neighbourhood of the cusp a reduction $=b(a+1)(c+1)\ldots[rm-(a+b+c\ldots)]^{t-1}$, &c., and hence when the given unicursal curve U^m has a cusp, the total reduction on account of the cusp in the number of the curves C^r which have with the given curve the t contacts of the orders $a, b, c, \ldots$ and besides pass through the requisite number of given points, is

$$=\{a(b+1)(c+1)\ldots+b(a+1)(c+1)+\&c.\}\,[rm-(a+b+c\ldots)]^{t-1},$$

which is the auxiliary theorem in question; some of the steps require however to be more completely made out.

90. I have calculated the following numerical results, wherein, as before, $\Delta=D-1=-m+\tfrac{1}{2}n+\tfrac{1}{2}\kappa$. I find also their values in the case where the curve C^r

is a conic (that is $r=2$) first in terms of m, n, κ, and finally in terms of m, n, α ($\alpha = 3n + \kappa$, as above). The results are

				for $r=2$, that is, curve C^r a conic.							
$[1]=$	$2rm+$	$2\Delta-$	κ	$=$	$2m+$	n		$=$	$2m+$	n	
$[2]=$	$3rm+$	$6\Delta-$	2κ	$=$		$3n+$	κ	$=$			α
$[3]=$	$4rm+$	$12\Delta-$	3κ	$=-$	$4m+$	$6n+$	3κ	$=-$	$4m-$	$3n+$	3α
$[4]=$	$5rm+$	$20\Delta-$	4κ	$=-$	$10m+$	$10n+$	6κ	$=-$	$10m-$	$8n+$	6α
$[5]=$	$6rm+$	$30\Delta-$	5κ	$=-$	$18m+$	$15n+$	10κ	$\doteq-$	$18m-$	$15n+$	10α
$[1,1]=-$	$12rm-$	$20\Delta+$	7κ	$=-$	$4m-$	$10n-$	3κ	$=-$	$4m-$	$n-$	3α
$[1,2]=-$	$24rm-$	$60\Delta+$	16κ	$=$	$12m-$	$30n-$	14κ	$=$	$12m+$	$12n-$	14α
$[1,3]=-$	$40rm-$	$136\Delta+$	29κ	$=$	$56m-$	$68n-$	39κ	$=$	$56m+$	$49n-$	39α
$[1,4]=-$	$60rm-$	$260\Delta+$	46κ	$=$	$140m-$	$130n-$	84κ	$=$	$140m+$	$122n-$	84α
$[2,2]=-$	$45rm-$	$144\Delta+$	32κ	$=$	$54m-$	$72n-$	$40\kappa.$	$=$	$54m+$	$48n-$	40α
$[2,3]=-$	$72rm-$	$288\Delta+$	54κ	$=$	$144m-$	$144n-$	90κ	$=$	$144m+$	$126n-$	90α
$[1,1,1]=$	$160rm+$	$352\Delta-$	98κ	$=-$	$32m+$	$176n+$	78κ	$=-$	$32m-$	$58n+$	78α
$[1,1,2]=$	$360rm+$	$1056\Delta-$	240κ	$=-$	$336m+$	$528n+$	288κ	$=-$	$336m-$	$336n+$	288α
$[1,1,3]=$	$672rm+$	$2528\Delta-$	478κ	$=-$	$1184m+$	$1264n+$	786κ	$=-$	$1184m-$	$1094n+$	786α
$[1,2,2]=$	$756rm+$	$2700\Delta+$	530κ	$=-$	$1188m+$	$1350n+$	820κ	$=-$	$1188m-$	$1110n+$	820α
$[1,1,1,1]=-$	$3360rm-$	$8928\Delta+$	2106κ	$=$	$2208m-$	$4464n-$	2358κ	$=$	$2208m+$	$2610n-$	2358α
$[1,1,1,2]=-$	$8064rm-$	$26784\Delta+$	5376κ	$=$	$10656m-$	$13392n-$	8016κ	$=$	$10656m+$	$10656n-$	8016α
$[1,1,1,1,1]=$	$96768rm+$	$296448\Delta-$	61464κ	$=-$	$102912m+$	$148224n+$	86760κ	$=-$	$10912m-$	$112056n+$	86760α

(It may be noticed as a curious circumstance that in the last column in the expressions of [2], [1, 2], [1, 1, 2] and [1, 1, 1, 2] respectively, the coefficients of m and n are in each case equal.)

91. In the case of the conic, (1), (2), &c. are the expressions denoted in the former part of this Memoir by (1 : :), (2 .·.), &c., the number of points being in each case such as to make in all five conditions; calculating these functions by means of the formulæ $(a)=[a]$, &c., the comparison of the resulting values with the values previously obtained will show *à posteriori* the limits within which the formulæ are applicable; where they cease to be applicable I find the difference, and annex it as a correction to the formula value: I have in some cases given what seems to be the proper theoretical form of this difference. We have

$$(1\ ::) \quad = \quad 2m + n;$$

$$(2\ .\!\cdot\!.) \quad = \quad \alpha;$$

$$(3\ :) \quad = -\ 4m - 3n + 3\alpha;$$

$$(4\ \cdot) \quad = -10m - 8n + 6\alpha;$$

$$(5) \quad = -18m - 15n + 10\alpha - [-3m + \alpha] \quad (= -[\iota]);$$

$$2\,(1,\ 1\ \therefore) \quad = \quad (2m+n)^2 \\ -\ 4m - n - 3\alpha\,;$$

$$(1,\ 2\ :) \quad = \quad (2m+n)\,\alpha \\ +\,12m + 12n - 14\alpha\,;$$

$$(1,\ 3\ \cdot) \quad = \quad (2m+n)(-\,4m - 3n + 3\alpha) \\ +\,56m + 49n - 39\alpha\,;$$

$$(1,\ 4) \quad = \quad (2m+n)(-\,10m - 8n + 6\alpha) \\ +\,140m + 122n - 84\alpha \\ -\,[(m-3)(-\,12m - 6n + 6\alpha)] \quad (= -\,[(m-3)(4\iota + 2\kappa)])\,;$$

$$2\,(2,\ 2\ \cdot) \quad = \quad \alpha^2 \\ +\,54m + 48n - 40\alpha\,;$$

$$(2,\ 3) \quad = \quad \alpha\,(-\,4m - 3n + 3\alpha) \\ +\,144m + 126n - 90\alpha \\ -\,[24m + 6n + (n-12)\,\alpha] \quad (= -\,[6\tau + (n-3)\,\kappa])\,;$$

$$6\,(1,\ 1,\ 1\ :) \quad = \quad (2m+n)^3 \\ +\,3\,(2m+n)(-\,4m - n - 3\alpha) \\ -\,32m - 58n + 78\alpha \\ -\,[4m\,(m-1)(m-2)]\,;$$

$$2\,(2,\ 1,\ 1\ \cdot) \quad = \quad (2m+n)^2\,\alpha \\ +\,2\,(2m+n)(12m + 12n - 14\alpha) + \alpha\,(-\,4m - n - 3\alpha) \\ -\,336m - 336n + 288\alpha \\ -\,[2\alpha\,(m-2)(m-3)] \quad (= -\,[6n\,(m-2)(m-3) + 2\kappa\,(m-2)(m-3)])\,;$$

$$2\,(3,\ 1,\ 1) \quad = \quad (2m+n)^2(-\,4m - 3n + 3\alpha) \\ +\,2\,(2m+n)(56m + 49n - 39\alpha) \\ +\,(-\,4m - n - 3\alpha)(-\,4m - 3n + 3\alpha) \\ -\,1184m - 1094n + 786\alpha \\ -\left[\begin{array}{l} -\,13m^3 - 8m^2n + 4mn^2 + 131m^2 + 92mn - 8n^2 - 316m - 226n \\ +\,\alpha\,(9m^2 - 87m - 3n + 204) \end{array}\right];$$

$$2\,(2,\ 2,\ 1) \quad = \quad (2m+n)\,\alpha^2 \\ +\,2\alpha\,(12m + 12n - 14\alpha) \\ +\,(2m+n)(54m + 48n - 40\alpha) \\ -\,1188m - 1110n - 820\alpha \qquad \left(= -\left[\begin{array}{l} \delta\,(2n + 10m - 38) \\ +\,\kappa\,(3m^2 - 19m + 30) \\ +\,\iota\,(6m^2 - 41m + 69) \\ +\,\tau\,(8m - 32) \end{array}\right]\right)$$

$$- \begin{bmatrix} 60m^2 + 42mn - 252m - 174n \\ + \alpha\,(-40m + 166) \\ + \alpha^2\,(m-4) \end{bmatrix} \left(= - \begin{bmatrix} (m-4)(\kappa^2 - \kappa) \\ + \;6\,(m-4)(n-3)\,\kappa \\ + 18\,(m-4)\,\tau \\ + \;(m-3)(4\iota + 2\kappa) \end{bmatrix}\right);$$

$$\begin{aligned} 24\,(1,\,1,\,1,\,1\cdot) = {} & (2m+n)^4 \\ & + 6\,(2m+n)^2\,(-4m-n-3\alpha) \\ & + 3\,(-4m-n-3\alpha)^2 \\ & + 4\,(2m+n)\,(-32m-58n+78\alpha) \\ & + 2208m + 2610n - 2358\alpha \\ & - \begin{bmatrix} 14m\,(m-1)(m-2)(m-3) \\ + 16n\;(m+2)(m-2)(m-3) \\ - 36\alpha\;(m-2)(m-3) \end{bmatrix} \\ & (= - [(m-2)(m-3)(14m^2 + 16mn - 14m - 76n - 36\kappa)]); \end{aligned}$$

$$\begin{aligned} 6\,(2,\,1,\,1,\,1) = {} & (2m+n)^3\,\alpha \\ & + 3\,(2m+n)^2\,(12m+12n-14\alpha) \\ & + 3\,(2m+n)\,\alpha\,(-4m-n-3\alpha) \\ & + 3\,(2m+n)\,(-336m-336n+288\alpha) \\ & + \alpha\,(-32m-58n+78\alpha) \\ & + 3\,(-4m-n-3\alpha)\,(12m+12n-14\alpha) \\ & + 10656m + 10656n - 8016\alpha \\ & - \begin{bmatrix} 108m^3 + 108m^2n - 1116m^2 - 1116mn + 2736m + 2736n \\ + \alpha\,(7m^3 + 6m^2n - 147m^2 - 30mn + 1040m + 24n - 2256) \\ + \alpha^2\,(-9m+36), \end{bmatrix}; \end{aligned}$$

where the correction is

$$= -(m-4)\begin{pmatrix} 108m^2 + 108mn - 684m - 684n \\ + \alpha\left(7m^2 - 119m + 564 + 6n\,(m-1)\right) \\ - 9\alpha^2 \end{pmatrix},$$

$$= -(m-4)\begin{pmatrix} 2\delta\,(21n-36) \\ + 2\tau\,(18m-126) \\ + \;\kappa\,(7m^2 + 6mn - 65m - 13n + 165) - 9\,(\kappa^2 - \kappa) \\ + \;\iota\,(+16n-96) \end{pmatrix};$$

$$
\begin{aligned}
120\,(1, 1, 1, 1, 1) = \;& (2m+n)^5 \\
& + 10\,(2m+n)^3\,(-4m-n-3\alpha) \\
& + 10\,(2m+n)^2\,(-32m-58n+78\alpha) \\
& + 10\,(-4m-n-3\alpha)\,(-32m-58n+78\alpha) \\
& + \;\;5\,(2m+n)\,(2208m+2610n-2358\alpha) \\
& + 15\,(2m+n)\,(-4m-n-3\alpha)^2 \\
& - 102912m - 112056n + 86760\alpha \\
& - \left(\begin{array}{l}
\qquad 31m^5 + \quad 70m^4n + \;40m^3n^2 \\
- \quad 310m^4 - \;\; 460m^3n - 120m^2n^2 \\
- \quad 235m^3 - \; 1030m^2n - 400m\,n^2 \\
+ \; 10690m^2 + 16060mn \; + 960 \quad n^2 \\
\; + \left(\begin{array}{l} - \quad 210m^3 - 180m^2n \\ + \;\; 2970m^2 + 900mn \\ - 15630m \; - 720n \\ + 28440 \end{array}\right) \\
+ \alpha^2\,(135m - 540),
\end{array}\right) ;
\end{aligned}
$$

where the correction is

$$
= -(m-4)\left(\begin{array}{l}
\;\; 31m^4 - 186m^3 - 979m^2 + 6774m \\
+ n\,(70m^3 - 180m^2 - 1750m + 9060) \\
+ 40n^2\,(m+3)\,(m-2) \\
+ \alpha \left(\begin{array}{l} -210m^2 + 2130m - 7110 \\ -180n\,(m-1) \end{array}\right) \\
+ \alpha^2 \,.\, 135,
\end{array}\right),
$$

which is

$$
= -(m-4)\left(\begin{array}{l}
31\,(2\delta+3\kappa)^2 + 110\,(2\delta+3\kappa)\,(2\tau+3\iota) \\
+ \;(\tfrac{70}{3}m^3 + 1142m + 3174 \quad)\,(\;\kappa - \;\; \iota) \\
+ \;(-14m - 638n - 1524 \quad)\,(2\delta+3\kappa) \\
+ \;(-390m + 110n + 4272 \;)\,(2\tau+3\iota\,) \\
+ \;(-210m^2 - 180mn + 2130m + 990n - 7110)\,\kappa + 135\kappa^2
\end{array}\right)
$$

but I have not sought to further reduce this expression, not knowing the proper form in which to present it.

92. The question which ought now to be considered is to determine the corrections or supplements which should be applied to the foregoing expressions (a), (a, b), &c., or to their equivalents $[a]$, $[a]\,[b] + [a,\; b]$, &c. in order to obtain formulæ for the cases

beyond the limits within which the present formulæ are applicable; but this I am not in a position to enter upon. If the extended formulæ were obtained, it would of course be an interesting verification or application of them to deduce from them the complete series of expressions (1 : :), (2 .·.) ... (1, 1, 1, 1, 1) for the number of the conics which satisfy given conditions of contact with a given curve, and besides pass through the requisite number of given points. It will be recollected that throughout these last investigations, I have put De Jonquières' $p=0$; that is, I have not considered the case of the curves C^r which (among the conditions satisfied by them) have with the curve U^m contacts of given orders at given points of the curve; it is probable that the general formulæ containing the number p admit of extensions and transformations analogous to the formulæ in which p is put $=0$, but this is a question which I have not considered.

93. The set of equations $(a)=[a]$, $(a, b)=[a][b]+[a, b]$, &c., considered irrespectively of the meaning of the symbols contained therein, gives rise to an analytical question which is considered in Annex No. 7.

The question of the conics satisfying given conditions of contact is considered from a different point of view in my Second Memoir above referred to.

Annex No. 1 (referred to in the notice of De Jonquières' memoir of 1861).—*On the form of the equation of the curves of a series of given index.*

To obtain the general form of the equation of the curves C^n of a series of the index N, it is to be observed that the equation of any such curve is always included in an equation of the order n in the coordinates, containing linearly and homogeneously certain parameters $a, b, c, ..$; this is universally the case, as we may, if we please, take the parameters $(a, b, c, ..)$ to be the coefficients of the general equation of the order n; but it is convenient to make use of any linear relations between these coefficients so as to reduce as far as possible the number of the parameters. Assume that the number of the parameters is $=\omega+1$, then in order that the curve should form a series (that is, satisfy $\tfrac{1}{2}n(n+3)-1$ conditions), we must have a $(\omega-1)$ fold relation between the parameters, or, what is the same thing, taking the parameters to be the coordinates of a point in ω-dimensional space, say the parametric point, the point in question must be situate on a $(\omega-1)$ fold locus. Moreover, the condition that the curve shall pass through a given point establishes between the parameters a linear relation (viz. that expressed by the original equation of the curve regarding the coordinates therein as belonging to the given point, and therefore as constants); that is, when the curve passes through a given point, the corresponding positions of the parametric point are given as the intersections of the $(\omega-1)$ fold locus by an omal onefold locus; the number of the curves is therefore equal to the number of these intersections, that is, to the order of the $(\omega-1)$ fold locus; or the index of the series being assumed to be $=N$, the order of the $(\omega-1)$ fold locus must be also $=N$. That is, the general form of the equation of the curves C^n which form a series of the index N, is that of an equation of the order n containing linearly and homogeneously the $\overline{\omega+1}$ coordinates of a certain $(\omega-1)$ fold locus of the order N. It is only in a particular case, viz. that in which the $(\omega-1)$ fold locus is unicursal, that the coordinates of a point of this locus can be expressed as rational and integral functions of the order N of a variable parameter θ; and consequently only in this same case that the equation of the curves C^n of the series of the index N can be expressed by an equation $(*\between x, y, z)^n=0$, or $(*\between x, y, 1)^n=0$, rational and integral of the degree N in regard to a variable parameter θ.

If in the general case we regard the coordinates of the parametric point as irrational functions of a variable parameter θ, then rationalising in regard to θ, we obtain an equation rational of the order N in θ, but the order in the coordinates instead of being $=n$, is equal to a multiple of n, say qn. Such an equation represents not a single curve but q distinct curves C^n, and it is to be observed that if we determine the parameter by substituting therein for the coordinates their values at a given point, then to each of the N values of the parameter there corresponds a system of q curves, only one of which passes through the given point, the other $\overline{q-1}$ curves are curves not passing through the given point, and having no proper connexion with the curves which satisfy this condition.

Returning to the proper representation of the series by means of an equation containing the coordinates of the parametric point, say an equation $(*\between x, y, 1)^n=0$, involving the two coordinates (x, y), it is to be noticed that forming the derived equation and eliminating the coordinates of the parametric point, we obtain an equation rational in the coordinates (x, y), and also rational of the degree N in the differential coefficient $\frac{dy}{dx}$; in fact since the number of curves through any given point (x_0, y_0) is $=N$, the differential equation must give this number of directions of passage from the point (x_0, y_0) to a consecutive point, that is, it must give this number of values of $\frac{dy}{dx}$, and must consequently be of the order N in this quantity.

Conversely, if a given differential equation rational in x, y, $\frac{dy}{dx}$, and of the degree N in the last-mentioned quantity $\frac{dy}{dx}$, admit of an algebraical general integral, the curves represented by this integral equation may be taken to be irreducible curves, and this being so they will be curves of a certain order n forming a series of the index N; whence the general integral (assumed to be algebraical) is given by an equation of the above-mentioned form, viz. an equation rational of a certain order n in the coordinates, and containing linearly and homogeneously the $\omega+1$ coordinates of a variable parametric point situate on an $(\omega-1)$ fold locus. The integral equation expressed in the more usual form of an equation rational of the order N in regard to the parameter or constant of integration, will be in regard to the coordinates of an order equal to a multiple of n, say $=qn$, and for any given value of the parameter will represent not a single curve C^n, but a system of q such curves: the first-mentioned form is, it is clear, the one to be preferred.

Annex No. 2 (referred to, No. 17).—*On the line-pairs which pass through three given points and touch a given conic.*

Taking the given points to be the angles of the triangle formed by the lines $(x=0, y=0, z=0)$, we have to find (f, g, h) such that the conic $(0, 0, 0, f, g, h \between x, y, z)^2 = 0$, or, what is the same thing, $fyz + gzx + hxy = 0$, shall reduce itself to a line-pair, and shall touch a given conic $(1, 1, 1, \lambda, \mu, \nu \between x, y, z)^2 = 0$. The condition for a line-pair is that one of the quantities f, g, h shall vanish, viz. it is $fgh = 0$; the condition for the contact of the two conics is found in the usual manner by equating to zero the discriminant of the function $1 - (\lambda + \theta f)^2 - (\mu + \theta g)^2 - (\nu + \theta h)^2 + 2(\lambda + \theta f)(\mu + \theta g)(\nu + \theta h) = (a, b, c, d \between \theta, 1)^3$ suppose; the values of a, b, c, d being

$$\begin{aligned} a &= \ 2fgh, \\ b &= -\tfrac{1}{3}(f^2 + g^2 + h^2 - 2\lambda gh - 2\mu hf - 2\nu fg), \\ c &= \ \tfrac{2}{3}\left((\mu\nu - \lambda) f + (\nu\lambda - \mu) g + (\lambda\mu - \nu) h\right), \\ d &= \ 1 - \lambda^2 - \mu^2 - \nu^2 + 2\lambda\mu\nu. \end{aligned}$$

Hence considering (f, g, h) as the coordinates of the parametric point, we have the discriminant-locus $a = 0$, and the contact-locus

$$a^2d^2 + 4ac^3 + 4b^3d - 3b^2c^2 - 6abcd = 0,$$

and at the intersection of the two loci, $a = 0$, $b^2(4bd - 3c^2) = 0$, equations breaking up into the system $(a = 0, b = 0)$ twice, and the system $a = 0$, $4bd - 3c^2 = 0$; the former of these is

$$fgh = 0, \quad f^2 + g^2 + h^2 - 2\lambda gh - 2\mu hf - 2\nu fg = 0,$$

which expresses that the intersection of the two lines of the line-pair intersect on the given conic; in fact the system is satisfied by $f = 0$, $g^2 + h^2 - 2\lambda gh = 0$, giving a line-pair $x(hy + gz) = 0$, the two lines whereof intersect on the conic $(1, 1, 1, \lambda, \mu, \nu \between x, y, z)^2 = 0$; and similarly, if $g = 0$, then $h^2 + f^2 - 2\mu hf = 0$, or if $h = 0$, then $f^2 + g^2 - 2\lambda fg = 0$. As noticed above this system occurs twice.

The second system is

$$\begin{aligned} fgh = 0, \qquad & (f^2 + g^2 + h^2 - 2\lambda gh - 2\mu hf - 2\nu fg)(1 - \lambda^2 - \mu^2 - \nu^2 + 2\lambda\mu\nu) \\ & \qquad + \left((\mu\nu - \lambda) f + (\nu\lambda - \mu) g + (\mu\lambda - \nu) h\right)^2 = 0, \end{aligned}$$

or, as the second equation may also be written,

$$\begin{aligned} & f^2(1 - \mu^2)(1 - \nu^2) + g^2(1 - \nu^2)(1 - \lambda^2) + h^2(1 - \lambda^2)(1 - \mu^2) \\ & \quad + 2gh(1 - \lambda^2)(\mu\nu - \lambda) + 2hf(1 - \mu^2)(\nu\lambda - \mu) + 2fg(1 - \nu^2)(\lambda\mu - \nu) = 0, \end{aligned}$$

which expresses that a line of the line-pair touches the conic; in fact the system is satisfied by $f = 0$, $g^2(1 - \nu^2) + h^2(1 - \mu^2) + 2gh(\mu\nu - \lambda) = 0$, viz. we have here the line-pair $x(hy + gz) = 0$, in which the line $hy + gz = 0$ touches the conic $(1, 1, 1, \lambda, \mu, \nu \between x, y, z)^2 = 0$ and the like if $g = 0$, or if $h = 0$. This system it has been seen occurs only once.

Annex No. 3 (referred to, No. 22).—*On the conics which pass through two given points and touch a given conic.*

Consider the conics which pass through two given points and touch a given conic. We may take $Z=0$ as the equation of the line through the two given points, and then taking the pole of this line in regard to the given conic and joining it with the two given points respectively, the equations of the joining lines may be taken to be $X=0$ and $Y=0$ respectively. This being so, we have for the given points $(X=0,\ Z=0)$ and $(Y=0,\ Z=0)$ respectively, and for the given conic

$$aX^2+bY^2+2hXY+cZ^2=0\,;$$

and since the required conic is to pass through the two given points its equation will be of the form

$$wX^2+2xYZ+2yZX+2zXY=0,$$

where $(x,\ y,\ z,\ w)$ are variable parameters which must satisfy a single condition in order that the last-mentioned conic may touch the given conic. The condition is at once seen to be that obtained by making the equation

$$\begin{aligned}&(a+\lambda w)\,bc\\ -\,&(a+\lambda w)\,(h+\lambda z)^2\\ -\,&b\lambda^2y^2\\ -\,&c\lambda^2z^2\\ +\,&2\lambda^2xy\,(h+\lambda z)=0,\end{aligned}$$

considered as a cubic equation in λ, have a pair of equal roots; or if we write

$$\begin{aligned}A&=3c\,(ab-h^2),\\ B&=(ab-h^2)\,w-2chz,\\ C&=-ax^2-by^2-cz^2+2h\,(xy-zw),\\ D&=3z\,(2xy-wz),\end{aligned}$$

then the required condition is

$$A^2D^2+4AC^3+4B^3D-6ABCD-3B^2C^2=0.$$

Hence the conic

$$wX^2+2xYZ+2yZX+2zXY=0$$

satisfies the prescribed conditions, if only the parameters $(x,\ y,\ z,\ w)$ satisfy the last-mentioned equation, that is, if $(x,\ y,\ z,\ w)$ are the coordinates of a point on the sextic surface represented by this equation.

The surface has upon it a cuspidal curve the equations whereof are

$$\left\|\begin{matrix}A, & B, & C\\ B, & C, & D\end{matrix}\right\|=0\,;$$

this may be considered as the intersection of the quadric surface $AC-B^2=0$ and the cubic surface $AD-BC=0$; and the cuspidal curve is consequently a sextic.

The surface has also a nodal curve made up of two conics; to prove this I write for shortness $k=h-\sqrt{ab}$, $k_1=h+\sqrt{ab}$; the values of A, B, C, D then are

$$\begin{aligned} A &= -3ckk_1,\\ B &= -kk_1w-2chz,\\ C &= -ax^2-by^2-cz^2+2h(xy-zw),\\ D &= 3z(2xy-zw); \end{aligned}$$

and it is in the first place to be shown that the surface contains the conic

$$x:y:z:w=\theta\sqrt{b}:\theta\sqrt{a}:1:-k\theta^2+\frac{c}{k},$$

where θ is a variable parameter. Substituting these values, we have

$$\begin{aligned} A &= -3ckk_1,\\ B &= k^2k_1\theta^2-c(3h+\sqrt{ab}),\\ C &= 2kk_1\theta^2-\frac{c}{k}(3h-\sqrt{ab}),\\ D &= 3\left(k_1\theta^2-\frac{c}{k}\right); \end{aligned}$$

and hence

$$\begin{aligned} AD-BC &= -2k\left(kk_1\theta^2+\frac{2c\sqrt{ab}}{k}\right)^2,\\ AC-B^2 &= -k^2\left(kk_1\theta^2+\frac{2c\sqrt{ab}}{k}\right)^2,\\ BD-C^2 &= -\left(kk_1\theta^2+\frac{2c\sqrt{ab}}{k}\right)^2, \end{aligned}$$

values which satisfy identically the equation of the surface written under the form

$$(AD-BC)^2-4(AC-B^2)(BD-C^2)=0.$$

Moreover, proceeding to form the derived equation, and to substitute therein the foregoing values of (x, y, z, w), we have

$$\partial A:\partial B:\partial C:\partial D=0:k^2:2k:3,$$

and then the derived equation is

$$\begin{aligned} &(AD-BC)(3A-2kB-k^2C)\\ &-2(AC-B^2)(3B-4kC+k^2D)\\ &-2(BD-C^2)(2kA-2k^2B)=0, \end{aligned}$$

that is,

$$\begin{aligned}&-k\,(\;3A-2kB-k^2C)\\&+k^2(\qquad\quad 3B-4kC+k^2D)\\&+\;\;(2kA-2k^2B),\end{aligned}$$

or finally

$$-k\,(A-3Bk+3k^2C-k^3D)=0,$$

which is satisfied by the foregoing values of A, B, C, D; hence the conic is a nodal curve on the sextic; and by merely changing the sign of one of the radicals $\sqrt{a}$, $\sqrt{b}$ (and therefore interchanging k, k_1) we obtain another conic which is also a nodal curve on the surface, that is, we have as nodal curves the two conics

$$\left\{\begin{aligned}&x:y:z:w=\theta\sqrt{b}:\;\;\theta\sqrt{a}:1:-k\,\theta^2+\frac{c}{k},\quad\text{and}\\&x:y:z:w=\theta\sqrt{b}:-\theta\sqrt{a}:1:-k_1\theta^2+\frac{c}{k_1}.\end{aligned}\right.$$

It is to be remarked that each of the nodal conics meets the cuspidal curve in two points, viz. writing for shortness $\Theta=\frac{1}{k}\sqrt{\frac{-2c\sqrt{ab}}{k_1}}$, $\Theta_1=\frac{1}{k_1}\sqrt{\frac{2c\sqrt{ab}}{k}}$, for the intersections of the first conic we have

$$x:y:z:w=\Theta\sqrt{a}:\;\;\Theta\sqrt{b}:1:\frac{c}{k_1}\quad\text{and}\quad=-\Theta\sqrt{a}:-\Theta\sqrt{b}:1:\frac{c}{k_1},$$

and for the intersections with the second conic

$$x:y:z:w=\Theta_1\sqrt{a}:-\Theta_1\sqrt{b}:1:\frac{c}{k}\quad\text{and}\quad=-\Theta_1\sqrt{a}:\;\;\Theta_1\sqrt{b}:1:\frac{c}{k}.$$

The condition of passing through any arbitrary point establishes a linear relation between the parameters (x, y, z, w). Hence, if the conic in addition to the prescribed conditions passes through two other given points, the point (x, y, z, w) is given as the intersection of a line with the sextic surface; the number of intersections is $=6$. If (x, y, z, w) is situate on the cuspidal curve, then the conic instead of simply touching the given conic will have with it a contact of the second order, and if we besides suppose that the conic passes through a given point, then the point (x, y, z, w) is given as the intersection of the cuspidal curve with a plane; the number is $=6$. Similarly, if the conic has two contacts with the given conic, and besides passes through a given point, then the point (x, y, z, w) is given as the intersection of the nodal curve by a plane; the number is $=4$. Finally (observing that in the case in question of the contacts of a conic with a conic we cannot have three simple contacts, or a simple contact and one of the second order), a point of intersection of the nodal and cuspidal curves answers to a contact of the third order; and the number is $=4$. That is, the theory of the sextic surface leads to the following values (agreeing with those obtained from the formulæ by writing therein $m = n = 2$, $\alpha = 6$), viz.

$$\begin{aligned}
&(1::) &&= 6, &&= 2m + n,\\
&(1,\ 1\ \therefore) &&= 4, &&= 2m^2 + 2mn + \tfrac{1}{2}n^2 - 2m - \tfrac{1}{2}n - \tfrac{3}{2}\alpha,\\
&(2\ \therefore) &&= 6, &&= \alpha,\\
&(3:) &&= 4, &&= -4m - 3n + 3\alpha.
\end{aligned}$$

I remark that the section by an arbitrary plane is a sextic curve having 6 cusps and 4 nodes; it is therefore a *unicursal* sextic; this suggests the theorem that the sextic surface is also *unicursal*, viz. that the coordinates are expressible rationally in terms of two parameters; I have found that this is in fact the case. In doing this there is no loss of generality in supposing that $a = b = c = 1$; and assuming that this is so, and putting also $-1 + h = k$, $1 + h = k_1$, and therefore $2h = k + k_1$, we have

$$\begin{aligned}
A &= -3kk_1,\\
B &= -kk_1 w - (k + k_1)\, z,\\
C &= -x^2 - y^2 - z^2 + (k + k_1)(xy - zw),\\
D &= 3z\,(2xy - zw).
\end{aligned}$$

The equation of the sextic surface being, as before,

$$A^2D^2 + 4AC^3 + 4B^3D - 3B^2C^2 - 6ABCD = 0,$$

I say that this equation is satisfied on writing therein

$$\begin{aligned}
x + y &= \sqrt{-\frac{2}{k_1}}\,(1 - k_1\alpha)\sin\phi,\\
x - y &= \sqrt{\frac{2}{k}}\,(1 + k\alpha)\cos\phi,\\
z &= 1,\\
w &= \left(2\alpha - \frac{1}{k}\right)\cos^2\phi + \left(2\alpha - \frac{1}{k_1}\right)\sin^2\phi,
\end{aligned}$$

where $(\alpha,\ \phi)$ are arbitrary. In fact these values give

$$\begin{aligned}
\tfrac{1}{3}A &= -kk_1 \cos^2\phi - kk_1 \sin^2\phi,\\
B &= -k\,(2\alpha k_1 + 1)\cos^2\phi - k_1\,(2\alpha k + 1)\sin^2\phi,\\
C &= -k\,(\alpha k_1 + 2)\cos^2\phi - k_1\alpha\,(\alpha k + 2)\sin^2\phi,\\
\tfrac{1}{3}D &= -k\alpha^2 \cos^2\phi - k_1\alpha^2 \sin^2\phi,
\end{aligned}$$

whence, ω being arbitrary, we have

$$\begin{aligned}
&\tfrac{1}{3}(A,\ B,\ C,\ D \between \omega,\ 1)^3\\
&\qquad = -\,[k\cos^2\phi\,(k_1\omega + 1) + k_1\sin^2\phi\,(k\omega + 1)]\,(\omega + \alpha)^2,
\end{aligned}$$

viz. the equation $(A, B, C, D \between \omega, 1)^3 = 0$, considered as a cubic equation in ω, has the twofold root $\omega = -\alpha$, that is, we have the above relation between (A, B, C, D). Whence also writing $\sin\phi = \frac{2\lambda}{1+\lambda^2}$, $\cos\phi = \frac{1-\lambda^2}{1+\lambda^2}$, the equation of the surface is satisfied by the values

$$\begin{aligned} x+y : x-y : z : w = & \sqrt{-\frac{2}{k_1}}(1-k_1\alpha)\, 2\lambda\, (1+\lambda^2) \\ : & \sqrt{\frac{2}{k}}(1+k\alpha) \quad (1-\lambda^4) \\ : & (1+\lambda^2)^2 \\ : & \left(2\alpha - \frac{1}{k}\right)(1-\lambda^2)^2 + \left(2\alpha - \frac{1}{k_1}\right)4\lambda^2, \end{aligned}$$

or the coordinates are expressed rationally in terms of α, λ.

Annex No. 4 (referred to, Nos. 22 and 71).—*On the Conics which touch a cuspidal cubic.*

In the cuspidal cubic, if $x=0$ be the equation of the tangent at the cusp, $y=0$ that of the line joining the cusp with the inflexion, and $z=0$ that of the tangent at the cusp, then the equation of the curve is $y^3 = x^2 z$; the coordinates of a point on the cubic are given by $x : y : z = 1 : \theta : \theta^3$, where θ is a variable parameter; and we have, at the cusp $\theta = \infty$, at the inflexion $\theta = 0$. In the cubic, $m = n = 3$, $\alpha\,(= 3n + \kappa) = 10$.

Considering now the conic

$$(a, b, c, f, g, h \between x, y, z)^2 = 0,$$

this meets the cubic in the 6 points the parameters of which are determined by the equation

$$(a, b, c, f, g, h \between 1, \theta, \theta^3)^2 = 0,$$

or, what is the same thing,

$$(c, 0, 2f, 2g, b, 2h, a \between \theta, 1)^6 = 0.$$

The discriminant of this sextic function contains the factor c, hence equating the residual factor to zero, we obtain the equation of the contact-locus in the form

$$(c, f, g, b, h, a)^9 = 0.$$

It follows that the number of the conics $(1 ::)$ is $=9$, which agrees with the general value $(1 ::) = 2m + n$. If the conic pass through the cusp we have $c = 0$, and the equation in θ is reduced to a quartic; it is convenient to alter the letters in such wise that the quartic equation may be obtained in the standard form $(a, b, c, d, e \between \theta, 1)^4 = 0$; viz. this will be the case if the equation of the conic is taken to be

$$(e, 6c, 0, \tfrac{1}{2}a, 2b, 2d \between x, y, z)^2 = 0,$$

and we then obtain the equation of the contact-locus in the form

$$(ae - 4bd + 3c^2)^3 - 27(ace + 2bcd - ad^2 - b^2e - c^3)^2 = 0,$$

which is a onefold locus of the order 6. It follows that we have

$$(\overline{1\kappa 1},\ 1\,\therefore) = 6, \text{ agreeing with } (\overline{1\kappa 1},\ 1\,\therefore) = n + 2m - 3.$$

The condition in order that the conic may touch a given line is given by an equation of the form

$$(*\between a^2,\ ab,\ b^2,\ 2ce - 3d^2,\ ae - 8bd,\ ad - 12bc)^1 = 0,$$

which is a onefold locus of the order 2; it at once follows that we have

$$(\overline{1\kappa 1},\ 1\ :/) = 12, \text{ agreeing with } (\overline{1\kappa 1},\ 1\ :/) = 2n + 4m - 6.$$

It is a matter of some difficulty to show that we have

$$(\overline{1\kappa 1},\ 1\ \cdot //) = 18, \text{ agreeing with } (\overline{1\kappa 1},\ 1\cdot //) = 4n + 4m - 6;$$

but I proceed to effect this, first remarking that I do not attempt to prove the remaining case

$$(\overline{1\kappa 1},\ 1\,///) = 15, \text{ agreeing with } (\overline{1\kappa 1},\ 1\,///) = 4n + 2m - 3.$$

Investigation of the value $(\overline{1\kappa 1},\ 1\cdot //) = 18$:

We have the sextic locus

$$(ae - 4bd + 3c^2)^3 - 27(ace + 2bcd - ad^2 - b^2e - c^3)^2 = 0,$$

and combined therewith two quadric loci,

$$(*\between a^2,\ ab,\ b^2,\ 2ce - 3d^2,\ ae - 8bd,\ ad - 12bc)^1 = 0,$$

$$(*'\between a^2,\ ab,\ b^2,\ 2ce - 3d^2,\ ae - 8bd,\ ad - 12bc)^1 = 0,$$

which intersect in a threefold locus of the order 24; it is to be shown that this contains as part of itself the quadric threefold locus $(a = 0,\ b = 0,\ 2ce - 3d^2 = 0)$ taken three times, leaving a residual locus of the order $24 - 6, = 18$.

We may imagine the coordinates a, b, c, d, e expressed as linear functions of any four coordinates, and so reduce the problem from a problem in 4-dimensional space to one in ordinary 3-dimensional space. We have thus a sextic surface, and two quadric surfaces; the sextic is a developable surface or torse, having for one of its generating lines the line $a = 0$, $b = 0$, and for the tangent plane along this line the plane $a = 0$; the two quadric surfaces meet in a quartic curve passing through the two points $(a = 0,\ b = 0,\ 3ce - 2d^2 = 0)$, which are points on the torse; it is to be shown that each of these points counts three times among the intersections of the torse with the quartic curve, the number of the remaining intersections being therefore $24 - 6, = 18$; and in order thereto it is to be shown that each of the points in question $(a = 0,\ b = 0,\ 3ce - 2d^2 = 0)$

is situate on the nodal line of the torse, and that the quartic curve touches there the sheet which is not touched by the tangent plane $a=0$; for this being so the quartic curve touching one sheet and simply meeting the other sheet meets the torse in three consecutive points, or the two points of intersection count each of them three times.

The torse has the cuspidal line

$$S = ae - 4bd + 3c^2 = 0, \quad T = ace + 2bcd - ad^2 - b^2e - c^3 = 0,$$

and the nodal line

$$\left\| \begin{matrix} 6(ac-b^2), & 3(ad-bc), & ae+2bd-3c^2, & 3(be-cd), & 6(ce-d^2) \\ a, & b, & c, & d, & e \end{matrix} \right\|$$

and the equations of the nodal line are satisfied by the values $(a=0,\ b=0,\ 3ce-2d^2=0)$ of the coordinates of the points in question. To find the tangent planes at these points, starting from the equation $S^3-27T^2=0$ of the torse, taking (A, B, C, D, E) as current coordinates, and writing

$$\partial = A\partial_a + B\partial_b + C\partial_c + D\partial_d + E\partial_e,$$

then the equation of the tangent plane is in the first instance given in the form $S^2\partial S - 18T\partial T = 0$, which writing therein $(a=0,\ b=0,\ 3ce-2d^2=0)$ assumes, as it should do, the form $0=0$; the left-hand side is in fact found to be $9c^3(3ce-2d^2)A$. Proceeding to the second derived equation, this is $S^2\partial^2 S + 2S(\partial S)^2 - 18T\partial^2 T - 18(\partial T)^2 = 0$, or substituting the values of the several terms, the equation is

$$\begin{aligned} &9c^4(AE-4BD+3C^2) \\ +\ &3c^2(\,eA-4dB\ +6cC)^2 \\ +\ &18c^3\{e(AC-B^2)+2d(BC-AD)+c(AE+2BD-3C^2)\} \\ -\ &9\ \{(ce-d^2)A+2cdB-3c^2C\}^2 = 0; \end{aligned}$$

the terms in BC, BD, C^2 vanish identically, that in B^2 is $(48-36=)\,12c^2d^2-18c^3e$, $=-6c^2(3ce-2d^2)B^2$, which also vanishes; hence there remain only the terms divisible by A, giving first the tangent plane $A=0$, and secondly the other tangent plane,

$$\begin{aligned} &A(-\ 6c^2e^2+18cd^2e-9d^4) \\ +\ &B(-\ 60c^2de+36cd^3) \\ +\ &C(\ 108c^3e-54c^2d^2) \\ +\ &D(-\ 36c^3d) \\ +\ &E\,.\,27c^4 \qquad\qquad = 0. \end{aligned}$$

Taking the equations of the quadric surfaces to be

$$\begin{aligned} &(\lambda, \mu, \nu, \rho, \sigma, \tau \between a^2, b^2, ab, 3ce-2d^2, ae-8bd, ad-12bc) = 0, \\ &(\lambda', \mu', \nu', \rho', \sigma', \tau' \between \quad " \quad " \quad " \quad " \quad) = 0, \end{aligned}$$

the equations of the tangent planes are

$$\rho\ (3cE+3eC-4dD)+\sigma\ (eA-8dB)+\tau\ (dA-12cB)=0,$$
$$\rho'\ (\quad\text{''}\quad)+\sigma'\ (\quad\text{''}\quad)+\tau'\ (\quad\text{''}\quad)=0,$$

in all which equations we have $3ce-2d^2=0$; and if to satisfy this equation we write $c:d:e=2:3\beta:3\beta^2$, then the equations of the tangent planes become

$$\beta^3(A\beta-8B)+8(3C\beta^2-4D\beta+2E)=0,$$
$$\rho\ (3CB^2-4D\beta+2E)+(\sigma\beta+\tau)(A\beta-8B)=0,$$
$$\rho'\ (\quad\text{''}\quad)+(\sigma'\beta-\tau)(\quad\text{''}\quad)=0,$$

or the three tangent planes intersect in the line $A\beta-8B=0$, $3C\beta^2-4D\beta+2E=0$, which completes the proof.

Reverting to the sextic locus,

$$(ae+4bd-3c^2)^2-27\,(ace+2bcd-ad^2-b^2e-c^3)^2=0,$$

considered as a locus in 4-dimensional space depending on the five coordinates (a, b, c, d, e), this has upon it the twofold locus

$$ae-4bd+3c^2=0,\quad ace+2bcd-ad^2-b^2e-c^3=0,$$

say the cuspidal locus, of the order 6, and the twofold locus

$$\left\Vert\begin{matrix} 6\,(ac-b^2), & 3\,(ad-bc), & ae+2bd-3c^2, & 3\,(be-cd), & 6\,(ce-d^2) \\ a\quad, & b\quad, & c\quad, & d\quad, & e \end{matrix}\right\Vert=0,$$

say the nodal locus, of the order 4: there is also a threefold locus,

$$\left\Vert\begin{matrix} a, & b, & c, & d \\ b, & c, & d, & e \end{matrix}\right\Vert=0,$$

say the supercuspidal locus, of the order 4. We thence at once infer

$(\overline{1\kappa1},\ 2\quad :)=6$, agreeing with $(\overline{1\kappa1},\ 2\quad :)=\alpha-4$,

$(\overline{1\kappa1},\ 1,\ 1:)=4$, " " $(\overline{1\kappa1},\ 1,\ 1:)=2m^2+2mn+\tfrac{1}{2}n^2-8m-\tfrac{7}{2}n+13-\tfrac{3}{2}\alpha$,

$(\overline{1\kappa1},\ 3\quad :)=4$, " " $(\overline{1\kappa1},\ 3\quad :)=-4m-3n-5+3\alpha$;

but I have not investigated the application to the symbols with $\cdot/$ or $//$.

If the conic, instead of simply passing through the cusp, touches the cuspidal tangent, then in the equation $(a, b, 0, f, g, h\between x, y, z)^2=0$ of the conic we have $f=0$, or, what is the same thing, in the equation $(e, 6c, 0, \tfrac{1}{2}a, 2b, 2d\between x, y, z)^2=0$ of the conic we have $a=0$. The equation in θ is thus reduced to $4b\theta^3+6c\theta^2+4d\theta+e=0$. For the independent discussion of this case it is convenient to alter the coefficients so that the equation in θ may be in the standard form $(a, b, c, d\between\theta, 1)^3=0$, viz. we assume the equation of the conic to be $(d,\ 3b,\ 0,\ 0,\ \tfrac{1}{2}a,\ \tfrac{3}{2}c)(x,\ y,\ z)^2 = 0$. The equation of the contact-locus then is

$$a^2d^2 + 4ac^3 + 4b^3d - 6abcd - 3b^2c^2 = 0,$$

viz. this is a developable surface, or torse, of the order 4, and we at once infer

$$(\overline{2\kappa 1},\ 1\,:) = 4, \text{ agreeing with } (\overline{2\kappa 1},\ 1\,:) = 2m + n - 5.$$

I will show also that we have

$$(\overline{2\kappa 1},\ 1\cdot/) = 6, \text{ agreeing with } (\overline{2\kappa 1},\ 1\cdot/) = 2m + 2n - 6,$$

and

$$(\overline{2\kappa 1},\ 1\ //) = 5,\quad\text{,,}\quad\text{,,}\quad (\overline{2\kappa 1},\ 1\ //) = m + 2n - 4.$$

The condition that the conic may touch an arbitrary line $\alpha x + \beta y + \gamma z = 0$, is in fact

$$(0,\ -\tfrac{1}{4}a^2,\ \tfrac{3}{4}(4bd - 3c^2),\ \tfrac{3}{4}ac,\ -\tfrac{3}{2}ab,\ 0)(\alpha,\ \beta,\ \gamma)^2 = 0,$$

which, considering therein $(a,\ b,\ c,\ d)$ as coordinates, is the equation of a quadric surface passing through the conic $a = 0$, $4bd - 3c^2 = 0$; the quartic torse also passes through this conic; hence the quadric surface and the torse intersect in this conic, which is of the order 2, and in a residual curve of the order 6; and the number of the conics $(\overline{2\kappa 1},\ 1\cdot/)$ is equal to the order of this residual curve, that is, it is $= 6$.

If the conic touch a second arbitrary line $\alpha' x + \beta' y + \gamma' z = 0$, then we have in like manner the quadric surface

$$(0,\ -\tfrac{1}{4}a^2,\ \tfrac{3}{4}(4bd - 3c^2),\ \tfrac{3}{4}ac,\ -\tfrac{3}{2}ab,\ 0)(\alpha',\ \beta',\ \gamma')^2 = 0;$$

that is, we have the quartic torse and two quadric surfaces, each passing through the conic $a = 0$, $4bd - 3c^2 = 0$, and it is to be shown that the number of intersections not on this conic is $= 5$. The two quadric surfaces intersect in the conic and in a second conic; this second conic meets the torse in 8 points, but 2 of these coincide with the point $a = 0$, $b = 0$, $c = 0$, which is one of the intersections of the two conics (the point $a = 0$, $b = 0$, $c = 0$ is in fact a point on the cuspidal edge of the torse, and, the conic passing through it, reckons for 2 intersections), and 1 of the 8 points coincides with the other of the intersections of the two conics; there remain therefore $8 - 2 - 1$, $= 5$ intersections, or we have $(\overline{2\kappa 1},\ 1\ //) = 5$.

Annex No. 5 (referred to, Nos. 22 and 71). *On the Conics which have contact of the third order with a given cuspidal cubic, and two contacts (double contact) with a given conic.*

Let the equation of the cuspidal cubic be $x^2z - y^3 = 0$ ($x = 0$ tangent at cusp, $z = 0$ tangent at inflexion, $y = 0$ line joining cusp and inflexion; equation satisfied by

$$x : y : z = 1 : \theta : \theta^3);$$

and let the equation of the given conic be

$$U = (a,\ b,\ c,\ f,\ g,\ h)(x,\ y,\ z)^3 = 0;$$

then writing

$$\begin{aligned}\Theta &= (a,\ b,\ c,\ f,\ g,\ h \between 1,\ \theta,\ \theta^3)^2 \\ &= c\theta^6 + 2f\theta^4 + 2g\theta^3 + b\theta^2 + 2h\theta + c,\end{aligned}$$

the equation of a conic having with the given cubic at a given point $(1, \theta, \theta^3)$ contact of the second order, and having double contact with the given conic, is

$$\left|\begin{array}{cccc} \sqrt{U}, & x, & y, & z \\ \sqrt{\Theta}, & 1, & \theta, & \theta^3 \\ (\sqrt{\Theta})' & . & 1, & 3\theta^2 \\ (\sqrt{\Theta})'' & . & . & 6\theta \end{array}\right| = 0,$$

viz. in the rational form this is

$$36\theta^2 U - \left|\begin{array}{cccc} & x, & y, & z \\ \sqrt{\Theta}, & 1, & \theta, & \theta^3 \\ (\sqrt{\Theta})' & . & 1, & 3\theta^2 \\ (\sqrt{\Theta})'' & . & . & 6\theta \end{array}\right|^2 = 0,$$

and this will have at the point $(1, \theta, \theta^3)$ a contact of the third order if θ be determined by

$$\left|\begin{array}{cccc} \sqrt{\Theta}, & 1, & \theta, & \theta^3 \\ (\sqrt{\Theta})' & . & 1, & 3\theta^2 \\ (\sqrt{\Theta})'' & . & . & 6\theta \\ (\sqrt{\Theta})''' & . & . & 6 \end{array}\right| = 0,$$

viz. this is

$$\Theta\,(\sqrt{\Theta})''' - (\sqrt{\Theta})'' = 0\,;$$

or developing and multiplying by $\Theta^{\frac{5}{2}}$, this is

$$\theta\,\{\Theta^2\Theta''' - \tfrac{3}{2}\Theta\Theta'\Theta'' + \tfrac{3}{4}\Theta'^3\} - (\Theta^2\Theta'' - \tfrac{1}{2}\Theta\Theta'^2) = 0,$$

or, what is the same thing,

$$\Theta^2\,(\theta\Theta''' - \Theta'') + \Theta\Theta'\,(-\tfrac{3}{2}\theta\Theta'' + \tfrac{1}{2}\Theta') + \Theta'^2\,.\,\tfrac{3}{4}\theta\Theta' = 0\,;$$

and substituting for Θ its value, this is

$$\begin{aligned} &(c\theta^6 + 2f\theta^4 + 2g\theta^3 + b\theta^2 + 2h\theta + a)^2\,(45c\theta^4 + 12f\theta^2 - b) \\ &+ (c\theta^6 + 2f\theta^4 + 2g\theta^3 + b\theta^2 + 2h\theta + a)\,(\ 3c\theta^5 + \ 4f\theta^3 + 3g\theta^2 + b\theta + h) \\ &\qquad\qquad (-\,42c\theta^5 - 32f\theta^3 - 15g\theta^2 - 2b\theta + h) \\ &+ 3\theta\,(3c\theta^5 + 4f\theta^3 + 3g\theta^2 + b\theta + h)^3 = 0. \end{aligned}$$

The coefficients of the powers 16, 15, 14, 13 of θ all vanish, so that this is in fact an equation of the twelfth order $(*\between\theta, 1)^{12}=0$; and putting, as usual,

$$(bc-f^2,\ ca-g^2,\ ab-h^2,\ gh-af,\ hf-bg,\ fg-ch)=(A,\ B,\ C,\ F,\ G,\ H),$$

the equation is found to be

$$\begin{array}{lll}
-\ 4cA\ \ \theta^{12} & +\ 72hA \\ & +\ \ 9gB \ \Big\}\ \theta^7 & +\ 45aB \\ & -\ 22bH & -\ 20fC\ \Big\}\ \theta^4 \\
+\ 30cH\ \ \theta^{11} & & +\ 10hG \\
-\ 36cB \\ +\ 16fA\ \Big\}\ \theta^{10} & +\ 40aA & +\ 5hF \\ & -\ 130hH & +\ 20aG\ \Big\}\ \theta^3 \\
-\ 10cG \\ +\ 40gA\ \Big\}\ \theta^9 & +\ 10gG\ \Big\}\ \theta^6 & -\ 4bC \\ & +\ 40fF & -\ 12aF\ \Big\}\ \theta^2 \\
+\ 20bA \\ -\ 60fB\ \Big\}\ \theta^8 & +\ 33hB & -\ 5hC\ \ \theta \\
-\ 90gH & +\ 2bG\ \Big\}\ \theta^5 & -\ aC\ \ =0, \\
& -\ 108aH
\end{array}$$

where the form of the coefficients may be modified by means of the identical equations

$$\begin{aligned}
(A,\ H,\ G\between a,\ h,\ g)&=K,\\
(H,\ B,\ F\between\ \ ,,\ \ )&=0,\\
(G,\ F,\ C\between\ \ ,,\ \ )&=0,\\
(A,\ H,\ G\between h,\ b,\ f)&=0,\\
(H,\ B,\ F\between\ \ ,,\ \ )&=K,\\
(G,\ F,\ C\between\ \ ,,\ \ )&=0,\\
(A,\ H,\ G\between g,\ f,\ c)&=0,\\
(H,\ B,\ F\between\ \ ,,\ \ )&=0,\\
(G,\ F,\ C\between\ \ ,,\ \ )&=K.
\end{aligned}$$

There is consequently a conic answering to each value of θ given by this equation, or we have in all 12 conics.

In the case where the given conic breaks up into a pair of lines, or say,

$$(a,\ b,\ c,\ f,\ g,\ h\between x,\ y,\ z)^2=2(\lambda x+\mu y+\nu z)(\lambda' x+\mu' y+\nu' z),$$

then, writing for shortness

$$\mu\nu'-\mu'\nu,\ \nu\lambda'-\nu'\lambda,\ \lambda\mu'-\lambda'\mu=X,\ Y,\ Z,$$

we have

$$(A,\ B,\ C,\ F,\ G,\ H)=(X^2,\ Y^2,\ Z^2,\ YZ,\ ZX,\ XY).$$

Substituting these values, but retaining (a, b, c, f, g, h) as standing for their values $a = 2\lambda\lambda'$, &c., the equation in θ is found to contain the cubic factor $2X\theta^3 - 3Y\theta^2 + Z$, where it is to be observed that this factor equated to zero determines the values of θ which correspond to the points of contact with the cuspidal cubic of the tangents from the point (X, Y, Z), which is the intersection of the lines $\lambda x + \mu y + \nu z = 0$, and $\lambda' x + \mu' y + \nu' z = 0$; and omitting the cubic factor, the residual equation is found to be

$$\left(\begin{array}{c|c|c|c|c|c|c|c|c|c} 2cX & -12cY & -8fX & \begin{matrix}-20gX\\-12fY\\+\ \ 4cZ\end{matrix} & \begin{matrix}-10bX\\+\ \ 3gY\\ \end{matrix} & \begin{matrix}-40hX\\-\ \ 8bY\\ \end{matrix} & \begin{matrix}-20aX\\+17hY\\+\ \ 7gZ\end{matrix} & \begin{matrix}+15aY\\+\ \ 4bZ\\ \end{matrix} & +5hZ & +aZ \end{array}\right)(\theta, 1)^9 = 0,$$

where the form of the coefficients may be modified by means of the identical equations

$$\begin{aligned} aX + hY + gZ &= 0, \\ hX + bY + fZ &= 0, \\ gX + fY + cZ &= 0. \end{aligned}$$

The equation is of the 9th order, and there are consequently 9 conics.

Annex No. 6 (referred to, No. 48).—*Containing, with the variation referred to in the text,* ZEUTHEN'S *forms for the characteristics of the conics which satisfy four conditions.*

(1)

$$\begin{aligned} (\ ::\ ) &= n + 2m, \\ (\ \therefore /) &= 2n + 4m, \\ (\ : //) &= 4n + 4m, \\ (\cdot ///) &= 4n + 2m, \\ (////) &= 2n + m; \end{aligned}$$

(1, 1)

$$\begin{aligned} (\therefore\ ) &= 2m(m + n - 3) + \tau, \\ (\ : /) &= 2m(m + 2n - 5) + 2\tau, \\ (\cdot //) &= 2n(2m + n - 5) + 2\delta, \\ (///\ ) &= 2n(m + n - 3) + \delta; \end{aligned}$$

(1, 1, 1)

$$\begin{aligned} (\ :\ ) &= \tfrac{1}{3}[2m^3 + 6m^2n - n^3 - 30m^2 - 18mn + 13n^2 + 84m - 42n + (6m + 3n - 26)\tau], \\ (\ \cdot /\ ) &= \tfrac{1}{6}[(m+n)\big(-(m+n)^2 - 7(m+n) + 48\big) + 4mn(3m + 3n - 13) + 2(3m + 3n - 20)(\delta + \tau)], \\ (\ //\ ) &= \tfrac{1}{3}[-m^3 + 6mn^2 + 2n^3 + 13m^2 - 18mn - 30n^2 - 42m + 84n + (3m + 6n - 26)\delta]; \end{aligned}$$

(1, 1, 1, 1)

$$(\cdot) = \tfrac{1}{6}\{\ 2(m-3)(m-4)(n^2-m-n)+(n-3)(n-4)(m^2-m-n)$$
$$+4(m^2-11m+28)\tau \qquad +2(n^2-11n+28)\delta$$
$$+\big(4(n-4)(m-4)-1\big)(2\delta+\tau)+2\delta^2+\tau^2\},$$

$$(/) = \tfrac{1}{6}\{\ (m-3)(m-4)(n^2-m-n)+2(n-3)(n-4)(m^2-m-n)$$
$$+2(m^2-11m+28)\tau \qquad +4(n^2-11n+28)\delta$$
$$+\big(4(n-4)(m-4)-1\big)(\delta+2\tau)+\delta^2+2\tau^2\};$$

(2)

$$(\therefore) = 3m+\iota,$$
$$(:/) = 2(3m+\iota),$$
$$(\cdot//) = 2(3m+\iota),$$
$$(///) = 3m+\iota,$$

(2, 1)

$$(:) = 3(2mn+n^2+4m-10n)+(2m+n-14)\kappa,$$
$$(\cdot/) = 2(3m+\iota)(m+n-12)+24(m+n),$$
$$(//) = 3(m^2+2mn-10m+4n)+(m+2n-14)\iota;$$

(2, 1, 1)

$$(\cdot) = (2m+n-7)\big(6\tau+(n-3)\kappa\big)$$
$$+\big((m-n)(m+n-5)+\tau\big)(3m+\iota-36)$$
$$+12(m-n)(m+n-3),$$

$$(/) = (m+2n-7)\big(6\delta+(m-3)\iota\big)$$
$$+\big((n-m)(m+n-5)+\delta\big)(3m+\iota-36)$$
$$+12(n-m)(m+n-3);$$

2, 2)

$$(\cdot) = \tfrac{1}{2}(3m+\iota)^2-3(3m+\iota)-9\tau-8\delta,$$
$$(/) = \tfrac{1}{2}(3m+\iota)^2-3(3m+\iota)-8\tau-9\delta;$$

(3)

$$(:) = 6n-4m+3\kappa = 5m-3n+3\iota,$$
$$(\cdot/) = 10n-8m+6\kappa = 10m-8n+6\iota,$$
$$(//) = 5n-3m+3\kappa = 6m-4n+3\iota;$$

(1, 3)

$$(\cdot) = 2(-4m^2+3mn+3n^2+28m-32n)+3(2m+n-13)\kappa,$$
$$(/) = 2(3m^2+3mn-4m^2-32m+28n)+3(m+2n-13)\iota;$$

(4)

$$(\cdot) = 10n-10m+6\kappa = 8m-8n+6\iota,$$
$$(/) = 8n-8m+6\kappa = 10m-10n+6\iota.$$

Annex No. 7 (referred to, No. 93).

In connexion with De Jonquières' formula, I have been led to consider the following question.

Given a set of equations:

$$\mathrm{a} = a \quad (\text{viz. } \mathrm{b} = b,\ \mathrm{c} = c,\ \text{\&c.}),$$

$$\begin{aligned}\mathrm{ab} &= ab \\ &\quad + (11)\, a\,.\,b\end{aligned} \quad \left(\begin{aligned}&\text{viz. } \mathrm{ac} = ac, \text{ \&c., and the like in all the subsequent equations} \\ &\qquad\quad + (11)\, a\,.\,c,\end{aligned}\right),$$

$$\begin{aligned}\mathrm{abc} &= abc \\ &\quad + (12)\,(a\,.\,bc + b\,.\,ac + c\,.\,ab) \\ &\quad + (111)\, a\,.\,b\,.\,c,\end{aligned}$$

$$\begin{aligned}\mathrm{abcd} &= abcd \\ &\quad + (13)\,(a\,.\,bcd + \text{\&c.}) \\ &\quad + (22)\,(ab\,.\,cd + \text{\&c.}) \\ &\quad + (112)\,(a\,.\,b\,.\,cd + \text{\&c.}) \\ &\quad + (1111)\, a\,.\,b\,.\,c\,.\,d,\end{aligned}$$

and so on indefinitely (where the ($\cdot$) is used to denote multiplication, and *ab*, *abc*, &c., and also ab, abc, &c. are so many separate and distinct symbols not expressible in terms of *a*, *b*, *c* &c., a, b, c &c.), then we have conversely a set of equations

$$a = \mathrm{a} \quad (\text{viz. } b = \mathrm{b},\ c = \mathrm{c}\ \text{\&c.}),$$

$$\begin{aligned}ab &= \mathrm{ab} \\ &\quad + [11]\, \mathrm{a}\,.\,\mathrm{b}\end{aligned} \quad \left(\begin{aligned}&\text{viz. } ac = \mathrm{ac} \text{ \&c., and the like in all the subsequent equations} \\ &\qquad\quad + [11]\, \mathrm{a}\,.\,\mathrm{c},\end{aligned}\right),$$

$$\begin{aligned}abc &= \mathrm{abc} \\ &\quad + [12]\,(\mathrm{a}\,.\,\mathrm{bc} + \mathrm{b}\,.\,\mathrm{ac} + \mathrm{c}\,.\,\mathrm{ab}) \\ &\quad + [111]\, \mathrm{a}\,.\,\mathrm{b}\,.\,\mathrm{c},\end{aligned}$$

$$\begin{aligned}abcd &= \mathrm{abcd} \\ &\quad + [13]\,(\mathrm{a}\,.\,\mathrm{bcd} + \text{\&c.}) \\ &\quad + [22]\,(\mathrm{ab}\,.\,\mathrm{cd} + \text{\&c.}) \\ &\quad + [112]\,(\mathrm{a}\,.\,\mathrm{b}\,.\,\mathrm{cd} + \text{\&c.}) \\ &\quad + [1111]\, \mathrm{a}\,.\,\mathrm{b}\,.\,\mathrm{c}\,.\,\mathrm{d},\end{aligned}$$

and so on; and it is required to find the relation between the coefficients ( ) and [ ]; we find, for example,

$$
\begin{aligned}
[11] &= - \quad (11),\\
[12] &= - \quad (12),\\
[111] &= \;\; 3 \; (11)(12)\\
&\quad - \quad (111),\\
[13] &= - \quad (13),\\
[22] &= - \quad (22),\\
[112] &= \;\; 2 \; (13)(12)\\
&\quad + \quad (22)(11)\\
&\quad - \quad (112),\\
[1111] &= - 12 \; (13)(12)(11)\\
&\quad + 4 \; (13)(111)\\
&\quad - 3 \; (22)(11)(11)\\
&\quad + 6 \; (112)(11)\\
&\quad - \quad (1111);
\end{aligned}
$$

and it is to be noticed that, conversely, the coefficients ( ) are given in terms of the coefficients [ ] by the like equations with the very same numerical coefficients; in fact from the last set of equations, this is at once seen to be the case as far as (112); and for the next term (1111) we have

$$
\begin{aligned}
(1111) = &+ 12\,[13]\,[12]\,[11] & &= (12-12-12=) - 12 \; [13][12][11]\\
&- 4\,[13]\,\{3\,[12]\,[11] - [111]\} & &\qquad\qquad\qquad\quad + 4 \; [13][111]\\
&+ 3\,[22]\,[11]\,[11] & &+ (3-6= \qquad) - 3 \; [22][11][11]\\
&- 6\,[11] \left\{\begin{array}{l} 2\,[13]\,[12] \\ +\,[22]\,[11] \\ -\,[112] \end{array}\right\} & &\qquad\qquad\qquad\quad + 6 \; [112][11]\\
&- [1111] & &\qquad\qquad\qquad\quad - \quad [1111]
\end{aligned}
$$

having the same coefficients -12, $+4$, -3, $+6$, -1 as in the formula for [1111] in terms of the coefficients ( ); it is easy to infer that the property holds good generally.

To explain the law for the expression of the coefficients of either set in terms of the other set, I consider, for example, the case where the sum of the numbers in the ( ), or [ ] is $=5$; and I form a kind of tree as follows:

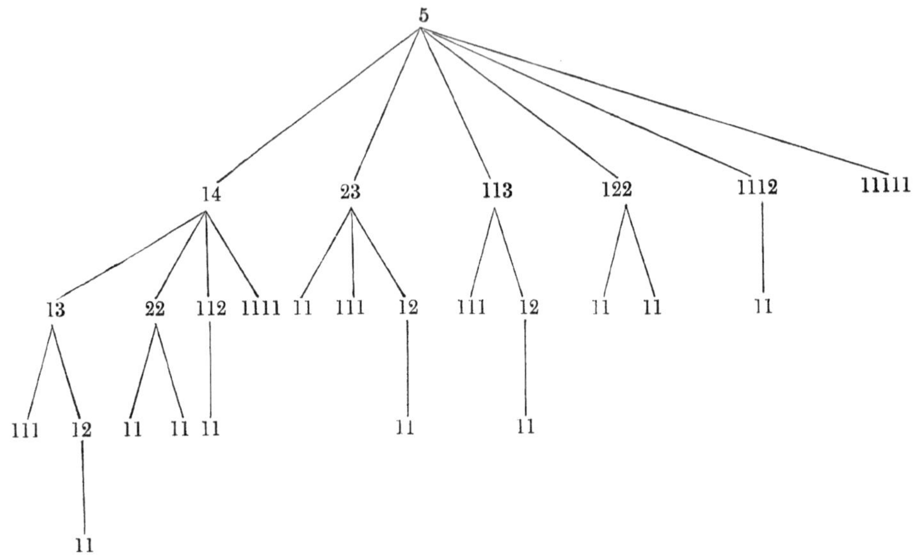

the formation of which is obvious; and I derive from it in the manner about to be explained the expressions for the coefficients [14], [23] &c. in terms of the corresponding coefficients in ( ); viz. we have

$$
\begin{aligned}
[14] &= - (14),\\
[23] &= - (23),\\
[113] &= 2\,(14)(13)\\
&\quad + (23)(11)\\
&\quad - (113),\\
[122] &= (14)(22)\\
&\quad + 2\,(23)(12)\\
&\quad - (112),\\
[1112] &= -6\,(14)(13)(12)\\
&\quad - 3\,(14)(22)(11)\\
&\quad + 3\,(14)(112)\\
&\quad - 6\,(23)(12)(11)\\
&\quad + 3\,(113)(12)\\
&\quad + 1\,(23)(111)\\
&\quad + 3\,(122)(11)\\
&\quad - 1\,(1112),
\end{aligned}
$$

$$
\begin{aligned}
[11111] = &+ 60 \quad (14)(13)(12)(11)\\
&- 20 \quad (14)(13)(111)\\
&+ 15 \quad (14)(22)(11)(11)\\
&- 30 \quad (14)(112)(11)\\
&+ 5 \quad (14)(1111)\\
&+ 30 \quad (23)(12)(11)(11)\\
&- 10 \quad (23)(111)(11)\\
&- 30 \quad (113)(12)(11)\\
&+ 10 \quad (113)(111)\\
&- 15 \quad (122)(11)(11)\\
&+ 10 \quad (1112)(11)\\
&- 1 \quad (11111).
\end{aligned}
$$

To form the symbolic parts, we follow each branch of the tree to each point of its course: thus from the branch 113 we have

(113)	belonging to	[113],
(113)(111)	„	[11111],
(113)(12)	„	[1112],
(113)(12)(11)	„	[11111];

viz. (113) belongs to [113]; (113)(111), read 11 (3 replaced by) 111, belongs to [11111]; (113)(12), read 11 (3 replaced by) 12, belongs to 1112; (113)(12)(11), read 11 (3 replaced by) 1 (2 replaced by) 11, belongs to [11111].

And observe that where (as, for example, with the symbol 122) there are branches derived from two or more figures, we pursue each such branch separately, and also all or any of them simultaneously to every point in the course of such branch or branches; thus for the branch 122 we have

(122)	belonging to	[122],
(122)(11) (122)(11) } (same twice)	„	[1112],
(122)(11)(11)	„	[11111].

Similarly for the branch 23 we have

(23)	belonging to	[23],
(23)(111)	„	[1112],
(23)(12)	„	[122],
(23)(12)(11) (same as *infrà*)	„	[1112],
(23)(11)(111)	„	[11111],
(23)(11)(12) (same as *suprà*)	„	[1112],
(23)(11)(12)(11)	„	[11111].

We thus obtain the symbolic parts of the several expressions for [14], [23]....[11111] respectively: the sign of each term is + or − according as the number of factors in ( ) is even or odd; thus in the expression for [11111], the term (14) (13) (12) (11) has four factors, and is therefore +, the term (113)(12)(11) has three factors, and is therefore −.

The numerical coefficients are obtained as follows. There is a common factor derived from the expression in [ ] on the left-hand side of the equation; viz. for [11111], which contains five equal symbols, this factor is $1.2.3.4.5, =120$; for [1112], which contains three equal symbols, it is $1.2.3, =6$; and so on (for a symbol such as [11222] containing two equal symbols, and three equal symbols, the factor would be $1.2.1.2.3, =12$, and so in other similar cases). In any term on the right-hand side of the equation, we must for a factor such as (11), which contains two equal symbols, multiply by $\frac{1}{2}$; for a factor such as (111), which contains three equal symbols, multiply by $\frac{1}{6}$, and so on. And in the case where a term (as, for example, the term (122)(11) or (23)(12)(11), *vide suprà*) occurs more than once, the term is to be taken account of each time that it occurs; or, what is the same thing, since the coefficient obtained as above is the same for each occurrence, the coefficient obtained as above is to be multiplied by the number of the occurrences of the term. For example, taking in order the several terms of the expression for [1112], the common factor is $=6$, and the several coefficients are

$$6,\quad 6.\tfrac{1}{2},\quad 6.\tfrac{1}{2},\quad 6.\tfrac{1}{2}\times 2,\quad 6.\tfrac{1}{2},\quad 6.\tfrac{1}{6},\quad 6.\tfrac{1}{2}.\tfrac{1}{2}\times 2,\quad 6.\tfrac{1}{6};$$

and similarly in the expression for [11111] the common factor is 120, and the coefficients taken in order are

$$120.\tfrac{1}{2},\quad 120.\tfrac{1}{6},\quad 110.\tfrac{1}{2}.\tfrac{1}{2}.\tfrac{1}{2},\ \&c.,$$

without there being in this case any coefficient with a factor arising from the plural occurrence of the term.

The foregoing result was established by induction, and I have not attempted a general proof.

I observe by way of a convenient numerical verification, that in each equation the sum of the coefficients (taken with their proper signs) is $(-)^{n-1}\,1.2..(n-1)$; if n be the number of parts in the [ ] ($n=5$ for [11111], $=4$ for [1112] &c.), and moreover, that the sum of these sums each multiplied by the proper polynomial coefficient and the whole increased by unity is $=0$; viz. for

[14], [23], [113], [122], [1112], [11111],

the sums of the coefficients are

$$-1,\quad -1,\quad +2,\quad +2,\quad -6,\quad +24 \text{ respectively,}$$

and we have

$$1+5(-1)+10(-1)+10(2)+15(2)+10(-6)+1(24),\ =75-75,\ =0.$$

If we have any five distinct things (a, b, c, d, e), then the polynomial coefficients 5, 10, 10, 15, 10, 1 denote respectively the number of ways in which these can be partitioned in the forms 14, 23, 113, 122, 1112, 11111 respectively, and the last-mentioned theorem is thus a theorem in the Partition of Numbers.

407.

SECOND MEMOIR ON THE CURVES WHICH SATISFY GIVEN CONDITIONS; THE PRINCIPLE OF CORRESPONDENCE.

[From the *Philosophical Transactions of the Royal Society of London*, vol. CLVIII. (for the year 1868), pp. 145—172. Received April 18,—Read May 2, 1867.]

In the present Memoir I reproduce with additional developments the theory established in my paper "On the Correspondence of two points on a Curve" (*London Math. Society*, No. VII., April 1866), [385]; and I endeavour to apply it to the determination of the number of the conics which satisfy given conditions; viz. these are conditions of contact with a given curve, or they may include arbitrary conditions Z, $2Z$, &c. If, for a moment, we consider the more general question where the Principle is to be applied to finding the number of the curves C^r of the order r, which satisfy given conditions of contact with a given curve, there are here two kinds of special solutions; viz., we may have proper curves C^r touching (specially) the given curve at a cusp or cusps thereof, and we may have improper curves, that is, curves which break up into two or more curves of inferior orders. In the case where the curves C^r are lines, there is only the first kind of special solution, where the sought for lines touch at a cusp or cusps. But in the case to which the Memoir chiefly relates, where the curves C^r are conics, we have the two kinds of special solutions, viz., proper conics touching at a cusp or cusps, and conics which are line-pairs or point-pairs. In the application of the Principle to determining the number of the conics which satisfy any given conditions, I introduce into the equation a term called the "Supplement" (denoted by the abbreviation "Supp."), to include the special solutions of both kinds. The expression of the Supplement should in every case be furnished by the theory; and this being known, we should then have an equation leading to the number of the conics which properly satisfy the prescribed conditions; but in thus finding the expression of the Supplements, there are difficulties which I am unable to overcome; and I have contented myself with the reverse course, viz., knowing in each case the number of the proper solutions, I use these results to determine *à posteriori* in each case the expression of the Supplement; the expression so obtained can in some cases be accounted for readily enough, and the knowledge of the whole series of them will be a convenient basis for ulterior investigations.

The Principle of Correspondence for points in a line was established by Chasles in the paper in the *Comptes Rendus*, June—July 1864, referred to in my First Memoir; it is extended to unicursal curves in a paper of the same series, March 1866, "Sur les courbes planes ou à double courbure dont les points peuvent se déterminer individuellement—Application du Principe de Correspondance dans la théorie de ces courbes," but not to the case of a curve of given deficiency D considered in my paper of April 1866 above referred to. The fundamental theorem in regard to unicursal curves, viz. that in a curve of the order m with $\frac{1}{2}(m-1)(m-2)$ double points (nodes or cusps) the coordinates (x, y, z) are proportional to rational and integral functions of a variable parameter θ,—as a case of a much more general theorem of Riemann's—dates from the year 1857, but was first explicitly stated by Clebsch in the paper "Ueber diejenigen ebenen Curven deren Coordinaten rationale Functionen eines Parameters sind," *Crelle*, t. LXIV. (1864), pp. 43—63. See also my paper "On the Transformation of Plane Curves," *London Mathematical Society*, No. III., Oct. 1865, [384].

The paragraphs of the present Memoir are numbered consecutively with those of the First Memoir.

Article Nos. 94 to 104.—*On the Correspondence of two points on a Curve.*

94. In a unicursal curve the coordinates (x, y, z) of any point thereof are proportional to rational and integral functions of a variable parameter θ. Hence if two points of the curve correspond in such wise that to a given position of the first point there correspond α' positions of the second point, and to a given position of the second point α positions of the first point, the number of points which correspond each to itself is $=\alpha+\alpha'$. For let the two points be determined by their parameters θ, θ' respectively, then to a given value of θ there correspond α' values of θ', and to a given value of θ' there correspond α values of θ; hence the relation between (θ, θ') is of the form $(\theta, 1)^{\alpha}(\theta', 1)^{\alpha'}=0$; and writing therein $\theta'=\theta$, then for the points which correspond each to itself, we have an equation $(\theta, 1)^{\alpha+\alpha'}=0$, of the order $\alpha+\alpha'$; that is, the number of these points is $=\alpha+\alpha'$.

Hence for a unicursal curve we have a theorem similar to that of M. Chasles' for a line, viz. the theorem may be thus stated:

If two points of a unicursal curve have an (α, α') correspondence, the number of united points is $=\alpha+\alpha'$. But a unicursal curve is nothing else than a curve with a deficiency $D=0$, and we thence infer:

THEOREM. If two points of a curve with deficiency D have an (α, α') correspondence, the number of united points is $=\alpha+\alpha'+2kD$; in which theorem $2k$ is a coefficient to be determined.

95. Suppose that the corresponding points are P, P' and imagine that when P is given the corresponding points P' are the intersections of the given curve by a curve Θ (the equation of the curve Θ will of course contain the coordinates of P as parameters, for otherwise the position of P' would not depend upon that of P). I find that if the curve Θ has with the given curve k intersections at the point P, then in the system of points (P, P') the number of united points is

$$\text{a} = \alpha + \alpha' + 2kD,$$

whence in particular if the curve Θ does not pass through the point P, then the number of united points is $= \alpha + \alpha'$, as in the case of a unicursal curve. (I have in the paper of April 1866 above referred to, proved this theorem in the particular case where the k intersections at the point P take place in consequence of the curve Θ having a k-tuple point at P, but have not gone into the more difficult investigation for the case where the k intersections arise wholly or in part from a contact of the curve Θ, or any branch or branches thereof, with the given curve at P.)

96. It is to be observed that the general notion of a united point is as follows: taking the point P at random on the given curve, the curve Θ has at this point k intersections with the given curve; the remaining intersections are the corresponding points P'; if for a given position of P one or more of the points P' come to coincide with P, that is, if for the given position of P the curve Θ has at this point more than k intersections with the given curve, then the point in question is a united point.

It might at first sight appear that if for a given position of P a number 2, 3,.. or j of the points P' should come to coincide with P, then that the point in question should reckon, for 2, 3,... or j (as the case may be) united points: but this is not so. This is perhaps most easily seen in the case of a unicursal curve; taking the equation of correspondence to be $(\theta, 1)^{\alpha} (\theta', 1)^{\alpha'} = 0$, then we have $\alpha + \alpha'$ united points corresponding to the values of θ which satisfy the equation $(\theta, 1)^{\alpha} (\theta, 1)^{\alpha'} = 0$; if this equation has a j-tuple root $\theta = \lambda$, the point P which answers to this value λ of the parameter is reckoned as j united points. But starting from the equation $(\theta, 1)^{\alpha} (\theta', 1)^{\alpha'} = 0$, if on writing in this equation $\theta = \lambda$, the resulting equation $(\lambda, 1)^{\alpha} (\theta', 1)^{\alpha'} = 0$ has a root $\theta' = \lambda$, it follows that the equation $(\theta, 1)^{\alpha} (\theta, 1)^{\alpha'} = 0$ has a root $\theta = \lambda$, and that the point which belongs to the value $\theta = \lambda$ is a united point; if on writing in the equation $\theta = \lambda$, the resulting equation $(\lambda, 1)^{\alpha} (\theta', 1)^{\alpha'} = 0$ has a j-tuple root $\theta' = \lambda$, *it does not follow* that the equation $(\theta, 1)^{\alpha} (\theta, 1)^{\alpha'} = 0$ has a j-tuple root $\theta = \lambda$, nor consequently that the point answering to $\theta = \lambda$ in anywise reckons as j united points.

97. This may be further illustrated by regarding the parameters θ, θ' as the coordinates of a point in a plane; the equation $(\theta, 1)^{\alpha} (\theta', 1)^{\alpha'} = 0$ is that of a curve of the order $\alpha + \alpha'$, having an α-tuple point at infinity on the axis $\theta = 0$, and an α'-tuple point at infinity on the axis $\theta' = 0$; the united points are given as the intersections of the curve with the line $\theta = \theta'$; a j-fold intersection, whether arising from a multiple point of the curve or from a contact of the line $\theta = \theta'$ with the curve,

gives a point which reckons as j united points. But if $\theta=\lambda$ gives the j-fold root $\theta'=\lambda$, this shows that the line $\theta=\lambda$ has with the curve j intersections at the point $\theta=\theta'=\lambda$; *not* that the line $\theta=\theta'$ has with the curve j intersections at the point in question.

98. Reverting to the notion of a united point as a point P which is such that one or more of the corresponding points P' come to coincide with P; in the case where P is at a node of the given curve, it is necessary to explain that the point P must be considered as belonging to one or the other of the two branches through the node, and that the point P is not to be considered as a united point unless we have *on the same branch of the curve* one or more of the corresponding points P' coming to coincide with the point P. If, to fix the ideas, $k=1$, that is, if the curve Θ simply pass through the point P, then if P be at a node the curve Θ passes through the node and has therefore at this point two intersections with the given curve; but the second intersection belongs to the other branch, and the node is not a united point; in order to make it so, it is necessary that the curve Θ should at the node touch the branch to which the point P is considered to belong. The thing appears very clearly in the case of a unicursal curve; we have here two values $\theta=\lambda$, $\theta=\lambda'$ answering to the node according as it is considered as belonging to one or the other branch of the curve; and in the equation of correspondence $(\theta,\ 1)^{a}(\theta',\ 1)^{a'}=0$, writing $\theta=\lambda$, we have an equation $(\lambda,\ 1)^{a}(\theta',\ 1)^{a'}=0$ satisfied by $\theta'=\lambda'$ but not by $\theta'=\lambda$, and the equation $(\theta,\ 1)^{a}(\theta,\ 1)^{a'}=0$ is thus not satisfied by the value $\theta=\lambda$. The conclusion is that a node *quâ* node is not a united point.

99. But it is otherwise as regards a cusp. When the point P is at a cusp, the curve Θ (which has in general with the given curve k intersections at P) has here more than k intersections, and (as in this case there is no distinction of branch) the cusp reckons as a united point. In the case of a unicursal curve, there is at the cusp a single value $\theta=\lambda$ of the parameter, and the equation $(\theta,\ 1)^{a}(\theta,\ 1)^{a'}=0$ is satisfied by the value $\theta=\lambda$. But for the very reason that the cusp *quâ* cusp reckons as a united point, the cusp is a united point only in an improper or special sense, and it is to be rejected from the number of true united points. We may include the cusps, along with any other special solutions which may present themselves, under a head "Supplement," and instead of writing as above $\mathrm{a}-\alpha-\alpha'=2kD$, write $\mathrm{a}-\alpha-\alpha'+\text{Supp.}=2kD$.

Before going further I apply the theorem to some examples in which the curve Θ is a system of lines.

100. Investigation of the class of a curve of the order m with δ nodes and κ cusps. Take as corresponding points on the given curve two points such that the line joining them passes through a fixed point O; the united points will be the points of contact of the tangents through O; that is, the number of the united points will be equal to the class of the curve. The curve Θ is here the line OP which has with the given curve a single intersection at P; that is, we have $k=1$. The points P' corresponding to a given position of P are the remaining $m-1$ intersections of OP with the curve, that is, we have $\alpha'=m-1$; and in like manner $\alpha=m-1$. Each of the

cusps is (specially) a united point, and counts once, whence the Supplement is $=\kappa$. Hence, writing n for the class, we have $n+2(m-1)+\kappa=2D$, or writing for $2D$ its value $=m^2-3m+2-2\delta-2\kappa$, we have $n=m^2-m-2\delta-3\kappa$, which is right.

101. Investigation of the number of inflexions. Taking the point P' to be a tangential of P (that is, an intersection of the curve by the tangent at P), the united points are the inflexions; and the number of the united points is equal to the number of the inflexions. The curve Θ is the tangent at P having with the given curve two intersections at this point; that is, $k=2$; P' is any one of the $m-2$ tangentials of P, that is, $\alpha'=m-2$; and P is the point of contact of any one of the $n-2$ tangents from P' to the curve, that is, $\alpha=n-2$. Each cusp is (specially) a united point, and counts once, whence the Supplement is $=\kappa$. Hence, writing ι for the number of inflexions, we have

$$\iota-(m-2)-(n-2)+\kappa=4D;$$

or substituting for $2D$ its value expressed in the form $n-2m+2+\kappa$, we have

$$\iota=3n-3m+\kappa,$$

which is right.

102. For the purpose of the next example it is necessary to present the fundamental equation under a more general form. The curve Θ may intersect the given curve in a system of points P', each p times, a system of points Q', each q times, &c. in such manner that the points (P, P'), the points (P, Q'), &c. are pairs of points corresponding to each other according to distinct laws; and we shall then have the numbers $(\mathrm{a}, \alpha, \alpha')$, $(\mathrm{b}, \beta, \beta')$, &c., corresponding to these pairs respectively, viz. (P, P') are points having an (α, α') correspondence, and the number of united points is $=\mathrm{a}$; (P, Q') are points having a (β, β') correspondence, and the number of united points is $=\mathrm{b}$, and so on. The theorem then is

$$p(\mathrm{a}-\alpha-\alpha')+q(\mathrm{b}-\beta-\beta')+\&\mathrm{c.}+\mathrm{Supp.}=2kD,$$

being in fact the most general form of the theorem for the correspondence of two points on a curve, and that which will be used in all the investigations which follow.

103. Investigation of the number of double tangents. Take P' an intersection of the curve with a tangent from P to the curve (or, what is the same thing, P, P' cotangentials of any point of the curve): the united points are here the points of contact of the several double tangents of the curve; or if τ be the number of double tangents, then the number of united points is $=2\tau$. The curve Θ is the system of the $n-2$ tangents from P to the curve; each tangent has with the curve a single intersection at P, that is, $k=n-2$; each tangent besides meets the curve in the point of contact Q' twice, and in $(m-3)$ points P'; hence if $(\mathrm{a}, \alpha, \alpha')$ refer to the points (P, Q'), and $(2\tau, \beta, \beta')$ to the points (P, P'), we have

$$2\{\mathrm{a}-\alpha-\alpha'\}+\{2\tau-\beta-\beta'\}+\mathrm{Supp.}=2(n-2)D.$$

From the foregoing example the value of $\mathrm{a}-\alpha-\alpha'$ is $=4D-\kappa$. In the case where the point P is at a cusp, then the $n-2$ tangents become the $n-3$ tangents from the cusp, and the tangent at the cusp; hence the curve Θ meets the given curve in $2(n-3)+3$, $=2n-3$ points, that is, $(n-2)+(n-1)$ points; this does not prove (*ante*, No. 96), but the fact is, that the cusp counts in the Supplement $(n-1)$ times, and the expression of the Supplement is $=(n-1)\kappa$. It is clear that we have $\beta=\beta'=(n-2)(m-3)$, so that the equation is

$$8D-2\kappa+2\tau-2(n-2)(m-3)+(n-1)\kappa=(n-2)2D,$$

that is

$$2\tau=2(n-2)(m-3)+(n-6)2D+(-n+3)\kappa;$$

or substituting for $2D$ its value $=n-2m+2+\kappa$ and reducing, this is

$$2\tau=n^2+8m-10n-3\kappa,$$

which is right.

104. As another example, suppose that the point P on a given curve of the order m and the point Q on a given curve of the order m' have an (α, α') correspondence, and let it be required to find the class of the curve enveloped by the line PQ. Take an arbitrary point O, join OQ, and let this meet the curve m in P'; then (P, P') are points on the curve m having a $(m'\alpha, m\alpha')$ correspondence; in fact to a given position of P there correspond α' positions of Q, and to each of these m positions of P'; that is, to each position of P there correspond $m\alpha'$ positions of P'; and similarly to each position of P' there correspond $m'\alpha$ positions of P. The curve Θ is the system of the lines drawn from each of the α' positions of Q to the point O, hence the curve Θ does not pass through P, and we have $k=0$. Therefore the number of the united points (P, P'), that is, the number of the lines PQ which pass through the point O, is $=m\alpha'+m'\alpha$, or this is the class of the curve enveloped by PQ.

It is to be noticed that if the two curves are curves in space (plane, or of double curvature), then the like reasoning shows that the number of the lines PQ which meet a given line O is $=m\alpha'+m'\alpha$, that is, the order of the scroll generated by the line PQ is $=m\alpha'+m'\alpha$.

Article Nos. 105 to 111.—*Application to the Conics which satisfy given conditions, one at least arbitrary.*

105. Passing next to the equations which relate to a conic, we seek for $(4Z)(1)$, the number of the conics which satisfy any four conditions $4Z$ and besides touch a given curve, $(3Z)(2)$ and $(3Z)(1, 1)$, the number of the conics which satisfy three conditions, and besides have with the given curve a contact of the second order, or (as the case may be) two contacts of the first order; and so on with the conditions $2Z$, Z, and then finally (5), $(4, 1)$, ... $(1, 1, 1, 1, 1)$, the numbers of the conics which have with the given curve a contact of the fifth order, or a contact of the fourth and also of the first order ..., or five contacts of the first order.

106. As regards the case $(4Z)(1)$, taking P an arbitrary point of the given curve m, and for the curve Θ the system of the conics $(4Z)(\bar{1})$ which pass through the given point P and besides satisfy the four conditions, then the curve Θ has with the given curve $(4Z)(\bar{1})$ intersections at P, and the points P' are the remaining $(2m-1)(4Z)(\bar{1})$ intersections: in the case of a united point (P, P'), some one of the system of conics becomes a conic $(4Z)(1)$; and the number of the united points is consequently equal to that of the conics $(4Z)(1)$; we have thus the equation

$$\{(4Z)(1) - 2(2m-1)(4Z)(\bar{1})\} + \text{Supp.}\,(4Z)(\bar{1}) = (4Z)(\bar{1})\,.\,2D.$$

107. It is in the present case easy to find *à priori* the expression for the Supplement. 1°. The system of conics $(4Z)$ contains $2(4Z\cdot) - (4Z/)$ point-pairs[1]; each of these, regarded as a line, meets the given curve in m points, and each of these points is (specially) a united point (P, P'); this gives in the Supplement the term $m\{2(4Z\cdot) - (4Z/)\}$. 2°. The number of the conics $(4Z)$ which can be drawn through a cusp of the given curve is $= (4Z\cdot)$; and the cusp is in respect of each of these conics a united point; we have thus the term $\kappa(4Z\cdot)$, and the Supplement is thus $= m\{2(4Z\cdot) - (4Z/)\} + \kappa(4Z\cdot)$. We have moreover $(4Z)(\bar{1}) = (4Z\cdot)$, $2D = n - 2m + 2 + \kappa$; and substituting these values, we find

$$\begin{aligned}(4Z)(1) = &\ (4m-2)(4Z\cdot)\\ &- m\{2(4Z\cdot) - (4Z/)\} - \kappa(4Z\cdot)\\ &+ (n - 2m + 2 + \kappa)(4Z\cdot)\\ = &\ n(4Z\cdot) + m(4Z/),\end{aligned}$$

which is right.

108. It is clear that if, instead of finding as above the expression of the Supplement, the value of $(4Z)(1)$, $= n(4Z\cdot) + m(4Z/)$, had been taken as known, then the equation would have led to

$$\text{Supp.}\,(4Z)(\bar{1}) = m\{2(4Z\cdot) - (4Z/)\} + \kappa(4Z\cdot);$$

and this, as in fact already remarked, is the course of treatment employed in the remaining cases. It is to be observed also that the equation may for shortness be written in the form

$$\begin{aligned}(4Z)\quad &\{(1) - 2(2m-1)(\bar{1})\}\\ &\qquad + \text{Supp.}\,(\bar{1}) = (\bar{1})\,2D;\end{aligned}$$

viz. the $(4Z)$ is to be understood as accompanying and forming part of each symbol; and the like in other cases.

109. We have the series of equations

$$\begin{aligned}(4Z)\quad &\{(1) - (\bar{1})(2m-1) - (\bar{1})(2m-1)\}\\ &+ \text{Supp.}\,(\bar{1}) \qquad\qquad = (\bar{1})\,2D;\end{aligned}$$

[1] The expression a point-pair is regarded as equivalent to and standing for that of a coincident line-pair: see First Memoir, No. 30.

$$(3Z)\quad \{(2)-(\bar{2})(2m-2)-(\bar{1},\ 1)\} + \text{Supp.}\ (\bar{2}) = 2(\bar{2})\,2D;$$

$$(3Z)\quad 2\{(2)-(\bar{1},\ 1)-(\bar{2})(2m-2)\} + \{2(1,\ 1)-(\bar{1},\ 1)(2m-3)-(\bar{1},\ 1)(2m-3)\} + \text{Supp.}\ (\bar{1},\ 1) = (\bar{1},\ 1)\,2D;$$

$$(2Z)\quad \{(3)-(\bar{3})(2m-3)-(\bar{1},\ 2)\} + \text{Supp.}\ (\bar{3}) = 3(\bar{3})\,2D;$$

$$(2Z)\quad 2\{(3)-(\bar{2},\ 1)-(\bar{2},\ 1)\} + \{(2,\ 1)-(\bar{2},\ 1)(2m-4)-(\bar{1},\ 1,\ 1)\,2\} + \text{Supp.}\ (\bar{2},\ 1) = 2(\bar{2},\ 1)\,2D;$$

$$(2Z)\quad 3\{(3)-(\bar{1},\ 2)-(\bar{3})(2m-3)\} + \{(1,\ 2)-(\bar{1},\ 2)(2m-4)-(\bar{1},\ 2)(2m-4)\} + \text{Supp.}\ (\bar{1},\ 2) = (\bar{1},\ 2)\,2D;$$

$$(2Z)\quad 2\{(2,\ 1)-(\bar{1},\ 1,\ 1)\,2-(\bar{2},\ 1)(2m-4)\} + \{3(1,\ 1,\ 1)-(\bar{1},\ 1,\ 1)(2m-5)-(\bar{1},\ 1,\ 1)(2m-5)\} + \text{Supp.}\ (1,\ 1,\ 1) = (\bar{1},\ 1,\ 1)\,2D;$$

$$(Z)\quad \{(4)-(\bar{4})(2m-4)-(\bar{1},\ 3)\} + \text{Supp.}\ (\bar{4}) = 4(\bar{4})\,2D;$$

$$(Z)\quad 2\{(4)-(\bar{3},\ 1)-(\bar{2},\ 2)\} + \{(3,\ 1)-(\bar{3},\ 1)(2m-5)-(\bar{1},\ 1,\ 2)\} + \text{Supp.}\ (\bar{3},\ 1) = 3(\bar{3},\ 1)\,2D;$$

$$(Z)\quad 3\{(4)-(\bar{2},\ 2)-(3,\ \bar{1})\} + \{2(2,\ 2)-(\bar{2},\ 2)(2m-5)-(\bar{1},\ 1,\ 2)\} + \text{Supp.}\ (\bar{2},\ 2) = 2(\bar{2},\ 2)\,2D;$$

$$(Z)\quad 2\{(3,\ 1)-(\bar{2},\ 1,\ 1)\,2-(\bar{2},\ 1,\ 1)\,2\} + \{(2,\ 1,\ 1)-(\bar{2},\ 1,\ 1)(2m-6)-(\bar{1},\ 1,\ 1,\ 1)\,3\} + \text{Supp.}\ (\bar{2},\ 1,\ 1) = 2(\bar{2},\ 1,\ 1)\,2D;$$

$$(Z)\quad 4\{(4)-(\bar{1},\ 3)-(\bar{4})(2m-4)\} + \{(1,\ 3)-(\bar{1},\ 3)(2m-5)-(\bar{1},\ 3)(2m-5)\} + \text{Supp.}\ (\bar{1},\ 3) = (\bar{1},\ 3)\,2D;$$

$$
\begin{aligned}
(Z)\quad & 3\{(3,\ 1)-(\bar{1},\ 1,\ 2)-(\bar{3},\ 1)(2m-5)\}\\
&+2\{2(2,\ 2)-(\bar{1},\ 1,\ 2)-(\bar{2},\ 2)(2m-5)\}\\
&+\ \{2(1,\ 1,\ 2)-(\bar{1},\ 1,\ 2)(2m-6)-(\bar{1},\ 1,\ 2)(2m-6)\}\\
&+\ \text{Supp.}\ (\bar{1},\ 1,\ 2) \qquad\qquad = (\bar{1},\ 1,\ 2)\,2D;
\end{aligned}
$$

$$
\begin{aligned}
(Z)\quad & 2\{(2,\ 1,\ 1)-(\bar{1},\ 1,\ 1,\ 1)\,3-(\bar{2},\ 1,\ 1)(2m-6)\}\\
&+\ \{4(1,\ 1,\ 1,\ 1)-(\bar{1},\ 1,\ 1,\ 1)(2m-7)-(\bar{1},\ 1,\ 1,\ 1)(2m-7)\}\\
&+\ \text{Supp.}\ (\bar{1},\ 1,\ 1,\ 1) \qquad\qquad = (\bar{1},\ 1,\ 1,\ 1)\,2D.
\end{aligned}
$$

110. I content myself with giving the expressions of only the following supplements.

$$
\begin{aligned}
\text{Supp.}\ (4Z)(\bar{1}) &= m\,[2\,(\cdot)-(/)]+\kappa\,(\cdot).\\
\text{Supp.}\ (3Z)(\bar{2}) &= \tfrac{1}{2}n\,[2\,(:)-(\cdot\,/)]+\tfrac{1}{2}\kappa\,(\cdot\,/).\\
\text{Supp.}\ (3Z)(\bar{1},\ 1) &= (\qquad 2mn-3n^2-\ n+n\alpha)\,(:)\\
&\quad +(2m^2-4mn \qquad -2m+2n+(m-\tfrac{1}{2})\,\alpha)\,(\cdot\,/)\\
&\quad +(-\,m^2 \qquad\qquad +\ m \qquad\qquad)\,(//).\\
\text{Supp.}\ (2Z)(\bar{3}) &= -\tfrac{1}{4}m\,[2\,(.\!\cdot\!.)-(:/)]\\
&\quad +\tfrac{1}{4}n\,[2\,(.\!\cdot\!.)-(:/)+2\,(2\,(:/)-(\cdot\,//))]\\
&\quad +\tfrac{1}{2}\kappa\,(:/).\\
\text{Supp.}\ (\ Z)(\bar{4}) &= a\kappa+b\,(2\kappa+2\iota),
\end{aligned}
$$

where a, b are the representatives of the condition Z.

It may be added that we have in general

$$\text{Supp.}\ (Z)(4X) \quad = a\,\text{Supp.}\ (4X\,\cdot)+b\,\text{Supp.}\ (4X\,/),$$

where $(4X)$ stands for any one of the symbols $(\bar{4})$, $(\bar{3},\ 1)\ldots.(\bar{1},\ 1,\ 1,\ 1)$.

111. The expression of Supp. $(4Z)(\bar{1})$ has been explained *suprà*, No. 108. That of Supp. $(3Z)(2)$ may also be explained. 1°. The point-pairs of the system of conics $(3Z)$, regarding each point-pair as a line, are a set of lines enveloping a curve; the class of this curve is equal to the number of the lines which pass through an arbitrary point, that is, as at first sight would appear, to the number of point-pairs in the system $(3Z\,\cdot)$, or to $2\,(3Z\,:)-(3Z\cdot/)$: it is, however, necessary to admit that the number of distinct lines, and therefore the class of the curve, is one-half of this, or $=\tfrac{1}{2}\,[2\,(3Z\,:)-(3Z\cdot/)]$; which being so, the number of the point-pairs $(3Z)$ which, regarded as lines, touch the given curve (of the order m and class n) is $=\tfrac{1}{2}n\,[2\,(3Z\,:)-(3Z\cdot/)]$. The point of contact of any one of these lines with the given curve is (specially) a united point, and we have thus the term $\tfrac{1}{2}n\,[2\,(3Z\,:)-(3Z\cdot/)]$ of the Supplement. 2°. The number of the conics $(3Z)$ which touch the given curve at a given cusp thereof, or, say, the conics $(3Z)(\overline{2\kappa 1})$, is $=\tfrac{1}{2}\,(3Z\cdot/)$, and the cusp is in respect of each of these conics a united point; we have thus the remaining term $\tfrac{1}{2}\kappa\,(3Z\cdot/)$ of the Supplement.

Article Nos. 112 to 135.—*Application to the Conics which satisfy five conditions of contact with a given Curve.*

112. We have twelve equations, which I first present in what I call their original forms; viz. these are—

First equation:

$$\{(5)-(\bar{5})(2m-5)-(\bar{1},\ 4)\}$$
$$+\ \text{Supp.}\ (\bar{5}) \qquad = 5\,(\bar{5})\,2D.$$

Second equation:

$$2\,\{(5)-(\bar{4},\ 1)-(\bar{2},\ 3)\}$$
$$+\ \{(4,\ 1)-(\bar{4},\ 1)(2m-6)-(\bar{1},\ 1,\ 3)\}$$
$$+\ \text{Supp.}\ (\bar{4},\ 1) \qquad = 4\,(\bar{4},\ 1)\,2D.$$

Third equation:

$$3\,\{(5)-2\,(\bar{3},\ 2)-2\,(\bar{3},\ 2)\}$$
$$+\ \{(3,\ 2)-(\bar{3},\ 2)\,(2m-6)-2\,(\bar{1},\ 2,\ 2)\}$$
$$+\ \text{Supp.}\ (\bar{3},\ 2) \qquad = 3\,(\bar{3},\ 2)\,2D.$$

Fourth equation:

$$2\,\{(4,\ 1)-2\,(\bar{3},\ 1,\ 1)-(\bar{2},\ 2,\ 1)\}$$
$$+\ \{(3,\ 1,\ 1)-(\bar{3},\ 1,\ 1)\,(2m-7)-(\bar{1},\ 1,\ 1,\ 2)\}$$
$$+\ \text{Supp.}\ (\bar{3},\ 1,\ 1) \qquad = 3\,(\bar{3},\ 1,\ 1)\,2D.$$

Fifth equation:

$$4\,\{(5)-(\bar{2},\ 3)-(\bar{4},\ 1)\}$$
$$+\ \{(3,\ 2)-(\bar{2},\ 3)\,(2m-6)-(\bar{1},\ 1,\ 3)\}$$
$$+\ \text{Supp.}\ (\bar{2},\ 3) \qquad = 2\,(\bar{2},\ 3)\,2D.$$

Sixth equation:

$$3\,\{(4,\ 1)-2\,(\bar{2},\ 2,\ 1)-2,\ (\bar{3},\ 1,\ 1)\}$$
$$+\,2\,\{(3,\ 2)-(\bar{2},\ 2,\ 1)-(\bar{2},\ 2,\ 1)\}$$
$$+\ \{2\,(2,\ 2,\ 1)-(\bar{2},\ 2,\ 1)\,(2m-7)-2\,(\bar{1},\ 1,\ 1,\ 2)\}$$
$$+\ \text{Supp.}\ (\bar{2},\ 2,\ 1) \qquad = 2\,(\bar{2},\ 2,\ 1)\,2D.$$

Seventh equation:

$$2\,\{(3,\ 1,\ 1)-3\,(\bar{2},\ 1,\ 1,\ 1)-3\,(\bar{2},\ 1,\ 1,\ 1)\}$$
$$+\ \{(2,\ 1,\ 1,\ 1)-(\bar{2},\ 1,\ 1,\ 1)\,(2m-8)-4\,(\bar{1},\ 1,\ 1,\ 1,\ 1)\}$$
$$+\ \text{Supp.}\ (\bar{2},\ 1,\ 1,\ 1) \qquad = 2\,(\bar{2},\ 1,\ 1,\ 1)\,2D.$$

Eighth equation:

$$\begin{aligned} &5\{(5)-(\bar{1},\ 4)-(\bar{5})(2m-5)\}\\ +\ &\{(4,\ 1)-(\bar{1},\ 4)(2m-5)-(\bar{1},\ 4)(2m-5)\}\\ +\ &\text{Supp.}\ (\bar{1},\ 4) \qquad\qquad =(\bar{1},\ 4)\,2D. \end{aligned}$$

Ninth equation:

$$\begin{aligned} &4\{(4,\ 1)-(\bar{1},\ 1,\ 3)-(\bar{4},\ 1)(2m-6)\}\\ +2&\{(3,\ 2)-(\bar{1},\ 1,\ 3)-(\bar{2},\ 3)(2m-6)\}\\ +\ &\{2(3,\ 1,\ 1)-(\bar{1},\ 1,\ 3)(2m-7)-(\bar{1},\ 1,\ 3)(2m-7)\}\\ +\ &\text{Supp.}\ (\bar{1},\ 1,\ 3) \qquad\qquad =(\bar{1},\ 1,\ 3)\,2D. \end{aligned}$$

Tenth equation:

$$\begin{aligned} &3\{(3,\ 2)-2(\bar{1},\ 2,\ 2)-(\bar{3},\ 2)(2m-6)\}\\ +\ &\{(2,\ 2,\ 1)-(\bar{1},\ 2,\ 2)(2m-7)-(\bar{1},\ 2,\ 2)(2m-7)\}\\ +\ &\text{Supp.}\ (\bar{1},\ 2,\ 2) \qquad\qquad =(\bar{1},\ 2,\ 2)\,2D. \end{aligned}$$

Eleventh equation:

$$\begin{aligned} &3\{(3,\ 1,\ 1)-(\bar{1},\ 1,\ 1,\ 2)-(\bar{3},\ 1,\ 1)(2m-7)\}\\ +2&\{2(2,\ 2,\ 1)-2(\bar{1},\ 1,\ 1,\ 2)-(\bar{2},\ 2,\ 1)(2m-7)\}\\ +\ &\{3(2,\ 1,\ 1,\ 1)-(\bar{1},\ 1,\ 1,\ 2)(2m-8)-(\bar{1},\ 1,\ 1,\ 2)(2m-8)\}\\ +\ &\text{Supp.}\ (\bar{1},\ 1,\ 1,\ 2) \qquad\qquad =(\bar{1},\ 1,\ 1,\ 2)\,2D. \end{aligned}$$

Twelfth equation:

$$\begin{aligned} &2\{(2,\ 1,\ 1,\ 1)-4(\bar{1},\ 1,\ 1,\ 1,\ 1)-(\bar{2},\ 1,\ 1,\ 1)(2m-8)\}\\ +\ &\{5(1,\ 1,\ 1,\ 1,\ 1)-(\bar{1},\ 1,\ 1,\ 1,\ 1)(2m-9)-(\bar{1},\ 1,\ 1,\ 1,\ 1)(2m-9)\}\\ +\ &\text{Supp.}\ (\bar{1},\ 1,\ 1,\ 1,\ 1) \qquad\qquad =(\bar{1},\ 1,\ 1,\ 1,\ 1)\,2D. \end{aligned}$$

113. I alter the forms of these equations by substituting for $2D$ its value $=n-2m+2+\kappa$, and by writing for the expressions with $(\bar{1})$ their values,

$$(\bar{1},\ 4)=(\cdot 4)-5(\bar{5}),\ \&\text{c.},$$

and except in the terms $\{\text{Supp.}(\bar{5})-\kappa(\bar{5})\}$, &c., by writing for κ its value $-3n+\alpha$. The resulting equations, if the Supplements were known, would serve to determine the values of (5), (4, 1), &c.; but I assume instead that the last-mentioned expressions are known (First Memoir, No. 50), and use the equations to determine the Supplements, or, what comes to the same thing, the values of the terms in { } which contain these Supplements. We have thus the twelve reduced equations, with resulting values of the supplements.

114. First equation:

$$\begin{array}{l|r}
 & =0= \\
(5) & -15m-15n+9\alpha \\
+\{\text{Supp. }(\bar{5})-\kappa(\bar{5})\} & -\ 3m\qquad +\ \alpha \\
+(\bar{5})(8m+7n-4\alpha) & 8m+\ 7n-4\alpha \\
-(\cdot 4) & +10m+\ 8n-6\alpha.
\end{array}$$

(that is, we have

$$\text{Supp. }(\bar{5})-\kappa(\bar{5})=-3m+\alpha,$$

and so in the subsequent cases, the equation gives the value of the term in { } which contains the Supplement).

115. Second equation:

$$\begin{array}{l|l}
 & =0= \\
2(5) & \qquad\qquad\qquad\qquad -\ 30m-\ 30n+\alpha(\qquad\qquad\qquad 18) \\
+(4,\ 1) & -8m^2-20mn-8n^2+104m+104n+\alpha(\ 6m+6n-66) \\
+\{\text{Supp. }(\bar{4},\ 1)-\kappa(\bar{4},\ 1)\} & -6m^2-\ 3mn\qquad\quad +\ 18m+\ \ 9n+\alpha(\ 3m\qquad -\ 9) \\
+(\bar{4},\ 1)(6m+5n-3\alpha) & +6m^2+11mn+5n^2-\ 36m-\ 30n+\alpha(-3m-3n+18) \\
-(\cdot 1,\ 3) & +8m^2+12mn+3n^2-\ 56m-\ 53n+\alpha(-6m-3n+39).
\end{array}$$

I stop for a moment to notice a very convenient verification of the term in { }; putting therein $\alpha=3n$, the term is

$$-6m^2-3mn+18m+9n+(9mn-27n);$$

and if in this we write $m=n=1$, $m^2=mn=n^2=2$, and when any higher terms enter $m^3=m^2n=mn^2=n^3=4$, $m^4=m^3n=m^2n^2=mn^3=n^4=8$, &c., the value is $-12-6+18+9+18-27$, $=0$, viz. we should always obtain a sum $=0$. The reason is that the term in question should always admit of being expressed in the form $p\delta+q\kappa+r\tau+s\iota$; the reduction to this form might be effected by the substitutions $m=\frac{1}{2}(m+n)+\frac{1}{6}(\kappa-\iota)$, $n=\frac{1}{2}(m+n)-\frac{1}{6}(\kappa-\iota)$, $m^2=2\,.\,\frac{1}{2}(m+n)+2\delta+3\kappa$, $n^2=2\,.\,\frac{1}{2}(m+n)+2\tau+3\iota$, giving a result $=A\,(m+n)+$ terms in $(\delta,\ \kappa,\ \tau,\ \iota)$, where A is a numerical coefficient calculable as above by simply writing $m=n=1$, $m^2=mn=n^2=2$, &c., and which is $=0$ when the term is of the proper form $p\delta+q\kappa+r\tau+s\iota$. The complete reduction to the form in question is material in the sequel, but I advert to the point here only for the sake of the numerical verification.

116. Third equation:

$$\begin{array}{l|l}
 & =0= \\
3(5) & -\ \ 45m-\ 45n+\alpha(\qquad\qquad\quad +27) \\
+(3,\ 2) & +120m+120n+\alpha(-4m-4n-78)+3\alpha^2 \\
+\{\text{Supp. }(\bar{3},\ 2)-\kappa(\bar{3},\ 2)\} & +\ \ 15m\qquad\quad +\alpha(\qquad\quad n-\ 7) \\
+(\bar{3},\ 2)(4m+3n-2\alpha) & -\ \ 36m-\ 27n+\alpha(\ 4m+3n+18)-2\alpha^2 \\
-2(\cdot 2,\ 2) & -\ \ 54m-\ 48n+\alpha(\qquad\qquad\quad +40)-\ \alpha^2.
\end{array}$$

Verification is $15+3(1\,.\,2-7)=0$.

117. Fourth equation:

$$= 0 =$$

$2\,(4,\ 1)$	$-\ 16m^2 -\ 40mn -\ 16n^2$	(1)
$+\,(3,\ 1,\ 1)$	$-\tfrac{3}{2}m^3 - 10m^2n - 10mn^2 - \tfrac{3}{2}n^3 + \tfrac{109}{2}m^2 + 116mn + \tfrac{109}{2}n^2$	(2)
$+\,\{\text{Supp.}\ (\bar{3},\ 1,\ 1) - \kappa\,(\bar{3},\ 1,\ 1)\}$	$-\tfrac{1}{2}m^3 + \tfrac{1}{2}m^2n + 2mn^2 + \tfrac{23}{2}m^2 + \tfrac{11}{2}mn - 7n^2$	(3)
$+\,(\bar{3},\ 1,\ 1)\,(4m + 3n - 2\alpha)$	$2m^3 + \tfrac{19}{2}m^2n + 8mn^2 + \tfrac{3}{2}n^3 - 26m^2 - \tfrac{91}{2}mn - \tfrac{39}{2}n^2$	(4)
$-\,(\cdot\,2,\ 1,\ 1)$	$-\ 24m^2 -\ 36mn -\ 12n^2$	(5)

(1) $+ 208m + 208n + \alpha\,(+ 12m + 12n - 132)$

(2) $- 434m - 434n + \alpha\,(\tfrac{3}{2}m^2 + 6mn + \tfrac{3}{2}n^2 - \tfrac{69}{2}m - \tfrac{69}{2}n + 291) - \tfrac{9}{2}\alpha^2$

(3) $-\ 50m -\ 23n + \alpha\,(\tfrac{1}{2}m^2 - \tfrac{19}{2}m - \tfrac{1}{2}n + 33)$

(4) $+ 108m + 81n + \alpha\,(- m^2 - 4mn - n^2 + 7m + \tfrac{17}{2}n - 54) + 3\alpha^2$

(5) $+ 168m + 168n + \alpha\,(- m^2 - 2mn - \tfrac{1}{2}n^2 + 25m + \tfrac{29}{2}n - 138) + \tfrac{3}{2}\alpha^2.$

Verification is $(-\tfrac{1}{2} + \tfrac{1}{2} + 2)\,4 + (\tfrac{23}{2} + \tfrac{11}{2} - 7)\,2 - 50 - 23 + 3\left(\tfrac{1}{2}\,.\,4 - (\tfrac{19}{2} + \tfrac{1}{2})\,2 + 33\right) = 0.$

118. Fifth equation:

$$= 0 =$$

$4\,(5)$	$-\ 60m -\ 60n + \alpha\,(36)$
$+\,(3,\ 2)$	$+ 120m + 120n + \alpha\,(- 4m - 4n - 78) + 3\alpha^2$
$+\,\{\text{Supp.}\ (\bar{2},\ 3) - \kappa\,(\bar{2},\ 3)\}$	$n^2 + 8m - n + \alpha\,(- 3)$
$+\,(\bar{2},\ 3)\,(2m + n - \alpha)$	$- 8m^2 - 12mn - 4n^2 - 12m - 6n + \alpha\,(10m + 7n + 6) - 3\alpha^2$
$-\,(\cdot\,1,\ 3)$	$+ 8m^2 + 12mn + 3n^2 - 56m - 53n + \alpha\,(- 6m - 3n + 39).$

Verification is $2 + 8 - 1 + 3\,(-3) = 0.$

119. Sixth equation:

$$= 0 =$$

$3\,(4,\ 1)$	$- 24m^2 - 60mn - 24n^2 + 312m + 312n$	(1)
$+\,2\,(3,\ 2)$	$+ 240m + 240n$	(2)
$+\,2\,(2,\ 2,\ 1)$	$+ 48m^2 + 108mn + 48n^2 - 936m - 936n$	(3)
$+\,\{\text{Supp.}\ (\bar{2},\ 2,\ 1) - \kappa\,(\bar{2},\ 2,\ 1)\}$	$+ 12m^2 + 6mn - 6n^2 - 60m - 6n$	(4)
$+\,(\bar{2},\ 2,\ 1)\,(2m + n - \alpha)$	$+ 12m^2 + 18mn + 6n^2 + 108m + 54n$	(5)
$-\,2\,(\cdot\,2,\ 1,\ 1)$	$- 48m^2 - 72mn - 24n^2 + 336m + 336n$	(6)

$$(1)\quad +\alpha(\qquad\qquad +18m+18n-198)$$

$$(2)\quad +\alpha(\qquad\qquad -8m-8n-156)+\alpha^2(\qquad +6)$$

$$(3)\quad +\alpha(\qquad\qquad -16m-16n+654)+\alpha^2(m+n-24)$$

$$(4)\quad +\alpha(\qquad mn\qquad -8m-2n+30)$$

$$(5)\quad +\alpha(2m^2+3mn+n^2-36m-21n-54)+\alpha^2(-m-n+15)$$

$$(6)\quad +\alpha(-2m^2-4mn-n^2+50m+29n-276)+\alpha^2(\qquad +3).$$

Verification is $(12+6-6)\,2-60-6+3\,\big(4-(8+2)\,2+30\big)=0.$

120. Seventh equation:

$$=0=$$

$2\,(3,\ 1,\ 1)$		(1)
$+(2,\ 1,\ 1,\ 1)$		(2)
$+\{\text{Supp.}\ (\overline{2},\ 1,\ 1,\ 1)-\kappa\,(\overline{2},\ 1,\ 1,\ 1)\}$	$\tfrac{1}{2}m^3n+m^2n^2$	(3)
$+(\overline{2},\ 1,\ 1,\ 1)\,(2m+n-\alpha)$	$\tfrac{1}{3}m^4+\tfrac{13}{6}m^3n+3m^2n^2+\tfrac{4}{3}mn^3+\tfrac{1}{6}n^4$	(4)
$-(\cdot 1,\ 1,\ 1,\ 1)$	$-\tfrac{1}{3}m^4-\tfrac{8}{3}m^3n-4m^2n^2-\tfrac{4}{3}mn^3-\tfrac{1}{6}n^4$	(5)

$$(1)\quad -3m^3-20m^2n-20mn^2-3n^3+109m^2+232mn+109n^2-868m-868n\quad(1)$$

$$(2)\quad 6m^3+30m^2n+30mn^2+6n^3-174m^2-348mn-174n^2+1320m\quad(2)$$

$$(3)\quad -\tfrac{7}{2}m^2n-5mn^2-\tfrac{3}{2}n^3-20m^2-5mn+\tfrac{29}{2}n^2+80m+26n\quad(3)$$

$$(4)\quad -5m^3-\tfrac{37}{2}m^2n-13mn^2-\tfrac{5}{2}n^3+\tfrac{74}{3}m^2+37mn+\tfrac{37}{2}n^2-150m-75n\quad(4)$$

$$(5)\quad +2m^3+12m^2n+8mn^2+n^3+\tfrac{181}{3}m^2+84mn+\tfrac{229}{6}n^2-382m-403n\quad(5)$$

$$(1)\quad +\alpha(\qquad 3m^2+12mn+3n^2-69m-69n+582)+\alpha^2(\qquad -9)$$

$$(2)\quad +\alpha(\tfrac{1}{6}m^3+m^2n+mn^2+\tfrac{1}{6}n^3-\tfrac{15}{2}m^2-26mn-\tfrac{15}{2}n^2+\tfrac{358}{3}m+\tfrac{358}{3}n-960)+\alpha^2(-\tfrac{3}{2}m-\tfrac{3}{2}n+28)$$

$$(3)\quad +\alpha(\qquad -m^2-\tfrac{3}{2}mn+\tfrac{1}{2}n^2+19m+\tfrac{5}{2}n-54)$$

$$(4)\quad +\alpha(-\tfrac{1}{6}m^3-m^2n-mn^2-\tfrac{1}{6}n^3-\tfrac{1}{2}m^2+\tfrac{7}{2}mn+n^2+\tfrac{50}{3}m+\tfrac{50}{3}n+75)+\alpha^2(\tfrac{3}{2}m+\tfrac{3}{2}n-\tfrac{29}{2})$$

$$(5)\quad +\alpha(\qquad 6m^2+12mn+3n^2-86m-55n+357)+\alpha^2(\qquad -\tfrac{9}{2})$$

Verification is

$$(\tfrac{1}{2}+1)\,8+(-\tfrac{7}{2}-5-\tfrac{3}{2})\,4+(-20-5+\tfrac{29}{2})\,2+80+26+3\,\big((-1-\tfrac{3}{2}+\tfrac{1}{2})\,4+(19+\tfrac{5}{2})\,2-54\big)=0.$$

121. Eighth equation:

$$=0=$$

$5\,(5)$	$-\ 75m-\ 75n+\alpha\,(\qquad\qquad 45)$
$+\ (4,\ 1)$	$-8m^2-20mn-8n^2+104m+104n+\alpha\,(6m+6n-66)$
$+\left\{\begin{array}{l}\text{Supp. }(\bar{1},\ 4)-\kappa\,(\bar{1},\ 4)\\ -\left(m-\frac{5}{3}\right)\big(2\,(\cdot\ 4)-(/\ 4)\big)\end{array}\right\}$	$+\ \ 1m-\ \ 4n+\alpha\,(\qquad\qquad 1)$
$+\ \left(-n+\frac{5}{3}\right)(\cdot\ 4)$	$10mn+8n^2-\frac{50}{3}m-\frac{40}{3}n+\alpha\,(\quad -6n+10)$
$+\ \left(-m+\frac{5}{3}\right)(/\ 4)$	$8m^2+10mn\qquad -\frac{40}{3}m-\frac{50}{3}n+\alpha\,(-6m\quad +10)$
$+\ 5n\,(\bar{5})$	$+5n.$

Verification is $1-4+3\,.\,1=0$.

122. Ninth equation:

$$=0=$$

$4\,(4,\ 1)$	$-\ 32m^2-\ 80mn-\ 32n^2$ (1)
$+\,2\,(3,\ 2)$	(2)
$+\,2\,(3,\ 1,\ 1)$	$-3m^3-20m^2n-20mn^2-3n^3+109m^2+232mn+109n^2$ (3)
$+\left\{\begin{array}{l}\text{Supp. }(\bar{1},\ 1,\ 3)-\kappa\,(\bar{1},\ 1,\ 3)\\ -(m-2)\big(2\,(\cdot\ 1,\ 3)-(/\ 1,\ 3)\big)\end{array}\right\}$	$-\ \ 2m^2+\ 12mn+\ \ 2n^2$ (4)
$+\ \ (-n+2)\,(\cdot\ 1,\ 3)$	$+\ \ 8m^2n+12mn^2+3n^3-\ 16m^2-\ 80mn-\ 59n^2$ (5)
$+\ \ (-m+2)\,(/\ 1,\ 3)$	$+3m^3+12m^2n+\ 8mn^2\qquad -\ 59m^2-\ 80mn-\ 16n^2$ (6)
$+\,n\big(4\,(\bar{4},\ 1)+2\,(\bar{2},\ 3)\big)$	$-\ \ 4mn-\ \ 4n^2$ (7)

(1)	$+416m+416n+\alpha\,(\qquad\qquad +24m+24n-264)$
(2)	$+240m+240n+\alpha\,(\qquad\qquad -\ 8m-\ 8n-156)+6\alpha^2$
(3)	$-868m-868n+\alpha\,(\ 3m^2+12mn+3n^2-69m-69n+582)-9\alpha^2$
(4)	$-\ \ 6m+\ 30n+\alpha\,(\qquad\qquad -\ 4m-10n-\ \ 6)+3\alpha^2$
(5)	$+112m+106n+\alpha\,(\quad -\ 6mn-3n^2+12m+45n-\ 78)$
(6)	$+106m+112n+\alpha\,(-3m^2-\ 6mn\qquad +45m+12n-\ 78)$
(7)	$-\ 36n+\alpha\,(\qquad\qquad +\ 6n\qquad).$

Verification is $(-2+12+2)\,2-6+30+3\,\big((-4-10)\,2-6\big)+3\,.\,9\,.\,2=0$.

123. Tenth equation:

$$=0=$$

$$\begin{array}{l|rr}
3\,(3,\ 2) & +360m+360n & (1)\\
+(2,\ 2,\ 1) & 24m^2+54mn+24n^2-468m-468n & (2)\\
+\left\{\begin{array}{l}\text{Supp. }(\bar{1},\ 2,\ 2)-\kappa\,(\bar{1},\ 2,\ 2)\\ -(m-2)\big(2\,(\cdot\,2,\ 2)-(/\,2,\ 2)\big)\end{array}\right\} & +\quad 6m+\ 33n & (3)\\
+\ (-n+2)\,(\cdot\,2,\ 2) & -27mn-24n^2+\ 54m+\ 48n & (4)\\
+\ (-m+2)\,(/\,2,\ 2) & -24m^2-27mn\qquad +\ 48m+\ 54n & (5)\\
+\ 3n\,(\bar{3},\ 2) & -\ 27n & (6)
\end{array}$$

$$\begin{array}{ll}
(1) & +\alpha\,(-12m-12n-234)+\alpha^2\,(\qquad\quad +\ 9)\\
(2) & +\alpha\,(-\ \ 8m-\ \ 8n+327)+\alpha^2\,(\tfrac{1}{2}m+\tfrac{1}{2}n-12)\\
(3) & +\alpha\,(\qquad\ \ -\ \ 3n-\ \ 13)+\alpha^2\,(\qquad\quad +\ 1)\\
(4) & +\alpha\,(\qquad\ +20n-\ \ 40)+\alpha^2\,(\quad -\tfrac{1}{2}n+\ 1)\\
(5) & +\alpha\,(\ \ 20m\qquad\ \ -\ \ 40)+\alpha^2\,(\tfrac{1}{2}m\qquad +\ 1)\\
(6) & +\alpha\,(\qquad\ +\ \ 3n\qquad\ ).
\end{array}$$

Verification is $+33+3\,(-3\,.\,2-13)+9\,.\,2=0$.

124. Eleventh equation:

$$=0=$$

$$\begin{array}{l|rr}
3\,(3,\ 1,\ 1) & -\ \tfrac{9}{2}m^3-30m^2n-30mn^2-\ \tfrac{9}{2}n^3 & (1)\\
+4\,(2,\ 2,\ 1) & & (2)\\
+3\,(2,\ 1,\ 1,\ 1) & +18m^3+90m^2n+90mn^2+18n^3 & (3)\\
+\left\{\begin{array}{l}\text{Supp. }(\bar{1},\ 1,\ 1,\ 2)-\kappa\,(\bar{1},\ 1,\ 1,\ 2)\\ -(m-\tfrac{7}{3})\big(2\,(\cdot\,1,\ 1,\ 2)-(/\,1,\ 1,\ 2)\big)\end{array}\right\} & -\ \tfrac{3}{2}m^3-\ \tfrac{3}{2}m^2n-\ \ 6mn^2-\ \ 3n^3 & (4)\\
+(-n+\tfrac{7}{3})\,(\cdot\,1,\ 1,\ 2) & -24m^2n-36mn^2-12n^3 & (5)\\
+(-m+\tfrac{7}{3})\,(/\,1,\ 1,\ 2) & -12m^3-36m^2n-24mn^2\qquad\quad & (6)\\
+n\,\big(3\,(\bar{3},\ 1,\ 1)+2\,(\bar{2},\ 2,\ 1)\big) & +\ \tfrac{3}{2}m^2n+\ \ 6mn^2+\ \tfrac{3}{2}n^3 & (7)
\end{array}$$

$$
\begin{array}{lll}
(1) & +\tfrac{327}{2}m^2+348mn+\tfrac{327}{2}n^2-1302m-1302n+\alpha\,(& (1)\\
(2) & +96m^2+216mn+96n^2-1872m-1872n+\alpha\,(& (2)\\
(3) & -522m^2-1044mn-522n^2+3960m+3960n+\alpha\,(\ \tfrac{1}{2}m^3+3m^2n+3mn^2+\tfrac{1}{2}n^3 & (3)\\
(4) & +\tfrac{21}{2}m^2-\tfrac{33}{2}mn+18n^2-2m-191n+\alpha\,(& (4)\\
(5) & +56m^2+252mn+196n^2-392m-392n+\alpha\,(\ -m^2n-2mn^2-\tfrac{1}{2}n^3 & (5)\\
(6) & +196m^2+252mn+56n^2-392m-392n+\alpha\,(-\tfrac{1}{2}m^3-2m^2n-mn^2 & (6)\\
(7) & -\tfrac{15}{2}mn-\tfrac{15}{2}n^2+189n+\alpha\,(& (7)
\end{array}
$$

$$
\begin{array}{ll}
(1) & \tfrac{9}{2}m^2+18mn+\tfrac{9}{2}n^2-\tfrac{207}{2}m-\tfrac{207}{2}n+873)+\alpha^2(-\tfrac{27}{2})\\
(2) & -32m-32n+1308)+\alpha^2(2m+2n-48)\\
(3) & -\tfrac{45}{2}m^2-78mn-\tfrac{45}{2}n^2+358m+358n-2880)+\alpha^2(-\tfrac{9}{2}m-\tfrac{9}{2}n+84)\\
(4) & -\tfrac{4}{3}mn-2n^2+\tfrac{23}{3}m+\tfrac{253}{6}n+55)+\alpha^2(m+n-\tfrac{31}{2})\\
(5) & +\tfrac{7}{3}m^2+\tfrac{89}{3}mn+\tfrac{47}{3}n^2-\tfrac{175}{3}m-\tfrac{1031}{6}n+322)+\alpha^2(\tfrac{3}{2}m-\tfrac{7}{2})\\
(6) & +\tfrac{47}{3}m^2+\tfrac{89}{3}mn+\tfrac{7}{3}n^2-\tfrac{1031}{6}m-\tfrac{175}{3}n+322)+\alpha^2(+\tfrac{3}{2}n-\tfrac{7}{2})\\
(7) & 2mn+2n^2\qquad \tfrac{69}{2}n\qquad).
\end{array}
$$

Verification is $4\,(-\tfrac{3}{2}-\tfrac{3}{2}-6-3)+2\,(\tfrac{21}{2}-\tfrac{33}{2}+18)-2-191$

$$+3\left((-\tfrac{4}{3}-2)\,4+(\tfrac{23}{3}+\tfrac{253}{6})\,2+55\right)+9\left((1+1)\,4-\tfrac{31}{2}\,.\,2\right)=0.$$

125. Twelfth equation:

$$=0=$$

$$
\begin{array}{l|ll}
2\,(\bar{2},\ 1,\ 1,\ 1) & & (1)\\
+\,5\,(1,\ 1,\ 1,\ 1,\ 1) & \tfrac{1}{24}m^5+\tfrac{5}{12}m^4n+\tfrac{5}{3}m^3n^2+\tfrac{5}{3}m^2n^3+\tfrac{5}{12}mn^4+\tfrac{1}{24}n^5 & (2)\\
+\left\{\begin{array}{l}\text{Supp.}\,(\bar{1},1,1,1,1)-\kappa\,(\bar{1},1,1,1,1)\\ -(m-\tfrac{8}{3})\left(2\,(\cdot\,1,1,1,1)-(/\,1,1,1,1)\right)\end{array}\right\} & & (3)\\
+\ (-n+\tfrac{8}{3})\,(\cdot\,1,\ 1,\ 1,\ 1) & -\tfrac{1}{12}m^4n-\tfrac{2}{3}m^3n^2-m^2n^3-\tfrac{1}{3}mn^4-\tfrac{1}{24}n^5 & (4)\\
+\ (-m+\tfrac{8}{3})\,(/\,1,\ 1,\ 1,\ 1) & -\tfrac{1}{24}m^5-\tfrac{1}{3}m^4n-m^3n^2-\tfrac{2}{3}m^2n^3-\tfrac{1}{12}mn^4 & (5)\\
+\,2n\,(\bar{2},\ 1,\ 1,\ 1) & & (6)
\end{array}
$$

(1) $+ 12m^3 + 60m^2n + 60mn^2 + 12n^3$ (1)

(2) $-\frac{5}{12}m^4 - \frac{25}{6}m^3n - 10m^2n^2 - \frac{25}{6}mn^3 - \frac{5}{12}n^4 - \frac{565}{24}m^3 - \frac{1045}{12}m^2n - \frac{1045}{12}mn^2 - \frac{565}{24}n^3$ (2)

(3) $-\frac{1}{6}m^4 - \frac{4}{3}m^3n - \frac{10}{3}m^2n^2 - 3mn^3 - \frac{1}{2}n^4 + 4m^3 + \frac{28}{3}m^2n + \frac{61}{3}mn^2 + 9n^3$ (3)

(4) $+\frac{2}{9}m^4 + \frac{41}{18}m^3n + \frac{17}{3}m^2n^2 + \frac{26}{9}mn^3 + \frac{13}{36}n^4 - \frac{4}{3}m^3 + \frac{85}{12}m^2n + \frac{47}{3}mn^2 + \frac{213}{24}n^3$ (4)

(5) $+\frac{13}{36}m^4 + \frac{26}{9}m^3n + \frac{17}{3}m^2n^2 + \frac{41}{18}mn^3 + \frac{2}{9}n^4 + \frac{213}{24}m^3 + \frac{47}{3}m^2n + \frac{85}{12}mn^2 - \frac{4}{3}n^3$ (5)

(6) $\frac{1}{3}m^3n + 2m^2n^2 + 2mn^3 + \frac{1}{3}n^4 - 5m^2n - 16mn^2 - 5n^3$ (6)

(1) $-348m^2 - 696mn - 348n^2 + 2640m + 2640n + \alpha\,(\frac{1}{3}m^3 + 2m^2n + 2mn^2 + \frac{1}{3}n^3$ (1)

(2) $+\frac{6335}{12}m^2 + \frac{2965}{3}mn + \frac{6335}{12}n^2 - 3159m - 3159n + \alpha\,(-\frac{5}{4}m^3 - \frac{15}{2}m^2n - \frac{15}{2}mn^2 - \frac{5}{4}n^3$ (2)

(3) $-\frac{27}{2}m^2 - 14mn - \frac{229}{6}n^2 - \frac{13}{3}m + \frac{437}{3}n + \alpha\,(\frac{1}{6}m^3 + m^2n + mn^2 + \frac{1}{6}n^3$ (3)

(4) $-\frac{362}{9}m^2 - \frac{302}{2}mn - \frac{4543}{36}n^2 + \frac{764}{3}m + \frac{806}{3}n + \alpha\,(\frac{3}{2}m^2n + 3mn^2 + \frac{3}{4}n^3$ (4)

(5) $-\frac{4543}{36}m^2 - \frac{303}{2}mn - \frac{362}{9}n^2 + \frac{806}{3}m + \frac{764}{3}n + \alpha\,(\frac{3}{4}m^3 + 3m^2n + \frac{2}{3}mn^2$ (5)

(6) $+\frac{74}{3}mn + \frac{74}{3}n^2 - 150n + \alpha\,($ (6)

(1) $-15m^2 - 52mn - 15n^2 + \frac{716}{3}m + \frac{716}{3}n - 1920) + \alpha^2(-3m - 3n + 56)$

(2) $+\frac{145}{4}m^2 + 115mn + \frac{145}{4}n^2 - \frac{1685}{4}m - \frac{1685}{4}n + 2430) + \alpha^2(\frac{45}{8}m + \frac{45}{8}n - 75)$

(3) $-\frac{3}{2}m^2 - mn + \frac{3}{2}n^2 - \frac{2}{3}m - \frac{89}{3}n - 34) + \alpha^2(-\frac{3}{2}m - \frac{3}{2}n + 13)$

(4) $-4m^2 - \frac{59}{2}mn - \frac{63}{4}n^2 + \frac{172}{3}m + \frac{1511}{12}n - 238) + \alpha^2(-\frac{9}{8}n + 3)$

(5) $-\frac{63}{4}m^2 - \frac{59}{2}mn - 4n^2 + \frac{1591}{12}m + \frac{172}{3}n - 238) + \alpha^2(-\frac{9}{8}m + 3)$

(6) $3mn - 3n^2 + 29n$).

Verification is

$$(-\tfrac{1}{6} - \tfrac{4}{3} - \tfrac{10}{3} - 3 - \tfrac{1}{2})\,8 + (4 + \tfrac{28}{3} + \tfrac{61}{3} + 9)\,4 + (-\tfrac{27}{2} - 14 - \tfrac{229}{6})\,2 + (-\tfrac{13}{3} + \tfrac{437}{3})$$
$$+3\left((\tfrac{1}{6} + 1 + 1 + \tfrac{1}{6})\,8 + (-\tfrac{3}{2} - 1 + \tfrac{3}{2})\,4 + (-\tfrac{2}{3} - \tfrac{89}{3})\,2 - 34\right) + 9\left((-\tfrac{3}{2} - \tfrac{3}{2})\,4 + 13\,.\,2\right) = 0.$$

126. It will be observed that in the eighth and following equations, viz. those wherein the expression of the Supplement contains the symbol $(\bar{1})$, I have included along with the Supplement within the { }, the terms $-(m - \frac{5}{3})\,\{2\,(\cdot\,4) - (/\,4)\}$, &c., viz. these are $-(m - \frac{5}{3})$ into number of point-pairs (4), &c.: this is for convenience only; it simplifies the calculation, both from the symmetrical form under which the remaining terms present themselves in the several equations, and because the expressions of the terms in question, (these terms being mere multiples of a number of point-pairs) are by Zeuthen's theory known in terms of the Capitals. It is to be noticed that for any equation, to find the system to which the Capitals belong, we diminish by unity the barred number and then remove the bar; thus for the seventh equation, where we have Supp. $(\bar{2}, 1, 1, 1)$, the Capitals belong to the system (1, 1, 1, 1).

127. Referring to Nos. 41 to 47 of the First Memoir, for convenience I collect the capitals which belong to a single curve, giving the values in terms of m, n, α as follows.

(1, 1, 1, 1)

(1) $A = \tfrac{1}{2}\delta(\delta-1) = \tfrac{1}{8}m^4 - \tfrac{1}{4}m^3 + 2m^2n - \tfrac{1}{8}m^2 - 2mn + 8n^2 + \tfrac{1}{4}m - 2n + \alpha(-\tfrac{2}{3}m^2 + \tfrac{3}{4}m - 6n + \tfrac{3}{4}) + \tfrac{9}{8}\alpha^2;$

(2) $B = \delta(n-4)(m-4) = \tfrac{1}{2}m^3n - 2m^3 - \tfrac{5}{2}m^2n + 4mn^2 + 10m^2 - 14mn - 16n^2 - 8m + 64n + \alpha(-\tfrac{3}{2}mn + 6m + 6n - 24);$

(4) $C = \tau.\tfrac{1}{2}(m-4)(m-5) = \tfrac{1}{4}m^2n^2 + 2m^3 - \tfrac{1}{4}m^2n - \tfrac{9}{4}mn^2 - 18m^2 + \tfrac{9}{4}mn + 5n^2 + 40m - 5n + \alpha(-\tfrac{3}{4}m^2 + \tfrac{27}{4}m - 15);$

(3) $D = \iota.\tfrac{1}{2}(m-3)(m-4) = -\tfrac{3}{2}m^3 + \tfrac{21}{2}m^2 - 18m + \alpha(\tfrac{1}{2}m^2 - \tfrac{7}{2}m + 6).$

(2, 1, 1)

(3) $E = \delta(n-4) = \tfrac{1}{2}m^2n - 2m^2 - \tfrac{1}{2}mn + 4n^2 + 2m - 16n + \alpha(-\tfrac{3}{2}n + 6);$

(3) $F = 2\delta(m-3) = m^3 - 4m^2 + 8mn + 3m - 24n + \alpha(-3m + 9);$

(6) $G = 2\tau(m-4) = mn^2 + 8m^2 - mn - 4n^2 - 32m + 4n + \alpha(-3m + 12);$

(2) $D\,[=\iota.\tfrac{1}{2}(m-3)(m-4)$ *suprà*$]$;

(1) $H = \delta\kappa = -\tfrac{3}{2}m^2n + \tfrac{3}{2}mn - 12n^2 + \alpha(\tfrac{1}{2}m^2 - \tfrac{1}{2}m + \tfrac{17}{2}n) - \tfrac{3}{2}\alpha^2;$

(2) $I = \kappa(n-3)(m-4) = -3mn^2 + 9mn + 12n^2 - 36n + \alpha(mn - 3m - 4n + 12);$

(5) $J = \iota(m-3) = -3m^2 + 9m + \alpha(m-3).$

(2, 2)

(9) $K = \tau = \tfrac{1}{2}n^2 + 4m - \tfrac{1}{2}n + \alpha(-\tfrac{3}{2});$

(3) $L = \kappa(n-3) = -3n^2 + 9n + \alpha(n-3);$

(1) $M = \tfrac{1}{2}\kappa(\kappa-1) = \tfrac{9}{2}n^2 + \tfrac{3}{2}n + \alpha(-3n - \tfrac{1}{2}) + \tfrac{1}{2}\alpha^2;$

(2) $N = \iota = -3m + \alpha(1);$

(1) $O = \kappa = -3n + \alpha(1).$

(3, 1)

$$\begin{array}{l|lll}
(2) & P = 2\delta & = m^2 & - m + 8n + \alpha(\quad - 3); \\
(2) & Q = 2\tau & = & n^2 + 8m - n + \alpha(\quad - 3); \\
(5) & J\ [= \iota(m-3)\ \textit{suprà}]; & & \\
(4) & R = \kappa(m-3) & = - 3mn & + 9n + \alpha(m-3).
\end{array}$$

(4)

$$\begin{array}{l|l}
(4) & N\ [= \iota\ \textit{suprà}]; \\
(2) & O\ [= \kappa\ \textit{suprà}].
\end{array}$$

128. I make the following calculations, serving to express in terms of Zeuthen's Capitals, the terms in { } contained in the twelve equations respectively.

$$N = -3m + \alpha$$

$$-3m + \alpha \text{ (first equation).}$$

$$\begin{aligned}
2J &= -6m^2 + 18m + \alpha(2m-6) \\
+R &= -3mn + 9n + \alpha(m-3)
\end{aligned}$$

$$-6m^2 - 3mn + 18m + 9n + \alpha(3m-9) \text{ (second equation).}$$

$$\begin{aligned}
6K &= 3n^2 + 24m - 3n + \alpha(\quad -9) \\
+L &= -3n^2 + 9n + \alpha(n-3) \\
+3N &= -9m + \alpha(\quad 3) \\
+2O &= -6n + \alpha(\quad 2)
\end{aligned}$$

$$15m + \alpha(n-7) \text{ (third equation).}$$

$$\begin{aligned}
E &= \tfrac{1}{2}m^2n - 2m^2 - \tfrac{1}{2}mn + 4n^2 + 2m - 16n + \alpha(\quad - \tfrac{3}{2}n + 6) \\
+F &= m^3 - 4m^2 + 8mn + 3m - 24n + \alpha(\quad - 3m + 9) \\
+2G &= 2mn^2 + 16m^2 - 2mn - 8n^2 - 64m + 8n + \alpha(\quad - 6m + 24) \\
+D &= -\tfrac{3}{2}m^3 + \tfrac{21}{2}m^2 - 18m + \alpha(\tfrac{1}{2}m^2 - \tfrac{7}{2}m + 6) \\
+3J &= -9m^2 + 27m + \alpha(\quad 3m - 9) \\
+J' &= -3n^2 + 9n + \alpha(\quad n - 3)
\end{aligned}$$

$$-\tfrac{1}{2}m^3 + \tfrac{1}{2}m^2n + 2mn^2 + \tfrac{23}{2}m^2 + \tfrac{11}{2}mn - 7n^2 - 50m - 23n + \alpha(\tfrac{1}{2}m^2 - \tfrac{19}{2}m - \tfrac{1}{2}n + 33)$$

(fourth equation)

$$Q = \underline{n^2 + 8m - n - 3\alpha}$$

$$n^2 + 8m - n - 3\alpha \text{ (fifth equation).}$$

$$\begin{array}{rl}
3G = & 3mn^2 + 24m^2 - 3mn - 12n^2 - 96m + 12n + \alpha(\quad -9m \quad +36) \\
+\ I = & -3mn^2 \quad + 9mn + 12n^2 \quad - 36n + \alpha(mn - 3m - 4n + 12) \\
+4J = & -12m^2 \quad + 36m \quad + \alpha(\quad 4m \quad - 12) \\
+2J' = & -\ 6n^2 \quad + 18n + \alpha(\quad 2n - \ 6) \\
\hline
& 12m^2 + 6mn - \ 6n^2 - 60m - \ 6n + \alpha(mn - 8m - 2n + 30)
\end{array}$$

(sixth equation).

$$\begin{array}{rll}
B = & \tfrac{1}{3}m^3n \quad - 2m^3 - \tfrac{5}{2}m^2n + 4mn^2 \quad + 10m^2 - 14mn - 16n^2 - \ 8m + 64n & (1) \\
+4C = & m^2n^2 + 8m^3 - \ m^2n - 9mn^2 \quad - 72m^2 + \ 9mn + 20n^2 + 160m - 20n & (2) \\
+4D = & -6m^3 \quad + 42m^2 \quad - \ 72m & (3) \\
+\ D' = & -\tfrac{3}{2}n^3 \quad + \tfrac{21}{2}n^2 \quad - 18n & (4) \\
\hline
& \tfrac{1}{2}m^3n + m^2n^2 \quad - \tfrac{7}{2}m^2n - 5mn^2 - \tfrac{3}{2}n^3 - 20m^2 - \ 5mn + \tfrac{29}{2}n^2 + \ 80m + 26n & (5)
\end{array}$$

$$\begin{array}{ll}
(1) & +\alpha(\quad -\tfrac{3}{2}mn \quad + \ 6m + 6n - 24) \\
(2) & +\alpha(-3m^2 \quad + 27m \quad - 60) \\
(3) & +\alpha(\ 2m^2 \quad - 14m \quad + 24) \\
(4) & +\alpha(\quad \tfrac{1}{2}n^2 \quad - \tfrac{7}{2}n + \ 6) \\
\hline
(5) & +\alpha(- \ m^2 - \tfrac{3}{2}mn + \tfrac{1}{2}n^2 + 19m + \tfrac{5}{2}n - 54) \text{ (seventh equation).}
\end{array}$$

$$\begin{array}{rl}
-\tfrac{1}{3}N = & m \quad - \tfrac{1}{3}\alpha \\
+\tfrac{4}{3}O = & -4n + \tfrac{4}{3}\alpha \\
\hline
& m - 4n + \ \alpha \text{ (eighth equation).}
\end{array}$$

$$\begin{array}{rl}
(\bar{2},\ 3) = & -4m - 4n - \ 6 + 3\alpha \\
+2(\bar{4},\ 1) = & 2m + 2n - 12 \\
\hline
& -2m - 2n - 18 + 3\alpha \text{ (used } \textit{infrà}\text{)}
\end{array}$$

$$\begin{array}{rl}
-2P = & -2m^2 \quad + 2m - 16n + \alpha(\quad 6) \\
-\ Q = & -\ n^2 - 8m + \ n + \alpha(\quad 3) \\
-2R = & 6mn \quad - 18n + \alpha(-2m \quad + \ 6) \\
+\ J' = & -3n^2 \quad + \ 9n + \alpha(\quad n - \ 3) \\
\kappa\{(\bar{2},\ 3) + 2(\bar{4},\ 1)\} = & 6mn + 6n^2 \quad + 54n + \alpha(-2m - 11n - 18) + 3\alpha^2 \\
\hline
& -2m^2 + 12mn + 2n^2 - 6m + 30n + \alpha(-4m - 10n - \ 6) + 3\alpha^2
\end{array}$$

(ninth equation).

$$\begin{aligned}
3L &= -9n^2 + 27n + \alpha(3n - 9)\\
+2M &= 9n^2 + 3n + \alpha(-6n - 1) + \alpha^2\\
-2N &= +6m + \alpha(\quad - 2)\\
-\,O &= 3n + \alpha(\quad - 1)\\
\hline
&\ 6m + 33n + \alpha(-3n - 13) + \alpha^2 \text{ (tenth equation).}
\end{aligned}$$

$$\begin{array}{lll}
\kappa(\bar{2},\,2,\,1) & = & -18mn - 18n^2 - 162n \qquad (1)\\
-\tfrac{1}{3}D & = & \tfrac{1}{2}m^3 - \tfrac{7}{2}m^2 + 6m \qquad (2)\\
-2E & = & -m^2n + 4m^2 + 1mn - 8n^2 - 4m + 32n \qquad (3)\\
-2F & = & -2m^3 + 8m^2 - 16mn - 6m + 48n \qquad (4)\\
-G & = & -mn^2 - 8m^2 + 1mn + 4n^2 + 32m - 4n \qquad (5)\\
+\tfrac{1}{3}H & = & -\tfrac{1}{2}m^2n + \tfrac{1}{2}mn - 4n^2 \qquad (6)\\
+\tfrac{5}{3}I & = & -5mn^2 + 15mn + 20n^2 - 60n \qquad (7)\\
-\tfrac{10}{3}J & = & +10m^2 - 30m \qquad (8)\\
+2D' & = & -3n^3 + 21n^2 - 36n \qquad (9)\\
-J' & = & 3n^2 - 9n \qquad (10)\\
\hline
 & & -\tfrac{3}{2}m^3 - \tfrac{3}{2}m^2n - 6mn^2 - 3n^3 + \tfrac{21}{2}m^2 - \tfrac{33}{2}mn + 18n^2 - 2m - 191n \qquad (11)
\end{array}$$

$$\begin{array}{ll}
(1) & +\alpha(-3mn - 3n^2 + 6m + 51n + 54) + \alpha^2(m + n - 15)\\
(2) & +\alpha(-\tfrac{1}{6}m^2 + \tfrac{7}{6}m - 2)\\
(3) & +\alpha(3n - 12)\\
(4) & +\alpha(6m - 18)\\
(5) & +\alpha(3m - 12)\\
(6) & +\alpha(\tfrac{1}{6}m^2 - \tfrac{1}{6}m + \tfrac{17}{6}n) + \alpha^2(\quad - \tfrac{1}{2})\\
(7) & +\alpha(\tfrac{5}{3}mn - 5m - \tfrac{20}{3}n + 20)\\
(8) & +\alpha(-\tfrac{10}{3}n + 10)\\
(9) & +\alpha(1n^2 - 7n + 12)\\
(10) & +\alpha(-n + 3)\\
\hline
(11) & +\alpha(-\tfrac{4}{3}mn - 2n^2 + \tfrac{23}{3}m + \tfrac{253}{6}n + 55) + \alpha^2(m + n - \tfrac{31}{2})
\end{array}$$

(eleventh equation).

$$\begin{array}{llllll}
\kappa(\bar{2},1,1,1)= & -\tfrac{1}{2}m^3n-3m^2n^2-3mn^3-\tfrac{1}{2}n^4 & +\tfrac{15}{2}m^2n+24mn^2+\tfrac{15}{2}n^3 & -37mn-37n^2 & (1)\\
-\tfrac{4}{3}A = -\tfrac{1}{6}m^4 & & +\tfrac{1}{3}m^3-\tfrac{3}{3}m^2n & +\tfrac{1}{6}m^2+\tfrac{8}{3}mn-\tfrac{32}{3}n^2 & (2)\\
-\tfrac{5}{3}B = & -\tfrac{5}{6}m^3n & +\tfrac{10}{3}m^3+\tfrac{25}{6}m^2n-\tfrac{20}{3}mn^2 & -\tfrac{50}{3}m^2+\tfrac{70}{3}mn+\tfrac{80}{3}n^2 & (3)\\
-\tfrac{4}{3}C = & -\tfrac{1}{3}m^2n^2 & -\tfrac{8}{3}m^3+\tfrac{1}{3}m^2n+3mn^2 & +24m^2-3mn-\tfrac{20}{3}n^2 & (4)\\
-2D = & & 3m^3 & -21m^2 & (5)\\
-D' = & & \tfrac{3}{2}n^3 & -\tfrac{21}{2}n^2 & (6)\\
\hline
-\tfrac{1}{6}m^4 & -\tfrac{4}{3}m^3n-\tfrac{10}{3}m^2n^2-3mn^3-\tfrac{1}{2}n^4 & +4m^3+\tfrac{28}{3}m^2n+\tfrac{61}{3}mn^2+9n^3 & -\tfrac{27}{2}m^2-14mn-\tfrac{229}{6}n^2 & (7)
\end{array}$$

$$\begin{array}{llllll}
(1) & +225n & +\alpha(\tfrac{1}{6}m^3+m^2n+mn^2+\tfrac{1}{6}n^3 & -\tfrac{5}{2}m^2-\tfrac{7}{2}mn+2n^2+\tfrac{37}{3}m-\tfrac{187}{6}n-75) & +\alpha^2(-\tfrac{3}{2}m-\tfrac{3}{2}n+\tfrac{29}{2})\\
(2) & -\tfrac{1}{3}m+\tfrac{8}{3}n & +\alpha(& m^2-m+8n-1) & +\alpha^2(-\tfrac{3}{2})\\
(3) & +\tfrac{40}{3}m-\tfrac{320}{3}n & +\alpha(& +\tfrac{5}{2}mn-10m-10n+40) & \\
(4) & -\tfrac{160}{3}m+\tfrac{20}{3}n & +\alpha(& m^2-9m+20) & \\
(5) & +36m & +\alpha(& -m^2+7m-12) & \\
(6) & +18n & +\alpha(& -\tfrac{1}{2}n^2+\tfrac{7}{2}n-6) & \\
\hline
(7) & -\tfrac{13}{3}m+\tfrac{437}{3}n & +\alpha(\tfrac{1}{6}m^3+m^2n+mn^2+\tfrac{1}{6}n^3 & -\tfrac{3}{2}m^2-mn+\tfrac{3}{2}n^2-\tfrac{2}{3}m-\tfrac{89}{3}n-34) & +\alpha^2(-\tfrac{3}{2}m-\tfrac{3}{2}n+13)
\end{array}$$

(twelfth equation).

129. We have consequently, by means of the results just obtained,

Supp. $(\bar{5})$ $= \kappa(\bar{5})$
$+N$ (first equation).

Supp. $(\bar{4}, 1)$ $= \kappa(\bar{4}, 1)$
$+2J+R$ (second equation).

Supp. $(\bar{3}, 2)$ $= \kappa(\bar{3}, 2)$
$+6K+L+3N+2O$ (third equation).

Supp. $(\bar{3}, 1, 1)$ $= \kappa(\bar{3}, 1, 1)$
$+D+E+F+2G+3J+J'$ (fourth equation).

Supp. $(\bar{2}, 3)$ $= \kappa(\bar{2}, 3)$
$+Q$ (fifth equation).

Supp. $(\bar{2}, 2, 1)$ $= \kappa(\bar{2}, 2, 1)$
$+3G+I+4J+2J'$ (sixth equation).

Supp. $(\bar{2}, 1, 1, 1) = \kappa(\bar{2}, 1, 1, 1)$
$+B+4C+4D+D'$ (seventh equation).

$$\text{Supp.}\ (\bar{1},\ 4) = \kappa(\bar{1},\ 4) + (m-\tfrac{5}{3})(4N+2O) - \tfrac{1}{3}N + \tfrac{4}{3}O \qquad \text{(eighth equation).}$$

$$\text{Supp.}\ (\bar{1},\ 1,\ 3) = \kappa(\bar{1},\ 1,\ 3) + \kappa(\bar{2},\ 3) + \kappa\,.\,2(\bar{4},\ 1) + (m-2)(2P+2Q+5J+4R) - 2P - Q - 2R + J' \qquad \text{(ninth equation).}$$

$$\text{Supp.}\ (\bar{1},\ 2,\ 2) = \kappa(\bar{1},\ 2,\ 2) + (m-2)(9K+3L+M+2N+O) + 3L + 2M - 2N - O \qquad \text{(tenth equation).}$$

$$\text{Supp.}\ (\bar{1},\ 1,\ 1,\ 2) = \kappa(\bar{1},\ 1,\ 1,\ 2) + \kappa(\bar{2},\ 2,\ 1) + (m-\tfrac{7}{3})(3E+3F+6G+2D+H+2I+5J) - 2E - 2F - G - \tfrac{1}{3}D + \tfrac{1}{3}H + \tfrac{5}{3}I - \tfrac{10}{3}J + 2D' - J' \qquad \text{(eleventh equation).}$$

Observe that

$$G - 2E' = 0, \quad G' - 2E = 0, \quad 3G + I + 8J = 3G' + I' + 8J',$$

relations which may be used to modify the form of the last preceding result.

$$\text{Supp.}\ (\bar{1},\ 1,\ 1,\ 1,\ 1) = \kappa(\bar{1},\ 1,\ 1,\ 1,\ 1) + \kappa(\bar{2},\ 1,\ 1,\ 1) + (m-\tfrac{8}{3})(A+2B+4C+3D) - \tfrac{4}{3}A - \tfrac{5}{3}B - \tfrac{4}{3}C - D' \qquad \text{(twelfth equation).}$$

130. We may in these equations introduce on the right-hand sides in place of a symbol such as $\bar{p}$ the symbol $\overline{p\kappa 1}$: for example, in the fifth equation, writing

$$(\bar{2},\ 3) = (\overline{2\kappa 1},\ 3) + [(\bar{2},\ 3) - (\overline{2\kappa 1},\ 3)],$$

and therefore also

$$\kappa(\bar{2},\ 3) = \kappa(\overline{2\kappa 1},\ 3) + \kappa[(\bar{2},\ 3) - (\overline{2\kappa 1},\ 3)],$$

the second term $\kappa[(\bar{2},\ 3) - (\overline{2\kappa 1},\ 3)]$ can be expressed in terms of Zeuthen's Capitals. The remark applies to all the twelve equations; only as regards the first four of them, inasmuch as $(\overline{5\kappa 1}) = 0, \ldots (\overline{3\kappa 1},\ 1,\ 1) = 0$, it is the whole original terms $\kappa(\bar{5}) \ldots \kappa(\bar{3},\ 1,\ 1)$ which are thus expressible by means of Zeuthen's Capitals. By the assistance of the formulæ (First Memoir, Nos. 69 and 73) we readily obtain

Referring to

$$\kappa(\bar{5}) = \kappa = O \qquad \text{(first equation).}$$

$$\kappa(\bar{4},\ 1) = \kappa(m+n-6) = R + J' \qquad \text{(second equation).}$$

$\kappa(\bar{3},\ 2) \quad = \kappa(-9+\alpha) = \kappa\big(3(n-3)+\kappa-1+1\big)$ Referring to

$= 3L + 2M + O$ (third equation).

$\kappa(\bar{3},\ 1,\ 1) \quad = \kappa(\tfrac{1}{2}m^2 + 2mn + \tfrac{1}{2}n^2 - \tfrac{13}{2}m - \tfrac{13}{2}n + 27 - \tfrac{3}{2}\alpha)$

$= H + 2I + D' + J'$ (fourth equation).

$$\begin{array}{lllllll}
\text{viz. } \kappa^{-1}\ H & = \tfrac{1}{2}m^2 & & & -\ \tfrac{1}{2}m & +\ 4n & -\ \tfrac{3}{2}\alpha \\
\kappa^{-1}.\ 2I & = & 2mn & & -\ 6m & -\ 8n + 24 & \\
\kappa^{-1}.\ D' & = & & \tfrac{1}{2}n^2 & & -\ \tfrac{7}{2}n +\ 6 & \\
\kappa^{-1}\ J' & = & & & & n -\ 3 & \\
\hline
 & \tfrac{1}{2}m^2 & + 2mn & + \tfrac{1}{2}n^2 & - \tfrac{13}{2}m & - \tfrac{13}{2}n + 27 & - \tfrac{3}{2}\alpha
\end{array}$$

$\kappa(\bar{2},\ 3) \quad = \kappa(\overline{2\kappa1},\ 3)$ (fifth equation).

$\kappa(\bar{2},\ 2,\ 1) \quad = \kappa(\overline{2\kappa1},\ 2,\ 1) + \kappa(n-3)$

$= \kappa(\overline{2\kappa1},\ 2,\ 1) + J'$ (sixth equation).

$\kappa(\bar{2},\ 1,\ 1,\ 1) = \kappa(\overline{2\kappa1},\ 1,\ 1,\ 1) + \kappa.\tfrac{1}{2}(n-3)(n-4)$

$= \kappa(\overline{2\kappa1},\ 1,\ 1,\ 1) + D'$ (seventh equation).

$\kappa(\bar{1},\ 4) \quad = \kappa(\overline{1\kappa1},\ 4) + \kappa$

$= \kappa(\overline{1\kappa1},\ 4) + O$ (eighth equation).

$\kappa(\bar{1},\ 1,\ 3) + \quad \kappa(\bar{2},\ 3) + \kappa\, 2(\bar{4},\ 1)$

$= \kappa(\overline{1\kappa1},\ 1,\ 3) + \kappa(\overline{2\kappa1},\ 3) \quad + \kappa(n-3)$

$+ \kappa(\overline{2\kappa1},\ 3)$

$+ \qquad \kappa(2m + 2n - 6)$

$= \kappa(\overline{1\kappa1},\ 1,\ 3) + 2\kappa(\overline{2\kappa1},\ 3) + 2R + 3J'$ (ninth equation).

$\kappa(\bar{1},\ 2,\ 2) \quad = \kappa(\overline{1\kappa1},\ 2,\ 2) + \kappa\{3(n-3)+\kappa-1\}$

$= \kappa(\overline{1\kappa1},\ 2,\ 2) + 3L + 2M$ (tenth equation).

$\kappa(\bar{1},\ 1,\ 1,\ 2) + \kappa(\bar{2},\ 2,\ 1)$

$= \kappa(\overline{1\kappa1},\ 1,\ 1,\ 2) + \kappa(\overline{2\kappa1},\ 2,\ 1) + \kappa\{\tfrac{1}{2}(n-3)(n-4) + \delta + 2\,\overline{n-3}\,\overline{m-4}\}$

$+ \kappa(\overline{2\kappa1},\ 2,\ 1) + \kappa(n-3)$

$= \kappa(\overline{1\kappa1},\ 1,\ 1,\ 2) + 2\kappa(\overline{2\kappa1},\ 1,\ 2)$

$+ D' + H + 2I + J'$ (eleventh equation).

$\kappa(\bar{1}, 1, 1, 1, 1) + \kappa(\bar{2}, 1, 1, 1)$ — Referring to

$= \kappa(\overline{1\kappa 1}, 1, 1, 1, 1) + \kappa(\overline{2\kappa 1}, 1, 1, 1)$

$+ \kappa(\overline{2\kappa 1}, 1, 1, 1) + \kappa \cdot \tfrac{1}{2}(n-3)(n-4)$

$= \kappa(\overline{1\kappa 1}, 1, 1, 1, 1) + 2\kappa(\overline{2\kappa 1}, 1, 1, 1) + D'$ (twelfth equation).

131. Hence, substituting in the expressions of the several Supplements, we have

Supp. $(\bar{5})$ $= O$
$+ N$ (first equation).

Supp. $(\bar{4}, 1)$ $= R + J'$
$+ 2J + R$ (second equation).

Supp. $(\bar{3}, 2)$ $= 3L + 2M + O$
$+ 6K + L + 3N + 2O$ (third equation).

Supp. $(\bar{3}, 1, 1)$ $= H + 2I + D' + J'$
$+ E + F + 2G + D + 3J + J'$ (fourth equation).

Supp. $(\bar{2}, 3)$ $= \kappa(\overline{2\kappa 1}, 3)$
$+ Q$ (fifth equation).

Supp. $(\bar{2}, 2, 1)$ $= \kappa(\overline{2\kappa 1}, 1, 1) + J'$
$+ 3G + I + 4J + 2J'$ (sixth equation).

Supp. $(\bar{2}, 1, 1, 1)$ $= \kappa(\overline{2\kappa 1}, 1, 1, 1) + D'$
$+ B + 4C + 4D + D'$ (seventh equation).

Supp. $(\bar{1}, 4)$ $= \kappa(\overline{1\kappa 1}, 4) + O$
$+ (m - \tfrac{5}{2})(4N + O)$
$- \tfrac{1}{3}N + \tfrac{4}{3}O$ (eighth equation).

Supp. $(\bar{1}, 1, 3)$ $= \kappa(\overline{1\kappa 1}, 1, 3) + 2\kappa(\overline{2\kappa 1}, 3) + 2R + 3J'$
$+ (m-2)(2P + 2Q + 5J + 4R)$
$- 2P - Q - 2R + J'$ (ninth equation).

Supp. $(\bar{1}, 2, 2)$ $= \kappa(\overline{1\kappa 1}, 2, 2) + 3L + 2M$
$+ (m-2)(9K + 3L + M + 2N + O)$
$+ 3L + 2M - 2N - O$ (tenth equation).

$$\text{Supp.}\ (\bar{1},\ 1,\ 1,\ 2) = \kappa(\overline{1\kappa 1},\ 1,\ 1,\ 2) + 2\kappa(\overline{2\kappa 1},\ 1,\ 2)$$
$$+ H + 2I + D' + J'$$
$$+ (m - \tfrac{7}{3})(3E + 3F + 6G + 2D + H + 2I + 5J)$$
$$- 2E - 2F - G - \tfrac{1}{3}D + \tfrac{1}{3}H + \tfrac{5}{3}I - \tfrac{10}{3}J + 2D' - J'$$

(eleventh equation).

$$\text{Supp.}\ (\bar{1},\ 1,\ 1,\ 1,\ 1) = \kappa(\overline{1\kappa 1},\ 1,\ 1,\ 1,\ 1) + 2\kappa(\overline{2\kappa 1},\ 1,\ 1,\ 1) + D'$$
$$+ (m - \tfrac{8}{3})(A + 2B + 4C + 3D)$$
$$- \tfrac{4}{3}A - \tfrac{5}{3}B - \tfrac{4}{3}C - 2D - D'.$$

(twelfth equation).

132. Hence finally, merely collecting the terms, we have the following expressions of the Supplements in the twelve equations respectively.

$\text{Supp.}\ (\bar{5}) = N + O$ (first equation).

$\text{Supp.}\ (\bar{4},\ 1) = 2J + 2R + J'$ (second equation).

$\text{Supp.}\ (\bar{3},\ 2) = 6K + 4L + 2M + 3N + 3O$ (third equation).

$\text{Supp.}\ (\bar{3},\ 1,\ 1) = D + E + F + 2G + H + 2I + 3J + D' + 2J'$ (fourth equation).

$\text{Supp.}\ (\bar{2},\ 3) = \kappa(\overline{2\kappa 1},\ 3) + Q$ (fifth equation).

$\text{Supp.}\ (\bar{2},\ 2,\ 1) = \kappa(\overline{2\kappa 1},\ 2,\ 1) + 3G + I + 4J + 3J'$ (sixth equation).

$\text{Supp.}\ (\bar{2},\ 1,\ 1,\ 1) = \kappa(\overline{2\kappa 1},\ 1,\ 1,\ 1) + B + 4C + 4D + 2D'$ (seventh equation).

$\text{Supp.}\ (\bar{1},\ 4) = \kappa(\overline{1\kappa 1},\ 4) + (4m - 7)N + (2m - 1)O$ (eighth equation).

$$\text{Supp.}\ (\bar{1},\ 1,\ 3) = \kappa(\overline{1\kappa 1},\ 1,\ 3) + 2\kappa(\overline{2\kappa 1},\ 3)$$
$$+ (2m-6)P + (2m-5)Q + (5m-10)J + (4m-8)R + 4J'$$ (ninth equation).

$$\text{Supp.}\ (\bar{1},\ 2,\ 2) = \kappa(\overline{1\kappa 1},\ 2,\ 2)$$
$$+ (9m-18)K + 3mL + (m+2)M + (2m-6)N + (m-3)O$$ (tenth equation).

$$\text{Supp.}\ (\bar{1},\ 1,\ 1,\ 2) = \kappa(\overline{1\kappa 1},\ 1,\ 1,\ 2) + 2\kappa(\overline{2\kappa 1},\ 1,\ 2)$$
$$+ (2m-5)D + (3m-9)E + (3m-9)F + (6m-15)G$$
$$+ (m-1)H + (2m-1)I + (5m-15)J + 3D'$$ (eleventh equation).

$$\text{Supp.}\ (\bar{1},\ 1,\ 1,\ 1,\ 1) = \kappa(\overline{1\kappa 1},\ 1,\ 1,\ 1,\ 1) + 2\kappa(\overline{2\kappa 1},\ 1,\ 1,\ 1)$$
$$+ (m-4)A + (2m-7)B + (4m-12)C + (3m-10)D,$$ (twelfth equation).

where I recall the remark, *ante*, No. 126, that in each equation the Capitals belong to the system obtained by diminishing the barred number by unity and removing the bar; (4) for the first equation, (3, 1) for the second, and so on.

133. These are, I think, the true theoretical forms of the Supplements, viz. (attending to the signification of the Capitals) the expressions actually exhibit how the Supplement arises, whether from proper conics passing through or touching at a cusp, or from point-pairs (coincident line-pairs) or line-pairs (including of course in these terms line-pair-points). Thus, for instance, Supp. $(\bar{5}) = N + O$. Referring to the explanations, First Memoir, Nos. 41 to 47, $N (= \iota)$ is the number of the line-pair-points described as "inflexion tangent terminated each way at inflexion," and O $(= \kappa)$ the number of the line-pair-points described as "cuspidal tangent terminated each way at cusp," or in what is here the appropriate point of view, we have as a coincident line-pair each inflexion tangent and each cuspidal tangent. Reverting to the generation of the first equation, when the point P is a point in general of the given curve, the curve Θ is the conic (5), having with the curve 5 intersections at P, and besides meeting it in the $2m-5$ points P'. When the point P is at an inflexion, the curve Θ becomes the coincident line-pair formed by the tangent taken twice, the number of intersections at P is therefore $=6$, and the inflexion is therefore (specially) a united point. Similarly, when the point P is at a cusp, the curve Θ becomes the coincident line-pair formed by the tangent taken twice, the number of intersections at P is therefore $=6$, and the cusp is thus (specially) a united point: we have thus the total number of special united points $= \kappa + \iota$, agreeing with the foregoing *à posteriori* result, Supp. $(\bar{5}) = N + O$.

134. Or to take another example; for the fifth equation we have

$$\text{Supp. } (\bar{2},\ 3) = \kappa\,(\overline{2\kappa 1},\ 3) + Q\,;$$

$Q\,(= 2\tau)$ is the number of the line-pair-points described as "double tangent terminated each way at point of contact," or, in the point of view appropriate for the present purpose, we have each double tangent as a coincident line-pair in respect to the one of its points of contact, and also as a coincident line-pair in respect to the other of its points of contact. Reverting to the generation of the equation, when the point P is a point in general on the given curve, the curve Θ is the system of conics $(\bar{2},\ 3)$ touching the curve at P, and having besides with it a contact of the third order; since for each conic the number of intersections at P is $=2$, the total number of intersections at P is $=2\,(\bar{2},\ 3)$, and the remaining $(2m-2)\,(\bar{2},\ 3)$ intersections are the points P'. Suppose that the point P is taken at the point of contact of a double tangent; of the $(\bar{2},\ 3)$ conics, 1 (I assume this is so) becomes the coincident line-pair formed by the double tangent taken twice, and gives therefore 4 intersections at P, the remaining $(\bar{2},\ 3)-1$ conics are proper conics, giving therefore $2\,(\bar{2},\ 3)-2$ intersections at P, or the total number of intersections at P is $2\,(\bar{2},\ 3)+2$ intersections; or there is a gain of 2 intersections. As remarked (No. 96), this does not of necessity imply that the point in question is to be considered as being (specially) 2 united points; I do not know how to decide *à priori* whether it is to be regarded as being 2 united points or as 1 united point, but it is in fact to be regarded as being (specially) only 1 united point; and as the points in question are the 2τ points of contact of the double tangents, we have thus the number 2τ of special united points. Again, when the point P is at a cusp, all the $(\bar{2},\ 3)$ conics remain proper conics $\big((\overline{2\kappa 1},\ 3) = (\bar{2},\ 3),$

First Memoir, No. 73), but each of these (*quà* conic touching the cuspidal tangent) has with the given curve at the cusp not 2 but 3 intersections, so that the total number of intersections at P is $3(\overline{2\kappa 1}, 3), = 3(\bar{2}, 3)$, and there is a gain of $(\bar{2}, 3) = (\overline{2\kappa 1}, 3)$ intersections. Each cusp counts (specially) as $(\overline{2\kappa 1}, 3)$ united points, and together the cusps count as $\kappa(\overline{2\kappa 1}, 3)$ united points; we have thus the total number $\kappa(\overline{2\kappa 1}, 3) + 2\tau$ of special united points, agreeing with the expression, Supp. $(\bar{2}, 3) = \kappa(\overline{2\kappa 1}, 3) + Q$.

135. As appears from the preceding example, or generally from the remark, *ante*, No. 96, I have not at present any *à priori* method of determining the proper numerical multipliers of the Capitals contained in the expressions of the several Supplements. I will only further remark, that the reason is obvious why (while in the first seven equations the multipliers are mere numbers) in the eighth and following equations the multipliers are linear functions of m; in fact in these last equations the barred symbol is $\bar{1}$, that is, when P is a point in general on the given curve, each of the conics which make up the curve Θ has with the given curve not a contact of any order, but an ordinary intersection at P. Imagine a position of P for which one of these conics becomes a coincident line-pair; this regarded as a single line has with the given curve $(m-\alpha)$ ordinary intersections (α a number, $=4$ at most, depending on the contacts which the line may have with the curve); for each of the $m-\alpha$ points, taken as a position of P, one of the conics which make up the curve Θ becomes the coincident line-pair, and there are in respect of this conic two intersections at P instead of one intersection only. We have thus in respect of the particular coincident line-pair a group of $(m-\alpha)$ special united points, viz. these are the $m-\alpha$ ordinary intersections of the coincident line-pair regarded as a single line with the given curve, and we thus understand in a general way how it is that the order m of the given curve enters into the expressions of the multipliers of the several Capitals in the last five equations. The object of the present Memoir was, however, the *à posteriori* derivation of the expressions (*ante*, No. 132) of the twelve Supplements; and having accomplished this, but being unable to discuss the results with any degree of completeness, I abstain from a further discussion of them.

408.

ADDITION TO MEMOIR ON THE RESULTANT OF A SYSTEM OF TWO EQUATIONS.

[From the *Philosophical Transactions of the Royal Society of London*, vol. CLVIII. (for the year 1868), pp. 173—180. Received August 6,—Read November 21, 1867.]

THE elimination tables in the Memoir on the Resultant of a System of two Equations, *Phil. Trans.* 1857, pp. 703—715, [148], relate to equations of the form $(a, b \ldots \between x, y)^m = 0$, *without* numerical coefficients; but it is, I think, desirable to give the corresponding tables for equations in the form $(a, b, \ldots \between x, y)^m = 0$ *with* numerical coefficients, which is the standard form in quantics. The transformation can of course be effected without difficulty, and the results are as here given. It is easy to see *à priori* that the sum of the numerical coefficients in each table ought to vanish; these sums do in fact vanish, and we have thus a verification as well of the tables of the present Addition as of the tables of the original memoir, by means whereof the present tables were calculated.

Table (2, 2).
Resultant of
$(a, b, c \between x, y)^2$,
$(p, q, r \between x, y)^2$.

Table (3, 2).
Resultant of
$(a, b, c, d \between x, y)^3$,
$(p, q, r \between x, y)^2$.

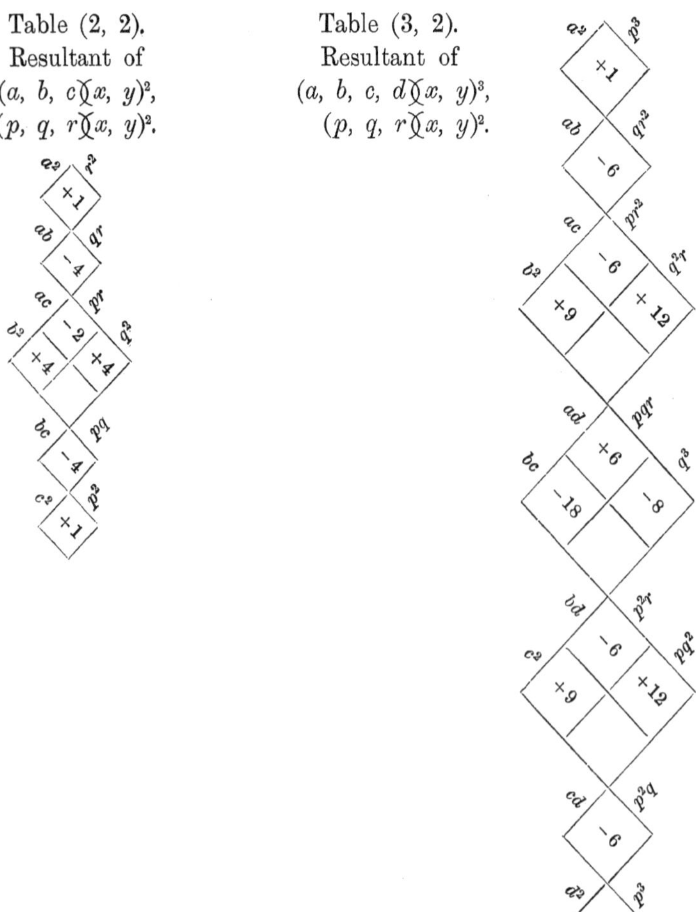

Table (4, 2).
Resultant of
$(a, b, c, d, e \between x, y)^4$,
$(p, q, r \between x, y)^2$.

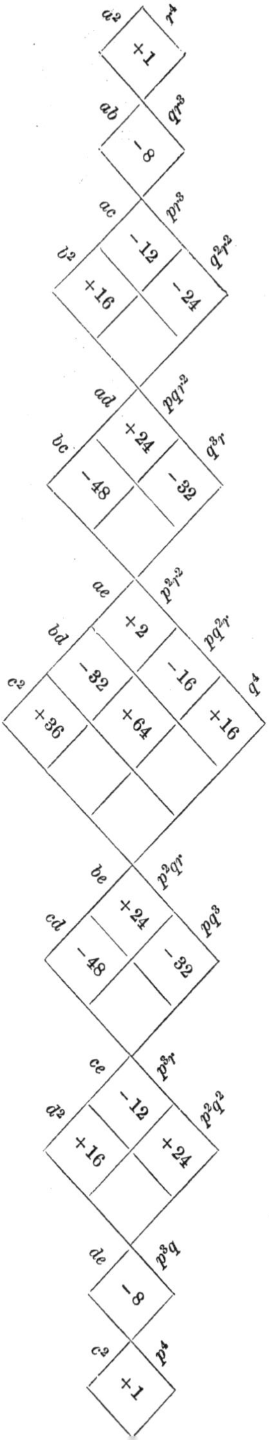

	r^4
a^2	+1

	qr^3
ab	−8

	pr^3	q^2r^2
ac	−12	−24
b^2	+16	

	pqr^2	q^3r
ad	+24	−32
bc	−48	

	p^2r^2	pq^2r	q^4
ae	+2	−16	+16
bd	−32	+64	
c^2	+36		

	p^2qr	pq^3
be	+24	−32
cd	−48	

	p^3r	p^2q^2
ce	−12	+24
d^2	+16	

	p^3q
de	−8

	p^4
e^2	+1

Table (3, 3)*.
Resultant of
$(a, b, c, d \between x, y)^3$,
$(p, q, r, s \between x, y)^3$.

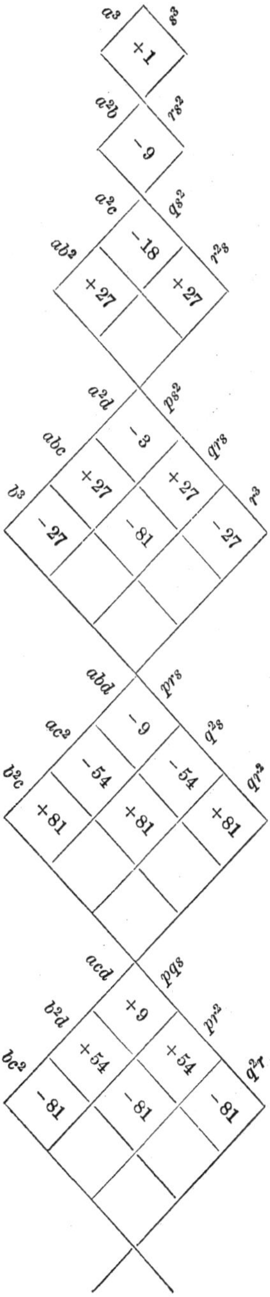

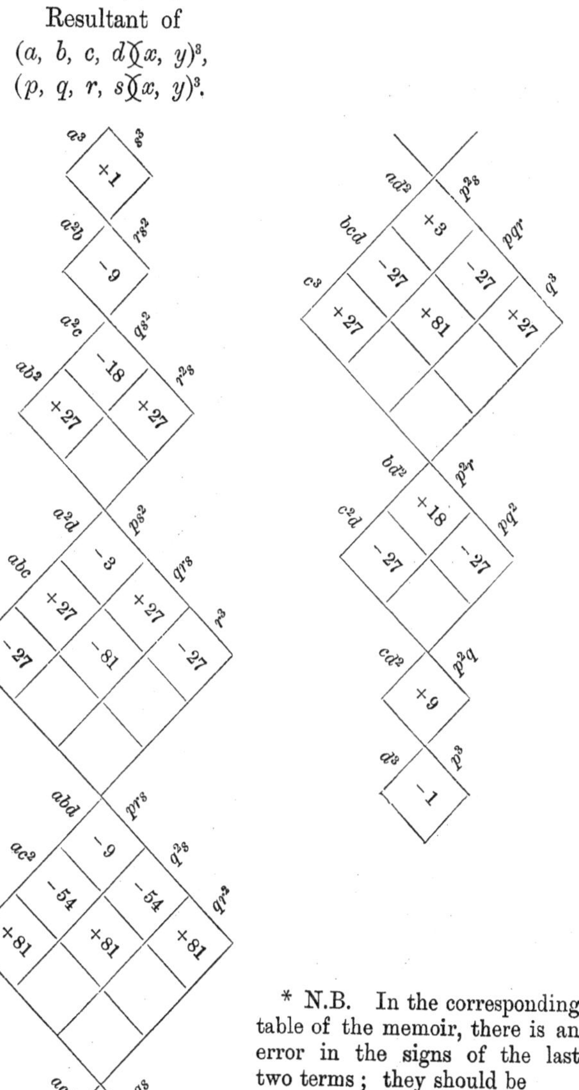

* N.B. In the corresponding table of the memoir, there is an error in the signs of the last two terms; they should be

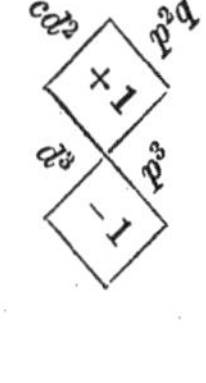

Table (4, 3).
Resultant of
$(a, b, c, d, e \between x, y)^4$,
$(p, q, r, s \between x, y)^3$.

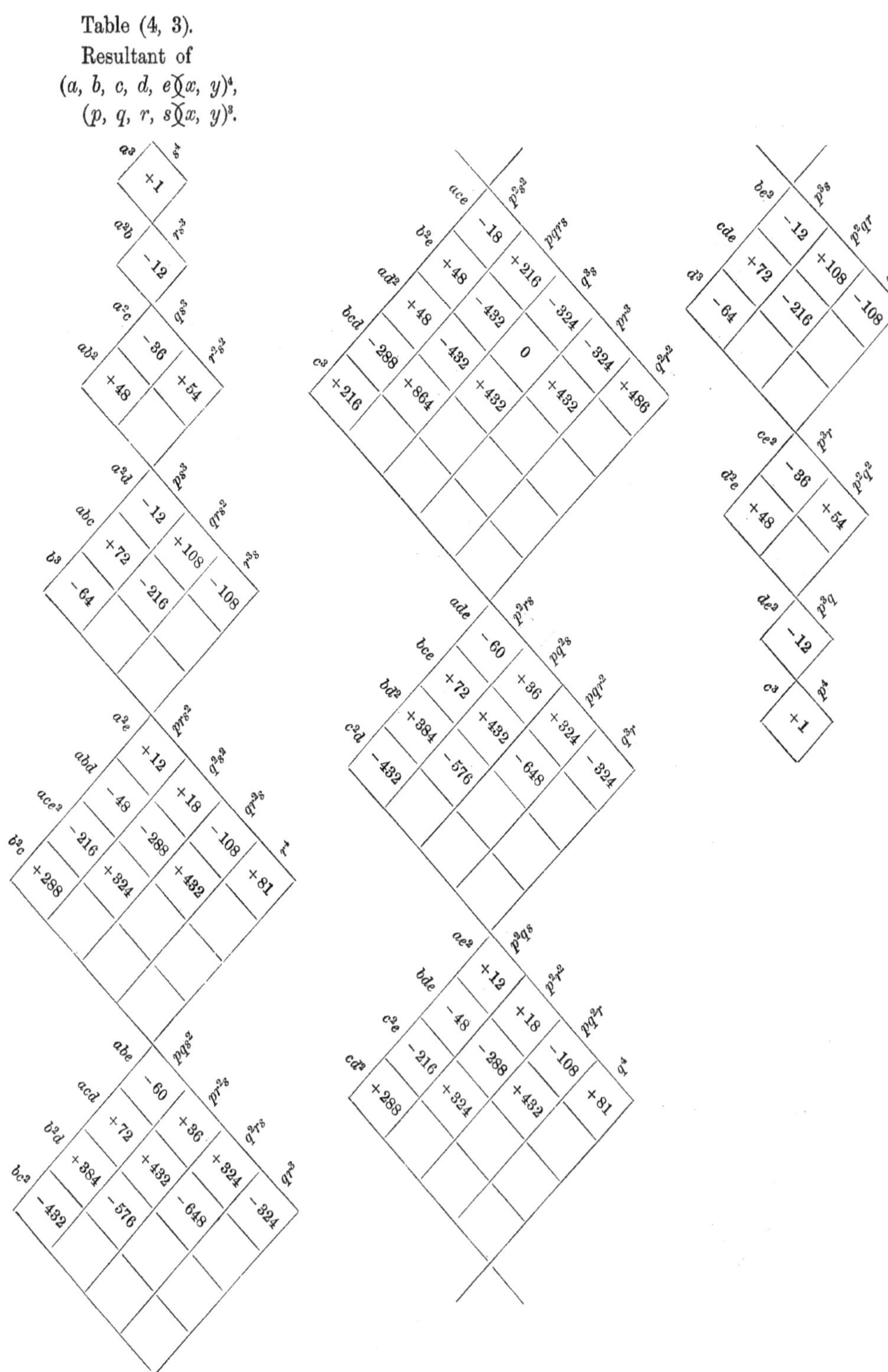

Table (4, 4).

Resultant of

$(a, b, c, d, e \between x, y)^4,$

$(p, q, r, s, t \between x, y)^4.$

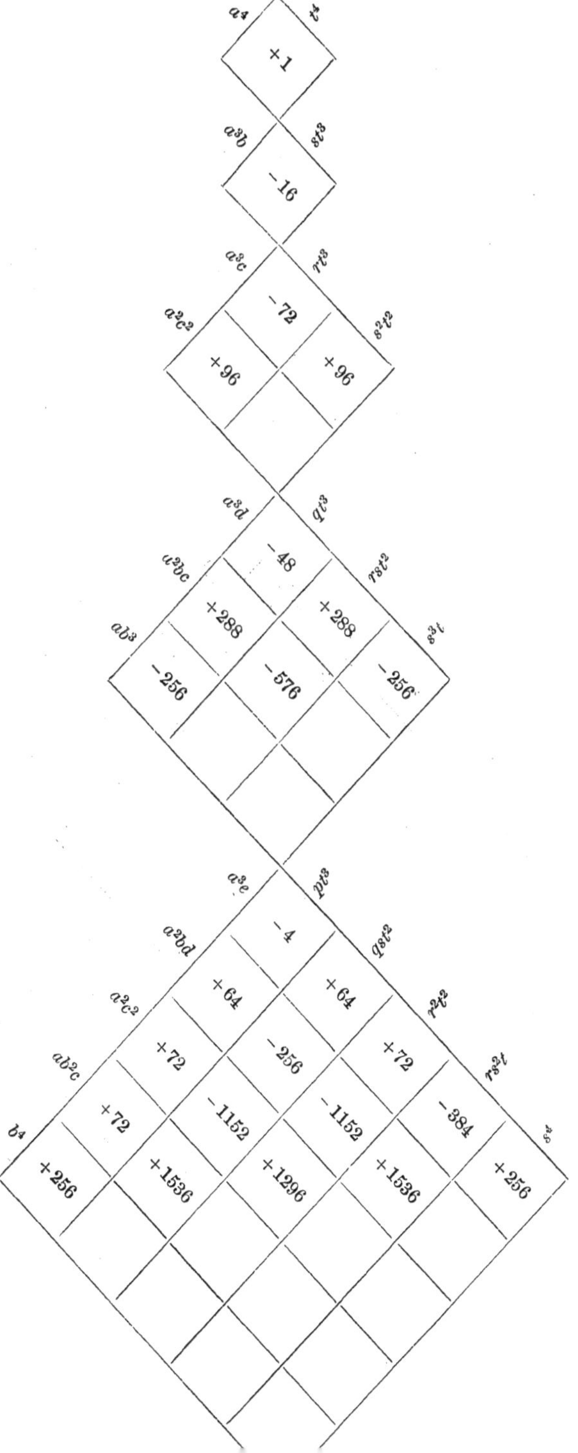

Table (4, 4) continued:

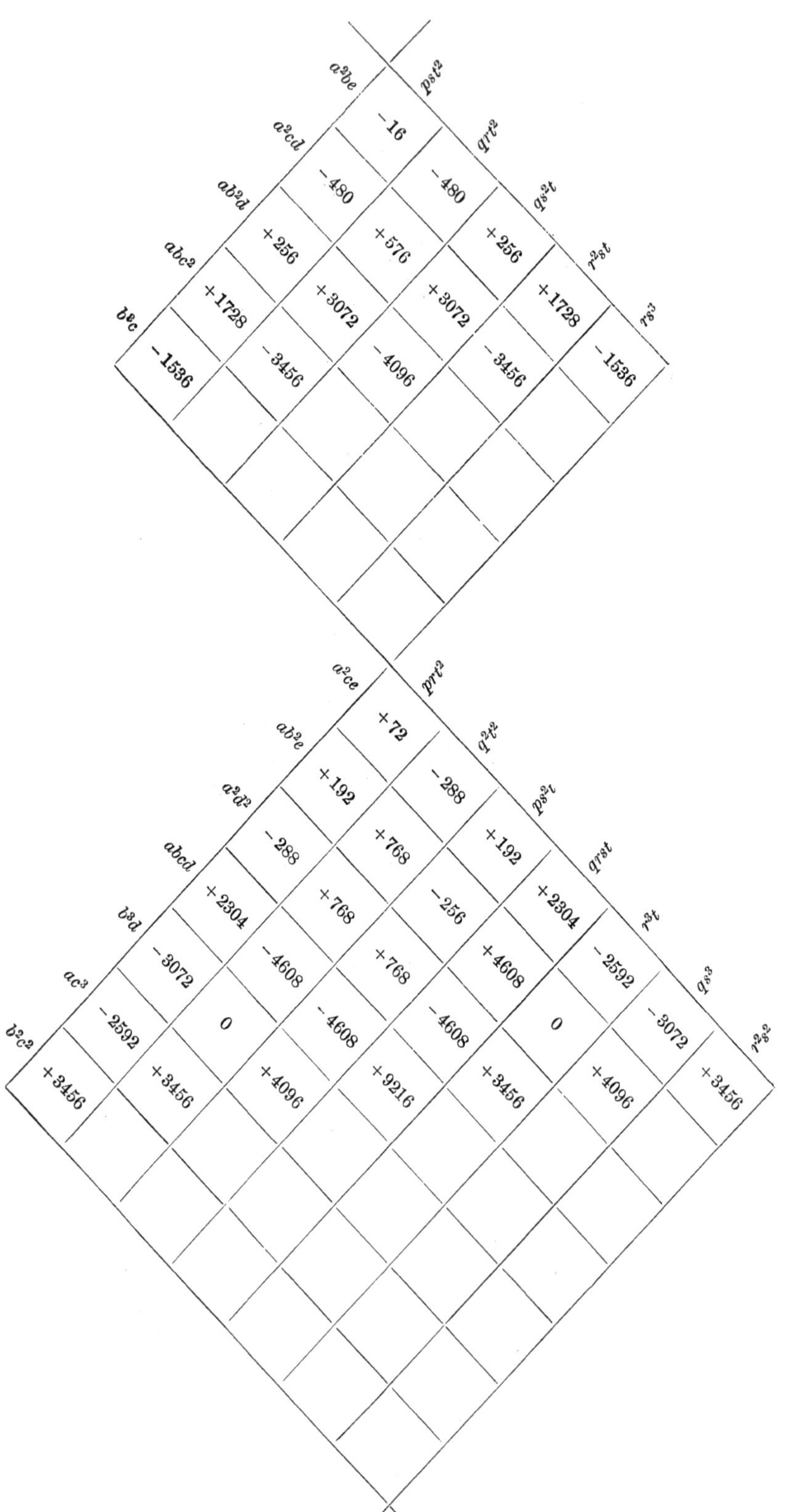

	pst^2	qrt^2	qs^2t	r^2st	rs^3
a^2be	−16	−480	+256	+1728	−1536
a^2cd	−480	+576	+3072	−3456	
ab^2d	+256	+3072	−4096		
abc^2	+1728	−3456			
b^3c	−1536				

	prt^2	q^2t^2	ps^2t	$qrst$	r^3t	qs^3	r^2s^2
a^2ce	+72	−288	+192	+2304	−2592	−3072	+3456
ab^2e	+192	+768	−256	+4608	0	+4096	
a^2d^2	−288	+768	+768	−4608	+3456		
$abcd$	+2304	−4608	−4608	+9216			
b^3d	−3072	0	+4096				
ac^3	−2592	+3456					
b^2c^2	+3456						

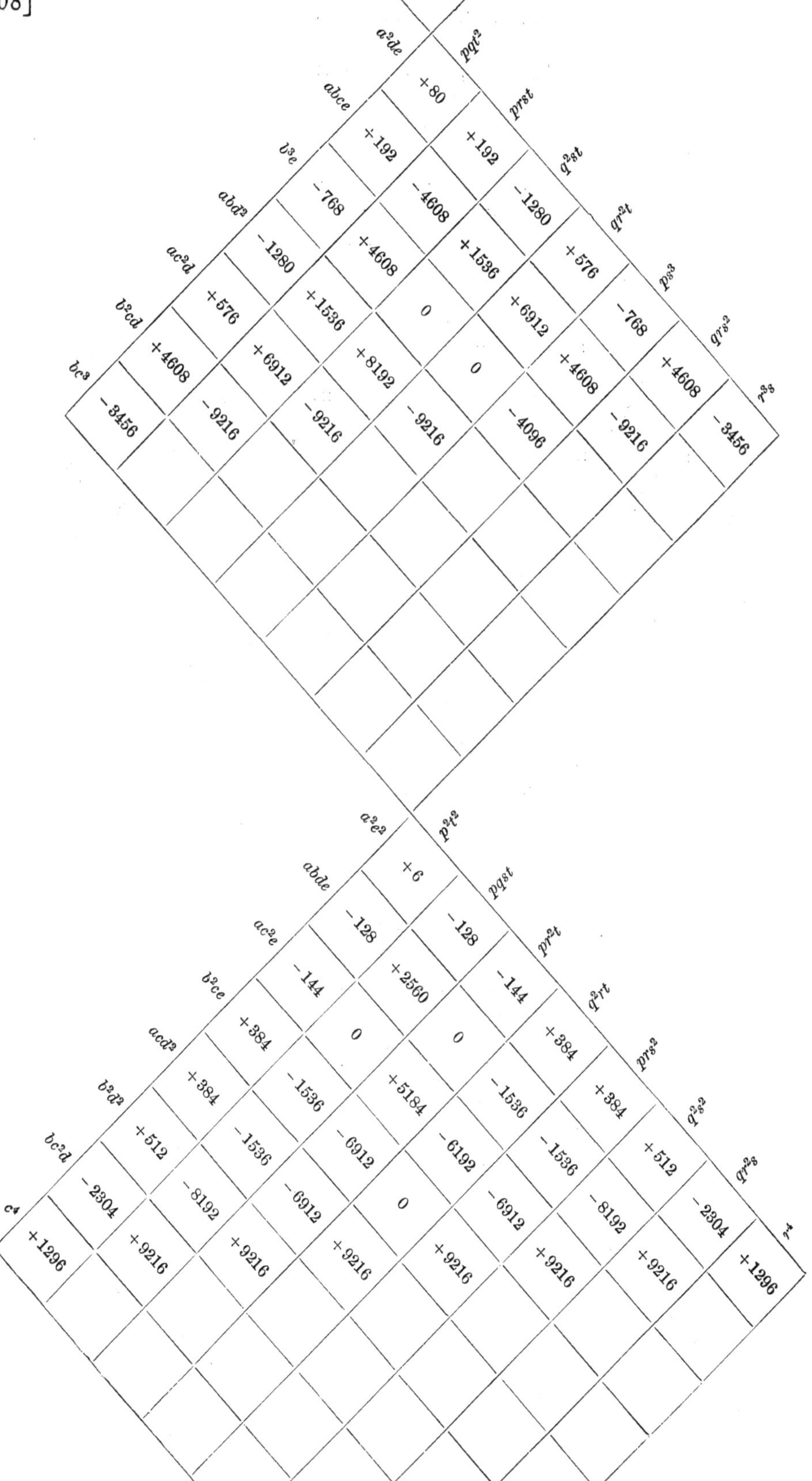

	pqt^2	$prst$	q^2st	qr^2t	ps^3	qrs^2	r^3s
a^2de	+80	+192	−1280	+576	−768	+4608	−3456
$abce$	+192	−4608	+1536	+6912	+4608	−9216	
b^3e	−768	+4608	0	0	−4096		
abd^2	−1280	+1536	+8192	−9216			
ac^2d	+576	+6912	−9216				
b^2cd	+4608	−9216					
bc^3	−3456						

	p^2t^2	$pqst$	pr^2t	q^2rt	prs^2	q^2s^2	qr^2s	r^4
a^2e^2	+6	−128	−144	+384	+384	+512	−2304	+1296
$abde$	−128	+2560	0	−1536	−1536	−8192	+9216	
ac^2e	−144	0	+5184	−6192	−6912	+9216		
b^2ce	+384	−1536	−6912	0	+9216			
acd^2	+384	−1536	−6912	+9216				
b^2d^2	+512	−8192	+9216					
bc^2d	−2304	+9216						
c^4	+1296							

Table (4, 4) concluded:

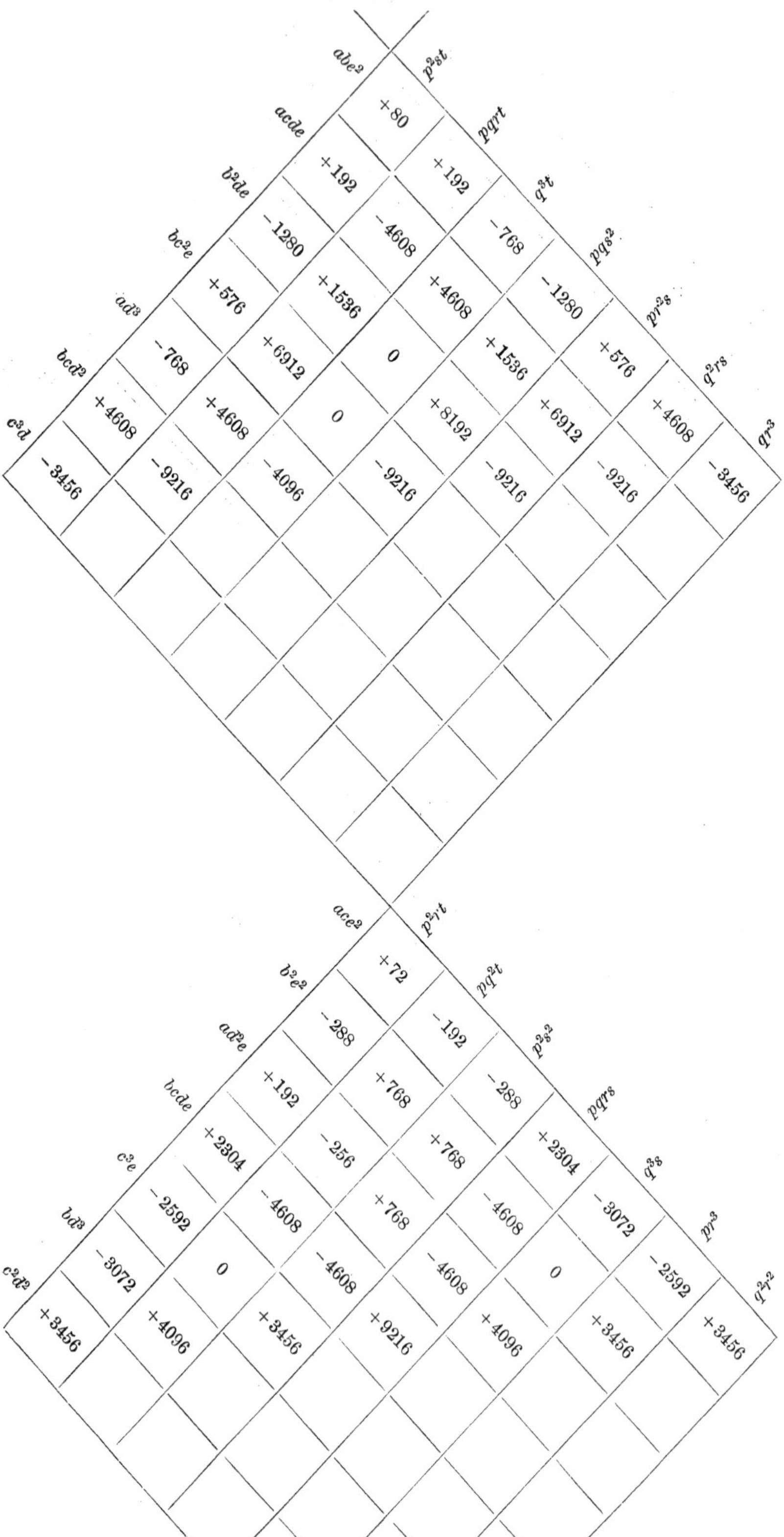

	p^2st	$pqrt$	q^3t	pqs^2	pr^2s	q^2rs	qr^3
abe^2	+80	+192	−768	−1280	+576	+4608	−3456
$acde$	+192	−4608	+4608	+1536	+6912	−9216	
b^2de	−1280	+1536	0	+8192	−9216		
bc^2e	+576	+6912	0	−9216			
ad^3	−768	+4608	−4096				
bcd^2	+4608	−9216					
c^3d	−3456						

	p^2rt	pq^2t	p^2s^2	$pqrs$	q^3s	pr^3	q^2r^2
ace^2	+72	−192	−288	+2304	−3072	−2592	+3456
b^2e^2	−288	+768	+768	−4608	0	+3456	
ad^2e	+192	−256	+768	−4608	+4096		
$bcde$	+2304	−4608	−4608	+9216			
c^3e	−2592	0	+3456				
bd^3	−3072	+4096					
c^2d^2	+3456						

	p^2qt	p^2rs	pq^2s	pqr^2	q^3r
ade^2	−16	−480	+256	+1728	−1536
bce^2	−480	+576	+3072	−3456	
bd^2e	+256	+3072	−4096		
c^2de	+1728	−3456			
cd^3	−1536				

	p^3t	p^2qs	p^2r	pq^2r	q^4
ae^3	−4	+64	+72	−384	+256
bde^2	+64	−256	−1152	+1536	
c^2e^2	+72	−1152	+1296		
cd^2e	−384	+1536			
d^4	+256				

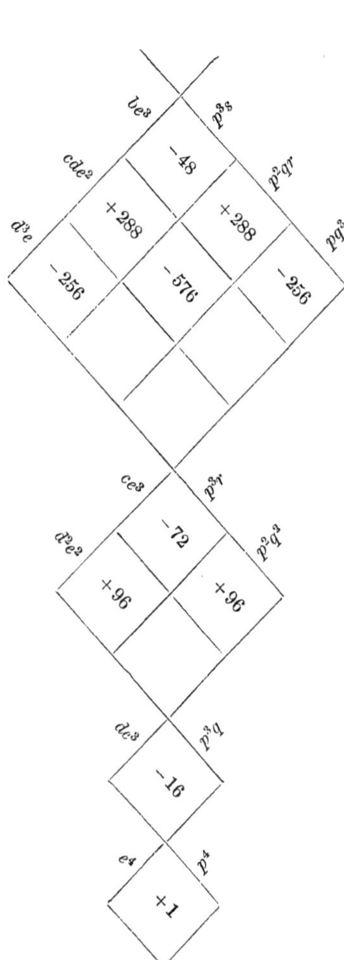

	p^3s	p^2qr	pq^3
be^3	−48	+288	−256
cde^2	+288	−576	
d^3e	−256		

	p^3r	p^2q^2
ce^3	−72	+96
d^2e^2	+96	

	p^3q
de^3	−16

	p^4
e^4	+1

409.

ON THE CONDITIONS FOR THE EXISTENCE OF THREE EQUAL ROOTS, OR OF TWO PAIRS OF EQUAL ROOTS, OF A BINARY QUARTIC OR QUINTIC.

[From the *Philosophical Transactions of the Royal Society of London*, vol. CLVIII. (for the year 1868), pp. 577—588. Received November 26, 1867,—Read January 9, 1868.]

[IT is remarked, *Proc. R. Soc.* vol. XVII. p. 314, that the above title is a misnomer: I had in fact in regard to the quintic considered *not* the twofold relations belonging to the root-systems 311 and 221 respectively, but the threefold relations belonging to the root-systems 41 and 32 respectively. The proper title would have been "On the conditions for the existence of certain systems of equal roots of a binary quartic or quintic."]

In considering the conditions for the existence of given systems of equalities between the roots of an equation, we obtain some very interesting examples of the composition of relations. A relation is either onefold, expressed by a single equation $U=0$, or it is, say k-fold, expressed by a system of k or more equations. Of course, as regards onefold relations, the theory of the composition is well known: the relation $UV=0$ is a relation compounded of the relations $U=0$, $V=0$; that is, it is a relation satisfied if, and not satisfied unless, one or the other of the two component relations is satisfied. The like notion of composition applies to relations in general; viz., the compound relation is a relation satisfied if, and not satisfied unless, one or the other of the two component relations is satisfied. I purposely refrain at present from any further discussion of the theory of composition. I say that the conditions for the existence of given systems of equalities between the roots of an equation furnish instances of such composition; in fact, if we express that the function $(*\between x, y)^n$, and its first-derived function in regard to x, or, what is the same thing, that the first-derived functions in regard to x, y respectively, have a common quadric factor, we obtain between the coefficients a certain twofold relation, which implies either that the equation $(*\between x, y)^n=0$ has three equal roots, or else that it has two pairs of equal roots; that is, the relation in question is satisfied if, and it is not satisfied unless, there is satisfied either the relation for the existence of three equal roots, or else the relation for the existence of two pairs of equal roots; or the relation for the existence of the quadric factor is compounded of the last-mentioned two relations. The relation for the quadric factor, for any value whatever of n, is at once seen to be expressible by means of an oblong matrix, giving rise to a series of determinants which are each to be put $=0$; the relation for three equal roots and that for two pairs of equal roots, in the particular cases $n=4$ and $n=5$, are given in my "Memoir on the Conditions for the existence of given Systems of Equalities between the roots of an Equation," *Phil. Trans.* vol. CXLVII. (1857), pp. 727—731, [150]; and I propose in the present Memoir to exhibit, for the cases in question $n=4$ and $n=5$, the connexion between the compound relation for the quadric factor with the component relations for the three equal roots and for the two pairs of equal roots respectively.

Article Nos. 1 to 8, *the Quartic.*

1. For the quartic function

$$(a, b, c, d, e \between x, y)^4,$$

the condition for three equal roots, or, say, for a root system 31, is that the quadrinvariant and the cubinvariant each of them vanish, viz. we must have

$$\begin{aligned} I &= ae - 4bd + 3c^2 = 0, \\ J &= ace - ad^2 - b^2e + 2bcd - c^3 = 0. \end{aligned}$$

2. The condition for two pairs of equal roots, or for a root system 22, is that the cubicovariant vanishes identically, viz. representing this by

$$(A, B, 5C, 10D, 5E, F, G \between x, y)^6 = 0,$$

we must have

$$\begin{aligned} A &= a^2d - 3abc + 2b^3 &&= 0, \\ B &= a^2e + 2abd - 9ac^2 + 6b^2c &&= 0, \\ C &= abe - 3acd + 2b^2d &&= 0, \\ D &= -ad^2 + b^2e &&= 0, \\ E &= -ade + 3bce - 2bd^2 &&= 0, \\ F &= -ae^2 - 2bde + 9c^2e - 6cd^2 &&= 0, \\ G &= -be^2 + 3cde - 2d^3 &&= 0. \end{aligned}$$

3. But the condition for the common quadric factor is

$$\left\| \begin{matrix} & a, & 3b, & 3c, & d \\ & b, & 3c, & 3d, & e \\ a, & 3b, & 3c, & d & \\ b, & 3c, & 3d, & e & \end{matrix} \right\| = 0,$$

and the determinants formed out of this matrix must therefore vanish for $(I, J)=0$, and also for $(A, B, C, D, E, F, G)=0$, that is, the determinants in question must be syzygetically related to the functions (I, J), and also to the functions (A, B, C, D, E, F, G).

4. The values of the determinants are

$1234 = 3\times$	$1235 = 3\times$	$1245 =$	$1345 = 3\times$	$2345 = 3\times$
$a^2ce\ +\ 1$	$a^2de\ -\ 1$	$a^2e^2\ -\ 1$	$abe^2\ -\ 1$	$ace^2\ +\ 1$
$a^2d^2\ -\ 3$	$abce\ +\ 4$	$abde\ +\ 2$	$acde\ +\ 4$	$ad^2e\ -\ 1$
$ab^2e\ -\ 1$	$abd^2\ +\ 1$	$ac^2e\ +\ 9$	$ad^3\ -\ 3$	$b^2e^2\ -\ 3$
$abcd\ +\ 14$	$ac^2d\ -\ 3$	$acd^2\ -\ 9$	$b^2de\ +\ 1$	$bcde\ +\ 14$
$ac^3\ -\ 9$	$b^3e\ -\ 3$	$b^2ce\ -\ 9$	$bc^2e\ -\ 3$	$bd^3\ -\ 8$
$b^3d\ -\ 8$	$b^2cd\ +\ 2$	$b^2d^2\ +\ 8$	$bcd^2\ +\ 2$	$c^3e\ -\ 9$
$b^2c^2\ +\ 6$				$c^2d^2\ +\ 6$

5. The syzygetic relation with $(I,\ J)$ is given by means of the identical equation

$$\begin{vmatrix} y^4, & -4xy^3, & 6x^2y^2, & -4x^3y, & x^4 \\ & a\ , & 3b\ , & 3c\ , & d \\ & b\ , & 3c\ , & 3d\ , & e \\ a\ , & 3b\ , & 3c\ , & d\ , & \\ b\ , & 3c\ , & 3d\ , & e\ , & \end{vmatrix} = -6I\,.\,\tilde{H}U + 9J\,.\,U,$$

or, as this may be written,

$$(1234,\ 1235,\ 1245,\ 1345,\ 2345\between x,\ y)^4 = -6I\,.\,\tilde{H}U + 9J\,.\,U,$$

where $\tilde{H}U$ is the Hessian of U,

$= ($

$ac\ +1$	$ad\ +2$	$ae\ +\ 1$	$be\ +\ 2$	$ce\ +1$
$b^2\ -1$	$bc\ -\ 2$	$bd\ +\ 2$	$cd\ -\ 2$	$d^2\ -1$
		$c^2\ -\ 3$		

$\between x,\ y)^4.$

6. That is, we have

$$\begin{aligned} 1234 &= (\ ac - b^2 &&,\ \ a\between -6I,\ 9J), \\ 4\,.\,1235 &= (2ad - 2bc &&,\ 4b\between -6I,\ 9J), \\ 6\,.\,1245 &= (\ ae + 2bd - 3c^2, &&\ \ 6c\between -6I,\ 9J), \\ 4\,.\,1345 &= (2be - 2cd &&,\ 4d\between -6I,\ 9J), \\ 2345 &= (\ ce - d^2 &&,\ \ e\between -6I,\ 9J). \end{aligned}$$

7. The determinants thus vanish if $(I,\ J)=0$, that is, for the root system 31; they will also vanish without this being so, if only

$$\left(\frac{3J}{2I}=\right)\frac{ac-b^2}{a}=\frac{ad-bc}{2b}=\frac{ae+2bd-3c^2}{6c}=\frac{be-cd}{2d}=\frac{ce-d^2}{e};$$

and we may omit the first member $\left(\dfrac{3J}{2I}=\right)$, since if the remaining terms are equal to each other they will also be $=\frac{3J}{2I}$. The equations may then be written

$$\left\| \begin{array}{ccccc} ac-b^2, & ad-bc, & ae+2bd-3c^2, & be-cd, & ce-d^2 \\ a\ \ , & 2b\ \ , & 6c\ \ , & 2d\ \ , & e \end{array} \right\| = 0,$$

and the ten equations of this system reduce themselves (as it is very easy to show) to the seven equations

$$(A,\ B,\ C,\ D,\ E,\ F,\ G) = 0,$$

which, as above mentioned, are the conditions for the root system 22.

8. It may be added that we have

	A	B	C	D	E	F	G
$\frac{1}{3}\cdot 1234 =$		c	$-4b$	$+3a$			
$\frac{1}{3}\cdot 1235 =$			c	$-3b$	$+a$		
$0 =$		d	$-3c$		$+a$		
$1245 =$		$-e$	$+4d$	$-3c$			
$0 =$		$-e$		$+6c$		$-a$	
$0 =$			$-d$	$+3c$	$-b$		
$\frac{1}{3}\cdot 1345 =$			$-e$	$+3d$	$-c$		
$0 =$			$-e$		$+3c$	$-b$	
$\frac{1}{3}\cdot 2345 =$				$-3e$	$+4d$	$-c$	

where it is to be noticed that the four equations having the left-hand side $=0$, give $B:C:D:E:F$ proportional to the determinants of the matrix

$$\left| \begin{array}{ccccc} d, & -3c, & \cdot\,, & a & \\ -e, & \cdot\,, & 6c, & \cdot\,, & -a \\ & -d\,, & 3c, & -b & \\ & -e\,, & \cdot\,, & +3c, & -b \end{array} \right|;$$

the determinants in question contain each the factor c, and omitting this factor, the system shows that B, C, D, E, F are proportional to their before-mentioned actual values.

Article Nos. 9 to 15, *the Quintic.*

9. For the quintic function

$$(a,\ b,\ c,\ d,\ e,\ f \between x,\ y)^5,$$

the condition of a root system 41 is that the covariant, $[B=]$ No. 14, shall vanish, viz. we must have

$$\begin{aligned} A &= 2\,(ae-4bd+3c^2\,)=0, \\ B &= \quad af-3be+2cd\ =0, \\ C &= 2\,(bf-4ce\ +3d^2)=0. \end{aligned}$$

10. The condition of a root system 32 is that the following covariant, viz.

$$[3A^2B-25C^2,\ =]\ 3\,(\text{No. }13)^2\,(\text{No. }14)-25\,(\text{No. }15)^2,$$

shall vanish, where

$[A =]$ No. 13 $= (a, b, c, d, e, f \between x, y)^5$, the quintic itself.

$[B =]$ No. 14 $= \left(\begin{array}{c|c|c} ae & af & bf \\ bd - 4 & be - 3 & ce - 4 \\ c^2 + 3 & cd + 2 & d^2 + 3 \end{array} \between x, y\right)^2$.

$[C =]$ No. 15 $= \left(\begin{array}{c|c|c|c|c|c|c} ac & ad \quad 3 & ae \quad 3 & af & bf \quad 3 & cf \quad 3 & df \\ b^2 - & bc - 3 & bd + 3 & be + 7 & ce + 3 & de - 3 & e^2 - \\ & & c^2 - 6 & cd - 8 & d^2 - 6 & & \end{array} \between x, y\right)^6$.

11. The developed expression of the foregoing function is as follows:

𝔄	𝔅	ℭ	𝔇	𝔈	𝔉	𝔊	ℌ	ℑ	𝔍	𝔎	𝔏	𝔐
(a^3e + 3	a^3f + 3	a^2bf + 33	a^2cf + 10	a^2df − 90	a^2ef − 114	a^2f^2 − 19	abf^2 − 114	acf^2 − 90	adf^2 + 10	aef^2 + 33	af^3 + 3	bf^3 + 3 $\between x, y)^{12}$
a^2bd − 12	a^2be + 21	a^2ce − 102	a^2de − 390	a^2e^2 − 195	$abdf$ − 264	$abef$ − 608	$acef$ − 264	$adef$ + 360	ae^2f + 155	bdf^2 − 102	bef^2 + 21	cef^3 − 12
a^2c^2 − 16	a^2cd − 144	a^2d^2 − 216	ab^2f + 155	$abcf$ + 360	abe^2 − 990	$acdf$ + 537	ad^2f + 468	ae^3 + 225	bcf^2 − 390	be^2f + 135	cdf^2 − 144	d^2f^2 − 16
ab^2c + 50	ab^2d + 30	ab^2e + 135	$abce$ + 100	$abde$ − 1500	ac^2f + 468	ace^2 − 245	ade^2 + 900	b^2f^2 − 195	$bdef$ + 100	c^2f^2 − 216	ce^2f + 30	de^2f + 50
b^4 − 25	abc^2 + 240	$abcd$ + 120	abd^2 − 600	ac^2e + 900	$acde$ + 1320	ad^2e + 1740	b^2ef − 990	$bcef$ − 1500	be^3 + 125	$cdef$ + 120	d^2ef + 240	e^4 − 25
	b^3c − 150	ac^3 + 480	ac^2d + 1600	acd^2 + 1800	ad^3 + 1080	b^2df − 245	$bcdf$ + 1320	bd^2f + 900	c^2ef − 600	ce^3 − 150	de^3 − 150	
		b^3d − 150	b^3e + 125	b^3f + 225	b^2cf + 900	b^2e^2 − 1700	bce^2 − 2700	c^2df + 1800	cd^2f + 1600	d^3f + 480		
		b^2c^2 − 300	b^2cd − 1000	b^2d^2 − 1500	b^2de − 2700	bc^2f + 1740	bd^2e + 900	c^2e^2 − 1500	cde^2 − 1000	d^2e^2 − 300		
					bc^2e + 900	$bcde$ − 2000	c^3f + 1080					
					bcd^2 − 600	bd^3 + 600	c^2de − 600					
						c^3e + 600						
						c^2d^2 − 400						

12. The conditions for the common [cubic] factor are

$$\left\|\begin{array}{cccccc} & a, & 4b, & 6c, & 4d, & e \\ a, & 4b, & 6c, & 4d, & e & \\ & b, & 4c, & 6d, & 4e, & f \\ b, & 4c, & 6d, & 4e, & f & \end{array}\right\| = 0,$$

the several determinants whereof are given in Table No. 27 of my "Third Memoir on Quantics," *Philosophical Transactions*, vol. CXLVI. (1856), pp. 627—647, [144].

13. These determinants must therefore vanish, for $(A, B, C)=0$, and also for $(\mathfrak{A}, \mathfrak{B}, \ldots \mathfrak{L}, \mathfrak{M})=0$, that is, they must be syzygetically connected with (A, B, C), and also with $(\mathfrak{A}, \mathfrak{B}, \ldots \mathfrak{L}, \mathfrak{M})$. The relation to (A, B, C) is in fact given in the Table appended to Table No. 27, viz. this is

	$C \times$	$+B \times$	$+A \times$
1234 =	$+\ 6\ a^2$	$-12\ ab$	$+16\ ac - 10\ b^2$
1235 =	$+\ 6\ ab$	$-\ 2\ ac - 10\ b^2$	$+\ 6\ ad$
1236 =	$-\ 2\ ac + 8\ b^2$	$+\ 6\ ad - 18\ bc$	$-\ 2\ df + 8\ e^2$
1245 =	$+18\ ac$	$-\ 6\ ad - 30\ bc$	$+\ 8\ ae + 10\ bd$
1246 =	$+12\ bc$	$+\ 4\ ae - 4\ bd - 24\ c^2$	$+\ 4\ be + 8\ cd$
1345 =	$+24\ ad$	$-\ 8\ ae - 40\ bd$	$+\ 4\ af + 20\ be$
1256 =	$-\ 1\ ae + 4\ bd + 3\ c^2$	$+\ 1\ af + 5\ be - 18\ cd$	$-\ 1\ bf + 4\ ce + 3\ d^2$
2345 =	$+20\ ae + 40\ bd - 30\ c^2$	$-80\ be + 20\ cd$	$+20\ bf + 40\ ce - 30\ d^2$
1346 =	$+\ 4\ ae + 8\ bd + 6\ c^2$	$-36\ cd$	$+\ 4\ bf + 8\ ce + 6\ d^2$
2346 =	$+\ 4\ af + 20\ be$	$-\ 8\ bf - 4\ ce$	$+24\ cf$
1356 =	$+\ 4\ be + 8\ cd$	$+\ 4\ bf - 4\ ce - 24\ d^2$	$+12\ de$
2356 =	$+\ 8\ bf + 10\ ce$	$-\ 6\ cf - 30\ de$	$+18\ df$
1456 =	$+\ 6\ ce$	$+\ 6\ cf - 18\ de$	$-\ 2\ df + 8\ e^2$
2456 =	$+\ 6\ cf$	$-\ 2\ df - 10\ e^2$	$+\ 6\ ef$
3456 =	$+16\ df - 10\ e^2$	$-12\ ef$	$+\ 6\ f^2$

14. Between the expressions $\mathfrak{A}$, $\mathfrak{B}$, &c., and 1234, 1235, &c., there exist relations the form of which is indicated by the following Table:

$\mathfrak{A}$	$\mathfrak{B}$	$\mathfrak{C}$	$\mathfrak{D}$	$\mathfrak{E}$	$\mathfrak{F}$	$\mathfrak{G}$	$\mathfrak{H}$	$\mathfrak{I}$	$\mathfrak{J}$	$\mathfrak{K}$	$\mathfrak{L}$	$\mathfrak{M}$	1234	1235	1236	1245	1246	1345	1256	2345	1346	2346	1356	2356	1456	2456	3456
c	b	a											a														
d	c	b	a										b	a													
e	d	c	b	a									c	b	a	a											
f	e	d	c	b	a								d	c	b	b	a	a									
	f	e	d	c	b	a							e	d	c	c	b	b	a	a	a						
		f	e	d	c	b	a						f	e	d	d	c	c	b	b	b	a	a				
			f	e	d	c	b	a						f	e	e	d	d	c	c	c	b	b	a	a		
				f	e	d	c	b	a						f	f	e	e	d	d	d	c	c	b	b	a	
					f	e	d	c	b	a							f	f	e	e	e	d	d	c	c	b	a
						f	e	d	c	b	a								f	f	f	e	e	d	d	c	b
							f	e	d	c	b	a										f	f	e	e	d	c
								f	e	d	c	b												f	f	e	d
									f	e	d	c														f	e
										f	e	d															f

viz. these relations are of the form

$$(\;)\,c\mathfrak{A}+(\;)\,b\mathfrak{B}+(\;)\,a\mathfrak{C}\qquad +(\;)\,a\,.\,1234 \qquad =0,$$

$$(\;)\,d\mathfrak{A}+(\;)\,c\mathfrak{B}+(\;)\,b\mathfrak{C}+(\;)\,a\mathfrak{D}\qquad +(\;)\,b\,.\,1234+(\;)\,a\,.\,1235=0,$$

&c.

where the brackets ( ) denote numerical coefficients, determinate as to their ratios.

15. Assuming the existence of these relations, we have for the determination of the numerical coefficients in each relation a set of linear equations, which are shown by the following Tables, viz. referring to the Table headed $c\mathfrak{A}$, $b\mathfrak{B}$, $a\mathfrak{C}$, $a\,.\,1234$, [first of the seven tables *infrà*] if the multipliers of the several terms respectively be A, B, C, X, then the Table denotes the system of linear equations

$$\begin{aligned}&0\ A\ +3\,B\ +\ 33\,C\ +\ 0\,X=0,\\&3\ A\ +0\,B\ -102\,C\ -16\,X=0,\\&\&c.,\end{aligned}$$

that is, nine equations to be satisfied by the ratios of the coefficients A, B, C, X, and which are in fact satisfied by the values at the foot of the Table, viz.

$$A : B : C : X = +66 : -11 : +1 : +6.$$

There would be in all fourteen Tables, but as those for the second seven would be at once deducible by symmetry from the first seven, I have only written down the seven Tables; the solutions for the first and second Tables were obtained without difficulty, but that for the third Table was so laborious to calculate, and contains such extraordinarily high numbers, that I did not proceed with the calculation, and it is accordingly only the first, second, and third Tables which have at the foot of them respectively the solutions of the linear equations.

16. The results given by these three Tables are, of course,

$$66\,c\mathfrak{A} - 11\,b\mathfrak{B} + 1\,a\mathfrak{C} + 6a\,.\,1234 = 0,$$

$$330\,d\mathfrak{A} + 110\,c\mathfrak{B} - 55\,b\mathfrak{C} + 9\,a\mathfrak{D} - 105\,a\,.\,1235 = 0,$$

$$\begin{aligned}&+266478575\ e\mathfrak{A}\\&-617359490\ d\mathfrak{B}\\&+144200810\ c\mathfrak{C}\\&+\ \ \ 9656911\ b\mathfrak{D}\\&+\ \ \ 9090785\ a\mathfrak{E}\\&-721004050\ c\,.\,1234\\&+\ \ 90914175\ b\,.\,1235\\&-160758675\ a\,.\,1245\\&+\ \ 11559295\ a\,.\,1236=0.\end{aligned}$$

It is to be noticed that the nine coefficients of this last equation were obtained from, and that they actually satisfy, a system of fourteen linear equations; so that the correctness of the result is hereby verified.

17. The seven Tables are

First Table.

	$c\mathfrak{A}$	$b\mathfrak{B}$	$a\mathfrak{C}$	$a.1234$
a^3bf		+ 3	+ 33	
a^3ce	+ 3		− 102	− 16
a^3d^2			− 216	+ 36
a^2be^2		+ 21	+ 135	+ 16
a^2bcd	− 12	− 144	+ 120	− 152
a^2c^3	− 16		+ 480	+ 96
ab^3d		+ 30	− 150	+ 80
ab^2c^2	+ 50	+ 240	− 300	− 60
b^4c	− 25	− 150		
	+ 66	− 11	+ 1	+ 6

Second Table.

	$d\mathfrak{A}$	$c\mathfrak{B}$	$b\mathfrak{C}$	$a\mathfrak{D}$	$a.1235$	$b.1234$
a^3cf		+ 3		+ 10	− 4	
a^3de	+ 3			− 390	+ 24	
a^2b^2f			+ 33	+ 155	+ 4	
a^2bce		+ 21	− 102	+ 100	− 84	− 16
a^2bd^2	− 12		− 216	− 600	− 24	+ 36
a^2c^2d	− 16	− 144		+ 1600	+ 64	+ 16
ab^3e			+ 135	+ 125	+ 60	
ab^2cd	+ 50	+ 30	+ 120	− 1000	− 40	− 152
abc^3		+ 240	+ 480			+ 96
b^4d	− 25		− 150			+ 80
b^3c^2		− 150	− 300			− 60
	+ 330	+ 110	− 55	+ 9	− 105	0

Third Table.

	$e\mathfrak{A}$	$d\mathfrak{B}$	$c\mathfrak{C}$	$b\mathfrak{D}$	$a\mathfrak{E}$	$c.1234$	$b.1235$	$a.1245$	$a.1236$
a^3df		+ 3			− 90			− 6	+ 6
a^3e^2	+ 3				− 195			+ 16	
a^2bcf			+ 33	+ 10	+ 360		− 4	+ 6	− 22
a^2bde	− 12	+ 21		− 390	− 1500		+ 24	− 26	− 6
a^2c^2e	− 16		− 102		+ 900	− 16		− 96	+ 16
a^2cd^2		− 144	− 216		+ 1800	+ 36		+ 96	
ab^3f				+ 155	+ 225		+ 4		+ 16
ab^2ce	+ 50		+ 135	+ 100		+ 16	− 84	+ 90	− 10
ab^2d^2		+ 30		− 600	− 1500		− 24	− 80	
abc^2d		+ 240	+ 120	+ 1600		− 152	+ 64		
ac^4			+ 480			+ 96			
b^4e	− 25			+ 125			+ 60		
b^3cd		− 150	− 150	− 1000		+ 80	− 40		
b^3c^3			− 300			− 60			
	+ 266478575	− 617359490	+ 144200810	+ 9656911	+ 9090785	− 721004050	+ 90914175	− 160758675	+ 11559295

Fourth Table.

	$f\mathfrak{A}$	$e\mathfrak{B}$	$d\mathfrak{C}$	$c\mathfrak{D}$	$b\mathfrak{E}$	$a\mathfrak{F}$	$d.1234$	$c.1235$	$b.1236$	$b.1245$	$a.1246$	$a.1345$
a^3ef	+ 3	+ 3				− 114					+ 4	
a^2bdf	− 12		+ 33		− 90	− 264			+ 6	− 6	− 4	− 24
a^2be^2		+ 21			− 195	− 990				+ 16	− 4	+ 64
a^2c^2f	− 16			+ 10		+ 468		− 4			− 24	+ 24
a^2cde		− 144	− 102	− 390		+ 1320	− 16	+ 24			+ 24	− 208
a^2d^3			− 216			+ 1080	+ 36					+ 144
ab^2cf	+ 50			+ 155	+ 360	+ 900		+ 4	− 22	+ 6	+ 24	
ab^2de		+ 30	+ 135		− 1500	− 2700	+ 16		− 6	− 26	− 20	− 40
abc^2e		+ 240		+ 100	+ 900	+ 900		− 84	+ 16	− 96		+ 60
$abcd^2$			+ 120	− 600	+ 1800	− 600	− 152	− 24		+ 96		− 40
ac^3d			+ 480	+ 1600			+ 96	+ 64				
b^4f	− 24				+ 225				+ 16			
b^3ce		− 150		+ 125				+ 60	− 10	+ 90		
b^3d^2			− 150		− 1500		+ 80			− 80		
b^2c^2d			− 300	− 1000			− 60	− 40				

Fifth Table.

	$f\mathfrak{B}$	$e\mathfrak{C}$	$d\mathfrak{D}$	$c\mathfrak{E}$	$b\mathfrak{F}$	$a\mathfrak{G}$	$e.1234$	$d.1235$	$c.1236$	$c.1245$	$b.1246$	$b.1345$	$a.1256$	$a.2345$	$a.1346$
f^2	+ 3					− 19							+ 1		
bef	+ 21	+ 33			− 114	− 608					+ 4		− 2		+ 16
cdf	− 144		+ 10	− 90		+ 537		− 4	+ 6	− 6			− 16	+ 20	− 36
ce^2		− 102		− 195		− 245	− 16			+ 16			+ 16	− 80	− 16
d^2e		− 216	− 390			+ 1740	+ 36	+ 24					+ 16	+ 60	+ 36
b^2df	+ 30		+ 155		− 264	− 245		+ 4			− 4	− 24	− 15	− 80	− 16
b^2e^2		+ 135			− 990	− 1700	+ 16				− 4	+ 64		+ 240	
bc^2f	+ 240			+ 360	+ 468	+ 1740			− 22	+ 6	− 24	+ 24		+ 60	+ 36
$bcde$		+ 120	+ 100	− 1500	+ 1320	− 2000	− 152	− 84	− 6	− 26	+ 24	− 208		− 860	− 20
bd^3			− 600		+ 1080	+ 600		− 24				+ 144		+ 960	
c^3e		+ 480		+ 900		+ 600	+ 96		+ 16	− 96				+ 960	
c^2d^2			+ 1600	+ 1800		− 400		+ 64		+ 96				− 320	
cf	− 150			+ 225	+ 900				+ 16		+ 24				
de		− 150	+ 125		− 2700		+ 80	+ 60			− 20	− 40			
c^2e		− 300			+ 900		− 60		− 10	+ 90		+ 60			
cd^2			− 1000	− 1500	− 600			− 40		− 80		− 40			

Sixth Table.

	$f\mathfrak{S}$	$e\mathfrak{D}$	$d\mathfrak{E}$	$c\mathfrak{F}$	$b\mathfrak{G}$	$a\mathfrak{H}$	$f.1234$	$e.1235$	$d.1236$	$d.1245$	$c.1246$	$c.1345$	$b.1256$	$b.2345$	$b.1346$	$a.2346$	$a.1356$
a^2bf^2	+ 33				− 19	− 114							+ 1				+ 4
a^2cef	− 102	+ 10		− 114		− 264	− 16	− 4			+ 4					− 24	− 4
a^2d^2f	− 216		− 90			+ 468	+ 36		+ 6	− 6						+ 24	− 24
a^2de^2		− 390	− 195			+ 900		+ 24		+ 16							+ 24
ab^2ef	+ 135	+ 155			− 698	− 990	+ 16	+ 4					− 2		+ 16	+ 64	− 4
$abcdf$	+ 120		+ 360	− 264	+ 537	+ 1320	− 152		− 22	+ 6	− 4	− 24	− 16	+ 20	− 36	− 208	+ 24
$abce^2$		+ 100		− 990	− 245	− 2700		− 84			− 4	+ 64	+ 16	− 80	− 16	− 40	− 20
abd^2e		− 600	− 1500		+ 1740	+ 900		− 24	− 6	− 26			+ 16	+ 60	+ 36	+ 60	
ac^3f	+ 480			+ 468		+ 1080	+ 96				− 24	+ 24				+ 144	
ac^2de		+ 1600	+ 900	+ 1320		− 600		+ 64	+ 16	− 96	+ 24	− 208				− 40	
acd^3			+ 1800	+ 1080						+ 96		+ 144					
b^3df	− 150		+ 225		− 245		+ 80		+ 16				− 80	− 80	− 16		
b^3e^2		+ 125			− 1700			+ 60						+ 240			
b^2c^2f	− 300			+ 900	+ 1740		− 60				+ 24			+ 60	+ 36		
b^2cde		− 1000		− 2700	− 2000			− 40	− 10	+ 90	− 20	− 40		− 860	− 20		
b^2d^3			− 1500		+ 600					− 80				+ 960			
bc^3e				+ 900	+ 600							+ 60		+ 960			
bc^2d^2				− 600	− 400							− 40		− 320			

Seventh Table.

	$f\mathfrak{D}$	$e\mathfrak{E}$	$d\mathfrak{F}$	$c\mathfrak{G}$	$b\mathfrak{H}$	$a\mathfrak{J}$	$f.1235$	$e.1236$	$e.1245$	$d.1246$	$d.1345$	$c.1256$	$c.2345$	$c.1346$	$b.2346$	$b.1356$	$a.2356$	$a.1456$
a^2cf^2	+ 10			− 19		− 90	− 4					+ 1					− 6	+ 6
a^2def	− 390	− 90	− 114			+ 360	+ 24	+ 6	− 6	+ 4							+ 6	− 22
a^2e^3		− 195				+ 225			+ 16									+ 16
ab^2f^2	+ 155				− 114	− 195	+ 4									+ 4	+ 16	
$abcef$	+ 100	+ 360		− 608	− 264	− 1500	− 84	− 22	+ 6			− 2		+ 16	− 24	− 4	− 26	− 6
abd^2f	− 600		− 264		+ 468	+ 900	− 24			− 4	− 24				+ 24	− 24	− 96	+ 16
$abde^2$		− 1500	− 990		+ 900			− 6	− 26	− 4	+ 64					+ 24	+ 90	− 10
ac^2df	+ 1600		+ 468	+ 537		+ 1800	+ 64			− 24	+ 24	− 16	+ 20	− 36			+ 96	
ac^2e^2		+ 900		− 245		− 1500		+ 16	− 96			+ 16	− 80	− 16			− 80	
acd^2e		+ 1800	+ 1320	+ 1740					+ 96	+ 24	− 208	+ 16	+ 60	+ 36				
ad^4			+ 1080								+ 144							
b^3ef	+ 125	+ 225			− 990		+ 60	+ 16							+ 64	− 4		
b^2cdf	− 1000		+ 900	− 245	+ 1320		− 40			+ 24		− 15	− 80	− 16	− 208	+ 24		
b^2ce^2				− 1700	− 2700			− 10	+ 90				+ 240		− 40	− 20		
b^2d^2e		− 1500	− 2700		+ 900				− 80	− 20	− 40				+ 60			
bc^3f				+ 1740	+ 1080								+ 60	+ 36	+ 144			
bc^2de			+ 900	− 2000	− 600						+ 60		− 860	− 20	− 40			
bcd^3			− 600	+ 600							− 40		+ 960					
c^4e				+ 600									+ 960					
c^3d^2				− 400									− 320					

And the remaining seven Tables might of course be deduced from these by writing (f, e, d, c, b, a) instead of (a, b, c, d, e, f), and making the corresponding alterations in the top line of each Table.

18. The equations $\mathfrak{A}=0$, $\mathfrak{B}=0, \ldots,$ $\mathfrak{M}=0$ consequently establish between the fifteen functions 1234, 1235, ... 3456 a system of fourteen equations, viz. the first and last three of these are

$$1234 = 0,$$

$$1235 = 0,$$

$$\begin{aligned} &-160758675 \,.\, 1245 \\ &+\ \ 11559295 \,.\, 1236 = 0, \end{aligned}$$

$$\vdots$$

$$\begin{aligned} &+\ \ 11559295 \,.\, 1456 \\ &-160758675 \,.\, 2356 = 0, \end{aligned}$$

$$2456 = 0,$$

$$3456 = 0.$$

To complete the proof that in virtue of the equations $\mathfrak{A}=0$, $\mathfrak{B}=0, \ldots,$ $\mathfrak{M}=0$ all the fifteen functions 1234, 1235, ... 3456 vanish, it is necessary to make use of the identical relations subsisting between these quantities 1234, &c.; thus we have

$$a \,.\, 1345 + 4b \,.\, 1245 + 6c \,.\, 1235 + 4d \,.\, 1234 = 0,$$

$$b \,.\, 1345 + 4c \,.\, 1245 + 6d \,.\, 1235 + 4e \,.\, 1234 = 0,$$

which, in virtue of the above equations $1234 = 0$ and $1235 = 0$, become

$$a \,.\, 1345 + 4b \,.\, 1245 = 0,$$

$$b \,.\, 1345 + 4c \,.\, 1245 = 0,$$

giving (unless indeed $ac - b^2 = 0$) $1245 = 0$, $1345 = 0$; the equation $1245 = 0$ then reduces the third of the above equations to $1236 = 0$, and so on until it is shown that the fifteen quantities all vanish.

410.

A THIRD MEMOIR ON SKEW SURFACES, OTHERWISE SCROLLS.

[From the *Philosophical Transactions of the Royal Society of London,* vol. CLIX. (for the year 1869), pp. 111—126. Received May 30,—Read June 18, 1868.]

THE present Memoir is supplementary to my "Second Memoir on Skew Surfaces, otherwise Scrolls," *Phil. Trans.* vol. CLIV. (1864), pp. 559—577, [340], and relates also to the theory of skew surfaces of the fourth order, or quartic scrolls. It was pointed out to me by Herr Schwarz(¹), in a letter dated Halle, June 1, 1867, that in the enumeration contained in my Second Memoir I have given only a particular case of the quartic scrolls which have a directrix skew cubic; viz. my eighth species, $S(1, 3^2)$, where there is also a directrix line. And this led me to observe that I had in like manner mentioned only a particular case of the quartic scrolls with a triple directrix line; viz. my third species, $S(1_3, 1, 4)$, where there is also a simple directrix line. The omitted species, say, *ninth species,* $S(1_3)$, *with a triple directrix line,* and *tenth species,* $S(3^2)$, *with a directrix skew* cubic, are considered in the present Memoir; and in reference to them I develope a theory of the reciprocal relations of these scrolls, which has some very interesting analytical features.

The paragraphs of the present Memoir are numbered consecutively with those of my Second Memoir above referred to.

Quartic Scroll, Ninth Species, $S(1_3)$, *with a triple directrix line.*

54. Consider a line the intersection of two planes, and let the equation of the one plane contain in the order 3, that of the second plane contain linearly, a variable parameter θ; the equations of the two planes may be taken to be

$$(p, q, r, s \between \theta, 1)^3 = 0, \quad (u, v \between \theta, 1) = 0,$$

¹ I take the opportunity of referring to his paper on Quintic Scrolls, Schwarz, "Ueber die geradlinigen Flächen fünften Grades," *Crelle,* t. LXVII. (1867), pp. 23—57.

where (p, q, r, s, u, v) are any linear functions whatever of the coordinates (x, y, z, w). Hence eliminating θ we have as the equation of the scroll generated by the line in question

$$(p, q, r, s \between v, -u)^3 = 0,$$

viz. this is a quartic scroll having the line $u=0$, $v=0$ for a triple line; that is, the line in question is a triple directrix line.

55. Taking $x=0$, $y=0$ for the equations of the directrix line, or writing $u=x$, $v=y$, and moreover expressing (p, q, r, s) as linear functions of the coordinates (x, y, z, w), the equation of the scroll takes the form

$$(* \between x, y)^4 + z(* \between x, y)^3 + w(*' \between x, y)^3 = 0;$$

and we may, by changing the values of z and w, make the term in $(x, y)^4$ to be

$$(* \between x, y)^4 + (\alpha x + \beta y)(* \between x, y)^3 + (\gamma x + \delta y)(*' \between x, y)^3,$$

where the arbitrary constants α, β, γ, δ may be so determined as to reduce this to a monomial kx^4, kx^3y, or kx^2y^2.

56. The coefficient k may vanish, and the equation of the scroll then is

$$z(* \between x, y)^3 + w(*' \between x, y)^3 = 0,$$

or, what is the same thing, it is

$$(* \between x, y)^3 (z, w) = 0,$$

viz. the scroll has in this particular case the simple directrix line $z=0$, $w=0$, thus reducing itself to the *third species*, $S(1_3, 1, 4)$, *with a triple directrix line and a single directrix line.* It is proper to exclude this, and consider the ninth species, $S(1_3)$, as having a triple directrix line, but no simple directrix line.

57. The scroll $S(1_3)$ may be considered as a scroll $S(m, n, p)$ generated by a line which meets each of three given directrices; viz. these may be taken to be the directrix line, and any two plane sections of the scroll. The section by any plane is a quartic curve having a triple point at the intersection with the directrix line; moreover the sections by any two planes meet in four points, the intersections of the scroll by the line of intersection of the two planes. Conversely, taking any line and two quartics related as above (that is, each quartic has a triple point at its intersection with the line, and the two quartics meet in four points lying in a line), the lines which meet the three curves generate a quartic scroll $S(1_3)$. This appears from the formula

$$S(m, n, p) = 2mnp - \alpha m - \beta n - \gamma p \text{ (Second Memoir, No. 5)};$$

we have in the present case

$$m=1,\ n=4,\ p=4,\quad \alpha=4,\ \beta=3,\ \gamma=3,$$

and the order of the scroll is $32-4-12-12$, $=4$, that is, the scroll is a quartic scroll; there is no difficulty in seeing that through each point of the line there pass three generating lines, but through each point of either of the plane quartics only a single generating line; that is, that the line is a triple directrix line, but each of the plane quartics a simple directrix curve.

58. We may instead of the section by any plane, consider the section by a plane through a generating line, or by a plane through two of the three generating lines which meet at any point of the directrix line; if (to consider only the most simple case) each of the planes be thus a plane through two generating lines, the section by either of these planes is made up of the two generating lines, and of a conic passing through the directrix line; the directrices are thus the line and two conics each of them meeting the line; we have therefore in the foregoing formula

$$m=1,\ n=2,\ p=2,\quad \alpha=0,\ \beta=1,\ \gamma=1,$$

and the order of the scroll is $8-2-2, =4$ as before.

Quartic Scroll, Tenth Species, (3^2), *with a directrix skew cubic met twice by each generating line*[1].

59. Consider a line, the intersection of two planes; and let the equation of each plane contain in the order 2 a variable parameter θ; the equations of the two planes may be taken to be

$$(p,\ q,\ r\between\theta,\ 1)^2=0,\quad (p',\ q',\ r'\between\theta,\ 1)^2=0,$$

where $(p,\ q,\ r,\ p',\ q',\ r')$ are linear functions of the coordinates $(x,\ y,\ z,\ w)$; hence eliminating θ, we have as the equation of the scroll generated by the line in question, $\square=0$, where $\square$ is the resultant of the two quadric functions. The equation may be written

$$4\,(pq'-p'q)\,(rq'-r'q)-(pr'-p'r)^2=0\,;$$

and the scroll has thus as a nodal (double) line the skew cubic determined by the equations

$$\left\|\begin{matrix} p\,, & q\,, & r \\ p', & q', & r' \end{matrix}\right\|=0.$$

It is easy to see (and indeed it will be shown presently) that this curve is met twice by each generating line of the scroll, and that the scroll is consequently a quartic scroll as described above.

[1] I have worded this heading in accordance with that of the eighth species, Second Memoir, No. 47, but the two headings might be expressed more completely thus:

Eighth Species, $S(1,\ 3_2{}^2)$, *with a directrix line and a double directrix skew cubic met twice by each generating line;*

Tenth Species, $S\,(3_2{}^2)$, *with a double directrix skew cubic met twice by each generating line;*

viz. the subscript 2 would indicate that the skew cubic is a nodal (double) line on the scroll, the exponent 2 indicating that it is met twice by each generating line.

60. The coordinates (x, y, z, w) may be fixed in such manner that the equations of the skew cubic shall be

$$\left\|\begin{matrix} x, & y, & z \\ y, & z, & w \end{matrix}\right\| = 0,$$

or, what is the same thing,

$$yw - z^2 = 0, \quad zy - xw = 0, \quad xz - y^2 = 0\,;$$

each of the equations $pq' - p'q = 0$, $rq' - r'q = 0$, $pr' - p'r = 0$ is then the equation of a quadric surface passing through the skew cubic, or, what is the same thing, each of the functions $pq' - p'q$, $rq' - r'q$, $pr' - p'r$ is a linear function of $yw - z^2$, $zy - xw$, $xz - y^2$; and the equation of the scroll is given as a quadric equation in the last-mentioned quantities. It will be convenient to represent the equation in the form

$$(H,\ F,\ C,\ B,\ A - F,\ -G \between yw - z^2,\ zy - xw,\ xz - y^2)^2 = 0,$$

or, writing for shortness

$$yw - z^2,\ zy - xw,\ xz - y^2 = p,\ q,\ r,$$

which letters (p, q, r) are used henceforward in this signification only, the equation will be

$$(H,\ F,\ C,\ B,\ A - F,\ -G \between p,\ q,\ r)^2 = 0,$$

viz. this is a quadric equation in (p, q, r), with arbitrary coefficients.

61. Comparing with the result, Second Memoir, Nos. 47 to 50, we see that in the particular case where the coefficients (A, B, C, F, G, H) satisfy the relation $AF + BG + CH = 0$, we have the *eighth species*, $S(1, 3^2)$, *with a directrix line and a directrix skew cubic met twice by each generating line.* We exclude this particular case, and in the tenth species consider the relation $AF + BG + CH = 0$ as not satisfied, and therefore the scroll as not having a directrix line.

62. I consider how the scroll may be obtained as a scroll $S(m^2, n)$ generated by a line meeting a curve of the order m twice and a curve of the order n once. The first curve will be the skew cubic, that is $m = 3$; the second curve may be any plane section of the scroll; such a section will be a quartic curve having three nodes, one at each intersection of its plane with the skew cubic. Conversely, if we have a skew cubic, and a plane quartic meeting the skew cubic in three points, each of them a node on the quartic, then the scroll generated by the lines which meet the skew cubic twice and the quartic once will be a quartic scroll. In fact (see First Memoir, No. 10, [339], and Second Memoir, No. 5) the order of the scroll is given by the formula $S(m^2, n) = n([m]^2 + M) -$ reduction, $= 16 -$ reduction. And in the present case the reduction arises (Second Memoir, No. 4) from the cones having their vertices at the intersections of the skew cubic and the quartic, and passing through the skew cubic. Each cone is of the order 2, and each intersection *quâ* double point on the quartic gives a reduction $2 \times$ order of cone, $= 4$; that is, the reduction arising from the three intersections is $= 12$; or the order of the scroll is $16 - 12$, $= 4$.

63. We may, instead of the section by a plane in general, consider the section by a plane through a generating line; the section is here made up of the generating line and of a plane cubic passing through each of the two points of intersection of the generating line with the skew cubic, and having a node at the remaining intersection of its plane with the skew cubic. Or we may consider the section by a plane through the two generating lines at any point of the skew cubic; the section is here made up of the two generating lines and of a conic passing through the second intersections of the two generating lines with the skew cubic; that is, meeting the skew cubic twice.

64. Conversely, consider a skew cubic, and a conic meeting it twice; the lines which meet the skew cubic twice, and also the conic, generate a quartic scroll; this appears by the before-mentioned formula $S(m^2,\ n) = n([m]^2 + M)$ − reduction; viz. we have $m = 3$, $n = 2$, and the order is $= 8$ − reduction; the reduction arises from the cones having their vertices at the intersections of the skew cubic and the conic. Each cone is of the order 2, and (*quà* simple point on the conic) each intersection gives a reduction = order of the cone; that is, the total reduction is $= 4$, and the order of the scroll is $8 - 4$, $= 4$ as above.

65. But a more elegant mode of generation of the scroll may be obtained by means of the skew cubic alone; viz. considering the system of lines which are in involution with five given lines, or say simply the lines which belong to an involution[1], I say that the locus of a line belonging to the involution, and meeting the skew cubic twice is the quartic scroll, *tenth species*, $S(3^2)$. In the particular case where the line (instead of belonging to a proper involution) meets a given line, the locus is a quartic scroll, *eighth species*, $S(1,\ 3^2)$.

66. The analysis is almost identical with that given (Second Memoir, Nos. 47 to 50) in regard to the scroll $S(1,\ 3^2)$. Considering a line defined by its "six coordinates" $(a,\ b,\ c,\ f,\ g,\ h)$, the condition which expresses that the line shall belong to an involution is

$$(A,\ B,\ C,\ F,\ G,\ H \between a,\ b,\ c,\ f,\ g,\ h) = 0,$$

where $(A,\ B,\ C,\ F,\ G,\ H)$ are arbitrary coefficients; if they are the coordinates of a line, that is, if $AF + BG + CH = 0$, then the condition expresses that the line $(a,\ b,\ c,\ f,\ g,\ h)$, instead of belonging to a proper involution, meets the line $(F,\ G,\ H,\ A,\ B,\ C)$.

[1] The theory is explained in my memoir "On the Six Coordinates of a Line," *Camb. Phil. Trans.* vol. XI. 1868, [348]. In explanation of the subsequent analytical investigations of the present memoir, it is convenient to remark that if on a given line we have the two points $(\alpha,\ \beta,\ \gamma,\ \delta)$ and $(\alpha',\ \beta',\ \gamma',\ \delta')$, and through the given line two planes $Ax + By + Cz + Dw = 0$ and $A'x + B'y + C'z + D'w = 0$; then we have

$$\begin{array}{l} \beta\gamma' - \beta'\gamma : \gamma\alpha' - \gamma'\alpha : \alpha\beta' - \alpha'\beta : \alpha\delta' - \alpha'\delta : \beta\delta' - \beta'\delta : \gamma\delta' - \gamma'\delta \\ = AD' - A'D : BD' - B'D : CD' - C'D : BC' - B'C : CA' - C'A : AB' - A'B; \end{array}$$

and denoting either of these sets of equal ratios by

$$a : b : c : f : g : h,$$

then $(a,\ b,\ c,\ f,\ g,\ h)$ satisfy identically the relation $af + bg + ch = 0$, and are said to be the six coordinates of the line.

We have to determine the locus of the line (a, b, c, f, g, h) the coordinates whereof satisfy the relation

$$(A, B, C, F, G, H \between a, b, c, f, g, h) = 0,$$

and which besides meets the skew cubic $yw - z^2 = 0$, $yz - xw = 0$, $xz - y^2 = 0$.

The equations of the skew cubic are satisfied by writing therein

$$x : y : z : w = 1 : t : t^2 : t^3;$$

and hence taking θ, ϕ for the parameters of the points of intersection of the line (a, b, c, f, g, h) with the skew cubic, we have

$$\begin{array}{llll} 1, & \theta, & \theta^2, & \theta^3; \\ 1, & \phi, & \phi^2, & \phi^3, \end{array}$$

as the coordinates of two points on the line in question; whence forming the expressions of the six coordinates of the line, and omitting the common factor $\phi - \theta$, these are

$$(a, b, c, f, g, h) = \theta\phi,\ -(\theta+\phi),\ 1,\ \theta^2+\theta\phi+\phi^2,\ \theta\phi(\theta+\phi),\ \theta^2\phi^2,$$

and hence the condition of involution gives between the parameters θ, ϕ the equation

$$(A, B, C, F, G, H \between \theta\phi,\ -\theta-\phi,\ 1,\ \theta^2+\theta\phi+\phi^2,\ \theta\phi(\theta+\phi),\ \theta^2\phi^2).$$

Moreover the coordinates of any point on the line in question are given by

$$x : y : z : w = l+m : l\theta+m\phi : l\theta^2+m\phi^2 : l\theta^3+m\phi^3;$$

and writing as above $p, q, r = yw - z^2,\ yz - xw,\ xz - y^2$, we thence find, omitting the common factor $(\theta-\phi)^2$,

$$p : q : r = \theta\phi : -(\theta+\phi) : 1;$$

and eliminating $\theta\phi$, $\theta+\phi$, we at once obtain

$$(A, B, C, F, G, H \between pr,\ qr,\ r^2,\ q^2-pr,\ -pq,\ p^2) = 0,$$

or, what is the same thing,

$$(H, F, C, B, A-F, -G \between p, q, r)^2 = 0$$

as the equation of the scroll generated by the line in involution which meets the given skew cubic twice.

Reciprocal of the Quartic Scroll $S(3^2)$.

67. I propose to reciprocate in regard to the quadric surface $x^2+y^2+z^2+t^2=0$ the foregoing scroll

$$(H, F, C, B, A-F, -G \between p, q, r)^2 = 0.$$

If the coordinates (a, b, c, f, g, h) of a line satisfy the condition of involution

$$(A, B, C, F, G, H \between a, b, c, f, g, h) = 0,$$

then the coordinates (a, b, c, f, g, h) of the reciprocal curve will satisfy the condition of involution

$$(F, G, H, A, B, C \between a, b, c, f, g, h) = 0.$$

The reciprocal of the before-mentioned skew cubic $x : y : z : w = 1 : t : t^2 : t^3$ is the quartic torse having for its edge of regression the skew cubic $3XZ - Y^2 = 0$, $YZ - 9XW = 0$, $3YW - Z^2 = 0$; or, what is the same thing, the skew cubic $X : Y : Z : W = 1 : 3t : 3t^2 : t^3$; see my paper "On the Reciprocation of a Quartic Developable," *Quart. Math. Journ.* vol. VII. (1866), pp. 87—92, [372].

68. Hence the reciprocal of the quartic scroll is the scroll generated by a line (a, b, c, f, g, h) the coordinates of which satisfy the condition of involution

$$(F, G, H, A, B, C \between a, b, c, f, g, h) = 0,$$

and which is moreover the intersection of two osculating planes of the skew cubic $X : Y : Z : W = 1 : 3t : 3t^2 : t^3$. For the point the parameter whereof is t, the equation of the osculating plane is

$$\begin{vmatrix} X, & Y, & Z, & W \\ 1, & 3t, & 3t^2, & t^3 \\ & 1, & 2t, & t^2 \\ & & 1, & t \end{vmatrix} = 0,$$

or, what is the same thing, the equation is

$$(t^3, -t^2, t, -1 \between X, Y, Z, W) = 0.$$

Hence for the line which is the intersection of the two osculating planes

$$(\theta^3, -\theta^2, \theta, -1 \between X, Y, Z, W) = 0,$$

$$(\phi^3, -\phi^2, \phi, -1 \between X, Y, Z, W) = 0,$$

forming the expressions of the six coordinates, but omitting the common factor $\phi - \theta$, these are

$$a, b, c, f, g, h = \theta^2 + \theta\phi + \phi^2, -\theta - \phi, 1, \theta\phi, \theta\phi(\theta + \phi), \theta^2\phi^2;$$

we have thus between the parameters θ, ϕ the relation

$$(F, G, H, A, B, C \between \theta^2 + \theta\phi + \phi^2, -\theta - \phi, 1, \theta\phi, \theta\phi(\theta + \phi), \theta^2\phi^2) = 0;$$

and the equation of the scroll is obtained by eliminating θ, ϕ between this equation and the last-mentioned two equations satisfied by θ, ϕ respectively.

69. We see that θ, ϕ are two of the roots of the equation

$$(X, -Y, Z, -W \between u, 1)^3 = 0;$$

let ρ be the third root, then we have

$$\theta+\phi+\rho \qquad =\frac{Y}{X},$$

$$\theta\phi+\rho(\theta+\phi)=\frac{Z}{X},$$

$$\theta\phi\,.\,\rho \qquad =\frac{W}{X},$$

and thence

$$\theta+\phi=\frac{1}{X}(Y-\rho X),\quad \theta\phi=\frac{1}{X}(Z-\rho Y+\rho^2 X)=\frac{1}{\rho X}W,$$

$$(X,\ -Y,\ Z,\ -W\between\rho,\ 1)^3=0.$$

Substituting for $\theta+\phi$ and $\theta\phi$ their values in terms of ρ, we find

$$\begin{aligned}F\rho\{Y^2-ZX-\rho XY\}-G\rho X(Y-\rho X)+H\rho X^2&\\ +W\{(AX+BY+CZ)-\rho(BX+CY)+\rho^2 CX\}&=0,\end{aligned}$$

or, what is the same thing,

$$\begin{aligned}&\rho^2 X(GX-FY+CW)\\ &-\rho\{F(Y^2-ZX)-GXY+HX^2-BXW-CYW\}\\ &+W(AX+BY+CZ)=0;\end{aligned}$$

from which and the equation

$$(X,\ -Y,\ Z,\ -W\between\rho,\ 1)^3=0,$$

we have to eliminate ρ.

70. Writing for shortness

$$\begin{aligned}(\ \ .\,,\quad H,\ -G,\ A\between X,\ Y,\ Z,\ W)&=\alpha,\\ (-H,\quad .\,,\quad F,\ B\between \qquad ,, \qquad)&=\beta,\\ (\ G,\ -F,\quad .\,,\ C\between \qquad ,, \qquad)&=\gamma,\\ (-A,\ -B,\ -C,\ .\ \between \qquad ,, \qquad)&=\delta,\end{aligned}$$

and therefore $\alpha X+\beta Y+\gamma Z+\delta W=0$: the two equations are

$$\begin{aligned}\rho^2 X\gamma+\rho(-\gamma Y-\beta X)-\delta W&=0,\\ \rho^3 X-\rho^2 Y+\rho Z \qquad\qquad -\ W&=0.\end{aligned}$$

Writing the first equation in the form

$$\gamma(\rho^2 X-\rho Y+Z)-\beta(\rho X-Y)+\alpha X=0,$$

multiplying by $-\rho$, and reducing by the other equation,

$$\beta(\rho^2 X-\rho Y)-\rho\alpha X-\gamma W=0,$$

or, as this may be written,

$$\beta(\rho^2 X-\rho Y+Z)-\alpha(\rho X-Y)-\alpha Y-\beta Z-\gamma W=0.$$

From this and the preceding equation we deduce the values of $\rho^2X-\rho Y+Z$ and $\rho X-Y$; viz. writing for shortness

$$\beta\delta-\gamma^2,\ \beta\gamma-\alpha\delta,\ \alpha\gamma-\beta^2=\mathrm{p},\ \mathrm{q},\ \mathrm{r},$$

we find

$$\rho^2X-\rho Y+Z : \rho X-Y : 1=-\mathrm{r}Z+\mathrm{q}W : \mathrm{r}Y-\mathrm{p}W : -\mathrm{r},$$

or, what is the same thing,

$$\rho^2X-\rho Y+Z \quad = \quad Z-\frac{\mathrm{q}}{\mathrm{r}}W,$$

$$\rho X \ - \ Y \quad =-Y+\frac{\mathrm{p}}{\mathrm{r}}W;$$

whence also

$$\rho^3X-\rho^2Y+\rho Z\ -W=\ 0,$$

$$\rho^2X-\rho Y \quad =-\frac{\mathrm{q}}{\mathrm{r}}W,$$

$$\rho X \quad = \ \frac{\mathrm{p}}{\mathrm{r}}W,$$

and thence

$$\rho\left(Z-\frac{\mathrm{q}}{\mathrm{r}}W\right) \quad =W,$$

$$\rho\left(Y-\frac{\mathrm{p}}{\mathrm{r}}W\right) \quad =\frac{\mathrm{q}}{\mathrm{r}}W,$$

$$\rho\ X \quad =\frac{\mathrm{p}}{\mathrm{r}}W,$$

and we have therefore

$$\left(\frac{W}{\rho}=\right)\quad Z-\frac{\mathrm{q}}{\mathrm{r}}W=\frac{\mathrm{r}}{\mathrm{q}}\left(Y-\frac{\mathrm{p}}{\mathrm{r}}W\right)=\frac{\mathrm{r}}{\mathrm{p}}X,$$

or omitting the first equation, we have (independent of ρ) a system which it is clear must be equivalent to a single equation.

71. I take any one of these equations, for instance the equation

$$Z-\frac{\mathrm{q}}{\mathrm{r}}W=\frac{\mathrm{r}}{\mathrm{q}}\left(Y-\frac{\mathrm{p}}{\mathrm{r}}W\right),$$

or, what is the same thing,

$$\mathrm{qr}Z-\mathrm{r}^2Y+(\mathrm{pr}-\mathrm{q}^2)\,W=0,$$

and I proceed to reduce it so as to obtain the result in a symmetrical form. For this purpose I observe that from the values of α, β, γ, δ, if only $AF+BG+CH$ not $=0$, we have

$$\begin{aligned} X : Y : Z : W = &\ (\ \ .\ ,\ \ -C,\quad B,\quad -F \between \alpha,\ \beta,\ \gamma,\ \delta)\\ &:(\ C,\quad .\ ,\quad -A,\quad -G \between \quad ,, \quad)\\ &:(-B,\quad A,\quad .\ ,\quad -H \between \quad ,, \quad)\\ &:(\ F,\quad G,\quad H,\quad .\ \between \quad ,, \quad); \end{aligned}$$

and substituting these values, the equation in question becomes

$$\begin{aligned} &\mathrm{qr}\ \ (-B\alpha + A\beta - H\delta) \\ &-\mathrm{r}^2\ (\ \ C\alpha - A\gamma - G\delta) \\ &+(\mathrm{pr}-\mathrm{q}^2)(\ \ F\alpha + G\beta + H\gamma) = 0. \end{aligned}$$

This becomes

$$\begin{array}{lcll} A\mathrm{r}(\mathrm{q}\beta + \mathrm{r}\gamma) & = & A\mathrm{r}(-\mathrm{p}\alpha) & = 0 \\ -B\mathrm{qr}\alpha & & -B\mathrm{qr}\alpha & \\ -C\mathrm{r}^2\alpha & & -C\mathrm{r}^2\alpha & \\ +F(\mathrm{pr}-\mathrm{q}^2)\alpha & & +F(\mathrm{pr}-\mathrm{q}^2)\alpha & \\ +G\{\mathrm{r}^2\delta + (\mathrm{pr}-\mathrm{q}^2)\beta\} & & +G\mathrm{pq}\alpha & \\ +H\{-\mathrm{qr}\delta + (\mathrm{pr}-\mathrm{q}^2)\gamma\} & & -H\mathrm{p}^2\alpha, & \end{array}$$

viz. the whole equation divides by α; and, omitting this factor, the equation is

$$A\mathrm{pr} + B\mathrm{qr} + C\mathrm{r}^2 + F(\mathrm{q}^2 - \mathrm{pr}) - G\mathrm{pq} + H\mathrm{p}^2 = 0,$$

or, what is the same thing, it is

$$(H,\ F,\ C,\ B,\ A-F,\ -G \between \mathrm{p},\ \mathrm{q},\ \mathrm{r})^2 = 0,$$

where I recall that we have

$$\mathrm{p},\ \mathrm{q},\ \mathrm{r} = \beta\delta - \gamma^2,\ \beta\gamma - \alpha\delta,\ \alpha\gamma - \beta^2,$$

α, β, γ, δ being linear functions of the current coordinates (X, Y, Z, W), viz. we have

$$\begin{aligned} \alpha &= (\ \ .\,,\quad H,\quad -G,\quad A \between X,\ Y,\ Z,\ W), \\ \beta &= (-H,\quad .\,,\quad F,\quad B \between \quad ,, \quad), \\ \gamma &= (\ \ G,\quad -F,\quad .\,,\quad C \between \quad ,, \quad), \\ \delta &= (-A,\quad -B,\quad -C,\quad .\ \between \quad ,, \quad). \end{aligned}$$

72. It thus appears that when $AF + BG + CH$ is not $= 0$, the reciprocal of the scroll

$$(H,\ F,\ C,\ B,\ A-F,\ -G \between p,\ q,\ r)^2 = 0$$

has an equation of the very same form,

$$(H,\ F,\ C,\ B,\ A-F,\ -G \between \mathrm{p},\ \mathrm{q},\ \mathrm{r})^2 = 0; \qquad \text{(Rec. I.)}$$

so that in fact the scroll, *tenth species*, $S(3^2)$, defined as the scroll generated by a line in involution which passes through two points of a skew cubic, may be reciprocally defined as the scroll generated by a line in involution which lies in two osculating planes of a skew cubic.

73. If for $(\alpha, \beta, \gamma, \delta)$ we substitute their values in terms of (X, Y, Z, W), the foregoing equation of the reciprocal scroll is obtained as an equation of the fourth order in the coordinates (X, Y, Z, W), and (in the first instance) of the fifth degree in the coefficients (A, B, C, F, G, H). It is a remarkable circumstance that the whole

equation contains the constant factor $AF + BG + CH$, so that throwing this out, the reduced equation will be only of the third degree in the coefficients.

74. The transformation is a very troublesome one, but I will indicate the steps by which I succeeded in accomplishing it. Each of the functions (p, q, r) is a quadric function of (X, Y, Z, W), say,

$$\begin{aligned} \mathrm{p} &= (a, b, c, d, f, g, h, l, m, n \between X, Y, Z, W)^2, \\ \mathrm{q} &= (a', \,.\,. \qquad \between \quad ,, \quad)^2, \\ \mathrm{r} &= (a'', \,.\,. \qquad \between \quad ,, \quad)^2; \end{aligned}$$

we have to form the value of

$$(H, F, C, B, A - F, -G \between \mathrm{p}, \mathrm{q}, \mathrm{r})^2,$$

viz. representing this for shortness by

$$(H, F, C, B, A - F, -G \between \begin{pmatrix} a, b, c, d, f, g, h, l, m, n \\ a', .. \\ a'', .. \end{pmatrix}^2,$$

the coefficient of X^4 is

$$(H, F, C, B, A - F, -G \between a^2, a'^2, a''^2, a'a'', a''a, aa''),$$

that of X^3Y is

$$(H, F, C, B, A - F, -G \between 2af, 2a'f', 2a''f'', a'f'' + a''f', a''f + af'', af' + a'f),$$

and so on, the successive terms a^2, a'^2, &c., $2af$, $2a'f'$, &c. being derived by an obvious law from the first terms a^2, $2af$, &c.; and these first terms are merely the coefficients of the terms X^4, X^3, Y, &c. in the development of

$$p^2, = \{(a, b, c, d, f, g, h, l, m, n \between X, Y, Z, W)^2\}^2;$$

viz. this is

X^4	X^3Y	X^3Z	X^3W	X^2Y^2	X^2YZ	X^2YW	X^2Z^2,	X^2ZW,	X^2W^2,	XY^3,	XY^2Z,	XY^2W,	XYZ^2
a^2	$2af$	$2ag$	$2al$	$2ab + h^2$	$2af + 2gh$	$2am + 2hl$	$2ac + g^2$	$2an + 2gl$	$2ad + l^2$	$2bh$	$2bg + 2fh$	$2bl + 2hm$	$2ch + 2fg$

$XYZW$,	XYW^2,	XZ^3,	XZ^2W,	XZW^2,	XW^3,	Y^4,	Y^3Z,	Y^3W,	Y^2Z^2,	Y^2ZW,	Y^2W^2,	YZ^3,
$2lf + 2mg + 2nh$	$2dh + 2lm$	$2cg$	$2cl + 2gn$	$2dg + 2ln$	$2dl$,	b^2,	$2bf$,	$2bm$,	$2bc, + f^2$	$2bn, + 2fm$	$2bd, + m^2$	$2cf$

YZ^2W,	YZW^2,	YW^3,	Z^4,	Z^3W,	Z^2W^2,	ZW^3,	W^4
$2cm + 2fn$	$+ 2df + 2mn$	$2dm$,	c^2,	$2cn$,	$2cd, + n^2$	$2dn$,	d^2

;

and the values of the coefficients $a, b, \ldots$ which enter into the formulæ are given by means of the following values of p, q, r; viz. these are

$$\begin{array}{lcccccccccc}
 & X^2 & Y^2 & Z^2 & W^2 & YZ & ZX & XY & XW & YW & ZW \\
\mathrm{p}=\Big(& AH, & -F^2, & -CF, & -C^2, & -BF, & -AF, & BH, & -AB, & -B^2, & -BC \between X, Y, Z, W)^2, \\
 & -G^2 & & & & & +CH & +2FG & -2CG & +2CF & \\
\mathrm{q}=(& -GH, & BH, & -CG, & BC, & -BG, & -AG, & AH, & A^2, & AB, & AC \between \quad " \quad)^2, \\
 & & & & & +CH & +FG & +FH & +BG & -BF & +CF \\
 & & & & & -F^2 & & & -CH & & \\
\mathrm{r}=(& -H^2, & -FH, & -F^2, & AC, & FG, & 2FH, & GH, & 2AG, & -AF, & -2BF \between \quad " \quad)^2. \\
 & & & & -B^2 & & -G^2 & & +BH & +CH & -CG
\end{array}$$

75. As an instance of the calculation of a single term, the coefficient of X^4 is

$$(H, F, C, B, A-F, -G \between AH-G^2, -GH, -H^2)^2;$$

viz. this is

$$\begin{array}{lcl}
H(AH-G^2)^2 & = & A^2H^3 - 2AG^2H^2 + G^4H^2 \\
+FH^2G^2 & = & FG^2H^2 \\
+CH^4 & = & CH^4 \\
+BGH^3 & = & BGH^3 \\
+(A-F)(-AH^3+G^2H^2) & = & -A^2H^3 + AG^2H^2 \\
 & & +AFH^3 - FG^2H^2 \\
-G(-AGH^2+G^3H) & = & AG^2H^2 - G^4H^2;
\end{array}$$

the whole term is thus $=(AF+BG+CH)H^3$, viz. there is the factor $AF+BG+CH$ as mentioned above.

76. Throwing out the factor in question, $AF+BG+CH$, the equation of the reciprocal scroll is found to be

$$\begin{array}{rll}
0 = & X^4 & .\ H^3 \qquad\qquad \text{(Rec. II.)} \\
 & +X^3Y & .\ -2GH^2 \\
 & +X^3Z & .\ AH^2 - 3FH^2 + G^2H \\
 & +X^3W & .\ -3AGH - 3BH^2 + G^3 \\
 & +X^2Y^2 & .\ 2FH^2 + G^2H \\
 & +X^2YZ & .\ -AGH + BH^2 + 2FGH - G^3 \\
 & +X^2YW & .\ A^2H + 3AFH + AG^2 + BGH - 2CH^2 - 3FG^2 \\
 & +X^2Z^2 & .\ -2AFH + AG^2 + CH^2 - FG^2 + 3F^2H \\
 & +X^2ZW & .\ -2A^2G - 2ABH + 3AFG + 6BFH - BG^2 - CGH \\
 & +X^2W^2 & .\ A^3 + 3ABG - 3ACH + 3B^2H + 3CG^2
\end{array}$$

$$\begin{array}{ll}
+XY^3 & -2FGH \\
+XY^2Z & AFH-BGH-3F^2H+2FG^2 \\
+XY^2W & 2ABH-2AFG-BFH+2CGH+3F^2G \\
+XYZ^2 & -AFG-2BFH+BG^2-CGH \\
+XYZW & A^2F-3AF^2-3ABG+ACH-2B^2H+BFG+5CFH-2CG^2 \\
+XYW^2 & 2A^2B-3ABF+2ACG+B^2G+BCH-6CFG \\
+XZ^3 & AF^2-2CFH+CG^2-F^3 \\
+XZ^2W & 2ABF-2ACG-2BCH-3BF^2+CFG \\
+XZW^2 & A^2C+AB^2+3ACF-3B^2F+BCG-2C^2H \\
+XW^3 & 3ABC-B^3+3C^2G \\
+Y^4 & F^2H \\
+Y^3Z & BFH-F^2G \\
+Y^3W & AF^2+B^2H-2CFH-F^3 \\
+Y^2Z^2 & -BFG+CFH+F^3 \\
+Y^2ZW & ABF-B^2G+BCH+2CFG \\
+Y^2W^2 & AB^2-2ACF-B^2F+C^2H+3CF^2 \\
+YZ^3 & BF^2-CFG \\
+YZ^2W & ACF+2B^2F-BCG-3CF^2 \\
+YZW^2 & ABC+B^3-2BCF-C^2G \\
+YW^3 & AC^2+B^2C-3C^2F \\
+Z^4 & CF^2 \\
+Z^3W & 2BCF \\
+Z^2W^2 & B^2C+2C^2F \\
+ZW^3 & 2BC^2 \\
+W^4 & C^3,
\end{array}$$

where, in regard to the symmetry of this equation, it is to be observed that we may interchange X and W, and Y and Z, leaving A, F unaltered but interchanging B and $-G$, and also C and H; thus the coefficient of X^3Z being $AH^2-3FH^2+G^2H$, that of YW^3 is $AC^2-3FC^2+B^2C$, $=AC^2+B^2C-3C^2F$. Or, again, the coefficient of Y^3Z being $BFH-F^2G$, that of YZ^3 is $-GFC+F^2B$, $=BF^2-CFG$.

77. But the equation may be written in the much more simple form

$$\begin{aligned}
&X(-\alpha^2\delta+3\alpha\beta\gamma-2\beta^3) \qquad \text{(Rec. III.)}\\
+&Y(-\alpha\beta\delta+2\alpha\gamma^2-\beta^2\gamma)\\
+&Z(\alpha\gamma\delta-2\beta^2\delta+\beta\gamma^2)\\
+&W(\alpha\delta^2-3\beta\gamma\delta+2\gamma^3)=0,
\end{aligned}$$

or, what is the same thing,

$$-\tfrac{1}{6}(3X\partial_\delta - Y\partial_\gamma + Z\partial_\beta - 3W\partial_\alpha)(\alpha^2\delta^2 - 6\alpha\beta\gamma\delta + 4\alpha\gamma^3 + 4\beta^3\delta - 3\beta^2\gamma^2) = 0, \qquad \text{(Rec. III.)}$$

as may be verified by actual substitution of the values of the coordinates.

78. By what precedes, substituting for p, q, r their values in terms of $\alpha, \beta, \gamma, \delta$, it appears that we have the remarkable identity

$$\begin{aligned}(H,\ F,\ C,\ B,\ A-F,\ -G\between \beta\delta-\gamma^2,\ \beta\gamma-\alpha\delta,\ \alpha\gamma-\beta^2)^2 \\ =(AF+BG+CH)\times\left\{\begin{array}{l} \ \ X(-\alpha^2\delta+3\alpha\beta\gamma-2\beta^3) \\ +Y(-\alpha\beta\delta+2\alpha\gamma^2-\beta^2\gamma) \\ +Z(\alpha\gamma\delta-2\beta^2\delta+\beta\gamma^2) \\ +W(\alpha\delta^2-3\beta\gamma\delta+2\gamma^3) \end{array}\right\}.\end{aligned}$$

79. In the case above considered of the tenth species, $S(3^2)$, for which $AF+BG+CH$ not $=0$, the three forms of the reciprocal equation are of course absolutely equivalent to each other. The first form has the advantage of putting in evidence the fact that the reciprocal scroll is also of the tenth species; the other two forms do not, at least obviously, put in evidence any special property of the reciprocal scroll.

Reciprocals of Eighth Species, $S(1, 3^2)$, and Ninth Species, $S(1^3)$.

80. If $AF+BG+CH=0$, then the equation

$$(H,\ F,\ C,\ B,\ A-F,\ -G\between p,\ q,\ r)^2=0$$

is a scroll of the eighth species, $S(1, 3^2)$. The first form of the reciprocal equation becomes identically $0=0$, on account of the evanescent factor $AF+BG+CH$, but the second and third forms continue to subsist, and either of them may be taken as the equation of the reciprocal scroll. Taking the third form, and calling to mind the significations of $(\alpha, \beta, \gamma, \delta)$, viz.

$$\begin{aligned}\alpha &= (\ .\ ,\ \ H,\ -G,\ A\between X,\ Y,\ Z,\ W),\\ \beta &= (-H,\ \ .\ ,\ \ F,\ B\between \quad ''\quad),\\ \gamma &= (\ G,\ -F,\ \ .\ ,\ C\between \quad ''\quad),\\ \delta &= (-A,\ -B,\ -C,\ .\between \quad ''\quad),\end{aligned}$$

it is to be observed that $\alpha=0$, $\beta=0$, $\gamma=0$, $\delta=0$ are the equations of four planes passing through a common line, viz. the line whose coordinates are (A, B, C, F, G, H), and the equation thus puts in evidence that this line is a triple line on the reciprocal scroll; that is, the reciprocal scroll is a scroll of the ninth species, $S(1^3)$. Or stating the theorem more completely: For the scroll, *eighth species*, $S(1, 3^2)$,

$$(H, F, C, B, A-F, -G \between \mathrm{p}, \mathrm{q}, \mathrm{r})^2 = 0,$$

generated by a line meeting the line (F, G, H, A, B, C), and the skew cubic $\mathrm{p}=0$ $\mathrm{q}=0$, $\mathrm{r}=0$ twice, the reciprocal scroll is of the *ninth species*, $S(1^3)$,

$$\begin{aligned} &X\,(-\alpha^2\delta + 3\alpha\beta\gamma - 2\beta^3\,) \\ +\,&Y\,(-\alpha\beta\delta + 2\alpha\gamma^2 - \beta^2\gamma) \\ +\,&Z\,(\alpha\gamma\delta - 2\beta^2\delta + \beta\gamma^2) \\ +\,&W(\alpha\delta^2 - 3\beta\gamma\delta + 2\gamma^3\,), \end{aligned}$$

having for its triple line the reciprocal line (A, B, C, F, G, H).

81. It should of course be possible, starting from the equation

$$(*\between X, Y)^4 + Z(*\between X, Y)^3 + W(*'\between X, Y)^3 = 0$$

of a scroll $S(1^3)$, to obtain the equation of the reciprocal scroll $S(1, 3^2)$. But I content myself with a very particular case. I consider the equation

$$Y^2Z^2 - Y^3W - Z^3X = 0,$$

which belongs to a scroll $S(1^3)$ having the line $Y=0$, $Z=0$ for its triple line. To find the equation of the reciprocal scroll, write

$$\begin{aligned} -\ Z^3 \quad &+ \lambda x = 0, \\ 2YZ^2 - 3Y^2W &+ \lambda y = 0, \\ 2Y^2Z - 3Z^2X &+ \lambda z = 0, \\ -\ Y^3 \quad &+ \lambda w = 0, \end{aligned}$$

we find without difficulty, reducing by means of the equation of the scroll,

$$\begin{aligned} \lambda^2(yw - z^2) &= -3Z^2\{Y^4 + 3XZ(XZ - Y^2)\}, \\ \lambda^2(xw - yz) &= 3Y^2Z^2\{YZ - 3WX\}, \\ \lambda^2(xz - y^2) &= -3Y^2\{Z^4 + 3YW(YW - Z^2)\}. \end{aligned}$$

Hence writing for a moment

$$\Omega = \{Y^4 + 3XZ(XZ - Y^2)\}\{Z^4 + 3YW(YW - Z^2)\} - Y^2Z^2(YZ - 3WX)^2,$$

we have

$$\begin{aligned} \Omega = \ &Y^4Z^4 + 3Y^5W(YW - Z^2) + 3Z^5X(XZ - Y^2) + 9XYZW(Y^2Z^2 - Y^3W - Z^3X + XYZW) \\ &- Y^4Z^4 + 6Y^3Z^3XW - 9Y^2Z^2X^2W^2, \end{aligned}$$

that is

$$\begin{aligned}\tfrac{1}{3}\Omega = & \quad Y^5W(YW-Z^2)+X^5X(XZ-Y^2)+2Y^3Z^3XW,\\ &= -Y^2Z^2(XZ-Y^2)(YW-Z^2)-Y^2Z^2(YW-Z^2)(XZ-Y^2)+2Y^3Z^3XW,\\ &= -2Y^2Z^2\{(XZ-Y^2)(YW-Z^2)-XYZW\},\\ &= -2Y^2Z^2\{Y^2Z^2-Y^3W-Z^3X\},\\ &= 0, \text{ by the equation of the scroll;}\end{aligned}$$

and we thus see that the equation of the reciprocal scroll is

$$(yw-z^2)(xz-y^2)-(yz-xw)^2=0,$$

or say $\mathrm{q}^2-\mathrm{pr}=0$, viz. it is a scroll $S(1,\ 3^2)$ generated by a line meeting the line $x=0$, $w=0$, and the cubic curve $\mathrm{p}=0$, $\mathrm{q}=0$, $\mathrm{r}=0$ twice. The equation is obviously included in the general equation

$$(H,\ F,\ C,\ B,\ A-F,\ -G\between \mathrm{p},\ \mathrm{q},\ \mathrm{r})^2=0,$$

where $AF+BG+CH=0$; viz. writing $A=B=C=G=H=0$, this becomes $F(\mathrm{q}^2-\mathrm{pr})=0$.

82. Returning to the general case of the scroll, *eighth species*, $S(1,\ 3^2)$, it is proper to show geometrically how it is that the reciprocal is a scroll, *ninth species*, $S(1^3)$. Consider in the scroll $S(1,\ 3^2)$ any plane through the directrix line; this contains three generating lines of the scroll, viz. these are the sides of the triangle formed by the three points of intersection of the plane with the skew cubic: hence in the reciprocal figure we have a directrix line such that at each point of it there are three generating lines; that is, we have a scroll $S(1^3)$ with a triple directrix line. Conversely, starting with the scroll $S(1^3)$, each plane through the triple directrix line meets the scroll in this line three times, and in a single generating line; whence there is in the reciprocal scroll a simple directrix line; but in order to show that it is a scroll $S(1,\ 3^2)$, we have yet to show that there is, as a nodal directrix, a skew cubic met twice by each generating line; this implies that, reciprocally, in the scroll $S(1^3)$ each generating line is the intersection of two osculating planes of a skew cubic (tangent planes of a quartic torse), each such plane containing two generating lines of the scroll—a geometrical property which is far from obvious; and similarly in the scroll, ninth species, $S(3^2)$, where the reciprocal scroll is of the same form, the property that each generating line is a line joining two points of a skew cubic leads to the property that each line is also the intersection of two osculating planes of a skew cubic (or, what is the same thing, two tangent planes of a quartic torse).

ADDITION, *May* 18, 1869.

Since the foregoing Memoir was written I received from Professor Cremona a letter dated Milan, November 20, 1868, in which (besides the ninth and tenth species considered above) he refers to two other species of quartic scrolls. He remarks that there is a bitangent torse which should in the classification be considered along with the nodal curve; and he enumerates in all 12 species as follows:

Deficiency.	No. of species.	Nodal curve.	Bitangent torse.	Corresponding to my species.
$p=0$	1	Γ_3	Σ_3	10
	2	H_2+R_1	K_2+R_1	7
	3	R_1^3	K_2+R_1	— (say, 12)
	4	H_2+R_1	R_1^3	— (say, 11)
	5	$R_1+R_1'+S_1$	$R_1+R_1'+S_1$	2
	6	$R_1^2+S_1$	$R_1^2+S_1$	5
	7	Γ_3	R_1^3	8
	8	R_1^3	Σ_3	9
	9	R_1^3	$R_1'^3$	3
	10	R_1^3	R_1^3	6
$p=1$	11	R_1+R_1'	R_1+R_1'	1
	12	R_1^2	R_1^2	4

where Γ_3 denotes a skew cubic, Σ_3 a torse of the 3rd class (or quartic torse), H_2 a conic, K_2 a quadric cone, R_1, R_1', S_1 different right lines, R_1^2, R_1^3 a line counted twice or three times, &c. I have in the last column added the references to my species 9 and 10; Professor Cremona notices (what I knew, but did not recollect) that the species 10 had been considered by M. Chasles, *Comptes Rendus*, June 3, 1861.

I have not yet examined the two new species mentioned in this enumeration; viz. these are (Cremona 3), say *twelfth species*, a scroll having a triple line, but a bitangent torse made up of a quadric cone and a line; and (Cremona 4), say *eleventh species*, a scroll having a nodal conic and line, but for its bitangent torse a triple line: the two species are, it is clear, reciprocal to each other; although properly treated as distinct, species 11 may be considered as a subform of 8, and species 12 as a subform of 9.

411.

A MEMOIR ON THE THEORY OF RECIPROCAL SURFACES.

[From the *Philosophical Transactions of the Royal Society of London,* vol. CLIX. (for the year 1869), pp. 201—229. Received November 12, 1868,—Read January 14, 1869.]

THE present Memoir contains some extensions of Dr Salmon's theory of Reciprocal Surfaces. I wish to put the formulæ on record, in order to be able to refer to them in a "Memoir on Cubic Surfaces," [412], but without at present attempting to completely develope the theory.

Article Nos. 1 to 5. *Extension of* SALMON'S *Fundamental Equations.*

1. The notation made use of is that of Salmon's *Geometry,* [2nd Ed.] pp. 450—459, [but reproduced in the later editions, see Ed. 4. (1882), pp. 580—592], with the additions presently referred to; the significations of all the symbols are explained by way of recapitulation at the end of the Memoir. I remark that my chief addition to Salmon's theory consists in a modification of his fundamental formulæ (A) and (B); these in their original form are

$$a(n-2) = \kappa + \rho + 2\sigma,$$

$$b(n-2) = \rho + 2\beta + 3\gamma + 3t,$$

$$c(n-2) = 2\sigma + 4\beta + \gamma,$$

$$a(n-2)(n-3) = 2\delta + 3[ac] + 2[ab],$$

$$b(n-2)(n-3) = 4k + [ab] + 3[bc],$$

$$c(n-2)(n-3) = 6h + [ac] + 2[bc],$$

where

$$[ab] = ab - 2\rho,$$

$$[ac] = ac - 3\sigma,$$

$$[bc] = bc - 3\beta - 2\gamma - i.$$

2. I take account of conical and biplanar nodes, or, as I call them, cnicnodes, and binodes; of pinch-points[1] on the nodal curve; and of close-points and off-points on the cuspidal curve: viz. I assume that there are

C, cnicnodes,

B, binodes,

j, pinch-points,

χ, close-points,

θ, off-points,

deferring for the present the explanation of these singularities. The same letters, accented, refer to the reciprocal singularities. Or using "trope" as the reciprocal term to node, these will be

C', cnictropes,

B', bitropes,

j', pinch-planes,

χ', close-planes,

θ', off-planes;

but these present themselves, not in the equations above referred to, but in the reciprocal equations.

3. The resulting alterations are that we must in the formulæ write $\kappa - B$, $\delta - C$ in place of κ, δ respectively; and change the formulæ for $c(n-2)$, $[ab]$, $[bc]$, into

$$\begin{aligned} c(n-2) &= 2\sigma + 4\beta + \gamma + \theta, \\ [ab] &= ab - 2\rho - j, \\ [ac] &= ac - 3\sigma - \chi, \end{aligned}$$

respectively.

4. Making these changes, and substituting for $[ab]$, $[ac]$, $[bc]$ their values, the formulæ become

$$\begin{aligned} a(n-2) &= \kappa - B + \rho + 2\sigma, \\ b(n-2) &= \rho + 2\beta + 3\gamma + 3t, \\ c(n-2) &= 2\sigma + 4\beta + \gamma + \theta, \\ a(n-2)(n-3) &= 2(\delta - C) + 3(ac - 3\sigma - \chi) + 2(ab - 2\rho - j), \\ b(n-2)(n-3) &= 4k + (ab - 2\rho - j) + 3(bc - 3\beta - 2\gamma - i), \\ c(n-2)(n-3) &= 6h + (ac - 3\sigma - \chi) + 2(bc - 3\beta - 2\gamma - i), \end{aligned}$$

which replace the original formulæ (A) and (B).

[1] This addition to the theory is in fact indicated in Salmon, see the note, p. 445; the i there employed, which is of course different from the i of his text, is the j of the present Memoir.

5. For convenience I annex the remaining equations; viz. these are

$$a' = n(n-1) - 2b - 3c,$$
$$\kappa' = 3n(n-2) - 6b - 8c,$$
$$\delta' = \tfrac{1}{2}n(n-2)(n^2-9) - (n^2-n-6)(2b+3c) + 2b(b-1) + 6bc + \tfrac{9}{2}c(c-1);$$

the equations

$$q = b^2 - b - 2k - 3\gamma - 6t,$$
$$r = c^2 - c - 2h - 3\beta,$$

(q, r in place of Salmon's R, S respectively); the equation

$$a = a';$$

and the corresponding equations, interchanging the accented and unaccented letters, in all 23 equations between the 42 quantities

$$n,\ a,\ \delta,\ \kappa;\ \ b,\ k,\ t,\ q,\ \rho,\ j;\ \ c,\ h,\ r,\ \sigma,\ \theta,\ \chi;\ \ \beta,\ \gamma,\ i;\ \ B,\ C,$$
$$n',\ a',\ \delta',\ \kappa';\ \ b',\ k',\ t',\ q',\ \rho',\ j';\ \ c',\ h',\ r',\ \sigma',\ \theta',\ \chi';\ \ \beta',\ \gamma',\ i';\ \ B',\ C'.$$

Article Nos. 6 to 12. *Developments.*

6. We have

$$(a - b - c)(n-2) = (\kappa - B - \theta) - 6\beta - 4\gamma - 3t,$$
$$(a - 2b - 3c)(n-2)(n-3) = 2(\delta - C)$$
$$- 8k - 18h - 6(bc - 3\beta - 2\gamma - i);$$

and substituting these values of δ, κ in the formula

$$n' = a(a-1) - 2\delta - 3\kappa,$$

and for a its value, $= n(n-1) - 2b - 3c$, we find

$$n' = n(n-1)^2 - n(7b + 12c) + 4b^2 + 8b + 9c^2 + 15c$$
$$- 8k - 18h + 18\beta + 12\gamma + 12i - 9t$$
$$- 2C - 3B - 3\theta,$$

where the foregoing equations for $a-b-c$ and $a-2b-3c$ show clearly the origin of the new terms $-2C-3B-3\theta$; these express that there is in the value of n' a reduction $=2$ for each cnicnode, $=3$ for each binode, and $=3$ for each off-point.

7. We have $(n-2)(n-3) = n^2 - n + (-4n+6) = a + 2b + 3c + (-4n+6)$; and making this substitution in the equations which contain $(n-2)(n-3)$, these become

$$a(-4n+6) = 2(\delta - C) - a^2 - 4\rho - 9\sigma - 2j - 3\chi,$$
$$b(-4n+6) = 4k - 2b^2 - 9\beta - 6\gamma - 3i + 2\rho - j,$$
$$c(-4n+6) = 6h - 3c^2 - 6\beta - 4\gamma - 2i - 3\sigma - \chi,$$

(Salmon's equations (C)); and adding to each equation 4 times the corresponding equation with the factor $(n-2)$, these become

$$\begin{aligned} a^2-2a &= 2(\delta-C)+4(\kappa-B)-\sigma-2j-3\chi, \\ 2b^2-2b &= 4k-\beta+6\gamma+12t-3i+2\rho-j, \\ 3c^2-2c &= 6h+10\beta+4\theta-2i+5\sigma-\chi. \end{aligned}$$

Writing in the first of these $a^2-2a=a(a-1)-a, =n'+2\delta+3\kappa-a$, and reducing the other two by means of the values of q, r, the equations become

$$\begin{aligned} n'-a &= -2C-4B+\kappa-\sigma-2j-3\chi, \\ 2q+\beta+3i+j &= 2\rho, \\ 3r+c+2i+\chi &= 5\sigma+\beta+4\theta, \end{aligned}$$

(Salmon's equations (D)).

I attend in particular to the first of these, or rather to the reciprocal equation, which will be

$$\sigma'=a-n+\kappa'-2j'-3\chi'-2C'-4B',$$

which, writing therein $a=n(n-1)-2b-3c$, and $\kappa=3n(n-2)-6b-8c$, becomes

$$\sigma'=4n(n-2)-8b-11c-2j'-3\chi'-2C'-4B'.$$

The singularity σ' is not *explicitly* defined in Salmon; σ' is the reciprocal of σ, and (as such) it denotes the number of common tangent planes of the spinode torse and of the torse generated by the tangent planes along a plane section of the surface; or, what is the same thing, it is the number of the spinode planes which touch the plane section; that is, it is equal to the number of points of intersection of the spinode curve and the plane section; or, finally, σ' is the order of the spinode curve. The spinode curve is in fact for a surface of the order n without singularities the intersection of the surface by the Hessian surface of the order $4(n-2)$, and is thus a curve of the order $4n(n-2)$, which agrees with the formula.

8. But the formula shows that there is in the order a reduction $8b+11c$ arising from the nodal and cuspidal curves of the surface, or, what is the same thing, that the Hessian surface meets the surface in the nodal curve taken 8 times, and in the cuspidal curve taken 11 times—a result which I had arrived at by other means, and also as appears *post*, No. 44. The formula shows further that there is a reduction $2j'+3\chi'+2C'+4B'$, or say there are reductions $=2, 3, 2, 4$, for the reciprocals of a pinch-point, a close-point, a cnicnode, and a binode respectively. Geometrically this must signify that the surface and its Hessian partially intersect in certain curves which are not regarded as belonging to the spinode curve. It will at once suggest itself that for the reciprocal of a cnicnode this curve is a conic, and for the reciprocal of a binode it is a line counting 4 times; while for the reciprocal of a pinch-point it is a line counting 2 times, and for the reciprocal of a close-point, a line counting 3 times.

9. It is clear that ρ' will in like manner denote the order of the node-couple curve.

10. I express in terms of

$$n,\ b,\ c,\ h,\ k,\ \beta,\ \gamma,\ j,\ \theta,\ \chi,\ C,\ B$$

such quantities and combinations of quantities as can be so expressed. We have

$$a = a' = n(n-1) - 2b - 3c,$$

$$\kappa' = 3n(n-2) - 6b - 8c,$$

$$\delta' = \tfrac{1}{2}n(n-2)(n^2-9) - (n^2-n-6)(2b+3c) + 2b(b-1) + 6bc + \tfrac{9}{2}c(c-1),$$

$$4i = 12h + c(5n-6) - 6c^2 - 5\gamma + 3\theta - 2\chi,$$

$$24t = (-8n+8)b + (15n-18)c + 8b^2 - 18c^2 - 2(8k-18h) + 20\beta - 15\gamma + 4j + 9\theta + 6\chi,$$

$$q = b^2 - b - 2k - 3\gamma - 6t, \quad (t \text{ suprà}),$$

$$r = c^2 - c - 2h - 3\beta,$$

$$2\sigma = c(n-2) - (4\beta+\gamma) - \theta,$$

$$8\rho = (16n-24)b + (-15n+18)c - 8b^2 + 18c^2 + 2(8k-18h) - 9(4\beta+\gamma) - 4j - 9\theta - 6\chi,$$

$$8\kappa = 8n(n-1)(n-2) + b(-32n+56) + c(-17n+46) + 8b^2 - 18c^2$$
$$- 2(8k-18h) + 17(4\beta+\gamma) + 4j + 17\theta + 6\chi + 8B,$$

$$2\delta = n(n-1)(n-2)(n-3) + b(-4n^2+20n-24) + c(-6n^2+15n-18) + 12bc + 18c^2$$
$$+ (8k-18h) - 9(4\beta+\gamma) - 9\theta + 2C,$$

$$8n' = 8n(n-1)^2 + (-32n+40)b + (-21n+30)c + 8b^2 - 18c^2$$
$$- 2(8k-18h) + 21(4\beta+\gamma) - 12j + 21\theta - 18\chi - 16C - 24B,$$

$$c' = 4n(n-1)(n-2) + (-16n+28)b + (-10n+26)c + 4b^2 - 9c^2$$
$$- (8k-18h) + 10(4\beta+\gamma) - 4j + 10\theta - 6\chi - 6C - 8B,$$

$$2b' = -a + n'(n'-1) - 3c', \quad (n',\ c' \text{ suprà}),$$

$$\sigma' + 2j' + 3\chi' + 2C' + 4B' = 4n(n-2) - 8b - 11c,$$

$$\rho' - 4j' - 6\chi' - 4C' - 9B' = -11n(n-2) + a(n'-2) + 22b + 30c, \quad (n',\ a \text{ suprà}),$$

$$2\sigma' + 4\beta' + \gamma' + \theta' = c'(n'-2), \quad (n',\ c' \text{ suprà}),$$

$$4k' - 3(i' + 3\beta' + 2\gamma') - 2\rho' - j' = (-4n'+6)b' + 2b'^2, \quad (n',\ b' \text{ suprà}),$$

$$6h - 2(i' + 3\beta' + 2\gamma') - 3\sigma' - \chi' = (-4n'+6)c' + 3c'^2, \quad (n',\ c' \text{ suprà});$$

{or in place of either of these,

$$8k' - 18h' - 4\rho' + 9\sigma' - 2j' + 3\chi' = (2b' - 3c')\{(n'-2)(n'-3) - a\}, \quad (n',\ b',\ c',\ a \text{ suprà})\},$$

$$\rho' + 2\beta' + 3\gamma' + 3t' = b'(n'-2), \quad (n',\ b' \text{ suprà}),$$

$$2q' + \beta' + 3i' + j' - 2\rho' = 0,$$

$$3r' + 2i' + \chi' - 5\sigma' - \beta' - 4\theta' = c', \quad (c' \text{ suprà}),$$

(twenty-three equations, being a transformation of the original system of twenty-three equations).

11. Forming the combinations $4i+6r$, $24t-8q+18r$ (the last of which introduces on the opposite side the term $+48t$), we obtain

$$4i+6r = c(5n-12)-5\gamma-18\beta+3\theta-2\chi,$$
$$-24t-8q+18r=-(8n-16)b+(15n-36)c-34\beta+9\gamma+4j+9\theta+6\chi,$$

equations which are used *post*, No. 53.

12. I remark that if there be on a surface a right line which is such that the tangent plane is different at different points of the line, the line is said to be *scrolar:* the section of the surface by any plane through the line contains the line *once.* But if there is at each point of the line one and the same tangent plane, then the section of the surface by the tangent plane contains the line at least twice; if it contain it twice only, the line is *torsal*; if three times the line is *oscular*; and the tangent plane containing the torsal or oscular line may in like manner be termed a torsal or an oscular tangent plane. These epithets, scrolar, torsal and oscular, will be convenient in the sequel.

Article Nos. 13 to 39. *Explanation of the New Singularities.*

I proceed to the explanation of the new singularities.

13. The cnicnode, or singularity $C=1$, is an ordinary conical point; instead of the tangent plane we have a proper quadricone.

14. The cnictrope, or reciprocal singularity $C'=1$, is also a well known one; it is in fact the conic of plane contact, or say rather the plane of conic contact, viz. the cnictrope is a plane touching a surface, not at a single point, but along a conic.

15. Consider a surface having the cnicnode $C=1$, and the reciprocal surface having the cnictrope $C'=1$. There are on the quadricone of the cnicnode six directions of closest contact[1], and reciprocal thereto we have six tangents of the cnictrope conic, touching it at six points. The plane of the cnictrope meets the surface in the conic twice, and in a residual curve which touches the conic at each of the six points. It would appear that these six contacts are part of the notion of the cnictrope.

16. We may of course have a surface with a conic of plane contact, but such that the residual curve of intersection in the plane of the conic does not touch the conic six times or at all; for instance the general equation of a surface with a conic of plane contact is $PM+V^2N=0$, where $P=0$ is a plane, $V=0$ a quadric surface; and here the conic $P=0$, $V=0$ does not touch the residual curve $P=0$, $N=0$. The reciprocal surface will in this case have a cnicnode, but there is some special circumstance doing away with the six directions of closest contact which in general belong thereto. I do not further pursue this inquiry.

[1] Taking for greater simplicity coordinates x, y, z, 1, then for a surface having a cnicnode at the origin, the equation is $U_2+U_3+\&c.=0$, the suffixes showing the degree in the coordinates; the equation of the quadricone is $U_2=0$, and the six directions are given as the lines of intersection of the two cones $U_2=0$, $U_3=0$.

17. For a surface having the cnictrope $C'=1$, the Hessian surface passes through the conic, which is thus thrown off from the spinode curve; or there is a reduction $=2$ in the order of the curve, which agrees with a foregoing result.

18. The binode, or singularity $B=1$, is a biplanar node, where instead of the proper quadricone we have two planes; these may be called the biplanes, and their line of intersection, the edge of the binode. The biplanes form a plane-pair.

19. The bitrope, or reciprocal singularity $B'=1$, is the plane of point-pair contact; but this needs explanation.

20. Consider a surface having a binode, and the reciprocal surface having a bitrope. We have the bitrope, a plane the reciprocal of the binode; in this plane a line, the reciprocal of the edge; in the line two points, or say a point-pair, the reciprocal of the biplanes: these points may be called the bipoints. There are in each biplane three directions of closest contact; the reciprocals of these are in the bitrope three directions through each of the two points. The section of the reciprocal surface by the bitrope is made up of the line counting three times (or the line is oscular), and of a curve passing in the three directions (having therefore a triple point) through each of the two bipoints. The bitrope contains thus an oscular line; but it is part of the notion that there are on this line two points each a triple point on the residual curve of intersection.

21. We may however have on a surface an oscular line without upon it two or any triple points of the residual curve of intersection. Such a surface is $Mx+Ny^3=0$; the intersections of the line $x=0$, $y=0$ with the curve $x=0$, $N=0$ will be all of them ordinary points. The reciprocal surface will have a binode, but there will be some special circumstance doing away with the existence of the directions of closest contact in the two biplanes respectively. I do not at present pursue the question.

22. For a surface having a bitrope $B'=1$, it appears from what precedes, that the oscular line must count 4 times in the intersection of the surface with the Hessian; for only in this way can the reduction 4 in the order of the spinode curve arise.

23. The pinch-point, or singularity $j=1$, is in fact mentioned in Salmon; it is a point on the nodal curve such that the two tangent planes coincide, or say it is a cuspidal point on the nodal curve. If, to fix the ideas, we take the nodal curve to be a complete intersection $P=0$, $Q=0$, then the equation of the surface is $(A, B, C \between P, Q)^2=0$ (A, B, C functions of the coordinates); we have a surface $AC-B^2=0$, which may be called the critic surface, intersecting the nodal curve in the points $P=0$, $Q=0$, $AC-B^2=0$, which are the pinch-points thereof; or if there be a cuspidal curve, then such of these points as are not situate on the cuspidal curve are the pinch-points: see my paper "On a Singularity of Surfaces," *Quart. Math. Journ.* vol. IX. (1868) pp. 332—338, [402]. The single tangent plane at the pinch-point meets the surface (see p. 338) in a curve having at the pinch-point a triple point, $=$ cusp $+ 2$ nodes, viz. there is a cuspidal branch the tangent to which coincides with that of the nodal curve; and there is a simple branch the tangent to which may be called the cotangent

at the pinch-point. In the particular case where the nodal curve is a right line the section is the line twice (representing the cuspidal branch), and a residual curve of the order $n-2$, the tangent to which is the cotangent.

24. The pinch-plane, or reciprocal singularity $j'=1$, is in fact a torsal plane touching the surface along a line, or meeting it in the line twice and in a residual curve. Let the line and curve meet in a point P; for the reason that the section by the plane is the line twice and the residual curve, the section has at P two coincident nodes; that is, the plane is a node-couple plane with two coincident nodes. The plane meets the consecutive node-couple plane in a line μ passing through P and touching at this point the residual curve. Considering now the reciprocal figure, the reciprocal of the pinch-plane is thus a point of the nodal curve, and is a pinch-point; the tangent plane at the pinch-point is the reciprocal of the point P; the tangent to the nodal curve is the reciprocal of the line μ, that is, of the tangent at P to the residual curve; and the cotangent at the pinch-point is the reciprocal of the torsal line.

25. There is in this theory the difficulty that for a surface of the order n, the torsal plane meets the residual curve of intersection in $(n-2)$ points P, and if each of these be a point on the node-couple curve, then in the reciprocal figure the pinch-point would be a multiple point on the nodal curve. I apprehend that starting with a pinch-point, a simple point on the nodal curve, we have in the reciprocal figure a pinch-plane or torsal plane as above, but with some speciality in virtue of which only one of the $(n-2)$ points of intersection of the torsal line with the residual plane curve is a point of the node-couple curve of the reciprocal surface. In the case of a pinch-plane or torsal plane of a cubic surface, $n-2$ is $=1$, and the question of multiplicity does not arise.

26. For a surface with a pinch-plane or torsal plane as above $(j'=1)$, the Hessian surface not only passes through the torsal line, but it touches the surface along this line, causing, as already mentioned, a reduction $=2$ in the order of the spinode curve. That the surfaces *touch* along the line is an important theorem[1], and I annex a proof.

27. Let $x=0$, $y=0$ be the torsal line, $x=0$ being the torsal plane; the equation of the surface therefore is $x\phi+y^2\psi=0$; and if A, B, C, D be the first derived functions of ϕ, $(a, b, c, d, f, g, h, l, m, n)$ the second derived functions, and if (A', B', C', D'), $(a', b', c', d', f', g', h', l', m', n')$ refer in like manner to ψ, then the equation of the Hessian is

$$0=\begin{vmatrix} 2A+xa+y^2a' , & B+xh+2yA'+y^2h', & C+xg+y^2g', & D+xl+y^2l' \\ B+xh+2yA'+y^2h', & xb+2\psi+4yB'+y^2b', & xf+2yC'+y^2f', & xm+2yD'+y^2m' \\ C+xg+y^2g' , & xf+2yC'+y^2f' , & xc+y^2c' , & xn+y^2n' \\ D+xl+y^2l' , & xm+y^2n'+2yD' , & xn+y^2n' , & xd+y^2d' \end{vmatrix};$$

[1] See Salmon, p. 218, where it is only stated that the Hessian passes through the line.

and representing this for a moment by

$$\begin{vmatrix} A, & H, & G, & L \\ H, & B, & F, & M \\ G, & F, & C, & N \\ L, & M, & N, & D \end{vmatrix} = 0,$$

then in the developed equation

$$D(ABC - AF^2 - BG^2 - CH^2 + 2FGH)$$
$$-(BC - F^2,\ CA - G^2,\ AB - H^2,\ GH - AF,\ HF - BG,\ FG - CH \between L,\ M,\ N)^2 = 0,$$

observing that C, F, M, N, D are of the first order in x, y, the only terms of the first order are contained in $B(-DG^2 - CL^2 + 2NGL)$; and since C, D, N are of the first order, we obtain all the terms of the first order by reducing B, G, L to the values 2ψ, C, D; viz. the terms of the first order are

$$2\psi(-C^2dx - D^2cx + 2CDnx), \quad = -2\psi(C^2d + D^2c - 2CDn)x.$$

Hence the complete equation is of the form

$$-2\psi(C^2d + D^2c - 2CDn)x + (x,\ y)^2 = 0,$$

or, what is the same thing, $x\Phi + y^2\Psi = 0$; the Hessian has therefore along the line $x = 0$, $y = 0$ the same tangent plane $x = 0$ as the surface; or it *touches* the surface along this line; that is, the line counts twice in the intersection of the two surfaces.

28. If instead of the right line we have a plane curve, say if the equation be $x\phi + P^2\psi = 0$, then the value of the Hessian is $x\Phi + P\Psi = 0$ (viz. the second term divides by P only, not by P^2), so that, as before mentioned in regard to a conic of contact, the surface and the Hessian merely cut (but do not touch) along the curve $x = 0$, $P = 0$. To show this in the most simple manner take the equation to be $x\phi + \frac{1}{2}P^2 = 0$; let A', B', C', D' be the first derived functions of ϕ, and (A, B, C, D), $(a, b, c, d, f, g, h, l, m, n)$ the first and second derived functions of P; then if in the equation of the Hessian we write for greater simplicity $x = 0$, the equation is

$$\begin{vmatrix} 2A' + Pa + A^2, & B' + Ph + AB, & C' + Pg + AC, & D' + Pl + AD \\ B' + Ph + AB, & Pb + B^2, & Pf + BC, & Pm + BD \\ C' + Pg + AC, & Pf + BC, & Pc + C^2, & Pn + CD \\ D' + Pl + AD, & Pm + BD, & Pn + CD, & Pd + D^2 \end{vmatrix} = 0.$$

The equation contains for example the term

$$-(D' + Pl + AD)^2\{P^2(bc - f^2) + P(bC^2 + cB^2 - 2fBC)\},$$

dividing as it should do by P, but not dividing by P^2; and considering the portion hereof $-D'^2P(bC^2 + cB^2 - 2fBC)$, there are no other terms in D'^2P which can destroy this, and to make the whole equation divide by P^2; which proves the required negative.

29. For the off-point or singularity $\theta = 1$; this is a point on the cuspidal curve at which the second derived functions all of them vanish. In further explanation hereof consider a surface $U = 0$, and the second polar of an arbitrary point $(\alpha, \beta, \gamma, \delta)$; viz. this is $(\alpha\partial_x + \beta\partial_y + \gamma\partial_z + \delta\partial_w)^2 U = 0$, or say for shortness $\Delta^2 U = 0$, where the coefficients of the powers and products of $(\alpha, \beta, \gamma, \delta)$ are of course the second derived functions of U; this equation, when reduced by means of the equations of the cuspidal curve, may acquire a factor Λ, thus assuming the form $\Lambda(\alpha P + \beta Q + \gamma R + \delta S)^2 = 0$, and if so the intersections of the cuspidal curve with the second polar ($= 2\sigma + \theta$, if, as for simplicity is supposed, there is no nodal curve) will be made up of the intersections of the cuspidal curve with the surface $\Lambda = 0$, and of those with the surface $\alpha P + \beta Q + \gamma R + \delta S = 0$ each twice; the latter of these, depending on the coordinates $(\alpha, \beta, \gamma, \delta)$ of the arbitrary points, are the points σ each twice; the former of them, or intersections of the cuspidal curve with the surface $\Lambda = 0$, are the points θ, or off-points of the cuspidal curve. If there is a nodal curve, the only difference is that the off-points are such of the above points as do not lie on the nodal curve.

30. As the most simple instance of the manner in which this singularity may present itself, consider a surface $FP^2 + GQ^3 = 0$, where the degrees of the functions are f, p, g, q, and therefore $n = f + 2p = g + 3q$, if n be the order of the surface. This has a cuspidal curve $P = 0, Q = 0$ of the order pq; the equation $\Delta^2(FP^2 + GQ^3) = 0$ of the second polar, when reduced by the equations $P = 0, Q = 0$ of the cuspidal curve, becomes simply $F(\Delta P)^2 = 0$; and we have thus the off-points $F = 0, P = 0, Q = 0$, consequently $\theta = fpq$.

31. But suppose, as before, the case of a surface $(A, B, C \between P, Q)^2 = 0$ having a cuspidal curve $P = 0, Q = 0$, and therefore $AC - B^2$ being $= 0$ for $P = 0, Q = 0$. The equation of the second polar, writing therein $P = 0, Q = 0$, becomes $(A, B, C \between \Delta P, \Delta Q)^2 = 0$, and if for any given surface this assumes the form $\Lambda(M\Delta P + N\Delta Q)^2 = 0$ (observe that M, N may be fractional provided only the $M\Delta P + N\Delta Q$ is integral), then there will be on the cuspidal curve the off-points $\Lambda = 0, P = 0, Q = 0$.

32. An interesting example is afforded by a surface which presents itself in the Memoir on Cubic Surfaces: the surface

$$\begin{aligned} &4y^6 \\ &- 4y^3x(x^2 + 3zw) \\ &+ \ zw(3x^2 + zw)^2 = 0 \end{aligned}$$

has the cuspidal conic $y = 0, 3x^2 + zw = 0$, and (as coming under the form $FP^2 + GQ^3 = 0$) has the off-points $zw = 0, y = 0, 3x^2 + zw = 0$; that is, the points $(x = 0, y = 0, z = 0)$, $(x = 0, y = 0, w = 0)$ each twice; $\theta = 4$.

But writing the same equation in the form

$$(4, 6x, 8x^2 + zw \between y^3 - 2x^3, x^2 - zw)^2 = 0,$$

where

$$4 \,.\, (8x^2 + zw) - (6x)^2 = -4(x^2 - zw),$$

it appears that there are also the three cuspidal conics $y^3 - 2x^3 = 0$, $x^2 - zw = 0$. Reducing by means of these two equations, the equation of the second polar is at first obtained in the form

$$(4,\ 6x,\ 8x^2 + zw \between 3y^2\Delta y - 6x^2\Delta x,\ 2x\Delta x - z\Delta\omega - w\Delta z)^2 = 0\,;$$

but further reducing by the same equations and writing for this purpose $y = \omega x$ $(\omega^3 = 2)$, the equation becomes

$$(4,\ 6x,\ 9x^2 \between x^2 (3\omega^2\Delta y - 6\Delta x),\ 2x\Delta x - z\Delta w - w\Delta z^2)^2 = 0,$$

that is

$$x^2 [2x (3\omega^2\Delta y - 6\Delta x) + 3 (2x\Delta x - z\Delta w - w\Delta z)]^2 = 0,$$

and we have thus the off-points $x^2 = 0$, $y^3 - 2x^3 = 0$, $x^2 - zw = 0$, in fact the before-mentioned two points each 6 times; and the complete value of θ is $\theta = (4 + 12 =) 16$; viz. the off-points are the points $(x = 0,\ y = 0,\ z = 0)$, $(x = 0,\ y = 0,\ w = 0)$ each 8 times. On account of this union of points the singularity is really one of a higher order, but equivalent to $\theta = 16$.

I am not at present able to explain the off-plane or reciprocal singularity $\theta' = 1$.

33. As to the close-point or singularity $\chi = 1$. I remark that at an ordinary point of the cuspidal curve the section by the tangent plane *touches*, at the point of contact, the cuspidal curve: the point of contact is on the curve of section a singular point {in the nature of a triple point, viz. taking the point of contact as origin, the form of the branch in the vicinity thereof is $y^3 - x^4 = 0$, where $y = 0$ is the equation of the tangent to the cuspidal curve}, such that the point of contact counts 4 times in the intersection of the cuspidal curve with the curve of section. At a close-point the form of the curve of section is altered; viz. the point of contact is here in the nature of a quadruple point with two distinct branches, one of them a triple branch of the form $y^3 = x^4$, but such that the tangent thereof, $y = 0$, is not the tangent of the cuspidal curve; the other of them a simple branch, the tangent of which is also distinct from the tangent of the cuspidal branch: the point of contact counts $3 + 1$ times, that is 4 times, as before, in the intersection of the cuspidal curve and the curve of section. The tangent to the simple branch may conveniently be termed the cotangent at the close-point; that of the other branch the cotriple tangent.

34. We may look at the question differently thus: to fix the ideas, let the cuspidal curve be a complete intersection $P = 0$, $Q = 0$; the equation of the surface is $(A,\ B,\ C \between P,\ Q)^2 = 0$, where $AC - B^2 = 0$, in virtue of the equations $P = 0$, $Q = 0$ of the cuspidal curve, that is, $AC - B^2$ is $= MP + NQ$ suppose. We have (as in the investigation regarding the pinch-point) a critic surface $AC - B^2 = 0$, this meets the surface in the cuspidal curve and in a residual curve of intersection; the residual curve by its intersection with the cuspidal curve determines the close-points; the tangent at the close-point is I believe the tangent of the residual curve. Analytically the close-points are given by the equations $P = 0$, $Q = 0$, $(A,\ B,\ C \between N,\ -M)^2 = 0$. It is proper to remark that if besides the cuspidal curve there be a nodal curve, only such of the points so determined as do not lie on the nodal curve are the close-points.

35. I take as an example a surface which is substantially the same as one which presents itself in the Memoir on Cubic Surfaces, viz. the surface $(1, w, xy \between w^2 - xy, z)^2 = 0$, having the cuspidal conic $w^2 - xy = 0$, $z = 0$. Since in the present case $AC - B^2 = P$, we have $M = 1$, $N = 0$, and the close-points are given by $P = 0$, $Q = 0$, $C = 0$; that is, they are the points $(z = 0, w = 0, x = 0)$ and $(z = 0, w = 0, y = 0)$.

36. I first however consider an ordinary point on the cuspidal curve, or conic $w^2 - xy = 0$, $z = 0$; the coordinates of any point on the conic are given by $x : y : z : w = 1 : \theta^2 : 0 : \theta$, where θ is an arbitrary parameter; we at once find $\theta^2 x + y - \theta(z + 2w) = 0$ for the equation of the tangent plane of the surface or cuspidal tangent plane at the point $(1, \theta^2, 0, \theta)$. Proceeding to find the intersection of this plane with the surface, the elimination of z gives

$$(\theta^2, \theta w, xy \between w^2 - xy, \theta^2 x + y, -2\theta w)^2 = 0,$$

which is of course the cone, vertex $(x = 0, y = 0, w = 0)$, which passes through the required curve of intersection. In place of the coordinates x, y take the new coordinates $\theta^2 x - y = 2p$, and $\theta^2 x + y - 2\theta w = 2q$; we have

$$\theta^2 x = \theta w + p - q,$$
$$-y = -\theta w + p - q,$$

and thence

$$-\theta^2 xy = p^2 - (q + \theta w)^2 = p^2 - q^2 - 2\theta qw - \theta^2 w^2,$$
$$\theta^2 (w^2 - xy) = p^2 - q^2 - 2\theta qw,$$

and the equation thus is

$$(\theta^4, \theta^3 w, -p^2 + q^2 + 2\theta qw + \theta^2 w^2 \between p^2 - q^2 - 2\theta qw, 2\theta^2 q) = 0,$$

or, what is the same thing,

$$(1, \theta w, -p^2 + q^2 + 2\theta qw + \theta^2 w^2 \between p^2 - q^2 - 2\theta qw, 2q)^2 = 0;$$

viz. this is

$$(p^2 - q^2 - 2\theta qw)^2 + 4\theta qw (p^2 - q^2 - 2\theta qw) + 4q^2 (-p^2 + q^2 + 2\theta qw + \theta^2 w^2) = 0;$$

or reducing, it is

$$(p^2 - q^2)(p^2 - 5q^2) + 8\theta q^3 w = 0,$$

the equation of the section in terms of the coordinates p, q, w. The equation is satisfied by the values $p = 0$, $q = 0$ which belong to the assumed point $(1, \theta^2, 0, \theta)$ of the conic, and in the vicinity of this point we have $p^4 + 8\theta q^3 w = 0$, which is a triple branch of the form $y^3 = x^4$, the tangent $q = 0$ being, it will be observed, the tangent of the conic. But at the close-points, or when $\theta = 0$ or $\theta = \infty$, the transformation fails; and these points must be considered separately.

37. At the first of these, viz. the point $z = 0$, $w = 0$, $x = 0$, the tangent plane of the surface or cuspidal tangent plane is $x = 0$, and this meets the surface in the curve $x = 0$,

$w^3(w+2z)=0$, that is in the line $x=0$, $w=0$ three times, and in the line $x=0$, $w+2z=0$ (that the section consists of right lines is of course a speciality, and it is clear that considering in a more general surface the section as defined by an equation in (w, z, y), the line $w=0$ represents the tangent to a triple branch $w^3=z^4+\&c.$, and the line $w+2z=0$ the tangent to a simple branch); these lines are each of them, it will be observed, distinct from the tangent to the cuspidal conic, which is $x=0$, $z=0$. And similarly the tangent plane at the other of the two points is $y=0$, meeting the surface in the curve $y=0$, $w^3(w+2z)=0$, that is in the line $y=0$, $w=0$ three times, and in the line $y=0$, $w+2z=0$.

38. The close-plane or reciprocal singularity $\chi'=1$ is (like the pinch-plane) a torsal plane, meeting the surface in a line twice and in a residual curve; the distinction is that the line and curve have an intersection P lying on the spinode curve; the close-plane is thus a spinode plane; it meets the consecutive spinode plane in a line μ passing through P, and which is *not* the tangent of the residual curve. In the reciprocal figure, the reciprocal of the close-plane is on the cuspidal curve, and is a close-point; the reciprocal of the point P is the cuspidal tangent plane; that of the line μ the tangent of the cuspidal curve; that of the tangent of the residual curve the cotriple tangent; that of the torsal line the cotangent.

39. The torsal line of a close-plane is not a mere torsal line; in fact by what precedes it appears that the surface and the Hessian intersect in this line, counting not twice but three times, and it is thus that the reduction in the order of the spinode curve caused by the close-plane is $=3$.

Article Nos. 40 and 41. *Application to a Class of Surfaces.*

40. Consider the surface $FP^2+GR^2Q^3=0$, where f, p, g, r, q being the degrees of the several functions, and n the order of the surface, we have of course $n=f+2p=g+2r+3q$.

There is here a nodal curve, the complete intersection of the two surfaces $P=0, R=0$; hence $b=pr$, $k=\tfrac{1}{2}pr(p-1)(r-1)$, $=\tfrac{1}{2}b(b-p-r+1)$; $t=0$; whence $(q)=pr(p+r-2)$. There is also a cuspidal curve the complete intersection of the two surfaces $P=0$, $Q=0$; hence $c=pq$, $h=\tfrac{1}{2}pq(p-1)(q-1)$, $=\tfrac{1}{2}c(c-p-q+1)$; whence $(r)=pq(p+q-2)$: I have written for distinction (q), (r), to denote the q, r of the fundamental equations. The two curves intersect in the pqr points $P=0$, $Q=0$, $R=0$, which are not stationary points on either curve; that is, $\beta=0$, $\gamma=0$, $i=pqr$.

There are on the nodal curve the $j=(f+g)pr$ pinch-points $F=0$, $P=0$, $R=0$, and $G=0$, $P=0$, $R=0$. There are on the cuspidal curve $\theta=fpq$ off-points $F=0$, $P=0$, $Q=0$; and there the gpq singular points $G=0$, $P=0$, $Q=0$. I find that these last, and also the θ points each three times, must be considered as close-points, that is, that we have $\chi=(g+3f)pq$.

41. We ought then to have

$$\begin{aligned} b(n-2) &= \rho, \\ c(n-2) &= 2\sigma+\theta; \\ 2(q)+3i+j &= 2\rho, \\ 3(r)+c+2i+\chi &= 5\sigma+4\theta; \end{aligned}$$

the first two of which give ρ, σ, and then, substituting their values, the other two equations should become identities. In fact, attending to the values $pr=b$, $pq=c$, the equations become

$$\begin{aligned} 2b(p+r-2)+3bq+b(f+g) &= 2b(n-2), \\ 3c(p+q-2)+c+2cr+b(g+3f) &= \tfrac{5}{2}\{c(n-2)-cf\}+4cf. \end{aligned}$$

The first of these is

$$2n=2p+2r+3q+f+g, \quad = \quad (2p+f)+(2r+3q+g),$$

and the second is

$$\tfrac{5}{2}n=3p+3q+2r+g+\tfrac{3}{2}f, \quad =\tfrac{3}{2}(2p+f)+(2r+3q+g),$$

so that the equations are satisfied.

Article No. 42. *The Flecnodal Curve.*

42. A point on a surface may be flecnodal, viz. the tangent plane may meet the surface in a curve having at the point a flecnode, that is, a node with an inflexion on one of the branches. Salmon has shown that, for a surface of the order n without singularities, the locus of the flecnodal points, or flecnodal curve, is the complete intersection of the surface by a surface of the order $11n-24$, which may be called the flecnodal surface, the order of the curve being thus $=n(11n-24)$. I have succeeded in showing, in a somewhat peculiar way by consideration of a surface of revolution, that if the surface of the order n has a nodal curve of the order b, and a cuspidal curve of the order c, then that the order of the flecnodal curve is $=n(11n-24)-22b-27c$; before giving this investigation, I will by the like principles demonstrate the above-mentioned theorem that the order of the spinode curve is $=4n(n-2)-8b-11c$.

Article Nos. 43 to 47. *Surfaces of Revolution, in connexion with the Spinode Curve and the Flecnodal Curve.*

43. Consider a plane curve of the order m with δ nodes and κ cusps, and let this be made to revolve about an axis in its own plane, so as to generate a surface of revolution. The complete meridian section is made up of the given curve and of an equal curve situate symmetrically therewith on the other side of the axis; the

order of the surface is thus $=2m$. The two curves intersect in m points on the axis and in m^2-m points, forming $\frac{1}{2}(m^2-m)$ pairs of points, situate symmetrically on opposite sides of the axes; these last generate $\frac{1}{2}(m^2-m)$ circles, nodal curves on the surface; the nodes generate δ circles, which are nodal curves on the surface, and the cusps generate κ circles, cuspidal curves on the surface. There are $m^2-m-2\delta-3\kappa$ circles of plane contact corresponding in the plane curve to the tangents perpendicular to the axis. Each of the m points on the axis gives in the surface a pair of (imaginary) lines; and we have thus two sets each of m lines, such that along the lines of each set the surface is touched by an (imaginary) meridian plane; viz. these are the circular planes $x+iy=0$, $x-iy=0$ passing through the axis. I assume without stopping to show it that these $2m$ lines are lines not j' but χ', that is, that they each reduce the order of the spinode curve by 3 [1]. The inflexions generate $3m^2-6m-6\delta-8\kappa$ circles which constitute the spinode curve on the surface.

44. And we can thus verify that the complete intersection of the surface with the Hessian is made up in accordance with the foregoing theory; viz.

Order of surface $=2m$,		
Order of Hessian $=4(2m-2)$,		
whence order of intersection		$=16m^2-16m$
Nodal curve, $\frac{1}{2}(m^2-m)+\delta$ circles,	8 times	$8m^2-\ \ 8m+16\delta$
Cuspidal curve, κ circles,	11 times	$+22\kappa$
Circles of contact $m^2-m-2\delta-3\kappa$,		$2m^2-\ \ 2m-\ \ 4\delta-\ \ 6\kappa$
Lines $2m$	, 3 times	$+\ 6m$
Spinode curve, $3m^2-6m-6\delta-8\kappa$ circles,		$6m^2-12m-12\delta-16\kappa$
		$16m^2-16m.$

45. We may by a similar reasoning show that the surface and the flecnode surface intersect in the nodal curve taken 22 times, and in the cuspidal curve taken 27 times; and consequently that the order of the residual intersection or flecnodal curve is

$$=n(11n-24)-22b-27c.$$

To effect this, observe that at any point whatever of a quadric surface the tangent plane meets the surface in a pair of lines, that is, in a curve having at the point of contact a node with an inflexion on each branch, or say, a fleflecnode. Imagine in the plane figure a conic having its centre on the axis of rotation and its axis coincident therewith, and the conic having with the curve of the order m a 4-pointic intersection at any point P; the point P generates a circle, such that along this circle the surface is osculated by a quadric surface of revolution in such wise that the meridian sections have a four-pointic contact; the circle in question is thus on the surface a fleflecnode circle; and I assume that it counts twice as a flecnode circle. Hence if the number of the points P be $=\theta$, we have on the surface θ fleflecnode circles, $=2\theta$ flecnode circles, that is, a flecnode curve of the order 4θ. I wish to show that we have $\theta=5m^2-9m-10\delta-12\kappa$.

[1] Observe that the terms in m cannot be got rid of in a different manner, by any alteration of the numbers 8 and 11 to which the present investigation relates.

46. The problem is as follows: given a curve of the order m with δ nodes and κ cusps; it is required to find the number of the conics, centre on a given line, and an axis coincident in direction with this line, which have with the given curve a 4-pointic intersection, or contact of the third order. This may be solved by means of formulæ contained in my "Memoir on the Curves which satisfy given Conditions," *Phil. Trans.* vol. CLVIII. (1868), pp. 75—144; see p. 88; [406].

Taking $x=0$ for the given line, the conic $(a, b, c, f, g, h\between x, y, 1)^2=0$ will have its centre on the given line and an axis coincident therewith, if only $h=0$, $g=0$; and denoting these two conditions by $2X$, it is easy to see that we have

$$(2X \therefore)=1, \quad (2X : /)=2, \quad (2X \cdot //)=2, \quad (2X\, ///)=1.$$

But in general if the conic satisfy any other three conditions $3Z$, then the number of the conics $(2X, 3Z)$ is

$$\begin{aligned} = \quad & \alpha'\,(\qquad\qquad\qquad\quad - \tfrac{1}{4}\gamma + \tfrac{1}{2}\delta) \\ & + \beta'\,(-\tfrac{3}{8}\alpha + \tfrac{5}{16}\beta + \tfrac{5}{16}\gamma - \tfrac{3}{8}\delta) \\ & + \gamma'\,(\ \tfrac{1}{2}\alpha - \tfrac{1}{4}\beta \qquad\qquad\qquad\ ), \end{aligned}$$

where α, β, γ, δ denote $(2X \therefore)$, $(2X : /)$, $(2X \cdot //)$, $(2X\, ///)$, viz. in the present case the values are 1, 2, 2, 1 respectively, and where α', β', γ' denote $(3Z :)$, $(3Z \cdot /)$, $(3Z\, //)$ respectively.

47. Substituting for α, β, γ, δ their values, the number of the conics in question is $=\tfrac{1}{2}\beta'$, that is $=\tfrac{1}{2}(3Z \cdot /)$. Suppose that $3Z$, or say 3, denotes the condition of a contact of the third order with a given curve (m, δ, κ), or say with a given curve (m, n, α) (m the order, n the class $=m^2-m-2\delta-3\kappa$, $\alpha=3n+\kappa$), then we have

$$\begin{aligned} (3 : \ ) &= -4m - 3n + 3\alpha, \\ (3 \cdot /) &= -8m - 8n + 6\alpha, \\ (3\, //\ ) &= -3m - 4n + 3\alpha; \end{aligned}$$

and from the second of these the number of the conics in question is $=-4m-4n+3\alpha$, that is, it is $=-4m+5n+3\kappa$, or finally it is $=5m^2-9m-10\delta-12\kappa$.

Hence, assuming that the $2m$ lines each counts 6 times[1],

Order of surface $=2m$

Order of flecnode surface $=11\,(2m-24)$ or $22m-24$

Order of intersection	$=44m^2-48m$
Nodal curve, $\tfrac{1}{2}(m^2-m)+\delta$ circles, 11 times	$22m^2-22m+44\delta$
Cuspidal curve κ circles, 27 times	$+54\kappa$
Circles of contact $m^2-m-2\delta-3\kappa$,	$2m^2-2m-4\delta-6\kappa$
Lines of contact $2m$, 6 times	$+12m$
Flecnodal curve, $5m^2-9m-10\delta-12\kappa$ circles each twice	$20m^2-36m-40\delta-48\kappa$
	$44m^2-48m.$

[1] See foot-note p. 343: the like remark applies to the present terms in m, which cannot be got rid of by an alteration of the numbers 22 and 27 to which the investigation relates.

Article Nos. 48 and 49. *The Flecnodal Torse.*

48. Starting from

$$\begin{aligned}22b' + 27c' &= 6(6b' + 8c') - 7(2b' + 3c')\\ &= 6(3n'^2 - 6n' - \kappa) - 7(n'^2 - n' - \delta)\\ &= 11n'^2 - 29n' + 7 - 6\kappa,\end{aligned}$$

that is

$$11n'^2 - 24n' - 22b' - 27c' = 5n' - 7\delta + 6\kappa,$$

I find

$$\begin{aligned}&n'(11n' - 24) - 22b' - 27c'\\ &\quad = n(n-1)(11n-24) + b(-59n+96) + c(-94n+156) + 26b^2 + 87c^2\\ &\qquad - 52k - 114h + 141\beta + 94\gamma + 77i + 3j + 4\chi - 15\theta - 45t - 10C - 9B.\end{aligned}$$

49. For a surface of the order n without singularities this equation is

$$n'(11n' - 24) - 22b' - 27c' = n(n-1)(11n-24);$$

to explain the meaning of it, I say that the reciprocal of a flecnode is a flecnodal plane, and *vice versâ:* the reciprocal of the flecnodal torse of the surface n (viz. the torse generated by the flecnodal planes of the surface) is thus the flecnodal curve of the reciprocal surface n'; and the class of the torse must therefore be equal to the order of the curve. The flecnodal torse is generated by the tangents of the surface n along the curve of intersection with a surface of the order $11n - 24$; the number of tangent planes which pass through an arbitrary point, or class of the torse, is at once found to be $n(n-1)(11n-24)$; for the reciprocal surface the order of the flecnodal curve is by what precedes $n'(11n' - 24) - 22b' - 27c'$; and the equation thus expresses that the order of the curve is equal to the class of the torse.

Article No. 50. *The general Surface of the Order n without Singularities.*

50. In the general surface of the order n without singularities, we have

$n = n,$

$a = n^2 - n,$

$\delta = \tfrac{1}{2}n(n-1)(n-2)(n-3),$

$\kappa = n(n-1)(n-2),$

$b = 0,$

$k = 0,$

$t = 0,$

$q = 0,$

$\rho = 0,$

$j = 0,$

$c = 0,$

$h = 0,$

$r = 0,$

$\sigma = 0,$

$\theta = 0,$

$\chi = 0,$

$C = 0,$

$B = 0,$

$\beta = 0,$

$\gamma = 0,$

$i = 0,$

$n' = n(n-1)^2,$

$a' = n(n-1),$

$\delta' = \tfrac{1}{2}n(n-2)(n^2-9),$

$\kappa' = 3n(n-2),$

$b' = \tfrac{1}{2}n(n-1)(n-2)(n^3-n^2+n-12),$

$k' = \tfrac{1}{8}n(n-2)(n^{10}-6n^9+16n^8-54n^7+164n^6-288n^5+547n^4-1058n^3+1068n^2-1214n+1464),$

$t' = \tfrac{1}{6}n(n-2)(n^7-4n^6+7n^5-45n^4+114n^3-111n^2+548n-960),$

$q' = n(n-2)(n-3)(n^2+2n-4),$

$\rho' = n(n-2)(n^3-n^2+n-12),$

$j' = 0,$

$c' = 4n(n-1)(n-2),$

$h' = \tfrac{1}{2}n(n-2)(16n^4-64n^3+80n^2-108n+156),$

$r' = 2n(n-2)(3n-4),$

$\sigma' = 4n(n-2),$

$\theta' = 0,$

$\chi' = 0,$

$C' = 0,$

$B' = 0,$

$\beta' = 2n(n-2)(11n-24),$

$\gamma' = 4n(n-2)(n-3)(n^3+3n-16),$

$i' = 0.$

Article Nos. 51 to 64. *Investigation of Formula for β'.*

51. The value $\beta' = 2n(n-2)(11n-24)$ for a surface without singularities was obtained by Salmon by independent geometrical considerations, viz. he obtains

$$2\beta' = 4n(n-2)(11n-24)$$

as the number of intersections of the spinode curve $\big(\text{order} = 4n(n-2)\big)$ by the flecnode surface of the order $11n-24$.

52. The value of β' must be obtainable in the case of a surface with singularities, and I have been led to conclude that we have

$$\begin{aligned}\beta' = \; & 2n(n-2)(11n-24) \\ & -(110n-272)\,b + 44q \\ & -(116n-303)\,c + \tfrac{63}{2}r \\ & +\tfrac{691}{2}\beta + 248\gamma + 198t \\ & + \text{linear function }(i, j, \theta, \chi, C, B, i', j', \theta', \chi', C', B'),\end{aligned}$$

but I have not yet completely determined the coefficients of the linear function. The reciprocal formula in the case of a surface of the order n without singularities, $i, j, \theta, \chi, C, B, i', j', \theta', \chi', C', B'$ then all vanishing, is the identity

$$\begin{aligned}0 = \; & 2n'(n'-2)(11n'-24) \\ & -(110n'-272)\,b' + 44q' \\ & -(116n'-303)\,c' + \tfrac{63}{2}r' \\ & +\tfrac{691}{2}\beta' + 248\gamma' + 198t'\end{aligned}$$

($n', b', q', c', r', \beta', \gamma', t'$ having the values in the foregoing Table). It was by assuming for β an expression of the above form but with indeterminate coefficients, and then determining these in such wise that the reciprocal equation should be an identity, that the foregoing formula for β' was arrived at.

53. I assume

$$\begin{aligned}\beta' = \; & 2n(n-2)(11n-24) \\ & -b(An-B) + Cq \\ & -c(Dn-E) + Fr \\ & -G\beta - H\gamma - It \\ & + \text{linear function }(i, j, \theta, \chi, C, B, i', j', \theta', \chi', C', B'),\end{aligned}$$

where it is to be remarked that, in virtue of the equations obtained No. 11, two of the coefficients of this form are really arbitrary: I cannot recall the considerations which led me to write $D = 116$, $E = 303$.

54. Forming the reciprocal equation

$$\begin{aligned}\beta = \;& 2n'(n'-2)(11n'-24)\\ & - b'(An'-B) + C'q\\ & - c'(Dn'-E) + F'r\\ & - G\beta' - H\gamma' - It'\\ & + \text{linear function } (i', j', \theta', \chi', C', B', i, j, \theta, \chi, C, B),\end{aligned}$$

and substituting herein the values which belong to the surface of the order n without singularities, we should have identically

$$\begin{aligned}0 = \;& 2n(n-1)^2(n-2)(n^2+1)(11n^3-22n^2+11n-24)\\ & - \tfrac{1}{2}n(n-1)(n-2)(n^3-n^2+n-12)[An(n-1)^2-B]\\ & + \;n(n-2)(n-3)(n^2+2n-4)\,C\\ & - 4n(n-1)(n-2)[Dn(n-1)^2-E]\\ & + 2n(n-2)(3n-4)\,F\\ & - 2n(n-2)(11n-24)\,G\\ & - 4n(n-2)(n-3)(n^3+3n-16)\,H\\ & - \tfrac{1}{6}n(n-2)(n^7-4n^6+7n^5-45n^4+114n^3-111n^2+548n-960)\,I\,;\end{aligned}$$

or dividing the whole by $n(n-2)$, this is

$$\begin{aligned}0 = \;& 2(n-1)^2(n^2+1)(11n^3-22n^2+11n-24)\\ & - \tfrac{1}{2}(n-1)(n^3-n^2+n-12)[An(n-1)^2-B]\\ & + (n-3)(n^2+2n-4)\,C\\ & - 4(n-1)[Dn(n-1)^2-E]\\ & + 2(3n-4)\,F\\ & - 2(11n-24)\,G\\ & - 4(n-3)(n^3+3n-16)\,H\\ & - \tfrac{1}{6}(n^7-4n^6+7n^5-45n^4+114n^3-111n^2+548n-960)\,I.\end{aligned}$$

55. And then, expanding in powers of n and equating to zero the coefficients of the several powers $n^7, \ldots n^0$, we obtain

$$\begin{array}{cccccccc}
22 & -88 & +154 & -224 & +250 & -184 & +118 & -48\\
-\tfrac{1}{2}A & +2A & -\tfrac{7}{2}A & +\tfrac{19}{2}A & -20A & +\tfrac{37}{2}A & -\;6A & \\
 & & & +\;\tfrac{1}{2}B & -\;B & +\;B & -\;\tfrac{13}{2}B & +\;6B\\
 & & & & +\;C & -\;C & -\;10C & +\;12C\\
 & & & -\;4D & +12D & -12D & +\;4D & \\
 & & & & & & +\;4E & -\;4E\\
 & & & & & & +\;6F & -\;8F\\
 & & & & & & -\;22G & +\;48G\\
 & & & -\;4H & +12H & -12H & +100H & -192H\\
-\tfrac{1}{6}I & +\tfrac{2}{3}I & -\tfrac{7}{6}I & +\;\tfrac{15}{2}I & -19\,I & +\tfrac{37}{2}\,I & -\tfrac{274}{3}\;I & +160I\\
\| & \| & \| & \| & \| & \| & \| & \|\\
0 & 0 & 0 & 0 & 0 & 0 & 0 & 0
\end{array}$$

viz. the equations are read vertically downwards. The first, second, and third equations, and the sum of the fourth and fifth, all give the same relation, $132 - 3A - I = 0$; there are consequently, inclusive of this, five independent relations. By combining the equations so as to simplify the numbers, I find these to be

$$\begin{aligned}
3A + I - 132 &= 0,\\
4A - B - 2C - 80 &= 0,\\
7A - B + 2E + 2F + 2G - 476 &= 0,\\
26A - B + 8D + 8H - 1532 &= 0,\\
6A - E - 2F + 12G - 48H + 40I - 132 &= 0.
\end{aligned}$$

56. I found, as presently mentioned, $A = 110$, $B = 272$, $C = 44$; values which satisfy (as they should do) the second equation; and then assuming $D = 116$ and $E = 303$, we have $F = \frac{63}{2}$, $G = -\frac{691}{2}$, $H = -248$, $I = -198$; and the formula is

$$\begin{aligned}
\beta' = {} & 2n(n-2)(11n-24)\\
& -(110n-272)\,b + 44q\\
& -(116n-303)\,c + \tfrac{63}{2}r\\
& + \tfrac{691}{2}\beta + 248\gamma + 198t\\
& + \text{linear function } (i, j, \theta, \chi, C, B, i', j', \theta', \chi', C', B'),
\end{aligned}$$

the process not enabling the determination of the coefficients of the linear function.

57. The values of A, B, C were found from the general theorem that if three surfaces of the orders μ, ν, ρ respectively intersect in a curve of the order m and class r which is α-tuple on μ, β-tuple on ν, and γ-tuple on ρ, then the number of the points of intersection of the three surfaces is

$$= \mu\nu\rho - m(\beta\gamma\mu + \gamma\alpha\nu + \alpha\beta\rho - 2\alpha\beta\gamma) + \alpha\beta\gamma r.$$

Apply this to the case of a surface of the order n with a nodal curve of the order b and class q, intersecting the Hessian and flecnodal surfaces, we have

	Order.	Passing through (b, q), times
Surface	n	2
Hessian	$4n-8$	4
Flecnodal	$11n-24$	11

whence number of intersections is

$$\begin{aligned}
&= 4n(n-2)(11n-24) - b\{n.4.11 + (4n-8)\,11.2 + (11n-24)\,2.4 - 2.2.4.11\}\\
&\qquad + 2.4.11q,
\end{aligned}$$

that is

$$= 4n(n-2)(11n-24) - (220n-544)\,b - 88q;$$

and the value of β' is one half of this,

$$= 2n(n-2)(11n-24) - (110n-272)b + 44q.$$

I have not succeeded in applying the like considerations to the cuspidal curve.

58. As regards the general theorem, we know (Salmon, p. 274) that if two surfaces of the orders μ, ν partially intersect in a curve of the order m and class r, and besides in a curve of the order m', then the curves m, m' meet in $m(\mu+\nu-2)-r$ points.

Suppose that the curve m is α-tuple on the surface μ; then to find the number I of the intersections of the curves m and m', we may imagine through m a surface of the order ρ; the surfaces μ, ν intersect in the curve m α times, and in a residual curve of the order $\mu\nu - m\alpha$, this last meets the surface ρ in $\rho(\mu\nu - m\alpha)$ points, and thence the three surfaces meet in $\mu\nu\rho - m\alpha\rho - I$ points. But since m is a simple curve on each of the surfaces ν, ρ, the three surfaces meet in $\mu(\nu\rho - m) - \alpha[m(\nu+\rho-2)-r]$ points, whence equating the two values

$$I = m(\mu + \alpha\nu - 2\alpha) - \alpha r.$$

Next, let the curve m be α-tuple on the surface μ, β-tuple on the surface ν. Considering the new surface ρ through m, then μ, ν intersect in the curve m $\alpha\beta$ times, and in a residual curve of the order $\mu\nu - m\alpha\beta$; this last meets the surface ρ in $\rho(\mu\nu - m\alpha\beta)$ points; whence the three surfaces meet in $\rho(\mu\nu - m\alpha\beta) - I$ points. But the curve m being a β-tuple curve on ν, and a simple curve on ρ, these meet in the curve m β times and in a residual curve of the order $\nu\rho - \beta m$, whence the three surfaces meet in

$$\mu(\nu\rho - \beta m) - \alpha[m(\nu + \beta\rho - 2\beta) - \beta r]$$

points; and equating the two values, we have

$$I = m(\beta\mu + \alpha\nu - 2\alpha\beta) - 2\alpha\beta r.$$

Lastly, if the curve m be γ-tuple on ρ, then the surfaces μ, ρ meet in m $\alpha\gamma$ times and in a residual curve of the order $\mu\rho - m\alpha\gamma$; this last meets ν in

$$\nu(\mu\rho - \alpha\gamma m) - \beta[m(\gamma\mu + \alpha\rho - 2\alpha\gamma) - \alpha\gamma r]$$

points, that is, the number of points of intersection of the three surfaces is

$$= \mu\nu\rho - m(\beta\gamma\mu + \gamma\alpha\nu + \alpha\beta\rho - 2\alpha\beta\gamma) + \alpha\beta\gamma r.$$

59. I represent the complete value of β' by

$$\begin{aligned}\beta' = \;& 2n(11n-24) \\ & -(110n-272)b + 44q \\ & -(116n-303)c + \tfrac{63}{2}r \\ & +\tfrac{691}{2}\beta + 248\gamma + 198t \\ & -hC - gB - xi - \lambda j - \mu\chi - \nu\theta \\ & -h'C' - g'B' - x'i' - \lambda' j' - \mu'\chi' - \nu'\theta',\end{aligned}$$

and (observing that the Table of Singularities in my Memoir on Cubic Surfaces was obtained without the aid of the formula now in question) I endeavour by means of the results therein contained to find the values of the unknown coefficients h, g, x, λ, μ, ν, h', g', x', λ', μ', ν'.

60. For a cubic surface $n=3$, and for a cubic surface without singular lines (in fact for all the cases except the cubic scrolls XXII and XXIII), the formula is

$$\beta' = 54 - hC - gB - \lambda' j' - \mu'\chi' - \nu'\theta' - h'C' - g'B';$$

and applying this to the several cases of cubic surfaces as grouped together in the Table, and referred to by the affixed roman numbers, the resulting equations are

$$\begin{array}{ll}
54 = 54, & \text{(I)}\\
30 = 54 - .\, h, & \text{(II)}\\
18 = 54 - g - 16\nu', & \text{(III)}\\
13 = 54 - 2h - \lambda', & \text{(IV)}\\
6 = 54 - h - g - \mu' - 8\nu', & \text{(VI)}\\
3 = 54 - 3h - 3\lambda', & \text{(VIII)}\\
0 = 54 - 2g - 16\nu' - g', & \text{(IX)}\\
1 = 54 - 2h - g - \lambda' - 2\mu', & \text{(XIII)}\\
0 = 54 - 4h - 6\lambda', & \text{(XVI)}\\
0 = 54 - h - 2g - 2\mu' - g', & \text{(XVII)}\\
0 = 54 - 3g - 3g', & \text{(XXI)}
\end{array}$$

which are all satisfied if only

$$\begin{array}{ll}
h & = 24,\\
g + 16\nu' & = 36,\\
g + 2\mu' & = 12,\\
g + g' & = 18,\\
\lambda' & = -7.
\end{array}$$

61. If we apply to the same surfaces the reciprocal equation for β, or, what is the same thing, apply the original equation to the reciprocal surfaces, as given by interchanging the upper and lower halves of the Table of Singularities, we have another series of equations, viz. this is

$$\begin{array}{llrl}
0 = 54432 - 54432, & & & \text{(I)}\\
0 = 27851 - 27846 - h', & h' & = 5, & \text{(II)}\\
0 = 18180 - 18318 - g' - 16\nu, & g' + 16\nu & = -138, & \text{(III)}\\
0 = 11765 - 11756 - 2h' - \lambda, & 2h' + \lambda & = 9, & \text{(IV)}\\
0 = 6917 - 6584 - h' - g' - \mu - 8\nu, & h' + g' + \mu + 8\nu & = 45, & \text{(VI)}\\
0 = 3534 - 3522 - 3h' - 3\lambda, & 3h' + 3\lambda & = 12, & \text{(VIII)}\\
0 = 3024 - 3144 - 2g' - 16\nu - g, & 2g' + 16\nu + g & = -120, & \text{(IX)}\\
0 = 1433 - 1386 - 2h' - g' - \lambda - 2\mu, & 2h' + g' + \lambda + 2\mu & = 47, & \text{(XIII)}\\
0 = 518 - 504 - 4h' - 6\lambda, & 4h' + 6\lambda & = 14, & \text{(XVI)}\\
0 = 383 - 322 - h' - 2g' - g - 2\mu, & h' + 2g' + g + 2\mu & = 61, & \text{(XVII)}\\
0 = 54 - 3g - 3g', & 3g + 3g' & = 54, & \text{(XXI)}
\end{array}$$

all satisfied if only

$$\begin{aligned} h' &= 5, \\ g' + 16\nu &= -138, \\ g' + 2\mu &= 38, \\ g + g' &= 18, \\ \lambda &= -1. \end{aligned}$$

62. I remark however that the cubic scroll XXII or XXIII gives

$$0 = 54 - (330 - 272) - 2(\lambda + \lambda'),$$

that is, $\lambda + \lambda' = -2$, instead of $\lambda + \lambda' = -8$. The investigation is in fact really inapplicable to a scroll, for every point of a scroll has the property of a flecnode; whence if $U = 0$ be the equation of the scroll, that of the flecnodal surface is $M.U = 0$, containing U as a factor, and there is not any definite curve of intersection constituting the flecnodal curve; but I am nevertheless surprised at the numerical contradiction.

63. Combining the two sets of results, we find

$$\begin{aligned} h &= 24, \\ g &= g, \\ x &= x, \\ \lambda &= -1, \\ \mu &= 10 + \tfrac{1}{2}g, \\ \nu &= -\tfrac{39}{4} + \tfrac{1}{16}g, \\ h' &= 5, \\ g' &= 18 - g, \\ x' &= x', \\ \lambda' &= -7, \\ \mu' &= 6 - \tfrac{1}{2}g, \\ \nu' &= \tfrac{9}{4} - \tfrac{1}{16}g; \end{aligned}$$

and the formula thus is

$$\begin{aligned} \beta' = {} & 2n(n-2)(11n-24) \\ & - (110n - 272)b + 44q \\ & - (116n - 303)c + \tfrac{63}{2}r \\ & + \tfrac{691}{2}\beta + 248\gamma + 198t \\ & - 24C + j - 10\chi + \tfrac{39}{4}\theta - 5C' - 18B' - 6\chi' - \tfrac{9}{4}\theta' \\ & - xi - x'i' + \tfrac{1}{16}g(-16B - 8\chi - \theta + 16B' - 8\chi' - \theta'), \end{aligned}$$

where x, x', g are constants which remain to be determined. The cubic surfaces fail to determine them, for the reason that in all of them we have $i=0$, $i'=0$; and $16B+8\chi+\theta=16B'+8\chi'+\theta'$: this last is a very remarkable relation, for the existence of which I do not perceive any *à priori* reason.

Substituting herein for q, r their values from No. 11, this may be written in the form

$$\begin{aligned}\beta' = {} & 2n(n-2)(11n-24)+b(-66n+184)+c(-\tfrac{369}{4}n+240)\\ & +141\beta+\tfrac{359}{4}\gamma+66t\\ & - \quad x'\,i'+7j'-6\chi'-\tfrac{9}{4}\theta'-5C'-18B'\\ & -(x+87)\,i-21j-\tfrac{41}{2}\chi+\tfrac{51}{2}\theta-24C\\ & +\tfrac{1}{16}g(-16B-8\chi-\theta+16B'+8\chi'+\theta').\end{aligned}$$

64. We have of course by interchanging the unaccented and accented letters, the reciprocal equation giving the value of β.

Article Nos. 65 to 68. *Recapitulation.*

65. In recapitulation, I say that we have between the 42 quantities

$$\begin{gathered} n,\ a,\ \delta,\ \kappa\ ;\ b,\ k,\ t,\ q,\ \rho,\ j\ ;\ c,\ h,\ r,\ \sigma,\ \theta,\ \chi\ ;\ \beta,\ \gamma,\ i\ ;\ B,\ C,\\ n',\ a',\ \delta',\ \kappa'\ ;\ b',\ k',\ t',\ q',\ \rho',\ j'\ ;\ c',\ h',\ r',\ \sigma',\ \theta',\ \chi'\ ;\ \beta',\ \gamma',\ i'\ ;\ B',\ C',\end{gathered}$$

in all 25 equations, viz. these are

$$\begin{aligned} & a = a',\\ & a' = n(n-1)-2b-3c,\\ & \kappa' = 3n(n-2)-6b-8c,\\ & \delta' = \tfrac{1}{2}n(n-2)(n^2-9)-(n^2-n-6)(2b+3c)+2b(b-1)+6bc+\tfrac{9}{2}c(c-1),\\ & a(n-2) = \kappa - B + \rho + 2\sigma,\\ & b(n-2) = \rho + 2\beta + 3\gamma + 3t,\\ & c(n-2) = 2\sigma + 4\beta + \gamma + \theta,\\ & a(n-2)(n-3) = 2(\delta - C) + 3(ac-3\sigma-\chi) + 2(ab-2\rho-j),\\ & b(n-2)(n-3) = 4k + (ab-2\rho-j) + 3(bc-3\beta-2\gamma-i),\\ & c(n-2)(n-3) = 6h + (ac-3\sigma-\chi) + 2(bc-3\beta-2\gamma-i),\\ & q = b^2-b-2k-3\gamma-6t,\\ & r = c^2-c-2h-3\beta,\\ & a = n'(n'-1)-2b'-3c',\\ & \kappa = 3n'(n'-2)-6b'-8c',\\ & \delta = \tfrac{1}{2}n'(n'-2)(n'^2-9)-(n'^2-n'-6)(2b'+3c')+2b'(b'-1)+6b'c'+\tfrac{9}{2}c'(c'-1),\end{aligned}$$

$$a'(n'-2) = \kappa' - B' + \rho' + 2\sigma',$$
$$b'(n'-2) = \rho' + 2\beta' + 3\gamma' + 3t',$$
$$c'(n'-2) = 2\sigma' + 4\beta' + \gamma' + \theta',$$
$$a'(n'-2)(n'-3) = 2(\delta' - C') + 3(a'c' - 3\sigma' - \chi') + 2(a'b' - 2\rho' - j'),$$
$$b'(n'-2)(n'-3) = 4k' + (a'b' - 2\rho' - j') + 3(b'c' - 3\beta' - 2\gamma' - i'),$$
$$c'(n'-2)(n'-3) = 6h' + (a'c' - 3\sigma' - \chi') + 2(b'c' - 3\beta' - 2\gamma' - i'),$$
$$q' = b'^2 - b' - 2k' - 3\gamma' - 6t',$$
$$r' = c'^2 - c' - 2h' - 3\beta',$$

together with the equations for β and β'.

66. The symbols signify as follows; viz.

n, order of the surface.

a, order of the tangent cone drawn from any point to the surface.

δ, number of nodal edges of the cone.

κ, number of its cuspidal edges.

b, order of nodal curve.

k, number of its apparent double points.

t, number of its triple points.

q, its class.

ρ, number of points where nodal curve is met by curve of contact of tangent cone.

j, number of pinch-points.

c, order of cuspidal curve.

h, number of its apparent double points.

r, its class.

σ, number of points where cuspidal curve is met by curve of contact of tangent cone.

θ, number of off-points.

χ, number of close-points.

β, number of intersections of nodal and cuspidal curves, stationary points on cuspidal curve.

γ, number of intersections, stationary points on nodal curve.

i, number of intersections, not stationary on either curve.

B, number of binodes of surface.

C, number of cnicnodes.

67. And the accented letters have the like significations in regard to the reciprocal surface; or, referring them to the original surface, we have

n', class of the surface.

a', class of curve of intersection by any plane.

δ', number of double tangents of curve of intersection.

κ', number of its inflexions.

b', class of node-couple torse.

k', number of its apparent double planes.

t', number of its triple planes.

q', its order.

ρ', order of node-couple curve.

j', number of pinch-planes.

c', class of spinode torse.

h', number of its apparent double planes.

r', its order.

σ', order of spinode curve.

θ', number of off-planes.

χ', number of close-planes.

β', number of common planes of node-couple and spinode torses, stationary planes of the spinode torse.

γ', number of common planes, stationary planes of node-couple torse.

i', number of common planes, not stationary planes of either torse.

B', number of bitropes of surface.

C', number of its cnictropes.

68. It is hardly necessary to recall that a spinode plane is a tangent plane meeting the surface in a curve having at the point of contact a spinode or cusp; the envelope of the spinode planes is the spinode torse, and the locus of their points of contact the spinode curve. And similarly a node-couple plane is a double tangent plane, or plane meeting the surface in a curve having two nodes; the envelope of the planes is the node-couple torse, and the locus of the points of contact the node-couple curve; the other terms made use of are all explained in the present Memoir.

ADDITION, *August* 3, 1869.

As in the theory of Curves, so in that of Surfaces, there are certain functions of the order, class, &c. and singularities which have the same values in the original and the reciprocal figures respectively; for convenience I represent any such identity by means of the symbol Σ, viz. $\phi(n, a, b, \ldots) = \Sigma$ denotes that the function $\phi(n, a, b, \ldots)$ is equal to the same function $\phi(n', a', b', \ldots)$ of the accented letters. By what precedes we have $a = \Sigma$; and it is moreover clear that any function of the unaccented letters which is $= 0$, or which is equal to a symmetrical function of any of the accented and unaccented letters, or to a function of a, is $= \Sigma$; for instance, from the equations of No. 5 we have $3a' - \kappa' = 3n - c$, and thence $3n - c - \kappa = 3a' - \kappa - \kappa'$, that is, $3n - c - \kappa = \Sigma$;

and from one of the equations of No. 11 we have $n-2C-4B+\kappa-\sigma-2j-3\chi=n+n'-a, =\Sigma$; we have thus the system of eight equations,

$$\begin{aligned}
&a && =\Sigma,\\
&3n-c-\kappa && =\Sigma,\\
&a(n-2)-\kappa+B-\rho-2\sigma && =\Sigma,\\
&b(n-2)-\rho-2\beta-3\gamma-3t && =\Sigma,\\
&c(n-2)-2\sigma-4\beta-\gamma-\theta && =\Sigma,\\
&n+\kappa-\sigma-2C-4B-2j-3\chi && =\Sigma,\\
&2q-2\rho+\beta+3i+j && =\Sigma,\\
&3r+c-5\sigma-\beta-4\theta+2i+\chi && =\Sigma;
\end{aligned}$$

or if from these we eliminate κ, ρ, σ, then the system of five equations,

$$\begin{aligned}
&a && =\Sigma,\\
&n(c-8)-4\beta-\gamma-\theta+4C+8B+6\chi+4j && =\Sigma,\\
&(a-b)(n-2)-11n+3c+4C+9B+2\beta+3\gamma+3t+6\chi+4j && =\Sigma,\\
&3r-20n+6c-\beta+2i+10C+20B+16\chi+10j-4\theta && =\Sigma,\\
&2q-2b(n-2)+5\beta+6\gamma+6t+3i+j && =\Sigma.
\end{aligned}$$

By means of a theorem of Dr Clebsch's I was led to the following expression for the "deficiency" of a surface of the order n having the singularities considered in the foregoing Memoir:

$$\text{Deficiency} =\tfrac{1}{6}(n-1)(n-2)(n-3)-(n-3)(b+c)+\tfrac{1}{2}(q+r)+2t+\tfrac{7}{2}\beta+\tfrac{5}{2}\gamma+i-\tfrac{1}{8}\theta.$$

This should be equal to the deficiency of the reciprocal surface, viz. we must have

$$2(n-1)(n-2)(n-3)-12(n-3)(b+c)+6q+6r+24t+42\beta+30\gamma+12i-\tfrac{3}{8}\theta=\Sigma;$$

but from a combination of the last-mentioned five equations we have

$$\begin{aligned}
-2n^3+6n^2+4n+(12n-36)b+(12n-48)c-6q-6r-24t&\\
-41\beta-30\gamma-13i-7j-8\chi+2\theta-4C-10B&=\Sigma;
\end{aligned}$$

and adding to the last preceding equation we have

$$26n-12c+\beta-i-7j-8\chi+\tfrac{1}{2}\theta-4C-10B=\Sigma.$$

Substituting for Σ its value in terms of the accented letters, we obtain for β' the value

$$\begin{aligned}
\beta'=\beta+26n-12c+i'+7j'+8\chi'-\tfrac{1}{2}\theta'+4C'+10B'&\\
-26n'+12c'-i-7j-8\chi+\tfrac{1}{2}\theta-4C-10B.&
\end{aligned}$$

We have

$$c'=-3a+\kappa+3n',$$

and thence

$$12c'-26n'=-36a+12\kappa+10n';$$

writing herein

$$n' = a + \kappa - \sigma - 2C - 4B - 2j - 3\chi - \sigma,$$

the value is

$$= -26a + 22\kappa - 20C - 40B - 20j - 30\chi - 10\sigma.$$

Substituting for κ its value $= a(n-2) + B - \rho - 2\sigma$, we have

$$12c' - 26n' = a(22n - 70) - 20C - 18B - 20j - 30\chi - 22\rho - 44\sigma\,;$$

or substituting for a, ρ, σ their values, this is

$$\begin{aligned} &= \{n(n-1) - 2b - 3c\}(22n - 70) - 20C - 18B - 20j - 30\chi \\ &\quad - 22b(n-2) + \ \ 44\beta + 66\gamma + 66t \\ &\quad - 27c(n-2) + 108\beta + 27\gamma \qquad + 27\theta, \end{aligned}$$

and adding hereto the remaining terms,

$$\begin{aligned} &\beta + 26n - 12c + i' + 7j' + 8\chi' - \tfrac{1}{2}\theta' + 4C' + 10B' \\ &\qquad\qquad - i \ - 7j \ - 8\chi \ + \tfrac{1}{2}\theta \ - 4C \ - 10B, \end{aligned}$$

we have

$$\begin{aligned} \beta' = 2n(n-2)(11n-24) + b(-66n+184) + c(-93n+252) + 153\beta + 93\gamma + 66t \\ + i' + \ 7j' + \ 8\chi' - \ \tfrac{1}{2}\theta' + \ 4C' + 10B' \\ - i \ - 27j \ - 38\chi \ + \tfrac{55}{2}\theta \ - 24C \ - 28B. \end{aligned}$$

Comparing this with the value of β', No. 63 of the foregoing Memoir, we should have

$$\begin{aligned} 0 &= \tfrac{13}{4}cn - 12c - 12\beta - \tfrac{13}{4}\gamma \\ &\quad - (x' + \ 1)\,i' \qquad - 14\chi' - \tfrac{7}{4}\theta' - 9C' - 28B' \\ &\quad - (x + 86)\,i \ + \ 6j + \tfrac{35}{2}\chi - 2\theta \qquad\quad + 28B \\ &\quad + \tfrac{1}{16}g(-16B - 8\chi - \theta + 16B' + 8\chi' + \theta'), \end{aligned}$$

or, what is the same thing,

$$0 = 13cn - 48c - 48\beta - 13\gamma + \Phi,$$

if for shortness

$$\begin{aligned} \Phi &= -(4x' + \quad 4)\,i' \qquad - 56\chi' - 7\theta' - 36C' - 112B' \\ &\quad - (4x + 344)\,i \ + 24j + 70\chi - 8\theta \qquad\quad + 112B \\ &\quad + \tfrac{1}{4}g(-16B - 8\chi - \theta + 16B' + 8\chi' + \theta'). \end{aligned}$$

I do not attempt to verify this equation, but I will partially verify a result deducible from it; viz. if Φ' is the like function of the accented letters, then we have

$$\Phi - \Phi' = \Pi - \Pi',$$

where

$$\begin{aligned} \Pi = (4x' - 4x - 340)\,i + 24j + 126\chi + 224B + 36C - \theta \\ - \tfrac{1}{2}g(16B + 8\chi + \theta)\,; \end{aligned}$$

and Π' is the like function of the accented letters. And this being so, we should have

$$13cn - 48c - 48B - 13\gamma + \Pi = 13c'n' - 48c' - 48\beta' - 13\gamma' + \Pi',$$

or, as this may be written,

$$13cn - 48c - 48\beta - 13\gamma + \Pi = \Sigma.$$

We have

$$26n - 12c + \beta - i - 7j - 8\chi + \tfrac{1}{2}\theta - 4C - 10B = \Sigma;$$

and multiplying by -4 and adding, the equation to be verified is

$$13n(c-8) - 13(4\beta + \gamma) + \Pi + 4i + 28j + 32\chi - 2\theta + 16C + 40B = \Sigma.$$

But we have from the Memoir

$$-13n(c-8) + 13(4\beta + \gamma) \qquad - 52j - 78\chi + 13\theta - 52C - 104B = \Sigma,$$

which reduces the equation to

$$\Pi + 4i - 24j - 46\chi + 11\theta - 36C - 64B = \Sigma;$$

or substituting for Π its value, this is

$$(4x' - 4x - 336)\,i + 80\chi + 10\theta + 160B - 2\lambda(16B + 8\chi + \theta) = \Sigma,$$

that is

$$4(x' - x - 84)\,i - (\tfrac{1}{2}g - 10)(16B + 8\chi + \theta) = \Sigma,$$

an equation which is satisfied if

$$i' = i, \quad x' = x,$$

and

$$g = 20, \text{ or else } 16B + 8\chi + \theta = 16B' + 8\chi' + \theta'.$$

412.

A MEMOIR ON CUBIC SURFACES.

[From the *Philosophical Transactions of the Royal Society of London*, vol. CLIX. (for the year 1869), pp. 231—326. Received November 12, 1868,—Read January 14, 1869.]

THE present Memoir is based upon, and is in a measure supplementary to that by Professor Schläfli, "On the Distribution of Surfaces of the Third Order into Species, in reference to the presence or absence of Singular Points, and the reality of their Lines," *Phil. Trans.* vol. CLIII. (1863), pp. 193—241. But the object of the Memoir is different. I disregard altogether the ultimate division depending on the reality of the lines, attending only to the division into (twenty-two, or as I prefer to reckon it) twenty-three cases depending on the nature of the singularities. And I attend to the question very much on account of the light to be obtained in reference to the theory of Reciprocal Surfaces. The memoir referred to furnishes in fact a store of materials for this purpose, inasmuch as it gives (partially or completely developed) the equations in plane-coordinates of the several cases of cubic surfaces, or, what is the same thing, the equations in point-coordinates of the several surfaces (orders 12 to 3) reciprocal to these respectively. I found by examination of the several cases, that an extension was required of Dr Salmon's theory of Reciprocal Surfaces in order to make it applicable to the present subject; and the preceding "Memoir on the Theory of Reciprocal Surfaces," [411], was written in connexion with these investigations on Cubic Surfaces. The latter part of the Memoir is divided into sections headed thus:— "Section I = 12, equation $(X, Y, Z, W)^3 = 0$" &c. referring to the several cases of the cubic surface; but the paragraphs are numbered continuously throughout the Memoir.

Article Nos. 1 to 13. *The twenty-three Cases of Cubic Surfaces—Explanations and Table of Singularities.*

1. I designate as follows the twenty-three cases of cubic surfaces, adding to each of them its equation:

$$\begin{array}{lll}
\text{I} & = 12, & (X, Y, Z, W)^3 = 0, \\
\text{II} & = 12 - C_2, & W(a, b, c, f, g, h \between X, Y, Z)^2 + 2kXYZ = 0, \\
\text{III} & = 12 - B_3, & 2W(X+Y+Z)(lX+mY+nZ) + 2kXYZ = 0
\end{array}$$

IV	$=12-2C_2$,	$WXZ+Y^2(\gamma Z+\delta W)+(a, b, c, d \between X, Y)^3=0$,
V	$=12-B_4$,	$WXZ+(X+Z)(Y^2-aX^2-bZ^2)=0$,
VI	$=12-B_3-C_2$,	$WXZ+Y^2Z+(a, b, c, d \between X, Y)^3=0$,
VII	$=12-B_5$,	$WXZ+Y^2Z+YX^2-Z^3=0$,
VIII	$=12-3C_2$,	$Y^3+Y^2(X+Z+W)+4aXZW=0$,
IX	$=12-2B_3$,	$WXZ+(a, b, c, d \between X, Y)^3=0$,
X	$=12-B_4-C_2$,	$WXZ+(X+Z)(Y^2-X^2)=0$,
XI	$=12-B_6$,	$WXZ+Y^2Z+X^3-Z^3=0$,
XII	$=12-U_6$,	$W(X+Y+Z)^2+XYZ=0$,
XIII	$=12-B_3-2C_2$,	$WXZ+Y^2(X+Y+Z)=0$,
XIV	$=12-B_5-C_2$,	$WXZ+Y^2Z+YX^2=0$,
XV	$=12-U_7$,	$WX^2+XZ^2+Y^2Z=0$,
XVI	$=12-4C_2$,	$W(XY+XZ+YZ)+XYZ=0$,
XVII	$=12-2B_3-C_2$,	$WXZ+XY^2+Y^3=0$,
XVIII	$=12-B_4-2C_2$,	$WXZ+(X+Z)Y^2=0$,
XIX	$=12-B_6-C_2$,	$WXZ+Y^2Z+X^3=0$,
XX	$=12-U_8$,	$WX^2+XZ^2+Y^3=0$,
XXI	$=12-3B_3$,	$WXZ+Y^3=0$,
XXII	$=3, S(1, 1)$,	$WX^2+ZY^2=0$,
XXIII	$=3, S(\overline{1, 1})$,	$X(WX+YZ)+Y^3=0$;

2. Where C_2 denotes a conic-node diminishing the class by 2; B_3, B_4, B_5, B_6 a biplanar node diminishing (as the case may be) the class by 3, 4, 5, or 6; and U_6, U_7, U_8 a uniplanar node diminishing (as the case may be) the class by 6, 7, or 8. The affixed explanation, which I shall usually retain in connexion with the Roman number, shows therefore in each case what the class is, and also the singularities which cause the reduction: thus XIII $=12-B_3-2C_2$ indicates that there is a biplanar node, B_3, diminishing the class by 3, and two conic-nodes, C_2, each diminishing the class by 2; and thus that the class is $12-3-2.2, =5$. As regards the cases XXII and XXIII, these are surfaces having a nodal right line, and are consequently scrolls, each of the class 3, viz. XXII is the scroll $S(1, 1)$ having a simple directrix right line distinct from the nodal line, and XXIII is the scroll $S(\overline{1, 1})$ having a simple directrix right line coincident with the nodal line: see as to this my "Second Memoir on Skew Surfaces, otherwise Scrolls," *Phil. Trans.* vol. CLIV. (1864), pp. 559—577, [340].

3. The nature of the points C_2, B_3, B_4, B_5, B_6, U_6, U_7, U_8 requires to be explained.

$C(=C_2)$ is a conic-node, where, instead of the tangent plane, we have a proper quadric cone.

$B(=B_3, B_4, B_5$ or $B_6)$ is a biplanar-node, where the quadric cone becomes a plane-pair (two distinct planes): the two planes are called the biplanes, and their line of intersection is the edge:

In B_3, the edge is not a line on the surface—in the other cases it is; this implies that the surface is touched along the edge by a plane, viz. in B_4, B_5 the edge is torsal, in B_6 it is oscular:

In B_4, the tangent plane is distinct from each of the biplanes:

In B_5, the tangent plane coincides with one of the biplanes; we have thus an ordinary biplane, and a torsal biplane:

In B_6, the tangent plane coinciding with one of the biplanes becomes oscular; we have thus an ordinary biplane, and an oscular biplane.

$U\,(=U_6,\ U_7$ or $U_8)$ is a uniplanar-node, where the quadric cone becomes a coincident plane-pair; say, the plane is the uniplane. It is to be observed that there is not in this case any edge. The uniplane meets the cubic surface in three lines, or say "rays," passing through the uniplanar-node, viz.

In U_6, the rays are three distinct lines:

In U_7, two of them coincide:

In U_8, they all three coincide.

4. To connect these singular points with the theory of the preceding Memoir, it is to be observed that they are respectively equivalent to a certain number of the cnicnodes $C\,(=C_2)$ and binodes $B\,(=B_3)$, viz. we have

$$\begin{array}{rcl} C_2 &=& C, \\ B_3 &=& \qquad B, \\ B_4 &=& 2C, \\ B_5 &=& C + \ B, \\ \left\{\begin{array}{l} B_6 \\ U_6 \end{array}\right. & \begin{array}{c} = \\ = \end{array} & \begin{array}{l} 3C, \\ 3C, \end{array} \\ U_7 &=& 2C + \ B, \\ U_8 &=& C + 2B. \end{array}$$

5. I take the opportunity of remarking that although the expressions cnicnode and binode properly refer to the simple singularities C and B, yet as $C_2 = C$, C_2 is properly spoken of as a cnicnode, and we may (using the term binode as an abbreviation for biplanar-node) speak of any of the singularities B_3, B_4, B_5, B_6 as a binode. Thus the surface $\mathrm{X} = 12 - B_4 - C_2$ has a binode B_4 and a cnicnode C_2; although theoretically the binode B_4 is equivalent to two cnicnodes, and the surface belongs to those with three cnicnodes, or for which $C = 3$. I use also the expression unode for shortness, instead of uniplanar-node, to denote any of the singularities U_6, U_7, U_8.

6. The foregoing equations (substantially the same as Schläfli's) are *Canonical forms*; the reduction of the equation of any case of surface to the above form is not always obvious. It would appear that each equation is from its simplicity in the form best adapted to the separate discussion of the surface to which it belongs; there is the disadvantage that the equations do not always (when from the geometrical connexion of the surfaces they ought to do so) lead the one to the other; for instance, $\mathrm{V} = 12 - B_4$ includes $\mathrm{VII} = 12 - B_5$, but we cannot from the equation $WXZ + (X+Z)(Y^2 - aX^2 - bZ^2) = 0$ of the former pass to the equation $WXZ + Y^2Z + YX^2 - X^3 = 0$ of the latter. This would be a serious imperfection if the object were to form a theory of the quaternary function $(X, Y, Z, W)^3$; but the equations are in the present Memoir used only as means to an end, the establishment of the geometrical theory of the surfaces to which they respectively belong, and the imperfection is not material.

7. I have used the capital letters (X, Y, Z, W) in place of Schläfli's (x, y, z, w), reserving these in place of his (p, q, r, s) for plane-coordinates of the cubic surfaces, or (what is the same thing) point-coordinates of the reciprocal surfaces; but I have in several cases interchanged the coordinates (X, Y, Z, W) so that they do not *in this order* correspond to Schläfli's (x, y, z, w): this has been done so as to obtain a greater uniformity in the representation of the surfaces. To explain this, let A, B, C, D be the vertices of the tetrahedron formed by the coordinate planes $A = YZW$, $B = ZWX$, $C = WXY$, $D = XYZ$; the coordinate planes have been chosen so that determinate vertices of the tetrahedron shall correspond to determinate singularities of the surface.

8. Consider first the surfaces which have no nodes B or U. It is clear that the nodes C_2 might have been taken at any vertices whatever of the tetrahedron; they are taken thus: there is always a node C_2 at D; when there is a second node C_2, this is at C, the third one is at A, and the fourth at B.

9. Consider next the surfaces which have a binode B_3, B_4, B_5, or B_6; this is taken to be at D, and the biplanes to be $X=0$, $Z=0$ [1] (the edge being therefore DB), viz. in B_5 or B_6, where the distinction arises, $X=0$ is the ordinary biplane, $Z=0$ the torsal or (as the case may be) oscular biplane. If there is a second node, this of necessity lies in an ordinary biplane; it may be and is taken to be in the biplane $X=0$, at C. I suppose for a moment that this is a node C_2. It is only when the binode is B_3 or B_4 that there can be a third node, for it is only in these cases that there is a second ordinary biplane $Z=0$; but in these cases respectively the third node, a C_2, may be and is taken to be in the biplane $Z=0$, at A.

10. The only case of two binodes is when each is a B_3. Here the first is as above at D, its biplanes being $X=0$, $Z=0$; and the second is as above in the biplane $X=0$, at C; the biplanes thereof are then $X=0$ (which is thus a biplane common to the two binodes, or say a common biplane), and a remaining biplane which may be and is taken to be $W=0$. If there is a third node, this may be either C_2 or B_3, but it will in either case lie in the biplane $Z=0$ of the first binode, and also in the biplane $W=0$ of the second binode, that is, in the line BA; and it may be and is taken to be at A; if a binode, then its biplanes are of necessity $Z=0$, $W=0$; and the plane $Y=0$ will be the plane through the three binodes D, C, A.

11. If there is a unode, then this may be and is taken to be at D, and its uniplane may be taken to be $X=0$; in the surface $\text{XII} = 12 - U_6$ the uniplane is, however, taken to be $X+Y+Z=0$. There is never, besides the unode, any other node.

12. The result is that the nodes, in the order of their speciality, are in the equations taken to be at D, C, A, B respectively; and that (except in the case $\text{III} = 12 - B_3$) the biplanes of the first binode are $X=0$, $Z=0$ (for a binode B_5 or B_6, $X=0$ being the ordinary biplane, $Z=0$ the special biplane), those of the second binode $X=0$, $W=0$, those of the third binode $Z=0$, $W=0$, and that (except in the case $\text{XII} = 12 - U_6$) the uniplane is $X=0$. For example, in the surface $\text{XVII} = 12 - 2B_3 - C_2$, as represented by its equation $WXZ + Y^2Z + X^3 = 0$, we have a B_3 at D, the biplanes being $X=0$, $Z=0$, a B_3 at C, the biplanes being $X=0$, $W=0$ (therefore $X=0$ the common biplane), and a C_2 at A.

[1] In the case, however, of a single B_3, $\text{III} = 12 - B_3$, the biplanes are taken to be $X+Y+Z=0$, $lX+mY+nZ=0$.

13. It will be convenient (anticipating the results of the investigations contained in the present Memoir) to give at once the following Table of Singularities; the several symbols have of course the significations explained in the former Memoir.

	I $=12.$	II $=12-C_2.$	III $=12-B_3.$	IV $=12-2C_2.$ V $=12-B_4.$	VI $=12-B_3-C_2.$ VII $=12-B_5.$	VIII $=12-3C_2.$ X $=12-B_4-C_2.$ XI $=12-B_6.$ XII $=12-U_6.$	IX $=12-2B_3.$	XIII $=12-B_3-2C_2.$ XIV $=12-B_5-C_2.$ XV $=12-U_7.$	XVI $=12-4C_2.$ XVIII $=12-B_4-2C_2.$ XIX $=12-B_6-C_2.$	XVII $=12-2B_3-C_2.$ XX $=12-U_8.$	XXI $=12-3B_3.$	XXII $=S(1, 1).$ XXIII $=S(\overline{1, 1}).$	
n	3	3	3	3	3	3	3	3	3	3	3	3	n
a	6	6	6	6	6	6	6	6	6	6	6	4	a
δ	0	1	0	2	1	3	0	2	4	1	0	0	δ
κ	6	6	7	6	7	6	8	7	6	8	9	3	κ
b	0	0	0	0	0	0	0	0	0	0	0	1	b
k	0	0	0	0	0	0	0	0	0	0	0	0	k
t	0	0	0	0	0	0	0	0	0	0	0	0	t
q	0	0	0	0	0	0	0	0	0	0	0	0	q
ρ	0	0	0	0	0	0	0	0	0	0	0	1	ρ
j	0	0	0	0	0	0	0	0	0	0	0	2	j
c	0	0	0	0	0	0	0	0	0	0	0	0	c
h	0	0	0	0	0	0	0	0	0	0	0	0	h
r	0	0	0	0	0	0	0	0	0	0	0	0	r
σ	0	0	0	0	0	0	0	0	0	0	0	0	σ
θ	0	0	0	0	0	0	0	0	0	0	0	0	θ
χ	0	0	0	0	0	0	0	0	0	0	0	0	χ
β	0	0	0	0	0	0	0	0	0	0	0	0	β
γ	0	0	0	0	0	0	0	0	0	0	0	0	γ
i	0	0	0	0	0	0	0	0	0	0	0	0	i
C	0	1	0	2	1	3	0	2	4	1	0	0	C
B	0	0	1	0	1	0	2	1	0	2	3	0	B
	I	II	III	IV V	VI VII	VIII X XI XII	IX	XIII XIV XV	XVI XVIII XIX	XVII XX	XXI	XXII XXIII	
n'	12	10	9	8	7	6	6	5	4	4	3	3	n'
a'	6	6	6	6	6	6	6	6	6	6	6	4	a'
δ'	0	0	0	0	0	0	0	0	0	0	0	0	δ'
κ'	9	9	9	9	9	9	9	9	9	9	9	3	κ'
b'	27	15	9	7	3	3	0	1	3	0	0	1	b'
k'	216	60	18	12	3	0	0	0	0	0	0	0	k'
t'	45	15	6	3	0	1	0	0	1	0	0	0	t'
q'	0	0	0	0	0	0	0	0	0	0	0	0	q'
ρ'	27	15	9	7	3	3	0	1	3	0	0	1	ρ'
j'	0	0	0	1	0	3	0	1	6	0	0	2	j'
c'	24	18	16	12	10	6	8	4	0	2	0	0	c'
h'	180	96	72	38	24	6	12	2	0	0	0	0	h'
r'	30	24	42	17	24	9	32	5	0	2	0	0	r'
σ'	12	12	12	10	9	6	8	4	0	2	0	0	σ'
θ'	0	0	16	0	8	0	16	0	0	0	0	0	θ'
χ'	0	0	0	0	1	0	0	2	0	2	0	0	χ'
β'	54	30	18	13	6	3	0	1	0	0	0	0	β'
γ'	0	0	0	0	0	0	0	0	0	0	0	0	γ'
i'	0	0	0	0	0	0	0	0	0	0	0	0	i'
C'	0	0	0	0	0	0	0	0	0	0	0	0	C'
B'	0	0	0	0	0	0	1	0	0	1	3	0	B

Article Nos. 14 to 19. *Explanation in regard to the Determination of the Number of certain Singularities.*

14. In the several cases I to XXI, we have a cubic surface ($n=3$), with singular points C and B but without singular lines. The section by an arbitrary plane is thus a curve, order $n=3$, that is, a cubic curve, without nodes or cusps, and therefore of the class $a'=6$, having $\delta'=0$ double tangents and $\kappa'=9$ inflexions. The tangent cone with an arbitrary point as vertex is a cone of the order $a=6$, having in the case $\text{I}=12$, $\delta=0$ nodal lines and $\kappa=6$ cuspidal lines, but with (in the several other cases) C nodal lines and B cuspidal lines (or rather singular lines tantamount to C double lines and B cuspidal lines): the class of the cone, or order of the reciprocal surface, is thus $n'=6\,.\,5-2(0+C)-3(6+B)=12-2B-3C$.

15. In the general case $\text{I}=12$, there are on the cubic surface 27 lines, lying by 3's in 45 planes; these 27 lines constitute the node-couple curve of the order $\rho'=27$, and the node-couple torse consists of the pencils of planes through these lines respectively, being thus of the class $\rho'=b'=27$; the 45 planes are triple tangent planes of the node-couple torse, which has thus $t'=45$ triple tangent planes. But in the other cases it is only certain of the 27 lines, say the "facultative lines" (as will be explained), which constitute the node-couple curve of the order ρ': the pencils of planes through these lines constitute the node-couple torse of the class $b'=\rho'$; the t' planes, each containing three facultative lines, are the triple tangent planes of the node-couple torse. Or if (as is somewhat more convenient) we refer the numbers b', t' to the reciprocal surface, then the lines, reciprocals of the facultative lines, constitute the nodal curve of the order b'; and the points t', each containing three of these lines, are the triple points of the nodal curve. Inasmuch as the nodal curve consists of right lines, the number k' of its apparent double points is given by the formula $2k'=b'^2-b'-6t'$; and comparing with the formula $q'=b'^2-b'-2k'-3\gamma'-6t'$, we have $q'+3\gamma'=0$, that is, $q'=0$ (q' the class of the nodal curve), and also $\gamma'=0$.

16. In the general case $\text{I}=12$, the spinode curve is the complete intersection of the cubic surface by the Hessian surface of the order 4, and it is thus of the order $\sigma'=12$; but in the other cases the complete intersection consists of the spinode curve together with certain right lines not belonging to the curve, and the spinode curve is of an order σ' less than 12: this will be further explained, and the reduction accounted for (see *post*, Nos. 24 *et seq.*).

17. Again, in the general case $\text{I}=12$, each of the 27 lines is a double tangent of the spinode curve, and the tangent planes of the surface at the points of contact are common tangent planes of the spinode torse and the node-couple torse, stationary planes of the spinode torse; or we have $\beta'=2\rho'=54$. In the other cases, however, instead of the 27 lines we must take only the facultative lines, each of which is or is not a double or a single tangent of the spinode curve; and the tangent planes of the surface at the points of contact are the common tangent planes as above—that is, the number of contacts gives β' not in general $=2\rho'$.

18. There are not, except as above, any common tangent planes of the two torses, that is, not only $\gamma'=0$ as already mentioned, but also $i'=0$. I do not at present account *à priori* for the values $\theta'=16$, 8, and 16, which present themselves in the Table. The cubic surface cannot have a plane of conic contact, and we have thus in every case $C'=0$; but the value of B' is not in every case $=0$.

19. In what precedes we see how a discussion of the equation of the cubic surface should in the several cases respectively lead to the values b', t', ρ', σ', β', j', χ', B', and how in the reciprocal surface the nodal curve of the order b' is known by means of the facultative lines of the original cubic surface. The cuspidal curve c' might also be obtained as the reciprocal of the spinode-torse; but this would in general be a laborious process, and it is the less necessary, inasmuch as the equation of the reciprocal surface is in each case obtained in a form putting in evidence the cuspidal curve.

Article Nos. 20 to 23. *The Lines and Planes of a Cubic Surface; Facultative Lines; Explanation of Diagrams.*

20. In the general surface I = 12, we have 27 lines and 45 triple-tangent planes, or say simply, planes: through each line pass 5 planes, in each plane lie 3 lines. For the surfaces II to XXI (the present considerations do not of course apply to the Scrolls) several of the lines come to coincide with each other, and several of the planes also come to coincide with each other; but the number of the lines is always reckoned as 27, and that of the planes as 45. If we attend to the distinct lines and the distinct planes, each line has a multiplicity, and the sum of these is $=27$; and so each plane has a multiplicity, and the sum of these is $=45$. Again, attending to a particular line in a particular plane, the line has a frequency 1, 2, or 3, that is, it represents 1, 2, or 3 of the 3 lines in the plane (this is in fact the distinction of a scrolar, torsal, or oscular line); and similarly, the plane has a frequency 1, 2, 3, 4, or 5, according to the number which it represents of the 5 planes through the line. It requires only a little consideration to perceive that the multiplicity of the plane into its frequency in regard to the line is equal to the multiplicity of the line into its frequency in regard to the plane. Observe, further, that if M be the multiplicity of the plane, then, considering it in regard to the lines contained therein, we get the products (M, M, M), $(2M, M)$, or $3M$, according as the three lines are or are not distinct, but that the sum of the products is always $=3M$, and that in regard to all the planes the total sum is 3×45, $=135$. And so if M' be the multiplicity of the line, then, considering it in regard to the planes which pass through it, we get the products (M', M', M', M', M'), $(2M', M', M', M')$, ...$(5M')$, as the case may be, but that the sum of the products is $=5M'$, and that in regard to all the lines the sum is 5×27, $=135$, as before.

21. The mode of coincidence of the lines and planes, and the several distinct lines and planes which are situate in or pass through the several distinct planes and lines respectively, are shown in the annexed diagrams I to XXI[1]: the multiplicity

[1] See the commencements of the several sections.

of each line appears by the upper marginal line, and that of each plane by the left-hand marginal column (thus in diagram I, $27 \times 1 = 27$ and $45 \times 1 = 45$, 1 is the multiplicity of each line, and it is also the multiplicity of each plane); the frequencies of a line and plane in regard to each other appear by the dots in the square opposite to the line and plane in question, these being read, for the frequency of the line vertically, and for the frequency of the plane horizontally; thus $\cdot\vdots$ indicates that the frequency of the line is $=3$, and the frequency of the plane is $=2$. There should be and are in every line of the diagram 3 dots, and in every column of the diagram 5 dots (a symbol $\cdot\vdots$ being read as just explained, 2 dots in the line, 3 dots in the column).

22. For the surface $\text{I} = 12$, there is of course no distinction between the lines, but these form only a single class, and the like for the planes; but for the other surfaces the lines and planes form separate classes, as shown in the diagrams by the lower marginal explanation of the lines, and the right-hand marginal explanation of the planes. I use here and elsewhere "ray" to denote a line passing through a single node; "axis" to denote a line joining two nodes; "edge" (as above) to denote the edge of a binode; any other line is a "mere line." An axis is always torsal or oscular; when it is torsal, the plane touching along the axis contains a third line which is the "transversal" of such axis; but a transversal may be a mere line, a ray, or an axis; in the case $\text{XVI} = 12 - 4C_2$, each transversal is a transversal in regard to two axes.

23. In the general case $\text{I} = 12$, each of the 27 lines is, as already mentioned, part of the node-couple curve; and the node-couple curve is made up of the 27 lines, and is thus a curve of the order 27. In fact each plane through a line meets the cubic surface in this line, and in a conic; the line and conic meet in two points, and the plane (that is in any plane) through the line is thus a double tangent plane touching the surface at the two points in question; the locus of the points of contact, that is the line itself, is thus part of the node-couple curve. But in the other cases, II to XXI, certain of the lines do not belong to the node-couple curve (this will be examined in detail in the several cases respectively); but I wish to show here how in a general way a line passing through a node, say a nodal ray, is not part of the node-couple curve. To fix the ideas, consider the surface $\text{II} = 12 - C_2$; there are here through C_2 six lines, or say rays: attending to any one of these, a plane through the ray meets the surface in the ray itself and in a conic; the ray and the conic meet as before in two points, one of them being the point C_2: the plane touches the surface at the other point, *but it does not touch the surface at* C_2. (I am not sure, and I leave it an open question, whether we ought to say that at a node C_2 there is *no* tangent plane, or to say that only the tangent planes of the nodal cone are tangent planes of the surface; but, at any rate, an arbitrary plane through C_2 is *not* a tangent plane.) The plane through the ray is only a single tangent plane, not a double tangent plane; and the ray is not part of the node-couple curve. We say that a line of the surface is or is not "facultative" according as it does or does not form part of the node-couple curve.

Article Nos. 24 to 26. *Axis; the different kinds thereof.*

24. A line joining two nodes is an axis; such a line is always a line, and it is a torsal or oscular line, of the surface. But some further distinctions are requisite; using the expressions in their strict sense, cnicnode $=C$, binode $=B$, an axis is a CC-axis joining two cnicnodes, or it is a CB-axis joining a cnicnode and a binode, or it is a BB-axis joining two binodes. A CC-axis is torsal, the transversal being a mere line, not a ray through either of the cnicnodes; a CB-axis is torsal, the transversal being a ray of the binode; a BB-axis is oscular. The distinction is of course carried through as regards the higher biplanar nodes B_4, B_5, B_6, and the uniplanar nodes U_6, U_7, U_8: thus $(B_3=B)$ the edge of a binode B_3 is not an axis at all, but $(B_4=2C)$ the edge of a binode B_4 is a CC-axis; $(B_5=B+C)$ the edge of a binode B_5 is a CB-axis; $(B_6=3C)$ the edge of a binode B_6 is a thrice-taken CC-axis; $(U_6=3C)$ each of the rays is regarded as a CC-axis; $(U_7=B+2C)$ the double ray is regarded as a twice-taken CB-axis, and the single ray as a CC-axis; $(U_8=2B+C)$ the ray is regarded as a BB-axis + a twice-taken CB-axis.

25. It has been mentioned that the intersection of the surface with the Hessian consists of the spinode curve, together with certain right lines; these lines are in fact the axes—viz. the examination of the several cases shows that in the complete intersection each CC-axis presents itself twice, each CB-axis 3 times, and each BB-axis 4 times. We thus see that a CC-axis, or rather the torsal plane along such axis, is the pinch-plane or singularity $j'=1$; the CB-axis, or rather the torsal plane along such axis, the close-plane or singularity $\chi'=1$; and the BB-axis, or oscular plane along such axis, the bitrope or singularity $B'=1$; for a cubic surface with singular lines the expression of σ' being in fact $\sigma'=12-2j'-3\chi'-4B'$. There are, however, some cases requiring explanation; thus for the case $\text{VIII}=12-B_5$, where the edge is by what precedes a CB-axis, the complete intersection is made up of the edge 4 times and of an octic curve; the consideration of the reciprocal surface shows, however, that the edge taken once is really part of the spinode curve (viz. that this curve is made up of the edge taken once and of the octic curve, its order being thus $\sigma'=9$); and the interpretation then of course is that the intersection is made up of the edge taken 3 times (as for a CB-axis it should be) and of the spinode curve.

26. I remark in further explanation, that in the several sections, in showing how the complete intersection of the cubic surface with the Hessian is made up, I have not referred to the axes in the above precise significations; thus $\text{XIV}=12-B_5-C_2$, the binode B_5 is $C+B$, and the edge is thus a CB-axis, while the axis B_5C_2 is a CB-axis + a CC-axis $(\chi'=1+1, =2, j'=1)$. The complete intersection should therefore consist of the spinode curve, + edge (as a CB-axis) 3 times + axis (as a CB-axis + a CC-axis) $2+3, =5$ times: it is in the section stated (in perfect consistency herewith, but without the full explanation) that the intersection is made up of the axis 5 times, the edge 4 times, and a cubic curve—which cubic curve together with the edge once constitutes the spinode curve; and so in other cases: this explanation will, I think, remove all difficulty.

Article Nos. 27 to 32. *On the Determination of the Reciprocal Equation.*

27. Consider in general the cubic surface $(*\between X, Y, Z, W)^3 = 0$, and in connexion therewith the equation $Xx + Yy + Zz + Ww = 0$, which regarding therein X, Y, Z, W as current coordinates, and x, y, z, w as constants, is the equation of a plane. If from the two equations we eliminate one of the coordinates, for instance W, we obtain

$$(*\between Xw,\ Yw,\ Zw,\ -(Xx + Yy + Zz))^3 = 0,$$

which, (X, Y, Z) being current coordinates, is obviously the equation of the cone, vertex $(X = 0,\ Y = 0,\ Z = 0)$, which stands on the section of the cubic surface by the plane. Equating to zero the discriminant of this function in regard to (X, Y, Z), we express that the cone has a nodal line; that is, that the section has a node, or, what is the same thing, that the plane $xX + yY + zZ + wW = 0$ is a tangent plane of the cubic surface; and we thus by the process in fact obtain the equation of the cubic surface in the reciprocal or plane coordinates (x, y, z, w). Consider in the same equation x, y, z, w as current coordinates, (X, Y, Z) as given parameters, the equation represents a system of three planes, viz. these are the planes $xX + yY + zZ + wW' = 0$, where W' has the three values given by the equation $(*\between X, Y, Z, W')^3 = 0$, or, what is the same thing, X, Y, Z, W' are the coordinates of any one of the three points of intersection of the cubic surface by the line $\dfrac{x}{X} = \dfrac{y}{Y} = \dfrac{z}{Z}$; (X, Y, Z, W') belongs to a point on the surface, and

$$xX + yY + zZ + wW' = 0$$

is the polar plane of this point in regard to a quadric surface $X^2 + Y^2 + Z^2 + W^2 = 0$; the equation

$$(*\between Xw,\ Yw,\ Zw,\ -(Xx + Yy + Zz))^3 = 0$$

is thus the equation of a system of 3 planes, the polar planes of three points of the cubic surface (which three points lie on an arbitrary line through the point $x = 0$, $y = 0$, $z = 0$). In equating to zero the discriminant in regard to (X, Y, Z), we find the envelope of the system of three planes, or say of a plane, the polar plane of an arbitrary point on the cubic surface,—or we have the equation of the reciprocal surface, being, as is known, the same thing as the equation of the cubic surface in the reciprocal or plane coordinates (x, y, z, w). In what precedes we have the explanation of an ordinary process of finding the equation of the reciprocal surface, this equation being thereby given by equating to zero the discriminant of a function $(*\between X, Y, Z)^3$, that is, of a ternary cubic function.

28. The process, as last explained, is a special one, viz. the position of a point on the surface is determined by means of certain two parameters, the ratios $X : Y : Z$ which fix the position of the line joining this point with the point $(x = 0,\ y = 0,\ z = 0)$. More generally we may consider the position of the point as determined by means of any two parameters; the equation of the polar plane then contains the two parameters, and by taking the envelope in regard to the two parameters considered as variable, we have the equation of the reciprocal surface.

29. But let the parameters, say θ, ϕ, be regarded as varying successively; if ϕ alone vary, we have on the surface a curve Θ, the equation whereof contains the parameter θ, and when θ varies this curve sweeps over the surface. The envelope in regard to ϕ of the polar plane of a point of the surface is a torse, the reciprocal of the curve Θ, and the envelope of the torse is the reciprocal surface. In particular the curve Θ may be the plane section by any plane through a fixed line, say, by the plane $P - \theta Q = 0$; the section is a cubic curve, the reciprocal is a sextic cone having its vertex in a fixed line (the reciprocal of the line $P=0$, $Q=0$), and the reciprocal surface is thus obtained as the envelope of this cone; assuming that the equation of the sextic cone has been obtained, this is an equation of a certain order in the parameter θ; or writing $\theta = P : Q$, we obtain the equation of the reciprocal surface by equating to zero the discriminant of a *binary* function of (P, Q).

30. With a variation, this process is a convenient one for obtaining the reciprocal of a cubic surface: we take the fixed line to be one of the lines on the cubic surface; the curve Θ is then a conic, its reciprocal is a quadricone, and the envelope of this quadricone is the required reciprocal surface. This is really what Schläfli does (but the process is not explained) in the several instances in which he obtains the equation of the reciprocal surface by means of a binary function. I remark that it would be very instructive, for each case of surface, to take the variable plane successively through the several kinds of lines on the particular surface; the equation of the reciprocal surface would thus be obtained under different forms, putting in evidence the relation to the reciprocal surface of the fixed line made use of. But this is an investigation which I do not enter upon: I adopt in each case Schläfli's process, without explanation, and merely write down the ternary or (as the case may be) binary function by means of which the equation of the reciprocal surface is obtained.

31. It is to be mentioned that there is a reciprocal process of obtaining the equation of the reciprocal surface; we may imagine, touching the cubic surface along any curve, a series of planes; that is, a torse circumscribed about the surface, and the equation whereof contains a variable parameter θ; the reciprocal figure is a curve, the equations whereof contain the parameter θ; the locus of this curve is the reciprocal surface; that is, the equation of the reciprocal surface is obtained by eliminating θ from the equations of the curve. In particular let the torse be the circumscribed cone having its vertex at any point of a fixed line; the reciprocal figure is then a plane curve, the plane of which passes through the line which is the reciprocal of the fixed line; it is moreover clear that if the position of the vertex on the fixed line be determined by the parameter θ *linearly* (for instance if the vertex be given as the intersection of the fixed line by a plane $P - \theta Q = 0$), then the equation of the plane of the curve will be of the form $P' = \theta Q'$, containing the parameter θ linearly; the other equation of the plane curve will contain θ rationally, and the elimination will be at once effected by substituting in this other equation for θ its value, $= P' \div Q'$. And observe moreover that if the fixed line be a line on the cubic surface, then the cone is a quadricone having for its reciprocal a conic; the reciprocal surface is thus given as the locus of a variable conic, the plane of which always passes through a fixed line; there are thus on the reciprocal surface

series of such conics. It would be very instructive and interesting to carry out the investigation in detail.

32. The equation of the reciprocal surface is found by equating to zero the discriminant of a ternary or a binary function[1], viz. this is a ternary cubic, or a binary quartic, cubic, or quadric. The equation as given in the form disct. $=0$, contains a factor which for the adopted forms of equations is always a power or product of powers of w, z, x[2] known *à priori*, and which is thrown out without difficulty, the equation being thereby reduced to the proper order. There is the singular advantage that the process puts in evidence the cuspidal curve of the resulting reciprocal surface, viz. for a ternary cubic, the form obtained is $S^3-T^2=0$, and for a binary quartic it is the equivalent form $I^3-27J^2=0$; but for the factor thrown out as just mentioned, we should have simply ($S=0$, $T=0$), or, as the case may be, ($I=0$, $J=0$) for equations of the cuspidal curve; the existence of the factor occasions however a modification, viz. the intersection of the two surfaces is not an indecomposable curve, and the cuspidal curve is in most cases, not the complete intersection, but a partial intersection of the two surfaces. In several cases it thus happens that the cuspidal curve is obtained as a curve $\left\|\begin{matrix} P, & Q, & R \\ P', & Q', & R' \end{matrix}\right\|=0$, without or with further speciality. Similarly when the equation of the reciprocal surface is obtained by means of a binary cubic; if the coefficients hereof (functions of course of the coordinates x, y, z, w) be A, B, C, D, then the surface is

$$(AD-BC)^2-4(AC-B^2)(BD-C^2)=0,$$

having the cuspidal curve $\left\|\begin{matrix} A, & B, & C \\ B, & C, & D \end{matrix}\right\|=0$, subject however to modification in the case of a thrown out factor.

Article Nos. 33 and 34. *Explanation as to the Sections of the Memoir.*

33. As regards the following Sections I to XXIII, it is to be observed that for the general surface $\mathrm{I}=12$, I do not attempt to form the equation of the reciprocal surface, and in some of the other cases, $\mathrm{II}=12-C_2$ &c., the equation of the reciprocal surface is either not obtained in a completely developed form, or it is too complicated to allow of its being dealt with, for instance so as to put in evidence the nodal curve of the surface. Portions of the theory given in the latter sections are consequently omitted in the earlier ones, and in particular in the Section I there is given only the diagram of the 27 lines and the 45 planes (with however developments as to notation and otherwise which have no place in the subsequent sections), and with the analytical expressions for the several lines and planes, although from the

[1] In some easy cases, for instance $\mathrm{XVI}=12-4C_2$, the equation of the reciprocal surface is obtained otherwise by a direct elimination.

[2] The factor is in general a power or product of powers of the linear functions which, equated to zero, give the equations of the planes reciprocal to the several nodes of the surface.

want of the equation of the reciprocal surface these analytical expressions have no present application. And so in some of the next following sections, no application is made of the analytical expressions of the lines and planes.

34. I call to mind that if a line be given as the intersection of the two planes

$$AX + BY + CZ + DW = 0, \quad A'X + B'Y + C'Z + D'W = 0,$$

then the six coordinates of the line are

$$\begin{array}{cccccc} a, & b, & c, & f, & g, & h \\ = AD' - A'D, & BD' - B'D, & CD' - C'D, & BC' - B'C, & CA' - C'A, & AB' - A'B, \end{array}$$

and that in terms of its six coordinates the line is given as the common intersection of the four planes

$$\left(\begin{array}{cccc} . & h, & -g, & a \\ -h, & . & f, & b \\ g, & -f, & . & c \\ -a, & -b, & -c, & . \end{array}\right)\!\!\left(X,\ Y,\ Z,\ W\right) = 0,$$

and that (reciprocating as usual in regard to $X^2 + Y^2 + Z^2 + W^2 = 0$) the coordinates of the reciprocal line are (f, g, h, a, b, c); that is, this is the common intersection of the four planes

$$\left(\begin{array}{cccc} . & c, & -b, & f \\ -c, & . & a, & g \\ b, & -a, & . & h \\ -f, & -g, & -h, & . \end{array}\right)\!\!\left(x,\ y,\ z,\ w\right) = 0.$$

It is in some cases more convenient to consider a line as determined as the intersection of two planes rather than by means of its six coordinates; thus, for instance, to speak of the line $X = 0$, $Y = 0$ rather than of the line (0, 0, 0, 1, 0, 0); and in some of the sections I have preferred not to give the expressions of the six coordinates of the several lines.

Article Nos. 35 to 46. § I = 12, Equation $(X, Y, Z, W)^3 = 0$.

35. There is in the system of the 27 lines and the 45 planes a complicated and many-sided symmetry which precludes the existence of any unique notation: the notation can only be obtained by starting from some arrangement which is not unique, but one of a system of several like arrangements. The notation employed in my original paper "On the Simple Tangent Planes of Surfaces of the Third Order," *Camb. and Dub. Math. Journ.* vol. IV. 1849, pp. 118—132, [76], and which is shown in the right hand and lower margins of the diagram, starts from such an arrangement; but

it is so complicated that it can hardly be considered as at all putting in evidence the relations of the lines and planes; that of Dr Hart (Salmon, "On the Triple Tangent Planes of a Surface of the Third Order," same volume, pp. 252—260), depending on an arrangement of the 27 lines according to a cube of 3 each way, is a singularly elegant one, and will be presently reproduced.

36. But the most convenient one is Schläfli's, starting from a double-sixer; viz. we can (and that in 36 different ways) select out of the 27 lines two systems each of six lines, such that no two lines of the same system intersect, but that each line of the one system intersects all but the corresponding line of the other system; or, say, if the lines are

$$\begin{array}{cccccc} 1, & 2, & 3, & 4, & 5, & 6 \\ 1', & 2', & 3', & 4', & 5', & 6', \end{array}$$

then these have the thirty intersections

	1′,	2′,	3′,	4′,	5′,	6′
1		.	.	.	.	.
2	.		.	.	.	.
3	.	.		.	.	.
4	.	.	.		.	.
5	.	.	.	.		.
6	.	.	.	.	.	

Any two lines such as 1, 2′ lie in a plane which may be called 12′; similarly the lines 1′, 2 lie in a plane which may be called 1′2; these two planes meet in a line 12; and any three lines such as 12, 34, 56 meet in pairs, lying in a plane 12.34.56. We have thus the entire system of the 27 lines and 45 planes, as in effect completely explained by what has been stated, but which is exhibited in full in the diagram.

37. The diagram of the lines and planes is

Planes.		56	46	45	36	35	34	26	25	24	23	16	15	14	13	12	6′	5′	4′	3′	2′	1′	6	5	4	3	2	1	
I=12		$\overline{27}$ $\overline{27}$														27 × 1 = 27													
12′																•					•							•	w
13′															•					•								•	ζ
14′														•					•									•	z
15′													•					•										•	z
16′												•					•											•	$\overline{\text{z}}$
21′																•						•					•		ξ
23′											•									•							•		θ
24′										•									•								•		$\overline{\text{f}}$
25′									•									•									•		$\overline{\text{p}}_1$
26′								•									•										•		p
31′															•							•				•			$\overline{\theta}$
32′											•										•					•			η
34′							•												•							•			g
35′						•												•								•			q
36′					•												•									•			q_1
41′														•								•			•				f
42′										•											•				•				y
43′							•													•					•				$\overline{\text{h}}$
45′				•														•							•				m
46′		•															•								•				m_1
51′													•									•		•					p
52′	45 × 1 = 45								•												•			•					y
53′						•														•				•					$\overline{\text{r}}$
54′				•															•					•					n_1
56′		•															•							•					l
61′												•										•	•						p_1
62′								•													•		•						$\overline{\text{y}}$
63′					•															•			•						$\overline{\overline{\text{r}}}$
64′			•																•				•						n
65′		•																•					•						l_1
12 . 34 . 56		•					•									•													x
12 . 35 . 46			•			•										•													x
12 . 36 . 45				•	•											•													x
13 . 24 . 56		•								•					•														h
13 . 25 . 46			•						•						•														r
13 . 26 . 45				•				•							•														$\overline{\text{r}}_1$
14 . 23 . 56			•								•			•															$\overline{\text{g}}$
14 . 25 . 36					•				•					•															$\overline{\text{n}}$
14 . 26 . 35						•		•						•															$\overline{\text{n}}_1$
15 . 23 . 46			•								•		•																$\overline{\text{q}}_1$
15 . 24 . 36					•					•			•																$\overline{\text{m}}_1$
15 . 26 . 34							•	•					•																$\overline{\text{l}}_1$
16 . 23 . 45				•							•	•																	$\overline{\text{q}}$
16 . 24 . 35						•				•		•																	$\overline{\text{m}}$
16 . 25 . 34	$\overline{45}$ $\overline{45}$						•		•			•																	$\overline{\overline{\text{l}}}_1$
		a_5	a_7	a_9	a_8	a_6	a_4	b_8	b_6	b_4	b_2	c_8	c_6	c_4	c_3	a_1	c_9	c_7	c_5	c_2	b_1	a_3	b_9	b_7	b_5	b_3	a_2	c_1	

Lines.

38. It has been mentioned that the number of double-sixers was $=36$, these are as follows:

1,	2,	3,	4,	5,	6	Assumed primitive	1
1′,	2′,	3′,	4′,	5′,	6′		
1,	1′,	23,	24,	25,	26	Like arrangements	15
2,	2′,	13,	14,	15,	16		
1,	2,	3,	56,	46,	45	Like arrangements	20
23,	13,	12,	4,	5,	6		36

where, if we take any column $\frac{1}{1'}$ of two lines, we have the complete number 216 of pairs of non-intersecting lines (each line meets 10 lines, there are therefore $27-1-10$, $=16$, which it does not meet, and the number of non-intersecting pairs is thus $\frac{1}{2}.27.16=216$).

39. We can out of the 45 planes select, and that in 120 ways, a trihedral-pair, that is, two triads of planes, such that the planes of the one triad, intersecting those of the other triad, give 9 of the 27 lines. Analytically if $X=0$, $Y=0$, $Z=0$ and $U=0$, $V=0$, $W=0$ are the equations of the six planes, then the equation of the cubic surface is $XYZ+kUVW=0$. See as to this *post*, No. 44.

The trihedral plane pairs are:

12′,	23′,	31′		
1′2,	2′3,	3′1	No. is =	20
12′,	34′,	14.23.56		
2′3,	4′1,	12.34.56	=	90
14.25.36,	35.16.24,	26.34.15		
14.35.26,	25.16.34,	36.24.15	=	10
				120

The construction of the last set is most easily effected by the diagram

$$\begin{array}{ccc}
1\ 2\ 3 & \times & 4\ 5\ 6 \\
3\ 1\ 2 & & 5\ 6\ 4 \\
2\ 3\ 1 & & 6\ 4\ 5
\end{array}$$

$$\underbrace{\qquad\qquad\qquad}$$

$$\Vert$$

$$\begin{array}{ccc}
14 & 25 & 36 \\
35 & 16 & 24 \\
26 & 34 & 15
\end{array}$$

It is immaterial how the two component triads 123 and 456 are arranged, we obtain always the same trihedral pair.

40. Dr Hart arranges the 27 lines, cubically, thus:

A_1	B_1	C_1	a_1	b_1	c_1	α_1	β_1	γ_1
A_2	B_2	C_2	a_2	b_2	c_2	α_2	β_2	γ_2
A_3	B_3	C_3	a_3	b_3	c_3	α_3	β_3	γ_3

where letters of the same alphabet denote lines in the same plane, if only the letters are the same or the suffixes the same; thus A_1, A_2, A_3 lie in a plane $A_1A_2A_3$; A_1, B_1, C_1 lie in a plane $A_1B_1C_1$. Letters of different alphabets denote lines which meet according to the Table

a_1	b_2	c_3	b_1	c_2	a_3	c_1	a_2	b_3
	A_1			B_1			C_1	
α_1	β_2	γ_3	β_1	γ_2	α_3	γ_1	α_2	β_3
c_2	a_3	b_1	a_2	b_3	c_1	b_2	c_3	a_1
	A_2			B_2			C_2	
β_3	γ_1	α_2	γ_3	α_1	β_2	α_3	β_1	γ_2
b_3	c_1	a_2	c_3	a_1	b_2	a_3	b_1	c_2
	A_3			B_3			C_3	
γ_2	α_3	β_1	α_2	β_3	γ_1	β_2	γ_3	α_1

where the letter in the centre of the square denotes a line lying in the same plane with the lines denoted by the letters of each vertical pair in the same square. Thus A_1 lies in the planes $A_1a_1\alpha_1$, $A_1b_2\beta_2$, $A_1c_3\gamma_3$ (and in the before-mentioned two planes $A_1A_2A_3$, $A_1B_1C_1$).

41. I find that one way in which this may be identified with the double-sixer notation is to represent the above arrangement by

1,	2′,	12	3′,	4,	34	13,	24,	56
14,	25,	36	2,	6′,	26	1′,	16,	6
4′,	5,	45	23,	46,	15	3,	35,	5′

and then the identification may apparently be effected in (720 × 36 =) 25920 ways, viz. we may first in any way permute the $\frac{1}{1'}$, $\frac{2}{2'}$, $\frac{3}{3'}$, $\frac{4}{4'}$, $\frac{5}{5'}$, $\frac{6}{6'}$, by this means not altering the double-sixer $\frac{1\ 2\ 3\ 4\ 5\ 6}{1'\ 2'\ 3'\ 4'\ 5'\ 6'}$, and then upon the arrangements so obtained make any of the substitutions which permute *inter se* the 36 double-sixers.

42. The equations of the 45 planes are obtained in my paper last referred to, viz. taking the equation of the surface to be

$$W\left(1,\ 1,\ 1,\ 1,\ mn+\frac{1}{mn},\ nl+\frac{1}{nl},\ lm+\frac{1}{lm},\ l+\frac{1}{l},\ m+\frac{1}{m},\ n+\frac{1}{n}\right)\!\!\left(X,\ Y,\ Z,\ W\right)^2+kXZY=0,$$

where
$$k=\frac{p^2-\beta^2}{2(p-\alpha)},\ \alpha=lmn+\frac{1}{lmn},\ \beta=lmn-\frac{1}{lmn},$$

then the equations of the planes are:

$W=0,$ [12′ = w]

$lX+mY+nZ+W\left[1+\frac{1}{k}\left(l-\frac{1}{l}\right)\left(m-\frac{1}{m}\right)\left(n-\frac{1}{n}\right)\right]=0,$ [23′ = θ]

$\frac{X}{l}+\frac{Y}{m}+\frac{Z}{n}+W\left[1-\frac{1}{k}\left(l-\frac{1}{l}\right)\left(m-\frac{1}{m}\right)\left(n-\frac{1}{n}\right)\right]=0,$ [31′ = $\bar{\theta}$]

$X=0,$ [12.34.56 = x]

$Y=0,$ [42′ = y]

$Z=0,$ [14′ = z]

$X+\frac{1}{k}\left(m-\frac{1}{m}\right)\left(n-\frac{1}{n}\right)W=0,$ [21′ = ξ]

$Y+\frac{1}{k}\left(n-\frac{1}{n}\right)\left(l-\frac{1}{l}\right)W=0,$ [32′ = η]

$Z+\frac{1}{k}\left(l-\frac{1}{l}\right)\left(m-\frac{1}{m}\right)W=0,$ [13′ = ζ]

$lX+\frac{Y}{m}+\frac{Z}{n}+W=0,$ [41′ = f]

$\frac{X}{l}+mY+\frac{Z}{n}+W=0,$ [34′ = g]

$\frac{X}{l}+\frac{Y}{m}+Z+W=0,$ [13.24.56 = h]

$\frac{X}{l}+mY+nZ+W=0,$ [24′ = $\bar{\mathrm{f}}$]

$lX+\frac{Y}{m}+nZ+W=0,$ [14.25.36 = $\bar{\mathrm{g}}$]

$lX+mY+\frac{Z}{n}+W=0,$ [43′ = $\bar{\mathrm{h}}$]

$X+\frac{l(p-\alpha)+2mn}{p+\beta}W=0,$ [12.35.46 = x]

$Y+\frac{m(p-\alpha)+2nl}{p+\beta}W=0,$ [52′ = y]

$Z+\frac{n(p-\alpha)+2lm}{p+\beta}W=0,$ [15′ = z]

$X+\frac{\frac{1}{l}(p-\alpha)+\frac{2}{mn}}{p+\beta}W=0,$ [12.36.45 = $\bar{\mathrm{x}}$]

$Y+\frac{\frac{1}{m}(p-\alpha)+\frac{2}{nl}}{p+\beta}W=0,$ [62′ = $\bar{\mathrm{y}}$]

$Z+\frac{\frac{1}{n}(p-\alpha)+\frac{2}{lm}}{p+\beta}W=0,$ [16′ = $\bar{\mathrm{z}}$]

$$-\frac{2n}{m(p-\alpha)}X+\frac{1}{m}Y+nZ+W=0, \qquad [56'=\mathrm{l}]$$

$$lX-\frac{2l}{n(p-\alpha)}Y+\frac{1}{n}Z+W=0, \qquad [45'=\mathrm{m}]$$

$$\frac{1}{l}X+mY-\frac{2m}{l(p-\alpha)}Z+W=0, \qquad [64'=\mathrm{n}]$$

$$-\frac{2m}{n(p-\alpha)}X+mY+\frac{1}{n}Z+W=0, \qquad [15.26.34=\bar{\mathrm{l}}]$$

$$\frac{1}{l}X-\frac{2n}{l(p-\alpha)}Y+nZ+W=0, \qquad [16.24.35=\bar{\mathrm{m}}]$$

$$lX+\frac{1}{m}Y-\frac{2l}{m(p-\alpha)}Z+W=0, \qquad [14.25.36=\bar{\mathrm{n}}]$$

$$-\frac{n(p-\alpha)}{2m}X+\frac{Y}{m}+nZ+W=0, \qquad [65'=\mathrm{l}_1]$$

$$lX-\frac{l(p-\alpha)}{2n}Y+\frac{1}{n}Z+W=0, \qquad [46'=\mathrm{m}_1]$$

$$\frac{1}{l}X+mY-\frac{m(p-\alpha)}{2l}Z+W=0, \qquad [54'=\mathrm{n}_1]$$

$$\frac{m(p-\alpha)}{2n}X+mY+\frac{1}{n}Z+W=0, \qquad [16.25.34=\bar{\mathrm{l}}_1]$$

$$\frac{1}{l}X-\frac{n(p-\alpha)}{2l}Y+nZ+W=0, \qquad [15.24.36=\bar{\mathrm{m}}_1]$$

$$lX+\frac{1}{m}Y-\frac{l(p-\alpha)}{2m}Z+W=0, \qquad [14.26.35=\bar{\mathrm{n}}_1]$$

$$-\frac{2X}{p-\alpha}+nY+mZ+\left(mn(p-\alpha)-2l\ (1-m^2-n^2)\right)\frac{W}{p+\beta}=0, \qquad [51'=\mathrm{p}]$$

$$nX-\frac{2Y}{p-\alpha}+lZ+\left(nl\ (p-\alpha)-2m(1-n^2-l^2)\right)\frac{W}{p+\beta}=0, \qquad [35'=\mathrm{q}]$$

$$mX+lY-\frac{2Z}{p-\alpha}+\left(lm\ (p-\alpha)-2n\ (1-l^2-m^2)\right)\frac{W}{p+\beta}=0, \qquad [13.25.46=\mathrm{r}]$$

$$-\frac{2X}{p-\alpha}+\frac{1}{n}Y+\frac{1}{m}Z+\left(\frac{1}{mn}(p-\alpha)-\frac{2}{l}\left(1-\frac{1}{m^2}-\frac{1}{n^2}\right)\right)\frac{W}{p-\beta}=0, \qquad [26'=\bar{\mathrm{p}}]$$

$$\frac{1}{n}X+\frac{2Y}{p-\alpha}+\frac{1}{l}Z+\left(\frac{1}{nl}(p-\alpha)-\frac{2}{m}\left(1-\frac{1}{n^2}-\frac{1}{l^2}\right)\right)\frac{W}{p-\beta}=0, \qquad [16.23.45=\bar{\mathrm{q}}]$$

$$\frac{1}{m}X+\frac{1}{l}Y-\frac{2Z}{p-\alpha}+\left(\frac{1}{lm}(p-\alpha)-\frac{2}{n}\left(1-\frac{1}{l^2}-\frac{1}{m^2}\right)\right)\frac{W}{p-\beta}=0, \qquad [36'=\bar{\mathrm{r}}]$$

$$-\frac{p-\alpha}{2}X+\frac{Y}{n}+\frac{Z}{m}-lmn\left(\frac{1}{l}\left(1-\frac{1}{m^2}-\frac{1}{n^2}\right)(p-\alpha)-\frac{2}{mn}\right)\frac{W}{p+\beta}=0, \quad [25'=\mathrm{p}_1]$$

$$\frac{X}{n}-\frac{p-\alpha}{2}Y+\frac{Z}{l}-lmn\left(\frac{1}{m}\left(1-\frac{1}{n^2}-\frac{1}{l^2}\right)(p-\alpha)-\frac{2}{nl}\right)\frac{W}{p+\beta}=0, \quad [15.23.46=\mathrm{q}_1]$$

$$\frac{X}{m}+\frac{Y}{l}-\frac{p-\alpha}{2}Z-lmn\left(\frac{1}{n}\left(1-\frac{1}{l^2}-\frac{1}{m^2}\right)(p-\alpha)-\frac{2}{lm}\right)\frac{W}{p+\beta}=0, \quad [53'=\mathrm{r}_1]$$

$$-\frac{p-\alpha}{2}X+nY+mZ-\frac{1}{lmn}\left(l(1-m^2-n^2)(p-\alpha)-2mn\right)\frac{W}{p-\beta}=0, \quad [61'=\bar{\mathrm{p}}_1]$$

$$nX-\frac{p-\alpha}{2}Y+lZ-\frac{1}{lmn}\left(m(1-n^2-l^2)(p-\alpha)-2nl\right)\frac{W}{p-\beta}=0, \quad [36'=\bar{\mathrm{q}}_1]$$

$$mX-lY-\frac{p-\alpha}{2}Z-\frac{1}{lmn}\left(n(1-l^2-m^2)(p-\alpha)-2lm\right)\frac{W}{p-\beta}=0, \quad [13.26.45=\bar{\mathrm{r}}_1]$$

43. The coordinates of the 27 lines are then found to be as follows:

(a)	(b)	(c)
1	0	0
0	1	0
0	0	1
$1-\frac{1}{kl}\left(m-\frac{1}{m}\right)\left(n-\frac{1}{n}\right)$	$-\frac{m}{k}\left(m-\frac{1}{m}\right)\left(n-\frac{1}{n}\right)$	$-\frac{n}{k}\left(m-\frac{1}{m}\right)\left(n-\frac{1}{n}\right)$
$-\frac{l}{k}\left(n-\frac{1}{n}\right)\left(l-\frac{1}{l}\right)$	$1-\frac{1}{km}\left(n-\frac{1}{n}\right)\left(l-\frac{1}{l}\right)$	$-\frac{n}{k}\left(n-\frac{1}{n}\right)\left(l-\frac{1}{l}\right)$
$-\frac{l}{k}\left(l-\frac{1}{l}\right)\left(m-\frac{1}{m}\right)$	$-\frac{m}{k}\left(l-\frac{1}{l}\right)\left(m-\frac{1}{m}\right)$	$1-\frac{1}{kn}\left(l-\frac{1}{l}\right)\left(m-\frac{1}{m}\right)$
$1-\frac{l}{k}\left(m-\frac{1}{m}\right)\left(n-\frac{1}{n}\right)$	$-\frac{1}{mk}\left(m-\frac{1}{m}\right)\left(n-\frac{1}{n}\right)$	$-\frac{1}{nk}\left(m-\frac{1}{m}\right)\left(n-\frac{1}{n}\right)$
$-\frac{1}{lk}\left(n-\frac{1}{n}\right)\left(l-\frac{1}{l}\right)$	$1-\frac{m}{k}\left(n-\frac{1}{n}\right)\left(l-\frac{1}{l}\right)$	$-\frac{1}{nk}\left(n-\frac{1}{n}\right)\left(l-\frac{1}{l}\right)$
$-\frac{1}{lk}\left(l-\frac{1}{l}\right)\left(m-\frac{1}{m}\right)$	$-\frac{1}{mk}\left(l-\frac{1}{l}\right)\left(m-\frac{1}{m}\right)$	$1-\frac{n}{k}\left(l-\frac{1}{l}\right)\left(m-\frac{1}{m}\right)$
1	0	0
0	1	0
0	0	1

(f)	(g)	(h)	
0	0	0	$(12 = a_1)$
0	0	0	$(\ 2' = b_1)$
0	0	0	$(\ 1 = c_1)$
0	$-n$	m	$(\ 2 = a_2)$
n	0	$-l$	$(23 = b_2)$
$-m$	l	0	$(\ 3' = c_2)$
0	$-\frac{1}{n}$	$\frac{1}{m}$	$(\ 1' = a_3)$
$\frac{1}{n}$	0	$-\frac{1}{l}$	$(\ 3 = b_3)$
$-\frac{1}{m}$	$\frac{1}{l}$	0	$(13 = c_3)$
0	$-\frac{1}{n}$	m	$(34 = a_4)$
n	0	$-\frac{1}{l}$	$(24 = b_4)$
$-\frac{1}{m}$	l	0	$(14 = c_4)$

(a)	(b)	(c)
1 0 0	0 1 0	0 0 1
$2m\left(l-\frac{1}{l}\right)$ $-m(p-\alpha)\left(1+\frac{2nl}{m(p-\alpha)}\right)$ $2\left(1+\frac{2lm}{n(p-\alpha)}\right)$	$2\left(1+\frac{2mn}{l(p-\alpha)}\right)$ $2n\left(m-\frac{1}{m}\right)$ $-n(p-\alpha)\left(1+\frac{2lm}{n(p-\alpha)}\right)$	$-l(p-\alpha)\left(1+\frac{2mn}{l(p-\alpha)}\right)$ $2\left(1+\frac{2nl}{m(p-\alpha)}\right)$ $2l\left(n-\frac{1}{n}\right)$
$2n\left(l-\frac{1}{l}\right)$ $2\left(1+\frac{2nl}{m(p-\alpha)}\right)$ $-n(p-\alpha)\left(1+\frac{2lm}{n(p-\alpha)}\right)$	$-l(p-\alpha)\left(1+\frac{2mn}{l(p-\alpha)}\right)$ $2l\left(m-\frac{1}{m}\right)$ $2\left(1+\frac{2lm}{n(p-\alpha)}\right)$	$2\left(1+\frac{2mn}{l(p-\alpha)}\right)$ $-m(p-\alpha)\left(1+\frac{2nl}{m(p-\alpha)}\right)$ $2m\left(n-\frac{1}{n}\right)$
$\frac{2}{n}\left(l-\frac{1}{l}\right)$ $-2\left(1+\frac{2m}{nl(p-\alpha)}\right)$ $\frac{p-\alpha}{n}\left(1+\frac{2n}{lm(p-\alpha)}\right)$	$\frac{p-\alpha}{l}\left(1+\frac{2l}{mn(p-\alpha)}\right)$ $\frac{2}{l}\left(m-\frac{1}{m}\right)$ $-2\left(1+\frac{2n}{lm(p-\alpha)}\right)$	$-2\left(1+\frac{2l}{mn(p-\alpha)}\right)$ $\frac{p-\alpha}{m}\left(1+\frac{2m}{nl(p-\alpha)}\right)$ $\frac{2}{m}\left(n-\frac{1}{n}\right)$
$\frac{2}{m}\left(l-\frac{1}{l}\right)$ $\frac{p-\alpha}{m}\left(1+\frac{2m}{nl(p-\alpha)}\right)$ $-2\left(1+\frac{2n}{lm(p-\alpha)}\right)$	$-2\left(1+\frac{2l}{mn(p-\alpha)}\right)$ $\frac{2}{n}\left(m-\frac{1}{m}\right)$ $\frac{p-\alpha}{n}\left(1+\frac{2n}{lm(p-\alpha)}\right)$	$\frac{p-\alpha}{l}\left(1+\frac{2l}{mn(p-\alpha)}\right)$ $-2\left(1+\frac{2m}{nl(p-\alpha)}\right)$ $\frac{2}{l}\left(n-\frac{1}{n}\right)$

(f)	(g)	(h)	
0	$-n$	$\frac{1}{m}$	$(56 = a_5)$
$\frac{1}{n}$	0	$-l$	$(4 = b_5)$
$-m$	$\frac{1}{l}$	0	$(4' = c_5)$
0	$-(p+\beta)$	$-\frac{2(p+\beta)}{l(p-\alpha)}$	$(35 = a_6)$
$-\frac{2(p+\beta)}{m(p-\alpha)}$	0	$-(p+\beta)$	$(25 = b_6)$
$-(p+\beta)$	$-\frac{2(p+\beta)}{n(p-\alpha)}$	0	$(15 = c_6)$
0	$\frac{2(p+\beta)}{l(p-\alpha)}$	$p+\beta$	$(46 = a_7)$
$p+\beta$	0	$\frac{2(p+\beta)}{m(p-\alpha)}$	$(5 = b_7)$
$\frac{2(p+\beta)}{n(p-\alpha)}$	$p+\beta$	0	$(5' = c_7)$
0	$-\frac{2l(p-\beta)}{p-\alpha}$	$-(p-\beta)$	$(36 = a_8)$
$-(p-\beta)$	0	$-\frac{2m(p-\beta)}{p-\alpha}$	$(26 = b_8)$
$-\frac{2n(p-\beta)}{p-\alpha}$	$-(p-\beta)$	0	$(16 = c_8)$
0	$p-\beta$	$\frac{2l(p-\beta)}{p-\alpha}$	$(45 = a_9)$
$\frac{2m(p-\beta)}{p-\alpha}$	0	$p-\beta$	$(6 = b_9)$
$p-\beta$	$\frac{2n(p-\beta)}{p-\alpha}$	0	$(6' = c_9)$

44. We have $X=0$, $Y=0$, $Z=0$, $W=0$ for the equations of the planes

$$(12.34.56=x),\quad (42'=y),\quad (14'=z),\quad (12'=w);$$

and representing by $\mathrm{f}=lX+\frac{1}{m}Y+\frac{1}{n}Z+W=0$ the equation of any other plane $(41'=\mathrm{f})$ the equation of the cubic surface may be presented in the several forms:

$$\begin{aligned}
0=U&=W\mathrm{f}\bar{\mathrm{f}} &&+\mathrm{k}\xi YZ,\\
&=W\mathrm{g}\bar{\mathrm{g}} &&+\mathrm{k}\eta ZX,\\
&=W\mathrm{h}\bar{\mathrm{h}} &&+\mathrm{k}\zeta XY,\\
&=W\theta\bar{\theta} &&+\mathrm{k}\xi\eta\zeta,\\
&=W\mathrm{l}\bar{\mathrm{l}}_1 &&+\mathrm{ky}\bar{\mathrm{z}}\mathrm{x},\\
&=W\mathrm{m}\bar{\mathrm{m}}_1 &&+\mathrm{kz}\bar{\mathrm{x}}\mathrm{y},\\
&=W\mathrm{n}\bar{\mathrm{n}}_1 &&+\mathrm{kx}\bar{\mathrm{y}}\mathrm{z},\\
&=W\mathrm{l}_1\bar{\mathrm{l}} &&+\mathrm{k}\bar{\mathrm{y}}\mathrm{zx},\\
&=W\mathrm{m}_1\bar{\mathrm{m}} &&+\mathrm{k}\bar{\mathrm{z}}\mathrm{xy},\\
&=W\mathrm{n}_1\bar{\mathrm{n}} &&+\mathrm{k}\bar{\mathrm{x}}\mathrm{yz},\\
&=W\mathrm{pp}_1 &&+\mathrm{k}\xi\mathrm{yz},\\
&=W\mathrm{qq}_1 &&+\mathrm{k}\eta\mathrm{zx},\\
&=W\mathrm{rr}_1 &&+\mathrm{k}\zeta\mathrm{xy},\\
&=W\bar{\mathrm{p}}\bar{\mathrm{p}}_1 &&+\mathrm{k}\xi\bar{\mathrm{y}}\bar{\mathrm{z}},\\
&=W\bar{\mathrm{q}}\bar{\mathrm{q}}_1 &&+\mathrm{k}\eta\bar{\mathrm{z}}\bar{\mathrm{x}},\\
&=W\bar{\mathrm{r}}\bar{\mathrm{r}}_1 &&+\mathrm{k}\zeta\bar{\mathrm{x}}\bar{\mathrm{y}},
\end{aligned}$$

which are the 16 forms containing W, out of the complete system of 120 trihedral-pair forms.

45. The 27 lines are each of them facultative; we have therefore $b'=\rho'=27$; $t'=45$; moreover each of the lines is a double tangent of the spinode curve, and therefore $\beta'(=2\rho')=54$.

46. The equation of the reciprocal surface is not here investigated; its form is

$$S^3-T^2=0,$$

where $S=(*\between x, y, z, w)^4$, $T=(*\between x, y, z, w)^6$; wherefore $n'=12$.

The nodal curve is composed of the lines which are the reciprocals of the original 27 lines ($b'=27$, $t'=45$ *ut suprà*). It may be remarked that the reciprocal of a double-sixer is a double-sixer. Hence the 27 lines of the reciprocal surface may be (and that in 36 different ways) represented by

$$\begin{matrix} 1, & 2, & 3, & 4, & 5, & 6 \\ 1', & 2', & 3', & 4', & 5', & 6' \end{matrix}$$

$$12,\ 13, \ldots .\ 56,$$

where 12 is now the line joining the points 12′ and 1′2; and so for the other lines. The lines 12, 34, 56 meet in a point 12.34.56; the 30 points 12′, 1′2 ... 56′, 5′6, and the fifteen points 12.34.56 make up the 45 points t'.

The above equation, $S^3 - T^2 = 0$, shows that the cuspidal curve is a complete intersection 6×4; $c' = 24$.

Section $\text{II} = 12 - C_2$.

Article Nos. 47 to 59. Equation $W(a, b, c, f, g, h \between X, Y, Z)^2 + 2kXYZ = 0$.

47. It may be remarked that the system of lines and planes is at once deduced from that belonging to I = 12, by supposing that in the double-sixer the corresponding lines 1 and 1′, &c. severally coincide; the line 12, instead of being given as the intersection of the planes 12′, 1′2, is given as the third line in the plane 12, which in fact represents the coincident planes 12′ and 1′2.

48. The diagram is

Planes.		Lines.																					
		56	46	45	36	35	34	26	25	24	23	16	15	14	13	12	6	5	4	3	2	1	
$\Pi = 12 - C_2$		$\overline{21}$ $\overline{27}$							15 × 1 = 15											6 × 2 = 12			
12																••					•	•	
13															••					•		•	
14														••					•			•	
15													••					•				•	
16												••					•					•	
23											••									•	•		
24										••									•		•		
25									••									•			•		Biradial planes, through each pair of rays.
26	15 × 2 = 30							••									•				•		
34							••												•	•			
35						••												•		•			
36					••												•			•			
45				••														•	•				
46			••														•		•				
56		••															•	•					
12 . 34 . 56		•					•									•							
12 . 35 . 46			•			•										•							
12 . 36 . 45				•	•											•							
13 . 24 . 56		•								•					•								
13 . 25 . 46			•						•						•								
13 . 26 . 45				•				•							•								
14 . 23 . 56		•									•			•									
14 . 25 . 36	15 × 1 = 15				•				•					•									Planes each containing three mere lines.
14 . 26 . 35						•		•						•									
15 . 23 . 46			•								•		•										
15 . 24 . 36					•					•			•										
15 . 26 . 34							•	•					•										
16 . 23 . 45				•							•	•											
16 . 24 . 35						•				•		•											
16 . 25 . 34	$\overline{30}$ $\overline{45}$						•		•			•											
									Mere lines, one in each biradial ray.											Rays, through the node situate in a quadric cone.			

49. Putting the equation of the surface in the form

$$W\left(1, 1, 1, l+\frac{1}{l}, m+\frac{1}{m}, n+\frac{1}{n}\right)(X, Y, Z)^2+\frac{\alpha\beta\gamma\delta}{p}XYZ=0,$$

where for shortness

$$\begin{aligned}\alpha &= mn - l,\\ \beta &= nl - m,\\ \gamma &= lm - n,\\ \delta &= lmn - 1,\\ p &= lmn,\end{aligned}$$

then taking $X=0$ as the equation of the plane [12], $Y=0$ as that of the plane [34], $Z=0$ as that of the plane [56], the equations of the 30 distinct planes are found to be

$$\begin{array}{rl}
X=0, & [12]\\
Y=0, & [34]\\
Z=0, & [56]\\
m\ X+l\ Y+Z=0, & [23]\\
m^{-1}X+l\ Y+Z=0, & [24]\\
m\ X+l^{-1}Y+Z=0, & [13]\\
m^{-1}X+l^{-1}Y+Z=0, & [14]\\
X+n\ Y+m\ Z=0, & [45]\\
X+n^{-1}Y+m\ Z=0, & [46]\\
X+n\ Y+m^{-1}Z=0, & [35]\\
X+n^{-1}Y+m^{-1}Z=0, & [36]\\
n\ X+Y+l\ Z=0, & [16]\\
n^{-1}X+Y+l\ Z=0, & [15]\\
n\ X+Y+l^{-1}Z=0, & [26]\\
n^{-1}X+Y+l^{-1}Z=0, & [25]\\
W=0, & [12.34.56]\\
X+\beta\gamma\ W=0, & [12.36.45]\\
X-\alpha\delta\ W=0, & [12.35.46]\\
Y+\alpha\gamma\ W=0, & [16.25.34]\\
Y-\beta\delta\ W=0, & [15.26.34]\\
Z+\alpha\beta\ W=0, & [14.23.56]\\
Z-\gamma\delta\ W=0, & [13.24.56]\\
mnX+nl\ Y+lm\ Z+\alpha\beta\gamma\delta\ W=0, & [16.23.45]\\
pX+n\ Y+m\ Z+\beta\gamma\delta\ W=0, & [13.26.45]\\
nX+p\ Y+l\ Z+\gamma\alpha\delta\ W=0, & [16.24.35]\\
mX+l\ Y+p\ Z+\alpha\beta\delta\ W=0, & [15.23.46]\\
X+lm\ Y+ln\ Z-\beta\gamma\delta\ W=0, & [15.24.36]\\
lmX+\ Y+mn\ Z-\gamma\alpha\delta\ W=0, & [13.25.46]\\
nlX+mn\ Y+\ Z-\alpha\beta\delta\ W=0, & [14.26.35]\\
lX+m\ Y+n\ Z-\alpha\beta\gamma\ W=0, & [14.25.36]
\end{array}$$

50. And the coordinates of the 21 distinct lines are

(a)	(b)	(c)	(f)	(g)	(h)	whence equations may be taken to be
l	0	0	0	-1	1	(1) $X=0,\ Y+lZ=0$
0	m	0	1	0	-1	(3) $Y=0,\ Z+mX=0$
0	0	n	-1	1	0	(5) $Z=0,\ X+nY=0$
l^{-1}	0	0	0	-1	1	(2) $X=0,\ Y+l^{-1}Z=0$
0	m^{-1}	0	1	0	-1	(4) $Y=0,\ Z+m^{-1}X=0$
0	0	n^{-1}	-1	1	0	(6) $Z=0,\ X+n^{-1}Y=0$
1	n	m	0	$\frac{m}{\beta\gamma}$	$-\frac{n}{\beta\gamma}$	(45) $X+nY+mZ=0,\ X+\beta\gamma W=0$
n	1	l	$-\frac{l}{\gamma\alpha}$	0	$\frac{n}{\gamma\alpha}$	(16) $Y+lZ+nX=0,\ Y+\gamma\alpha W=0$
m	l	1	$\frac{l}{\alpha\beta}$	$-\frac{m}{\alpha\beta}$	0	(23) $Z+mX+lY=0,\ Z+\alpha\beta W=0$
1	$\frac{1}{n}$	m	0	$-\frac{m}{\alpha\delta}$	$\frac{1}{n\alpha\delta}$	(46) $X+n^{-1}Y+mZ=0,\ X-\alpha\delta W=0$
n	1	$\frac{1}{l}$	$\frac{1}{l\beta\delta}$	0	$-\frac{n}{\beta\delta}$	(26) $Y+l^{-1}Z+nX=0,\ Y-\beta\delta W=0$
$\frac{1}{m}$	l	1	$-\frac{l}{\gamma\delta}$	$\frac{1}{m\gamma\delta}$	0	(24) $Z+m^{-1}X+lY=0,\ Z-\gamma\delta W=0$
1	n	$\frac{1}{m}$	0	$-\frac{1}{m\alpha\delta}$	$\frac{n}{\alpha\delta}$	(35) $X+nY+m^{-1}Z=0,\ X-\alpha\delta W=0$
$\frac{1}{n}$	1	l	$\frac{l}{\beta\delta}$	0	$-\frac{1}{n\beta\delta}$	(15) $Y+lZ+n^{-1}X=0,\ Y-\beta\delta W=0$
m	$\frac{1}{l}$	1	$-\frac{1}{l\gamma\delta}$	$\frac{m}{\gamma\delta}$	0	(13) $Z+mX+l^{-1}Y=0,\ Z-\gamma\delta W=0$
1	$\frac{1}{n}$	$\frac{1}{m}$	0	$\frac{1}{m\beta\gamma}$	$-\frac{1}{n\beta\gamma}$	(36) $X+n^{-1}Y+m^{-1}Z=0,\ X+\beta\gamma W=0$
$\frac{1}{n}$	1	$\frac{1}{l}$	$-\frac{1}{l\gamma\alpha}$	0	$\frac{1}{n\gamma\alpha}$	(25) $Y+l^{-1}Z+n^{-1}X=0,\ Y+\gamma\alpha W=0$
$\frac{1}{m}$	$\frac{1}{l}$	1	$\frac{1}{l\alpha\beta}$	$-\frac{1}{m\gamma\alpha}$	0	(14) $Z+m^{-1}X+l^{-1}Y=0,\ Z+\alpha\beta W=0$
1	0	0	0	0	0	(12) $X=0,\ W=0$
0	1	0	0	0	0	(34) $Y=0,\ W=0$
0	0	1	0	0	0	(56) $Z=0,\ W=0$

51. The six nodal rays are not, the fifteen mere lines are, facultative. Hence

$$b' = \rho' = 15\,;\ t' = 15.$$

52. Resuming the equation $W(a, b, c, f, g, h \between X, Y, Z)^2 + 2kXYZ = 0$, the equation of the Hessian surface is found to be

$$\begin{aligned} &KW^2(a, b, c, f, g, h \between X, Y, Z)^2 \\ &+ 2kW\{(a, b, c, f, g, h \between X, Y, Z)^2 (FX + GY + HZ) - 3KXYZ\} \\ &- k^2\{a^2X^4 + b^2Y^4 + c^2Z^4 - 2bcY^2Z^2 - 2caZ^2X^2 - 2abX^2Y^2 \\ &\qquad - 4XYZ[(af + gh)X + (bg + hf)Y + (ch + fg)Z]\} = 0, \end{aligned}$$

where

$$(A, B, C, F, G, H) = (bc - f^2,\ ca - g^2,\ ab - h^2,\ gh - af,\ hf - bg,\ fg - ch),$$

$$K = abc - af^2 - bg^2 - ch^2 + 2fgh.$$

The Hessian and the cubic intersect in an indecomposable curve, which is the spinode curve; that is, spinode curve is a complete intersection 3×4; $\sigma' = 12$.

The equations of the spinode curve may be written in the simplified form

$$\begin{aligned} &W(a, b, c, f, g, h \between X, Y, Z)^2 + 2kXYZ = 0, \\ &- 8KXYZW \\ &+ 8kXYZ(afX + bgY + chZ) \\ &- k^2\{a^2X^4 + b^2Y^4 + c^2Z^4 - 2bcY^2Z^2 - 2caZ^2X^2 - 2abX^2Y^2\} = 0\,; \end{aligned}$$

and it appears hereby that the node C_2 is a sixfold point on the curve, the tangents of the curve in fact coinciding with the six rays.

Each of the 15 lines touches the spinode curve twice; in fact, for the line 12 we have $X = 0$, $W = 0$; and substituting in the equations of the spinode curve, we have $(bY^2 - cZ^2) = 0$; that is, we have the two points of contact $X = 0$, $W = 0$, $Y\sqrt{b} = \pm Z\sqrt{c}$. Hence $\beta' = 30$.

Reciprocal Surface.

53. The equation is found by equating to zero the discriminant of the ternary cubic function

$$(Xx + Yy + Zz)(a, b, c, f, g, h \between X, Y, Z)^2 - 2kwXYZ,$$

viz. the discriminant contains the factor w^2 which is to be thrown out, thus reducing the order to $n' = 10$.

The ternary cubic, multiplying by 3 to avoid fractions, is

$$\begin{array}{cccccccccc} X^3, & Y^3, & Z^3, & 3Y^2Z, & 3Z^2X, & 3X^2Y, & 3YZ^2, & 3ZX^2, & 3XY^2, & 6XYZ, \\ 3ax, & 3by, & 3cz, & bz + 2fy, & cx + 2gz, & ay + 2hx, & cy + 2fz, & az + 2gx, & bx + 2hy, & fx + gy + hz - kw. \end{array}$$

Write as before (A, B, C, F, G, H) for the inverse coefficients $(A = bc - f^2$, &c.), and $K = abc - af^2 - bg^2 - ch^2 + 2fgh$; and moreover

$$\begin{aligned}
\Phi &= (A, B, C, F, G, H \between x, y, z)^2, \\
P &= Ax + Hy + Gz, \\
Q &= Hx + By + Fz, \\
R &= Gx + Fy + Cz, \\
t &= fx + gy + hz, \\
U &= afyz + bgzx + chxy, \\
V &= 2Kxyz - aPyz - bQzx - cRxy \\
&= -a\,Hy^2z - b\,Fz^2x - c\,Gx^2y \\
&\quad - a\,Gyz^2 - b\,Hzx^2 - c\,Fxy^2 \\
&\quad + (-abc - af^2 - bg^2 - ch^2 + 4fgh)\,xyz, \\
W &= (A, B, C, F, G, H \between ayz, bzx, cxy)^2, \\
L &= k^2w^2 - 2ktw - \Phi, \\
M &= kw\,U + V, \\
N &= 2kabc\,xyzw + W:
\end{aligned}$$

54. Then the invariants of the ternary cubic are

$$\begin{aligned}
S &= L^2 - 12kw\,M, \\
T &= L^3 - 18kw\,LM - 54k^2w^2N;
\end{aligned}$$

and the required equation of the reciprocal surface is

$$\frac{1}{108w^2}\{(L^2 - 12kwM)^3 - (L^3 - 18kwLM - 54k^2w^2N)^2\} = 0,$$

viz. this is

$$\begin{aligned}
0 = \quad & L^3N && = (k^2w^2 - 2ktw - \Phi)^3\,(2kabc\,xyzw + W) \\
& + L^2M^2 && + (k^2w^2 - 2ktw - \Phi)^2\,(kw\,U + V)^2 \\
& - 18kwLMN && - 18kw\,(k^2w^2 - 2ktw - \Phi)\,(kw\,U + V)\,(2kabc\,xyzw + W) \\
& - 16kwM^3 && - 16kw\,(kw\,U + V)^3 \\
& - 27k^2w^2N^2 && - 27k^2w^2\,(2kabc\,xyzw + W)^2,
\end{aligned}$$

which, arranged in powers of kw, is as follows; viz. we have

$$\begin{aligned}
\text{Coeff. } (kw)^7 &= 2abc\,xyz, \\
(kw)^6 &= 2abc\,xyz\,(-6t) + W \\
&\quad + U^2, \\
(kw)^5 &= 2abc\,xyz\,(-3\Phi + 12t^2) + W\,(-6t) \\
&\quad + U^2\,(-4t) + 2\,UV \\
&\quad - 36abc\,xyz\,U,
\end{aligned}$$

$$
\begin{aligned}
(kw)^4 = \; & 2abc\,xyz\,(12t\Phi - 8t^3) + W(-3\Phi + 12t^2) \\
& + U^2(-2\Phi + 4t^2) + 2UV(-4t) + V^2 \\
& - 36abc\,xyz\,V - 18UW + 72abc\,xyzt\,U \\
& - 16U^3 \\
& - 108a^2b^2c^2\,x^2y^2z^2,
\end{aligned}
$$

$$
\begin{aligned}
(kw)^3 = \; & 2abc\,xyz\,(3\Phi^2 - 12t^2\Phi) + W(12t\Phi - 8t^3) \\
& + U^2 4t\Phi + 2UV(-2\Phi + 4t^2) + V^2(-4t) \\
& - 18VW + 72abc\,xyzt\,V + 36tUW - 36abc\,xyz\Phi U \\
& - 48U^2V \\
& - 108abc\,xyz\,W,
\end{aligned}
$$

$$
\begin{aligned}
,, \quad (kw)^2 = \; & 2abc\,xyz\,(-6t\Phi^2) + W(3\Phi^2 - 12t^2\Phi) \\
& + U^2\Phi^2 + 2UV(4t\Phi) + V^2(-2\Phi + 4t^2) \\
& + 36tVW - 36abc\,xyz\Phi V - 18\Phi UW \\
& - 48UV^2 \\
& - 27W^2,
\end{aligned}
$$

$$
\begin{aligned}
,, \quad (kw)^1 = \; & 2abc\,xyz\,(-\Phi^3) + W(-6t\Phi)^2 \\
& + 2UV\Phi^2 + V^2(4t\Phi) \\
& - 18\Phi VW \\
& - 16V^3,
\end{aligned}
$$

$$
\begin{aligned}
,, \quad (kw)^0 = \; & W(-\Phi^3) \\
& + V^2\Phi^2;
\end{aligned}
$$

but I have not carried the ultimate reduction further than in Schläfli, viz. I give only the terms in $(kw)^7$, $(kw)^6$, $(kw)^5$, and $(kw)^0$.

55. I present the result as follows; the coefficients deducible from those which precede, by mere cyclical permutations of the letters a, b, c and f, g, h, are indicated by (,,).

$0 = (kw)^7 \,.\, 2abc\,xyz$

	y^2z^2	z^2x^2	x^2y^2	x^2yz	xy^2z	xyz^2
$+ (kw)^6\,.$	$a^2bc + 1$	,,	,,	$abcf - 14$ $gcbh + 4$	,,	,,

	y^3z^2	y^2z^3	z^3x^2	z^2x^3	x^3y^2	x^2y^3	x^3yz	xy^3z	xyz^3	xy^2z^2	x^2yz^2	x^2y^2z
$+ (kw)^5\,.$	$a^2bcg - 6$ $a^2cfh + 2$	$a^2bch - 6$ $a^2bfg + 2$	,,	,,	,,	,,	$ab^2c^2 - 6$ $abcf^2 + 42$ $b^2cg^2 + 2$ $bc^2h^2 + 2$ $bcfgh - 24$	,,	,,	$a^2bcf - 32$ $abcfgh + 64$ $abfg^2 - 24$ $acfh^2 - 24$ $af^2gh + 8$	,,	,,

⋮

$+ (kw)^0 \,.\, -K\,[(A, B, C, F, G, H \between x, y, z)^2]^2\,(cy^2 - 2fyz + bz^2)\,(az^2 - 2gzx + bx^2)\,(bx^2 - 2hxy + ay^2).$

56. In explanation of the discussion of the reciprocal surface, it is convenient to remark that we have

Node C_2, $X=0$, $Y=0$, $Z=0$.

Tangent cone is

$$(a, b, c, f, g, h \between X, Y, Z)^2 = 0.$$

Nodal rays are sections of cone by planes $X=0$, $Y=0$, $Z=0$ respectively, viz. equations of the rays are

$$X=0,\quad bY^2+2fYZ+cZ^2=0,$$
$$Y=0,\quad cZ^2+2gZX+aX^2=0,$$
$$Z=0,\quad aX^2+2hXY+bY^2=0.$$

Reciprocal plane is $w=0$.

Conic of contact is

$$(A, B, C, F, G, H \between x, y, z)^2=0,\ w=0.$$

Lines are tangents of this conic from points $(y=0, z=0)$, $(z=0, x=0)$, $(x=0, y=0)$ respectively, viz. equations are

$$w=0,\quad cy^2-2fyz+bz^2=0,$$
$$w=0,\quad az^2-2gzx+cx^2=0,$$
$$w=0,\quad bx^2-2hxy+ay^2=0.$$

57. The equation shows that the section by the plane $w=0$ is made up of the conic $(A, B, C, F, G, H \between x, y, z)^2=0$, twice, and of the six lines, tangents to this conic, viz. the lines

$$w=0,\quad cy^2-2fyz+bz^2=0,$$
$$w=0,\quad az^2-2gzx+cx^2=0,$$
$$w=0,\quad bx^2-2hxy+ay^2=0,$$

each once; the lines in question (reciprocals of the nodal rays) are thus mere scrolar lines on the reciprocal surface.

58. I do not attempt to put in evidence the nodal curve of the surface; by what precedes it is made up of 15 lines, intersecting 3 together in 15 points; and if we denote the six tangents of the conic just referred to by

1, 2, 3, 4, 5, 6,

then the fifteen lines are respectively lines passing through the intersections of each pair of these tangents; viz. through the intersection of the tangents 1 and 2, we have a line 12; and so in other cases; that is, the 15 lines are 12, 13 .... 56. The lines 12 and 34 meet; and the lines 12, 34, 56 meet in a point; we have thus the 15 points 12.34.56, triple points of the nodal curve.

59. As regards the cuspidal curve, the equation of the surface may be written

$$(L^2-12kwM)(4M^2+3LN)-(LM+9kwN)^2$$
$$=3(L^2M^2+L^3N-18kwLMN-16kwM^3-27k^2w^2N^2)=0,$$

and we thus have

$$4M^2+3LN=0,$$
$$LM+9kwN=0,$$
$$L^2-12kwM=0,$$

or, what is the same thing,

$$\left\|\begin{matrix} L, & 12M, & -9N \\ kw, & L, & M \end{matrix}\right\|=0$$

(equivalent to two equations) for the equations of the cuspidal curve. Attending to the second and third equations, the cuspidal curve may be considered as the residual intersection of the quartic and quintic surfaces $L^2-12kwM=0$, $LM+9kwN=0$, which partially intersect in the conic $w=0$, $L=0$; or say it is a curve $4\times5-2$; $c'=18$.

Section III $=12-B_3$.

Article Nos. 60 to 72. Equation $2W(X+Y+Z)(lX+mY+nZ)+2kXYZ=0$.

60. The system of lines and planes is at once deduced from that belonging to II $=12-C_2$, by supposing the tangent cone to reduce itself to the pair of biplanes; 3 of the planes (a) of II $=12-C_2$ thus coming to coincide with the one biplane, and three of them with the other biplane.

61. The diagram is

III $=12-B_3$		Lines. 36	35	34	26	25	24	16	15	14	6	5	4	3	2	1	
		$\overline{15}$ $\overline{27}$				$9\times1=9$							$6\times3=18$				
123	$2\times6=12$													••	••	••	Biplanes.
456											••	••	••				
14	$9\times3=27$									•••			•			•	Biradial planes each containing a ray 1, 2, or 3 of the one biplane, and a ray 4, 5, or 6 of the other biplane.
15									•••			•				•	
16								•••			•					•	
24							•••						•		•		
25						•••						•			•		
26					•••						•				•		
34				•••									•	•			
35			•••									•		•			
36		•••									•			•			
14.25.36	$6\times1=6$	•				•				•							Planes each containing three mere lines.
14.26.35			•		•					•							
15.26.34				•	•				•								
15.24.36		•					•		•								
16.24.35			•				•	•									
16.25.34	$\overline{17}$ $\overline{45}$			•		•		•									
						Mere lines, in each biradial plane, one.							Rays 1, 2, 3 and 4, 5, 6, in the two biplanes respectively.				

Planes.

62. Taking $X+Y+Z=0$ for the biplane that contains the rays 1, 2, 3, and $lX+mY+nZ=0$ for that which contains the rays 4, 5, 6, we may take $X=0$, $Y=0$, $Z=0$ for the equations of the planes [14], [25], [36] respectively; and then writing for shortness

$$m-n,\ n-l,\ l-m=\lambda,\ \mu,\ \nu,$$

and assuming, as we may do, $k=\lambda\mu\nu$, so that the equation of the surface is

$$W(X+Y+Z)(lX+mY+nZ)+(m-n)(n-l)(l-m)XYZ=0,$$

the equations of the 17 distinct planes are

$X=0,$	[14]
$Y=0,$	[25]
$Z=0,$	[36]
$X+\ \ Y+\ \ Z=0,$	[123]
$lX+mY+nZ=0,$	[456]
$lX+\ nY+nZ=0,$	[15]
$lX+\ nY+nZ=0,$	[16]
$lX+mY+\ lZ=0,$	[25]
$nX+mY+nZ=0,$	[26]
$mX+mY+nZ=0,$	[35]
$lX+\ lY+nZ=0,$	[36]
$W=0,$	[14 . 25 . 36]
$W+\ l\lambda X=0,$	[14 . 26 . 35]
$W+m\mu Y=0,$	[16 . 25 . 34]
$W+\ n\nu Z=0,$	[15 . 24 . 36]
$lmX+mnY+\ nlZ+W=0,$	[15 . 26 . 34]
$nlX+lmY+mnZ-W=0,$	[16 . 24 . 35]

63. And the coordinates of the fifteen distinct lines are

(a)	(b)	(c)	(f)	(g)	(h)	whence equations may be written
0	0	0	0	-1	1	(1) $X=0,\ Y+Z=0$
0	0	0	1	0	-1	(2) $Y=0,\ Z+X=0$
0	0	0	-1	1	0	(3) $Z=0,\ X+Y=0$
0	0	0	0	$-n$	m	(4) $X=0,\ mY+nZ=0$
0	0	0	n	0	$-l$	(5) $Y=0,\ nZ+lX=0$
0	0	0	$-m$	l	0	(6) $Z=0,\ lX+mY=0$
1	0	0	0	0	0	(14) $X=0,\ W=0$
0	1	0	0	0	0	(25) $Y=0,\ W=0$
0	0	1	0	0	0	(36) $Z=0,\ W=0$
l	n	n	$n^2\nu$	$-nl\nu$	0	(15) $lX+nY+nZ=0,\ W+n\nu Z=0$
l	m	m	$-m^2\mu$	0	$lm\mu$	(16) $lX+mY+mZ=0,\ W+m\mu Y=0$
l	m	l	0	$l^2\lambda$	$-lm\mu$	(26) $lX+mY+lZ=0,\ W+l\lambda X=0$
n	m	n	$mn\nu$	$-n^2\nu$	0	(24) $nX+mY+nZ=0,\ W+n\nu Z=0$
m	m	n	$-mn\mu$	0	$m^2\mu$	(34) $mX+mY+nZ=0,\ W+m\mu Y=0$
l	l	n	0	$nl\lambda$	$-l^2\lambda$	(35) $lX+lY+nZ=0,\ W+l\lambda X=0$

64. The rays are not, the mere lines are, facultative; hence $b'=\rho'=9:\ t'=6$.

65. The equation of the Hessian surface is

$$\begin{aligned}&-W(X+Y+Z)(lX+mY+nZ)(\mu\nu X+\nu\lambda Y+\lambda\mu Z)\\&-k(l^2X^4+m^2Y^4+n^2Z^4-2mnY^2Z^2-2nlZ^2X^2-2lmX^2Y^2)\\&+kXYZ\{(l^2+3lm+3ln+mn)X+(m^2+3mn+3ml+nl)Y+(n^2+3nl+3nm+lm)Z\}=0.\end{aligned}$$

The Hessian and cubic surfaces intersect in an indecomposable curve, which is the spinode curve; that is, spinode curve is a complete intersection 3×4; $\sigma'=12$.

The equations may be written in the simplified form

$$\begin{aligned}&W(X+Y+Z)(lX+mY+nZ)+kXYZ=0,\\&l^2X^4+m^2Y^4+n^2Z^4-2mnY^2Z^2-2nlZ^2X^2-2lmX^2Y^2\\&\qquad-4XYZ\{l(m+n)X+m(n+l)Y+n(l+m)Z\}=0.\end{aligned}$$

We may also obtain the equation

$$\begin{aligned}&k^2(X+Y+Z)(lX+mY+nZ)\{lX^2+mY^2+nZ^2-(m+n)YZ-(n+l)ZX-(l+m)XY\}\\&\qquad+\lambda^2Y^2Z^2+\mu^2Z^2X^2+\nu^2X^2Y^2-2XYZ(\mu\nu X+\nu\lambda Y+\lambda\mu Z)=0,\end{aligned}$$

which shows that there is at B_3 an eightfold point, the tangents being given by

$$(X+Y+Z)(lX+mY+nZ)=0,$$

$$(\lambda^2,\ \mu^2,\ \nu^2,\ -\mu\nu,\ -\nu\lambda,\ -\lambda\mu \between YZ,\ ZX,\ XY)^2=0.$$

Each of the facultative lines is a double tangent of the spinode curve; whence $\beta'=18$.

Reciprocal Surface.

66. The equation may be deduced from that for $\text{II}=12-C_2$, viz. writing

$$(a,\ b,\ c,\ f,\ g,\ h \between X,\ Y,\ Z)^2=2(X+Y+Z)(lX+mY+nZ),$$

that is

$$(a,\ b,\ c,\ f,\ g,\ h) \quad = \quad (2l,\ 2m,\ 2n,\ m+n,\ n+l,\ l+m),$$

we have

$$(A,\ B,\ C,\ F,\ G,\ H)=-(\lambda^2,\ \mu^2,\ \nu^2,\ \mu\nu,\ \nu\lambda,\ \lambda\mu);\ K=0.$$

Writing also

$$\lambda,\ \mu,\ \nu=m-n,\ n-l,\ l-m \text{ as before,}$$

$$\lambda x+\mu y+\nu z=\sigma,$$

$$lmn\,xyz=\theta,$$

$$l(m+n)\,yz+m(n+l)\,zx+n(l+m)\,xy=\upsilon,$$

$$l\lambda\,yz+\quad m\mu\,zx+\quad n\nu\,xy=\psi,$$

$$(m+n)\,x+\quad(n+l)\,y+\quad(l+m)\,z=t,$$

we have

$$U=2\upsilon,\quad V=2\sigma\psi,\quad W=-4\psi^2,$$

and then

$$L=k^2w^2-2ktw+\sigma^2,\quad M=2(kw\upsilon+\sigma\psi),\quad N=4(4lmnkxyzw-\psi^2);$$

so that the equation is

$$\begin{array}{lll} 0= & L^3N= & 4(k^2w^2-2kwt+\sigma^2)^3(4kw\theta-\psi^2) \\ & +L^2M^2 & +\ 4(k^2w^2-2kwt+\sigma^2)^2(kw\upsilon+\sigma\psi)^2 \\ & -18kwLMN & -144\ kw(k^2w^2-2kwt+\sigma^2)(kw\upsilon+\sigma\psi)(kw\theta-\psi^2) \\ & -16kw\ \ M^3 & -128\ kw(kw\upsilon+\sigma\psi)^3 \\ & -27k^2w^2\ N^2 & -432\ k^2w^2(kw\theta-\psi^2); \end{array}$$

or reducing the first two terms so as to throw out from the whole equation the factor kw, the equation is

$$4L^2\{\theta L+(\upsilon^2-\psi^2)kw+2\psi(t\psi+\upsilon\sigma)\}-18LMN-16M^3-27kwN^2=0$$

or, what is the same thing, it is

$$(k^2w^2 - 2kwt + \sigma^2)^2 \{k^2w^2\theta + kw(-2t\theta + v^2 - \psi^2) + \sigma^2\theta + 2\sigma v\psi + 2t\psi^2\}$$
$$- \;\; 36\,(k^2w^2 - 2kwt + \sigma^2)(kwv + \sigma\psi)(4kw\theta - \psi^2)$$
$$- \;\; 32\,(kwv + \sigma\psi)^3$$
$$- 108kw\,(4kw\theta - \psi^2)^2 = 0.$$

67. This is

$$\begin{aligned}
&(kw)^6 . \quad \theta \\
&+ (kw)^5 . - \psi^2 - 6t\theta + v^2 \\
&+ (kw)^4 . \quad \sigma^2 . 3\theta + \sigma\psi . 2v + \psi^2 . 6t + 12t^2\theta - 4tv^2 - 144\theta v \\
&+ (kw)^3 . - 2\sigma^2\psi^2 + \sigma^2(2v^2 - 10t\theta) + \sigma\psi(-8tv - 144\theta) + \psi^2(-12t^2 + 36v) \\
&\qquad - 8t^3\theta + 4t^2v^2 + 288tv\theta - 32v^3 - 1728\theta^2 \\
&+ (kw)^2 . \quad \sigma^4 . 3\theta + \sigma^3\psi . 4v + \sigma^2\psi^2 . 12t + \sigma\psi^3 . 37 \\
&\qquad + \sigma^2(12t^2\theta - 4tv^2 - 144\theta v) + \sigma\psi(+8t^2v + 288t\theta - 96v^2) + \psi^2(8t^3 - 72tv + 864\theta) \\
&+ (kw) \;. - \sigma^4\psi^2 + \sigma^4(-6t\theta + v^2) + \sigma^3\psi(-8tv - 144\theta) \\
&\qquad + \sigma^2\psi^2(-8t - 90v) + \sigma\psi^3 . -72 + \psi^4 . -108 \\
&+ (kw)^0 . \quad \sigma^3(\theta,\ 2v,\ 2t,\ 4 \between \sigma,\ \psi)^3 = 0,
\end{aligned}$$

which, reducing the last term, is

$$\begin{aligned}
&(kw)^6\, lmnxyz \\
&\quad \vdots \\
&- 4\sigma^3\lambda\mu\nu\,(y - z)(z - x)(x - y)(ny - mz)(lz - nx)(mx - ly) = 0.
\end{aligned}$$

68. I verify the last term in the particular case $z = 0$ as follows: the coefficient of σ^3 is

$$(0,\ 2n(l + m)xy,\ 2(m + n)x + 2(n + l)y,\ 4 \between \lambda x + \mu y,\ n\nu xy)^3,$$

which is

$$\begin{aligned}
&= 2n^2\nu x^2y^2 \{(l + m)(\lambda x + \mu y)^2 + [(m + n)x + (n + l)y](\nu\lambda x + \mu\nu y) + 2n\nu^2 xy\} \\
&= 2n^2\nu x^2y^2 \{[(l + m)\lambda + (m + n)\nu]\lambda x^2 \\
&\qquad + [2(l + m)\lambda\mu + (m + n)\mu\nu + (n + l)\nu\lambda + 2n\nu]\,xy \\
&\qquad + [(l + m)\mu + (n + l)\nu]\mu y^2\},
\end{aligned}$$

which, substituting for λ, μ, ν their values $m - n,\ n - l,\ l - m$, is

$$= 2n^2\nu x^2y^2 . - 2\lambda\mu\,(x - y)(mx - ly);$$

or for $z = 0$ the coefficient of σ^3 is

$$= - 4\lambda\mu\nu\, n^2x^2y^2(x - y)(mx - ly),$$

agreeing with the general value

$$- 4\lambda\mu\nu\,(y - z)(z - x)(x - y)(ny - mz)(lz - nx)(lx - my).$$

69. In the discussion of the equation it is convenient to write down the relations of the two surfaces, thus:

Cubic surface.	Reciprocal surface.
B_3, $X=0,\ Y=0,\ Z=0$	Plane $w=0$,
Biplanes $X+\ Y+\ Z=0$	Points in $w=0$, viz.
$lX+mY+nZ=0$,	$x=y=z$ and $x:y:z=l:m:n$,
intersecting in edge.	in line $(m-n)x+(n-l)y+(l-m)z=0$,
	that is, $\lambda x+\mu y+\nu z=0$, or $\sigma=0$.
Rays in first biplane,	Lines in plane $w=0$, and through first point, viz.
$X=0,\ Y+Z=0;\ Y=0,\ Z+X=0,$ $Z=0,\ X+Y=0;$	$y-z=0,\ z-x=0,\ x-y=0;$
rays in second biplane,	lines through second point, viz.
$X=0,\ mY+nZ=0;\ Y=0,\ nZ+lX=0,$ $Z=0,\ lX+mY=0.$	$ny-mz=0,\ nz-lx=0,\ lx-my=0.$

70. The equation puts in evidence the section by the plane $w=0$, viz. this is the line $\sigma=0$ (reciprocal of the edge) three times, and the six lines (reciprocals of the rays) each once. Observe that the edge is *not* a line on the cubic; but its reciprocal is a line, and that an oscular line on the reciprocal surface; the six lines (reciprocals of the rays) are mere scrolar lines on the reciprocal surface; they pass, three of them, through the point $x=y=z$, and the other three through the point $x:y:z=l:m:n$; that is, they are six tangents of the point-pair (reciprocal of the pair of biplanes) formed by these two points.

71. I do not attempt to put in evidence the nodal curve on the surface; by what precedes it consists of 9 lines, reciprocals of the mere lines. If we denote by 1, 2, 3 and 4, 5, 6 the lines which pass through the points $x=0,\ y=0,\ z=0$ and through the point $x:y:z=l:m:n$ respectively, then these intersect in the nine points 14, 15, 16, 24, 25, 26, 34, 35, 36; and through each of these there passes a nodal line which may be represented by the same symbol; that is, we have the nodal lines 14,.... 36. Two lines such as 14, 25 meet; and three lines such as 14, 25, 36 meet in a point; we have thus the six points 14.25.36 &c. triple points on the nodal curve; as before, $b'=9,\ t'=6$.

72. The cuspidal curve is given by the equations

$$\left\| \begin{array}{ccc} k^2w^2-2kwt+\sigma^2, & 24(kwv+\sigma\psi), & -36(4lmnk\,xyzw-\psi^2) \\ kw \quad , & k^2w^2-2kwt+\sigma^2, & 2(kwv+\sigma\psi) \end{array} \right\| = 0.$$

Writing down the two equations,

$$(k^2w^2-2kwt+\sigma^2)^2-24kw(kwv+\sigma\psi)=0,$$
$$(k^2w^2-2kwt+\sigma^2)(kwv+\sigma\psi)+18w(lmnk\,xyzw-\psi^2)=0,$$

these are respectively of the orders 4 and 5; but they intersect in the line $w=0$, $\sigma=0$ taken four times, or say, the cuspidal curve is a partial intersection $4.5-4$; $c'=16$.

Section $IV = 12 - 2C_2$.

Article Nos. 73 to 84. Equation $WXZ + Y^2(\gamma Z + \delta W) + (a, b, c, d \between X, Y)^3 = 0$.

73. The diagram of the lines is

Planes.	$IV = 12 - 2C_2$.	34 . 1′2′	24 . 1′3′	23 . 1′4′	14 . 2′3′	13 . 2′4′	12 . 3′4′	4′	3′	2′	1′	4	3	2	1	5	0	
		$\overline{16}$		$6 \times 1 = 6$			$\overline{27}$				$8 \times 2 = 16$					$1 \times 1 = 1$	$1 \times 4 = 4$	
[0]	$1 \times 2 = 2$															••	••	Plane touching along axis.
11′	$4 \times 4 = 16$										••				••		•	Planes through axis, each containing a ray of the one node and a ray of the other node.
22′										••				••			•	
33′									••				••				•	
44′								••				••					•	
12	$12 \times 2 = 24$						••							•	•			Biradial planes, each containing two rays of the one node or two rays of the other node.
13						••							•		•			
14					••							•			•			
23				••									•	•				
24			••									•		•				
34		••										•	•					
1′2′		••								•	•							
1′3′			••						•		•							
1′4′				••				•			•							
2′3′					••				•	•								
2′4′						••		•		•								
3′4′							••	•	•									
12 . 34	$3 \times 1 = 3$	•					•									•		Planes each containing three mere lines.
13 . 24			•			•										•		
14 . 23	$\overline{20}$ $\overline{45}$			•	•											•		
		Mere lines, each the intersection of a biradial plane of the one node and a biradial plane of the other node.						Rays 1, 2, 3, 4, and 1′, 2′, 3′, 4′ through the two nodes respectively.								Transversal, in the tangent plane along the axis.	Axis, through the two nodes.	

74. Writing $X(a, b, c, d\between X, Y)^3 - \gamma\delta Y^4 = -\gamma\delta(f_1X - Y)(f_2X - Y)(f_3X - Y)(f_4X - Y)$, the 20 planes are

$$\begin{array}{ll}
X = 0, & [\ 0] \\
X - f_1Y = 0, & [11'] \\
X - f_2Y = 0, & [22'] \\
X - f_3Y = 0, & [33'] \\
X - f_4Y = 0, & [44'] \\
\delta\{X - (\mathrm{f}_1 + \mathrm{f}_2)Y\} - \mathrm{f}_1\mathrm{f}_2Z = 0, & [12] \\
\delta\{X - (\mathrm{f}_1 + \mathrm{f}_3)Y\} - \mathrm{f}_1\mathrm{f}_3Z = 0, & [13] \\
\delta\{X - (\mathrm{f}_1 + \mathrm{f}_4)Y\} - \mathrm{f}_1\mathrm{f}_4Z = 0, & [14] \\
\delta\{X - (\mathrm{f}_2 + \mathrm{f}_3)Y\} - \mathrm{f}_2\mathrm{f}_3Z = 0, & [23] \\
\delta\{X - (\mathrm{f}_2 + \mathrm{f}_4)Y\} - \mathrm{f}_2\mathrm{f}_4Z = 0, & [24] \\
\delta\{X - (\mathrm{f}_3 + \mathrm{f}_4)Y\} - \mathrm{f}_3\mathrm{f}_4Z = 0, & [34] \\
\gamma\{X - (\mathrm{f}_1 + \mathrm{f}_2)Y\} - \mathrm{f}_1\mathrm{f}_2W = 0, & [1'2'] \\
\gamma\{X - (\mathrm{f}_1 + \mathrm{f}_3)Y\} - \mathrm{f}_1\mathrm{f}_3W = 0, & [1'3'] \\
\gamma\{X - (\mathrm{f}_1 + \mathrm{f}_4)Y\} - \mathrm{f}_1\mathrm{f}_4W = 0, & [1'4'] \\
\gamma\{X - (\mathrm{f}_2 + \mathrm{f}_3)Y\} - \mathrm{f}_2\mathrm{f}_3W = 0, & [2'3'] \\
\gamma\{X - (\mathrm{f}_2 + \mathrm{f}_4)Y\} - \mathrm{f}_2\mathrm{f}_4W = 0, & [2'4'] \\
\gamma\{X - (\mathrm{f}_3 + \mathrm{f}_4)Y\} - \mathrm{f}_3\mathrm{f}_4W = 0, & [3'4'] \\
-\gamma\delta\left(\frac{1}{\mathrm{f}_1\mathrm{f}_2} + \frac{1}{\mathrm{f}_3\mathrm{f}_4}\right)X + dy + \gamma Z + \delta W = 0, & [12.34] \\
-\gamma\delta\left(\frac{1}{\mathrm{f}_1\mathrm{f}_3} + \frac{1}{\mathrm{f}_2\mathrm{f}_4}\right)X + dy + \gamma Z + \delta W = 0, & [13.24] \\
-\gamma\delta\left(\frac{1}{\mathrm{f}_1\mathrm{f}_4} + \frac{1}{\mathrm{f}_2\mathrm{f}_3}\right)X + dy + \gamma Z + \delta W = 0, & [14.23]
\end{array}$$

75. And the 16 lines are

(a)	(b)	(c)	(f)	(g)	(h)	whence equations may be written
0	0	0	0	0	1	(0) $X = 0,\ Y = 0$
δ	0	0	0	$-\gamma$	d	(5) $X = 0,\ dY + \gamma Z + \delta W = 0$
0	0	0	f_1^2	f_1	$-\delta$	(1) $X = \mathrm{f}_1 Y = 0,\ \delta Y + \mathrm{f}_1 Z = 0$
0	0	0	f_2^2	f_2	$-\delta$	(2) " "
0	0	0	f_3^2	f_3	$-\delta$	(3) " "
0	0	0	f_4^2	f_4	$-\delta$	(4) " "

(a)	(b)	(c)	(f)	(g)	(h)		
f_1	$-f_1^2$	0	0	0	γ	(1′)	$X-f_1Y=0,\ \gamma Y+f_1W=0$
f_2	$-f_2^2$	0	0	0	γ	(2′)	,, ,,
f_3	$-f_3^4$	0	0	0	γ	(3′)	,, ,,
f_4	$-f_4^2$	0	0	0	γ	(4′)	,, ,,
$-\frac{\delta}{f_1f_2}$	$\delta\left(\frac{1}{f_1}+\frac{1}{f_2}\right)$	1	$-\gamma\left(\frac{1}{f_3}+\frac{1}{f_4}\right)$	$-\frac{\gamma}{f_3f_4}$	$\frac{1}{f_3f_4}\left(\frac{1}{f_1}+\frac{1}{f_2}\right)-\frac{1}{f_1f_2}\left(\frac{1}{f_3}+\frac{1}{f_4}\right)$	(12 . 3′4′)*	
$-\frac{\delta}{f_1f_3}$	$\delta\left(\frac{1}{f_1}+\frac{1}{f_3}\right)$	1	$-\gamma\left(\frac{1}{f_2}+\frac{1}{f_4}\right)$	$-\frac{\gamma}{f_2f_4}$	$\frac{1}{f_2f_4}\left(\frac{1}{f_1}+\frac{1}{f_3}\right)-\frac{1}{f_1f_3}\left(\frac{1}{f_2}+\frac{1}{f_4}\right)$	(13 . 2′4′) ,,	
$-\frac{\delta}{f_1f_4}$	$\delta\left(\frac{1}{f_1}+\frac{1}{f_4}\right)$	1	$-\gamma\left(\frac{1}{f_2}+\frac{1}{f_3}\right)$	$-\frac{\gamma}{f_2f_3}$	$\frac{1}{f_2f_3}\left(\frac{1}{f_1}+\frac{1}{f_4}\right)-\frac{1}{f_1f_4}\left(\frac{1}{f_2}+\frac{1}{f_3}\right)$	(14 . 2′3′) ,,	
$-\frac{\delta}{f_2f_3}$	$\delta\left(\frac{1}{f_2}+\frac{1}{f_3}\right)$	1	$-\gamma\left(\frac{1}{f_1}+\frac{1}{f_4}\right)$	$-\frac{\gamma}{f_1f_4}$	$\frac{1}{f_1f_4}\left(\frac{1}{f_2}+\frac{1}{f_3}\right)-\frac{1}{f_2f_3}\left(\frac{1}{f_1}+\frac{1}{f_4}\right)$	(23 . 1′4′) ,,	
$-\frac{\delta}{f_2f_4}$	$\delta\left(\frac{1}{f_2}+\frac{1}{f_4}\right)$	1	$-\gamma\left(\frac{1}{f_1}+\frac{1}{f_3}\right)$	$-\frac{\gamma}{f_1f_3}$	$\frac{1}{f_1f_3}\left(\frac{1}{f_2}+\frac{1}{f_4}\right)-\frac{1}{f_2f_4}\left(\frac{1}{f_1}+\frac{1}{f_3}\right)$	(24 . 1′3′) ,,	
$-\frac{\delta}{f_3f_4}$	$\delta\left(\frac{1}{f_3}+\frac{1}{f_4}\right)$	1	$-\gamma\left(\frac{1}{f_1}+\frac{1}{f_2}\right)$	$-\frac{\gamma}{f_1f_2}$	$\frac{1}{f_1f_2}\left(\frac{1}{f_3}+\frac{1}{f_4}\right)-\frac{1}{f_3f_4}\left(\frac{1}{f_1}+\frac{1}{f_2}\right)$	(34 . 1′2′) ,,	

*equations are

$$\delta\{X-(f_1+f_2)Y\}-f_1f_2Z=0,\quad \gamma\{X-(f_3+f_4)Y\}-f_3f_4W=0,$$

[and similarly for each of the remaining five lines].

76. To verify the equations of the line 12 . 3′4′, observe that the two equations give

$$\gamma Z+\delta W=\gamma\delta\left\{X\left(\frac{1}{f_1f_2}+\frac{1}{f_3f_4}\right)-Y\left(\frac{1}{f_1}+\frac{1}{f_2}+\frac{1}{f_3}+\frac{1}{f_4}\right)\right\},$$

$$ZW=\frac{\gamma\delta}{f_1f_2f_3f_4}\{X-(f_1+f_2)YX-(f_3+f_4)Y\}:$$

the equation of the surface, multiplying by X and observing that $-\gamma\delta=af_1f_2f_3f_4$, becomes

$$X^2ZW+XY^2(\gamma Z+\delta W)+\gamma\delta Y^4-\frac{\gamma\delta}{f_1f_2f_3f_4}(X-f_1Y)(X-f_2Y)(X-f_3Y)(X-f_4Y)=0;$$

and substituting the values just obtained, this is

$$X^2[X-(f_1+f_2)Y][X-(f_3+f_4)Y]+XY^2[X(f_1f_2+f_3f_4)-Y(f_1f_2f_3+f_1f_2f_4+f_1f_3f_4+f_2f_3f_4)]$$
$$+f_1f_2f_3f_4Y_4-(X-f_1Y)(X-f_2Y)(X-f_3Y)(X-f_4Y)=0,$$

which is in fact an identity.

77. The facultative lines are the transversal and the six mere lines; $b'=\rho'=7$; $t'=3$.

78. The equation of the Hessian surface is found to be

$$\begin{aligned}&(\gamma Z+\delta W)XZW+Y^2(\gamma Z-\delta W)^2+3(cX+dY)XZW+12\gamma\delta XY^2(aX+bY)\\&-(\gamma Z+\delta W)(3aX^3+9bX^2Y+6cXY^2)\\&-9X^2\{(ac-b^2)X^2+(ad-bc)XY+(bd-c^2)Y^2\}=0.\end{aligned}$$

79. Combining with the foregoing the equation of the surface

$$XZW+Y^2(\gamma Z+\delta W)+(a,\ b,\ c,\ d\between X,\ Y)^3=0,$$

it appears that these have along the line $X=0$, $Y=0$ the common tangent plane $X=0$, or, what is the same thing, that they meet in the line $X=0$, $Y=0$ (the axis) twice, and in a residual curve of the tenth order, which is the spinode curve; the equations may be presented in the somewhat more simple form

$$\begin{aligned}&XZW+Y^2(\gamma Z+\delta W)+(a,\ b,\ c,\ d\between X,\ Y)^3=0,\\&-4\gamma\delta Y^2ZW-4(\gamma Z+\delta W)(a,\ b,\ c,\ d\between X,\ Y)^3+12\gamma\delta XY^2(aX+bY)\\&\qquad+X^4(-12ac+9b^2)-3d(4aX^3Y+6bX^2Y^2+4cXY^3+dY^4)=0,\end{aligned}$$

which, however, still contain the line $X=0$, $Y=0$ twice. The spinode curve, as just mentioned, is of the tenth order; that is, we have $\sigma'=10$.

Each of the 6 mere lines is a double tangent to the spinode curve, but the transversal is only a single tangent: to show this, observe that the equations of the transversal are $X=0$, $\gamma Z+\delta W+dY=0$; substituting in the equations of the curve the first equation, that of the cubic surface is of course satisfied identically; for the second equation, writing $X=0$, this becomes $Y^2\{-4\gamma\delta ZW-4dY(\gamma Z+\delta W)-3d^2Y^2\}=0$; or writing herein $dY=-(\gamma Z+\delta W)$, it becomes $Y^2(\gamma Z-\delta W)^2=0$. The value $Y^2=0$ gives $X=0$, $Y=0$, $\gamma Z+\delta W=0$, viz. this is a point on the axis $X=0$, $Y=0$ not belonging to the spinode curve; the value $(\gamma Z-\delta W)^2=0$ gives a point of contact $X=0$, $\gamma Z+\delta W+dY=0$, $\gamma Z-\delta W=0$; and the transversal is thus a single tangent. Hence the number of contacts is $2.6+1$, $=13$; that is, we have $\beta'=13$.

Reciprocal Surface.

80. The equation is found by equating to zero the discriminant of the binary quartic

$$\{xX^2+yXY-(\delta z+\gamma w)Y^2\}+4Zw\{X(a,\ b,\ c,\ d\between X,\ Y)^3-\gamma\delta Y^4\},$$

or say this is $(*\between X,\ Y)^4$, where the coefficients are

$$\begin{aligned}&6x^2 &&+24azw,\\&3xy &&+18bzw,\\&y^2-2(\delta z+\gamma w)x &&+12czw,\\&-3(\delta z+\gamma w)y &&+\ 6dzw,\\&6(\delta z-\gamma w)^2.\end{aligned}$$

81. Forming the invariants, these are

$$\tfrac{1}{3}I = \Lambda^2 + \quad 24\,Uzw + 144\mu z^2w^2,$$
$$-J = \Lambda^3 + 36\Lambda\,Uzw + 216\,Vz^2w^2 + 864\nu z^3w^3,$$

where

$$\begin{aligned}
\Lambda &= y^2 + 4\,(\delta z + \gamma w)\,x,\\
U &= 2\gamma\delta x^2 + 2a\,(\delta z - \gamma w)^2 + 3by\,(\delta z + \gamma w) + c\,[y^2 - 2\,(\delta z + \gamma w)\,x] - dxy,\\
V &= \quad (-8ac + 9b^2)\,(\delta z - \gamma w)^2\\
&\quad + (2c^2 - bd)\,[y^2 - 2\,(\delta z + \gamma w)\,x]\\
&\quad + (-4ad + 6bc)\,y\,(\delta z + \gamma w)\\
&\quad - 2cd\,xy\\
&\quad + \quad d^2x^2\\
&\quad + 4\gamma\delta\,(2cx^2 - 3bxy + ay^2),\\
\mu &= c^2 - bd,\\
\nu &= ad^2 - 2bcd + 2c^3,
\end{aligned}$$

and the equation is

$$\frac{1}{432z^2w^2}\{(\Lambda^2 + 24\,Uzw + 144\mu z^2w^2)^3 - (\Lambda^3 + 36\Lambda\,Uzw + 216\,Vz^2w^2 + 864\nu z^3w^3)^2\} = 0;$$

or, expanding, this is

$$\begin{aligned}
&\Lambda^4\mu - \Lambda^3V + \Lambda^2U^2\\
+ \quad &4zw\,(\quad - \Lambda^3\nu + 12\Lambda^2U\mu - 9\Lambda UV + 8U^3)\\
+ \quad &36z^2w^2\,(\qquad 4\Lambda^2\mu^2 - 4\Lambda U\nu + 16U^2\mu - 3V^2)\\
+ \quad &864z^3w^3\,(\qquad 4U\mu^2 - \;V\nu)\\
+ \; &1728z^4w^4\,(\qquad 4\mu^3 - \nu^2) = 0,
\end{aligned}$$

where observe that the value of

$$4\mu^3 - \nu^2, \; = 4\,(bd - c^2)^3 - (ad^2 - 3bcd + 2c^3)^2 \text{ is } = - d^2\,(a^2d^2 + 4ac^3 + 4b^3d - 3b^2c^2 - 6abcd).$$

82. It is convenient to modify the form of the equation as follows; write

$$U_1 = U + 8a\gamma\delta zw, \;\; V_1 = V + (-8ac + 9b^2)\,\gamma\delta zw,$$

so that

$$\begin{aligned}
\Lambda \; &= y^2 + 4\,(\delta z + \gamma w)\,x,\\
U_1 &= -\,2\gamma\delta x^2 + 2a\,(\delta z + \gamma w)^2 + 3by\,(\delta z + \gamma w) + c\,[y^2 - 2\,(\delta z + \gamma w)\,x] - dxy,\\
V_1 &= \quad (-8ac + 9b^2)\,(\delta z + \gamma w)^2\\
&\quad + (2c^2 - bd)\,[y^2 - 2\,(\delta z + \gamma w)\,x]\\
&\quad + (-4ad + 6bc)\,y\,(\delta z + \gamma w)\\
&\quad - 2cdxy\\
&\quad + d^2x^2\\
&\quad + 4\gamma\delta\,(2cx^2 - 3bxy + ay^2),\\
\mu &= \quad c^2 - bd,\\
\nu &= \quad ad^2 - 2bcd + 2c^3,
\end{aligned}$$

Λ, U_1, V_1 being, it will be observed, functions of x, y, $\delta z+\gamma w$. The transformed equation is

$$\Lambda^2(\Lambda^2\mu-\Lambda V_1+U_1^2)+\Omega zw=0,$$

where the term Ω may be calculated without difficulty: the first term of this is

$$=\{y^2+4(\delta z+\gamma w)x\}^2.4\gamma^2\delta^2[x+\mathrm{f}_1y-\mathrm{f}_1^2(\delta z+\gamma w)]..[x+\mathrm{f}_4y-\mathrm{f}_4^2(\delta z+\gamma w)],$$

the developed expressions of $\frac{1}{4}(\Lambda^2\mu-\Lambda V_1+U_1^2)$ and of $\gamma^2\delta^2$ into the product of the linear factors being in fact each

$$\begin{aligned}
= \quad & x^4.\gamma^2\delta^2+x^3y.d\gamma\delta+x^2y^2.-3c\gamma\delta+xy^3.3b\gamma\delta+y^4.-a\gamma\delta\\
&+[x^3(-d^2-6c\gamma\delta)+x^2y(3cd+9b\gamma\delta)+xy^2(-3bd-4a\gamma\delta)+y^3.ad](\delta z+\gamma w)\\
&+[x^2(9c^2-6bd-2a\gamma)+xy(3ad-9bc)+y^2.3ac](\delta z+\gamma w)^2\\
&+[x(6ac-9b^2)+y.3ab](\delta z+\gamma w)^3\\
&+a^2\delta^4.(\delta z+\gamma w)^4.
\end{aligned}$$

The form puts in evidence the section by the plane $w=0$, which is the reciprocal of the node D, viz. this is a conic (the reciprocal of the tangent cone) twice, and four lines, the reciprocals of the nodal rays, each once. And similarly for the section by the plane $z=0$.

83. The nodal curve is made up of the lines which are the reciprocals of the six mere lines and the transversal; viz. we have three pairs of lines and a seventh line, the lines of each pair intersecting at a point of the seventh line, and these three points being the triple points of the nodal curve; $t'=3$ as before.

84. The equations of the cuspidal curve are at once reduced to the form

$$\begin{aligned}
&\Lambda^2 \quad +24Uzw+144\mu z^2w^2=0,\\
&\Lambda U \ +(18V-12\mu\Lambda)zw+72\nu z^2w^2=0,
\end{aligned}$$

which are two quartic surfaces having in common the conics $z=0$, $\Lambda=0$, and $w=0$, $\Lambda=0$; or we may say that the cuspidal curve is a curve $4.4-2-2$; that is $c'=12$.

Section $\mathrm{V} = 12 - B_4$.

Article Nos. 85 to 94. Equation $WXZ + (X+Z)(Y^2 - aX^2 - bZ^2) = 0$.

85. The diagram of the lines and planes is

$\mathrm{V}=12-B_4$.	Lines.	22′	21′	12′	11′	2′	1′	2	1	4	3	
		$\overline{10}$ 4×1= 4 $\overline{27}$				4×4=16				1×1= 1	1×6= 6	
Planes. 12	2×12=24							•••	•••		••	Biplanes containing rays 1, 2 and 1′, 2′ respectively.
1′2′						•••	•••				••	
0	1× 3= 3									•••	••	Plane touching along edge and containing the transversal.
11′	4× 4=16				••••		•		•			Biradial planes each containing a ray of the one and a ray of the other biplane.
12′				••••		•			•			
21′			••••				•	•				
22′		••••				•		•				
11′. 22′	2× 1= 2	•			•					•		Planes each through the transversal.
12′. 21′	$\overline{9}$ $\overline{45}$		•	•						•		
		Mere lines.				Rays 1, 2 and 1′, 2′ in the two biplanes respectively.				Transversal in the plane touching along the edge.	Edge.	

86. The planes are

$$
\begin{aligned}
X &= 0, & [12]\\
Z &= 0, & [1'2']\\
X + Z &= 0, & [0]\\
-X\sqrt{a} + Y - Z\sqrt{b} &= 0, & [11']\\
X\sqrt{a} + Y - Z\sqrt{b} &= 0, & [12']\\
-X\sqrt{a} + Y + Z\sqrt{b} &= 0, & [21']\\
X\sqrt{a} + Y + Z\sqrt{b} &= 0, & [22']\\
\sqrt{ab}\,(X+Z) + W &= 0, & [11'.22']\\
-2\sqrt{ab}\,(X+Z) + W &= 0, & [12'.21'].
\end{aligned}
$$

87. And the lines are

a	b	c	f	g	h	equations may be written	
0	0	0	0	1	0	(3)	$X=0,\ Z=0$
1	0	1	0	0	0	(4)	$X+Z=0,\ W=0$
0	0	0	0	$\sqrt{b}$	1	(1)	$X=0,\quad Y-Z\sqrt{b}=0$
0	0	0	0	$-\sqrt{b}$	1	(2)	$X=0,\quad Y+Z\sqrt{b}=0$
0	0	0	1	$\sqrt{a}$	0	(1′)	$Z=0,\ -X\sqrt{a}+Y=0$
0	0	0	1	$-\sqrt{a}$	0	(2′)	$Z=0,\quad X\sqrt{a}+Y=0$
$-\frac{1}{\sqrt{b}}$	$\frac{1}{\sqrt{ab}}$	$-\frac{1}{\sqrt{a}}$	2	$2(\ \sqrt{a}-\sqrt{b})$	-2	(11′)	but for the other lines the coordinate expressions are the more convenient.
$-\frac{1}{\sqrt{b}}$	$-\frac{1}{\sqrt{ab}}$	$\frac{1}{\sqrt{a}}$	2	$2(-\sqrt{a}-\sqrt{b})$	-2	(12′)	
$\frac{1}{\sqrt{b}}$	$-\frac{1}{\sqrt{ab}}$	$-\frac{1}{\sqrt{a}}$	2	$2(\ \sqrt{a}+\sqrt{b})$	-2	(21′)	
$\frac{1}{\sqrt{b}}$	$\frac{1}{\sqrt{ab}}$	$\frac{1}{\sqrt{a}}$	2	$2(-\sqrt{a}+\sqrt{b})$	-2	(22′)	

88. The four mere lines and the transversal are each facultative; *the edge is also facultative, counting twice;* $\rho'=b'=7,\ t'=3$.

That the edge is as stated a facultative line counting twice, I discovered, and accept, *à posteriori*, from the circumstance that on the reciprocal surface the reciprocal of the edge is (as will be shown) a tacnodal line, that is, a double line with coincident tangent planes, counting twice as a nodal line. Reverting to the cubic surface, I notice that the section by an arbitrary plane through the edge consists of the edge and of a conic touching the edge at the biplanar point; by what precedes it appears that the arbitrary plane is to be considered, and that twice, as a node-couple plane of the surface: I do not attempt to further explain this.

89. Hessian surface. The equation is

$$(X+Z)XZW+(X-Z)^2Y^2+(X+Z)(3a,\ -a,\ -b,\ 3b\between X,\ Z)^3=0.$$

Combining with the equation

$$XZW+(X+Z)(Y^2-aX^2-bZ^2)=0,$$

and observing that from the two equations we deduce

$$-XZY^2+(X+Z)(aX^3+bZ^3)=0,$$

it appears that the complete intersection of the Hessian and the surface is made up of the line $X=0,\ Z=0$ (the edge) twice (that is, the two surfaces touch along the edge), and of a curve of the tenth order, which is the spinode curve; $c'=10$.

The equations of the spinode curve may be presented in the form

$$\left\| \begin{matrix} XZ, & aX^2+bZ^2-Y^2, & aX^3+bZ^3 \\ X+Z, & W \quad , & Y^2 \end{matrix} \right\| = 0\,;$$

it is a curve $3.4-2$, the partial intersection of a quartic and a cubic surface which touch along a line.

The binode is on the spinode curve a singular point; through it we have two branches represented in the vicinity thereof by the equations

$$\left(\frac{X}{W}=-\tfrac{1}{2}\left(\frac{Y}{W}\right)^2,\ \frac{Z}{W}=-\left(\frac{1}{2b}\right)^{\frac{1}{3}}\left(\frac{Y}{W}\right)^{\frac{4}{3}}\right) \text{ and } \left(\frac{Z}{W}=-\tfrac{1}{2}\left(\frac{Y}{W}\right)^2,\ \frac{X}{W}=-\left(\frac{1}{2a}\right)^{\frac{1}{3}}\left(\frac{Z}{W}\right)^{\frac{4}{3}}\right)$$

respectively.

90. The edge counted once is regarded as a double tangent of the spinode curve (I do not understand this, there is apparently a higher tangency); each of the four mere lines is a double tangent; the transversal is a single tangent; hence $\beta'=2.2+2.4+1,\ =13$.

Reciprocal Surface.

91. The equation is found by equating to zero the discriminant of the binary quartic

$$y^2X^2Z^2+4w\,(Xx+Zz)\,XZ\,(X+Z)+4w^2\,(aX^2+bZ^2)\,(X+Z)^2,$$

viz. multiplying by 6 to avoid fractions, and calling the function $(*\between X,\ Z)^4$, the coefficients are

$$\begin{aligned}&24aw^2,\\ &6w\,(x+2aw),\\ &y^2+4\,(x+z)\,w+4\,(a+b)\,w^2,\\ &6w\,(z+2bw),\\ &24bw^2\,;\end{aligned}$$

and then writing

$$\begin{aligned}L&=y^2+4\,(x+z)\,w+4\,(a+b)\,w^2,\\ M&=\ 4\,\{xz+2\,(bx+az)\,w\},\\ N&=16aby^2-bx^2-ay^2,\end{aligned}$$

we find

$$\begin{aligned}\tfrac{1}{3}I&=L^2-12w^2M,\\ -J&=L^3-18w^2LM-54w^4N,\end{aligned}$$

and then the equation is

$$\frac{1}{108w^4}\{(L^2-12w^2M)^3-(L^3-18w^2LM-54w^4N)^2\}=0,$$

viz. it is

$$L^3N+L^2M^2-18w^2LMN-16w^2M^3-27w^4N^2=0.$$

92. This, completely developed, is

$$\begin{aligned}
&64w^6 . ab(a+b)^2\{(a+b)y^2-(x-z)^2\}\\
&+32w^5 . 2ab\begin{Bmatrix} 3(a+b)[(a-2b)x+(-2a+b)z]y^2 \\ +(x-z)^2[(-3a+5b)x+(5a-3b)z]\end{Bmatrix}\\
&+16w^4\begin{Bmatrix} 3ab(a^2-7ab+b^2)y^4 \\ +[b(9a^2+26ab-b^2)x^2-26ab(a+b)xz+a(-a^2+26ab+9b^2)z^2]y^2 \\ +(x-z)^2[b(-12a+b)x^2+22abxz+a(a-12b)z^2]\end{Bmatrix}\\
&+\ 8w^3\begin{Bmatrix} 3ab[(2a-b)x+(-a+2b)z]y^4 \\ +[b(-2a+5b)x^3+b(3a-2b)x^2z+a(-2a+3b)xz^2+a(5a-2b)z^3]y^2 \\ +2(x-z)^2[-2bx^3+bx^2z+axz^2-2ay^3]\end{Bmatrix}\\
&+\ 4w^2\begin{Bmatrix} 3ab(a+b)y^6 \\ +\ [b(9a-2b)x^2+8abxz+a(-2a+9b)z^2]y^4 \\ +2[-6bx^4+bx^3z-(a+b)x^2z^2+axz^3-6az^4]y^2 \\ +4x^2z^2(x-z)^2\end{Bmatrix}\\
&+\ 2w\begin{Bmatrix} 2ab(x+z)y^6 \\ -[3bx^3+2bx^2z+2axz+3az^3]y^4 \\ +4x^2z^2(x+z)y^2\end{Bmatrix}\\
&+\qquad y^4(ay^2-x^2)(cy^2-z^2)=0,
\end{aligned}$$

where we see that the section by the plane $w=0$ (reciprocal of B_4) is made up of the line $w=0$, $y=0$ (reciprocal of the edge) four times, and of the lines $w=0$, $ay^2-x^2=0$; $w=0$, $by^2-z^2=0$ (reciprocals of the rays) each once.

93. The surface contains the line $y=0$, $w=0$ (reciprocal of the edge); and if we attend only to the terms of the lowest order in y, w, viz.

$$x^2z^2\{16(x-z)^2w^2+8(x+z)y^2w+y^4\},$$

which terms equated to zero give

$$w=-\tfrac{1}{4}\frac{1}{(\sqrt{x}\pm\sqrt{z})^2}y^2,$$

we see that the line in question ($y=0$, $w=0$) is a tacnodal line on the surface, the tacnodal plane being $w=0$, *a fixed plane for all points of the line*: it has already been seen that this plane meets the surface in the line taken 4 times; every other plane through the line meets the surface in the line taken twice. We have in what precedes the *à posteriori* proof that in the cubic surface the edge is a facultative line to be counted twice.

94. Cuspidal curve. The equation of the surface may be written

$$(L^2-12w^2M)(4M^2+3LN)-(LM+9w^2N)^2=0,$$

and we thus have

$$\begin{aligned} 4M^2 + 3LN &= 0, \\ LM + w^2 N &= 0, \\ L^2 - 12w^2 M &= 0, \end{aligned}$$

or, what is the same thing,

$$\left\| \begin{matrix} L, & 12M, & -9N \\ w^2, & L, & M \end{matrix} \right\| = 0$$

for the equation of the cuspidal curve. Attending to the second and third equations, these are quartics having in common $w^2 = 0$, $L = 0$, that is, the line $y = 0$, $w = 0$ four times; or the cuspidal curve is a partial intersection $4 \times 4 - 4$: $c' = 12$.

Section $\text{VI} = 12 - B_3 - C_2$.

Article Nos. 95 to 102. Equation $WXZ + Y^2Z + (a, b, c, d)(X, Y)^3 = 0$.

95. The diagram of the lines and planes is

$\text{VI} = 12 - B_3 - C_2$. Planes.		Lines. 14 . 2′3′	13 . 2′4′	12 . 3′4′	4′	3′	2′	4	3	2	1	0	
		$\overline{11}$ $\overline{27}$	$3 \times 1 = 3$			$3 \times 2 = 6$			$3 \times 3 = 9$		$1 \times 3 = 3$	$1 \times 6 = 6$	
0	$1 \times 6 = 6$										••	•••	Biplane touching along axis, and containing transversal ray.
00	$1 \times 6 = 6$							••	••	••			Other biplane.
22′							•••			••		•	Planes each through the axis and containing a ray of the binode and a ray of the cnicnode.
33′	$3 \times 6 = 18$					•••			••			•	
44′					•••			••				•	
12				•••						•	•		Biradial planes of the binode, each containing ray of axial biplane and a ray of other biplane.
13	$3 \times 3 = 9$		•••						•		•		
14		•••						•			•		
2′3′		••				•	•						Biradial planes of the cnicnode.
2′4′	$3 \times 2 = 6$		••		•		•						
3′4′	$\overline{11}$ $\overline{45}$			••	•	•							
		Mere lines.			Cnicnodal rays.			Biplanar rays of other biplane.			Biplanar ray of axial biplane, being a transversal ray.	Axis, joining the two nodes.	

96. Writing $(a, b, c, d \between X, Y)^3 = -d(\theta_2 X - Y)(\theta_3 X - Y)(\theta_4 X - Y)$, the planes are

$$
\begin{array}{rl}
X = 0, & [0] \\
Z = 0, & [00] \\
\theta_2 X - Y = 0, & [22'] \\
\theta_3 X - Y = 0, & [33'] \\
\theta_4 X - Y = 0, & [44'] \\
d(\theta_2 X - Y) - Z = 0, & [12] \\
d(\theta_3 X - Y) - Z = 0, & [13] \\
d(\theta_4 X - Y) - Z = 0, & [14] \\
X\theta_2\theta_3 - Y(\theta_2 + \theta_3) - W = 0, & [2'3'] \\
X\theta_2\theta_4 - Y(\theta_2 + \theta_4) - W = 0, & [2'4'] \\
X\theta_3\theta_4 - Y(\theta_3 + \theta_4) - W = 0, & [3'4']
\end{array}
$$

97. And the lines are

a	b	c	f	g	h		equations may be written
0	0	0	0	0	1	(0)	$X = 0,\ Y = 0$
0	0	0	0	-1	d	(1)	$X = 0,\ dY + Z = 0$
0	0	0	1	θ_2	0	(2)	$\theta_2 X - Y = 0,\ Z = 0$
0	0	0	1	θ_3	0	(3)	$\theta_3 X - Y = 0,\ Z = 0$
0	0	0	1	θ_4	0	(4)	$\theta_4 X - Y = 0,\ Z = 0$
θ_2	-1	0	0	0	θ_2^2	(2′)	$\theta_2 X - Y = 0,\ \theta_2^2 X + W = 0$
θ_3	-1	0	0	0	θ_3^2	(3′)	$\theta_3 X - Y = 0,\ \theta_3^2 X + W = 0$
θ_4	-1	0	0	0	θ_4^2	(4′)	$\theta_4 X - Y = 0,\ \theta_4^2 X + W = 0$
$-d\theta_2$	d	1	$-(\theta_3 + \theta_4)$	$-\theta_3\theta_4$	$d(\theta_3\theta_4 - \theta_2\theta_3 - \theta_2\theta_4)$	(12 . 3′4′)	but for the remaining lines
$-d\theta_3$	d	1	$-(\theta_2 + \theta_4)$	$-\theta_2\theta_4$	$d(\theta_2\theta_4 - \theta_3\theta_2 - \theta_3\theta_4)$	(13 . 2′4′)	the coordinate expressions
$-d\theta_4$	d	1	$-(\theta_2 + \theta_3)$	$-\theta_2\theta_3$	$d(\theta_2\theta_3 - \theta_4\theta_2 - \theta_4\theta_3)$	(14 . 2′3′)	are more convenient.

The mere lines are each of them facultative; $b' = \rho' = 3$; $t' = 0$.

98. Hessian surface. The equation is

$$
\begin{aligned}
&\{Z + 3(cX + dY)\}\{XZW + Y^2 Z + (a, b, c, d \between X, Y)^3\} \\
&\quad - 4Z(a, b, c, d \between X, Y)^3 \\
&\quad - 3(4ac - 3b^2, ad, bd, cd, d^2 \between X, Y)^4 = 0;
\end{aligned}
$$

and it is thence easy to see that the complete intersection is made up of the line $X = 0$, $Y = 0$ (the axis) three times, and of a curve of the ninth order, which is the spinode curve; $\sigma' = 9$.

99. The equations of the spinode curve may be written in the simplified form

$$XZW + Y^2Z + (a,\ b,\ c,\ d \between X,\ Y)^3 = 0,$$

$$4Z(a,\ b,\ c,\ d \between X,\ Y)^3 + 3(4ac - 3b^2,\ ad,\ bd,\ cd,\ d^2 \between X,\ Y)^4 = 0,$$

the line $X=0$, $Y=0$ here appearing as a triple line on the second surface; the curve is a partial intersection, $3 \times 4 - 3$.

The node C_2 is a triple point on the curve, the tangents being the nodal rays.

The node B_3 is a quintuple point, one tangent being $X=0$, $3dY+4Z=0$, and the other tangents being given by $Z=0$, $(4ac-3b^2,\ ad,\ bd,\ cd,\ d^2 \between X,\ Y)^4 = 0$.

Each of the facultative lines is a double tangent to the curve, or we have $\beta' = 6$.

Reciprocal Surface.

100. Comparing the equation of the cubic surface with that for $\text{IV} = 12 - 2C_2$, it appears that the equation of $\text{VI} = 12 - B_3 - C_2$ is obtained by substituting in that equation the values $\delta = 0$, $\gamma = 1$. But instead of making this substitution in the final formula, it is convenient to make it in the binary quartic $(* \between X,\ Y)^4$, thus in fact working out the reciprocal surface by means of the function

$$(xX^2 + yXY - wY^2)^2 + 4zwX(a,\ b,\ c,\ d \between X,\ Y)^3,$$

the coefficients whereof (multiplying by 6 to avoid fractions) are

$$\begin{aligned} &6x^2 + 24azw,\\ &3xy + 18bzw,\\ y^2 - {}&2xw + 12czw,\\ -{}&3yw + 6dzw,\\ &6w^2. \end{aligned}$$

We find

$$\begin{aligned} \tfrac{1}{3}I &= L^2 - 12zwM,\\ -J &= L^3 - 18zwLM - 54z^2w^2N, \end{aligned}$$

where

$$\begin{aligned} L &= y^2 + 6(x+3cz)w,\\ M &= 2dxy + 6(2cx - by + 2bdz)w - 4aw^2,\\ N &= -4d^2x^2 - 8d(3bx - 2ay + 2adz)w - 12(3b^2 - 4ac)w^2. \end{aligned}$$

The equation is

$$\frac{1}{108z^2w^2}\{(L^2 - 12zwM)^3 - (L^3 - 18zwLM - 54z^2w^2N)^2\} = 0,$$

viz. it is

$$L^2(LN + M^2) - 18zwLMN - 16zwM^3 - 27z^2w^2N^2 = 0,$$

where however $LN + M^2$ contains the factor w, $= wP$ suppose; the equation thus is

$$L^2P - 18zLMN - 16zM^3 - 27z^2wN^2 = 0.$$

Write

$$\begin{aligned} A &= 4x + 12cz,\\ B &= 6cx - 3by + 6bdz - 2aw,\\ C &= 6bdx - 4ady + 4ad^2z + 3(3b^2 - 4ac)w, \end{aligned}$$

and therefore

$$L = \quad y^2 + \ Aw,$$
$$M = \ 2dxy + 2Bw,$$
$$N = -4d^2x^2 - 4Cw,$$

then we have

$$P = \frac{1}{w}\{-(y^2 + Aw)(4d^2x^2 + 4Cw) + (2dxy + 2Bw)^2\}$$
$$= -4\{Cy^2 - 2Bdxy + Ad^2x^2 + w(AC - B^2)\},$$

or the equation is

$$4L^2\{Cy^2 - 2Bdxy + Ad^2x^2 + w(AC - B^2)\} + 18zLMN + 16zM^3 + 27z^2wN^2 = 0.$$

101. Consider the section by the plane $w = 0$, we have $L = y^2$, $M = 2dxy$, $N = -4d^2x^2$, and the equation becomes $4y^4(Cy^2 - 2Bdxy + Ad^2x^2) + (128 - 144 =) - 16d^3x^3y^3z = 0$; which substituting for A, B, C the values

$$A = 4x + 12cz,$$
$$B = 6cx - 3by + 6bdz,$$
$$C = 6bdx - 4ady + 4ad^2z,$$

becomes $16dy^3(y - dz)(dx^3 - 3cx^2y + 3cxy^2 - ay^3) = 0$; which is in fact the line $w = 0$, $y = 0$ (reciprocal of the edge) three times, and the lines $w = 0$, $(y - dz)(d, -c, b, -a\between x, y)^3 = 0$ (reciprocals of the biplanar rays) each once. Observe that the edge $(X = 0, Z = 0)$ is not a line of the cubic surface, but the reciprocal line $y = 0$, $w = 0$ presents itself as an oscular line of the reciprocal surface.

102. The equations of the cuspidal curve are in the first instance obtained in the form

$$\left\| \begin{matrix} L, & M, & 3N \\ 12zw, & L, & -4M \end{matrix} \right\| = 0.$$

Consider the two equations

$$L^2 \ - 12zwM = 0,$$
$$LM + \ 9zwN = 0,$$

each of the fourth order, but which are satisfied by $zw = 0$, $L = 0$; that is, by $(w = 0, y^2 = 0)$, $(z = 0, y^2 + 4xw = 0)$. The line $(w = 0, y = 0)$ however presents itself in the intersection of the two surfaces, not twice only, but 4 times. To show this, observe that the line in question is a nodal line on the surface $L^2 - 12zwM = 0$; in fact, attending only to the terms of the second order in y, w, we find

$$\{(4x + 12cz)^2 - 144cxz - 144bdz^2\}w^2 - 24dxzyw = 0,$$

giving the two sheets

$$\{(4x + 12cz)^2 - 144cxz - 144bdz^2\}w - 24dxzy \quad = 0 \text{ and } w = 0;$$

in regard to the last-mentioned sheet the form in the vicinity thereof is given by $w = Ay^3$, viz. we have approximately $L = y^2$, $M = 2dxy$, and thence $y^4 - 12z \,.\, Ay^3 \,.\, 2dxy = 0$, that is, $A = \frac{1}{24dxz}$ or $w = \frac{1}{24dxz}y^3$; the line is thus a flecnodal line on the surface

$L^2 - 12zwM = 0$. Next as regards the surface $LM + 9zwN = 0$; the line $y = 0$, $w = 0$ is a simple line on the surface, the terms of the lowest order being $9zw(-4d^2x^2) = 0$; that is, we have $w = 0$, and for a next approximation $w = Ay^3$, viz. $L = y^2$, $M = -2dxy$, $N = -4d^2x^2$, and therefore $-2dxy^3 + 9z \,.\, Ay^3(-4d^2x^2) = 0$, that is, $A = -\dfrac{1}{18dxz}$, or $w = -\dfrac{1}{18dxz}y^3$; there is thus a threefold intersection with one sheet and a simple intersection with the other sheet of the surface $L^2 - 12zwM = 0$. The surfaces intersect, as has been mentioned in the conic $z = 0$, $y^2 + 4xw = 0$; or we have the line $y = 0$, $w = 0$ four times, the conic once, and a residual cuspidal curve of the order $4 \,.\, 4 - 4 - 2, = 10$; that is, $c' = 10$.

Section VII $= 12 - B_5$.

Article Nos. 103 to 116. Equation $WXZ + Y^2Z + YX^2 - Z^3 = 0$.

103. The diagram of lines and planes[1] is

$\text{VII} = 12 - B_5$.		Lines: 13′	12′	3′	2′	1	0	
		$\overline{6}$	2× 1= 2 $\overline{27}$	2× 5=10		1× 5= 5	1×10=10	
Planes. 01	1×15=15					· · ·	· · · · ·	Torsal biplane.
00	1×20=20			· · · ·	· · · ·		· ·	Ordinary biplane.
12′	2× 5=10		· · · · ·		·	·		Planes each containing a mere line.
13′	$\overline{4}$ $\overline{45}$	· · · · ·		·		·		
		Mere lines.		Ray of ordinary biplane.		Ray of torsal biplane.	Edge.	

[1] The marginal symbols in the preceding diagrams constitute a real notation of the lines and planes; but here, and still more so in some of the following diagrams, they are mere marks of reference, showing which are the lines and planes to which the several equations respectively belong.

104. The planes are

$Z=0,$	[10]
$X=0,$	[00]
$Y+Z=0,$	[12′]
$Y-Z=0,$	[13′]

The lines are

$X=0,$	$Z=0,$	(0)
$Y=0,$	$Z=0,$	(1)
$X=0,$	$Y+Z=0,$	(2′)
$X=0,$	$Y-Z=0,$	(3′)
$X-W=0,$	$Y+Z=0,$	(12′)
$X+W=0,$	$Y-Z=0,$	(13′).

105. The two mere lines are facultative, and the edge is also facultative; $\rho'=b'=3$; $t'=0$.

106. Hessian surface. The equation is

$$Z(WXZ+Y^2Z+YX^2-Z^3)-4X^2YZ+X^4+4Z^4=0.$$

The complete intersection with the surface is thus given by the equations

$$WXZ+Y^2Z+YX^2-Z^3=0,\quad -4X^2YZ+X^4+4Z^4=0,$$

which is made up of the line $X=0$, $Z=0$ (the edge) four times and a curve of the eighth order. To see this, observe that the last-mentioned surfaces have in common the line $X=0$, $Z=0$, which is on the first surface a torsal line (equation in vicinity being $Z=-\frac{1}{Y}X^2$), and on the second surface a triple line (equations in vicinity being $Z=\frac{1}{Y}X^2$ and $X^2=\frac{1}{Y}Z^3$). But $Z=-\frac{1}{Y}X^2$ *touches* $Z=\frac{1}{Y}X^2$, and the line counts thus $(2+2=)$ 4 times.

107. I say that the complete intersection is the line $(X=0,\ Z=0)$ three times together with a spinode curve made up of this same line once and of the curve of the eighth order; and that thus $\sigma'=9$.

The discussion of the reciprocal surface in fact shows that the reciprocal of the edge is a singular line thereof, counting once as a nodal and twice as a cuspidal line thereof; the cuspidal tangent planes are the reciprocals of the several points of the edge, and the edge is thus part of the spinode curve. The reasoning may appear to show that the edge should be counted twice, but it must be counted once only, making the order $=9$ as mentioned.

108. I find that the octic component of the spinode curve is a unicursal curve, the equations of which may be written

$$X:Y:Z:W=16\theta^2:4\theta+16\theta^5:16\theta^3:-5-8\theta^4-16\theta^8;$$

the values of θ at the binode B_5 are $\theta=0$, $\theta=\infty$, and we thus obtain in the neighbourhood thereof the two branches

$$\frac{Y}{W}=-5\left(\frac{Z}{W}\right)^2,\quad \frac{X}{W}=\tfrac{25}{4}\left(\frac{Z}{W}\right)^3 \text{ and } \frac{X}{W}=\left(\frac{Z}{W}\right)^{\frac{5}{3}},\quad \frac{Y}{W}=-\left(\frac{Z}{W}\right)^2.$$

109. Each of the lines $(X-W=0,\ Y+Z=0)$ and $(X+W=0,\ Y-Z=0)$ is a double tangent of the spinode octic; in fact for the first of these lines we have

$$16\theta^8+8\theta^4+16\theta^2+5=0,\quad 16\theta^5+16\theta^3+4\theta=0,$$

that is,

$$(2\theta^2+1)^2(4\theta^4-4\theta^2+5)=0,\quad 4\theta(2\theta^2+1)^2=0,$$

so that the line touches at the two points given by $2\theta^2+1=0$; and similarly the other line touches at the two points given by $2\theta^2-1=0$.

The edge $X=0$, $Z=0$ has apparently a higher contact with the spinode octic, viz. the equations $X=0$, $Z=0$ are satisfied by $\theta=0$ twice, $\theta=\infty$ five times; but it must be reckoned only as a double tangent. Hence $\beta'=2\,.\,2+2,\ =6$.

Reciprocal Surface.

110. The equation is obtained by equating to zero the discriminant of the binary quartic

$$X^2(yZ-wX)^2+4wZ^2(wZ^2+zZX+xX^2),$$

viz. calling this $(*\between X,\ Z)^4$, the coefficients (multiplying by 6) are

$$(6w^2,\ -3yw,\ y^2+4xw,\ 6zw,\ 24w^2);$$

and then writing

$$\begin{aligned} L &= \quad y^2+4xw,\\ M &= -2yz-4w^2,\\ N &= -4z^2+16xw,\end{aligned}$$

we have

$$\begin{aligned} \tfrac{1}{3}I &= L^2-12w^2M,\\ -J &= L^3-18w^2LM-54w^4N,\end{aligned}$$

and the equation is, as in former cases,

$$L^2(LN+M^2)-18w^2LMN-16w^2M^3-27w^4N^2=0;$$

but $LN+M^2$ and therefore the whole equation divides by w, and we thus obtain

$$16L^2\left(-xz^2+y^2x+w(yz+4x^2)+w^3\right)-18wLMN-16wM^3-27w^3N^2=0;$$

or, completely developed, this is

$$\begin{aligned}
&w^7\,.\,64\\
+\,&w^5\,.\,32\ (\ 3yz-4x^2)\\
+\,&w^4\,.\,16x\,(\ 5y^2+9z^2)\\
+\,&w^3\,.\qquad (\ y^4+30y^2z^2+160yzx^2-27z^4+64x^4)\\
+\,&w^2\,.\ \ 4x\,(11y^3z+12y^2x^2-9yz^3-4z^2x^2)\\
+\,&w\ \ .\ \ y^2\,(\ \ y^3z+12y^2x^2-yz^3-8z^2x^2)\\
+\,&\qquad\ y^4x\,(\ \ y^2-z^2)=0.
\end{aligned}$$

111. To transform the equation so as to put in evidence the nodal curve, I collect the terms according to their degrees in (y, z) and (x, w); viz. the equation thus becomes

$$\begin{aligned}
&64x^4w^3 - 128x^2w^5 + 64w^7\\
&+ z^2(-16x^3w^2 + 144xw^4)\\
&+ zy(160x^2w^3 + 96w^5)\\
&+ y^2(48x^3w^2 + 80xw^4)\\
&+ z^4 \,.\, -27w^3\\
&+ z^3y \,.\, -36xw^2\\
&+ z^2y^2 \,.\, -8x^2w + 30w^3\\
&+ zy^3 \,.\, 44xw^2\\
&+ y^4 \,.\, 12x^2w + w^3\\
&+ z^3y^3 \,.\, -w\\
&+ z^2y^4 \,.\, -x\\
&+ zy^5 \,.\, w\\
&+ y^6 \,.\, x \qquad = 0;
\end{aligned}$$

and if for a moment we write $z = \alpha + \gamma$, $y = \alpha - \gamma$ and collect, ultimately replacing α, γ by their values $\frac{1}{2}(z+y)$, $\frac{1}{2}(z-y)$, the equation can be expressed in the form

$$\begin{aligned}
&64w^3(x^2-w^2)^2\\
&+ 8w^2(z+y)^2(x+w)^2(x+3w)\\
&+ 8w^2(z-y)^2(x-w)^2(x-3w)\\
&- 32w^2(z^2-y^2)(x^2-w^2)x\\
&+ \tfrac{1}{4}w(z+y)^4(x+w)^2\\
&- w(z+y)^3(z-y)(x+w)(3x+7w)\\
&+ \tfrac{1}{2}w(z^2-y^2)^2(11x^2-27w^2)\\
&- w(z+y)(z-y)^3(x-w)(3x-7w)\\
&+ \tfrac{1}{4}w(z-y)^4(x-w)^2\\
&- y^3(z^2-y^2)(zw+xy) = 0,
\end{aligned}$$

and observing that we have

$$\begin{aligned}
zw + xy &= -z(x-w) + x(z+y)\\
&= z(x+w) - x(z-y),
\end{aligned}$$

we see that every term of the equation is at least of the second order in $z+y$ and $x-w$ conjointly; and also at least of the second order in $z-y$ and $x+w$ conjointly;

that is, the surface has the nodal lines $(z+y=0,\ x-w=0)$ and $(z-y=0,\ x+w=0)$, which are the reciprocals of the lines 12′ and 13′ respectively. The nodal curve is made up of these two lines and of the line $y=0$, $w=0$ (reciprocal of edge), as will presently appear; so that we have $b'=3$.

112. The equations of the cuspidal curve are

$$\begin{aligned} L^2 - 12w^2M &= 0, \\ LM + \ \ 9w^2N &= 0, \\ 4M^2 + \ \ 3LN \ &= 0. \end{aligned}$$

Attending to the two equations

$$\begin{aligned} L^2 - 12w^2M &= y^4 + 8y^2xw + 16x^2w^2 + 24yzw^2 + 48w^4 = 0, \\ LM + 9w^2N &= y^3z + 2y^2w^2 + 4xyzw + (8 - 72 =) - 64xw^3 + 18z^2w^2 = 0, \end{aligned}$$

these surfaces are each of the order 4, and the order of their intersection is $=16$. But the two surfaces contain in common the line $(y=0,\ w=0)$ 7 times; in fact on the first surface this is a cusp-nodal line $4xw+y^2+Ay^{\frac{5}{2}}=0$; and on the second surface it is a nodal line $w(4xy+18zw)=0$; the sheet $w=0$ is more accurately $4xw+y^2+By^3\ldots=0$; whence in the intersection with the first surface the line counts 5 times in respect of the first sheet and 2 times in respect of the second sheet; together $(5+2=)$ 7 times, and the residual curve is of the order $(16-7=)$ 9.

113. I say that the cuspidal curve is made up of this curve of the 9th order, and of the line $y=0$, $w=0$ (reciprocal of the edge) once; so that $c'=10$. In fact, considering the line in question $y=0$, $w=0$ in relation to the surface, the equation of the surface (attending only to the lowest terms in y, w) may be written

$$- xz^2(y^2+4xw)^2 + w(-y^3z^2) + w^2(-36xyz^3) + \&\text{c.} = 0,$$

giving in the vicinity of the line

$$4xw + y^2 = Ay^{\frac{5}{2}},$$

and then

$$- xz^2A^2 + \frac{z^3}{x}\left(\tfrac{1}{4} - \tfrac{36}{16}\right) = 0,$$

that is, $A^2 = -2\dfrac{z}{x^2}$ or $4xw+y^2 = \sqrt{-2}\,.\,\dfrac{z^{\frac{1}{2}}}{x}y^{\frac{5}{2}}$; wherefore the line is a cusp-nodal line, counting once as a nodal and once as a cuspidal line; and so giving the foregoing results $b'=3$, $c'=10$.

114. I revert to the equation which exhibits the nodal lines $(x-w=0,\ y+z=0)$, $(x+w=0,\ y-z=0)$ for the purpose of showing that they have respectively no pinch-points; that is, that in regard to each of them we have $j'=0$. In fact for the first

of these lines, neglecting the terms which contain $x-w$, $y+z$ conjointly in an order above the second, the equation may be written

$$\begin{aligned}
& 64w^3(x+w)^2 \qquad\qquad (x-w)^2\\
+\ & 8w^2(x+w)^2(x+3w)(z+y)^2\\
+\ & 8w^2(z-y)^2(x-3w)(x-w)^2\\
-\ & 32w^2(z-y)\ (x+\ w)\,x \qquad (x-w)(z+y)\\
+\ & \tfrac{1}{2}w\ (z-y)^2(11x^2-27w^2)(z+y)^2\\
-\ & w\ (z-y)^3(3x-7w) \qquad (x-w)(z+y)\\
+\ & \tfrac{1}{4}w\ (z-y)^4 \qquad\qquad (x-w)^2\\
+\ & y^3z\ (z-y) \qquad\qquad (x-w)(z+y)\\
-\ & y^3x\ (z-y) \qquad\qquad (z+y)^2=0,
\end{aligned}$$

viz. this is

$$(A,\ B,\ C\between x-w,\ z+y)^2=0,$$

where, collecting the terms and reducing the values by means of the equations $x-w=0$, $z+y=0$, or say by writing $x=w$, $-y=z$, we have

$$\begin{aligned}
A = \ & 64w^3(x+w)^2 && = \ 256w^5\\
+\ & 8w^2(z-y)^2(x-3w) && -\ 64w^3z^2\\
+\ & \tfrac{1}{4}w^2(z-y)^4 && +\ 4wz^4\\
& && = \ 4w(z^2-8w^2)^2,\\
B = & -32(z-y)(x+w)\,xw^2 && = -128w^4z\\
& -w(z-y)^3(3x-7w) && +\ 32w^2z^3\\
& +y^3z(z-y) && -\ 2z^5\\
& && = -\ 2z(z^2-8w^2)^2,\\
C = \ & 8w^2(x+w^2)(x+3w) && = \ 128w^5\\
+\ & \tfrac{1}{2}w(z-y)^2(11x^2-27w^2) && -\ 32w^3z^2\\
-\ & xy^3(z-y) && +\ 2wz^4\\
-\ & xy^3(z-y) && +\ 2wz^4\\
& && = \ 2w(z^2-8w^2)^2.
\end{aligned}$$

Hence the condition $4AC-B^2=0$ of a pinch-point is $(z^2-8w^2)^5=0$, so that the pinch-points (if any) would be at the points $x-w=0$, $y+z=0$, $z^2-8w^2=0$; or say at $x, y, z, w =1,\ -2\sqrt{2},\ 2\sqrt{2},\ 1$. But these values give $L, M, N=12,\ 12,\ -16$; values which satisfy the equations $L^2-12w^2M=0$, $LM+9w^2N=0$, $4M^2+3LN=0$, and as the points in question are obviously not on the line $y=0$, $w=0$, they lie on the ninthic component of the cuspidal curve, being in fact points β', and not pinch-points.

The line $y=0$, $w=0$ *quà* nodal line would have every point a pinch-point, but being part of the cuspidal curve, no point thereof is regarded as a pinch-point; that is, in regard to this line also we have $j'=0$. And therefore for the entire nodal curve $j'=0$.

115. The cuspidal ninthic curve is a unicursal curve, the equations of which can be very readily obtained by considering it as the reciprocal of the spinode torse; we in fact have

$$x : y : z : w = ZW + 2XY : 2YZ + X^2 : WX + Y^2 - 3Z^2 : ZX,$$

or substituting for X, Y, Z, W their values $(=16\theta^2,\ 4\theta + 16\theta^5,\ 16\theta^3,\ -5 - 8\theta^4 - 16\theta^8)$ and omitting a common factor $16\theta^2$, we find for the cuspidal curve

$$x : y : z : w = 3\theta + 24\theta^5 - 16\theta^9 : 24\theta^2 + 32\theta^6 : -4 - 48\theta^4 : 16\theta^3$$

(values which verify the equation $Xx + Yy + Zz + Ww = 0$); the spinode curve being thus of the order $=9$ as mentioned.

For $\theta = \infty$ we have the singular point $(y=0,\ z=0,\ w=0)$ (reciprocal of torsal biplane), and in the vicinity thereof $x : y : z : w = 1 : -2\theta^{-3} : 3\theta^{-5} : -\theta^{-6}$, therefore

$$\left(\frac{y}{x}\right)^2 = -4\frac{w}{x}, \quad \left(\frac{y}{x}\right)^5 = -\tfrac{32}{27}\left(\frac{z}{x}\right)^3.$$

For $\theta = 0$ we have the singular point $x=0,\ y=0,\ w=0$ (reciprocal of the other biplane), and in the vicinity thereof $x : y : z : w = -\tfrac{3}{4}\theta : -6\theta^2 : 1 : -4\theta^3$, therefore

$$\frac{y}{z} = -\tfrac{32}{3}\left(\frac{x}{z}\right)^2, \quad \frac{w}{z} = \tfrac{256}{27}\left(\frac{x}{z}\right)^3.$$

116. The section of the surface by the plane $z=0$ is an interesting curve. Writing $z=0$ in the equation of the surface, I find that the resulting equation may be written

$$(64w^3,\ 144xw^2,\ w^3 + 76x^2w + xy^2 \between w^2 + 27x^2,\ y^2 - 32xw)^2 = 0,$$

where observe that

$$64w^3(w^3 + 76x^2w + xy^2) - (72xw^2)^2 = 64w^3[w(w^2 + 27x^2) + x(y^2 - 32xw)];$$

so that the curve has the four cusps $w^2 + 27x^2 = 0$, $y^2 - 32xw = 0$; the plane $z=0$ intersects the cuspidal ninthic curve in the point $(y=0,\ z=0,\ w=0)$ counting 5 times, and in the last-mentioned four points: in fact, writing in the equations of the ninthic curve $z=0$, that is $1 + 12\theta^4 = 0$, we find $x, y, w = \tfrac{8}{9}\theta,\ \tfrac{64}{3}\theta^2,\ 16\theta^3$, and thence $w^2 + 27x^2 = \tfrac{64}{3}\theta^2(1 + 12\theta^4) = 0$, $y^2 - 32xw = 0$.

The curve has also nodes at the points $(y=0,\ x+w=0;\ y=0,\ x-w=0)$, viz. these are the intersections of the plane $z=0$ with the nodal lines $(y-z=0,\ x+w=0)$ and $(y+z=0,\ x-w=0)$, and it has at the point $(y=0,\ w=0)$ (intersection of its plane with the cusp-nodal line $y=0,\ w=0$, and quintic intersection with the cuspidal ninthic) a singular point $=2$ cusps $+\,7$ nodes; hence the curve has cusps $=(4+2=)\,6$; nodes $(2+7=)\,9$; or 2 nodes $+\,3$ cusps $=36$; class $=6$, as it should be.

Section VIII $= 12 - 3C_2$.

Article Nos. 117 to 125. Equation $Y^3 + Y(X+Z+W) + 4aXZW = 0$.

117. The diagram of the lines and planes is

		Lines.												
		6	5	4	3	2	1	$\bar{9}$	$\bar{8}$	$\bar{7}$	9	8	7	
VIII $= 12 - 3C_2$.		$\overline{12}$ $\overline{27}$			$6 \times 2 = 12$				$3 \times 1 = 3$			$3 \times 4 = 12$		
Planes. 7	$3 \times 2 = 6$									••			••	Planes each touching along an axis and containing the corresponding transversal.
8									••			••		
9								••			••			
12	$3 \times 2 = 6$					•	•			••				Biradial planes each containing two rays of the same node.
34				•	•				••					
56		•	•					••						
13	$6 \times 4 = 24$				••		••					•		Planes each containing an axis, and two rays through the terminal nodes respectively.
24				••		••						•		
16		••					••				•			
25			••			••					•			
46		••		••									•	
35			••		••								•	
789	$1 \times 8 = 8$										••	••	••	Plane through the three axes.
$\overline{789}$	$1 \times 1 = 1$ $\overline{14}$ $\overline{45}$							•	•	•				Plane through the three transversals.
					Rays, 1, 2 through node 89. 3, 4 " " 97. 5, 6 " " 78.				Transversals, meeting the three axes respectively.			Axes, each joining two nodes.		

118. Take m_1, m_2 as the roots of the equation $(m-1)^2 = 4am$, so that $m_1 + m_2 = 2 + 4a$, $m_1 m_2 = 1$, then the planes are

$X = 0,$	[7]
$Y = 0,$	[8]
$Z = 0,$	[9]
$Y + Z + X = 0,$	[12]
$Y + X + W = 0,$	[34]
$Y + Z + W = 0,$	[56]
$Y = (m_1 - 1) X,$	[13]
$Y = (m_2 - 1) X,$	[24]
$Y = (m_2 - 1) Z,$	[16]
$Y = (m_1 - 1) Z,$	[25]
$Y = (m_1 - 1) W,$	[46]
$Y = (m_2 - 1) W,$	[35]
$Y = 0,$	[789]
$Y + X + Z + W = 0,$	[$\overline{789}$]

119. And the lines are

a	b	c	f	g	h	equations may be written
0	0	0	0	0	1	(7) $X = 0,\ Y = 0$
0	0	0	1	0	0	(8) $Z = 0,\ Y = 0$
0	1	0	0	0	0	(9) $W = 0,\ Y = 0$
1	1	1	0	0	0	$(\bar{7})$ $Y + Z + X = 0,\ W = 0$
0	0	1	-1	1	0	$(\bar{8})$ $Y + X + W = 0,\ Z = 0$
1	0	0	0	-1	1	$(\bar{9})$ $Y + Z + W = 0,\ X = 0$
0	0	0	$\dfrac{1}{m_1 - 1}$	1	$\dfrac{1}{m_2 - 1}$	(1) $Y = (m_1 - 1) X = (m_2 - 1) Z$
0	0	0	$\dfrac{1}{m_2 - 1}$	1	$\dfrac{1}{m_1 - 1}$	(2) $Y = (m_2 - 1) X = (m_1 - 1) Z$
-1	$\dfrac{1}{m_1 - 1}$	0	0	0	$\dfrac{1}{m_2 - 1}$	(3) $Y = (m_2 - 1) W = (m_1 - 1) X$
-1	$\dfrac{1}{m_2 - 1}$	0	0	0	$\dfrac{1}{m_1 - 1}$	(4) $Y = (m_1 - 1) W = (m_2 - 1) X$
0	$-\dfrac{1}{m_1 - 1}$	1	$\dfrac{1}{m_2 - 1}$	0	0	(5) $Y = (m_1 - 1) Z = (m_2 - 1) W$
0	$-\dfrac{1}{m_2 - 1}$	1	$\dfrac{1}{m_1 - 1}$	0	0	(6) $Y = (m_2 - 1) Z = (m_1 - 1) W$

120. The three transversals are each facultative; $\rho' = b' = 3$; $t' = 0$.

121. Hessian surface. The equation is

$$4aXZW(3Y+X+Z+W)+Y^2(X^2+Z^2+W^2-2XZ-2XW-2ZW)=0.$$

The complete intersection with the cubic surface is made up of the lines ($Y=0$, $X=0$), ($Y=0$, $Z=0$), ($Y=0$, $W=0$) (the axes) each twice, and of a sextic curve which is the spinode curve; $\sigma' = 6$.

The spinode curve is a complete intersection 2×3; the equations may in fact be written

$$Y^3 + Y^2(X+Z+W) + 4aXZW = 0,$$
$$3Y^2 + 4Y(X+Z+W) + 4(XZ+XW+ZW) = 0;$$

the nodes D, C, A are nodes (double points) of the curve, the tangents at each node being the nodal rays.

Each of the transversals is a single tangent of the spinode curve; in fact for the transversal $Y+Z+X=0$, $W=0$, these equations of course satisfy the equation of the cubic surface; and substituting in the equation of the Hessian, we have $Y^2(X-Z)^2=0$. But $Y+Z+X=0$, $W=0$, $Y=0$ is a point on the axis $W=0$, $Y=0$, not belonging to the spinode curve; we have only the point of contact $Y+X+Z=0$, $W=0$, $X-Z=0$. Hence $\beta' = 3$.

Reciprocal Surface.

122. The equation is found by means of the binary cubic,

$$aT(T-yU)^2+(T-xU)(T-zU)(T-wU),$$

viz. writing for shortness

$$\beta = x+z+w,$$
$$\gamma = xz+xw+zw,$$
$$\delta = xzw;$$

this is a binary cubic $(*\between T, U)^3$, the coefficients whereof are

$$3(a+1),\ -2ay-\beta,\ ay^2+\gamma,\ -3\delta,$$

and the equation is hence found to be

$$\begin{aligned}&4a^3y^3(y^3-\beta y^2+\gamma y-\delta)\\ &+a^2\{(12\gamma-\beta^2)y^4-(8\beta\gamma+36\delta)y^3+(30\beta\delta+8\gamma^2)y^2-36\gamma\delta y+27\delta^2\}\\ &+2a\{(6\gamma^2-\beta^2\gamma-9\beta\delta)y^2+(12\beta^2\delta-2\beta\gamma^2-18\gamma\delta)y+2\gamma^3+27\delta^2-9\beta\gamma\delta\}\\ &-(\beta^2\gamma^2+18\beta\gamma\delta-4\beta^3\delta-4\gamma^3-27\delta^2)=0;\end{aligned}$$

or substituting for β, γ, δ in the first and last lines their values

$$(= x + z + w,\ xz + xw + zw,\ xzw),$$

this is

$$\begin{aligned}
&4a^3y^3(y-x)(y-z)(y-w)\\
&+a^2\{(12\gamma-\beta^2)y^4-(8\beta\gamma+36\delta)y^3+(30\beta\delta+8\gamma^2)y^2-36\gamma\delta y+27\delta^2\}\\
&+2a\{(6\gamma^2-\beta^2\gamma-9\beta\delta)y^2+(12\beta^2\delta-2\beta\gamma^2-18\gamma\delta)y+2\gamma^3+27\delta^2-9\beta\gamma\delta\}\\
&-(x-z)^2(x-w)^2(z-w)^2=0.
\end{aligned}$$

123. The nodal curve is made up of the lines $(y=x=z)$, $(y=x=w)$, $(y=z=w)$, reciprocals of the three transversals.

To show this I remark that, writing

$$\begin{aligned}
\beta' &= (x-y)+(z-y)+(w-y),\\
\gamma' &= (x-y)(z-y)+(x-y)(w-y)+(z-y)(w-y),\\
\delta' &= (x-y)(z-y)(w-y),
\end{aligned}$$

the equation of the surface may be written

$$\begin{aligned}
&4a^3y^3(y-x)(y-z)(y-w)\\
&+a^2\{y^2(12\beta'\delta'-\gamma'^2)+x\,.\,18\gamma'\delta'+27\delta'^2\}\\
&+2a\{y(-6\beta'^2\delta'+2\beta'\gamma'^2+9\gamma'\delta')+2\gamma'^3+27\delta'^2-9\beta'\gamma'\delta'\}\\
&-(x-z)^2(x-w)^2(z-w)^2=0,
\end{aligned}$$

whence observing that γ' is of the order 1 and δ' of the order 2 in $(x-y)$, $(z-y)$ conjointly, each term of the equation is at least of the second order in $(x-y)$, $(z-y)$ conjointly; or we have $y=x=z$, a nodal line; and similarly the other two lines are nodal lines.

124. The foregoing transformed equation is most readily obtained by reverting to the cubic in T, U, viz. writing $p=x-y$, $r=z-y$, $s=w-y$, and therefore $x=y+p$, $z=y+r$, $w=y+s$, the cubic function (putting therein $T=V+yU$) becomes

$$a(V+yU)V^2+(V-pU)(V-rU)(V-sU);$$

writing $\beta', \gamma', \delta' = p+r+s,\ pr+ps+rs,\ prs$, the coefficients are $\big(3(a+1),\ ay-\beta',\ \gamma',\ -3\delta'\big)$, and the equation of the surface is thus obtained in the form

$$\begin{aligned}
&27(a+1)^2\delta'^2\\
&+18(a+1)(ay-\beta')\gamma'\delta'\\
&+\ \ 4(a+1)\gamma'^3\\
&-\ \ 4(ay-\beta')^3\delta'\\
&-\ \ \ \ (ay-\beta')^2\gamma'^2=0,
\end{aligned}$$

which, arranging in powers of a, and reversing the sign, is the foregoing transformed result.

125. The cuspidal curve is given by the equations

$$\left\| \begin{matrix} 3(a+1), & -2ay-\beta, & ay^2+\gamma \\ -2ay-\beta, & ay^2+\gamma, & -\delta \end{matrix} \right\| = 0,$$

or say by the equations

$$3(a+1)(ay^2+\gamma)-(2ay+\beta)^2=0,$$

that is

$$a(a-3)y^2+4a\beta y-3(a+1)\gamma=0,$$

and

$$-3(a+1)\delta+(2ay+\beta)(ay^2+\gamma)=0,$$

consequently $c'=6$. It is to be added that the cuspidal curve is a complete intersection, 2×3.

Section $\text{IX} = 12 - 2B_3$.

Article Nos. 126 to 136. Equation $WXZ + (a, b, c, d \between X, Y)^3 = 0$.

126. The diagram of the lines and planes is

$\text{IX}=12-2B_3$.	Lines.	6	5	4	3	2	1	0	
		7 ... 27			6×3=18			1×9= 9	
Planes. 0	1×6= 6							• • • •	Common biplane, oscular along the axis.
7	2×6=12				• •	• •	• •		Other biplanes of the two binodes respectively.
8		• •	• •	• •					
14	3×9=27			• • •			• • •	•	Planes each through the axis and containing rays of the two binodes respectively.
25			• • •			• • •		•	
36		• • •			• • •			•	
6	45								
		Rays, 1, 2, 3 in the non-axial biplane 7 of the one binode, and 4, 5, 6 in the non-axial biplane 8 of the other binode.						Axis joining the two binodes.	

127. Writing $(a, b, c, d \between X, Y)^3 = -d(f_1X - Y)(f_2X - Y)(f_3X - Y)$, the planes are

$$\begin{array}{ll} X = 0, & [\ 0] \\ Z = 0, & [\ 7] \\ W = 0, & [\ 8] \\ f_1X - Y = 0, & [14] \\ f_2X - Y = 0, & [25] \\ f_3X - Y = 0, & [36]\,; \end{array}$$

and the lines are

$$\begin{array}{lll} X = 0, & Y = 0, & (0) \\ f_1X - Y = 0, & Z = 0, & (1) \\ f_2X - Y = 0, & Z = 0, & (2) \\ f_3X - Y = 0, & Z = 0, & (3) \\ f_1X - Y = 0, & W = 0, & (4) \\ f_2X - Y = 0, & W = 0, & (5) \\ f_3X - Y = 0, & W = 0, & (6). \end{array}$$

128. There is no facultative line; $\rho' = b' = 0$, $t' = 0$; and hence also $\beta' = 0$.

129. Hessian surface. The equation is

$$X\{ZW(cX + dY) - 3X(ac - b^2,\ ad - bc,\ bd - c^2 \between X, Y)^2\} = 0,$$

so that the Hessian breaks up into the plane $X = 0$ (axial or common biplane) and into a cubic surface.

The complete intersection of the Hessian with the cubic surface is made up of the line $X = 0$, $Y = 0$ (the axis) four times; and of a system of four conics, which is the spinode curve; $c' = 8$.

In fact combining the equations

$$WXZ + (a, b, c, d \between X, Y)^3 = 0$$

and

$$ZW(cX + dY) - 3X(ac - b^2,\ ad - bc,\ bd - c^2 \between X, Y)^2 = 0,$$

these intersect in the axis once, and in a curve of the eighth order which breaks up into four conics; for we can from the two equations deduce

$$(a, b, c, d \between X, Y)^3 (cX + dY) + 3X^2 (ac - b^2,\ ad - bc,\ bd - c^2 \between X, Y)^2 = 0,$$

that is

$$(4ac - 3b^2,\ ad,\ bd,\ cd,\ d^2 \between X, Y)^4 = 0,$$

a system of four planes each intersecting the cubic $XZW + (a, b, c, d \between X, Y)^3 = 0$ in the axis and a conic; whence, as above, spinode curve is four conics.

It is easy to see that the tangent planes along any conic on the surface pass through a point, and form therefore a quadric cone; hence in particular the spinode torse is made up of the quadric cones which touch the surface along the four conics respectively.

Reciprocal Surface.

130. The equation is obtained by means of the binary cubic

$$X(xX+yY)^2+4zw(a,\ b,\ c,\ d\between X,\ Y)^3,$$

viz. calling this $(*\between X,\ Y)^3$ the coefficients are

$$(3x^2+12azw,\ 2xy+12bzw,\ y^2+12czw,\ 12dzw).$$

The equation is found to be

$$\begin{aligned}&432\,(a^2d^2-6abcd+4ac^3+4b^3d-3b^2c^2)\,z^3w^3\\ &+216\,[(ad^2-3bcd+2c^3)\,x^2+(-2acd+4b^2d-2bc^2)\,xy+(-abd+2ac^2-b^2c)\,y^2]\,z^2w^2\\ &+\quad 9\,[3d^2x^4-12cdx^3y+(10bd+8c^2)\,x^2y^2-(4ad+8bc)\,xy^3+(4ac-b^2)\,y^4]\,zw\\ &-\quad y^3\,(dx^3-3cx^2y+3bxy^2-ay^3)=0.\end{aligned}$$

The section by the plane $w=0$ (reciprocal of $B_3=D$) is the line $w=0$, $y=0$ (reciprocal of edge) three times, and the lines $w=0$, $dx^3-3cx^2y+3bxy^2-ay^3=0$ (reciprocals of the biplanar rays). And similarly for the section by the plane $z=0$ (reciprocal of $B_3=C$).

The section by the plane $y=0$ is made up of the lines $(y=0,\ z=0)$, $(y=0,\ w=0)$ each once, and of two conics, $y=0$,

$$\begin{aligned}&16\,(a^2d^2-6abcd+4ac^3+4b^3d-3b^2c^2)\,z^2w^2\\ &+\ 8\,(ad^2-3bcd+2c^3)\,x^2zw\\ &+\quad d^2x^4=0.\end{aligned}$$

131. There is not any nodal curve; $b'=0$.

132. Cuspidal curve. The equations may be written

$$\left\|\begin{matrix}3x^2+12azw, & 2xy+12bzw, & y^2+12czw\\ 2xy+12bzw, & y^2+12czw, & 12dzw\end{matrix}\right\|=0.$$

Forming the equations

$$\begin{aligned}&(bd-c^2)\,.\,144z^2w^2+2\,(dxy-cy^2)\,.\,12zw\ -y^4=0,\\ &(ad-bc)\,.\,144z^2w^2+(3dx^2-2cxy-by^2)\,.\,12zw\ -2xy^3=0,\end{aligned}$$

these are two quartic surfaces having in common the lines $(y=0,\ w=0)$, $(y=0,\ z=0)$: attending to the line $(y=0,\ z=0)$, this is on the second surface an oscular line, $z=\dfrac{1}{18dxw}y^3$; on the first surface it is a nodal line, the one tangent plane being $6\,(bd-c^2)\,w\,.\,z+dxz\,.\,y=0$, the other tangent plane being $z=0$, but the line being in regard to this sheet an oscular line, $z=\dfrac{1}{24dxw}y^3$. Hence in the intersection of the two surfaces the line counts $(1+3=)$ 4 times; similarly the line $y=0$, $w=0$ counts $(1+3=)$ four times; and there is a residual intersection of the order $(16-4-4=)$ 8, which is the cuspidal curve; $c'=8$.

133. The cuspidal curve is a system of four conics; in fact from the preceding equations written in the forms

$$(bd - c^2,\ 2(dxy - cy^2),\ -y^4 \between 12zw,\ 1)^2 = 0,$$
$$(ad - bc,\ 3dx^2 - 2cxy - by^2,\ -2xy^3 \between 12zw,\ 1)^2 = 0,$$

eliminating zw, we obtain

$$\left\{\begin{matrix} 3(bd - c^2), \\ 2(-ad^2 - 3bcd + 4c^3), \\ 6(acd + b^2d - 2bc^2), \\ 6(bc - ad)\,b, \\ a^2d - b^3, \end{matrix}\right\} \between x,\ y)^4 = 0,$$

which shows that the cuspidal curve lies in four planes, and it hence consists of four conics; these are of course the reciprocals of the quadric cones which touch the cubic surface along the four conics which make up the spinode curve.

134. The equation of the surface, attending only to the terms of the second order in y, z, w, is $27d^2x^4zw = 0$; it thus appears that the point $y = 0$, $z = 0$, $w = 0$ (reciprocal of the plane $X = 0$) (which is oscular along the axis joining the two binodes, or BB-axis) is a binode on the reciprocal surface, the biplanes being $z = 0$, $w = 0$, viz. these are the planes reciprocal to the binodes ($X = 0$, $Y = 0$, $W = 0$) and ($X = 0$, $Y = 0$, $Z = 0$) of the cubic surface; we have thus $B' = 1$.

It is proper to remark that the binode $y = 0$, $z = 0$, $w = 0$ is not on the cuspidal curve, as its being so would probably imply a higher singularity.

135. A simple case, presenting the same singularities as the general one, is when $a = d$, $b = c = 0$: to diminish the numerical coefficients assume $a = d = \frac{1}{12}$, the cubic surface is thus $12XZW + X^3 + Y^3 = 0$, and the equation of the reciprocal surface, multiplying it by 4, becomes

$$\begin{aligned} &z^3w^3 \\ &+ 6x^2z^2w^2 \\ &+ (9x^4 - 12x^3y)\,zw \\ &- 4y^3(x^3 - y^3) = 0, \end{aligned}$$

viz. this is the surface

$$\begin{aligned} &4y^6 \\ &- 4y^3x(x^2 + 3zw) \\ &+ zw(3x^2 + zw)^2 = 0 \end{aligned}$$

considered in the Memoir "On the Theory of Reciprocal Surfaces." The cuspidal curve is, as there shown, composed of the four conics $y = 0$, $3x^2 + zw = 0$ and $y^3 - 2x^3 = 0$, $x^2 - zw = 0$; and it is there shown that the two points ($x = 0$, $y = 0$, $z = 0$), ($x = 0$, $y = 0$, $w = 0$), each reckoned eight times, are to be considered as off-points of the reciprocal surface.

136. The like investigation applies to the general surface, and we have thus $\theta' = 16$; the points in question are still the points $(x=0, y=0, z=0)$, $(x=0, y=0, w=0)$; viz. these are the points of intersection of the surface by the line $(x=0, y=0)$, which points are also the common points of intersection of the four conics which compose the cuspidal curve, that is, they are quadruple points on the cuspidal curve; it does not appear that the points are on this account, viz. *quà* quadruple points of the cuspidal curve, off-points of the surface, nor does this even show that the points should be reckoned each eight times. As already remarked, the singularity requires a more complete investigation.

Section $X = 12 - B_4 - C_2$.

Article Nos. 137 to 143. Equation $WXZ + (X+Z)(Y^2 - X^2) = 0$.

137. The diagram of the lines and planes is

$X = 12 - B_4 - C_2$.		Lines. 12: $1\times1=1$	2′ 1′: $2\times2=4$	2 1: $2\times4=8$	3: $1\times6=6$	0: $1\times8=8$	
	$\overline{6}$	$\overline{7}$				$\overline{27}$	
Planes. 0	$1\times12=12$				• •	• • • •	Biplane touching along axis, and containing edge.
3	$1\times12=12$			• • • • • •	• •		Other biplane.
11′ 22′	$2\times\ 8=16$		• • • • • • • •	• • • •		• •	Planes each through the axis and containing a biplanar ray and a cnicnodal ray.
3′	$1\times\ 3=\ 3$	• • •			• •		Plane touching along the edge and containing the mere line.
1′2′	$1\times\ 2=\ 2$	• •	• •				Biradial plane through the two cnicnodal rays.
	$\overline{45}$						
		Mere line, being a transversal.	Cnicnodal rays.	Biplanar rays in the non-axial biplane.	Edge of binode, being a transversal.	Axis, through the two nodes.	

138. The planes are and the lines are

$X = 0,$	[0]	$X = 0,\ Y = 0,$	(0)
$Z = 0,$	[3]	$X = 0,\ Z = 0,$	(3)
$X - Y = 0,$	[11′]	$X - Y = 0,\ Z = 0,$	(1)
$X + Y = 0,$	[22′]	$X + Y = 0,\ Z = 0,$	(2)
$W = 0,$	[3′]	$X - Y = 0,\ W = 0,$	(1′)
$X + Z = 0,$	[1′2′]	$X + Y = 0,\ W = 0,$	(2′)
		$X + Z = 0,\ W = 0,$	(12).

139. The facultative lines are the edge counting twice, and the mere line;

$$\rho = b' = 3;\ t' = 1.$$

140. Hessian surface. The equation is

$$X(X+Z)(ZW+3X^2-XZ)+Y^2(X-Z)^2=0.$$

The complete intersection with the surface consists of the line ($X=0$, $Y=0$), the axis, four times; the line ($X=0$, $Z=0$), the edge, twice; and a sextic curve, which is the spinode curve; $c'=6$.

Writing the equations of the surface and the Hessian in the form

$$X(ZW+Y^2)-X^3+Z(Y^2-X^2)=0,$$
$$X(X+Z)(ZW+Y^2)+(Z-3X)\{-X^3+Z(Y^2-X^2)\}=0,$$

we see that the equations of the spinode curve may be written

$$ZW+Y^2=0,$$
$$-X^3+Z(Y^2-X^2)=0,$$

viz. the curve is a complete intersection, 2×3.

There is at B_4 a triple point $\frac{Y}{W}=-\left(\frac{Z}{W}\right)^2,\ \frac{X}{W}=-\left(\frac{Z}{W}\right)^{\frac{4}{3}}$; and at C_2 a double point, the tangents coinciding with the nodal rays $W=0$, $Y^2-X^2=0$.

The edge and the mere line are each of them single tangents of the spinode curve. But the edge counting twice in the nodal curve, its contact with the spinode curve will also count twice, that is, we have $\beta'=2.1+1,\ =3$.

Reciprocal Surface.

141. The equation is obtained by means of the binary cubic

$$4w^2X(X+Z)^2+4wZ(X+Z)(xX+zZ)+y^2XZ^2;$$

or calling this $(*\between X, Z)^3$, the coefficients are

$$(12w^2,\ 8w^2+4wx,\ 4w^2+4wx+4wz+y^2,\ 12wz),$$

and thence the equation is found to be

$$\begin{aligned}&16w^4[y^2-(x-z)^2]\\&+16w^3[(2x-5z)y^2-2(x-2z)(x-z)^2]\\&+\ 8w^2[y^4+(x^2-xz+6z^2)y^2-2x^2(x-z)^2]\\&+\ 4w[(2x+3z)y^4-2x^2(x+z)y^2]\\&+\ y^4(y^2-x^2)=0,\end{aligned}$$

where the section by the plane $w=0$ (reciprocal of binode) is $y^4(y^2-x^2)=0$, viz. this is the line $w=0$, $y=0$ (reciprocal of the edge) four times, and the lines $w=0$, $y^2-x^2=0$ (reciprocals of the biplanar rays).

The section by the plane $z=0$ is found to be $(y^2-x^2)(y^2+4xw+4w^2)^2=0$, viz. this is the two lines $z=0$, $y^2-x^2=0$ (reciprocals of the nodal rays), and the conic $z=0$, $y^2+4xw+4w^2=0$ (reciprocal of the nodal cone $WX+Y^2-X^2=0$) twice.

142. Nodal curve. The equation shows that the line $y=0$, $x-z=0$ (reciprocal of the line $W=0$, $X+Z=0$) is a nodal line on the surface.

It also shows that the line $y=0$, $w=0$ (reciprocal of the edge) is a tacnodal line (= 2 nodal lines) on the surface; in fact attending only to the lowest terms in y, w, we have

$$-x^2[16(x-z)^2w^2+8(x+z)wy^2+y^4]=0,$$

that is,

$$4(x-z)w+\frac{\sqrt{x}\pm\sqrt{z}}{\sqrt{x}\mp\sqrt{z}}y^2=0,$$

two values, $w=Ay^2$, $w=By^2$, which indicates a tacnodal line.

The nodal curve is thus made up of the line $y=0$, $x-z=0$ once, and the line $y=0$, $w=0$ twice; $b'=3$.

143. Cuspidal curve. The equations

$$\left\|\begin{matrix}12w^2, & 8w^2+4wx, & 4w^2+4wx+4wz+y^2\\ 8w^2+4wx, & 4w^2+4wx+4wz+y^2, & 12wz\end{matrix}\right\|=0$$

give

$$\begin{aligned}&(4w+2x)^2-3(4w^2+4wx+4wz+y^2)=0,\\&-36w^2z+(2w+x)(4w^2+4wx+4wz+y^2)=0,\end{aligned}$$

or, as these are more simply written,

$$\begin{aligned}&4w^2+\ 4wx-12wz+4x^2-3y^2=0,\\&8w^3+12w^2x-28w^2z+w(4x^2+4xz+2y^2)+xy^2=0,\end{aligned}$$

so that the cuspidal curve is a complete intersection 2×3; $c'=6$.

Section $\text{XI} = 12 - B_6$.

Article Nos. 144 to 149. Equation $WXZ + Y^2Z + X^3 - Z^3 = 0$.

144. The diagram of the lines and planes is

$\text{XI} = 12 - B_6$. Planes are			Lines are: $X=0,\ Y+Z=0$ 1; $X=0,\ Y-Z=0$ 2 — $2 \times 6 = 12$; $\overline{3}$ $\overline{27}$	$X=0,\ Z=0$ 0 — $1 \times 15 = 15$	
$X=0$	0	$1 \times 15 = 15$		⋯	Oscular biplane.
$X=0$	3	$1 \times 30 = 30$; $\overline{2}$ $\overline{45}$	⋯ ⋯	⋯	Ordinary biplane.
			Rays in the ordinary biplane.	Edge.	

where the equations of the lines and planes are shown in the margins of the diagram.

145. The edge is a facultative line counting three times; this will appear from the discussion of the reciprocal surface. Therefore $\rho' = b' = 3$; $t' = 1$.

146. Hessian surface. This is

$$Z(WXZ + Y^2Z - 3X^3 - 3Z^3) = 0,$$

breaking up into $Z = 0$, the oscular biplane, and into a cubic surface (itself a surface $\text{XI} = 12 - B_6$). The complete intersection with the cubic surface is made up of the line $X = 0$, $Z = 0$ (the edge) six times, and of a residual sextic ($= 3$ conics), which is the spinode curve; $c' = 6$.

The equations of the sextic are in fact $XZ + Y^2 = 0$, $X^3 + Z^3 = 0$, so that this consists of three conics, each in a plane passing through the edge.

The edge touches each of the three conics at the point $X = 0$, $Z = 0$, $Y = 0$; but it must be reckoned as a single tangent of the spinode curve, and then counting it three times, $\beta' = 3$.

Reciprocal Surface.

147. The equation is obtained by means of the binary cubic

$$(12w^2,\ 4zw,\ y^2+4xw,\ -12w^2 \between Z,\ X)^3,$$

viz. it is

$$\begin{aligned} & 432w^6 \\ + & \ 72w^3z\,(4xw+y^2) \\ - & \ 64w^3z^3 \\ + & \qquad (4xw+y^2)^3 \\ - & \quad z^2\,(4xw+y^2)^3 = 0, \end{aligned}$$

or, completely developed, it is

$$\begin{aligned} & w^6\,.\,432 \\ & +w^4\,.\,288xz \\ & +w^3\,.\,72y^2z+64x^3-64z^3 \\ & +w^2\,.\,48x^2y^2-16x^2z^2 \\ & +w\ \,.\,12xy^4-8xy^2z^2 \\ & + \quad y^4\,(y^2-z^2)=0\,; \end{aligned}$$

the section by the plane $w=0$ (reciprocal of B_6) is $w=0$, $y=0$ (reciprocal of edge) four times, together with $w=0$, $y^2-z^2=0$, reciprocals of the two rays.

148. The nodal curve is the line $y=0$, $w=0$ (reciprocal of edge counting as three lines); $b'=3$. In fact the form of the surface in the vicinity is given by $w=-\frac{1}{4x}y^2 \pm \frac{1}{4}\sqrt{\frac{z}{x^5}}\,y^3$, viz. there are two sheets osculating along the line in question, that is intersecting in this line taken three times.

149. For the cuspidal curve we have

$$\left\| \begin{matrix} 12w^2, & 4zw, & y^2+4xw \\ 4zw, & y^2+4xw, & -12w^2 \end{matrix} \right\| = 0,$$

giving

$$\begin{aligned} 12xw+3y^2 \quad -4z^2 &= 0, \\ 36w^3\ +4wxz+y^2z &= 0\,; \end{aligned}$$

or multiplying the first by $3z$ and subtracting the second, we have $108w^3+4z^3=0$. Hence the equations are

$$\begin{aligned} z^3+27w^3 &= 0, \\ 12xw+3y^2-4z^2 &= 0, \end{aligned}$$

viz. the cuspidal curve is made up of three conics lying in planes through the line $z=0$, $w=0$.

The curve may be put in evidence by writing the equation of the surface in the form

$$(3y^2+5z^2+12xw,\ 24z,\ 16 \between 3y^2-4z^2+12xw,\ z^3+27w^3)^2=0,$$

where

$$16\,(3y^2+5z^2+12xw)-144z^2=16\,(3y^2-4z^2+12xw).$$

Section XII = 12 − U_6.

Article Nos. 150 to 156. Equation $W(X+Y+Z)^2 + XYZ = 0$.

150. The diagram of the lines and planes is

XII=12 − U_6.		Lines. 3′	2′	1′	3	2	1	
		$\overline{6}$ $\overline{27}$	3×1= 3			3×8=24		
0	1×32=32				· · ·	· · ·	· · ·	Uniplane.
1				· · · ·			· ·	
2	3× 4=12		· · · ·			· ·		Planes each touching along a ray, and containing a mere line.
3		· · · ·			· ·			
1′2′3′	1× 1= 1 $\overline{5}$ $\overline{45}$							Plane through the three mere lines.
Planes.			Mere lines.			Rays in the plane.		

151. The planes are

$X + Y + Z = 0$, [0]
$X = 0$, [1]
$Y = 0$, [2]
$Z = 0$, [3]
$W = 0$, [1′2′3′]

The lines are

$X = 0,\ Y + Z = 0,$ (1)
$Y = 0,\ Z + X = 0,$ (2)
$Z = 0,\ X + Y = 0,$ (3)
$X = 0,\ W = 0,$ (1′)
$Y = 0,\ W = 0,$ (2′)
$Z = 0,\ W = 0,$ (3′).

152. The three mere lines are each facultative: $\rho' = b' = 3$; $t' = 1$.

153. Hessian surface. The equation is

$$(X+Y+Z)^2 (X^2 + Y^2 + Z^2 - 2YZ - 2ZX - 2XY) = 0,$$

viz. the surface consists of the uniplane $X + Y + Z = 0$ twice, and of a quadric cone having its vertex at U_6, and touching each of the planes $X = 0$, $Y = 0$, $Z = 0$.

The complete intersection with the cubic surface is made up of the rays each twice and of a residual sextic, which is the spinode curve; $\sigma'=6$.

The equations of the spinode curve are

$$W(X+Y+Z)^2+XYZ=0,$$
$$X^2+Y^2+Z^2-2YZ-2ZX-2XY=0,$$

viz. the curve is a complete intersection, 2×3.

Each of the mere lines is a single tangent (as at once appears by writing for instance $W=0$, $X=0$, which gives $(Y-Z)^2=0$); that is, $\beta'=3$.

Reciprocal Surface.

154. The equation is found by means of the binary cubic

$$4(T-xU)(T-yU)(T-zU)+wT^2U,$$

viz. writing for shortness

$$\beta=x+y+z,$$
$$\gamma=yz+zx+xy,$$
$$\delta=xyz,$$

then the cubic function is

$$(12,\ w-4\beta,\ 4\gamma,\ -12\delta \between T,\ U)^3,$$

and the equation of the reciprocal surface is found to be

$$\begin{aligned}&432\,\delta^2\\ +\ &64\,\gamma^3\\ -\ &\quad(w-4\beta)^3\,\delta\\ +\ &72\,(w-4\beta)\,\gamma\delta\\ -\ &\quad(w-4\beta)^2\,\gamma^2=0\,;\end{aligned}$$

expanding, this is

$$\begin{aligned}&\quad w^3\,.-\delta\\ +\ &\quad w^2\,.\,12\beta\delta-\gamma^2\\ +\ &\quad 8w\ .-6\beta^2\delta+\beta\gamma^2+9\gamma\delta\\ +\ &\qquad 16\,(4\beta^3\delta-\beta^2\gamma^2-18\beta\gamma\delta+4\gamma^3+27\delta^2)=0\,;\end{aligned}$$

or substituting for β, γ, δ in the first and last lines, this is

$$\begin{aligned}&\quad w^3\,.-xyz\\ +\ &\quad w^2.(12\beta\delta-\gamma^2)\\ +\ &\quad 8w\ .-6\beta^2\delta+\beta\gamma^2+9\gamma\delta\\ +\ &\qquad 16\,(y-z)^2\,(z-x)^2\,(x-y)^2=0\end{aligned}$$

(where $\beta,\ \gamma,\ \delta=x+y+z,\ yz+zx+xy,\ xyz$). The section by the plane $w=0$ (reciprocal of the unode) is made up of the lines $w=0$, $y-z=0$; $w=0$, $z-x=0$; $w=0$, $x-y=0$ (reciprocals of the rays) each twice.

155. The nodal curve is at once seen to consist of the lines $(y=0,\ z=0)$, $(z=0,\ x=0)$, $(x=0,\ y=0)$, reciprocals of the facultative lines; in fact, in regard to $(y,\ z)$ conjointly γ is of the order 1, and δ is of the order 2; hence every term of the equation is of the order 2 in y, z; and the like as to the other two lines: $b'=3$ as above.

156. For the cuspidal curve we have

$$\left\| \begin{matrix} 12\ , & w-4\beta, & 4\gamma \\ w-4\beta, & 4\gamma\ , & -12\delta \end{matrix} \right\| = 0,$$

or say

$$48\gamma - (w-4\beta)^2 = 0,$$
$$36\delta + \gamma(w-4\beta) = 0,$$

whence the cuspidal curve is a complete intersection 2×3; $c'=6$.

Section XIII $= 12 - B_3 - 2C_2$.

Article Nos. 157 to 164. Equation $WXZ + Y^2(Y+X+Z) = 0$.

157. The diagram of the lines and planes is

XIII $=12-B_3-2C_2$.		Lines. 012	4	3	2	1	0	6	5	
		$1\times 1=1$	$2\times 2=4$		$2\times 3=6$		$1\times 4=4$	$2\times 6=12$		
		$\overline{8}$ $\overline{27}$								
Planes. 1	$2\times\ 6=12$					· ·		·	· ·	Biplanes.
2					· ·			· · ·		
056	$1+12=12$						· · ·	· ·	· ·	Plane through the three axes.
5	$2\times\ 6=12$			· · ·		· ·			·	Planes each through an axis joining the binode with a cnicnode.
6			· · ·		· ·			·		
34	$1\times\ 4=\ 4$		· ·	· ·			·			Plane through the axis joining the two cnicnodes.
12	$1\times\ 3=\ 3$	· · ·			·	·				Planes through the biplanar rays.
0	$1\times\ 2=\ 2$	· ·					· ·			Plane touching along the axis which joins the two cnicnodes.
	$\overline{8}$ $\overline{45}$									
		Transversal.	Cnicnodal rays, one through each cnicnode.		Biplanar rays, one in each biplane, and being each a transversal.		Axis joining the two cnicnodes.	Axes, each joining the binode with a cnicnode.		

158. The planes are

The planes are		The lines are	
$X=0$,	[1]	$X=0,\ Y=0$,	(5)
$Z=0$,	[2]	$Z=0,\ Y=0$,	(6)
$Y=0$,	[056]	$Y=0,\ W=0$,	(0)
$Y+X=0$,	[5]	$X=0,\ Y+Z=0$,	(1)
$Y+Z=0$,	[6]	$Z=0,\ Y+X=0$,	(2)
$Y-W=0$,	[34]	$W=Y=-Z$,	(3)
$X+Y+Z=0$,	[12]	$W=Y=-X$,	(4)
$W=0$,	[0]	$W=0;\ X+Y+Z=0$,	(012)

159. The transversal is facultative; $\rho'=b'=1,\ t'=0$.

160. The Hessian surface is

$$WXZ(3Y+X+Z)+Y^2(Z-X)^2=0.$$

The complete intersection with the surface is made up of the line $Y=0,\ X=0$ (CB-axis) three times; the line $Y=0,\ Z=0$ (CB-axis) three times; line $Y=0,\ W=0$ (CC-axis) twice, and of a residual quartic, which is the spinode curve; $\sigma'=4$.

161. Representing the two equations by $U=0,\ H=0$, we have

$$(3Y+X+Z)\,U-H=Y^2(3Y^2+4YX+4YZ+4XZ),\ =MY^2 \text{ suppose},$$

and

$$27(X+Z)\,U+9H=9WXZ(3Y+4X+4Z)+36Y^2(X^2+XZ+Z^2)+27Y^3(X+Z);$$

but

$$\left(-9(X+Z)\,Y+16XZ\right)M=$$
$$64X^2Z^2+28YXZ(X+Z)-Y^2(36X^2+28XZ+36Z^2)-27Y^3(X+Z),$$

whence

$$27(X+Z)\,U+9H+(-9XY-9ZY+16XZ)\,M$$
$$=ZX\left\{12Y^2+28Y\overline{X+Z}+64XZ+9W(3Y+4X+4Z)\right\};$$

or, as this may also be written,

$$\begin{aligned}&27Y^2(X+Z)\,U && +9Y^2H\\ +&(-9YX-9YZ+16XZ)(3Y+X+Z)\,U && +(9Y\overline{X+Z}-16XZ)\,H,\end{aligned}$$

that is,

$$\{-9Y(X+Z)^2+48YXZ+16XZ(X+Z)\}\,U+\{9Y^2+9Y\overline{X+Z}-16XZ\}\,H$$
$$=Y^2ZX\,\{12Y^2+28Y(Z+X)+64XZ+9W(3Y+4X+4Z)\}=0;$$

and we thus obtain the equation of the residual quartic, or spinode curve, in the form

$$3Y^2+\ \ 4Y(X+Z)+\ \ 4XZ=0,$$
$$12Y^2+28Y(X+Z)+64XZ+9W(3Y+4X+4Z)=0.$$

The spinode curve is thus a complete intersection, 2×2; and since the first surface is a cone having its vertex on the second surface, we see moreover that the spinode curve is a nodal quadriquadric. Instead of the last equation we may write more simply

$$4Y(X+Z)+16XZ+3W(3Y+4X+4Z)=0.$$

The equations of the transversal are $W=0$, $X+Y+Z=0$, and substituting in the equations of the spinode curve we obtain from each equation $(X-Z)^2=0$, that is, the transversal is a single tangent of the spinode curve; $\beta'=1$.

Reciprocal Surface.

162. The equation of the cubic is derived from that belonging to $\text{VI}=12-B_3-C_2$ by writing therein $a=b=0$, $c=\tfrac{1}{3}$, $d=1$. Making this change in the formulæ for the reciprocal surface of the case just referred to, we have

$$\begin{aligned}
L &= y^2+4(x+z)w,\\
M &= 2x(y+2w),\\
N &= -4x^2,\\
P &= 16x^2(y+w-x-z);
\end{aligned}$$

and substituting in the equation

$$L^2P+8zM^3-9zLMN-27z^2wN^2=0,$$

the equation divides by x^2; or throwing this out, the equation is

$$\begin{aligned}
&(y^2+4xw+4zw)^2(y+w-x-z)\\
&-8xz(y+2w)^3\\
&+9xz(y^2+4xw+4zw)(y+2w)\\
&-27x^2z^2w=0;
\end{aligned}$$

reducing, this is

$$\begin{aligned}
&w^3.\,16(x-z)^2\\
&+w^2\left\{\begin{aligned}&y^2(x+z)\\&+2y(x^2-4xz+z^2)\\&+(x+z)(2x-z)(-x+2z)\end{aligned}\right\}\\
&+w\left\{\begin{aligned}&y^4\\&+8y^3(x+z)\\&-2y^2(4x^2+23xz+4z^2)\\&+36xyz(x+z)\\&-27x^2z^2\end{aligned}\right\}\\
&+y^3(y-x)(y-z)=0.
\end{aligned}$$

The section by the plane $w=0$ (reciprocal of B_3) is $w=0$, $y=0$ (the edge) three times; and $w=0$, $y-x=0$; $w=0$, $y-z=0$ (reciprocals of the CB-axes).

163. Nodal curve. This is the line $y=x=z$; wherefore $b'=1$. To put the line in evidence, write for a moment $x=y+\alpha$, $z=y+\gamma$, then the equation is readily converted into

$$\begin{aligned}
&w^3 . 16(\alpha-\gamma)^2\\
&+ w^2 \begin{Bmatrix} -y(\alpha^2-4\alpha\gamma+\gamma^2) \\ +(\alpha+\gamma)(2\alpha-\gamma)(-\alpha+2\gamma) \end{Bmatrix}\\
&+ w \begin{Bmatrix} y^2(\alpha^2-10\alpha\gamma+\gamma^2) \\ -18y\alpha\gamma(\alpha+\gamma) \\ -27\alpha^2\gamma^2 \end{Bmatrix}\\
&+ y^3\alpha\gamma=0,
\end{aligned}$$

which, each term being at least of the second order in α, $\gamma\,(=x-y,\ z-y)$, exhibits the nodal line in question.

164. Cuspidal curve. Multiplying by 27, the equation may be written

$$\begin{aligned}(7y-3x-3z-5w,\ -y+6w,\ -w\between y^2+16yw-12xw-12zw+16w^2,&\\ -20y^2+24yx+24yz-27xz-8yw+16w^2)^2=0,&\end{aligned}$$

where

$$4w(7y-3x-3z-5w)+(-y+6w)^2=y^2+16yw-12(x+z)w+16w^2;$$

and we have thus in evidence the cuspidal curve,

$$\begin{aligned}y^2+16yw-12(x+z)w+16w^2=0,&\\ -20y^2+24y(x+z)-27xz-8yw+16w^2=0,&\end{aligned}$$

which is a complete intersection, 2×2, or quadriquadric curve; $c'=4$.

Section XIV $= 12 - B_5 - C_2$.

Article Nos. 165 to 171. Equation $WXZ + Y^2Z + YX^2 = 0$.

165. The diagram of the lines and planes is

$\text{XIV} = 12 - B_5 - C_2$.			$W=0,\ Y=0$ — 3 — $1 \times 2 = 2$, $\overline{4}$ $\overline{27}$	$Z=0,\ Y=0$ — 2 — $1 \times 5 = 5$	$X=0,\ Z=0$ — 1 — $1 \times 10 = 10$	$X=0,\ Y=0$ — 0 — $1 \times 10 = 10$	Lines are
Planes are							
$Z=0$	12	$1 \times 15 = 15$		• • •	• • • •		Torsal biplane.
$X=0$	01	$1 \times 20 = 20$			• •	• • • • •	Ordinary biplane.
$Y=0$	023	$1 \times 10 = 10$, $\overline{3}$ $\overline{45}$	• • • • •	• •		•	Plane through axis and the two rays.
			Cnicnodal ray.	Biplanar ray in torsal biplane.	Edge.	Axis.	

where the equations of the planes and lines are shown in the margins.

166. The edge is a facultative line, as will appear from the discussion of the reciprocal surface: $\rho' = b' = 1$; $t' = 0$.

167. Hessian surface. The equation is

$$WXZ^2 + Y^2Z^2 - 3X^2YZ + X^4 = 0.$$

The complete intersection with the surface is made up of the line $X=0$, $Y=0$ (the axis) five times, the line $X=0$, $Z=0$ (the edge) four times, and a skew cubic, the equations of which may be written

$$\left\| \begin{matrix} X, & Y, & W \\ 4Z, & X, & -5Y \end{matrix} \right\| = 0.$$

In fact from the equations $U=0$, $H=0$ we deduce $H - ZU = X^2(X^2 + 4YZ) = 0$; and if in $U=0$ we write $X^2 = 4YZ$, it becomes $Z(XW + 5Y^2) = 0$; and then in $5U = 0$, writing $5Y^2 = -XW$, we have

$$5WXZ + Z(-XW) + 5X^2Y = X(5XY + 4ZW) = 0.$$

168. I say that the spinode curve is made up of the edge $X=0$, $Z=0$ once, and of the cubic curve; and therefore $\sigma'=4$.

In fact in the reciprocal surface the cuspidal curve is made up of the skew cubic, and of a line the reciprocal of the axis, being a cusp-nodal line, and so counting once as part of the cuspidal curve: the pencil of planes through the line is thus part of the cuspidal torse; and reverting to the original cubic surface, we have the axis as part of the spinode curve: I assume that it counts *once*.

The edge is a single tangent of the spinode curve; $\beta'=1$.

Reciprocal Surface.

169. The equation is obtained by means of the binary cubic

$$4wZ^2(Xx+Zz)+X(YZ-wX)^2,$$

or, as this may be written,

$$(3w^2,\ -2yw,\ y^2+4xw,\ 12zw \between X,\ Z)^3.$$

The equation is in the first instance obtained in the form

$$\begin{aligned}&108w^6z^2\\ -\ &32w^4y^3z\\ +\ &36w^4yz(y^2+4xw)\\ +\ &w^2\ (y^2+4xw)^3\\ -\ &w^2y^2(y^2+4xw)^2=0;\end{aligned}$$

but the last two terms being together $=4w^3x(y^2+4xw)^2$, the whole divides by $4w^3$, and it then becomes

$$\begin{aligned}&27w^3z^2\\ -\ &8wy^3z\\ +\ &9wyz(y^2+4xw)\\ +\ &x(y^2+4xw)^2=0;\end{aligned}$$

or, expanding, it is

$$\begin{aligned}&w^3\,.\,27z^2\\ +\ &w^2\,.\,36xyz+16x^3\\ +\ &w\ .\quad y^3z+\ 8x^2y^2\\ +\ &\qquad xy^4=0.\end{aligned}$$

The section by the plane $w=0$ (reciprocal of B_5) is $w=0$, $y=0$ (reciprocal of edge) four times, together with $w=0$, $x=0$ (reciprocal of biplanar ray).

The section by the plane $z=0$ (reciprocal of C_2) is $x(y^2+4xw)^2=0$, viz. this is $z=0$, $y^2+4xw=0$ (reciprocal of nodal cone) twice, together with $z=0$, $x=0$ (reciprocal of nodal ray).

170. Nodal curve. This is the line $w=0$, $y=0$, reciprocal of edge. The equation in the vicinity is $y=-\frac{1}{4x}w\pm\sqrt{-\frac{z}{8x^4}}\,w^{\frac{5}{2}}$, showing that the line is a cusp-nodal line counting once in the nodal and once in the cuspidal curve: wherefore $b'=1$.

171. Cuspidal curve. The equation of the surface may be written

$$(x,\ -y,\ 3w \between 12xw - y^2,\ 9zw + 8xy)^2 = 0,$$

where $4x \,.\, 3w - y^2 = 12xw - y^2$. This exhibits the cuspidal curve $12xw - y^2 = 0$, $9zw + 8xy = 0$, breaking up into the line $w = 0$, $y = 0$ (reciprocal of edge) and a skew cubic; the line is really part of the cuspidal curve, or $c' = 4$.

The equations of the cuspidal cubic may be written in the more complete form

$$\left\| \begin{matrix} 12x, & y, & z \\ y, & w, & -8x \end{matrix} \right\| = 0.$$

Section XV $= 12 - U_7$.

Article Nos. 172 to 176. Equation $WX^2 + XZ^2 + Y^2Z = 0$.

172. The diagram of the lines and planes is

XV $= 12 - U_7$.			3: $Z=0,\ W=0$	2: $X=0,\ Z=0$	1: $X=0,\ Y=0$	Lines are
			$1\times\ 1=\ 1$	$1\times 10 = 10$	$1\times 16 = 16$	
			$\overline{3}$ $\overline{27}$			
Planes are						
$X=0$	21	$1\times 40 = 40$		· · · ·	· · · · · ·	Uniplane.
$Z=0$	23	$1\times\ 5=\ 5$ $\overline{2}$ $\overline{45}$	· · · · ·	· ·		Plane touching along the single ray.
			Mere line.	Single ray.	Torsal ray.	

where the equations of the lines and planes are shown in the margins.

173. The mere line is facultative: $\rho' = b' = 1$; $t' = 0$.

174. The Hessian surface is

$$X^2(XZ - Y^2) = 0,$$

viz. this is the uniplane $X = 0$ twice, and a quadric cone having its vertex at U_7.

The complete intersection with the surface is made up of $X=0$, $Y=0$ (torsal ray) six times; $X=0$, $Z=0$ (single ray) twice; and of a residual quartic, which is the spinode curve; $\sigma'=4$.

The equations of the spinode curve are $XZ-Y^2=0$, $XW+2Z^2=0$; the first surface is a cone having its vertex on the second surface; and the curve is thus a nodal quadriquadric.

The mere line is a single tangent of the spinode curve; $\beta'=1$.

Reciprocal Surface.

175. The equation is obtained by means of the binary cubic

$$(-3y^2,\ 2yz,\ 4xw,\ 6yw \between X,\ Y)^3,$$

viz. throwing out the factor y, the equation is

$$w^2(-64x^3)+w(-16x^2z^2+72xy^2z+27y^4)+16y^2z^3=0.$$

The section by the plane $w=0$ (reciprocal of U_7) is $w=0$, $z=0$ (reciprocal of torsal ray) three times, and $w=0$, $y=0$ (reciprocal of single ray) twice.

Nodal curve. This is the line $x=0$, $y=0$, reciprocal of the mere line: $b'=1$.

Cuspidal curve. The equation of the surface may be written

$$(64x,\ -16z,\ -3w \between z^2+3xw,\ 9y^2+4zx)^2=0,$$

where

$$4\,.\,64x(-3w)-256z^2=-256(z^2+3xw).$$

This exhibits the cuspidal curve $z^2+3xw=0$, $9y^2+4zx=0$, where the surfaces are each of them cones; the vertex of the second cone is on the first cone, and the two cones have at this point a common tangent plane; the curve is thus a cuspidal quadriquadric.

176. {The equation

$$(64x,\ -16z,\ -3w \between z^2+3xw,\ 9y^2+4zx)^2=0$$

resembles that of a quintic torse, viz. the equation of a quintic torse is

$$(\ x,\ -4z,\ 8w \between z^2-2wx,\ y^2-2zx)^2=0,$$

which equation, writing $9y$ for y, $-2x$ for x, and $\tfrac{3}{4}w$ for w, becomes

$$(-2x,\ -4z,\ 6w \between z^2+3xw,\ 9y^2+4zx)^2=0,$$

or, what is the same thing,

$$(\ x,\ 2z,\ -3w \between z^2+3xw,\ 9y^2+4zx)^2=0;$$

and developing, this is

$$\begin{aligned} & x^3\,.\,w^2 \\ & +x^2\,.-2z^2w \\ & +x\ .-18y^2zw+z^4 \\ & \qquad -27y^4w+2y^2z^3=0, \end{aligned}$$

which, however, differs from the equation of the reciprocal surface, not only in the numerical coefficients, but by the presence of a term xz^4.}

Section XVI $= 12 - 4C_2$.

Article No. 177 to 180. Equation $W(XY + XZ + YZ) + XYZ = 0$.

177. The diagram of the lines and planes is

			$X+W=0,\ Y+Z=0$	$X+Z=0,\ Y+W=0$	$X+Y=0,\ Z+W=0$	$X=0,\ Y=0$	$X=0,\ Z=0$	$X=0,\ W=0$	$Y=0,\ Z=0$	$Y=0,\ W=0$	$Z=0,\ W=0$	Lines are
			14.23	13.24	12.34	34	24	23	14	13	12	
XVI $= 12 - 4C_2$.			$\overline{9}$ $3\times1=3$ $\overline{27}$			$6\times4=24$						
Planes are												
$Z+W=0$	12	$6\times2=12$			·						··	Planes each touching along an axis, and containing a transversal.
$Y+W=0$	13			·						··		
$Y+Z=0$	14		·						··			
$X+W=0$	23		·					··				
$X+Z=0$	24			·			··					
$X+Y=0$	34				·	··						
$X=0$	1	$4\times8=32$				:	:	:				Planes each through three axes.
$Y=0$	2					:			:	:		
$Z=0$	3						:		:		:	
$W=0$	4							:		:	:	
$X+Y+Z+W=0$	1234	$1\times1=1$ $\overline{11}$ $\overline{45}$	⋮	⋮	⋮							Plane through the three transversals.
			Transversals, each a transversal in respect of two opposite axes.			Axes each through two nodes.						

where the equations of the lines and planes are shown in the margins.

178. The transversals are each facultative: $\rho' = b' = 3$; $t' = 1$.

179. Hessian surface. The equation is

$$4XYZW - (X + Y + Z + W)(WXY + WXZ + WYZ + XYZ) = 0,$$

or, what is the same thing,

$$\begin{aligned} &X^2 (YZ + YW + ZW) \\ + &Y^2 (ZW + ZX + WX) \\ + &Z^2 (WX + WY + XY) \\ + &W^2 (XY + XZ + YZ) = 0. \end{aligned}$$

The complete intersection with the cubic surface is made up of the six axes each twice, and there is no spinode curve; $\sigma' = 0$, whence also $\beta' = 0$.

Reciprocal Surface.

180. The equation is immediately obtained in the irrational form

$$\sqrt{x} + \sqrt{y} + \sqrt{z} + \sqrt{w} = 0,$$

or rationalizing, it is

$$(x^2 + y^2 + z^2 + w^2 - 2yz - 2zx - 2xy - 2xw - 2yw - 2zw)^2 - 64xyzw = 0;$$

so that this is in fact Steiner's quartic surface.

Nodal curve. This consists of the lines $x - y = 0$, $z - w = 0$; $x - z = 0$, $y - w = 0$; $x - w = 0$, $y - z = 0$; so that $b' = 3$.

To put any one of these, for instance the line $x - y = 0$, $z - w = 0$, in evidence, we may write the equation of the surface in the form

$$[(x-y)^2 + (z-w)^2 - 2(x+y)(z+w)]^2 - 64xyzw = 0,$$

that is

$$\begin{aligned} &\{(x-y)^2 + (z-w)^2\}\{(x-y)^2 + (z-w)^2 - 4(x+y)(z+w)\} \\ &\quad + 4[(x+y)^2(z+w)^2 - 16xyzw] = 0, \end{aligned}$$

or finally

$$\begin{aligned} &\{(x-y)^2 + (z-w)^2\}\{(x-y)^2 + (z-w)^2 - 4(x+y)(z+w)\} \\ &\quad + 4\{(x-y)^2(z-w)^2 + 4xy(z-w)^2 + 4zw(x-y)^2\} = 0, \end{aligned}$$

where each term is at least of the second order in $x - y$, $z - w$.

There is no cuspidal curve; $c' = 0$.

Section XVII $= 12 - 2B_3 - C_2$.

Article Nos. 181 to 185. Equation $WXZ + XY^2 + Y^3 = 0$.

181. The diagram of the lines and planes is

XVII $= 12 - 2B_3 - C_2$.

Planes are			Lines are: 0 — $X=0, Y=0$	1 — $Y=0, Z=0$	2 — $Y=0, W=0$	3 — $Z=0, X+Y=0$	4 — $W=0, X+Y=0$	
			$1\times 9 = 9$	$2\times 6 = 12$		$2\times 3 = 6$		
							$\overline{5}$ $\overline{27}$	
$X=0$	0	$1\times 6 = 6$	· · · ·					Common biplane, through the axis joining the two binodes.
$Z=0$	13			· · · · ·		· · · ·		Remaining biplanes, one for each binode.
$W=0$	24	$2\times 6 = 12$			· · · · ·		· · · ·	
$Y=0$	012	$1\times 18 = 18$	· ·	·	·			Plane through the three axes.
$X+Y=0$	034	$1\times 9 = 9$	·			·	·	Plane through the axis joining the two binodes.
	$\overline{5}$	$\overline{45}$						
			Axis joining the two binodes.	Axes each through the cnicnode and a binode.		Biplanar rays, one for each binode.		

where the equations of the lines and planes are shown in the margins.

182. There is no facultative line; $b' = \rho' = 0$, $t' = 0$.

183. The Hessian surface is

$$X\,(WXZ + 3YZW + XY^2) = 0,$$

viz. this breaks up into $X = 0$ (the common biplane), and into a cubic surface.

The complete intersection with the cubic surface is made up of $X=0$, $Y=0$ (BB-axis) four times, of $Y=0$, $Z=0$ and $Y=0$, $W=0$ (CB-axes) each three times; and of a residual conic, which is the spinode curve; $\sigma'=2$. The equations of the spinode curve are $Y^2-3ZW=0$, $4X+3Y=0$; viz. it lies in a plane passing through the BB-axis; since there is no facultative line, $\beta'=0$.

Reciprocal Surface.

184. The equation is found to be

$$(y^2+4zw)^2-xy^3-36xyzw+27x^2zw=0,$$

or say this is

$$16z^2w^2+(8y^2-36xy+27x^2)\,zw+y^3\,(y-x)=0.$$

The section by plane $w=0$ (reciprocal of $B_3=D$) is $w=0$, $y^3\,(y-x)=0$, viz. this is the line $w=0$, $y=0$ (reciprocal of edge) three times, and the line $w=0$, $y-x=0$ (reciprocal of ray) once; and the like as to section by plane $z=0$.

The section by plane $x=0$ (reciprocal of $C_2=A$) is $x=0$, $(y^2+4zw)^2=0$, viz. this is the conic (reciprocal of nodal cone) twice.

There is no nodal curve; $b'=0$.

185. Cuspidal curve. The equation of the surface may be written

$$(1,\ -y,\ 3zw\between y^2-12zw,\ 9x-8y)^2=0,$$

where $4\,.\,1\,.\,3zw-y^2=-(y^2-12zw)$; and there is thus a cuspidal conic $y^2-12zw=0$, $9x-8y=0$: wherefore $c'=2$.

Attending only to the terms of the second order in y, z, w, the equation becomes $x^2zw=0$; that is, the point $y=0$, $z=0$, $w=0$ (reciprocal of the common biplane) is a binode of the surface; or there is the singularity $B'=1$.

Section XVIII $= 12 - B_4 - 2C_2$.

Article Nos. 186 to 189. Equation $WXZ + Y^2(X+Z) = 0$.

186. The diagram of the lines and planes are

XVIII $= 12 - B_4 - 2C_2$.			$W=0,\ X+Z=0$	$X=0,\ Z=0$	$Y=0,\ W=0$	$Y=0,\ Z=0$	$Y=0,\ X=0$	Lines are
			4	3	0	2	1	
			$1\times1=1$ $\overline{5}$ $\overline{27}$	$1\times6=6$	$1\times4=4$	$2\times8=16$		
Planes are								
$Y=0$	0	$1\times16=16$			· · · ·	· ·	· ·	Plane through the three axes.
$X=0$	01	$2\times12=24$		· ·			· · · ·	Biplanes.
$Z=0$	02			· ·		· · · ·		
$X+Z=0$	34	$1\times\ 3=\ 3$	· · ·	· ·				Plane touching along edge and containing the mere line.
$W=0$	04	$1\times\ 2=\ 2$ $\overline{5}$ $\overline{45}$	· ·		· ·			Plane touching along axis through the two cnicnodes and containing the mere line.
			Mere line, being also a transversal.	Edge of the binode.	Axis through the two cnicnodes.	Axes, each through the binode and a cnicnode.		

where the equations of the lines and planes are shown in the margins.

187. The mere line is facultative; the edge is also facultative counting twice (this will appear from the discussion of the reciprocal surface): $b' = \rho' = 3,\ t' = 1$.

188. The Hessian surface is

$$(X+Z)\,WXZ+(X-Z)^2\,Y^2=0.$$

The complete intersection with the cubic surface is $Y=0$, $Z=0$ and $Y=0$, $X=0$ (the CB-axes) each four times; $Y=0$, $W=0$ (BB-axis) twice; and $X=0$, $Z=0$ (the edge) twice. There is no spinode curve, $\sigma'=0$; wherefore also $\beta'=0$.

Reciprocal Surface.

189. The equation is obtained from the binary quadric $4w(X+Z)(Xx+Zz)+y^2XZ$, or say

$$(8wx,\ 4w(x+z)+y^2,\ 8wz\between X,\ Z)^2.$$

The equation is thus

$$(y^2+4wx+4wz)^2-64w^2xz=0,$$

or in an irrational form

$$iy+2\sqrt{wx}+2\sqrt{wz}=0.$$

The section by the plane $w=0$ (reciprocal of B_4) is $w=0$, $y=0$ (reciprocal of edge) four times.

The section by the plane $z=0$ (reciprocal of $C_2=C$) is $z=0$, $y^2+4wx=0$ (reciprocal of nodal cone) twice; and similarly for the section by $x=0$ (reciprocal of $C_2=A$).

Nodal curve. Writing the equation in the form

$$y^4+8wy^2(z+x)+16w^2(x-z)^2=0,$$

we have a nodal line $y=0$, $x-z=0$, reciprocal of the mere line:

and writing the equation in the form

$$w=\frac{1}{4(\sqrt{x}\pm\sqrt{z})^2}\,y^2,$$

we have $y=0$, $w=0$ (reciprocal of edge), a tacnodal line counting as two lines; $b'=3$.

There is no cuspidal curve; $c'=0$.

Section XIX $= 12 - B_6 - C_2$.

Article Nos. 190 to 193. Equation $WXZ + Y^2Z + X^3 = 0$.

190. The diagram of the lines and planes is

		$X=0,\ Z=0$	$X=0,\ Z=0$	Lines are
XIX $=12 - B_6 - C_2$. Planes are		$1 \times 15 = 15$ $\overline{2}$ $\overline{27}$	$1 \times 12 = 12$	
$Z=0$	$1 \times 15 = 15$	· · · · ·		Oscular biplane.
$X=0$	$1 \times 30 = 30$ $\overline{2}$ $\overline{45}$	· ·	· · · · · · ·	Ordinary biplane.
		Edge of the binode.	Axis joining the binode and the cnicnode.	

where the equations of the lines and planes are shown in the margins.

191. The axis is a facultative line counting three times (as will appear from the reciprocal surface); $\rho' = b' = 3$, $t' = 1$.

192. The Hessian surface is

$$Z(WXZ + Y^2Z - 3X^3) = 0,$$

viz. this is the oscular biplane $Z = 0$ and a cubic surface.

The complete intersection with the cubic surface is made up of $X = 0$, $Z = 0$ (the edge) six times, and $X = 0$, $Y = 0$ (the axis) six times. There is no spinode curve, $\sigma' = 0$; whence also $\beta' = 0$.

Reciprocal Surface.

193. The equation is at once found to be

$$64zw^3 + (y^2 + 4xw)^2 = 0.$$

The section by the plane $w=0$ (reciprocal of B_6) is $w=0$, $y=0$ (reciprocal of edge) four times. The section by the plane $z=0$ (reciprocal of C_2) is $z=0$, $y^2+4xw=0$ (reciprocal of nodal cone) twice.

Nodal curve. The equation gives

$$w=-\frac{1}{4x}y^2\pm\frac{i\sqrt{z}}{x^{\frac{5}{2}}}y^3+\&c.,$$

showing that the line $w=0$, $y=0$ (reciprocal of edge) is an oscnodal line counting as three lines; $b'=3$.

There is no cuspidal curve; $c'=0$.

Section $XX=12-U_8$.

Article Nos. 194 to 197. Equation $X^2W+XZ^2+Y^3=0$.

194. The diagram of the lines and planes is

Line is $X=0$, $Y=0$ 1

$XX=12-U_8$.

$\overline{1}$ $1\times27=27$ $\overline{27}$

Plane is $X=0$ 0

$1\times45=45$

$\overline{1}$ $\overline{45}$

Uniplane.

Triple ray.

where the equations of the line and plane are shown in the margins.

195. There is no facultative line; $b'=\rho'=0$, $t'=0$.

196. The Hessian surface is $X^3Y=0$, viz. this is the uniplane $X=0$, three times, and the plane $Y=0$ through the ray. The complete intersection with the cubic surface is made up of $X=0$, $Y=0$ (the ray) ten times, and of a residual conic, which is the spinode curve; $\sigma'=2$.

The equations of the spinode conic are $Y=0$, $XW+Z^2=0$, viz. the plane of the conic passes through the ray. Since there is no facultative line, $\beta'=0$.

Reciprocal Surface.

197. The equation is at once found to be

$$27\,(z^2+4xw)^2-64w^3y=0.$$

The section by the plane $w=0$ (reciprocal of the Unode) is $w=0$, $z=0$ (reciprocal of ray) four times.

There is no nodal curve; $b'=0$. But there is a cuspidal conic, $y=0$, $z^2+4xw=0$.

The point $y=0$, $z=0$, $w=0$ (reciprocal of the uniplane $X=0$) is a point which must be considered as uniting the singularities $B'=1$, $\chi'=2$.

I give in an Annex a further investigation in reference to this case of the cubic surface.

Section XXI $=12-3B_3$.

Article Nos. 198 to 201. Equation $WXZ+Y^3=0$.

198. The diagram of the lines and planes is

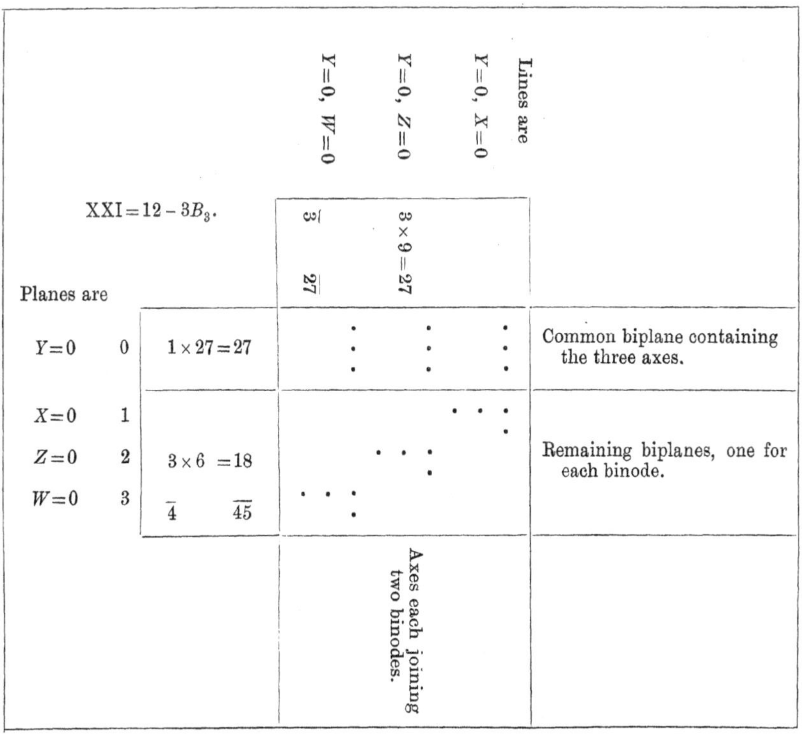

$\text{XXI}=12-3B_3$.			Lines are $Y=0,\ W=0$	$Y=0,\ Z=0$	$Y=0,\ X=0$	
Planes are			$3 \times 9 = 27$; total $\overline{3}$, $\overline{27}$			
$Y=0$	0	$1\times 27=27$	⋮	⋮	⋮	Common biplane containing the three axes.
$X=0$	1				· · ·	Remaining biplanes, one for each binode.
$Z=0$	2	$3\times 6\ =18$		· · ·		
$W=0$	3	$\overline{4}$ $\overline{45}$	· · ·			
			Axes each joining two binodes.			

where the equations of the lines and planes are shown in the margins.

199. There is no facultative line; $\rho'=b'=0$, $t'=0$.

200. The Hessian surface is $XYZW=0$, the common biplane and the other biplanes each once. The complete intersection with the surface consists of the axes each four times; there is no spinode curve, $\sigma'=0$; whence also $\beta'=0$.

Reciprocal Surface.

201. This is $27xzw - y^3 = 0$, viz. it is a cubic surface of the form $\text{XXI} = 12 - 3B_3$. There is no nodal curve, $b' = 0$, and no cuspidal curve, $c' = 0$. Moreover $B' = 3$.

Article No. 202. *Synopsis for the foregoing sections.*

202. I annex the following synopsis, for the several cases, of the facultative lines (or node-couple curve) and of the spinode curve of the cubic surface; also of the nodal curve and the cuspidal curve of the reciprocal surface. It is to be observed that in designating a curve, for instance, as $18 = 4 \times 5 - 2$, this means that it is a curve of the order 18, the partial intersection of a quartic surface and a quintic surface, but without any explanation of the nature of the common curve 2 which causes the reduction, viz. without explaining whether this is a conic or a pair of lines, and so in other cases; this may be seen by reference to the proper section of the Memoir.

	Facultative lines.	Nodal curve.	Spinode curve.	Cuspidal curve.
$\text{I} = 12$	27	27	$12 = 3 \times 4$	$24 = 6 \times 4$
$\text{II} = 12 - C_2$	15	15	$12 = 3 \times 4$	$18 = 4 \times 5 - 2$
$\text{III} = 12 - B_3$	9	9	$12 = 3 \times 4$	$16 = 4 \times 5 - 4$
$\text{IV} = 12 - 2C_2$	7	7	$10 = 3 \times 4 - 2$	$12 = 4 \times 4 - 2 - 2$
$\text{V} = 12 - B_4$	$7 = 5 +$ edge twice	$7 = 5 +$ rec. of edge twice, rec. of edge tacnodal	$10 = 3 \times 4 - 2$	$12 = 4 \times 4 - 4$
$\text{VI} = 12 - B_3 - C_2$	3	3	$9 = 3 \times 4 - 3$	$10 = 4 \times 4 - 4 - 2$
$\text{VII} = 12 - B_5$	$3 = 2 +$ edge	$3 = 2 +$ rec. of edge, rec. of edge is cuspnodal	$9 =$ edge + unicursal 8-thic	$10 =$ rec. of edge + unicursal 9-thic, rec. of edge is cuspnodal
$\text{VIII} = 12 - 3C_2$	3	3	$6 = 2 \times 3$	$6 = 2 \times 3$
$\text{IX} = 12 - 2B_3$	none	none	$8 = 4$ conics	$8 = 4$ conics
$\text{X} = 12 - B_4 - C_2$	$3 = 1 +$ edge twice	$3 = 1 +$ rec. of edge twice, rec. of edge is tacnodal	$6 = 2 \times 3$	$6 = 2 \times 3$
$\text{XI} = 12 - B_6$	$3 =$ edge 3 times	$3 =$ rec. of edge 3 times, rec. of edge is oscnodal	$6 = 3$ conics	$6 = 3$ conics
$\text{XII} = 12 - U_6$	3	3	$6 = 2 \times 3$	$6 = 2 \times 3$
$\text{XIII} = 12 - B_3 - 2C_2$	1	1	$4 = 2 \times 2$, nodal quadriquadric	$4 = 2 \times 2$ quadriquadric
$\text{XIV} = 12 - B_5 - C_2$	$1 =$ edge	$1 =$ rec. of edge, rec. of edge is cuspnodal	$4 = 3 +$ edge	$4 = 3 +$ rec. of edge, rec. of edge is cuspnodal
$\text{XV} = 12 - U_7$	1	1	$4 = 2 \times 2$, nodal quadriquadric	$4 = 2 \times 2$ cuspidal quadriquadric
$\text{XVI} = 12 - 4C_2$	3	3	none	none
$\text{XVII} = 12 - 2B_3 - C_2$	none	none	$2 =$ conic	$2 =$ conic
$\text{XVIII} = 12 - B_4 - 2C_2$	$3 = 1 +$ edge twice	$1 +$ rec. of edge twice, rec. of edge tacnodal	none	none
$\text{XIX} = 12 - B_6 - C_2$	$3 =$ axis 3 times	$3 =$ rec. of axis 3 times, rec. of axis oscnodal	none	none
$\text{XX} = 12 - U_8$	none	none	$2 =$ conic	$2 =$ conic
$\text{XXI} = 12 - 3B_3$	none	none	none	none

I pass now to the two cases of cubic scrolls.

Article No. 203. Section XXII $= S(1, 1)$. Equation $X^2W + Y^2Z = 0$.

203. As this is a scroll there is here no question of the 27 lines and 45 planes; there is a nodal line $X = 0$, $Y = 0$, $(b = 1)$ and a single directrix line, $Z = 0$, $W = 0$.

The Hessian surface is $X^2Y^2 = 0$; the complete intersection with the cubic surface is made up of $X = 0$, $Y = 0$ (the nodal line) eight times, and of the lines $X = 0$, $Z = 0$, and $Y = 0$, $W = 0$, each twice.

The reciprocal surface is $x^2z - y^2w = 0$; viz. this is a like scroll, XXII $= S(1, 1)$; $c' = 0$, $b' = 1$.

Article No. 204. Section XXIII $= S(\overline{1, 1})$. Equation $X(XW + YZ) + Y^3 = 0$.

204. This is also a scroll; there is a nodal line $X = 0$, $Y = 0$, and a single directrix line united therewith.

The Hessian surface is $X^4 = 0$; the complete intersection with the cubic surface is $X = 0$, $Y = 0$ (the nodal line) twelve times.

The reciprocal surface is $w(xw + yz) - z^3 = 0$; viz. this is a like scroll, XXIII $= S(\overline{1, 1})$; $c' = 0$, $b' = 1$.

Annex containing Additional Researches in regard to the case XX $= 12 - U_8$; *equation* $WX^2 + XZ^2 + Y^3 = 0$.

Let the surface be touched by the line (a, b, c, f, g, h), that is, the line the equations whereof are

$$\left(\begin{array}{cccc} 0, & h, & -g, & a \\ -h, & 0, & f, & b \\ g, & -f, & 0, & c \\ -a, & -b, & -c, & 0 \end{array}\right)(X, Y, Z, W) = 0.$$

Writing the equation in the form $cW \,.\, cX^2 + X(cZ)^2 + c^2Y^3 = 0$, and substituting for cW, cZ their values in terms of X, Y, we have

$$(-gX + fY)\,cX^2 + X(aX + bY)^2 + c^2Y^3 = 0,$$

that is

$$(a^2 - cg,\ 2ab + cf,\ b^2,\ c^2)(X, Y)^3 = 0,$$

or say

$$(3(a^2 - cg),\ 2ab + cf,\ b^2,\ 3c^2)(X, Y)^3 = 0,$$

viz. the condition of contact is obtained by equating to zero the discriminant of the cubic function. We have thus

$$\begin{aligned} & 27c^4(a^2 - cg)^2 \\ + \ & 4b^6(a^2 - cg) \\ + \ & 4c^2(2ab + cf)^3 \\ - \ & b^4(2ab + cf)^2 \\ - \ & 18b^2c^2(a^2 - cg)(2ab + cf) = 0, \end{aligned}$$

viz. this is

$$
\begin{aligned}
&+ 27\, a^4c^2 \\
&- \ \ 4\, a^3b^3 \\
&+ 30\, a^2b^2cf \\
&- 54\, a^2c^3g \\
&+ 36\, ab^3cg \\
&+ 24\, abc^2f^2 \\
&+ \ \ 4\, b^5h \\
&- \ \ 1\, b^4f^2 \\
&+ 18\, b^2c^2fg \\
&+ 27\, c^4g^2 \\
&+ \ \ 4\, c^3f^3 = 0,
\end{aligned}
$$

which is the condition in order that the line (a, b, c, f, g, h) may touch the surface $X^2W + XZ^2 + Y^3 = 0$; and if we unite thereto the conditions that the line shall pass through a given point $(\alpha, \beta, \gamma, \delta)$, we have in effect the equation of the circumscribed cone, vertex $(\alpha, \beta, \gamma, \delta)$.

Writing (f, g, h, a, b, c) in place of (a, b, c, f, g, h), we obtain

$$
\begin{aligned}
&\ \ \ 27\, f^4h^2 \\
&- \ \ 4\, f^3g^3 \\
&+ 30\, f^2g^2ha \\
&- 54\, f^2h^3b \\
&+ 36\, fg^3hb \\
&+ 24\, fgh^2a^2 \\
&+ \ \ 4\, g^5c \\
&- \ \ 1\, g^4a^2 \\
&+ 18\, g^2h^2ab \\
&+ 27\, h^4f^2 \\
&+ \ \ 4\, h^3a^3 = 0
\end{aligned}
$$

as the condition that the line (a, b, c, f, g, h) shall touch the reciprocal surface

$$27\,(4\,xw + z^2)^2 + 64\,y^3w = 0\,;$$

and if we consider a, b, c, f, g, h as standing for

$$\gamma y - \beta z,\ \ \alpha z - \gamma x,\ \ \beta x - \alpha y,\ \ \delta x - \alpha w,\ \ \delta y - \beta w,\ \ \delta z - \gamma w,$$

values which satisfy the relation

$$\left(\begin{array}{cccc} 0, & h, & -g, & a \\ -h, & 0, & f, & b \\ g, & -f, & 0, & c \\ -a, & -b, & -c, & 0 \end{array}\right\rangle\!\!\!\left(\alpha,\ \beta,\ \gamma,\ \delta\right)=0,$$

then the equation in (a, b, c, f, g, h) is that of the circumscribed cone, vertex $(\alpha, \beta, \gamma, \delta)$; the order being (as it should be) $a'=6$.

The cuspidal conic is $y=0$, $4xw+z^2=0$, and we at once obtain $a^2-4cg=0$ as the condition that the line (a, b, c, f, g, h) shall pass through the cuspidal cone. Hence attributing to (a, b, c, f, g, h) the foregoing values, we have

$$a^2-4cg=0$$

for the equation of the cone, vertex $(\alpha, \beta, \gamma, \delta)$, which passes through the cuspidal conic; this is of course a quadric cone, $c'=2$. I proceed to determine the intersections of the two cones.

Representing by $\Theta=0$ the foregoing equation of the circumscribed cone, and putting for shortness

$$X=27h^2(f^2-bh)-2g^2(2fg+ah),$$

I find that we have identically

$$\Theta-(f^2-bh)X+(g^4-4ah^3-8fgh^2)(a^2-4cg)-(32fg^2h+16agh^2)(af+bg+ch)=0:$$

whence in virtue of the relation $af+bg+ch=0$, we see that the equations $\Theta=0$, $a^2-4cg=0$, are equivalent to

$$(f^2-bh)X=0, \quad a^2-4cg=0,$$

or the twelve lines of intersection break up into the two systems

$$f^2-bh=0, \quad a^2-4cg=0,$$

and

$$(X=)\quad 27h^2(f^2-bh)-2g^2(2fg+ah)=0, \quad a^2-4cg=0.$$

To determine the lines in question, observe that we have

$$\left(\begin{array}{cccc} 0, & h, & -g, & a \\ -h, & 0, & f, & b \\ g, & -f, & 0, & c \\ -a, & -b, & -c, & 0 \end{array}\right\rangle\!\!\!\left(\alpha,\ \beta,\ \gamma,\ \delta\right)=0;$$

and we can by the first three of these express a, b, c linearly in terms of f, g, h; the equations $f^2 - bh = 0$, $a^2 - 4cg = 0$, $27h^2(f^2 - bh) - 2g^2(2fg + ah) = 0$ become thus homogeneous equations in (f, g, h); the equations may in fact be written

$$\begin{aligned}
\delta^2(a^2 - 4cg) &= (\gamma^2 + 4\alpha\delta) g^2 + \beta^2 h^2 - 2\beta\gamma gh - 4\beta\delta hf = 0,\\
\delta\ (f^2 - bh) &= \delta f^2 - \alpha h^2 + \gamma hf = 0,\\
\delta X\quad &= 27h^2(\delta f^2 - \alpha h^2 + \gamma hf) + 2g^2(\beta h^2 - \gamma gh - 2\delta fg) = 0,
\end{aligned}$$

viz. interpreting (f, g, h) as coordinates *in plano*, the first equation represents a conic, the second a pair of lines, and the third a quartic.

We have identically

$$\begin{aligned}
&\{2\beta\delta f - (\gamma^2 + 4\alpha\delta) g + \beta\gamma h\}^2 - 4\beta^2\delta(\delta f^2 - \alpha h^2 + \gamma hf)\\
&\qquad = (\gamma^2 + 4\alpha\delta)\{(\gamma^2 + 4\alpha\delta) g^2 + \beta^2 h^2 - 2\beta\gamma gh - 4\beta\delta hf\};
\end{aligned}$$

and it thus appears that the two conics touch at the points given by the equations

$$\begin{aligned}
\delta f^2 - \alpha h^2 + \gamma hf &= 0,\\
2\beta\delta f - (\gamma^2 + 4\alpha\delta) g + \beta\gamma h &= 0:
\end{aligned}$$

we have moreover

$$\begin{aligned}
-(\gamma^2 + 4\alpha\delta)(\beta h^2 - \gamma gh - 2\delta fg) &= 4\beta\delta(\delta f^2 - \alpha h^2 + \gamma hf)\\
&\quad + (-2\delta f - \gamma h)[2\beta\delta f - (\gamma^2 + 4\alpha\delta) g + \beta\gamma h],
\end{aligned}$$

hence at the last-mentioned two points $-\beta h^2 + \gamma gh + 2\delta fg$ is $=0$; and the quartic $X = 0$ thus passes through these two points.

The conic $(a^2 - cg) = 0$ and the quartic $X = 0$ intersect besides (as is evident) in the point $g = 0$, $h = 0$ reckoning as two points, since it is a node of the quartic; and they must consequently intersect in four more points: to obtain these in the most simple manner, write for a moment

$$\Omega = -(\gamma^2 + 4\alpha\delta) g^2 + \beta^2 h^2,$$

then we have identically

$$\begin{aligned}
16\beta^2\delta g^2(\delta f^2 - \alpha h^2 + \gamma hf) - \Omega^2 &= -[(\gamma^2 + 4\alpha\delta) g^2 + \beta^2 h^2] + 4\beta^2 g^2(\gamma h + 2\delta f)^2,\\
&= -\{(\gamma^2 + 4\alpha\delta) g^2 + \beta^2 h^2 - 2\beta\gamma gh - 2\beta\delta fg\}\{(\gamma^2 + 4\alpha\delta) g^2 + \beta^2 h^2 + 2\beta\gamma gh + 4\beta\delta fg\},\\
&= -\delta(a^2 - 4cg)\{(\gamma^2 + 4\alpha\delta) g^2 + \beta^2 h^2 + 2\beta\gamma gh + 4\beta\delta fg\};
\end{aligned}$$

and moreover

$$2\beta(\beta h^2 - 2\delta fg - \gamma gh) - \Omega = (\gamma^2 + 4\alpha\delta) g^2 + \beta^2 h^2 - 2\beta\gamma gh - 4\beta\delta fg = \delta(a^2 - 4cg).$$

Hence when $a^2 - 4cg = 0$, we have

$$\delta f^2 - \alpha h^2 + \gamma hf = \frac{\Omega^2}{16\beta^2\delta g^2},\quad \beta h^2 - 2\delta fg - \gamma gh = \frac{\Omega}{2\beta};$$

and substituting these values in the equation $X=0$, it becomes

$$27h^2 \cdot \frac{\Omega^2}{16\beta^2\delta} + 2g^2 \cdot \frac{\Omega}{2\beta} = 0,$$

viz. multiplying by $16\beta^2\delta$, and omitting the factor Ω, this is

$$27h^2\Omega + 16\beta\delta g^4 = 0,$$

or finally

$$16\beta\delta g^4 - 27(\gamma^2 + 4\alpha\delta) g^2h^2 + 27\beta^2h^4 = 0,$$

a pencil of four lines, each passing through the point $g=0$, $h=0$, and therefore intersecting the conic

$$(\gamma^2 + 4\alpha\delta) g^2 + \beta^2h^2 - 2\beta\gamma gh - 4\beta\delta hf = 0$$

at that point and at one other point; and we have thus four points of intersection, which are the required four points.

Recapitulating, the conic $a^2 - 4cg = 0$ meets the sextic $(f^2 - bh) X = 0$ in the two points

$$\begin{cases} \delta f^2 - \alpha h^2 + \gamma hf = 0, \\ 2\beta\delta f - (\gamma^2 + 4\alpha\delta) g + \beta\gamma h = 0 \end{cases}$$

each three times, in the point $g=0$, $h=0$ twice, and in the four points

$$\begin{cases} 16\beta\delta g^4 - 27(\gamma^2 + 4\alpha\delta) g^2h^2 + 27\beta^2h^4 = 0, \\ (\gamma^2 + 4\alpha\delta) g^2 + \beta^2h^2 - 2\beta\gamma gh - 4\beta\delta hf = 0 \end{cases}$$

each once. Or reverting to the proper significations of (a, b, c, f, g, h), instead of points we have 2 lines each three times, a line twice, and 4 lines each once; the line $g=0$, $h=0$, that is, $g=0$, $h=0$, $a=0$, being, it will be observed, the line $\frac{y}{\beta} = \frac{z}{\gamma} = \frac{w}{\delta}$ drawn from $(\alpha, \beta, \gamma, \delta)$ to the point $y=0$, $z=0$, $w=0$, which is the reciprocal of the uniplane $X=0$: the twelve lines are the $a'c'$ lines of intersection of the circumscribed cone a' with the cuspidal cone c', viz. $a'c' = [a'c'] + 3\sigma' + \chi'$; $[a'c'] = 4$ referring to the last-mentioned four lines; $\sigma' = 2$ to the two lines; and $\chi' = 2$ to the line $g=0$, $h=0$, $a=0$, which it thus appears must in the present case be reckoned twice.

413.

A MEMOIR ON ABSTRACT GEOMETRY.

[From the *Philosophical Transactions of the Royal Society of London*, vol. CLX. (for the year 1870), pp. 51—63. Received October 14,—Read December 16, 1869.]

I SUBMIT to the Society the present exposition of some of the elementary principles of an Abstract m-dimensional Geometry. The science presents itself in two ways,—as a legitimate extension of the ordinary two- and three-dimensional geometries; and as a need in these geometries and in analysis generally. In fact whenever we are concerned with quantities connected together in any manner, and which are, or are considered as variable or determinable, then the nature of the relation between the quantities is frequently rendered more intelligible by regarding them (if only two or three in number) as the coordinates of a point in a plane or in space: for more than three quantities there is, from the greater complexity of the case, the greater need of such a representation; but this can only be obtained by means of the notion of a space of the proper dimensionality; and to use such representation, we require the geometry of such space. An important instance in plane geometry has actually presented itself in the question of the determination of the number of the curves which satisfy given conditions: the conditions imply relations between the coefficients in the equation of the curve; and for the better understanding of these relations it was expedient to consider the coefficients as the coordinates of a point in a space of the proper dimensionality.

A fundamental notion in the general theory presents itself, slightly in plane geometry, but already very prominently in solid geometry; viz. we have here the difficulty as to the form of the equations of a curve in space, or (to speak more accurately) as to the expression by means of equations of the twofold relation between the coordinates of a point of such curve. The notion in question is that of a k-fold relation,—as distinguished from any system of equations (or onefold relations) serving for the expression of it, and as giving rise to the problem how to express such relation by means of a system of equations (or onefold relations). Applying to the case of solid geometry my conclusion in the general theory, it may be mentioned that I regard the twofold relation of a curve in space as being completely and precisely expressed by means of a system of equations ($P=0$, $Q=0$, ... $T=0$), when no one of the functions P, Q, ... T is a linear function, with constant or variable *integral* coefficients, of the others of them, and when *every surface whatever* which passes through the curve has its equation expressible in the form $U = AP + BQ \ldots + KT$, with constant or variable integral coefficients, A, B, ... K. It is hardly necessary to remark that all the functions and coefficients are taken to be rational functions of the coordinates, and that the word integral has reference to the coordinates.

Article Nos. 1 to 36. *General Explanations; Relation, Locus, &c.*

1. Any m quantities may be represented by means of $m+1$ quantities as the ratios of m of these to the remaining $(m+1)$th quantity, and thus in place of the absolute values of the m quantities we may consider the ratios of $m+1$ quantities.

2. It is to be noticed that we are *throughout* concerned with the ratios of the $m+1$ quantities, not with the absolute values; this being understood, any mention of the ratios is in general unnecessary; thus I shall speak of a relation between the $m+1$ quantities, meaning thereby a relation as regards the ratios of the quantities; and so in other cases. It may also be noticed that in many instances a limiting or extreme case is sometimes included, sometimes not included, under a general expression; the general expression is intended to include whatever, having regard to the subject-matter and context, can be included under it.

3. Postulate. We may conceive between the $m+1$ quantities a *relation*[1].

4. A relation is either *regular*, that is, it has a definite manifoldness, or, say, it is a k-fold relation; or else it is *irregular*, that is, composed of relations not all of the same manifoldness. As to the word "composed," see *post*, No. 14.

5. The ratios are determined (not in general uniquely) by means of a m-fold relation; and a relation cannot really be more than m-fold. But the notion of a more than m-fold relation has nevertheless to be considered. A relation may be, either in mere appearance or else according to a provisional conception thereof, more than m-fold, and be really m-fold or less than m-fold. Thus a relation expressed by $m+1$ or more

[1] The whole difficulty of the subject is (so to speak) in the analytical representation of a relation; without solving it, the theories of the text cannot be exhibited analytically with equivalent generality; and I have for this reason presented them in an abstract form without analytical expression or commentary. But it is perfectly easy to obtain analytical illustrations; a onefold relation is expressed by an equation $P=0$; and (although a k-fold relation is not in general expressible by k equations) any k independent equations $P=0$, $Q=0$, &c. constitute a k-fold relation. Thus, No. 4, an instance of an irregular relation is $MP=0$, $MQ=0$, viz. this is satisfied by the satisfaction either of the onefold relation $M=0$, or of the twofold relation $P=0$, $Q=0$. And *post*, Nos. 14 and 21, the relation composed of the two onefold relations $P=0$ and $Q=0$ is the onefold relation $PQ=0$; the relation aggregated of the same two relations is the twofold relation $P=0$, $Q=0$.

equations is in general and *primâ facie* more than m-fold; but if the equations are not independent, and equivalent to m or fewer equations, then the relation will be m-fold or less than m-fold. Or the given relation may depend on parameters, and so long as these are arbitrary be really more than m-fold; but the parameters may have to be, and be accordingly, so determined that the relation shall be m-fold or less than m-fold. A more than m-fold relation is said to be superdeterminate.

6. A system of any number of onefold relations, whether independent or dependent, and if more than m of them, whether compatible or incompatible, is termed a 'Plexus,' viz. if the number of onefold relations be $=\theta$, then the plexus is θ-fold. A θ-fold plexus constitutes a relation which is at most θ-fold, but which may be less than θ-fold.

7. Every relation whatever is expressible, and that *precisely*, by means of a plexus; but for the expression of a k-fold relation we may require a more than k-fold plexus.

8. Postulate. We may conceive a m-dimensional space, the indetermination of the ratios of $m+1$ coordinates, and *locus in quo* of the point, the unique determination of these ratios. More generally we may conceive any number of spaces, each of its own dimensionality, and existing apart by itself.

9. Conversely, any $m+1$ quantities may be taken as the coordinates of a point in a m-dimensional space.

10. The $m+1$ coordinates may have a k-fold relation; it appears (*ante*, No. 5) that the case $k>m$, or where the relation is more than m-fold, is not altogether excluded; but this is not now under consideration. The two limiting cases $k=0$ and $k=m$ will be presently mentioned; the remaining case is $k>0<m$; the system of points the coordinates of which satisfy such a relation constitutes a k-fold or $(m-k)$-dimensional locus. And k is the manifoldness, $m-k$ the dimensionality, of the locus.

11. If $k=m$, that is, if the ratios are determined, we have the point-system, which, if the determination be unique, is a single point. The expression "a locus" may extend to include the point-system, and therefore also the point. If $k=0$, that is, if the coordinates are not connected by any relation, we have the original m-dimensional space.

12. We may say that the m-dimensional space is the *locus in quo* not only of the points in such space, but of the locus determined by any relation whatever between the coordinates; and in like manner that any $(m-k)$dimensional locus in such space is a $(m-k)$dimensional space, a *locus in quo* of the points thereof, and of every locus determined by a relation between the coordinates, implying the k-fold relation which corresponds to the $(m-k)$dimensional locus.

13. There is not any locus corresponding to a relation which is really more than m-fold; hence in speaking of the locus corresponding to a given relation, we either assume that the relation is not more than m-fold, or we mean the locus, *if any*, corresponding to such relation.

14. Any two or more relations may be *composed* together, and they are then factors of a single *composite* relation; viz. the composite relation is a relation satisfied if, and not satisfied unless, some one of the component relations be satisfied.

15. The foregoing notion of composition is, it will be noticed, altogether different from that which would at first suggest itself. The definition is defective as not explaining the composition of a relation any number of times with itself, or elevation thereof to power; which however must be admitted as part of the notion of composition.

16. A k-fold relation which is not satisfied by any other k-fold relation, and which is not a power, is a *prime* relation. A relation which is not prime is composite, viz. it is a relation composed of prime factors each taken once or any other number of times; in particular, it may be the power of a single prime factor. Any prime factor is single or multiple according as it occurs once or a greater number of times.

17. A relation which is either prime, or else composed of prime factors each of the same manifoldness, is a regular relation; a k-fold relation is *ex vi termini* regular. An irregular relation is a composite relation the prime factors whereof are not all of the same manifoldness.

18. A prime k-fold relation cannot be implied in any prime k-fold relation different from itself. But a prime k-fold relation may be implied in a prime more-than-k-fold relation,—or in a composite relation, regular or irregular, each factor whereof is more than k-fold; and so also a composite relation, regular or irregular, each factor whereof is at most k-fold, may be implied in a composite relation, regular or irregular, each factor whereof is more than k-fold. In a somewhat different sense, each factor of a composite relation implies the composite relation.

19. A composite relation is satisfied if any particular one of the component relations is satisfied; but in order to exclude this case we may speak of a composite relation as being satisfied *distributively*; viz. this will be the case if, in order to the satisfaction of the composite relation, it is *necessary* to consider *all* the factors thereof, or, what is the same thing, when the reduced relation obtained by the omission of any one factor whatever is *not always satisfied*. And when the composite relation is satisfied distributively, the several factors thereof are satisfied *alternatively*; viz. there is no one which is throughout unsatisfied.

20. A composite onefold relation is never distributively implied in a prime k-fold relation—that is, a prime k-fold relation implies only a prime onefold relation, or at least only implies a composite onefold relation improperly, in the sense that it implies a certain prime factor of such composite onefold relation. Conversely, every k-fold relation which implies distributively a composite onefold relation is composite.

21. Any two or more relations may be *aggregated* together into, and they are then constituents of, a single *aggregate* relation; viz. the aggregate relation is only satisfied when all the constituent relations are satisfied. The aggregate relation implies each of the constituent relations.

22. There is no meaning in aggregating a relation with itself; such aggregation only occurs accidentally when two relations aggregated together become one and the same relation; and the aggregate of a relation with itself is nothing else than the original relation.

23. A onefold relation is not an aggregate, but is its own sole constituent; a more than onefold relation may always be considered as an aggregate of two or more constituent relations. The constituent relations determine, they in fact constitute, the aggregate relation; but the aggregate relation does not in any wise determine the constituent relations. Any relation implied in a given relation may be considered as a constituent of such given relation.

24. The aggregate of a k-fold and a l-fold relation is in general and at most a $(k+l)$fold relation; when it is a $(k+l)$fold relation, the constituent relations are independent, but otherwise, viz. if the aggregate relation is, or has for factor, a less than $(k+l)$fold equation, the constituent relations are dependent or interconnected.

25. Passing from relations to loci, we may say that the composition of relations corresponds to the *congregation* of loci, and the aggregation of relations to the *intersection* of loci.

26. For, first, the locus (if any) corresponding to a given composite relation is the congregate of the loci corresponding to the several prime factors of the given relation, the locus corresponding to a single factor being taken once, and the locus corresponding to a multiple factor being taken a number of times equal to the multiplicity of the factor.

27. And, secondly, the locus (if any) corresponding to a given aggregate relation is the locus common to and contained in each of the loci corresponding to the several constituent relations respectively; or, what is the same thing, it is the intersection of these several loci.

28. It may be remarked that a k-fold locus and a l-fold locus where $k+l>m$ (or where the aggregate relation is more than m-fold) have not in general any common locus.

29. Any onefold relation implied in a given k-fold relation is said to be in *involution* with the k-fold relation, and so in a system of onefold relations, if any relation be implied in the other relations, or, what is the same thing, in the relation aggregated of the other relations, then the system is said to be in *involution*; a system not in involution is said to be *asyzygetic.*

30. Consider a given k-fold relation, and, in conjunction therewith, a system of any number of onefold relations each implied in the given k-fold relation. We may omit from the system any relation implied in the remaining relations, and so successively until we arrive at an asyzygetic system. Consider now any other onefold relation implied in the given k-fold relation; this is either implied in the system of onefold relations, and it is then to be rejected, or if it is not implied in the system, it is to

be added on to and made part of the system. It may happen that, in the system thus obtained, some one relation of the original system is implied in the remaining relations of the new system; but if this is so the implied relation is to be rejected; the new system will in this case contain only as many relations as the original system, and in any case the new system will be asyzygetic. Treating in the same manner every other onefold relation implied in the given k-fold relation, we ultimately arrive at an asyzygetic system of onefold relations, such that every onefold relation implied in the given k-fold relation is implied in the asyzygetic system. The number of onefold relations will be at least equal to k (for if this were not so we should have the given k-fold relation as an aggregate of less than k onefold relations); but it may be greater than k, and it does not appear that there is any [assignable] superior limit to the number of onefold relations of the asyzygetic system.

31. The system of onefold relations is a precise equivalent of the given k-fold relation. Every set of values of the coordinates which satisfies the given k-fold relation satisfies the system of onefold relations; and reciprocally every set of values which satisfies the system of onefold relations satisfies the given k-fold relation. But if we omit any one or more of the onefold relations, then the reduced system so obtained is not a precise equivalent of the given k-fold relation; viz. there exist sets of values satisfying the reduced system, but not satisfying the given k-fold relation.

32. In fact consider a k-fold relation the aggregate of less than all of the onefold relations of the asyzygetic system, and in connexion therewith an omitted onefold relation; this omitted relation is not implied in the aggregate, and it constitutes with the aggregate not a $(k+1)$fold, but only a k-fold relation. This happens as follows, viz. the omitted relation is a factor of a composite onefold relation distributively implied in the aggregate; hence the aggregate is composite, and it implies distributively a composite onefold relation composed of the omitted relation and of an associated onefold relation; that is, the aggregate will be satisfied by values which satisfy the omitted relation, and also by values which (not satisfying the omitted relation) satisfy the associated relation just referred to.

33. Selecting at pleasure any k of the onefold relations of the asyzygetic system, being such that the aggregate of the k relations is a k-fold relation, we have a composite k-fold relation wherein each of the remaining onefold relations is alternatively implied; viz. each remaining onefold relation is a factor of a composite onefold relation implied distributively in the composite k-fold relation. Hence considering the $k+1$ onefold relations, viz. any $k+1$ relations of the asyzygetic system, each one of these is implied alternatively in the aggregate of the remaining k relations; and we may say that the $k+1$ onefold relations are in *convolution*.

34. More generally any $k+1$ or more, or all the relations of the asyzygetic system are in *convolution*, that is, any relation of the system is alternatively implied in the aggregate of the remaining relations, or indeed in the aggregate of any k relations (not being themselves in convolution) of the remaining relations of the asyzygetic

system. It may be added that, besides the relations of the system, there is not any onefold relation alternatively implied in the asyzygetic system.

35. The foregoing theory has been stated without any limitation as to the value of k, and it has I think a meaning even when k is $>m$; but the ordinary case is $k \not> m$. Considering the theory as applying to this case, I remark that the last proposition, viz. that no reduced system is a precise equivalent of the given k-fold relation, is generally true only on the assumption of the existence or quasi-existence of sets of values satisfying a more than m-fold relation. For let k be $\not> m$, and, on the contrary, assume, as we usually do, that it is not in general possible to satisfy a more than m-fold relation between the coordinates; the number of relations in the system may be $>m+1$; and if this is so, then selecting any $m+1$ relations of the system, it may very well happen that the given k-fold relation is not satisfied by any sets of values other than those which satisfy the $m+1$ relations,—that is, that the $m+1$ relations are a precise equivalent of the given k-fold relation. But even in this case the consideration of the entire system of the onefold relations is not the less advantageous; and I say in general that the given k-fold relation has for its precise and complete equivalent the asyzygetic system of onefold relations.

36. {In illustration of the foregoing Nos. 29 to 35, I remark that, for the functions or equations $P=0$, $Q=0$, $R=0$, &c., if we have identically $AP+BQ+CR+\ldots=0$, where the factors $A, B, C, \ldots$ are integral functions of the coordinates, and where some one of these factors, say, A, is a constant (or if we please $=1$), then the system of functions or equations is in involution; or, to speak more accurately, the function or equation $P=0$ is in involution with the remaining functions or equations $Q=0$, $R=0, \ldots$. But when the factors $A, B, C, \ldots$ are no one of them constant, then we have a convolution. If $P=0$ is in involution with the remaining equations $Q=0$, $R=0, \ldots$, then $P=0$ is implied in these equations, and the relations $(Q=0, R=0, \ldots)$ and $(P=0, Q=0, R=0, \ldots)$ are equivalent to each other. But in the case of a convolution where

$$AP+BQ+CR+\ldots=0,$$

then the relation the equations $Q=0$, $R=0, \ldots$ imply $AP=0$, that is, $A=0$ or else $P=0$; or, what is the same thing, the relation $(Q=0, R=0, \ldots)$ is a relation composed of the two relations $(A=0, Q=0, R=0, \ldots)$ and $(P=0, Q=0, R=0, \ldots)$. In the k-fold relation expressed by the more than k equations $(P=0, Q=0, R=0, \ldots)$, selecting any k of these equations which are not in convolution, and uniting thereto any one of the remaining equations, we have a convolution of $k+1$ equations; and when a k-fold relation is precisely expressed by means of a system of k or more equations $(P=0, Q=0, \ldots)$, then every equation $\Omega=0$ implied in the given relation, or, what is the same thing, the equation of any onefold locus passing through the locus given by the k-fold relation is in involution with the equations $P=0$, $Q=0, \ldots$, that is, we have identically $\Omega=AP+BQ+CR+\ldots$, $A, B, C, \ldots$ being integral functions of the coordinates.}

Article Nos. 37 to 42. *Omal Relation; Order.*

37. A k-fold relation may be linear or omal. If $k=m$, the corresponding locus is a point; if $k<m$ the locus is a k-fold, or $(m-k)$dimensional omaloid; the expression omaloid used absolutely denotes the onefold or $(m-1)$dimensional omaloid; the point may be considered as a m-fold omaloid.

38. A m-fold relation which is not linear or omal is of necessity composite, composed of a certain number M of m-fold linear or omal relations; viz. the m-fold locus corresponding to the m-fold relation is a point-system of M points, each of which may be considered as given by a separate m-fold linear or omal relation; each which relation is a factor of the original m-fold relation. The given m-fold relation, and the point-system corresponding thereto, are respectively said to be of the order M.

39. The order of a point-system of M points is thus $=M$, but it is of course to be borne in mind that the points may be single or multiple points; and that if the system consists of a point taken α times, another point taken β times, &c., then the number of points and therefore the order M of the system is considered to be $=\alpha+\beta+\ldots$.

40. If to a given k-fold relation $(k<m)$ we unite an absolutely arbitrary $(m-k)$fold linear relation, so as to obtain for the aggregate a m-fold relation, then the order M of this m-fold relation (or, what is the same thing, the number M of points in the corresponding point-system) is said to be the order of the given k-fold relation. The notion of order does not apply to a more than m-fold relation.

41. The foregoing definition of order may be more compendiously expressed as follows: viz.

Given between the $m+1$ coordinates a relation which is at most m-fold; then if it is not m-fold, join to it an arbitrary linear relation so as to render it m-fold; we have a m-fold relation giving a point-system; and the order of the given relation is equal to the number of points of the point-system.

42. The relation aggregated of two or more given relations, when the notion of order applies to the aggregate relation, that is, when it is not more than m-fold, is of an order equal to the product of the orders of the constituent relations; or, say, the orders of the given relations being μ, μ', ..., the order of the aggregate relation is $=\mu\mu'\ldots$.

Article Nos. 43 and 44. *Parametric Relations.*

43. We have considered so far relations which involve only the coordinates $(x, y, \ldots)$ [1]; the coefficients are purely numerical, or, if literal, they are absolute constants, which either do or do not satisfy certain conditions; if they do not, the relation assumed in the first instance to be k-fold is really k-fold, or, as we may express it, the relation is

[1] The only exception is *ante*, No. 5, where, in illustration of the notion of a more than m-fold relation, mention is made of "parameters."

really as well as formally k-fold; if they do satisfy certain relations in virtue whereof the formally k-fold relation is really less than k-fold, say, it is $(k-l)$fold, then the relation is in fact to be considered *ab initio* as a $(k-l)$fold relation: there is no question of a relation being in general k-fold and becoming less than k-fold, or suffering any other modification in its form; and the notion of a more than m-fold relation is in the preceding theory meaningless.

44. But a relation between the coordinates $(x, y, \ldots)$ may involve parameters, and so long as these remain arbitrary it may be really as well as formally k-fold; but when the parameters satisfy certain conditions, it may become $(k-l)$fold, or may suffer some other modification in its form. And we have to consider the theory of a relation between the coordinates $(x, y, \ldots)$, involving besides parameters which may satisfy certain conditions, or, say simply, a relation involving variable parameters. If the number of the parameters be m', then these parameters may be regarded as the ratios of m' quantities to a remaining $m'+1$ th quantity, and the relation may be considered as involving *homogeneously* the $m'+1$ parameters $(x', y', \ldots)$. And these may, if we please, be regarded as coordinates of a point in their own m'-dimensional space, or we have to consider relations between the $m+1$ coordinates $(x, y, \ldots)$ and the $m'+1$ (parameters or) coordinates $(x', y', \ldots)$. It is to be added that a relation may involve distinct sets of parameters, say, we have besides the original set of parameters, a set of $m''+1$ parameters $(x'', y'', \ldots)$ involved homogeneously. But this is a generalization the necessity for which has hardly arisen.

Article Nos. 45 to 55. *Quantics, Notation, &c.*

45. A homogeneous function of the coordinates $(x, y, \ldots)$ is represented by a notation such as

$$(*\between x, y, \ldots)^{(\cdot)}$$

(where $(*)$ indicates the coefficients and $(\cdot)$ the degree), and it is said to be a quantic; and in reference to the quantic the quantities or coordinates $(x, y, \ldots)$ are also termed *facients*. More generally a quantic involving two or more sets of coordinates, or facients, is represented by the similar notation

$$(*\between x, y, \ldots)^{(\cdot)}(x', y', \ldots)^{(:)}\ldots.$$

46. The quantic is unipartite, bipartite, tripartite, &c., according as the number of sets is one, two, three, &c.; and with respect to any set of coordinates, it is binary, ternary, quaternary, ... $(m+1)$ary, according as the number of the coordinates is two, three, four, or $m+1$; and it is linear, quadric, cubic, quartic, ..., according as the degree in regard to the coordinates in question is 1, 2, 3, 4, ...

47. A quantic involving two or more sets of coordinates, and linear in regard to each of them, is said to be tantipartite; or, in particular, when there are only two sets, it is said to be lineo-linear; we may even extend the epithet lineo-linear to the case of any number of sets.

48. Instead of the general notation

$$(* \between x, y, \ldots)^{(\cdot)} (x', y', \ldots)^{(:)} \ldots$$

we may write

$$(a, \ldots \between x, y, \ldots)^{\mu} (x', y', \ldots)^{\mu'}, \ldots,$$

where the coefficients are now indicated by $(a, \ldots)$, and the degrees are $\mu, \mu', \ldots$

49. In the cases where the particular values of the coefficients have to be attended to, we write down the entire series of coefficients, or at least refer thereto by the notation $(a, \ldots)$; and it is to be understood that the coefficients expressed or referred to are each to be multiplied by the appropriate numerical coefficient, viz. for the term $x^{\alpha} y^{\beta} \ldots x'^{\alpha'} y'^{\beta'} \ldots$ this numerical coefficient is

$$= \frac{[\mu]^{\mu} [\mu']^{\mu'} \ldots}{[\alpha]^{\alpha} [\beta]^{\beta} \ldots [\alpha']^{\alpha'} [\beta']^{\beta'} \ldots}.$$

50. It is sometimes convenient not to introduce these numerical multipliers, and we then use the notation

$$(a, \ldots \widetilde{\between} x, y, \ldots)^{\mu} (x', y', \ldots)^{\mu'} \ldots,$$

or

$$(a, \ldots \between x, y, \ldots \widetilde{)}^{\mu} (x', y', \ldots \widetilde{)}^{\mu'} \ldots.$$

In particular $(a, b, c \between x, y)^2$, $(a, b, c, d \between x, y)^3$ &c. denote respectively

$$ax^2 + 2bxy + cy^2,$$
$$ax^3 + 3bx^2y + 3cxy^2 + dy^3,$$
&c.;

but $(a, b, c \widetilde{\between} x, y)^2$, $(a, b, c, d \widetilde{\between} x, y)^2$, &c. denote

$$ax^2 + bxy + cy^2,$$
$$ax^3 + bx^2y + cxy^2 + dy^3,$$
&c.,

and so $(a, b, c, f, g, h \between x, y, z)^2$ and $(a, b, c, f, g, h \widetilde{\between} x, y, z)^2$ denote respectively

$$ax^2 + by^2 + cz^2 + 2fyz + 2gzx + 2hxy,$$

and

$$ax^2 + by^2 + cz^2 + fyz + gzx + hxy.$$

51. To show which are the coefficients that belong to the several terms respectively, it is obviously proper that the quantic should be once written out at full length; thus, in speaking of a ternary cubic function, we say let $U = (a, \ldots \between x, y, z)^3$

$$\begin{aligned} &= (a, b, c, f, g, h, i, j, k, l \between x, y, z)^3 \\ &= ax^3 + by^3 + cz^3 \\ &\quad + 3(fy^2z + gz^2x + hx^2y + lyz^2 + jzx^2 + kxy^2) \\ &\quad + 6lxyz, \end{aligned}$$

and the like in other cases.

52. A onefold relation between the coordinates is expressible by means of an equation of the form

$$(*\between x,\ y, \ldots)^{(\cdot)} = 0.$$

53. The expression "an equation" used without explanation may be taken to mean an equation of the form in question, viz. the equation obtained by putting a quantic equal to zero; the quantic is said to be the *nilfactum* of the equation. We may consequently say simply that a onefold relation between the coordinates is always expressible by an equation.

54. It is frequently convenient to denote the quantic or nilfactum by a single letter, and to use a locution such as "the equation $U = (*\between x,\ y \ldots,)^{(\cdot)} = 0$," which really means that the single letter U stands for the quantic $(*\between x,\ y, \ldots)^{(\cdot)}$, so that we are afterwards at liberty to write $U = 0$ as an abbreviated expression for $(*\between x,\ y, \ldots)^{\mu} = 0$. We may also speak of the equation or function $U = 0$, meaning thereby the equation $U = 0$, or the function U.

55. A k-fold relation between the coordinates is (as has been shown) equivalent to a system of k or more onefold relations; each of these is expressible by an equation $U = 0$, and the k-fold relation is thus expressible by a system of k or more such equations. Representing by $((U))$ the system of functions which are the nilfacta of these equations respectively, the k-fold relations may be represented thus, $((U)) = 0$; or more completely, the relation being k-fold, and the number of equations being $= s$, by the notation

$$((U)\ s)\ (k\text{-fold}) = 0.$$

We may also speak of the system or relation $((U)) = 0$, meaning thereby the system of functions $((U))$, or the relation $((U)) = 0$.

Article Nos. 56 to 62. *Resultant, Discriminant, &c.*

56. In the case $k > m$, a given k-fold relation between the $m+1$ coordinates $(x,\ y, \ldots)$ and the parameters $(x',\ y', \ldots)$ leads to a $(k-m)$fold relation between the parameters. This is termed the *resultant relation* of the given k-fold relation, or when the additional specification is necessary, the *resultant relation* obtained by elimination of the coordinates $(x,\ y, \ldots)$.

57. Consider a k-fold relation between the $m+1$ coordinates $(x,\ y, \ldots)$ and the $m'+1$ coordinates $(x',\ y', \ldots)$. If $k \not> m$, then, considering the $(x,\ y, \ldots)$ as coordinates and the $(x',\ y', \ldots)$ as parameters, we have corresponding to the given relation a k-fold locus in the m-space; and so if $k \not> m'$, then, considering the $(x',\ y', \ldots)$ as coordinates, but the $(x,\ y, \ldots)$ as parameters, we have corresponding to the given relation a k-fold locus in the m'-space.

58. If $k > m$, but if the $(k-m)$fold resultant relation is satisfied, then the given k-fold relation becomes a m-fold linear relation between the coordinates $(x, y, \ldots)$, and is consequently satisfied by a single set of values of the coordinates. Hence, considering the given k-fold relation as implying the $(k-m)$fold resultant relation, the k-fold relation will represent a single point in the m-space, say, the *common point*.

59. A m-fold relation, or the locus, or point-system thereby represented, may have a *double* or *nodal* point, viz. two of the points of the point-system may be coincident. More generally a k-fold relation $(k \ngtr m)$, or the locus thereby represented, may have a *double* or *nodal* point; for let the relation if less than m-fold be made m-fold by adjoining to it a linear $(m-k)$fold relation satisfied by the coordinates of the point in question but otherwise arbitrary, then, if the point in question be a double or nodal point of the m-fold relation, or of the point-system thereby represented, the point is said to be a double or nodal point of the original k-fold relation, or of the locus thereby represented.

60. A given-fold relation $(k \ngtr m)$ between the $m+1$ coordinates, or the locus thereby represented, has not in general a nodal point. But if the relation involve the $m'+1$ parameters $(x', y', \ldots)$, then, if a certain onefold relation be satisfied between the parameters, there will be a nodal point. The onefold relation between the parameters is the *discriminant relation* of the given k-fold relation.

61. In the case in question, $k \ngtr m$, the discriminant relation is the resultant relation of a $(m+1)$fold relation which is the aggregate of the given k-fold relation with a certain relation called the *Jacobian relation*, or when the distinction is required, the *Jacobian relation* in regard to the $(x, y, \ldots)$.

62. Consider a k-fold relation $(k \ngtr m, \ngtr m')$ between the $m+1$ coordinates $(x, y, \ldots)$ and the $m'+1$ coordinates $(x', y', \ldots)$. It has been seen that to a given set of values of the $(x', y', \ldots)$ or, say, to a given point in the m'-space, there corresponds a k-fold locus in the m-space, and that to a given set of values of the $(x, y, \ldots)$, or to a given point in the m-space, there corresponds a k-fold locus in the m'-space. The k-fold locus in the m'-space may have a nodal point; this will be the case if there is satisfied between the $(x, y, \ldots)$ a certain one-fold relation, the discriminant relation of the given k-fold relation in regard to the $(x', y', \ldots)$. This onefold relation represents in the m-space a onefold locus, the *envelope* of the k-fold loci in the m-space corresponding to the several points of the m'-space. The property of the envelope is that to each point thereof there corresponds in the m'-space a k-fold locus having a nodal point.

Article Nos. 63—69. *Consecutive Points; Tangent Omals.*

63. As the notions of proximity and remoteness have been thus far altogether ignored, it seems necessary to make the following

Postulate. We may conceive a point consecutive (or indefinitely near) to a given point.

64. If the coordinates of the given point are $(x, y, \ldots)$, those of the consecutive point may be assumed to be $(x+\delta x, y+\delta y, \ldots)$, where $\delta x, \delta y, \ldots$ are indefinitely small in regard to $(x, y, \ldots)$.

65. It may be remarked that, taking the coordinates to be $(x+X, y+Y, \ldots)$, there is no obligation to have $(X, Y, \ldots)$ indefinitely small; in fact whatever the magnitudes of these quantities are, if only $X : Y : \ldots = x : y : \ldots$, then the point $(x+X, y+Y, \ldots)$ will be the very same with the original point, and it is therefore clear that a consecutive point may be represented in the same manner with magnitudes, however large, of $X, Y, \ldots$ But we *may* assume them indefinitely small, that is, the ratios $x+\delta x : y+\delta y, \ldots$, where $\delta x, \delta y, \ldots$ are indefinitely small in regard to $(x, y, \ldots)$, will represent any set of ratios indefinitely near to the ratios $(x : y, \ldots)$.

The foregoing quantities $(\delta x, \delta y, \ldots)$ are termed the increments.

66. Consider a k-fold relation between the $m+1$ coordinates $(x, y, \ldots)$, $k \not> m$; the increments $(\delta x, \delta y, \ldots)$ are connected by a linear k-fold relation.

The linear k-fold relation is satisfied if we assume the increments proportional to the coordinates—this is, in fact, assuming that the point remains unaltered. We may write $(\delta x, \delta y, \ldots) = (x, y, \ldots)$, since in such an equation only the ratios are attended to. But it may be preferable to write $(\delta x, \delta y, \ldots) = \lambda (x, y, \ldots)$. In particular if $k = m$, then the increments are connected by a linear m-fold relation; that is, the ratio of the increments is uniquely determined; and as the relation is satisfied by taking the increments proportional to the coordinates, it is clear that the values which the linear m-fold relation gives for the increments are in fact proportional to the coordinates: viz. there is not in this case any consecutive point.

67. Considering the k-fold relation as belonging to a k-fold locus in the m-space, so that $(x, y, \ldots)$ are the coordinates of a point on this locus, then if in the linear k-fold relation between the increments these increments are replaced by the coordinates $(\mathrm{x}, \mathrm{y}, \ldots)$ of a point in the m-space, then considering the original coordinates $(x, y, \ldots)$ as parameters, the locus of the point $(\mathrm{x}, \mathrm{y}, \ldots)$ is a k-fold omal locus: it is to be observed that, by what precedes, the linear k-fold relation is satisfied by writing therein the values $\mathrm{x} : \mathrm{y}, \ldots = x : y, \ldots$, that is, the k-fold omal locus passes through the original point $(x, y, \ldots)$; the k-fold omal locus is said to be the *tangent-omal* of the original k-fold locus at the point $(x, y, \ldots)$, which point is said to be the *point of contact.*

68. If in the original k-fold locus we replace $(x, y, \ldots)$ by $(\mathrm{x}, \mathrm{y}, \ldots)$, and combine therewith the k-fold linear relation, we have between the coordinates $(\mathrm{x}, \mathrm{y}, \ldots)$ a $2k$-fold relation (containing as parameters the coordinates $(x, y, \ldots)$); these parameters satisfy the original k-fold relation, and in virtue hereof the $2k$-fold relation (whether $2k$ is or is not greater than m) is satisfied by the values $\mathrm{x}, \mathrm{y}, \ldots = x : y : \ldots$; and not only so, but the point in question is a nodal or double point on the $2k$-fold locus. It also follows that the tangent-omal locus, considering in the k-fold linear relation $(x, y, \ldots)$ as parameters satisfying the original k-fold relation, has for its envelope the k-fold locus.

69. We thus arrive at the notion of the double generation of a k-fold locus, viz. such locus is the locus of the points, or, say, of the *ineunt-points* thereof; and it is also the envelope of the tangent-omals thereof. We have thus a theory of duality; I do not at present attempt to develope the theory, but it is necessary to refer to it, in order to remark that this theory is essential to the systematic development of a m-dimensional geometry; the original classification of loci as onefold, twofold,... $(m-1)$fold is incomplete, and must be supplemented with the loci reciprocally connected with these loci respectively. And moreover the theory of the singularities of a locus can only be systematically established by means of the same theory of duality; the singularities in regard to the ineunt-point must be treated of in connexion with the singularities in regard to the tangent-omal. These theories (that is, the classification of loci, and the establishment and discussion of the singularities of each kind of locus), vast as their extent is, should in the logical order precede that for which other reasons it may be expedient next to consider, the theory of Transformation, as depending on relations involving simultaneously the $m+1$ coordinates $(x, y, \ldots)$ and the $m'+1$ coordinates $(x' y', \ldots)$.

414.

ON POLYZOMAL CURVES, OTHERWISE THE CURVES $\sqrt{U}+\sqrt{V}+\&c.=0$.

[From the *Transactions of the Royal Society of Edinburgh,* vol. XXV. (1868), pp. 1—110. Read 16th December 1867.]

If U, V, &c., are rational and integral functions $(*\between x, y, z)^r$, all of the same degree r, in regard to the coordinates (x, y, z), then $\sqrt{U}+\sqrt{V}+\&c.$ is a polyzome, and the curve $\sqrt{U}+\sqrt{V}+\&c.=0$ a polyzomal curve. Each of the curves $\sqrt{U}=0$, $\sqrt{V}=0$, &c. (or say the curves $U=0$, $V=0$, &c.) is, on account of its relation of circumscription to the curve $\sqrt{U}+\sqrt{V}+\&c.=0$, considered as a girdle thereto (ζῶμα), and we have thence the term "zome" and the derived expressions "polyzome," "zomal," &c. If the number of the zomes $\sqrt{U}$, $\sqrt{V}$, &c. be $=\nu$, then we have a ν-zome, and corresponding thereto a ν-zomal curve; the curves $U=0$, $V=0$, &c., are the zomal curves or zomals thereof. The cases $\nu=1$, $\nu=2$, are not, for their own sake, worthy of consideration; it is in general assumed that ν is $=3$ at least. It is sometimes convenient to write the general equation in the form $\sqrt{lU}+\&c.=0$, where l, &c. are constants. The Memoir contains researches in regard to the general ν-zomal curve; the branches thereof, the order of the curve, its singularities, class, &c.; also in regard to the ν-zomal curve $\sqrt{l(\Theta+L\Phi)}+\&c.=0$, where the zomal curves $\Theta+L\Phi=0$, all pass through the points of intersection of the same two curves $\Theta=0$, $\Phi=0$ of the orders r and $r-s$ respectively; included herein we have the theory of the depression of order as arising from the ideal factor or factors of a branch or branches. A general theorem is given of "the decomposition of a tetrazomal curve," viz. if the equation of the curve be $\sqrt{lU}+\sqrt{mV}+\sqrt{nW}+\sqrt{pT}=0$; then if U, V, W, T are in involution, that is, connected by an identical equation $\mathrm{a}U+\mathrm{b}V+\mathrm{c}W+\mathrm{d}T=0$, and if l, m, n, p, satisfy the condition $\frac{l}{\mathrm{a}}+\frac{m}{\mathrm{b}}+\frac{n}{\mathrm{c}}+\frac{p}{\mathrm{d}}=0$, the tetrazomal curve breaks up into two trizomal curves, each expressible by means of any three of the four functions U, V, W, T; for example, in the form $\sqrt{l'U}+\sqrt{m'V}+\sqrt{p'T}=0$. If, in this theorem, we take $p=0$, then the original curve is the trizomal $\sqrt{lU}+\sqrt{mV}+\sqrt{nW}=0$, T is any function $=-\frac{1}{\mathrm{d}}(\mathrm{a}U+\mathrm{b}V+\mathrm{c}W)$, where, considering l, m, n as given, a, b, c are quantities subject only to the condition $\frac{l}{\mathrm{a}}+\frac{m}{\mathrm{b}}+\frac{n}{\mathrm{c}}=0$, and we have the theorem of "the variable zomal of a trizomal curve," viz. the equation of the trizomal $\sqrt{lU}+\sqrt{mV}+\sqrt{nW}=0$, may be expressed by means of any two of the three functions U, V, W, and of a function T determined as above, for example in the form $\sqrt{l'U}+\sqrt{m'V}+\sqrt{n'T}=0$; whence also it may be expressed in terms of three new functions T, determined as above. This theorem, which occupies a prominent position in the whole theory, was suggested to me by Mr Casey's theorem, presently referred to, for the construction of a bicircular quartic as the envelope of a variable circle.

In the ν-zomal curve $\sqrt{l(\Theta+L\Phi)}+\&c.=0$, if $\Theta=0$ be a conic, $\Phi=0$ a line, the zomals $\Theta+L\Phi=0$, &c. are conics passing through the same two points $\Theta=0$, $\Phi=0$, and there is no real loss of generality in taking these to be the circular points at infinity—that is, in taking the conics to be circles. Doing this, and using a special notation $\mathfrak{A}^\circ=0$ for the equation of a circle having its centre at a given point A, and similarly $\mathfrak{A}=0$ for the equation of an evanescent circle, or say of the point A, we have the ν-zomal curve $\sqrt{l\mathfrak{A}^\circ}+\&c.=0$, and the more special form $\sqrt{l\mathfrak{A}}+\&c.=0$. As regards the last-mentioned curve, $\sqrt{l\mathfrak{A}}+\&c.=0$, the point A to which the equation $\mathfrak{A}=0$ belongs, is a focus of the curve, viz. in the case $\nu=3$, it is an ordinary focus, and in the case $\nu>3$, it is a special kind of focus, which, if the term were required, might be called a foco-focus; the Memoir contains an explanation of the general theory of the foci of plane curves. For $\nu=3$, the equation $\sqrt{l\mathfrak{A}}+\sqrt{m\mathfrak{B}}+\sqrt{n\mathfrak{C}}=0$ is really equivalent to the apparently more general form $\sqrt{l\mathfrak{A}^\circ}+\sqrt{m\mathfrak{B}^\circ}+\sqrt{n\mathfrak{C}^\circ}=0$. In fact, this last is in general a bicircular quartic, and, in regard to it, the before-mentioned theorem of the variable zomal becomes Mr Casey's theorem, that "the bicircular quartic (and, as a particular case thereof, the circular cubic) is the envelope of a variable circle having its centre on a given conic and cutting at right angles a given circle." This theorem is a sufficient basis for the complete theory of the trizomal curve $\sqrt{l\mathfrak{A}^\circ}+\sqrt{m\mathfrak{B}^\circ}+\sqrt{n\mathfrak{C}^\circ}=0$; and it is thereby very easily seen that the curve $\sqrt{l\mathfrak{A}^\circ}+\sqrt{m\mathfrak{B}^\circ}+\sqrt{n\mathfrak{C}^\circ}=0$ can be represented by an equation $\sqrt{l'\mathfrak{A}'}+\sqrt{m'\mathfrak{B}'}+\sqrt{n'\mathfrak{C}'}=0$. But for $\nu>3$ this is not so, and the curve $\sqrt{l\mathfrak{A}}+\&c.=0$ is only a particular form of the curve $\sqrt{l\mathfrak{A}^\circ}+\&c.=0$; and the discussion of this general form is scarcely more difficult than that of the special form $\sqrt{l\mathfrak{A}}+\&c.=0$, included therein. The investigations in relation to the theory of foci, and in particular to that of the foci of the circular cubic and bicircular quartic, precede in the Memoir the theories of the trizomal curve $\sqrt{l\mathfrak{A}^\circ}+\sqrt{m\mathfrak{B}^\circ}+\sqrt{n\mathfrak{C}^\circ}=0$, and the tetrazomal curve $\sqrt{l\mathfrak{A}^\circ}+\sqrt{m\mathfrak{B}^\circ}+\sqrt{n\mathfrak{C}^\circ}+\sqrt{p\mathfrak{D}^\circ}=0$, to which the concluding portions relate. I have accordingly divided the Memoir into four parts, viz. these are—Part I., On Polyzomal Curves in general; Part II., Subsidiary

Investigations; Part III., On the Theory of Foci; and Part IV., On the Trizomal and Tetrazomal Curves where the zomals are circles. There is, however, some necessary intermixture of the theories treated of, and the arrangement will appear more in detail from the headings of the several articles. The paragraphs are numbered continuously through the Memoir. There are four Annexes, relating to questions which it seemed to me more convenient to treat of thus separately.

It is right that I should explain the very great extent to which, in the composition of the present Memoir, I am indebted to Mr Casey's researches. His Paper "On the Equations and Properties (1) of the System of Circles touching three circles in a plane; (2) of the System of Spheres touching four spheres in space; (3) of the System of Circles touching three circles on a sphere; (4) on the System of Conics inscribed in a conic and touching three inscribed conics in a plane," was read to the Royal Irish Academy, April 9, 1866, and is published in their "Proceedings." The fundamental theorem for the equation of the pairs of circles touching three given circles was, previous to the publication of the paper, mentioned to me by Dr Salmon, and I communicated it to Professor Cremona, suggesting to him the problem solved in his letter of March 3, 1866, as mentioned in my paper, "Investigations in connexion with Casey's Equation," *Quarterly Math. Journ.* vol. VIII. 1867, pp. 334—341, [395], and as also appears, Annex No. IV of the present Memoir.

In connexion with this theorem, I communicated to Mr Casey, in March or April 1867, the theorem No. 164 of the present Memoir, that for any three given circles, centres A, B, C, the equation $\overline{BC}\sqrt{\mathfrak{A}^\circ}+\overline{CA}\sqrt{\mathfrak{B}^\circ}+\overline{AB}\sqrt{\mathfrak{C}^\circ}=0$ (where $\overline{BC}$, $\overline{CA}$, $\overline{AB}$, denote the mutual distances of the points A, B, C) belongs to a Cartesian. Mr Casey, in a letter to me dated 30th April, 1867, informed me of his own mode of viewing the question as follows:—"The general equation of the second order $(a, b, c, f, g, h\between\alpha, \beta, \gamma)^2=0$, where α, β, γ are circles, is a bicircular quartic. If we take the equation $(a, b, c, f, g, h\between\lambda, \mu, \nu)^2=0$ in tangential coordinates (that is, when λ, μ, ν are perpendiculars let fall from the centres of α, β, γ on any line), it denotes a conic; denoting this conic by F, and the circle which cuts α, β, γ orthogonally by J, I proved that, if a variable circle moves with its centre on F, and if it cuts J orthogonally, its envelope will be the bicircular quartic whose equation is that written down above;" and among other consequences, he mentions that the foci of F are the double foci of the quartic, and the points in which J cuts F single foci of the quartic, and also the theorem which I had sent him as to the Cartesian, and he refers to his Memoir on Bicircular Quartics as then nearly finished. An Abstract of the Memoir as read before the Royal Irish Academy, 10th February, 1867, and published in their *Proceedings*, pp. 44, 45, contains the theorems mentioned in the letter of 30th April, and some other theorems. It is not necessary that I should particularly explain in what manner the present Memoir has been, in the course of writing it, added to or altered in consequence of the information which I have thus had of Mr Casey's researches; it is enough to say that I have freely availed myself of such information, and that there is no question as to Mr Casey's priority in anything which there may be in common in his memoir on Bicircular Quartics and in the present Memoir.

Part I. (Nos. 1 to 55).—On Polyzomal Curves in General.

Article Nos. 1 to 4. *Definition and Preliminary Remarks.*

1. As already mentioned, U, V, &c. denote rational and integral functions $(*\between x, y, z)^r$, all of the same degree r in the coordinates (x, y, z), and the equation

$$\sqrt{U} + \sqrt{V} + \&c. = 0$$

then belongs to a polyzomal curve, viz., if the number of the zomes $\sqrt{U}$, $\sqrt{V}$, &c. is $= \nu$, then we have a ν-zomal curve. The radicals, or any of them, may contain rational factors, or be of the form $P\sqrt{Q}$; but in speaking of the curve as a ν-zomal, it is assumed that any two terms, such as $P\sqrt{Q} + P'\sqrt{Q}$, involving the same radical $\sqrt{Q}$, are united into a single term, so that the number of distinct radicals is always $= \nu$; in particular (r being even), it is assumed that there is only one rational term P. But the ordinary case, and that which is almost exclusively attended to, is that in which the radicals $\sqrt{U}$, $\sqrt{V}$, &c. are distinct irreducible radicals without rational factors.

2. The curves $U = V = 0$, &c. are said to be the zomal curves, or simply the zomals of the polyzomal curve $\sqrt{U} + \sqrt{V} + \&c. = 0$; more strictly, the term zomal would be applied to the functions U, V, &c. It is to be noticed, that although the form $\sqrt{U} + \sqrt{V} + \&c. = 0$ is equally general with the form $\sqrt{lU} + \sqrt{mV} + \&c. = 0$ (in fact, in the former case, the functions U, V, &c. are considered as implicitly containing the constant factors l, m, &c., which are expressed in the latter case), yet it is frequently convenient to express these factors, and thus write the equation in the form $\sqrt{lU} + \sqrt{mV} + \&c.$ For instance, in speaking of any given curves $U = 0$, $V = 0$, &c., we are apt, disregarding the constant factors which they may involve, to consider U, V, &c. as given functions; but in this case the general equation of the polyzomal with the zomals $U = 0$, $V = 0$, &c., is of course $\sqrt{lU} + \sqrt{mV} + \&c. = 0$.

3. Anticipating in regard to the cases $\nu = 1$, $\nu = 2$, the remark which will be presently made in regard to the ν-zomal, that $\sqrt{U} + \sqrt{V} + \&c. = 0$ is the curve represented by the rationalised form of this equation, the monozomal curve $\sqrt{U} = 0$ is merely the curve $U = 0$, viz., this is any curve whatever $U = 0$ of the order r; and similarly, the bizomal curve $\sqrt{U} + \sqrt{V} = 0$ is merely the curve $U - V = 0$, viz. this is any curve whatever $\Omega = 0$, of the order r; the zomal curves $U = 0$, $V = 0$, taken separately, are not curves standing in any special relation to the curve in question $\Omega = 0$, but $U = 0$ may be any curve whatever of the order r, and then $V = 0$ is a curve of the same order r, in involution with the two curves $\Omega = 0$, $U = 0$; we may, in fact, write the equation $\Omega = 0$ under the bizomal form $\sqrt{U} + \sqrt{\Omega + U} = 0$. In the case r even, we may, however, notice the bizomal curve $P + \sqrt{U} = 0$ (P a rational function of the degree $\tfrac{1}{2}r$); the rational equation is here $\Omega = U - P^2 = 0$, that is $U = \Omega + P^2$, viz., P is any curve whatever of the order $\tfrac{1}{2}r$, and $U = 0$ is a curve of the order r, touching the given curve $\Omega = 0$ at each of its $\tfrac{1}{2}r^2$ intersections with the curve $P = 0$. I further

remark that the order of the ν-zomal curve $\sqrt{V}+\&c.=0$ is $=2^{\nu-2}r$; this is right in the case of the bizomal curve $\sqrt{U}+\sqrt{V}=0$, the order being $=r$, but it fails for the monozomal curve $\sqrt{U}=0$, the order being in this case r, instead of $\frac{1}{2}r$, as given by the formula. The two unimportant and somewhat exceptional cases $\nu=1$, $\nu=2$, are thus disposed of, and in all that follows (except in so far as this is in fact applicable to the cases just referred to), ν may be taken to be $=3$ at least.

4. It is to be throughout understood that by the curve $\sqrt{U}+\sqrt{V}+\&c.=0$ is meant the curve represented by the rationalised equation

$$\text{Norm}(\sqrt{U}+\sqrt{V}+\&c.)=0,$$

viz. the Norm is obtained by attributing to all but one of the zomes $\sqrt{U}$, $\sqrt{V}$, &c., each of the two signs $+$, $-$, and multiplying together the several resulting values of the polyzome; in the case of a ν-zomal curve, the number of factors is thus $=2^{\nu-1}r$ (whence, as each factor is of the degree $\frac{1}{2}r$, the order of the curve is $2^{\nu-1}\cdot\frac{1}{2}r$, $=2^{\nu-2}r$, as mentioned above). I expressly mention that, as regards the polyzomal curve, we are not in any wise concerned with the signs of the radicals, which signs are and remain essentially indeterminate; the equation $\sqrt{U}+\sqrt{V}+\&c.=0$, is a mere symbol for the rationalised equation, Norm $(\sqrt{U}+\sqrt{V}+\&c.)=0$.

Article Nos. 5 to 12. *The Branches of a Polyzomal Curve.*

5. But we may in a different point of view attend to the signs of the radicals: if for all values of the coordinates we take the symbol $\sqrt{\overline{\overline{\ ,,}}}$, and consider $\sqrt{\overline{\overline{U}}}$, $\sqrt{\overline{\overline{V}}}$, &c. as signifying determinately, say the *positive* values of $\sqrt{U}$, $\sqrt{V}$, &c.; then each of the several equations $\pm\sqrt{\overline{\overline{U}}}\pm\sqrt{\overline{\overline{V}}}+\&c.=0$, or, fixing at pleasure one of the signs, suppose that prefixed to $\sqrt{\overline{\overline{U}}}$, then each of the several equations $\sqrt{\overline{\overline{U}}}\pm\sqrt{\overline{\overline{V}}}\pm\&c.=0$, will belong to a *branch* of the polyzomal curve: a ν-zomal curve has thus $2^{\nu-1}$ branches corresponding to the $2^{\nu-1}$ values respectively of the polyzome. The separation of the branches depends on the precise fixation of the significations of $\sqrt{\overline{\overline{U}}}$, $\sqrt{\overline{\overline{V}}}$, &c., and in regard hereto some further explanation is necessary.

6. When U is real and positive, $\sqrt{\overline{\overline{U}}}$ may be taken to be, in the ordinary sense, the positive value of $\sqrt{U}$, and so when U is real and negative, $\sqrt{\overline{\overline{U}}}$ may be taken to be $=i$ into the positive value of $\sqrt{-U}$; and the like as regards $\sqrt{\overline{\overline{V}}}$, &c. The functions U, V, &c. are assumed to be real functions of the coordinates; hence, for any real values of the coordinates, U, V, &c. are real positive or negative quantities, and the significations of $\sqrt{\overline{\overline{U}}}$, $\sqrt{\overline{\overline{V}}}$, &c. are completely determined.

7. But the coordinates may be imaginary. In this case the functions U, V, &c. will for any given values of the coordinates acquire each of them a determinate, in general imaginary, value. If for all real values whatever of α, β, we select once for

all one of the two opposite values of $\sqrt{\alpha+\beta i}$, *calling* it the *positive* value, and representing it by $\sqrt{\overline{\overline{\alpha+\beta i}}}$, then, for any particular values of the coordinates, U being $=\alpha+\beta i$, the value of $\sqrt{\overline{\overline{U}}}$ may be taken to be $=\sqrt{\overline{\overline{\alpha+\beta i}}}$; and the like as regards $\sqrt{\overline{\overline{V}}}$, &c. $\sqrt{\overline{\overline{U}}}$, $\sqrt{\overline{\overline{V}}}$, &c. have thus each of them a determinate signification for any values whatever, real or imaginary, of the coordinates. The coordinates of a given point on the curve $\sqrt{U}+\sqrt{V}+\&c.=0$, will in general satisfy only one of the equations $\sqrt{\overline{\overline{U}}}\pm\sqrt{\overline{\overline{V}}}\pm\&c.=0$; that is, the point will belong to one (but in general only one) of the $2^{\nu-1}$ branches of the curve; the entire series of points the coordinates of which satisfy any one of the $2^{\nu-1}$ equations, will constitute the branch corresponding to that equation.

8. The signification to be attached to the expression $\sqrt{\overline{\overline{\alpha+\beta i}}}$ should agree with that previously attached to the like symbol in the case of a positive or negative real quantity; and it should, as far as possible, be subject to the condition of continuity, viz., as $\alpha+\beta i$ passes continuously to $\alpha'+\beta' i$, so $\sqrt{\overline{\overline{\alpha+\beta i}}}$ should pass continuously to $\sqrt{\overline{\overline{\alpha'+\beta' i}}}$; but (as is known) it is not possible to satisfy universally this condition of continuity; viz., if for facility of explanation we consider (α, β) as the coordinates of a point in a plane, and imagine this point to describe a closed curve surrounding the origin or point $(0, 0)$, then it is not possible so to define $\sqrt{\overline{\overline{\alpha+\beta i}}}$ that this quantity, varying continuously as the point moves along the curve, shall, when the point has made a complete circuit, resume its original value. The signification to be attached to $\sqrt{\overline{\overline{\alpha+\beta i}}}$ is thus in some measure arbitrary, and it would appear that the division of the curve into branches is affected by a corresponding arbitrariness, but this arbitrariness relates only to the imaginary branches of the curve: the notion of a real branch is perfectly definite.

9. It would seem that a branch may be impossible for any series whatever of points real or imaginary. Thus, in the bizomal curve $\sqrt{U}+\sqrt{V}=0$, the branch $\sqrt{\overline{\overline{U}}}+\sqrt{\overline{\overline{V}}}=0$ is impossible. In fact, for any point whatever, real or imaginary, of the curve, we have $U=V$, and therefore $\sqrt{\overline{\overline{U}}}=\sqrt{\overline{\overline{V}}}$; the point thus belongs to the other branch $\sqrt{\overline{\overline{U}}}-\sqrt{\overline{\overline{V}}}=0$, not to the branch $\sqrt{\overline{\overline{U}}}+\sqrt{\overline{\overline{V}}}=0$; the only points belonging to the last-mentioned branch are the isolated points for which simultaneously $\sqrt{\overline{\overline{U}}}=0$, $\sqrt{\overline{\overline{V}}}=0$; viz., the points of intersection of the two curves $U=0$, $V=0$.

10. It is not clear to me whether the case is the same in regard to the branch $\sqrt{\overline{\overline{U}}}+\sqrt{\overline{\overline{V}}}+\sqrt{\overline{\overline{W}}}=0$ of a trizomal curve. In fact, for each point of the curve $\sqrt{\overline{\overline{U}}}+\sqrt{\overline{\overline{V}}}+\sqrt{\overline{\overline{W}}}=0$ we have $(U-V-W)^2=4VW$, and therefore, $U-V-W=\pm 2\sqrt{\overline{\overline{V}}}\sqrt{\overline{\overline{W}}}$; there may very well be points for which the sign is $+$; that is, points for which $U=V+W+2\sqrt{\overline{\overline{V}}}\sqrt{\overline{\overline{W}}}$, and for these points we have $\pm\sqrt{\overline{\overline{U}}}=\sqrt{\overline{\overline{V}}}+\sqrt{\overline{\overline{W}}}$; for real values of the coordinates the sign on the left hand must be $+$ (for otherwise the two sides

would have opposite signs), but there is no apparent reason, or at least no obviously apparent reason, why this should be so for imaginary values of the coordinates, and if the sign be in fact $-$, then the point will belong to the branch $\sqrt{U}+\sqrt{V}+\sqrt{W}=0$.

11. But the branch in question is clearly impossible for any series of real points; so that, leaving it an open question whether the epithet "impossible" is to be understood to mean impossible for any series of real points (that is, as a mere synonym of imaginary), or whether it is to mean impossible for any series of points, real or imaginary, whatever, I say that in a ν-zomal curve some of the branches are or may be impossible, and that there is at least one impossible branch; viz., the branch $\sqrt{U}+\sqrt{V}+\text{\&c.}=0$.

12. For the purpose of referring to any branch of a polyzomal curve it will be convenient to consider $\sqrt{U}$ as signifying determinately $+\sqrt{U}$, or else $-\sqrt{U}$; and the like as regards $\sqrt{V}$, &c., but without any identity or relation between the signs prefixed to the $\sqrt{U}$, $\sqrt{V}$, &c., respectively; the equation $\sqrt{U}+\sqrt{V}+\text{\&c.}=0$, so understood, will denote determinately some one (that is, any one at pleasure) of the equations $\sqrt{U}\pm\sqrt{V}\pm\text{\&c.}=0$, and it will thus be the equation of some one (that is, any one at pleasure) of the branches of the polyzomal curve – all risk of ambiguity which might otherwise exist will be removed if we speak either of the *curve* $\sqrt{U}+\sqrt{V}$, $\text{\&c.}=0$, or else of the *branch* $\sqrt{U}+\sqrt{V}+\text{\&c.}=0$. Observe that by the foregoing convention, when only one branch is considered, we avoid the necessity of any employment of the sign $\pm$, or of the sign $-$; but when two or more branches are considered in connection with each other, it is necessary to employ the sign $-$ with one or more of the radicals $\sqrt{U}$, $\sqrt{V}$, &c.; thus in the trizomal curve $\sqrt{U}+\sqrt{V}+\sqrt{W}=0$, we may have to consider the branches $\sqrt{U}+\sqrt{V}+\sqrt{W}=0$, $\sqrt{U}+\sqrt{V}-\sqrt{W}=0$; viz., either of these equations apart from the other denotes any one branch at pleasure of the curve, but when the branch represented by the one equation is fixed, then the branch represented by the other equation is also fixed.

Article Nos. 13 to 17. *The Points common to Two Branches of a Polyzomal Curve.*

13. I consider the points which are situate simultaneously on two branches of the ν-zomal curve $\sqrt{U}+\sqrt{V}+\text{\&c.}=0$. The equations of the two branches may be taken to be

$$\sqrt{U}+\text{\&c.}+(\sqrt{W}+\text{\&c.})=0,$$
$$\sqrt{U}+\text{\&c.}-(\sqrt{W}+\text{\&c.})=0,$$

viz., fixing the significations of $\sqrt{U}$, $\sqrt{V}$, $\sqrt{W}$, &c. in such wise that in the equation of one branch these shall each of them have the sign $+$, we may take $\sqrt{U}$, &c. to be those radicals which, in the equation of the other branch, have the sign $+$, and

$\sqrt{W}$, &c. to be those radicals which have the sign $-$. The foregoing equations break up into the more simple equations

$$\sqrt{U} + \&c. = 0, \quad \sqrt{W} + \&c. = 0,$$

which are the equations of certain *branches* of the *curves* $\sqrt{U} + \&c. = 0$, and $\sqrt{W} + \&c. = 0$, respectively, and conversely each of the intersections of these two curves is a point situate simultaneously on some two branches of the original ν-zomal curve $\sqrt{U} + \sqrt{V} + \&c. = 0$. Hence, partitioning in any manner the ν-zome $\sqrt{U} + \sqrt{V} + \&c.$ into an α-zome, $\sqrt{U} = \&c.$ and a β-zome $\sqrt{W} + \&c.$ $(\alpha + \beta = \nu)$, and writing down the equations

$$\sqrt{U} + \&c. = 0, \quad \sqrt{W} + \&c. = 0$$

of an α-zomal curve and a β-zomal curve respectively, each of the intersections of these two curves is a point situate simultaneously on two branches of the ν-zomal curve; and the points situate simultaneously on two branches of the ν-zomal curve are the points of intersection of the several pairs of an α-zomal curve and a β-zomal curve, which can be formed by any bipartition of the ν-zome.

14. There are two cases to be considered:—First, when the parts are $1, \nu - 1$ ($\nu - 1$ is > 1, except in the case $\nu = 2$, which may be excluded from consideration), or say when the ν-zome is partitioned into a *zome* and *antizome*. Secondly, when the parts α, β, are each > 1 (this implies $\nu = 4$ at least), or say when the ν-zome is partitioned into a pair of *complementary parazomes*.

15. To fix the ideas, take the tetrazomal curve $\sqrt{U} + \sqrt{V} + \sqrt{W} + \sqrt{T} = 0$, and consider first a point for which $\sqrt{U} = 0$, $\sqrt{V} + \sqrt{W} + \sqrt{T} = 0$. The Norm is the product of $(2^3 =)$ 8 factors; selecting hereout the factors

$$\sqrt{U} + \sqrt{V} + \sqrt{W} + \sqrt{T},$$

$$\sqrt{U} - \sqrt{V} - \sqrt{W} - \sqrt{T},$$

let the product of these

$$= U - (\sqrt{V} + \sqrt{W} + \sqrt{T})^2$$

be called F, and the product of the remaining six factors be called G; the rationalised equation of the curve is therefore $FG = 0$. The derived equation is $GdF + FdG = 0$; at the point in question $\sqrt{U} = 0$, $\sqrt{V} + \sqrt{W} + \sqrt{T} = 0$; G and dG are each of them finite (that is, they neither vanish nor become infinite), but we have

$$F = 0,\ dF = dU - (\sqrt{V} + \sqrt{W} + \sqrt{T})(dV \div \sqrt{V} + dW \div \sqrt{W} + dT \div \sqrt{T}),\ = dU,$$

and the derived equation is thus $GdU = 0$, or simply $dU = 0$. It thus appears that the point in question is an ordinary point on the tetrazomal curve; and, further, that the tetrazomal curve is at this point touched by the zomal curve $U = 0$. And similarly, each of the points of intersection of the two curves $\sqrt{U} = 0$, $\sqrt{V} + \sqrt{W} + \sqrt{T} = 0$, is an ordinary point on the tetrazomal curve; and the tetrazomal curve is at each of these points touched by the zomal curve $U = 0$.

16. Consider, secondly, a point for which $\sqrt{U}+\sqrt{V}=0$, $\sqrt{W}+\sqrt{T}=0$; to form the Norm, taking in this case the two factors

$$\sqrt{U}+\sqrt{V}+\sqrt{W}+\sqrt{T},$$
$$\sqrt{U}+\sqrt{V}-\sqrt{W}-\sqrt{T},$$

let their product

$$=(\sqrt{U}+\sqrt{V})^2-(\sqrt{W}+\sqrt{T})^2$$

be called F, and the product of the remaining six factors be called G; the rationalised equation is $FG=0$, and the derived equation is $FdG+GdF=0$. At the point in question G and dG are each of them finite (that is, they neither vanish nor become infinite), but we have

$$F=0,\ dF=(\sqrt{U}+\sqrt{V})(dU\div\sqrt{U}+dV\div\sqrt{V})-(\sqrt{W}+\sqrt{T})(dW\div\sqrt{W}+dT\div\sqrt{T}),\ =0,$$

that is, the derived equation becomes identically $0=0$; the point in question is thus a singular point, and it is easy to see that it is in fact a node, or ordinary double point, on the tetrazomal curve. And similarly, each of the points of intersection of the two curves $\sqrt{U}+\sqrt{V}=0$, $\sqrt{W}+\sqrt{T}=0$ is a node on the tetrazomal curve.

17. The proofs in the foregoing two examples respectively are quite general, and we may, in regard to a ν-zomal curve, enunciate the results as follows, viz., in a ν-zomal curve, the points situate simultaneously on two branches are either the intersections of a zomal curve and its antizomal curve, or else they are the intersections of a pair of complementary parazomal curves. In the former case, the points in question are ordinary points on the ν-zomal, but they are points of contact of the ν-zomal with the zomal; it may be added, that the intersections of the zomal and antizomal, each reckoned twice, are *all* the intersections of the ν-zomal and zomal. In the latter case, the points in question are nodes of the ν-zomal; it may be added, that the ν-zomal has not, *in general*, any nodes other than the points which are thus the intersections of a pair of complementary parazomals, and that it has not *in general* any cusps.

Article Nos. 18 to 21. *Singularities of a ν-zomal Curve.*

18. It has been already shown that the order of the ν-zomal curve is $=2^{\nu-2}r$. Considering the case where ν is $=3$ at least, the curve, as we have just seen, has contacts with each of the zomal curves, and it has also nodes. I proceed to determine the number of these contacts and nodes respectively.

19. Consider first the zomal curve $U=0$, and its antizomal $\sqrt{V}+\sqrt{W}+\text{\&c.}=0$, these are curves of the orders r and $2^{\nu-3}r$ respectively, and they intersect therefore in $2^{\nu-3}r^2$ points. Hence the ν-zomal touches the zomal in $2^{\nu-3}r^2$ points, and reckoning each of these twice, the number of intersections is $=2^{\nu-2}r^2$, viz., these are all the intersections of the ν-zomal with the zomal $U=0$. The number of contacts of the ν-zomal with the several zomals $U=0$, $V=0$, &c., is of course $=2^{\nu-3}r^2\nu$.

20. Considering next a pair of complementary parazomal curves, an α-zomal and a β-zomal respectively $(\alpha+\beta=\nu)$, these are of the orders $2^{\alpha-2}r$ and $2^{\beta-2}r$ respectively, and they intersect therefore in $2^{\alpha+\beta-4}r^2=2^{\nu-4}r^2$ points, nodes of the ν-zomal. This number is independent of the particular partition (α, β), and the ν-zomal has thus this same number, $2^{\nu-4}r^2$, of nodes in respect of each pair of complementary parazomals; hence the total number of nodes is $=2^{\nu-4}r^2$ into the number of pairs of complementary parazomals. For the partition (α, β) the number of pairs is $=[\nu]^\nu \div [\alpha]^\alpha [\beta]^\beta$, or when $\alpha=\beta$, which of course implies ν even, it is one-half of this; extending the summation from $\alpha=2$ to $\alpha=\nu-2$, each pair is obtained twice, and the number of pairs is thus $=\frac{1}{2}\Sigma\{[\nu]^\nu \div [\alpha]^\alpha [\beta]^\beta\}$; the sum extended from $\alpha=0$ to $\alpha=\nu$ is $(1+1)^\nu$, $=2^\nu$, but we thus include the terms 1, ν, ν, 1, which are together $=2\nu+2$, hence the correct value of the sum is $=2^\nu-2\nu-2$, and the number of pairs is the half of this $=2^{\nu-1}-\nu-1$. Hence the number of nodes of the ν-zomal curve is $=(2^{\nu-1}-\nu-1)\,2^{\nu-4}r^2$.

21. The ν-zomal is thus a curve of the order $2^{\nu-2}r$, with $(2^{\nu-1}-\nu-1)\,2^{\nu-4}r^2$ nodes, but without cusps; the class is therefore

$$=2^{\nu-3}r\,[(\nu+1)\,r-2],$$

and the deficiency is

$$=2^{\nu-4}r\,[(\nu+1)\,r-6]+1.$$

These are the general expressions, but even when the zomal curves $U=0$, $V=0$, &c., are given, then writing the equation of the ν-zomal under the form $\sqrt{lU}+\sqrt{mV}+\text{\&c.}=0$, the constants $l : m :$ &c., may be so determined as to give rise to nodes or cusps which do not occur in the general case; the formulæ will also undergo modification in the particular cases next referred to.

Article Nos. 22 to 27. *Special Case where all the Zomals have a Common Point or Points.*

22. Consider the case where the zomals $U=0$, $V=0$ have all of them any number, say k, of common intersections—these may be referred to simply as the common points. Each common point is a $2^{\nu-2}$-tuple point on the ν-zomal curve; it is on each zomal an ordinary point, and on each antizomal a $2^{\nu-3}$-tuple point, and on any α-zomal parazomal a $2^{\alpha-2}$-tuple point. Hence, considering first the intersections of any zomal with its antizomal, the common point reckons as $2^{\nu-3}$ intersections, and the k common points reckon as $2^{\nu-3}k$ intersections; the number of the remaining intersections is therefore $=2^{\nu-3}(r^2-k)$, and the zomal touches the ν-zomal in each of these points. The intersections of the zomal with the ν-zomal are the k-common points, each of them a $2^{\nu-2}$-tuple point on the ν-zomal, and therefore reckoning together as $2^{\nu-2}k$ intersections; and the $2^{\nu-3}(r^2-k)$ points of contact, each reckoning twice, and therefore together as $2^{\nu-2}(r^2-k)$ intersections $\left(2^{\nu-2}k+2^{\nu-2}(r^2-k)=2^{\nu-2}r^2,\ =r\,.\,2^{\nu-2}r\right)$; the total number of contacts with the zomals $U=0$, $V=0$, &c., is thus $=2^{\nu-3}(r^2-k)\,\nu$.

23. Secondly, considering any pair of complementary parazomals, an α-zomal and a β-zomal, each of the common points, being a $2^{\alpha-2}$-tuple point and a $2^{\beta-2}$-tuple point on the two curves respectively, counts as $2^{\alpha+\beta-4}$, $=2^{\nu-4}$ intersections, and the k common points count as $2^{\nu-4}k$ intersections; the number of the remaining intersections is therefore $=2^{\nu-4}(r^2-k)$, each of which is a node on the ν-zomal curve; and we have thus in all $2^{\nu-4}(2^{\nu-1}-\nu-1)(r^2-k)$ nodes.

24. There are, besides, the k common points, each of them a $2^{\nu-2}$-tuple point on the ν-zomal, and therefore each reckoning as $\frac{1}{2}2^{\nu-2}(2^{\nu-2}-1)$, $=2^{2\nu-5}-2^{\nu-3}$ double points, or together as $(2^{2\nu-5}-2^{\nu-3})k$ double points. Reserving the term node for the above-mentioned nodes or proper double points, and considering, therefore, the double points (dps.) as made up of the nodes and of the $2^{\nu-2}$-tuple points, the total number of dps. is thus

$$2^{\nu-4}(2^{\nu-1}-\nu-1)(r^2-k)+(2^{2\nu-5}-2^{\nu-3})k,$$

$$=2^{\nu-4}(2^{\nu-1}-\nu-1)\,r^2+\{(\nu+1)\,2^{\nu-4}-2^{\nu-3}\}\,k\,;$$

or finally this is

$$=2^{\nu-4}\{(2^{\nu-1}-\nu-1)\,r^2+\ (\nu-1)\}\,;$$

so that there is a gain $=2^{\nu-4}(\nu-1)k$ in the number of dps. arising from the k common points. There is, of course, in the class a diminution equal to twice this number, or $2^{\nu-3}(\nu-1)k$; and in the deficiency a diminution equal to this number, or $2^{\nu-4}(\nu-1)k$.

25. The zomal curves $U=0$, $V=0$, &c., may all of them pass through the same ν^2 points; we have then $k=r^2$, and the expression for the number of dps. is $=(2^{2\nu-5}-2^{\nu-3})r^2$, viz., this is $=\frac{1}{2}2^{\nu-2}(2^{\nu-2}-1)r^2$. But in this case the dps. are nothing else than the r^2 common points, each of them a $2^{\nu-2}$-tuple point, the ν-zomal curve in fact breaking up into a system of $2^{\nu-2}$ curves of the order r, each passing through the r^2 common points. This is easily verified, for if $\Theta=0$, $\Phi=0$ are some two curves of the order r, then, in the present case, the zomal curves are curves in involution with these curves; that is, they are curves of the form $l\Theta+l'\Phi=0$, $m\Theta+m'\Phi=0$, &c., and the equation of the ν-zomal curve is

$$\sqrt{l\Theta+l'\Phi}+\sqrt{m\Theta+m'\Phi}+\text{\&c.}=0.$$

The rationalised equation is obviously an equation of the degree $2^{\nu-2}$ in Θ, Φ, giving therefore a constant value for the ratio $\Theta:\Phi$; calling this q, or writing $\Theta=q\Phi$, we have

$$\sqrt{lq+l'}+\sqrt{mq+m'}+\text{\&c.}=0,$$

viz., the rationalised equation is an equation of the degree $2^{\nu-2}$ in q, and gives therefore $2^{\nu-2}$ values of q. And the ν-zomal curve thus breaks up into a system of $2^{\nu-2}$ curves each of the form $\Theta-q\Phi=0$, that is, each of them in involution with the curves $\Theta=0$, $\Phi=0$. The equation in q may have a multiple root or roots, and the system of curves so contain repetitions of the same curve or curves; an instance of this (in relation to the trizomal curve) will present itself in the sequel; but I do not at present stop to consider the question.

26. A more important case is when the zomal curves are each of them in involution with the same two given curves, one of them of the order r, the other of an inferior order. Let $\Theta=0$ be a curve of the order r, $\Phi=0$ a curve of an inferior order $r-s$; $L=0$, $M=0$, &c., curves of the order s; then the case in question is when the zomal curves are of the form $\Theta+L\Phi=0$, $\Theta+M\Phi=0$, &c., the equation of the ν-zomal is

$$\sqrt{l(\Theta+L\Phi)}+\sqrt{m(\Theta+M\Phi)}+\&c.=0,$$

where l, m, &c. are constants. This is the most convenient form for the equation, and by considering the functions L, M, &c. as containing implicitly the factors l^{-1}, m^{-1}, &c. respectively, we may take it to include the form $\sqrt{l\Theta+L\Phi}+\sqrt{m\Theta+M\Phi}+\&c.=0$, which last has the advantage of being immediately applicable to the case where any one or more of the constants l, m, &c. may be $=0$.

27. In the case now under consideration we have the $r(r-s)$ points of intersection of the curves $\Theta=0$, $\Phi=0$ as common points of all the zomals. Hence, putting in the foregoing formula $k=r(r-s)$, we have a ν-zomal curve of the order $2^{\nu-2}r$, having with each zomal $2^{\nu-2}rs$ contacts, or with all the zomals $2^{\nu-2}rs\nu$ contacts, having a node at each of the $2^{\nu-4}rs$ intersections (not being common points $\Theta=0$, $\Phi=0$) of each pair of complementary parazomals; that is, together $2^{\nu-4}(2^{\nu-1}-\nu-1)rs$ nodes, and having, besides, at each of the $r(r-s)$ common points, a $2^{\nu-2}$-tuple point, counting as $2^{2\nu-5}-2^{\nu-3}$ dps., together as $(2^{2\nu-5}-2^{\nu-3})r(r-s)$ dps.; whence, taking account of the nodes, the total number of dps. is $=2^{\nu-4}r[(2^{\nu-1}-2)r-(\nu-1)s]$.

Article Nos. 28 to 37. *Depression of Order of the ν-zomal Curve from the Ideal Factor of a Branch or Branches.*

28. In the case of the $r(r-s)$ common points as thus far considered, the order of the ν-zomal curve has remained throughout $=2^{\nu-2}r$, but the order admits of depression, viz., the constants l, m, &c., and those of the functions L, M, &c., may be such that the Norm contains the factor Φ^ω; the ν-zomal curve then contains as part of itself ($\Phi^\omega=0$) the curve $\Phi=0$ taken ω times, and this being so, if we discard the factor in question, and consider the residual curve as being the ν-zomal, the order of the ν-zomal will be $=2^{\nu-2}r-\omega(r-s)$.

29. To explain how such a factor Φ^ω presents itself, consider the polyzome $\sqrt{l(\Theta+L\Phi)}+\&c.$, or, what is the same thing, $\sqrt{l}\sqrt{\Theta+L\Phi}+\&c.$, belonging to any particular branch of the curve, we may, it is clear, take $\sqrt{\Theta+L\Phi}$, &c. each in a fixed signification as equivalent to $\overline{\sqrt{\Theta+L\Phi}}$, &c., respectively, and the particular branch will then be determined by means of the significations attached to $\sqrt{l}$, $\sqrt{m}$, &c. Expanding the several radicals, the polyzome is

$$\sqrt{l}\left\{\sqrt{\Theta}+\tfrac{1}{2}L\frac{\Phi}{\sqrt{\Theta}}-\tfrac{1}{8}L^2\frac{\Phi^2}{\Theta\sqrt{\Theta}}+\&c.,\right\}+\&c.;$$

or, what is the same thing, it is

$$\sqrt{\Theta}(\sqrt{l}+\&c.)+\tfrac{1}{2}\frac{\Phi}{\sqrt{\Theta}}(L\sqrt{l}+\&c.)-\tfrac{1}{8}\frac{\Phi^2}{\Theta\sqrt{\Theta}}(L^2\sqrt{l}+\&c.)+\&c.$$

which expansion may contain the factor Φ, or a higher power of Φ. For instance, if we have $\sqrt{l}+\&c.=0$, the expansion will then contain the factor Φ; and if we also have $L\sqrt{l}+\&c.=0$ (observe this implies as many equations as there are asyzygetic terms in the whole series of functions L, M, &c.; thus, if L, M, &c., are each of them of the form $aP+bQ+cR$, with the same values of P, Q, R, but with different values of the coefficients a, b, c, then it implies the three equations $a\sqrt{l}+\&c.=0$, $b\sqrt{l}+\&c.=0$, $c\sqrt{l}+\&c.=0$; and so in other cases), if I say $L\sqrt{l}+\&c.$ be also $=0$, then the expansion will contain the factor Φ^2, and so on; the most general supposition being, that the expansion contains as factor a certain power Φ^α of Φ. Imagine each of the polyzomes expanded in this manner, and let certain of the expansions contain the factors Φ^α, Φ^β, &c., respectively. The produce of the expansions is identically equal to the product of the unexpanded polyzomes—that is, it is equal to the Norm; hence, if $\alpha+\beta+\&c.=\omega$, the Norm will contain the factor Φ^ω.

30. It has been mentioned that the form $\sqrt{l(\Theta+L\Phi)}$ is considered as including the form $\sqrt{l\Theta+L\Phi}$, that is, when $l=0$, the form $\sqrt{L\Phi}$. If in the equation of the ν-zomal curve there is any such term—for instance, if the equation be $\sqrt{L\Phi}+\sqrt{m(\Theta+M\Phi)}+\&c.=0$, the radical $\sqrt{L\Phi}$ contains the factor $\Phi^{\frac{1}{2}}$; but if L contains as factor an odd or an even power of Φ, then $\sqrt{L\Phi}$ will contain the factor Φ^α where α is either an integer, or an integer $+\frac{1}{2}$. Consider the polyzome $\sqrt{L\Phi}+\sqrt{m(\Theta+M\Phi)}+\&c.$, belonging to any particular branch of the curve; the radical $\sqrt{L\Phi}$ contains, as just mentioned, the factor Φ^α, and if the remaining terms $\sqrt{m(\Theta+M\Phi)}+\&c.$, are such that the expansion contains as factor the same or any higher power of Φ, then the expansion of the polyzome $\sqrt{L\Phi}+\sqrt{m(\Theta+M\Phi)}+\&c.$, belonging to the particular branch will contain the factor Φ^α; and similarly we may have branches containing the factors Φ^α, Φ^β, &c., whence, as before, if $\omega=\alpha+\beta+\&c.$, the Norm will contain the factor Φ^ω; the only difference is, that now α, β, &c., instead of being of necessity all integers, are each of them an integer, or an integer $+\frac{1}{2}$; of course, in the latter case the integer may be zero, or the index be $=\frac{1}{2}$. It is clear that ω must be an integer, and it is, in fact, easy to see that the fractional indices occur in pairs; for observe that α being fractional, the expansion of $\sqrt{m(\Theta+M\Phi)}+\&c.$, will contain not Φ^α, but a higher power, $\Phi^{\alpha+q}$, where $\alpha+q$ is an integer; whence *each* of the polyzomes $\sqrt{L\Phi}\pm\left(\sqrt{m(\Theta+M\Phi)}+\&c.\right)$ will contain the factor Φ^α.

31. Observe that in every case the factor Φ^α presents itself as a factor of the expansion of the polyzome corresponding to a particular branch of the curve; the polyzome itself does not contain the factor Φ^α, and we cannot in anywise say that the corresponding branch contains as factor the curve $\Phi^\alpha=0$; but we may, with great propriety of expression, say that *the branch ideally contains the curve* $\Phi^\alpha=0$; and this

being so, the general theorem is, that if we have branches ideally containing the curves $\Phi^\alpha = 0$, $\Phi^\beta = 0$, &c. respectively, then the ν-zomal curve contains not ideally but actually the factor $\Phi^\omega = 0$ $(\omega = \alpha + \beta + \&c.)$, the order of the ν-zomal being thus reduced from $2^{\nu-2}r$ to $2^{\nu-2}r - \omega(r-s)$; and conversely, that any such reduction in the order of the ν-zomal arises from the factors $\Phi^\alpha = 0$, $\Phi^\beta = 0$, &c., ideally contained in the several branches of the ν-zomal.

32. It is worth while to explain the notion of an ideal factor somewhat more generally; an irrational function, taking the irrationalities thereof in a determinate manner, may be such that, as well the function itself as all its differential coefficients up to the order $\alpha - 1$, vanish when a certain parameter Φ contained in the function is put $= 0$; this is only saying, in other words, that the function expanded in ascending powers of Φ contains no power lower than Φ^α; and, in this case, we say that the irrational function contains *ideally* the factor Φ^α. The rationalised expression, or Norm, in virtue of the irrational function (taken determinately as above) thus ideally containing Φ^α, will actually contain the factor Φ^α; and if any other values of the irrational function contain respectively Φ^β, &c., then the Norm will contain the factor $\Phi^{\alpha+\beta+\&c.}$.

33. A branch ideally containing $\Phi^\alpha = 0$ may for shortness be called integral or fractional, according as the index α is an integer or a fraction; by what precedes the fractional branches present themselves in pairs. If for a moment we consider integral branches only, then if the ν-zomal contain $\Phi = 0$, this can happen in one way only, there must be some one branch ideally containing $\Phi = 0$; but if the ν-zomal contain $\Phi^2 = 0$, then this may happen in two ways,—either there is a single branch ideally containing $\Phi^2 = 0$, or else there are two branches, each of them ideally containing $\Phi = 0$. And generally, if the ν-zomal contain $\Phi^\omega = 0$, then forming any partition $\omega = \alpha + \beta + \&c.$ (the parts being integral), this may arise from there being branches ideally containing $\Phi^\alpha = 0$, $\Phi^\beta = 0$, &c. respectively. The like remarks apply to the case where we attend also to fractional branches,—thus, if the ν-zomal contain $\Phi = 0$, this may arise (not only, as above mentioned, from a branch ideally containing $\Phi = 0$, but also) from a pair of branches, each ideally containing $\Phi^{\frac{1}{2}} = 0$. And so in general, if the ν-zomal contain $\Phi^\omega = 0$, the partition $\omega = \alpha + \beta + \&c.$ is to be made with the parts integral or fractional ($= \frac{1}{2}$ or integer $+ \frac{1}{2}$ as above), but with the fractional terms in pairs; and then the factor $\Phi^\omega = 0$ may arise from branches ideally containing $\Phi^\alpha = 0$, $\Phi^\beta = 0$, &c. respectively.

34. Any zomal, antizomal, or parazomal of a ν-zomal curve, $\sqrt{l(\Theta + L\Phi)} + \&c. = 0$, is a polyzomal curve (including in the term a monozomal curve) of the same form as the ν-zomal; and may in like manner contain $\Phi = 0$, or more generally, $\Phi^\omega = 0$, viz., if $\omega = \alpha + \beta + \&c.$ be any partition of ω as above, this will be the case if the zomal, antizomal, or parazomal has branches ideally containing $\Phi^\alpha = 0$, $\Phi^\beta = 0$, &c. respectively. It is to be observed that if a zomal, antizomal, or parazomal contain $\Phi = 0$, or any higher power $\Phi^\omega = 0$, this does not in anywise imply that the zomal contains even $\Phi = 0$. But if (attending only to the most simple case) a zomal and its antizomal, or a pair of complementary parazomals, each contain $\Phi = 0$ inseparably (that

is, through a single branch ideally containing $\Phi=0$), then the ν-zomal will have two branches, each ideally containing $\Phi=0$, and it will thus contain $\Phi^2=0$. In fact, if in the zomal and antizomal, or in the complementary parazomals, the branches which ideally contain $\Phi=0$ are

$$\sqrt{l(\Theta+L\Phi)}+\&c.=0,\quad \sqrt{n(\Theta+N\Phi)}+\&c.=0$$

respectively (for a zomal, the $+\&c.$ should be omitted, and the first equation be written $\sqrt{l(\Theta+L\Phi)}=0$), then in the ν-zomal there will be the two branches

$$\left(\sqrt{l(\Theta+L\Phi)}+\&c.\right)\pm\left(\sqrt{n(\Theta+N\Phi)}+\&c.\right)=0,$$

each ideally containing $\Phi=0$.

Conversely, if a ν-zomal contain $\Phi^2=0$ by reason that it has two branches each ideally containing $\Phi=0$, then either a zomal and its antizomal will each of them, or else a pair of complementary parazomals will each of them, inseparably contain $\Phi=0$.

35. Reverting to the case of the ν-zomal curve

$$\sqrt{l(\Theta+L\Phi)}+\sqrt{m(\Theta+M\Phi)}+\&c.=0,$$

which does not contain $\Phi=0$, each of the $r(r-s)$ common points $\Theta=0$, $\Phi=0$ is a $2^{\nu-2}$-tuple point on the ν-zomal; each of these counts therefore for $2^{\nu-2}$ intersections of the ν-zomal with the curve $\Phi=0$, and we have thus the complete number $2^{\nu-2}r(r-s)$ of intersections of the two curves, viz., the curve $\Phi=0$ meets the ν-zomal in the $r(r-s)$ common points, each of them a $2^{\nu-2}$-tuple point on the ν-zomal, and in no other point.

36. But if the ν-zomal contains $\Phi^\omega=0$, then each of the $r(r-s)$ common points is still a $2^{\nu-2}$-tuple point on the aggregate curve; the aggregate curve therefore passes $2^{\nu-2}$ times through each common point; but among these passages are included ω passages of the curve $\Phi=0$ through the common point. The residual curve—say the ν-zomal—passes therefore only $2^{\nu-2}-\omega$ times through the common point; that is, each of the $r(r-s)$ common points is a $(2^{\nu-2}-\omega)$ tuple point on the ν-zomal. The curve $\Phi=0$ meets the ν-zomal in $\{2^{\nu-2}r-\omega(r-s)\}(r-s)$ points, viz., these include the $r(r-s)$ common points, each of them a $(2^{\nu-2}-\omega)$ tuple point on the ν-zomal, and therefore counting together as $(2^{\nu-2}-\omega)r(r-s)$ intersections; there remain consequently $\omega s(r-s)$ other intersections of the curve $\Theta=0$ with the ν-zomal.

37. In the case where the ν-zomal contains the factor $\Phi^\omega=0$, then throughout excluding from consideration the $r(r-s)$ common points $\Theta=0$, $\Phi=0$, the *remaining* intersections of any zomal with its antizomal are points of contact of the zomal with the ν-zomal, and the *remaining* intersections of each pair of complementary parazomals are nodes of the ν-zomal, it being understood that if any zomal, antizomal, or parazomal contain a power of $\Phi=0$, such powers of $\Phi=0$ are to be discarded, and only the residual curves attended to. The number of contacts and of nodes may in any particular case be investigated without difficulty, and some instances will present themselves in the sequel, but on account of the different ways in which the factor

$\Phi^{\omega} = 0$ may present itself, ideally in a single branch, or in several branches, and the consequent occurrence in the latter case of powers of $\Phi = 0$ in certain of the zomals, antizomals, or parazomals, the cases to be considered would be very numerous, and there is no reason to believe that the results could be presented in any moderately concise form; I therefore abstain from entering on the question.

Article Nos. 38 and 39. *On the Trizomal Curve and the Tetrazomal Curve.*

38. The trizomal curve

$$\sqrt{U} + \sqrt{V} + \sqrt{W} = 0$$

has for its rationalised form of equation

$$U^2 + V^2 + W^2 - 2VW - 2WU - 2UV = 0;$$

or as this may also be written,

$$(1,\ 1,\ 1,\ -1,\ -1,\ -1)(U,\ V,\ W)^2 = 0;$$

and we may from this rational equation verify the general results applicable to the case in hand, viz., that the trizomal is a curve of the order $2r$, and that

$U = 0$, at each of its r^2 intersections with $V - W = 0$,
$V = 0$, " " $W - U = 0$,
$W = 0$, " " $U - V = 0$,

respectively touch the trizomal. There are not, in general, any nodes or cusps, and the order being $= 2r$, the class is $= 2r(2r-1)$.

39. The tetrazomal curve

$$\sqrt{U} + \sqrt{V} + \sqrt{W} + \sqrt{T} = 0$$

has for its rationalised form of equation

$$(U^2 + V^2 + W^2 + T^2 - 2UV - 2UW - 2UT - 2VW - 2VT - 2WT)^3 - 64UVWT = 0,$$

and we may hereby verify the fundamental properties, viz., that the tetrazomal is a curve of the order $4r$, touched by each of the zomals $U = 0$, $V = 0$, $W = 0$, $T = 0$ in $2r^2$ points, viz., by $U = 0$ at its intersections with $\sqrt{U} + \sqrt{W} + \sqrt{T} = 0$, that is, $V^2 + W^2 + T^2 - 2VW - 2VT - 2WT = 0$; (and the like as regards the other zomals), and having $3r^2$ nodes, viz., these are the intersections of $(\sqrt{U} + \sqrt{V} = 0,\ \sqrt{W} + \sqrt{T} = 0)$, $(\sqrt{U} + \sqrt{W} = 0,\ \sqrt{V} + \sqrt{T} = 0)$, $(\sqrt{U} + \sqrt{T} = 0,\ \sqrt{V} + \sqrt{W} = 0)$, or, what is the same thing, the intersections of $(U - V = 0,\ W - T = 0)$, $(U - W = 0,\ V - T = 0)$, $(U - T = 0,\ V - W = 0)$. There are not in general any cusps, and the class is thus $= 4r(4r-1) - 6r^2, = 10r^2 - 4r$.

Article Nos. 40 and 41. *On the Intersection of two ν-Zomals having the same Zomal Curves.*

40. Without going into any detail, I may notice the question of the intersection of two ν-zomals which have the same zomal curves—say the two trizomals $\sqrt{U}+\sqrt{V}+\sqrt{W}=0$, $\sqrt{lU}+\sqrt{mV}+\sqrt{nW}=0$, or two similarly related tetrazomals. For the trizomals, writing the equations under the form

$$\sqrt{U}+\sqrt{V}+\sqrt{W}=0, \quad \sqrt{l}\sqrt{U}+\sqrt{m}\sqrt{V}+\sqrt{n}\sqrt{W}=0,$$

then, when these equations are considered as existing simultaneously, we may, without loss of generality, attribute to the radicals $\sqrt{U}$, $\sqrt{V}$, $\sqrt{W}$, the same values in the two equations respectively; but doing so, we must in the second equation successively attribute to all but one of the radicals $\sqrt{l}$, $\sqrt{m}$, $\sqrt{n}$, each of its two opposite values. For the intersections of the two curves we have thus

$$\sqrt{U} : \sqrt{V} : \sqrt{W} = \sqrt{m}-\sqrt{n} : \sqrt{n}-\sqrt{l} : \sqrt{l}-\sqrt{m},$$

viz., this is one of a system of four equations, obtained from it by changes of sign, say in the radicals $\sqrt{m}$ and $\sqrt{n}$. Each of the four equations gives a set of r^2 points; we have thus the complete number, $=4r^2$, of the points of intersection of the two curves.

41. But take, in like manner, two tetrazomal curves; writing their equations in the form

$$\begin{gathered}\sqrt{U}+\quad\sqrt{V}+\quad\sqrt{W}+\quad\sqrt{T}=0,\\ \sqrt{l}\sqrt{U}+\sqrt{m}\sqrt{V}+\sqrt{n}\sqrt{W}+\sqrt{p}\sqrt{T}=0,\end{gathered}$$

then $\sqrt{U}$, $\sqrt{V}$, $\sqrt{W}$, $\sqrt{T}$ may be considered as having the same values in the two equations respectively, but we must in the second equation attribute successively, say to $\sqrt{m}$, $\sqrt{n}$, $\sqrt{p}$, each of their two opposite values. For the intersections of the two curves we have

$$\begin{aligned} &\quad\cdot\quad (\sqrt{m}-\sqrt{l})\sqrt{V}+(\sqrt{n}-\sqrt{l}\,)\sqrt{W}+(\sqrt{p}-\sqrt{l}\,)\sqrt{T}=0,\\ &(\sqrt{l}-\sqrt{m})\sqrt{U}\quad\cdot\quad +(\sqrt{n}-\sqrt{m})\sqrt{W}+(\sqrt{p}-\sqrt{m})\sqrt{T}=0,\end{aligned}$$

viz., this is one of a system of eight similar pairs of equations, obtained therefrom by changes of sign of the radicals $\sqrt{m}$, $\sqrt{n}$, $\sqrt{p}$. The equations represent each of them a trizomal curve, of the order $2r$; the two curves intersect therefore in $4r^2$ points, and if each of these was a point of intersection of the two tetrazomals, we should have in all $8\times 4r^2=32r^2$ intersections. But the tetrazomals are each of them a curve of the order $4r$, and they intersect therefore in only $16r^2$ points. The explanation is, that not all the $4r^2$ points, but only $2r^2$ of them are intersections of the tetrazomals. In fact, to find *all* the intersections of the two trizomals, it is necessary in their two equations to attribute opposite signs to one of the radicals $\sqrt{W}$, $\sqrt{T}$; we obtain $2r^2$ intersections from the equations as they stand, the remaining $2r^2$ intersections from the two equations after we have in the second equation reversed the sign, say of $\sqrt{T}$.

Now, from the two equations as they stand we can pass back to the two tetrazomal equations, and the first-mentioned $2r^2$ points are thus points of intersection of the two tetrazomal curves—from the two equations after such reversal of the sign of $\sqrt{T}$, we cannot pass back to the two tetrazomal equations, and the last-mentioned $2r^2$ points are thus not points of intersection of the two tetrazomal curves. The number of intersections of the two curves is thus $8 \times 2r^2$, $=16r^2$, as it should be.

Article Nos. 42 to 45. *The Theorem of the Decomposition of a Tetrazomal Curve.*

42. I consider the tetrazomal curve—

$$\sqrt{lU} + \sqrt{mV} + \sqrt{nW} + \sqrt{pT} = 0,$$

where the zomal curves are in involution,—that is, where we have an identical relation,

$$\mathrm{a}U + \mathrm{b}V + \mathrm{c}W + \mathrm{d}T = 0;$$

and I proceed to show that if l, m, n, p satisfy the relation

$$\frac{l}{\mathrm{a}} + \frac{m}{\mathrm{b}} + \frac{n}{\mathrm{c}} + \frac{p}{\mathrm{d}} = 0,$$

the curve breaks up into two trizomals. In fact, writing the equation under the form

$$(\sqrt{lU} + \sqrt{mV} + \sqrt{nW})^2 - pT = 0,$$

and substituting for T its value, in terms of U, V, W, this is

$$(l\mathrm{d} + p\mathrm{a})\, U + (m\mathrm{d} + p\mathrm{b})\, V + (n\mathrm{d} + p\mathrm{c})\, W$$

$$+ 2\sqrt{mn\mathrm{d}}\,\sqrt{VW} + 2\sqrt{nl\mathrm{d}}\,\sqrt{WU} + 2\sqrt{lm\mathrm{d}}\,\sqrt{UV} = 0;$$

or, considering the left-hand side as a quadric function of $(\sqrt{U}, \sqrt{V}, \sqrt{W})$, the condition for its breaking up into factors is

$$\begin{vmatrix} l\mathrm{d} + p\mathrm{a}, & \mathrm{d}\sqrt{lm}, & \mathrm{d}\sqrt{ln} \\ \mathrm{d}\sqrt{ml}, & m\mathrm{d} + p\mathrm{b}, & \mathrm{d}\sqrt{mn} \\ \mathrm{d}\sqrt{nl}, & \mathrm{d}\sqrt{nm}, & n\mathrm{d} + p\mathrm{c} \end{vmatrix} = 0,$$

that is

$$p^2 (l\mathrm{bcd} + m\mathrm{cda} + n\mathrm{dab} + p\mathrm{abc}) = 0,$$

or finally, the condition is

$$\frac{l}{\mathrm{a}} + \frac{m}{\mathrm{b}} + \frac{n}{\mathrm{c}} + \frac{p}{\mathrm{d}} = 0.$$

43. Multiplying by $l\mathrm{d} + p\mathrm{a}$, and observing that in virtue of the relation we have

$$(l\mathrm{d} + p\mathrm{a})(m\mathrm{d} + p\mathrm{b}) = lm\mathrm{d}^2 - \frac{\mathrm{abd}}{\mathrm{c}} pn,$$

$$(l\mathrm{d} + p\mathrm{a})(n\mathrm{d} + p\mathrm{c}) = ln\mathrm{d}^2 - \frac{\mathrm{acd}}{\mathrm{b}} pm,$$

the equation becomes

$$\left((l\mathrm{d}+p\mathrm{a})\sqrt{U}+\mathrm{d}\sqrt{lm}\sqrt{V}+\mathrm{d}\sqrt{ln}\sqrt{W}\right)^2=\frac{\mathrm{ad}}{\mathrm{bc}}p\left(\mathrm{b}\sqrt{n}\sqrt{V}-\mathrm{c}\sqrt{m}\sqrt{W}\right)^2,$$

or as this is more conveniently written

$$\left(\left(\sqrt{l}+\frac{\mathrm{a}p}{\mathrm{d}\sqrt{l}}\right)\sqrt{U}+\sqrt{mV}+\sqrt{nW}\right)^2=\frac{\mathrm{a}}{\mathrm{bcd}}\frac{p}{l}\left(\mathrm{b}\sqrt{nV}-\mathrm{c}\sqrt{mV}\right)^2,$$

an equation breaking up into two equations, which may be represented by

$$\sqrt{l_1U}+\sqrt{m_1V}+\sqrt{n_1W}=0,\quad \sqrt{l_2U}+\sqrt{m_2V}+\sqrt{n_2W}=0,$$

where

$$\sqrt{l_1}=\sqrt{l}+\frac{\mathrm{a}}{\mathrm{d}}\frac{p}{\sqrt{l}}\quad,\quad \sqrt{l_2}=\sqrt{l}+\frac{\mathrm{a}}{\mathrm{d}}\frac{p}{\sqrt{l}}$$

$$\sqrt{m_1}=\sqrt{m}-\sqrt{\frac{\mathrm{a}}{\mathrm{bcd}}\frac{p}{l}}\,\mathrm{b}\sqrt{n}\quad,\quad \sqrt{m_2}=\sqrt{m}+\sqrt{\frac{\mathrm{a}}{\mathrm{bcd}}\frac{p}{l}}\,\mathrm{b}\sqrt{n}$$

$$\sqrt{n_1}=\sqrt{n}+\sqrt{\frac{\mathrm{a}}{\mathrm{bcd}}\frac{p}{l}}\,\mathrm{c}\sqrt{m}\quad,\quad \sqrt{n_2}=\sqrt{n}-\sqrt{\frac{\mathrm{a}}{\mathrm{bcd}}\frac{p}{l}}\,\mathrm{c}\sqrt{m},$$

where, in the expressions for $\sqrt{l}$, &c., the signs of the radicals

$$\sqrt{l},\ \sqrt{m},\ \sqrt{n},\ \sqrt{\frac{\mathrm{a}}{\mathrm{bcd}}\frac{p}{l}},$$

may be taken determinately in any way whatever at pleasure; the only effect of an alteration of sign would in some cases be to interchange the values of $(\sqrt{l_1}, \sqrt{m_1}, \sqrt{n_1})$ with those of $(\sqrt{l_2}, \sqrt{m_2}, \sqrt{n_2})$. The tetrazomal curve thus breaks up into two trizomals.

44. It is to be noticed that we have

$$\begin{aligned}\frac{l_1}{\mathrm{a}}+\frac{m_1}{\mathrm{b}}+\frac{n_1}{\mathrm{c}}= &\ \frac{l}{\mathrm{a}}+\frac{\mathrm{a}p^2}{\mathrm{d}^2l}+2\frac{p}{\mathrm{d}}\\ &+\frac{m}{\mathrm{b}}+\frac{\mathrm{a}}{\mathrm{cd}}\frac{np}{l}\\ &+\frac{n}{\mathrm{c}}+\frac{\mathrm{a}}{\mathrm{bd}}\frac{mp}{l}\\ = &\ \left(l+\frac{\mathrm{a}p}{\mathrm{d}l}\right)\left(\frac{l}{\mathrm{a}}+\frac{m}{\mathrm{b}}+\frac{n}{\mathrm{c}}+\frac{p}{\mathrm{d}}\right);\end{aligned}$$

that is

$$\frac{l_1}{\mathrm{a}}+\frac{m_1}{\mathrm{b}}+\frac{n_1}{\mathrm{c}}=0;$$

and that similarly we have

$$\frac{l_2}{\mathrm{a}}+\frac{m_2}{\mathrm{b}}+\frac{n_2}{\mathrm{c}}=0.$$

The meaning is, that, taking the trizomal curve $\sqrt{l_1 U}+\sqrt{m_1 V}+\sqrt{n_1 W}=0$, this regarded as a tetrazomal curve, $\sqrt{l_1 U}+\sqrt{m_1 V}+\sqrt{n_1 W}+\sqrt{0T}=0$, satisfies the condition $\frac{l_1}{\mathrm{a}}+\frac{m_1}{\mathrm{b}}+\frac{n_1}{\mathrm{c}}+\frac{0}{\mathrm{d}}=0$; and the like as to the trizomal curve $\sqrt{l_2 U}+\sqrt{m_2 V}+\sqrt{n_2 W}=0$.

45. The equation by which the decomposition was effected is, it is clear, one of twelve equivalent equations; four of these are

$$\left(\sqrt{l}+\frac{\mathrm{a}p}{\mathrm{d}\sqrt{l}},\ \sqrt{m},\ \sqrt{n},\ 0\right)\left(\sqrt{U},\ \sqrt{V},\ \sqrt{W},\ \sqrt{T}\right)^2=$$
$$\frac{\mathrm{a}}{\mathrm{bcd}}\,\frac{p}{l}\left(\mathrm{b}\sqrt{nW}-\mathrm{c}\sqrt{mW}\right)^2,$$

$$\left(0,\ \sqrt{m}+\frac{\mathrm{b}l}{\mathrm{a}\sqrt{m}},\ \sqrt{n},\ \sqrt{p}\right)\left(\quad ,, \quad\right)^2=$$
$$\frac{\mathrm{b}}{\mathrm{cda}}\,\frac{l}{m}\left(\mathrm{c}\sqrt{pW}-\mathrm{d}\sqrt{nT}\right)^2,$$

$$\left(\sqrt{l},\ 0,\ \sqrt{n}+\frac{\mathrm{c}m}{\mathrm{b}\sqrt{n}},\ \sqrt{p}\right)\left(\quad ,, \quad\right)^2=$$
$$\frac{\mathrm{c}}{\mathrm{dab}}\,\frac{m}{n}\left(\mathrm{d}\sqrt{lT}-\mathrm{a}\sqrt{pU}\right)^2,$$

$$\left(\sqrt{l},\ \sqrt{m},\ 0,\ \sqrt{p}+\frac{\mathrm{d}n}{\mathrm{c}\sqrt{p}}\right)\left(\quad ,, \quad\right)^2=$$
$$\frac{\mathrm{d}}{\mathrm{abc}}\,\frac{n}{p}\left(\mathrm{a}\sqrt{mU}-\mathrm{b}\sqrt{lV}\right)^2,$$

and the others may be deduced from these by a cyclical permutation of (U, V, W), (a, b, c), (l, m, n), leaving T, d, p unaltered.

Article Nos. 46 to 51. *Application to the Trizomal; the Theorem of the Variable Zomal.*

46. I take the last equation written under the form

$$(\mathrm{a}\sqrt{mU}-\mathrm{b}\sqrt{lV})^2=\frac{\mathrm{abc}}{\mathrm{d}n}\left(\sqrt{lpU}+\sqrt{mpV}+\left(p+\frac{\mathrm{d}n}{\mathrm{c}}\right)\sqrt{T}\right)^2,$$

which, putting therein $p=0$, is

$$(\mathrm{a}\sqrt{mU}-\mathrm{b}\sqrt{lV})^2=\frac{\mathrm{abd}}{\mathrm{c}}\,nT,$$

which is in fact the trizomal curve,

$$\mathrm{a}\sqrt{mU}-\mathrm{b}\sqrt{lV}+\sqrt{\frac{\mathrm{abd}}{\mathrm{c}}\,nT}=0,$$

viz., the trizomal curve $\sqrt{lU}+\sqrt{mV}+\sqrt{nW}=0$,—if a, b, c be any quantities connected by the equation

$$\frac{l}{a}+\frac{m}{b}+\frac{n}{c}=0,$$

(the ratios a, b, c thus involving a single arbitrary parameter); and if we take T a function such that $aU+bV+cW+dT=0$; that is, $T=0$, any one of the series of curves $aU+bV+cW=0$, in involution with the given curves $U=0$, $V=0$, $W=0$,—has its equation expressible in the form

$$a\sqrt{mU}-b\sqrt{lV}+\sqrt{\frac{abd}{c}nT}=0;$$

that is, we have the curve $T=0$ (the equation whereof contains a variable parameter) as a zomal of the given trizomal curve $\sqrt{lU}+\sqrt{mV}+\sqrt{nW}=0$; and we have thus from the theorem of the decomposition of a tetrazomal deduced the theorem of the variable zomal of a trizomal. The analytical investigation is somewhat simplified by assuming $p=0$ *ab initio*, and it may be as well to repeat it in this form.

47. Starting, then, with the trizomal curve

$$\sqrt{lU}+\sqrt{mV}+\sqrt{nW}=0,$$

and writing

$$aU+\ bV+\ cW+dT=0$$

as the definition of T, the coefficients being connected by

$$\frac{l}{a}\ +\ \frac{m}{b}\ +\ \frac{n}{c}=0,$$

the equation gives

$$lU+\ mV+2\sqrt{lmUV}-nW=0;$$

or substituting in this equation for W its value in terms of U, V, T, we have

$$(an+cl)\,U+(bn+cm)\,V+2c\sqrt{lmUV}+dnT=0,$$

which by the given relation between a, b, c, is converted into

$$-\frac{ac}{b}mU-\frac{bc}{a}lV+2c\sqrt{lmUV}+dnT=0;$$

that is

$$a^2mU+b^2lV-2ab\sqrt{lmUV}\ =\frac{abd}{c}nT,$$

viz., this is

$$(a\sqrt{mU}-b\sqrt{lV})^2\qquad=\frac{abd}{c}nT,$$

or finally

$$a\sqrt{mU}-b\sqrt{lV}+\sqrt{\frac{abd}{c}nT}=0.$$

48. The result just obtained of course implies that when as above

$$\mathrm{a}U + \mathrm{b}V + \mathrm{c}W + \mathrm{d}T = 0, \quad \frac{l}{\mathrm{a}} + \frac{m}{\mathrm{b}} + \frac{n}{\mathrm{c}} = 0,$$

the trizomal curve $\sqrt{lU} + \sqrt{mV} + \sqrt{nW} = 0$ can be expressed by means of any three of the four zomals U, V, W, T, and we may at once write down the four forms

$$\left(\begin{array}{cccc} \cdot\ , & \sqrt{\frac{n}{\mathrm{c}^2}}\ , & -\sqrt{\frac{m}{\mathrm{b}^2}}\ , & -\sqrt{\frac{l\mathrm{d}}{\mathrm{abc}}} \\ -\sqrt{\frac{n}{\mathrm{c}^2}}\ , & \cdot\ , & \sqrt{\frac{l}{\mathrm{a}^2}}\ , & -\sqrt{\frac{m\mathrm{d}}{\mathrm{abc}}} \\ \sqrt{\frac{m}{\mathrm{b}^2}}\ , & -\sqrt{\frac{l}{\mathrm{a}^2}}\ , & \cdot\ , & -\sqrt{\frac{n\mathrm{d}}{\mathrm{abc}}} \\ \sqrt{\frac{l\mathrm{d}}{\mathrm{abc}}}\ , & \sqrt{\frac{m\mathrm{d}}{\mathrm{abc}}}\ , & \sqrt{\frac{n\mathrm{d}}{\mathrm{abc}}}\ , & \cdot \end{array}\right)\left(\sqrt{U}, \sqrt{V}, \sqrt{W}, \sqrt{T}\right) = 0,$$

the last of which is the original equation $\sqrt{lU} + \sqrt{mV} + \sqrt{nW} = 0$. It may be added that if the first equation be represented by $\sqrt{m_1 V} + \sqrt{n_1 W} + \sqrt{p_1 T} = 0$,—that is, if we have

$$\sqrt{m_1} = \sqrt{\frac{n}{\mathrm{c}^2}}, \quad \sqrt{n_1} = -\sqrt{\frac{m}{\mathrm{b}^2}}, \quad \sqrt{p_1} = \sqrt{\frac{l\mathrm{d}}{\mathrm{abc}}},$$

and therefore

$$\frac{m_1}{\mathrm{b}} + \frac{n_1}{\mathrm{c}} + \frac{p_1}{\mathrm{d}} = \frac{l}{\mathrm{bc}}\left(\frac{l}{\mathrm{a}} + \frac{m}{\mathrm{b}} + \frac{n}{\mathrm{c}}\right), \ = 0;$$

or if the second equation be represented by $\sqrt{l_2 U} + \sqrt{n_2 W} + \sqrt{p_2 T} = 0$,—that is, if we have

$$\sqrt{l_2} = -\sqrt{\frac{n}{\mathrm{c}^2}}, \quad \sqrt{n_2} = \sqrt{\frac{l}{\mathrm{a}^2}}, \quad \sqrt{p_2} = \sqrt{\frac{m\mathrm{d}}{\mathrm{abc}}},$$

and therefore

$$\frac{l_2}{\mathrm{a}} + \frac{n_2}{\mathrm{c}} + \frac{p_2}{\mathrm{d}} = 0;$$

or if the third equation be represented by $\sqrt{l_3 U} + \sqrt{m_3 V} + \sqrt{p_3 T} = 0$,—that is, if we have

$$\sqrt{l_3} = \sqrt{\frac{m}{\mathrm{b}^2}}, \quad \sqrt{m_3} = -\sqrt{\frac{l}{\mathrm{a}^2}}, \quad \sqrt{p_3} = \sqrt{\frac{m\mathrm{d}}{\mathrm{abc}}},$$

and therefore

$$\frac{l_3}{\mathrm{a}} + \frac{m_3}{\mathrm{b}} + \frac{p_3}{\mathrm{d}} = 0,$$

then the equation of the trizomal may also be expressed in the forms

$$\left(\begin{array}{cccc} \cdot\;, & \sqrt{m_1}\;, & \sqrt{n_1}\;, & \sqrt{p_1} \\ -\sqrt{m_1}, & \cdot\;, & \sqrt{\frac{p_1\mathrm{bc}}{\mathrm{ad}}}, & -\sqrt{\frac{n_1\mathrm{bd}}{\mathrm{ac}}} \\ -\sqrt{n_1}, & -\sqrt{\frac{p_1\mathrm{bc}}{\mathrm{ad}}}, & \cdot\;, & \sqrt{\frac{m_1\mathrm{cd}}{\mathrm{ab}}} \\ -\sqrt{p_1}, & \sqrt{\frac{n_1\mathrm{bd}}{\mathrm{ac}}}, & -\sqrt{\frac{m_1\mathrm{cd}}{\mathrm{ab}}}, & \cdot \end{array}\right)(\sqrt{U},\ \sqrt{V},\ \sqrt{W},\ \sqrt{T})=0,$$

$$\left(\begin{array}{cccc} \cdot\;, & -\sqrt{l_2}\;, & -\sqrt{\frac{p_2\mathrm{ac}}{\mathrm{bd}}}, & \sqrt{\frac{n_2\mathrm{ad}}{\mathrm{bc}}} \\ \sqrt{l_2}\;, & \cdot\;, & \sqrt{n_2}\;, & \sqrt{p_2} \\ \sqrt{\frac{p_2\mathrm{ac}}{\mathrm{bd}}}, & -\sqrt{n_2}\;, & \cdot\;, & -\sqrt{\frac{l_2\mathrm{cd}}{\mathrm{ab}}} \\ -\sqrt{\frac{n_2\mathrm{ad}}{\mathrm{bc}}}, & -\sqrt{p_2}\;, & -\sqrt{\frac{l_2\mathrm{cd}}{\mathrm{ab}}}, & \cdot \end{array}\right)\left(\sqrt{U},\ \sqrt{V},\ \sqrt{W},\ \sqrt{T}\right)=0,$$

and

$$\left(\begin{array}{cccc} \cdot\;, & \sqrt{\frac{p_3\mathrm{ab}}{\mathrm{cd}}}, & -\sqrt{l_3}\;, & -\sqrt{\frac{m_3\mathrm{ad}}{\mathrm{bc}}} \\ -\sqrt{\frac{p_3\mathrm{ab}}{\mathrm{cd}}}, & \cdot\;, & -\sqrt{m_3}, & -\sqrt{\frac{l_3\mathrm{bd}}{\mathrm{ac}}} \\ \sqrt{l_3}\;, & \sqrt{m_3}\;, & \cdot\;, & \sqrt{p_3} \\ \sqrt{\frac{m_3\mathrm{ad}}{\mathrm{bc}}}, & -\sqrt{\frac{l_3\mathrm{bd}}{\mathrm{ac}}}, & -\sqrt{p_3}\;, & \cdot \end{array}\right)\left(\sqrt{U},\ \sqrt{V},\ \sqrt{W},\ \sqrt{T}\right)=0.$$

49. These equations may, however, be expressed in a much more elegant form. Write

$$\mathrm{a}'=\frac{a}{(\beta\gamma\delta)},\quad \mathrm{b}'=-\frac{b}{(\gamma\delta\alpha)},\quad \mathrm{c}'=\frac{c}{(\delta\alpha\beta)},\quad \mathrm{d}'=\frac{-d}{(\alpha\beta\gamma)},$$

where, for shortness, $(\beta\gamma\delta)=(\beta-\gamma)(\gamma-\delta)(\delta-\beta)$, &c.; (α, β, γ) being arbitrary quantities: or, what is the same thing,

$$\mathrm{a} : \mathrm{b} : \mathrm{c} : \mathrm{d}=\mathrm{a}'(\beta\gamma\delta) : -\mathrm{b}'(\gamma\delta\alpha) : \mathrm{c}'(\delta\alpha\beta) : -\mathrm{d}'(\alpha\beta\gamma).$$

Assume

$$l : m : n \quad =\rho\mathrm{a}'(\beta-\gamma)^2 : \sigma\mathrm{b}'(\gamma-\alpha)^2 : \tau\mathrm{c}'(\alpha-\beta)^2;$$

then the equation $\frac{l}{\mathrm{a}}+\frac{m}{\mathrm{b}}+\frac{n}{\mathrm{c}}=0$ takes the form

$$\rho(\beta-\gamma)(\alpha-\delta)+\sigma(\gamma-\alpha)(\beta-\delta)+\tau(\alpha-\beta)(\gamma-\delta),$$

and the four forms of the equation are found to be

$$\left(\begin{array}{cccc} . & \sqrt{\tau}\,(\delta-\gamma), & \sqrt{\sigma}\,(\beta-\delta), & \sqrt{\rho}\,(\gamma-\beta) \\ \sqrt{\tau}\,(\gamma-\delta), & . & \sqrt{\rho}\,(\delta-\alpha), & \sqrt{\sigma}\,(\alpha-\gamma) \\ \sqrt{\sigma}\,(\delta-\beta), & \sqrt{\rho}\,(\alpha-\delta), & . & \sqrt{\tau}\,(\beta-\alpha) \\ \sqrt{\rho}\,(\beta-\gamma), & \sqrt{\sigma}\,(\gamma-\alpha), & \sqrt{\tau}\,(\alpha-\beta), & . \end{array}\right)\left(\sqrt{\mathrm{a}'U},\ \sqrt{\mathrm{b}'V},\ \sqrt{\mathrm{c}'W},\ \sqrt{\mathrm{d}'T}\right)=0,$$

viz., these are the equivalent forms of the original equation assumed to be

$$(\beta-\gamma)\sqrt{\rho\mathrm{a}'U}+(\gamma-\alpha)\sqrt{\sigma\mathrm{b}'V}+(\alpha-\beta)\sqrt{\tau\mathrm{c}'W}=0.$$

50. I remark that the theorem of the variable zomal may be obtained as a transformation theorem—viz., comparing the equation $\sqrt{lU}+\sqrt{mV}+\sqrt{nW}=0$ with the equation $\sqrt{lx}+\sqrt{my}+\sqrt{nz}=0$; this last belongs to a conic touched by the three lines $x=0$, $y=0$, $z=0$; the equation of the same conic must, it is clear, be expressible in a similar form by means of any other three tangents thereof, but the equation of any tangent of the conic is $\mathrm{a}x+\mathrm{b}y+\mathrm{c}z=0$, where a, b, c are any quantities satisfying the condition $\frac{l}{\mathrm{a}}+\frac{m}{\mathrm{b}}+\frac{n}{\mathrm{c}}=0$; whence, writing $\mathrm{a}x+\mathrm{b}y+\mathrm{c}z+\mathrm{d}w=0$, we may introduce $w=0$ along with any two of the original zomals $x=0$, $y=0$, $z=0$, or, instead of them, any three functions of the form w; and then the mere change of x, y, z, w into U, V, W, T gives the theorem. But it is as easy to conduct the analysis with (U, V, W, T) as with (x, y, z, w), and, so conducted, it is really the same analysis as that whereby the theorem is established *ante*, No. 47.

51. It is worth while to exhibit the equation of the curve

$$\sqrt{lU}+\sqrt{mV}+\sqrt{nW}=0,$$

in a form containing three new zomals. Observe that the equation $\frac{l}{\mathrm{a}}+\frac{m}{\mathrm{b}}+\frac{n}{\mathrm{c}}=0$ is satisfied by $\mathrm{a}=l\phi\chi$, $\mathrm{b}=m\chi\theta$, $\mathrm{c}=n\theta\phi$, if only $\theta+\phi+\chi=0$; or say, if $\theta=a'-a''$, $\phi=a''-a$, $\chi=a-a''$. The equation

$$\lambda\sqrt{(a-a')(a-a'')\,lU+(a'-a'')(a'-a)\,mV+(a''-a)(a''-a')\,nW}$$
$$+\mu\sqrt{(b-b')(b-b'')\,lU+(b'-b'')(b'-b)\,mV+(b''-b)(b''-b')\,nW}$$
$$+\nu\sqrt{(c-c')(c-c'')\,lU+(c'-c'')(c'-c)\,mV+(c''-c)(c''-c')\,nW}=0$$

is consequently an equation involving three zomals of the proper form; and we can determine λ, μ, ν in suchwise as to identify this with the original equation $\sqrt{lU}+\sqrt{mV}+\sqrt{nW}$, viz., writing successively $U=0$, $V=0$, $W=0$, we find

$$(a'-a'')\lambda+(b'-b'')\mu+(c'-c'')\nu=0,$$
$$(a''-a)\lambda+(b''-b)\mu+(c''-c)\nu=0,$$
$$(a-a')\lambda+(b-b')\mu+(c-c')\nu=0,$$

equations which are, as they should be, equivalent to two equations only, and which give

$$\lambda : \mu : \nu = \begin{vmatrix} 1, & 1, & 1 \\ b, & b', & b'' \\ c, & c', & c'' \end{vmatrix} : \begin{vmatrix} 1, & 1, & 1 \\ c, & c', & c'' \\ a, & a', & a'' \end{vmatrix} : \begin{vmatrix} 1, & 1, & 1 \\ a, & a', & a'' \\ b, & b', & b'' \end{vmatrix},$$

and the equation, with these values of λ, μ, ν substituted therein, is in fact the equation of the trizomal curve $\sqrt{lU}+\sqrt{mV}+\sqrt{nW}=0$ in terms of three new zomals. It is easy to return to the forms involving one new zomal and any two of the original three zomals.

Article No. 52. *Remark as to the Tetrazomal Curve.*

52. I return for a moment to the case of the tetrazomal curve, in order to show that there is not, in regard to it in general, any theorem such as that of the variable zomal. Considering the form $\sqrt{lx}+\sqrt{my}+\sqrt{nz}+\sqrt{pw}=0$ (the coordinates x, y, z, w are of course connected by a linear equation, but nothing turns upon this), the curve is here a quartic touched twice by each of the lines $x=0$, $y=0$, $z=0$, $w=0$ (viz., each of these is a double tangent of the curve), and having besides the three nodes $(x=y,\ z=w)$, $(x=z,\ y=w)$, $(x=w,\ y=z)$. But a quartic curve with three nodes, or trinodal quartic, has only four double tangents—that is, besides the lines $x=0$, $y=0$, $z=0$, $w=0$, there is no line $\alpha x+\beta y+\gamma z+\delta w=0$ which is a double tangent of the curve; and writing U, V, W, T in place of x, y, z, w, then if U, V, W, T are connected by a linear equation (and, *à fortiori*, if they are not so connected), there is not any curve $\alpha U+\beta V+\gamma W+\delta T=0$ which is related to the curve in the same way with the lines $U=0$, $V=0$, $W=0$, $T=0$; or say there is not (besides the curves $U=0$, $V=0$, $W=0$, $T=0$), any other zomal $\alpha U+\beta V+\gamma W+\delta T=0$, of the tetrazomal curve. The proof does not show that for special forms of U, V, W, T there may not be zomals, not of the above form $\alpha U+\beta V+\gamma W+\delta T=0$, but belonging to a separate system. An instance of this will be mentioned in the sequel.

Article Nos. 53 to 56. *The Theorem of the Variable Zomal of a Trizomal Curve resumed.*

53. I resume the foregoing theorem of the variable zomal of the trizomal curve $\sqrt{lU}+\sqrt{mV}+\sqrt{nW}=0$. The variable zomal $T=0$ is the curve $\mathrm{a}U+\mathrm{b}V+\mathrm{c}W=0$, where a, b, c are connected by the equation $\frac{l}{\mathrm{a}}+\frac{m}{\mathrm{b}}+\frac{n}{\mathrm{c}}=0$; that is, it belongs to a single series of curves selected in a certain manner out of the double series $\mathrm{a}U+\mathrm{b}V+\mathrm{c}W=0$ (a double series, as containing the *two* variable parameters a : b : c). These are the whole series of curves in involution with the given curves $U=0$, $V=0$, $W=0$, or being such that the Jacobian of any three of them is identical with the Jacobian of the three given curves; in particular, the Jacobian of any one of the curves $\mathrm{a}U+\mathrm{b}V+\mathrm{c}W=0$,

and of two of the three given curves, is identical with the Jacobian of the three given curves. I call to mind that, by the Jacobian of the curves $U=0$, $V=0$, $W=0$, is meant the curve

$$J(U,\ V,\ W)=\frac{d(U,\ V,\ W)}{d(x,\ y,\ z)}=\begin{vmatrix} d_xU, & d_yU, & d_zU \\ d_xV, & d_yV, & d_zV \\ d_xW, & d_yW, & d_zW \end{vmatrix}=0,$$

viz., the curve obtained by equating to zero the Jacobian or functional determinant of the functions U, V, W. Some properties of the Jacobian, which are material as to what follows, are mentioned in the Annex No. I.

For the complete statement of the theorem of the variable zomal, it would be necessary to interpret geometrically the condition $\frac{l}{\mathrm{a}}+\frac{m}{\mathrm{b}}+\frac{n}{\mathrm{c}}=0$, thereby showing how the single series of the variable zomal is selected out of the double series of the curves $\mathrm{a}U+\mathrm{b}V+\mathrm{c}W=0$ in involution with the given curves. Such a geometrical interpretation of the condition may be sought for as follows, but it is only in a particular case, as afterwards mentioned, that a convenient geometrical interpretation is thereby obtained.

54. Consider the fixed line $\Omega=px+qy+rz=0$, and let it be proposed to find the locus of the $(r-1)^2$ poles of the line $\Omega=0$ in regard to the series of curves $\mathrm{a}U+\mathrm{b}V+\mathrm{c}W=0$, where $\frac{l}{\mathrm{a}}+\frac{m}{\mathrm{b}}+\frac{n}{\mathrm{c}}=0$. Take $(x,\ y,\ z)$ as the coordinates of any one of the poles in question, then in order that $(x,\ y,\ z)$ may belong to one of the $(r-1)^2$ poles of the line $\Omega=px+qy+rz=0$ in regard to the curve $\mathrm{a}U+\mathrm{b}V+\mathrm{c}W=0$, we must have

$$d_x(\mathrm{a}U+\mathrm{b}V+\mathrm{c}W)\ :\ d_y(\mathrm{a}U+\mathrm{b}V+\mathrm{c}W)\ :\ d_z(\mathrm{a}U+\mathrm{b}V+\mathrm{c}W)=p\ :\ q\ :\ r;$$

or, what is the same thing,

$$=d_x\Omega\ :\ d_y\Omega\ :\ d_z\Omega;$$

and these equations give without difficulty

$$\mathrm{a}:\mathrm{b}:\mathrm{c}=J(V,\ W,\ \Omega)\ :\ J(W,\ U,\ \Omega)\ :\ J(U,\ V,\ \Omega),$$

whence, substituting in the equation $\frac{l}{\mathrm{a}}+\frac{m}{\mathrm{b}}+\frac{n}{\mathrm{c}}=0$, we have

$$\frac{l}{J(V,\ W,\ \Omega)}+\frac{m}{J(W,\ U,\ \Omega)}+\frac{n}{J(U,\ V,\ \Omega)}=0$$

as the locus of the $(r-1)^2$ poles in question. Each of the Jacobians is a function of the order $2r-2$, and the order of the locus is thus $=4r-4$. As the given curves $U=0$, $V=0$, $W=0$ belong to the single series of curves, it is clear that the locus passes through the $3(r-1)^2$ points which are the $(r-1)^2$ poles of the fixed line in regard to the curves $U=0$, $V=0$, $W=0$ respectively.

55. In the case where the given trizomal is

$$\sqrt{l(\Theta+L\Phi)}+\sqrt{m(\Theta+M\Phi)}+\sqrt{n(\Theta+N\Phi)}=0,$$

$s=r-1$, that is, where the zomals $\Theta+L\Phi=0$, $\Theta+M\Phi=0$, $\Theta+N\Phi=0$ are each of them curves of the order r, passing through the r intersections of the line $\Phi=0$ with the curve $\Theta=0$, then, taking this line $\Phi=0$ for the fixed line $\Omega=0$, we have

$$J(V,\ W,\ \Omega)=J(\Theta+M\Phi,\ \Theta+N\Phi,\ \Phi)=\Phi\,\{M,\ N\},$$

if, for shortness, $\{M,\ N\}=J(M-N,\ \Theta,\ \Phi)+\Phi\,J(M,\ N,\ \Phi)$, and the like as to the other two Jacobians, so that, attaching the analogous significations to $\{N,\ L\}$ and $\{L,\ M\}$, the equation of the locus is

$$\frac{l}{\{M,\ N\}}+\frac{m}{\{N,\ L\}}+\frac{n}{\{L,\ M\}}=0,$$

where observe that each of the curves $\{M,\ N\}=0$, $\{N,\ L\}=0$, $\{L,\ M\}=0$ is a curve of the order $2r-3$; the order of the locus is thus $=4r-6$, and (as before) this locus passes through the $3(r-1)^2$ points which are the $(r-1)^2$ poles of the line $\Phi=0$ in regard to the curves $\Theta+L\Phi=0$, $\Theta+M\Phi=0$, $\Theta+N\Phi=0$ respectively.

56. In the case $r=2$, the trizomal is

$$\sqrt{l(\Theta+L\Phi)}+\sqrt{m(\Theta+M\Phi)}+\sqrt{n(\Theta+N\Phi)}=0,$$

where the zomals are the conics $\Theta+L\Phi=0$, $\Theta+M\Phi=0$, $\Theta+N\Phi=0$, each passing through the same two points $\Theta=0$, $\Phi=0$; the locus of the pole of the line $\Phi=0$, in regard to the variable zomal, is the conic

$$\frac{l}{\{M,\ N\}}+\frac{m}{\{N,\ L\}}+\frac{n}{\{L,\ M\}}=0,$$

viz., $\{M,\ N\}=0$, $\{N,\ L\}=0$, $\{L,\ M\}=0$, are here the lines passing through the poles of the line $\Phi=0$ in regard to the second and third, the third and first, and the first and second of the given conics respectively: treating l, m, n as arbitrary, the locus is clearly *any* conic through the poles of the line $\Phi=0$ in regard to the three conics respectively. The Jacobian of the three given conics is a conic related in a special manner to the three given conics, and which might be called the Jacobian conic thereof, and it would be easy to give a complete enunciation of the theorem for the case in hand. (See as to this, Annex No. I, above referred to.) But if, in accordance with the plan adopted in the remainder of the memoir, we at once assume that the points $\Theta=0$, $\Phi=0$ are the circular points at infinity, then the theorem can be enunciated under a more simple form—viz., if $\mathfrak{A}^\circ=0$, $\mathfrak{B}^\circ=0$, $\mathfrak{C}^\circ=0$ are the equations of any three circles, then in the trizomal

$$\sqrt{l\mathfrak{A}^\circ}+\sqrt{m\mathfrak{B}^\circ}+\sqrt{n\mathfrak{C}^\circ}=0,$$

the variable zomal is any circle whatever of the series of circles cutting at right angles the orthotomic circle of the three given circles, and having its centre on a certain conic which passes through the centres of the given circles. Moreover, if the coefficients l, m, n are not given in the first instance, but are regarded as arbitrary, then the last-mentioned conic is any conic whatever through the three centres, and there belongs to such conic and the series of zomals derived therefrom as above, a trizomal curve $\sqrt{l\mathfrak{A}^\circ}+\sqrt{m\mathfrak{B}^\circ}+\sqrt{n\mathfrak{C}^\circ}=0$. This is obviously the theorem, that if a variable circle has its centre on a given conic, and cuts at right angles a given circle, then the envelope of the variable circle is a trizomal curve $\sqrt{l\mathfrak{A}^\circ}+\sqrt{m\mathfrak{B}^\circ}+\sqrt{n\mathfrak{C}^\circ}$, where $\mathfrak{A}^\circ=0$, $\mathfrak{B}^\circ=0$, $\mathfrak{C}^\circ=0$ are any three circles, positions of the variable circle, and l, m, n are constant quantities depending on the selected three circles.

PART II. (Nos. 57 to 104). SUBSIDIARY INVESTIGATIONS.

Article Nos. 57 and 58. *Preliminary Remarks.*

57. We have just been led to consider the conics which pass through two given points. There is no real loss of generality in taking these to be the circular points at infinity, or say the points I, J—viz., every theorem which in anywise explicitly or implicitly relates to these two points, may, without the necessity of any change in the statement thereof, be understood as a theorem relating instead to any two points P, Q. I call to mind that a circle is a conic passing through the two points I, J, and that lines at right angles to each other are lines harmonically related to the pair of lines from their intersection to the points I, J respectively, so that when (I, J) are replaced by any two given points whatever, the expression a circle must be understood to mean a conic passing through the two given points; and in speaking of lines at right angles to each other, it must be understood that we mean lines harmonically related to the pair of lines from their intersection to the two given points respectively. For instance, the theorem that the Jacobian of any three circles is their orthotomic circle, will mean that the Jacobian of any three conics which each of them passes through the two given points is the orthotomic conic through the same two points, that is, the conic such that at each of its intersections with any one of the three conics, the two tangents are harmonically related to the pair of lines from this intersection to the two given points respectively. Such extended interpretation of any theorem is applicable even to the theorems which involve distances or angles—viz., the terms "distance" and "angle" have a determinate signification when interpreted in reference (not to the circular points at infinity, but instead thereof) to any two given points whatever (see as to this my "Sixth Memoir on Quantics," Nos. 220, *et seq.*). *Phil. Trans.*, vol. CXLIX. (1859), pp. 61—90; see p. 86; [158]. And this being so, the theorem can, without change in the statement thereof, be understood as referring to the two given points.

58. I say then that any theorem (referring explicitly or implicitly) to the circular points at infinity I, J, may be understood as a theorem referring instead to any two given points. We might of course give the theorems in the first instance in terms explicitly referring to the two given points—(viz., instead of a circle, speak of a conic through the two given points, and so in other instances); but, as just explained, this is not really more general, and the theorems would be given in a less concise and

familiar form. It would not, on the face of the investigations, be apparent that in treating of the polyzomal curves

$$\sqrt{l(\Theta + L\Phi)} + \sqrt{m(\Theta + M\Phi)} + \&c. = 0,$$

($\Theta = 0$ a conic, $\Phi = 0$ a line, as above), that we were really treating of the curves the zomals whereof are circles, and therein of the theories of foci and focofoci as about to be explained. And for these reasons I shall consider the two points $\Theta = 0$, $\Phi = 0$, to be the circular points at infinity I, J, and in the investigations, &c., make use of the terms circle, right angles, &c., which, in their ordinary significations, have implicit reference to these two points.

The present Part does not explicitly relate to the theory of polyzomal curves, but contains a series of researches, partly analytical and partly geometrical, which will be made use of in the following Parts III. and IV. of the Memoir.

Article Nos. 59 to 62. *The Circular Points at Infinity; Rectangular and Circular Coordinates.*

59. The coordinates made use of (except in the cases where the general trilinear coordinates (x, y, z), or any other coordinates, are explicitly referred to), will be either the ordinary rectangular coordinates x, y, or else, as we may term them, the circular coordinates ξ, η $(= x + iy,\ x - iy$ respectively, $i = \sqrt{-1}$ as usual), but in either case I shall introduce for homogeneity the coordinate z, it being understood that this coordinate is in fact $= 1$, and that it may be retained or replaced by this its value, in different investigations or stages of the same investigation, as may for the time being be most convenient. In more concise terms, we may say that the coordinates are either the rectangular coordinates x, y, and z $(= 1)$, or else the circular coordinates ξ, η, and $z\,(= 1)$. The equation of the line infinity is $z = 0$; the points I, J are given by the equations $(x + iy = 0,\ z = 0)$ and $(x - iy = 0,\ z = 0)$, or, what is the same thing, by the equations $(\xi = 0,\ z = 0)$ and $(\eta = 0,\ z = 0)$ respectively; or in the rectangular coordinates the coordinates of these points are $(-i, 1, 0)$ and $(i, 1, 0)$ respectively, and in the circular coordinates they are $(1, 0, 0)$ and $(0, 1, 0)$ respectively. It is, of course, only for points at infinity that the coordinate z is $= 0$ (and observe that for any such point the x and y or ξ and η coordinates may be regarded as finite); for every point whatever not at infinity the coordinate z is, as stated above, $= 1$.

60. Consider a point A, whose coordinates (rectangular) are $(a, a', 1)$ and (circular) $(\alpha, \alpha', 1)$, viz., $\alpha = a + a'i$, $\alpha' = a - a'i$; then the equations of the lines through A to the points I, J, are

$$x - az + i(y - a'z) = 0, \quad x - az - i(y - a'z) = 0$$

respectively, or they are

$$\xi - \alpha z = 0 \qquad , \quad \eta - \alpha' z = 0$$

respectively. These equations, if (a, a') or (α, α') are arbitrary, will, it is clear, be the equations of any two lines through the points I, J, respectively.

61. We have from either of the equations in (x, y, z)

$$(x-az)^2+(y-a'z)^2=0,$$

that is, the distance from each other of any two points $(x, y, 1)$, and $(a, a', 1)$ in a line through I or J is $=0$. And in particular, if $z=0$, then $x^2+y^2=0$; that is, the distance of the point $(a, a', 1)$ from I or J is in each case $=0$.

62. Consider for a moment any three points P, Q, A; the perpendicular distance of P from QA is $=2$ triangle $PQA \div$ distance QA; if Q be any point on the line through A to either of the points I, J, and in particular if Q be either of the points I, J, then the triangle PQA is finite, but the distance QA is $=0$: that is, the perpendicular distance of P from the line through A to either of the points I, J, that is, from any line through either of these points, is $=\infty$. But, as just stated, the triangle PQA is finite, or say the triangles PIA, PJA are each finite; viz., the coordinates (rectangular) of P, A being $(x, y, z=1)$, $(a, a', 1)$ or (circular) $(\xi, \eta, z=1)$, $(\alpha, \alpha', 1)$, the expressions for the doubles of these triangles respectively are

$$\begin{vmatrix} x, & y, & z \\ -i, & 1, & 0 \\ a, & a', & 1 \end{vmatrix}, \quad \begin{vmatrix} x, & y, & z \\ i, & 1, & 0 \\ a, & a', & 1 \end{vmatrix}$$

that is, they are (rectangular coordinates) $x-az+i(y-a'z)$, $x-az-i(y-a'z)$, or (circular coordinates) $\xi-\alpha z$, $\eta-\alpha' z$.

Representing the double areas by PIA, PJA, respectively, and the squared distance of the points A, P, by $\mathfrak{A}$, we have—

$$\begin{aligned} \mathfrak{A} &= (x-az)^2+(y-a'z)^2 \\ &= (\xi-\alpha z)(\eta-\alpha' z), \;= PIA\,.\,PJA. \end{aligned}$$

Article No. 63. *Antipoints; Definition and Fundamental Properties.*

63. Two pairs of points (A, B) and (A_1, B_1) which are such that the lines AB, A_1B_1 bisect each other at right angles in a point O in such wise that $OA=OB=i\,OA_1=i\,OB_1$, are said to be antipoints, each of the other. In rectangular coordinates, taking the coordinates of $(AB,)$ to be $(a, 0, 1)$ and $(-a, 0, 1)$, those of (A_1, B_1) will be $(0, ai, 1)$ and $(0, -ai, 1)$ respectively, whence joining the points (A, B) with the points (I, J), the points A_1, B_1 are given as the intersections of the lines AI and BJ, and of the lines AJ and BI respectively. Or, what is the same thing, in any quadrilateral wherein I, J are opposite angles, the remaining pairs (A, B) and (A_1, B_1) are antipoints each of the other.

64. In circular coordinates, if the coordinates of A are $(\alpha, \alpha', 1)$, and those of B are $(\beta, \beta', 1)$, then the equations of

$$\begin{aligned} &AI,\ AJ \text{ are } \xi-\alpha z=0, \quad \eta-\alpha' z=0, \\ &BI,\ BJ \;\;\text{,,}\;\; \xi-\beta z=0, \quad \eta-\beta' z=0, \end{aligned}$$

whence the equations of

$$A_1I,\ A_1J \text{ are } \xi - \alpha z = 0, \quad \eta - \beta' z = 0,$$
$$B_1I,\ B_1J \quad ,, \quad \xi - \beta z = 0, \quad \eta - \alpha' z = 0.$$

65. Considering any point P the coordinates of which are ξ, η, z $(=1)$, let $\mathfrak{A}$, $\mathfrak{B}$, $\mathfrak{A}_1$, $\mathfrak{B}_1$ be its squared distances from the points A, B, A_1, B_1 respectively; then by what precedes

$$\mathfrak{A} = (\xi - \alpha z)(\eta - \alpha' z),$$
$$\mathfrak{B} = (\xi - \beta z)(\eta - \beta' z),$$
$$\mathfrak{A}_1 = (\xi - \alpha z)(\eta - \beta' z),$$
$$\mathfrak{B}_1 = (\xi - \beta z)(\eta - \alpha' z),$$

and thence

$$\mathfrak{A} . \mathfrak{B} = \mathfrak{A}_1 . \mathfrak{B}_1;$$

that is, the product of the squared distances of a point P from any two points A, B, is equal to the product of the squared distances of the same point P from the two antipoints A_1, B_1. This theorem, which was, I believe, first given by me in the *Educational Times* (see reprint, vol. VI. 1866, p. 81), is an important one in the theory of foci. It is to be further noticed that we have

$$\mathfrak{A} + \mathfrak{B} - \mathfrak{A}_1 - \mathfrak{B}_1 = (\alpha - \beta)(\alpha' - \beta') z^2, \ = Kz^2 = K,$$

if K, $=(\alpha - \alpha')(\beta - \beta')$, be the squared distance of the points A, B, $=-$ squared distance of points A_1, B_1.

Article No. 66. *Antipoints of a Circle.*

66. A similar notion to that of two pairs of antipoints is as follows, viz., if from the centre of a circle perpendicular to its plane and in opposite senses, we measure off two distances each $=i$ into the radius, the extremities of these distances are antipoints of the circle. It is clear that the antipoints of the circle and the extremities of any diameter thereof are (in the plane of these four points) pairs of antipoints. It is to be added that each antipoint is the centre of a sphere radius zero, or say of a cone sphere, passing through the circle: the circle is thus the intersection of the two cone spheres having their centres at the two antipoints respectively.

Article No. 67. *Antipoints in relation to a Pair of Orthotomic Circles.*

67. It is a well-known property that if any circle pass through the points (A, B), and any other circle through the antipoints (A_1, B_1), then these two circles cut at right angles. Conversely if a circle pass through the points A, B, then all the orthotomic circles which have their centres on the line AB pass through the antipoints A_1, B_1. In particular, if on AB as diameter we describe a circle and on A_1B_1 as diameter a circle, then these two circles—being, it is clear, concentric circles with their radii in the ratio $1 : i$, and as concentric circles touching each other at the points (I, J)—cut each other at right angles; or say they are concentric orthotomic circles.

Article Nos. 68 to 71. *Forms of the Equation of a Circle.*

68. In rectangular coordinates the equation of a circle, coordinates of centre $(a, a', 1)$ and radius $= a''$, is

$$\mathfrak{A}^\circ = (x - az)^2 + (y - a'z)^2 - a''^2z^2 = 0\ ;$$

and in circular coordinates, the coordinates of the centre being $(\alpha, \alpha', 1)$, and radius $= a''$ as before, the equation is

$$\mathfrak{A}^\circ = (\xi - \alpha z) \quad (\eta - \alpha' z) - a''^2z^2 = 0.$$

69. I observe in passing, that the origin being at the centre and the radius being $= 1$, then writing also $z = 1$, the equation of the circle is $\xi\eta = 1$, that is the circular coordinates of any point of the circle, expressed by means of a variable parameter θ, are $\left(\theta, \dfrac{1}{\theta}, 1\right)$.

70. Consider a current point P, the coordinates of which (rectangular) are $x, y, z (= 1)$, and (circular) are $\xi, \eta, z (= 1)$, then the foregoing expression

$$\begin{aligned}\mathfrak{A}^\circ &= (x - az)^2 + (y - a'z)^2 - a''^2 z^2 \\ &= (\xi - \alpha z)(\eta - \alpha' z) - a''^2 z^2\end{aligned}$$

denotes, it is clear, the square of the tangential distance of the point P from the circle $\mathfrak{A}^\circ = 0$.

71. But there is another interpretation of this same function $\mathfrak{A}^\circ$, viz., writing therein $z = 1$, and then

$$\mathfrak{A}^\circ = (x - a)^2 + (y - a')^2 + (a''i)^2,$$

we see that $\mathfrak{A}^\circ$ is the squared distance of P from either of the antipoints of the circle (points lying, it will be recollected, out of the plane of the circle), and we have thus the theorem that the square of the tangential distance of any point P from the circle is equal to the square of its distance from either antipoint of the circle.

Article Nos. 72 to 77. *On a System of Sixteen Points.*

72. Take (A, B, C, D) any four concyclic points, and let the antipoints of

$$\begin{array}{llcll} (B, C), & (A, D) & \text{be} & (B_1, C_1), & (A_1, D_1), \\ (C, A), & (B, D) & ,, & (C_2, A_2), & (B_2, D_2), \\ (A, B), & (C, D) & ,, & (A_3, B_3), & (C_3, D_3), \end{array}$$

then each of the three new sets (A_1, B_1, C_1, D_1), (A_2, B_2, C_2, D_2), (A_3, B_3, C_3, D_3) will be a set of four concyclic points.

73. Let O be the centre of the circle through (A, B, C, D), say of the circle O, and then, the lines BC, AD meeting in R, the lines CA, BD in S, and the lines AD, CD in T, let each of these points be made the centre of a circle orthotomic to O, viz., let these new circles be called the circles R, S, T respectively.

As regards the circle R, since its centre lies in BC, the circle passes through (B_1, C_1); and since the centre lies in AD, the circle passes through (A_1, D_1), that is, the four points (A_1, B_1, C_1, D_1) lie in the circle R. Similarly (A_2, B_2, C_2, D_2) lie in the circle S, and (A_3, B_3, C_3, D_3) in the circle T.

74. The points R, S, T are conjugate points in relation to the circle O; that is, ST, TR, RS are the polars of R, S, T respectively in regard to this circle; and they are, consequently, at right angles to the lines OR, OS, OT respectively; viz., the four centres O, R, S, T are such that the line joining any two of them cuts at right angles the line joining the other two of them, and we see that the relation between the four sets is in fact a symmetrical one; this is most easily seen by consideration of the circular points at infinity I, J, the four sets of points may be arranged thus:

$$\begin{array}{cccc} A\,, & A_3, & A_2, & A_1, \\ B_3, & B\,, & B_1, & B_2, \\ C_2, & C_1\,, & C\,, & C_3\,, \\ D_1, & D_2, & D_3, & D\,, \end{array}$$

in such wise that any four of them in the same vertical line pass through I, and any four in the same horizontal line pass through J; and this being so, starting for instance with (A_3, B_3, C_3, D_3) we have antipoints

$$\begin{array}{llcll} \text{of } (B_3\,, C_3\,), & (A_3, D_3) & \text{are} & (B_2\,, C_2\,), & (A_2, D_2), \\ \;\;,,\; (C_3\,, A_3), & (B_3, D_3) & ,, & (C_1, A_1), & (B_1, D_1), \\ \;\;,,\; (A_3, B_3\,), & (C_3, D_3) & ,, & (A\,, B\,), & (C\,, D\,), \end{array}$$

and similarly if we start from (A_1, B_1, C_1, D_1) or (A_2, B_2, C_2, D_2).

75. I return for a moment to the construction of (A_1, B_1, C_1, D_1); these are points on the circle R, and (B_1, C_1) are the antipoints of (B, C); that is, they are the intersections of the circle R by the line at right angles to BC from its middle point, or, what is the same thing, by the perpendicular on BC from O. Similarly (A_1, D_1) are the antipoints of (A, D); that is, they are the intersections of the circle R by the perpendicular on AD from O. And the like as to (A_2, B_2, C_2, D_2) and (A_3, B_3, C_3, D_3) respectively.

76. Hence, starting with the points A, B, C, D on the circle O, and constructing as above the circles P, Q, R, and constructing also the perpendiculars from O on the six chords AB, AC, &c.,

$$\begin{array}{lllllll} \text{the perpendiculars on} & BC,\ AD & \text{meet circle} & R & \text{in} & (B_1, C_1), & (A_1, D_1), \\ \quad ,, & CA,\ BD & ,, \quad ,, & S & ,, & (C_2, A_2), & (B_2, D_2), \\ \quad ,, & AB,\ CD & ,, \quad ,, & T & ,, & (A_3, B_3), & (C_3, D_3), \end{array}$$

so that the whole system is given by means of the circles P, Q, R, and the six perpendiculars.

77. If to fix the ideas (A, B, C, D) are real points taken in order on the real circle O, then the points R, S, T are each of them real; but R and T lie outside, S inside the circle O. The circles R and T are consequently real, but the circle S imaginary, viz., its radius is $=i$ into a real quantity; the imaginary points (A_1, B_1, C_1, D_1) are thus given as the intersections of a real circle by a pair of real lines, and the like as to the imaginary points (A_3, B_3, C_3, D_3); but the imaginary points (A_2, B_2, C_2, D_2) are only given as the intersections of an imaginary circle (centre real and radius a pure imaginary) by a pair of real lines. The points (C_2, A_2) *quâ* antipoints of (C, A) are easily constructed as the intersections of a real circle by a real line, and the like as to the points (B_2, D_2) *quâ* antipoints of (B, D), but the construction for the two pairs of points cannot be effected by means of the same real circle.

Article Nos. 78 to 80. *Property in regard to Four Confocal Conics.*

78. All the conics which pass through the four concyclic points A, B, C, D, have their axes in fixed directions; but three such conics are the line-pairs (BC, AD), (CA, BD), and (AB, CD), whence the directions of the axes are those of the bisectors of the angles formed by any one of these pairs of lines; hence, in particular, considering either axis of a conic through the four points, the lines AB and CD are equally inclined on opposite sides to this axis, and this leads to the theorem that the antipoints $(A_3, B_3)(C_3, D_3)$ are in a conic confocal to the given conic through (A, B, C, D); whence, also, considering any given conic whatever through (A, B, C, D), the points (A_1, B_1, C_1, D_1), (A_2, B_2, C_2, D_2), (A_3, B_3, C_3, D_3) lie severally in three conics, each of them confocal with the given conic.

79. To prove this, consider any two confocal conics, say an ellipse and a hyperbola, and let F be one of their four intersections; join F with the common centre O, and let OT, ON be parallel to the tangent and normal respectively of the ellipse at the point F. OF, OT are in direction conjugate axes of the ellipse, and OF, ON are in direction conjugate axes of the hyperbola; and if they are also the axes in magnitude, that is, if the points T, N are the intersections of OT with the ellipse and of ON with the hyperbola respectively, then it is easy to show that $\overline{OT}^2 + \overline{ON}^2 = 0$. And this being so, imagine on the ellipse any two points A, B such that the chord AB is parallel to OT, that is conjugate to OF; AB is bisected by OF, say in a point K, or we have parallel to OT the semichords or ordinates $KA = KB$; and we may, perpendicularly to this or parallel to ON, draw through K in the hyperbola a chord A_3B_3, which chord will be bisected in K, or we shall have $KA_3 = KB_3$. Hence KA, KA_3 are in the ellipse and the hyperbola respectively ordinates conjugate to the same diameter OF, and the semi-diameters conjugate to OF being OT, ON respectively, we have $KA^2 (= KB^2) : \overline{KA_3}^2 (= KB_3^2) = \overline{OT}^2 : \overline{ON}^2$, this is, $\overline{KA}^2 = \overline{KB}^2 = -\overline{KA_3}^2 = -\overline{KB_3}^2$; or (A_3, B_3) will be the antipoints of (A, B).

80. Conversely, if in the ellipse we have the two points (A, B), then drawing the diameter OF conjugate to AB, and through its extremity F, the confocal hyperbola, then the antipoints (A_3, B_3) will lie on the hyperbola. And similarly, if on the

ellipse we have the two points (C, D), then drawing the diameter OG conjugate to CD, and through its extremity G a confocal hyperbola, the antipoints (C_3, D_3) will lie on the hyperbola. Suppose (A, B, C, D) are concyclic, then, as noticed, AB and CD will be equally inclined on opposite sides to the transverse axis of the ellipse—the conjugate diameters OF, OG will therefore be equally inclined on opposite sides of the transverse axis—and the points F and G will therefore be situate symmetrically on opposite sides of the transverse axis, that is, the points F and G will respectively determine the same confocal hyperbola, and we have thus the required theorem, viz., if (A, B, C, D) are any four concyclic points on an ellipse, or say on a conic, and if (A_3, B_3) are the antipoints of (A, B), and (C_3, D_3) the antipoints of (C, D), then (A_3, B_3, C_3, D_3) will lie on a conic confocal with the given conic.

Article Nos. 81 to 85. *System of the Sixteen Points, the Axial Case.*

81. The theorems hold good when the four points A, B, C, D are in a line; the antipoints (B_1, C_1) of (B, C), &c., are in this case situate symmetrically on opposite sides of the line, so that it is evident at sight that we have (A_1, B_1, C_1, D_1), (A_2, B_2, C_2, D_2), (A_3, B_3, C_3, D_3), each set in a circle; and that the centres R, S, T of these circles lie in the line. The construction for the general case becomes, however, indeterminate, and must therefore be varied. If in the general case we take any circle through (B, C), and any circle through (A, D), then the circle R cuts at right angles these two circles, and has, consequently, its centre R in the radical axis of the two circles; whence, when the four points are in a line, taking any circle through (B, C), or in particular the circle on BC as diameter, and any circle through (A, D), or in particular the circle on AD as diameter,—the radical axis of these two circles intersects the line in the required centre R, and the circle R is the circle with this centre cutting at right angles the two circles respectively; the circles S and T are, of course, obtained by the like construction in regard to the combinations $(C, A;\ B, D)$ and $(A, B;\ C, D)$, respectively. It may be added, that we have

$$\left.\begin{matrix}R\\ S\\ T\end{matrix}\right\}\text{ centre and }\left\{\begin{matrix}\text{extremities } R\\ \text{of diameter } S\\ \text{of circles } T\end{matrix}\right\}\text{ sibiconjugate points of involutions }\left\{\begin{matrix}B,\ C;\ A,\ D,\\ C,\ A;\ B,\ D,\\ A,\ B;\ C,\ D,\end{matrix}\right.$$

and that (as in the general case) the circles R, S, T intersect each pair of them at right angles; and they are evidently each intersected at right angles by the line $ABCD$ (or axis of the figure), which replaces the circle O in the general case.

82. If the points A, B, C, D are taken in order on the line, then the points R, S, T are all real, viz., the point R is situate, on one side or the other, outside AD, but the points S and T are each of them situate between B and C; the circles R and T are real, but the circle S has its radius a pure imaginary quantity.

83. If one of the four points, suppose D, is at infinity on the line, then the antipoints of (A, D), of (B, D), and of (C, D) are each of them the two points (I, J).

It would at first sight appear that the only conditions for the circles R, S, T were the conditions of passing through the antipoints of (B, C), of (C, A), and of (A, B) respectively, and that these circles thus became indeterminate; but in fact the definition of the circles is then as follows, viz., R has its centre at A, and passes through the antipoints of (B, C): (whence squared radius $= AB \,.\, AC$). And similarly, S has its centre at B, and passes through antipoints of (C, A), (squared radius $= BA \,.\, BC$); and T has its centre at C, and passes through antipoints of (A, B). (squared radius $= CA \,.\, CB$); these three circles cut each other at right angles. As before, A, B, C being in order on the line, the circles R, T are real, but the circle S has its radius a pure imaginary quantity.

84. That the circles are as just mentioned appears as follows: taking the line as axis of x, and a, b, c, d for the x coordinates of the four points respectively, then the coordinates of A_1, D_1 are

$$\tfrac{1}{2}(a+d),\ \pm \tfrac{1}{2}i(a-d);$$

whence, m being arbitrary, the general equation of a circle through A_1, D_1 is

$$x^2 + y^2 - 2mxz + [m(a+d) - ad]\, z^2 = 0,$$

writing herein

$$m = a - \frac{k^2}{d}$$

this becomes

$$x^2 + y^2 - 2\left(a - \frac{k^2}{d}\right) xz + \left(a^2 - k^2 - \frac{ak^2}{d}\right) z^2 = 0,$$

viz., for $d = \infty$ it is

$$(x - az)^2 + y^2 - k^2 z^2 = 0,$$

which is a circle having A for its centre, and its radius an arbitrary quantity k. If the circle passes through the antipoints of B, C, the coordinates of these are

$$\tfrac{1}{2}(b+c),\ \pm \tfrac{1}{2}i(b-c),$$

and we find

$$k^2 = [\tfrac{1}{2}(b+c) - a]^2 - \tfrac{1}{4}(b-c)^2 = (a-b)(a-c).$$

85. Reverting to the general case of four points A, B, C, D on a line, the theorem as to the confocal conics holds good under the form that, drawing any conic whatever through (A_1, B_1, C_1, D_1), the points (A_2, B_2, C_2, D_2), and (A_3, B_3, C_3, D_3) lie in confocal conics, these conics have their centre on the line, and axes in the direction of and perpendicular to the line. When D is at infinity, the confocal conics become any three concentric circles through (B_1, C_1), (C_2, A_2), and (A_3, B_3) respectively.

Article Nos. 86 to 91. *The Involution of Four Circles.*

86. Consider any four points A, B, C, D, the centres of circles denoted by these same letters, and let $\mathfrak{A}^\circ$, $\mathfrak{B}^\circ$, $\mathfrak{C}^\circ$, $\mathfrak{D}^\circ$ signify as usual, viz., if (in orthogonal coordinates) $(a, a', 1)$ are the coordinates of the centre, and a'' the radius of the circle A, then $\mathfrak{A}^\circ$ stands for $(x - az)^2 + (y - a'z)^2 - a''^2 z^2$, and the like for $\mathfrak{B}^\circ$, $\mathfrak{C}^\circ$, $\mathfrak{D}^\circ$. Write also

$$\mathrm{a} : \mathrm{b} : \mathrm{c} : \mathrm{d} = BCD : -CDA : DAB : -ABC,$$

where BCD, &c., are the triangles formed by the points (B, C, D), &c.; the analytical expressions are

$$\mathrm{a} : \mathrm{b} : \mathrm{c} : \mathrm{d} = \begin{vmatrix} b, & b', & 1 \\ c, & c', & 1 \\ d, & d', & 1 \end{vmatrix} : -\begin{vmatrix} c, & c', & 1 \\ d, & d', & 1 \\ a, & a', & 1 \end{vmatrix} : \begin{vmatrix} d, & d', & 1 \\ a, & a', & 1 \\ b, & b', & 1 \end{vmatrix} : -\begin{vmatrix} a, & a', & 1 \\ b, & b', & 1 \\ c, & c', & 1 \end{vmatrix}$$

so that

$$\begin{aligned} \mathrm{a} + \mathrm{b} + \mathrm{c} + \mathrm{d} &= 0, \\ \mathrm{a}a + \mathrm{b}b + \mathrm{c}c + \mathrm{d}d &= 0, \\ \mathrm{a}a' + \mathrm{b}b' + \mathrm{c}c' + \mathrm{d}d' &= 0; \end{aligned}$$

this being so, it is clear that we have

$$\mathrm{a}\mathfrak{A}^\circ + \mathrm{b}\mathfrak{B}^\circ + \mathrm{c}\mathfrak{C}^\circ + \mathrm{d}\mathfrak{D}^\circ =$$

$$z^2\left[\mathrm{a}(a^2 + a'^2 - a''^2) + \mathrm{b}(b^2 + b'^2 - b''^2) + \mathrm{c}(c^2 + c'^2 - c''^2) + \mathrm{d}(d^2 + d'^2 - d''^2)\right] = Kz^2, \ = K,$$

a constant.

87. I am not aware that in the general case there is any convenient expression for this constant K; it is $=0$ when the four circles have the same orthotomic circle; in fact, taking as origin the centre of the orthotomic circle, and its radius to be $=1$, we have

$$a^2 + a'^2 - a''^2 = 1, \text{ \&c.},$$

whence

$$K = \mathrm{a} + \mathrm{b} + \mathrm{c} + \mathrm{d} = 0;$$

that is, if the circles A, B, C, D have the same orthotomic circle, then $\mathfrak{A}^\circ, \mathfrak{B}^\circ, \mathfrak{C}^\circ, \mathfrak{D}^\circ$, a, b, c, d, signifying as above, we have

$$\mathrm{a}\mathfrak{A}^\circ + \mathrm{b}\mathfrak{B}^\circ + \mathrm{c}\mathfrak{C}^\circ + \mathrm{d}\mathfrak{D}^\circ = 0,$$

and, in particular, if the circles reduce themselves to the points A, B, C, D respectively, then (writing as usual $\mathfrak{A}, \mathfrak{B}, \mathfrak{C}, \mathfrak{D}$ in place of $\mathfrak{A}^\circ, \mathfrak{B}^\circ, \mathfrak{C}^\circ, \mathfrak{D}^\circ$) if the four points A, B, C, D are on a circle, we have

$$\mathrm{a}\mathfrak{A} + \mathrm{b}\mathfrak{B} + \mathrm{c}\mathfrak{C} + \mathrm{d}\mathfrak{D} = 0.$$

88. This last theorem may be regarded as a particular case of the theorem

$$\mathrm{a}\mathfrak{A} + \mathrm{b}\mathfrak{B} + \mathrm{c}\mathfrak{C} + \mathrm{d}\mathfrak{D} = Kz^2 = K,$$

viz., the four circles reducing themselves to the points A, B, C, D, we can find for the constant K an expression which will of course vanish when the points are on a circle. For this purpose, let the lines BC, AD meet in R, the lines CA, BD in S, and the lines AB, CD in T; we may, to fix the ideas, consider $ABCD$ as forming a convex quadrilateral, R and T will then be the exterior centres, S the interior centre; a, b, c, d, may be taken *equal* to $BCD, -CDA, DAB, -ABC$, where the areas BCD, &c., are each taken positively. The expression $\mathrm{a}\mathfrak{A} + \mathrm{b}\mathfrak{B} + \mathrm{c}\mathfrak{C} + \mathrm{d}\mathfrak{D}$ has the same value, whatever is the position of the point P $(x, y, z=1)$; taking this point at R, and writing for a moment

$$RA = \alpha, \quad RB = \beta, \quad RC = \gamma, \quad RD = \delta,$$

then

$$BCD = (RCD - RBD) = \tfrac{1}{2}RD\,(RC - RB)\sin R = (\gamma - \beta)\,\delta \sin R,$$

with similar expressions for the other triangles; and we thus have

$$\mathrm{a}\mathfrak{A} + \mathrm{b}\mathfrak{B} + \mathrm{c}\mathfrak{C} + \mathrm{d}\mathfrak{D} = \tfrac{1}{2}z^2 . \sin R \left\{\begin{matrix} \alpha^2(\gamma-\beta)\,\delta \\ -\beta^2(\delta-\alpha)\,\gamma \\ +\gamma^2(\delta-\alpha)\,\beta \\ -\delta^2(\gamma-\beta)\,\alpha \end{matrix}\right\} = \tfrac{1}{2}z^2 \sin R\,(\beta\gamma - \alpha\delta)(\gamma-\beta)(\delta-\alpha),$$

that is, replacing α, β, γ, δ, by their values, and writing also $z = 1$, we have

$$\mathrm{a}\mathfrak{A} + \mathrm{b}\mathfrak{B} + \mathrm{c}\mathfrak{C} + \mathrm{d}\mathfrak{D} = \tfrac{1}{2}\sin R . (RB . RC - RA . RD)\, BC . AD,$$

where $\tfrac{1}{2}\sin R . BC . AD$ is in fact the area of the quadrilateral $ABCD$; we have thus

$$\begin{aligned} \mathrm{a}\mathfrak{A} + \mathrm{b}\mathfrak{B} + \mathrm{c}\mathfrak{C} + \mathrm{d}\mathfrak{D} &= (RB . RC - RA . RD)\,\square \\ &= (SC . SA - SB . SD)\,\square \\ &= (TA . TB - TC . TD)\,\square \end{aligned}$$

where it is to be observed that SA, SC being measured in opposite directions from S, must be considered, one as positive, the other as negative, and the like as regards SB, SD. This expression for the value of the constant is due to Mr Crofton. In the particular case where A, B, C, D, are on a circle, we have as before

$$\mathrm{a}\mathfrak{A} + \mathrm{b}\mathfrak{B} + \mathrm{c}\mathfrak{C} + \mathrm{d}\mathfrak{D} = 0.$$

89. If the four points A, B, C, D, are on a circle, then, taking as origin the centre of this circle and its radius as unity, the circular coordinates of the four points will be

$$\left(\alpha, \frac{1}{\alpha}, 1\right), \quad \left(\beta, \frac{1}{\beta}, 1\right), \quad \left(\gamma, \frac{1}{\gamma}, 1\right), \quad \left(\delta, \frac{1}{\delta}, 1\right),$$

the corresponding forms of $\mathfrak{A}^\circ$, &c., being

$$\mathfrak{A}^\circ = (\xi - \alpha z)\left(\eta - \frac{1}{\alpha}z\right) - a''^2 z^2, \text{ \&c.}$$

the expressions for a, b, c, d, observing that we have

$$\begin{vmatrix} \beta, & \beta^{-1}, & 1 \\ \gamma, & \gamma^{-1}, & 1 \\ \delta, & \delta^{-1}, & 1 \end{vmatrix} = \frac{1}{\beta\gamma\delta}\begin{vmatrix} 1, & \beta, & \beta^2 \\ 1, & \gamma, & \gamma^2 \\ 1, & \delta, & \delta^2 \end{vmatrix} = \frac{1}{\beta\gamma\delta}(\beta\gamma\delta), \text{ \&c.}$$

if $(\beta\gamma\delta)$, &c. denote $(\beta-\gamma)(\gamma-\delta)(\delta-\beta)$, &c., become

$$\mathrm{a} : \mathrm{b} : \mathrm{c} : \mathrm{d} = \alpha(\beta\gamma\delta) : -\beta(\gamma\delta\alpha) : \gamma(\delta\alpha\beta) : -\delta(\alpha\beta\gamma),$$

which are convenient formulæ for the case in question.

90. If the points A, B, C, D, are on a line, then taking this line for the axis of x, we may write $\mathfrak{A}^\circ = (x - az)^2 + y^2 - a''^2 z^2$, &c. It is to be remarked here that we can, without any relation whatever between the radii of the circles, satisfy the equation

$$\mathrm{a}\mathfrak{A}^\circ + \mathrm{b}\mathfrak{B}^\circ + \mathrm{c}\mathfrak{C}^\circ + \mathrm{d}\mathfrak{D}^\circ = 0;$$

in fact this will be the case if we have

$$\begin{aligned} &\mathrm{a} + \mathrm{b} + \mathrm{c} + \mathrm{d} = 0,\\ &\mathrm{a}a + \mathrm{b}b + \mathrm{c}c + \mathrm{d}d = 0,\\ &\mathrm{a}(a^2 - a''^2) + \mathrm{b}(b^2 - b''^2) + \mathrm{c}(c^2 - c''^2) + \mathrm{d}(d^2 - d''^2) = 0, \end{aligned}$$

equations which determine the ratios a : b : c : d. In the case where the circles reduce themselves to the points A, B, C, D, these equations become

$$\begin{aligned} &\mathrm{a} + \mathrm{b} + \mathrm{c} + \mathrm{d} = 0,\\ &\mathrm{a}a + \mathrm{b}b + \mathrm{c}c + \mathrm{d}d = 0,\\ &\mathrm{a}a^2 + \mathrm{b}b^2 + \mathrm{c}c^2 + \mathrm{d}d^2 = 0, \end{aligned}$$

giving

$$\mathrm{a} : \mathrm{b} : \mathrm{c} : \mathrm{d} = (bcd) : -(cda) : (dab) : -(abc);$$

if for shortness (bcd), &c. stand for $(b-c)(c-d)(d-b)$, &c.; and for these values, we have

$$\mathrm{a}\mathfrak{A} + \mathrm{b}\mathfrak{B} + \mathrm{c}\mathfrak{C} + \mathrm{d}\mathfrak{D} = 0.$$

91. A very noticeable case is when the four circles are such that the foregoing values of (a, b, c, d) also satisfy the equation

$$\mathrm{a}\mathfrak{A}^\circ + \mathrm{b}\mathfrak{B}^\circ + \mathrm{c}\mathfrak{C}^\circ + \mathrm{d}\mathfrak{D}^\circ = 0;$$

the condition for this is obviously

$$\mathrm{a}a''^2 + \mathrm{b}b''^2 + \mathrm{c}c''^2 + \mathrm{d}d''^2 = 0;$$

or, as it may also be written,

$$\frac{a''^2}{(a-b)(a-c)(a-d)} + \frac{b''^2}{(b-c)(b-d)(b-a)} + \frac{c''^2}{(c-d)(c-a)(c-b)} + \frac{d''^2}{(d-a)(d-b)(d-c)} = 0.$$

Article No. 92. *On a Locus connected with the foregoing Properties.*

92. If, as above, A, B, C, D are any four points, and $\mathfrak{A}$, $\mathfrak{B}$, $\mathfrak{C}$, $\mathfrak{D}$ are the squared distances of a current point P from the four points respectively, then the locus of the foci of the conics which pass through the four points is the tetrazomal curve

$$\mathrm{a}\sqrt{\mathfrak{A}} + \mathrm{b}\sqrt{\mathfrak{B}} + \mathrm{c}\sqrt{\mathfrak{C}} + \mathrm{d}\sqrt{\mathfrak{D}} = 0.$$

In fact the sum $\mathrm{a}\mathfrak{A} + \mathrm{b}\mathfrak{B} + \mathrm{c}\mathfrak{C} + \mathrm{d}\mathfrak{D}$ has, it has been seen, a constant value for all positions of the point P; taking P to be the other focus, its squared distances are $(k - \sqrt{A})^2$, &c., whence for the first-mentioned focus we have

$$\mathrm{a}\mathfrak{A} + \mathrm{b}\mathfrak{B} + \mathrm{c}\mathfrak{C} + \mathrm{d}\mathfrak{D} = \mathrm{a}(k - \sqrt{\mathfrak{A}})^2 + \mathrm{b}(k - \sqrt{\mathfrak{B}})^2 + \mathrm{c}(k - \sqrt{\mathfrak{C}})^2 + \mathrm{d}(k - \sqrt{\mathfrak{D}})^2;$$

or recollecting that $\mathrm{a}+\mathrm{b}+\mathrm{c}+\mathrm{d}=0$, it follows that we have for the locus in question $\mathrm{a}\sqrt{\mathfrak{A}}+\mathrm{b}\sqrt{\mathfrak{B}}+\mathrm{c}\sqrt{\mathfrak{C}}+\mathrm{d}\sqrt{\mathfrak{D}}=0$; this locus will be discussed in the sequel. I remark here, that in the case where the four points are on a circle, then (as mentioned above), the axes of the several conics are in the same fixed directions; there are thus two sets of foci, those on the axis in one direction, and those on the axis in the other direction; it might therefore be anticipated, and it will appear, that in this case the tetrazomal breaks up into two trizomal curves.

Article Nos. 93 to 98. *Formulæ as to the two Sets (A, B, C, D), and (A_1, B_1, C_1, D_1), each of four Concyclic Points.*

93. Consider the four points A, B, C, D on a circle, then taking, as before, their circular coordinates to be $(\alpha, \alpha', 1)$, $(\beta, \beta', 1)$, $(\gamma, \gamma', 1)$, $(\delta, \delta', 1)$, the condition that the points may be on a circle is

$$\begin{vmatrix} 1, & \alpha, & \alpha', & \alpha\alpha' \\ 1, & \beta, & \beta', & \beta\beta' \\ 1, & \gamma, & \gamma', & \gamma\gamma' \\ 1, & \delta, & \delta', & \delta\delta' \end{vmatrix} = 0,$$

viz., this equation may be written

$$\begin{aligned} &(\beta-\gamma)(\alpha-\delta) : (\gamma-\alpha)(\beta-\delta) : (\alpha-\beta)(\gamma-\delta) \\ =&(\beta'-\gamma')(\alpha'-\delta') : (\gamma'-\alpha')(\beta'-\delta') : (\alpha'-\beta')(\gamma'-\delta'); \end{aligned}$$

or if, for shortness, we take

$$\begin{array}{llll} a=\beta-\gamma, & f=\alpha-\delta, & a'=\beta'-\gamma', & f'=\alpha'-\delta', \\ b=\gamma-\alpha, & g=\beta-\delta, & b'=\gamma'-\alpha', & g'=\beta'-\delta', \\ c=\alpha-\beta, & h=\gamma-\delta, & c'=\alpha'-\beta', & h'=\gamma'-\delta', \end{array}$$

and consequently

$$\begin{array}{ll} af+bg+ch=0, & a'f'+b'g'+c'h'=0, \\ a \qquad\qquad =g-h, & a' \qquad\qquad =g'-h', \\ b \qquad\qquad =h-f, & b' \qquad\qquad =h'-f', \\ c \qquad\qquad =f-g, & c' \qquad\qquad =f'-g', \\ a+b+c=0, & a'+b'+c'=0, \end{array}$$

then the equation is

$$af : bg : ch = a'f' : b'g' : c'h'.$$

94. Let a, b, c, d, denote as before $(\mathrm{a} : \mathrm{b} : \mathrm{c} : \mathrm{d} = BCD : -CDA : DAB : -ABC)$, then we have

$$\mathrm{a} : \mathrm{b} : \mathrm{c} : \mathrm{d} = \begin{vmatrix} \beta, & \beta', & 1 \\ \gamma, & \gamma', & 1 \\ \delta, & \delta', & 1 \end{vmatrix} : - \begin{vmatrix} \gamma, & \gamma', & 1 \\ \delta, & \delta', & 1 \\ \alpha, & \alpha', & 1 \end{vmatrix} : \begin{vmatrix} \delta, & \delta', & 1 \\ \alpha, & \alpha', & 1 \\ \beta, & \beta', & 1 \end{vmatrix} : - \begin{vmatrix} \alpha, & \alpha', & 1 \\ \beta, & \beta', & 1 \\ \gamma, & \gamma', & 1 \end{vmatrix}$$

and we may write

$$\begin{array}{llll} \mathrm{a} = \quad . \quad , & ah' - a'h, & ag' - a'g, & gh' - g'h, \\ \mathrm{b} = bh' - b'h, & \quad . \quad , & bf' - b'f, & hf' - h'f, \\ \mathrm{c} = cg' - c'g, & cf' - c'f, & \quad . \quad , & fg' - f'g, \\ \mathrm{d} = cb' - c'b, & ac' - a'c, & ba' - b'a, & \quad . \quad , \end{array}$$

viz., the expressions in the same horizontal line are equal, and a, b, c, d are proportional to the expressions in the four lines respectively.

95. I say that we have

$$\frac{c'f'}{ah}\,\mathrm{a} = \frac{c'g'}{bh}\,\mathrm{b} = \frac{a'f'}{af}\,\mathrm{c} = \frac{f'g'}{ab}\,\mathrm{d},$$

viz., this will be the case if

$$\begin{aligned} bc'\mathrm{a} &= hg'\,\mathrm{d}, \\ ac'\mathrm{b} &= hf'\mathrm{d}, \\ a'b\mathrm{c} &= fg'\,\mathrm{d}, \end{aligned}$$

and selecting the convenient expressions for a, b, c, d, these equations become

$$\begin{aligned} bc'\,(gh' - g'h) &= g'h\,(cb' - c'b), \\ ac'\,(hf' - h'f) &= f'h\,(ac' - a'c), \\ a'b\,(fg' - f'g) &= fg'\,(ba' - b'a), \end{aligned}$$

viz., these equations are respectively $bgc'h' = b'g'ch$, $cha'f' = c'h'af$, $afb'g' = a'f'bg$, and are consequently satisfied. It thus appears that the equation

$$\frac{l}{\mathrm{a}} + \frac{m}{\mathrm{b}} + \frac{n}{\mathrm{c}} + \frac{p}{\mathrm{d}} = 0$$

is transformable into

$$\frac{c'f'}{ah}\,l + \frac{c'g'}{bh}\,m + \frac{a'f'}{af}\,n + \frac{f'g'}{ab}\,p = 0,$$

which is of course one of a system of similar forms.

96. Take (A_1, D_1) the antipoints of (A, D); (B_1, C_1) the antipoints of (B, C); or say that the circular coordinates of A_1, B_1, C_1, D_1 are $(\alpha, \delta', 1)$, $(\beta, \gamma', 1)$, $(\gamma, \beta', 1)$, $(\delta, \alpha', 1)$ respectively; the points A_1, B_1, C_1, D_1 are, as above mentioned, on a circle, the condition that this may be so being in fact

$$\begin{vmatrix} 1, & \alpha, & \delta', & \alpha\delta' \\ 1, & \beta, & \gamma', & \beta\gamma' \\ 1, & \gamma, & \beta', & \gamma\beta' \\ 1, & \delta, & \alpha', & \delta\alpha' \end{vmatrix} = 0,$$

equivalent to

$$af : bg : ch = a'f' : b'g' : c'h'.$$

97. Let $(\mathrm{a}_1, \mathrm{b}_1, \mathrm{c}_1, \mathrm{d}_1)$ be the corresponding quantities to $(\mathrm{a}, \mathrm{b}, \mathrm{c}, \mathrm{d})$, viz., $\mathrm{a}_1 : \mathrm{b}_1 : \mathrm{c}_1 : \mathrm{d}_1 = B_1C_1D_1 : -C_1D_1A_1 : D_1A_1B_1 : -A_1B_1C_1$; we have

$$\mathrm{a}_1 : \mathrm{b}_1 : \mathrm{c}_1 : \mathrm{d}_1 = \begin{vmatrix} \beta, & \gamma', & 1 \\ \gamma, & \beta', & 1 \\ \delta, & \alpha', & 1 \end{vmatrix} : - \begin{vmatrix} \gamma, & \beta', & 1 \\ \delta, & \alpha', & 1 \\ \alpha, & \delta', & 1 \end{vmatrix} : \begin{vmatrix} \delta, & \alpha', & 1 \\ \alpha, & \delta', & 1 \\ \beta, & \gamma', & 1 \end{vmatrix} : - \begin{vmatrix} \alpha, & \delta', & 1 \\ \beta, & \gamma', & 1 \\ \gamma, & \beta', & 1 \end{vmatrix}$$

giving rise to a similar set of forms

$$\begin{array}{llll} \mathrm{a}_1 = \quad . & , -ac' + ha', & a'g + b'a, & -c'g - b'h, \\ \mathrm{b}_1 = -c'b - g'h, & \quad . & , -f'b - g'f, & -f'h + c'f, \\ \mathrm{c}_1 = \ b'c + h'g, & -f'c + h'f, & \quad . & , \ f'g + g'f, \\ \mathrm{d}_1 = \ g'c + h'b, & -h'a + a'c, & -a'b - g'a, & \quad . \quad , \end{array}$$

and leading to

$$\frac{cf}{a'c'}\mathrm{a}_1 = -\frac{cg}{c'g'}\mathrm{b}_1 = \frac{af}{a'f'}\mathrm{c}_1 = -\frac{fg}{a'g'}\mathrm{d}_1,$$

so that the equation

$$\frac{l_1}{\mathrm{a}_1} + \frac{m_1}{\mathrm{b}_1} + \frac{n_1}{\mathrm{c}_1} + \frac{p_1}{\mathrm{d}_1} = 0,$$

is transformable into

$$\frac{cf}{a'c'}l_1 - \frac{cg}{c'g'}m_1 + \frac{af}{a'f'}n_1 - \frac{fg}{a'g'}p_1 = 0.$$

98. Let A, B, C, D, be, as above, points on a circle; (A_1, D_1) and (B_1, C_1) the antipoints of (A, D), (B, C) respectively. Write

$$\begin{array}{ll} \mathfrak{A} = (\xi - \alpha z)(\eta - \alpha' z), & \mathfrak{A}_1 = (\xi - \alpha z)(\eta - \delta' z), \\ \mathfrak{B} = (\xi - \beta z)(\eta - \beta' z), & \mathfrak{B}_1 = (\xi - \beta z)(\eta - \gamma' z), \\ \mathfrak{C} = (\xi - \gamma z)(\eta - \gamma' z), & \mathfrak{C}_1 = (\xi - \gamma z)(\eta - \beta' z), \\ \mathfrak{D} = (\xi - \delta z)(\eta - \delta' z), & \mathfrak{D}_1 = (\xi - \delta z)(\eta - \alpha' z); \end{array}$$

then we have identically

$$\begin{aligned} (\delta-\alpha)(\delta'-\alpha')\mathfrak{B} &= (\beta-\delta)(\beta'-\delta')\mathfrak{A} + (\beta-\alpha)(\beta'-\alpha')\mathfrak{D} - (\beta-\delta)(\beta'-\alpha')\mathfrak{A}_1 - (\beta-\alpha)(\beta'-\delta')\mathfrak{D}_1, \\ (\delta-\alpha)(\delta'-\alpha')\mathfrak{C} &= (\gamma-\delta)(\gamma'-\delta')\mathfrak{A} + (\gamma-\alpha)(\gamma'-\alpha')\mathfrak{D} - (\gamma-\delta)(\gamma'-\alpha')\mathfrak{A}_1 - (\gamma-\alpha)(\gamma'-\delta')\mathfrak{D}_1, \\ (\delta-\alpha)(\delta'-\alpha')\mathfrak{B}_1 &= (\beta-\delta)(\gamma'-\delta')\mathfrak{A} + (\beta-\alpha)(\gamma'-\alpha')\mathfrak{D} - (\beta-\delta)(\gamma'-\alpha')\mathfrak{A}_1 - (\beta-\alpha)(\gamma'-\delta')\mathfrak{D}_1, \\ (\delta-\alpha)(\delta'-\alpha')\mathfrak{C}_1 &= (\gamma-\delta)(\beta'-\delta')\mathfrak{A} + (\gamma-\alpha)(\beta'-\alpha')\mathfrak{D} - (\gamma-\delta)(\beta'-\alpha')\mathfrak{A}_1 - (\gamma-\alpha)(\beta'-\delta')\mathfrak{D}_1, \end{aligned}$$

or, in the foregoing notation,

$$\begin{aligned} ff'\mathfrak{B} &= gg'\mathfrak{A} + cc'\mathfrak{D} + gc'\mathfrak{A}_1 + cg'\mathfrak{D}_1, \\ ff'\mathfrak{C} &= hh'\mathfrak{A} + bb'\mathfrak{D} - hb'\mathfrak{A}_1 - bh'\mathfrak{D}_1, \\ ff'\mathfrak{B}_1 &= gh'\mathfrak{A} - cb'\mathfrak{D} - gb'\mathfrak{A}_1 + ch'\mathfrak{D}_1, \\ ff'\mathfrak{C}_1 &= hg'\mathfrak{A} - bc'\mathfrak{D} + hc'\mathfrak{A}_1 - bg'\mathfrak{D}_1. \end{aligned}$$

Article Nos. 99 to 104. *Further Properties in relation to the same Sets* (A, B, C, D) *and* (A_1, B_1, C_1, D_1).

99. It is to be shown that in virtue of these equations, and if moreover $\frac{l}{\mathrm{a}}+\frac{m}{\mathrm{b}}+\frac{n}{\mathrm{c}}+\frac{p}{\mathrm{d}}=0$, then it is possible to find l_1, m_1, n_1, p_1, such that we have identically

$$-l\mathfrak{A}+m\mathfrak{B}+n\mathfrak{C}-p\mathfrak{D}+l_1\mathfrak{A}_1-m_1\mathfrak{B}_1-n_1\mathfrak{C}_1+p_1\mathfrak{D}_1=0.$$

This equation will in fact be identically true if only

$$\begin{array}{llll} -ff'l+gg'm+hh'n & . & -gh'm_1-g'hn_1 & =0, \\ \quad cc'm+bb'n-ff'p & . & +cb'm_1+bc'n_1 & =0, \\ \quad gc'm-hb'n & +ff'l_1 & +gb'm_1-hc'n_1 & =0, \\ \quad cg'm-bh'n & . & +ch'm_1+bg'n_1+ff'p_1 & =0. \end{array}$$

From the first and second equations eliminating m_1 or n_1, the other of these quantities disappears of itself, and we thus obtain two equations which must be equivalent to a single one, viz., we have

$$bc'ff'l+c'g'afm+bha'f'n+g'hff'p=0,$$
$$b'cff'l+cga'f'm+b'h'afn+gh'ff'p=0;$$

which equations may also be written

$$\frac{c'f'}{ah}l+\frac{c'g'}{bh}m+\frac{a'f'}{af}n+\frac{f'g'}{ab}p=0,$$

$$\frac{cf}{a'h'}l+\frac{cg}{b'h'}m+\frac{af}{a'f'}n+\frac{fg}{a'b'}p=0;$$

and it thus appears that the equations are equivalent to each other, and to the assumed relation

$$\frac{l}{\mathrm{a}}+\frac{m}{\mathrm{b}}+\frac{n}{\mathrm{c}}+\frac{p}{\mathrm{d}}=0.$$

100. Similarly, from the third and fourth equations eliminating m or n, the other of these quantities disappears of itself, and we find

$$cg'ff'l_1-cga'f'm_1+afc'g'n_1-c'gff'p_1=0,$$
$$bh'ff'l_1-afb'h'm_1+bha'f'n_1-b'hff'p_1=0,$$

equations which may be written

$$\frac{cf}{a'c'}l-\frac{cg}{c'g'}m+\frac{af}{a'f'}n-\frac{fg}{g'a'}p=0,$$

$$\frac{f'h'}{ah}l-\frac{b'h'}{bg}m+\frac{a'f'}{af}n-\frac{b'g'}{ab}p=0,$$

where we see that the two equations are equivalent to each other and to the equation

$$\frac{l_1}{\mathrm{a}_1}+\frac{m_1}{\mathrm{b}_1}+\frac{n_1}{\mathrm{c}_1}+\frac{p_1}{\mathrm{d}_1}=0.$$

It thus appears that the quantities l_1, m_1, n_1, p_1, must satisfy this last equation. It is to be observed that the first and second equations being, as we have seen, equivalent to a single equation, either of the quantities m_1, n_1, may be assumed at pleasure, but the other is then determined; the third and fourth equations then give l_1, p_1; and the quantities l_1, m_1, n_1, p_1, so obtained, satisfy *identically* the equation $\frac{l_1}{\mathrm{a}_1}+\frac{m_1}{\mathrm{b}_1}+\frac{n_1}{\mathrm{c}_1}+\frac{p_1}{\mathrm{d}_1}=0$.

101. Now writing

$$\begin{aligned} ff'l_1 &= -g(c'm+b'm_1)+h(b'n+c'n_1),\\ ff'p_1 &= -c(g'm-h'm_1)+b(h'n-g'n_1), \end{aligned}$$

and

$$\begin{aligned} ff'p &= \quad c(c'm+b'm_1)+b(b'n+c'n_1),\\ ff'l &= \quad g(g'm-h'm_1)+h(h'n-g'n_1), \end{aligned}$$

we find

$$\begin{aligned} f^2f'^2(l_1p_1-lp) &= -(bg+ch)\left[(c'm+b'm_1)(h'n-g'n_1)+(g'm-h'm_1)(b'n+c'n_1)\right],\\ &= \quad (bg+ch)\;(b'g'+c'h')\;(m_1n_1-mn),\\ &= \quad aa'ff'(m_1n_1-mn), \end{aligned}$$

that is

$$ff'(l_1p_1-lp)=aa'(m_1n_1-mn)$$

viz., this equation is satisfied identically by the values of l_1, m_1, n_1, p_1 determined as above.

102. Hence if $m_1n_1=mn$, we have also $l_1p_1=lp$, and we can determine m_1, n_1, so that m_1n_1 shall $=mn$, viz., in the first or second of the four equations (these two being equivalent to each other, as already mentioned), writing $m_1=\theta n$, and therefore $n_1=\frac{1}{\theta}m$, we have

$$-ff'l+gg'm+hh'n-gh'n\theta-g'hm\frac{1}{\theta}=0,$$

$$cc'm+bb'n-ff'p+cb'n\theta+bc'm\frac{1}{\theta}=0,$$

which are, in fact, the same quadric equation in θ, viz., we have

$$\frac{-ff'l+gg'm+hh'n}{cc'm+bb'n-ff'p}=-\frac{gh'}{cb'}=-\frac{g'h}{bc'}\,.$$

The final result is that there are two sets of values of l_1, m_1, n_1, p_1, each satisfying the identity

$$-l\mathfrak{A}+m\mathfrak{B}+n\mathfrak{C}-p\mathfrak{D}+l_1\mathfrak{A}_1-m_1\mathfrak{B}_1-n_1\mathfrak{C}_1+p_1\mathfrak{D}_1=0,$$

and for each of which we have

$$\frac{l_1}{\mathrm{a}_1}+\frac{m_1}{\mathrm{b}_1}+\frac{n_1}{\mathrm{c}_1}+\frac{p_1}{\mathrm{d}_1}=0, \quad l_1p_1=lp, \ m_1n_1=mn.$$

103. Consider, in particular, the case where $p=0$; the relation

$$\frac{l}{\mathrm{a}}+\frac{m}{\mathrm{b}}+\frac{n}{\mathrm{c}}+\frac{p}{\mathrm{d}}=0,$$

here becomes

$$l=-\frac{ag'}{bf'}m-\frac{a'h}{c'f}n.$$

The equation in θ is

$$(cc'm+bb'n)\,\theta+cb'n\theta^2+bc'm=0,$$

viz., this is

$$(c\theta+c'm)(b'n\theta+b)=0,$$

giving

$$\theta=-\frac{b}{c}\ , \quad m_1=-\frac{bn}{c}\ , \quad n_1=-\frac{cm}{b},$$

or else

$$\theta=-\frac{c'm}{b'n}, \quad m_1=-\frac{c'm}{b'}, \quad n_1=-\frac{b'n}{c'}.$$

Since in the present case $l_1p_1=0$, we have either $l_1=0$, or else $p_1=0$, and as might be anticipated, the two values of θ correspond to these two cases respectively, viz., proceeding to find the values of l_1, p_1, the completed systems are

$$\theta=-\frac{b}{c}\ , \quad l_1=\frac{a}{bcf'}\left(cc'm-bb'n\right), \quad m_1=-\frac{bn}{c}\ , \quad n_1=-\frac{cm}{b}, \quad p_1=0,$$

$$\theta=-\frac{c'm}{b'n}, \quad l'_1=0 \qquad , \quad m'_1=-\frac{c'm}{b'}, \quad n'_1=-\frac{b'n}{c'}, \quad p'_1=\frac{a'}{b'c'f}\left(cc'm-bb'n\right),$$

so that for the first system we have

$$\frac{l_1}{\mathrm{a}_1}+\frac{m_1}{\mathrm{b}_1}+\frac{n_1}{\mathrm{c}_1}=0, \quad m_1n_1=mn, \quad -l\mathfrak{A}+m\mathfrak{B}+n\mathfrak{C}=-l_1\mathfrak{A}_1+m_1\mathfrak{B}_1+n_1\mathfrak{C}_1,$$

and for the second system

$$\frac{m'_1}{\mathrm{b}_1}+\frac{n'_1}{\mathrm{c}_1}+\frac{p'_1}{\mathrm{d}_1}=0, \quad m'_1n'_1=mn, \quad -l\mathfrak{A}+m\mathfrak{B}+n\mathfrak{C}=-p'_1\mathfrak{D}_1+m'_1\mathfrak{B}_1+n'_1\mathfrak{C}_1.$$

104. The whole of the foregoing investigation would have assumed a more simple form if the circular coordinates had been taken with reference to the centre of the circle $ABCD$ as origin, and the radius of this circle been put $=1$; we should then have $\alpha'=\frac{1}{\alpha}$, &c., and consequently

$$a'=-\frac{1}{\beta\gamma}a, \quad b'=-\frac{1}{\gamma\alpha}b, \quad c'=-\frac{1}{\alpha\beta}c, \quad f'=-\frac{1}{\alpha\delta}f, \quad g'=-\frac{1}{\beta\delta}g, \quad h'=-\frac{1}{\gamma\delta}h;$$

but the symmetrical relation of the circles $ABCD$ and $A_1B_1C_1D_1$ would not have been so clearly shown.

I will however give the investigation in this simplified form, for the identity $-l\mathfrak{A}+m\mathfrak{B}+n\mathfrak{C}=-l_1\mathfrak{A}+m_1\mathfrak{B}+n_1\mathfrak{C}$; viz., in this case we have

$$\frac{l}{\alpha}=-\frac{m}{\beta}\frac{(\beta-\gamma)(\beta-\delta)}{(\gamma-\alpha)(\alpha-\delta)}-\frac{n}{\gamma}\frac{(\beta-\gamma)(\gamma-\delta)}{(\alpha-\beta)(\alpha-\gamma)},$$

and the identity to be satisfied is

$$\begin{aligned}-l\ (\xi-\alpha z)\left(\eta-\frac{1}{\alpha}z\right) &= -l_1\ (\xi-\alpha z)\left(\eta-\frac{1}{\delta}z\right)\\ +m\,(\xi-\beta z)\left(\eta-\frac{1}{\beta}z\right) &\quad +m_1\,(\xi-\beta z)\left(\eta-\frac{1}{\gamma}z\right)\\ +n\ (\xi-\gamma z)\left(\eta-\frac{1}{\gamma}z\right) &\quad +n_1\ (\xi-\gamma z)\left(\eta-\frac{1}{\beta}z\right);\end{aligned}$$

writing $\xi=\alpha z$, $\eta=\frac{1}{\beta}z$, we find m_1, and writing $\xi=\alpha z$, $\eta=\frac{1}{\gamma}z$, we find n_1, and it is then easy to obtain the value of l_1, viz., the results are

$$\frac{l_1}{\delta}=\frac{m}{\beta}\frac{(\alpha-\beta)(\beta-\gamma)}{(\gamma-\alpha)(\alpha-\delta)}+\frac{n}{\gamma}\frac{(\beta-\gamma)(\gamma-\alpha)}{(\alpha-\beta)(\alpha-\delta)},\quad m_1=-n\frac{\gamma-\alpha}{\alpha-\beta},\quad n_1=-m\frac{\alpha-\beta}{\gamma-\alpha},$$

and therefore $m_1n_1=mn$; it may be added that we have

$$-\frac{l_1}{\delta}=\frac{\beta-\gamma}{\alpha-\delta}\left(\frac{m_1}{\gamma}+\frac{n_1}{\delta}\right),$$

viz., this is the form assumed by the equation $\frac{l_1}{\mathrm{a}_1}+\frac{m_1}{\mathrm{b}_1}+\frac{n_1}{\mathrm{c}_1}=0.$

PART III. (Nos. 105 to 157). ON THE THEORY OF FOCI.

Article Nos. 105 to 110. *Explanation of the General Theory.*

105. If from a focus of a conic we draw two tangents to the curve, these pass respectively through the two circular points at infinity, and we have thence the generalised definition of a focus as established by Plücker, viz., in any curve a focus is a point such that the lines joining it with the two circular points at infinity are respectively tangents to the curve; or, what is the same thing, if from each of the circular points at infinity, say from the points I, J, tangents are drawn to the curve, the intersections of each tangent from the one point with each tangent from the other point are the foci of the curve. A curve of the class n has thus in general n^2 foci. It is to be added that, as in the conic the line joining the points of contact of the two tangents from a focus is the directrix corresponding to that focus, so in general the line joining the points of contact of the tangents from the focus through the points I, J respectively is the directrix corresponding to the focus in question.

106. A circular point at infinity I or J, may be an ordinary or a singular point on the curve, and the tangent at this point then counts, or, in the case of a multiple point, the tangents at this point count a certain number of times, say q times, among the tangents which can be drawn to the curve from the point; the number of the remaining tangents is thus $=n-q$. In particular, if the circular point at infinity be an ordinary point, then the tangent counts twice, or we have $q=2$; if it be a node, each of the tangents counts twice, or $q=4$; if it be a cusp, the tangent counts three times, or $q=3$. Similarly, if the other circular point an infinity be an ordinary or a singular point on the curve, the tangent or tangents there count a certain number of times, say q' times, among the tangents to the curve from this point; the number of the remaining tangents is thus $=n-q'$. And if as usual we disregard the tangents at the two points I, J respectively, and attend only to the remaining tangents, the number of the foci is $=(n-q)(n-q')$.

107. Among the tangents from the point I or J there may be a tangent which, either from its being a multiple tangent (that is, a tangent having ordinary contact at two or more distinct points), or from being an osculating tangent at one or more points, counts a certain number of times, say r, among the tangents from the point in question. Similarly, if among the tangents from the other point J or I, there is a tangent which counts r' times, then the foci are made up as follows, viz. we have

Intersections of the two singular tangents counting as .	$r'r$ foci.
Intersections of the first singular tangent with each of the ordinary tangents from the other circular point at infinity, as	$(n-q'-r')r$ „
Do. for second singular tangent, ...	$(n-q-r)r'$ „
Intersections of the ordinary tangents	$(n-q-r)(n-q'-r')$ „
Giving together the	$(n-q)(n-q')$ foci:

and the like observation applies to the more general case where the tangents from each of the points I, J include more than one singular tangent.

108. There is yet another case to be considered; the line infinity may be an ordinary or a singular tangent to the curve: assuming that it counts s times among the tangents from either of the circular points at infinity, the numbers of the remaining tangents are $n-q-s$, $n-q'-s$ from the two points I, J respectively, and the number of foci is $=(n-q-s)(n-q'-s)$.

109. In the case of a real curve the two points I, J are related in the same manner to the curve, and we have therefore $q=q'$; the singular tangents (if any) from the two points respectively being the same as well in character as in number. Writing $n-q-s=n-q'-s$, $=p$, and not for the present attending to the case of singular tangents, I shall assume that the number of tangents to the curve from each of the two points is $=p$; the number of foci is thus $=p^2$; and to each focus there corresponds a directrix, viz., this is the line through the points of contact of the tangents from the focus to the two points I, J respectively.

110. Consider any two foci A, B not *in lineâ* with either of the points I, J, then joining these with the points I, J, and taking A_1, B_1 the intersections of AI, BJ and of AJ, BI (A_1, B_1 being therefore by a foregoing definition the antipoints of (A, B)), then A_1, B_1 are, it is clear, foci of the curve. We may out of the p^2 foci select, and that in $1.2..p$ different ways, a system of p foci such that no two of them lie *in lineâ* with either of the points I, J; and this being so, taking the antipoints of each of the $\frac{1}{2}p(p-1)$ pairs out of the p foci, we have, inclusively of the p foci, in all $p+2.\frac{1}{2}p(p-1)$, that is p^2 foci, the entire system of foci.

Article Nos. 111 to 117. *On the Foci of Conics.*

111. A conic is a curve of the class 2, and the number of foci is thus $=4$. Taking as foci any two points A, B, the remaining two foci will be the antipoints A_1, B_1. In order that a given point A may be a focus, the conic must touch the lines AI, AJ; similarly, in order that a given point B may be a focus, the conic must touch the lines BI, BJ; the equation of a conic having the given points A, B for foci contains therefore a single arbitrary parameter.

112. In the case, however, of the parabola the curve touches the line infinity; there is consequently from each of the points I, J only a single tangent to the curve, and consequently only one focus: the parabola having a given point A for its focus is a conic touching the line infinity and the lines AI, AJ, or say the three sides of the triangle AIJ; its equation contains therefore two arbitrary parameters.

113. Returning to the general conic, there are certain trizomal forms of the focal equation, not of any great interest, but which may be mentioned. Using circular coordinates, and taking $(\alpha, \alpha', 1)$ and $(\beta, \beta', 1)$ for the coordinates of the given foci A, B respectively, the conic touches the lines $\xi-\alpha z=0$, $\eta-\alpha' z=0$, $\xi-\beta z=0$, $\eta-\beta' z=0$; the equation of a conic touching the first three lines is

$$\sqrt{l(\xi-\alpha z)}+\sqrt{m(\xi-\beta z)}+\sqrt{n(\eta-\alpha' z)}=0,$$

where l, m, n are arbitrary, and it is easy to obtain, in order that the conic may touch the fourth line $\eta-\beta' z=0$, the condition

$$n=-\frac{\beta-\alpha}{\beta'-\alpha'}(m-l).$$

114. In fact, n having this value, the equation gives

$$l(\xi-\alpha z)+m(\xi-\beta z)+2\sqrt{lm(\xi-\alpha z)(\xi-\beta z)}=-\frac{\beta-\alpha}{\beta'-\alpha'}(m-l)\big(\eta-\beta' z+(\beta'-\alpha')z\big),$$

and taking over the term

$$\frac{\beta-\alpha}{\beta'-\alpha'}(m-l)(\beta'-\alpha')z, \quad =(\beta-\alpha)(m-l)z,$$

this gives

$$l(\xi-\beta z)+m(\xi-\alpha z)+2\sqrt{lm(\xi-\alpha z)(\xi-\beta z)}=-\frac{\beta-\alpha}{\beta'-\alpha'}(m-l)(\eta-\beta' z),$$

which puts in evidence the tangent $\eta-\beta' z$. It is easy to see that the equation may be written in any one of the four forms

$$\sqrt{l\,(\xi-\alpha z)}+\sqrt{m\,(\xi-\beta z)}+\sqrt{-\frac{\beta-\alpha}{\beta'-\alpha'}(m-l)(\eta-\alpha' z)}=0,$$

$$\sqrt{m\,(\xi-\alpha z)}+\sqrt{l\,(\xi-\beta z)}+\sqrt{-\frac{\beta-\alpha}{\beta'-\alpha'}(m-l)(\eta-\beta' z)}=0,$$

$$\sqrt{l\,(\eta-\alpha' z)}+\sqrt{m(\eta-\beta' z)}+\sqrt{-\frac{\beta'-\alpha'}{\beta-\alpha}(m-l)(\xi-\alpha z)}=0,$$

$$\sqrt{m\,(\eta-\alpha' z)}+\sqrt{l\,(\eta-\beta' z)}+\sqrt{-\frac{\beta'-\alpha'}{\beta-\alpha}(m-l)(\xi-\beta z)}=0,$$

viz., in forms containing any three of the four radicals $\sqrt{\xi-\alpha z}$, $\sqrt{\xi-\beta z}$, $\sqrt{\eta-\alpha' z}$, $\sqrt{\eta-\beta' z}$. The conic is thus expressed as a trizomal curve, the zomals being each a line, viz., they are any three out of the four focal tangents; the order of the curve, as deduced from the general expression $2^{\nu-2}r$, is $=2$; so that there is here no depression of order.

115. But the ordinary form of the focal equation is a more interesting one; viz., $\mathfrak{A}$, $\mathfrak{B}$ being as usual the squared distances of the current point from the two given foci respectively, say

$$\mathfrak{A}=(\xi-\alpha z)(\eta-\alpha' z),$$

$$\mathfrak{B}=(\xi-\beta z)(\eta-\beta' z),$$

then $2a$ being an arbitrary parameter, the equation is

$$2az+\sqrt{\mathfrak{A}}+\sqrt{\mathfrak{B}}=0,$$

viz., the equation is here that of a trizomal curve, the zomals being curves of the second order, that is, the zomals are $(z^2=0)$ the line infinity twice, and the line-pairs AI, AJ and BI, BJ respectively: the general expression $2^{\nu-2}r$ gives therefore the order $=4$; but in the present case there are two branches, viz., the branches

$$2az+\sqrt{\mathfrak{A}}-\sqrt{\mathfrak{B}}=0,\quad 2az-\sqrt{\mathfrak{A}}+\sqrt{\mathfrak{B}}=0,$$

each ideally containing $(z=0)$ the line infinity; the curve contains therefore $(z^2=0)$ the line infinity twice, and omitting this factor the order is $=2$, as it should be.

116. To express the equation by means of the other two foci A_1, B_1, writing the equation under the form

$$\mathfrak{A}+\mathfrak{B}+2\sqrt{\mathfrak{A}\mathfrak{B}}-4a^2z^2=0,$$

and then if $\mathfrak{A}_1$, $\mathfrak{B}_1$ are the squared distances of the current point from A_1, B_1 respectively, we have (*ante*, No. 65),

$$\mathfrak{A}\mathfrak{B}=\mathfrak{A}_1\mathfrak{B}_1,$$

$$\mathfrak{A}+\mathfrak{B}-\mathfrak{A}_1-\mathfrak{B}_1=kz^2,$$

where k is the squared distance of the foci A, B, $=4a^2e^2$ suppose: whence putting $a^2(1-e^2)=b^2$, the equation becomes

$$\mathfrak{A}_1+\mathfrak{B}_1+2\sqrt{\mathfrak{A}_1\mathfrak{B}_1}-4b^2z^2=0,$$

that is

$$\sqrt{\mathfrak{A}_1}+\sqrt{\mathfrak{B}_1}+2bz=0,$$

which is the required new form. It is hardly necessary to remark that the equation $2az+\sqrt{\mathfrak{A}}+\sqrt{\mathfrak{B}}=0$, putting therein $z=1$, and expressing $\mathfrak{A}$, $\mathfrak{B}$ in rectangular coordinates measured along the axes, is the ordinary focal equation $2a=\sqrt{(x-ae)^2+y^2}+\sqrt{(x+ae)^2+y^2}$.

117. I remark that the equation $2az+\sqrt{\mathfrak{A}}+\sqrt{\mathfrak{B}}=0$ gives rise to $4a^2z^2+\mathfrak{A}-\mathfrak{B}+4az\sqrt{\mathfrak{A}}=0$, but here $\mathfrak{A}-\mathfrak{B}=-4aexz$, so that the equation contains $z=0$, and omitting this it becomes $(az-ex)+\sqrt{\mathfrak{A}}=0$, a bizomal form, being a curve of the order $=2$, as it should be; this is in fact the ordinary equation in regard to a focus and its directrix.

Article Nos. 118 to 123. *Theorem of the Variable Zomal as applied to a Conic.*

118. The equation $2kz+\sqrt{\mathfrak{A}^\circ}+\sqrt{\mathfrak{B}^\circ}=0$ is in like manner that of a conic; in fact, this would be a curve of the order $=4$, but there are as before the two branches $2kz+\sqrt{\mathfrak{A}^\circ}-\sqrt{\mathfrak{B}^\circ}=0$, $2kz-\sqrt{\mathfrak{A}^\circ}+\sqrt{\mathfrak{B}^\circ}=0$, each ideally containing ($z=0$) the line infinity, and the order is thus reduced to be $=2$. Each of the circles $\mathfrak{A}^\circ=0$, $\mathfrak{B}^\circ=0$ is a circle having double contact with the conic (this of course implies that the centre of the circle is on an axis of the conic). We may if we please start from the form $2kz+\sqrt{\mathfrak{A}}+\sqrt{\mathfrak{B}}=0$, and then by means of the theorem of the variable zomal introduce into the equation one, two, or three such circles.

119. It is in this point of view that I will consider the question, viz., adapting the formula to the case of the ellipse, and starting from the form

$$2az+\sqrt{(x-aez)^2+y^2}+\sqrt{(x+aez)^2+y^2}=0,$$

the equation of the variable zomal or circle of double contact may be taken to be

$$\frac{4a^2z^2}{-2}+\frac{(x-aez)^2+y^2}{1-q}+\frac{(x+aez)^2+y^2}{1+q}=0,$$

where q is an arbitrary parameter; writing for greater simplicity $z=1$, and reducing, the equation is

$$(x-qae)^2+y^2=b^2(1-q^2).$$

120. If $q<1$, then writing $q=\sin\theta$, we obtain the ellipse

$$\frac{x^2}{a^2}+\frac{y^2}{b^2}=1,$$

as the envelope of the variable circle

$$(x-ae\sin\theta)^2+y^2=b^2\cos^2\theta,$$

viz., of a circle having its centre on the major axis at a distance $=ae\sin\theta$ from the centre, and its radius $=b\cos\theta$. (I notice, in passing, that this gives in practice a very convenient graphical construction of the ellipse.) It may be remarked that for $\theta=\pm\sin^{-1}e$, the circle becomes

$$\left(x\pm\left(a-\frac{b^2}{a}\right)\right)^2+y^2=\frac{b^4}{a^2},$$

viz., this is the circle of curvature at one or other of the extremities of the major axis; as θ passes from 0 to $\pm\sin^{-1}e$ we have a series of real circles, which, by their continued intersection, generate the ellipse; as θ increases from $\theta=\pm\sin^{-1}e$ to $\pm 90°$, the circles continue real, but the consecutive circles no longer intersect in any real point,—and ultimately for $\theta=\pm 90°$, the circles become evanescent at the two foci respectively.

121. In the case $q>1$, we have a real representation of

$$(x-qae)^2+y^2+b^2(q^2-1),$$

as the squared distance of the point (x, y) from a point $(X, 0, Z)$ out of the plane of the figure, viz., putting this $=(x-X)^2+y^2+Z^2$, we have

$$qae=X,\quad Z^2=b^2(q^2-1),$$

whence

$$Z^2=b^2\left(\frac{X^2}{a^2e^2}-1\right),$$

or, what is the same thing,

$$\frac{X^2}{a^2-b^2}-\frac{Z^2}{b^2}=1\,;$$

that is, the locus is the focal hyperbola, viz., a hyperbola in the plane of zx, having its vertices at the foci, and its foci at the vertices of the ellipse.

122. If instead of the form first considered, we start from the trizomal form

$$2bz+\sqrt{x^2+(y-aeiz)^2}+\sqrt{x^2+(y+aeiz)^2}=0,$$

then we have the zomal or circle of double contact under the form

$$x^2+(y-qaei)^2=a^2(1-q^2)\,;$$

or putting herein $q=-i\tan\phi$, this is

$$x^2+(y-ae\tan\phi)^2=a^2\sec^2\phi\,;$$

so that we have the ellipse as the envelope of a variable circle having its centre on the minor axis of the ellipse, distance from the centre $=ae\tan\phi$, and radius $=a\sec\phi$. This is, in fact, Gergonne's theorem, according to which the ellipse is the secondary caustic or orthogonal trajectory of rays issuing from a point and refracted at a right line into a rarer medium. It is to be remarked that for $\tan\phi=\pm\frac{ae}{b}$, the equation of the circle is

$$x^2+\left(y\pm\left(b-\frac{a^2}{b}\right)\right)^2=\frac{a^4}{b^2},$$

viz., this is the circle of curvature at one or other extremity of the minor axis; from $\phi = 0$ to $\phi = \pm \tan^{-1}\frac{ae}{b}$, the intersections of the consecutive circles are real, and give the entire real ellipse; from $\phi = \pm \tan^{-1}\frac{ae}{b}$ to $\phi = \pm 90°$, the circles are still real, but the intersections of consecutive circles are imaginary.

123. If in the equation of the generating circle we interchange x, y, a, b, the equation becomes

$$(x - aei \tan \phi)^2 + y^2 = b^2 \sec^2 \phi,$$

which is (as it should be) equivalent to the former equation

$$(x - ae \sin \theta)^2 + y^2 = b^2 \cos \theta,$$

the identity being established by means of the equation

$$\cos \theta = \frac{1}{\cos \phi}, \text{ and therefore } \sin \theta = i \tan \phi,\ \tan \theta = i \sin \phi,$$

which is Jacobi's imaginary transformation in the theory of Elliptic Functions.

Article Nos. 124 to 126. *Foci of the Circular Cubic and the Bicircular Quartic.*

124. For a cubic curve, the class is in general $= 6$, and the number of the foci is $= 36$. But a specially interesting case is that of a circular cubic, viz., a cubic passing through each of the circular points at infinity. Here, at each of the circular points at infinity, the tangent at this point reckons twice among the tangents to the curve from the point; the number of the remaining tangents is thus $= 4$, and the number of the foci is $= 16$. If from any two points whatever on the curve tangents be drawn to the curve, then the two pencils of tangents are, and that in four different ways, homologous to each other, viz., if the tangents of the first pencil are (1, 2, 3, 4), and those of the second pencil, taken in a proper order, are (1′, 2′, 3′, 4′), then we have (1, 2, 3, 4) homologous with each of the arrangements (1′, 2′, 3′, 4′), (2′, 1′, 4′, 3′), (3′, 4′, 1′, 2′), (4′, 3′, 2′, 1′). And in each case the intersections of the four corresponding tangents lie on a conic passing through the two given points on the curve (1).

1 It may be remarked that if the equation of the first pencil of lines be

$$(x - ay)(x - by)(x - cy)(x - dy) = 0,$$

and that of the second pencil

$$(z - aw)(z - bw)(z - cw)(z - dw) = 0,$$

then the equations of four conics are

$$xw - yz = 0,$$

$$(a + d - b - c)\, xz + (bc - ad)(xw + yz) + \big(ad\,(b + c) - bc\,(a + d)\big)\, yw = 0,$$

$$(b + d - c - a)\, xz + (ca - bd)(xw + yz) + \big(bd\,(c + a) - ca\,(b + d)\big)\, yw = 0,$$

$$(c + d - a - b)\, xz + (ab - cd)(xw + yz) + \big(cd\,(a + b) - ab\,(c + d)\big)\, yw = 0.$$

125. Hence taking the points on the curve to be the circular points at infinity, we have the sixteen foci lying in fours upon four different circles—that is, we have four tetrads of concyclic foci. Let any one of these tetrads be A, B, C, D, then if

$$\begin{array}{lll} \text{Antipoints of } (B, C)(A, D) & \text{are} & (B_1, C_1), \ (A_1, D_1), \\ \quad ,, \quad (C, A)(B, D) & ,, & (C_2, A_2), \ (B_2, D_2), \\ \quad ,, \quad (A, B)(C, D) & ,, & (A_3, B_3), \ (C_3, D_3), \end{array}$$

the four tetrads of concyclic foci are

$$\begin{array}{llll} A, & B, & C, & D; \\ A_1, & B_1, & C_1, & D_1; \\ A_2, & B_2, & C_2, & D_2; \\ A_3, & B_3, & C_3, & D_3. \end{array}$$

It is to be observed that if A, B, C, D are any four points on a circle, then if, as above, we pair these in any manner, and take the antipoints of each pair, the four antipoints lie on a circle, and thus the original system A, B, C, D, of four points on a circle, leads to the remaining three systems of four points on a circle. The theory is in fact that already discussed *ante*, No. 72 *et seq.*

126. The preceding theory applies without alteration to the bicircular quartic, viz., the quartic curve which has a node at each of the circular points at infinity. The class is here $=8$, but among the tangents from a node each of the two tangents at the node is to be reckoned twice, and the number of the remaining tangents is $=4$: the number of foci is $=16$. And, by the general theorem that in a binodal quartic the pencils of tangents from the two nodes respectively are homologous, the sixteen foci are related to each other precisely in the manner of the foci of the circular cubic. The latter is in fact a particular case of the former, viz., the bicircular quartic may break up into the line infinity, and a circular cubic.

Article Nos. 127 to 129. *Centre of the Circular Cubic, and Nodo-Foci, &c. of the Bicircular Quartic.*

127. The tangents at I, J have not been recognised as tangents from I, J, giving by their intersection a focus, but it is necessary in the theory to pay attention to the tangents in question. It is clear that these tangents are in fact asymptotes—viz., in the case of the circular cubic they are the two imaginary asymptotes of the curve, and in the case of a bicircular quartic, the two pairs of imaginary parallel asymptotes; but it is convenient to speak of them as the tangents at I, J.

128. In the case of a circular cubic, the tangents at I and J meet in a point which I call the centre of the curve, viz., this is the intersection of the two imaginary asymptotes.

129. In the case of a bicircular quartic, the two tangents at I and the two tangents at J meet in four points, which (although not recognising them as foci) I call the nodo-foci; these lie in pairs on two lines, diagonals of the quadrilateral formed by the four tangents (the third diagonal is of course the line IJ), which diagonals I call the "nodal axes;" and the point of intersection of the two nodal axes is the "centre" of the curve. The nodo-foci are four points, two of them real, the other two imaginary, viz., they are two pairs of antipoints, the lines through the two pairs respectively being, of course, the nodal axes; these are consequently real lines bisecting each other at right angles in the centre (with the relation $1 : i$ between the distances). The centre may also be defined as the intersection of the harmonic of IJ in regard to the tangents at I, and the harmonic of this same line in regard to the tangents at J. Speaking of the tangents as asymptotes, the nodo-foci are the angles of the rhombus formed by the two pairs of parallel asymptotes; the nodal axes are the diagonals of this rhombus, and the centre is the point of intersection of the two diagonals; as such it is also the intersection of the two lines drawn parallel to and midway between the lines forming each pair of parallel asymptotes.

Article No. 130. *Circular Cubic and Bicircular Quartic; the Axial or Symmetrical Case.*

130. In a circular cubic or bicircular quartic, the pencil of the tangents from I and that of the tangents through J, considered as corresponding to each other in some one of the four arrangements, may be such that the line IJ considered as belonging to the two pencils respectively shall correspond to itself, and when this is so, the four foci, A, B, C, D, which are the intersections of the corresponding tangents in question, will lie in a line (viz., the conic which exists in the general case will break up into a line-pair consisting of the line IJ and another line). The line in question may be called the focal axis; it will presently be shown that in the case of the circular cubic it passes through the centre, and that in the case of the bicircular quartic it not only passes through the centre, but coincides with one or other of the nodal axes, viz., with that passing through the real or the imaginary nodo-foci; that is, the curve may have on the focal axis two real or else two imaginary nodo-foci. The focal axis contains, as has been mentioned, four foci—the remaining twelve foci are situate symmetrically, six on each side of the focal axis, the arrangement of the sixteen foci being as mentioned *ante*, No. 81 *et seq.*; the focal axis is in fact an axis of symmetry of the curve, and if preferred it may be named the axis of symmetry, transverse axis, or simply the axis. And the curve (circular cubic, or bicircular quartic) is in this case a "symmetrical" or "axial" curve.

Article Nos. 131 to 140. *Circular Cubic and Bicircular Quartic: Singular Forms.*

131. The circular cubic may have a node or a cusp. If this were at one of the points I, J the curve would be imaginary, and I do not attend to the case; and for the same reason, for the bicircular quartic I do not attend to the case where *one* of

the points I, J is a cusp. There remain then for the circular cubic and for the bicircular quartic the cases where there is a node or a cusp at a real point of the curve; and for the bicircular quartic the case where each of the points I, J is a cusp—in general the curve has no other node or cusp, but it may besides have a node or cusp at a real point thereof.

132. I consider first the case of the bicircular quartic where each of the points I, J is a cusp. The curve is in this case of necessity symmetrical[1]—it is in fact a Cartesian; viz., the Cartesian may be taken by definition to be a quartic curve having a cusp at each of the circular points at infinity. But in this case, as distinguished from the general case of the bicircular quartic, there is an essential degeneration of all the focal properties, and it is necessary to explain what these become. The centre is evidently the intersection of the cuspidal tangents; the nodo-foci (so far as they can be said to exist) coalesce with the centre, and they do not in so coalescing determine any definite directions for the nodal axes; that is, there are no nodal axes, and the only theorem in regard to the focal axis or axis of symmetry is, that it passes through the centre. Of the four tangents through the point I, one has come to coincide with the line IJ; and similarly, of the four tangents through the point J one has come to coincide with the line JI: there remain only three tangents through I and three tangents through J, and these by their intersections determine nine foci—viz., three foci A, B, C on the axis, and besides (B_1, C_1) the antipoints of (B, C): (C_2, A_2) the antipoints of (C, A) and (A_3, B_3) the antipoints of (A, B).

133. The remaining seven foci have disappeared, viz., we may consider that one of them has gone off to infinity on the focal axis, and that three pairs of foci have come to coincide with the points I, J respectively. The circle O (as in the general case of a symmetrical quartic) has become a line, the focal axis; the circles R, S, T (contrary to what might at first sight appear) continue to be determinate circles, viz., these have their centres at A, B, C respectively, and pass through the points (B_1, C_1), (C_2, A_2), and (A_3, B_3) respectively, see *ante*, No. 83. But on each of these circles we have not more than two proper foci, and it is only on the axis as representing the circle O that we have three proper foci, the axial foci A, B, C: in regard hereto it is to be remarked that the equation of the curve can be expressed not only by means of these three foci in the form $\sqrt{l\mathfrak{A}} + \sqrt{m\mathfrak{B}} + \sqrt{n\mathfrak{C}} = 0$; but by means of any two of them in the form $\sqrt{l\mathfrak{A}} + \sqrt{m\mathfrak{B}} + K = 0$, where K is a constant, or, what is the same thing (z being introduced for homogeneity in the expressions of $\mathfrak{A}$ and $\mathfrak{B}$ respectively), in the form $\sqrt{l\mathfrak{A}} + \sqrt{m\mathfrak{B}} + Kz^2 = 0$.

134. Using for the moment the expression "twisted" as opposed to symmetrical—

[1] It will appear, *post* Nos. 161—164, that if starting with three given points as the foci of a bicircular quartic, we impose the condition that the nodes at I, J shall be each of them a cusp, then either the quartic will be the circle through the three points taken twice, in which case the assumed focal property of the given three points disappears altogether, or else the three points must be *in lineâ*, and thus the curve be symmetrical, that is, a Cartesian.

(viz., the curve is twisted when there is not any axis of symmetry, but the foci lie only on circles)—then the classification is

Circular Cubics, twisted,
" " symmetrical,
Bicircular Quartics, twisted,
" " symmetrical, {Ordinary, Bicuspidal = Cartesian,

and each of these kinds may be general, nodal, or cuspidal—viz., for the two last mentioned kinds there may be a node or a cusp at a real point of the curve.

135. In the case of a node, say the point N; first if the curve (circular cubic or bicircular quartic) be twisted—then of the four foci A, B, C, D we have two, suppose B and C, coinciding with N; and the sixteen foci are as follows, viz.

B, C, A, D are N, N, A, D;
B_1, C_1, A_1, D_1 " N, N, Antipoints of (A, D);
C_2, A_2, B_2, D_2 " Antipoints of (N, A), Antipoints of (N, D);
A_3, B_3, C_3, D_3 " Do. do.

viz., we have the points (A, D) each once, the node N four times, the antipoints of (A, D) once, and the antipoints of (N, A) and of (N, D), each pair twice. But properly there are only four foci, viz., the points A, D and their antipoints. The circle O subsists as in the general case, and so does the circle $R(BC, AD)$, viz., this has for centre the intersection of the line AD by the tangent at N to the circle O, and it passes through the point N, of course cutting the circle O at right angles: the circles S and T each reduce themselves each to the point N considered as an evanescent circle, or what is the same thing to the line-pair NI, NJ.

136. The case is nearly the same if the curve be symmetrical, but in the case of the bicircular quartic excluding the Cartesian: viz., we have on the axis the foci B, C coinciding at N, and the other two foci A, D; the sixteen foci are as above—and the circle R is determined by the proper construction as applied to the case in hand, viz., the centre R is the intersection of the axis by the radical axis of the point N (considered as an evanescent circle) and the circle on AD as diameter; that is $\overline{RN}^2 = RA \, . \, RD$. And the circles S and T reduce themselves each to the point N considered as an evanescent circle.

137. Next if we have a cusp, say the point K: first if the curve (circular cubic or bicircular quartic) be twisted—then of the four foci A, B, C, D, three, suppose A, B, C, coincide with K; and the sixteen foci are as follows, viz.,

B, C, A, D are K, K, K, D,
B_1, C_1, A_1, D_1 " K, K, Antipoints of (K, D),
C_2, A_2, B_2, D_2 " Do. do.
A_3, B_3, C_3, D_3 " Do. do.

viz., we have the point D once, the point K nine times, and the antipoints of K, D three times. But properly the point D is the only focus. The circle O is, it would appear, *any* circle through K, D, but possibly the particular circle which touches the cuspidal tangent may be a better representative of the circle O of the general case—the circles R, S, T reduce themselves each to the point K considered as an evanescent point.

138. The like is the case if the curve be symmetrical, but in the case of the bicircular quartic excluding the Cartesian; the circle O is here the axis, which is in fact the cuspidal tangent.

139. For the Cartesian, if there is a node N; then of the three foci A, B, C, two, suppose B and C, coincide with N; the nine foci are A once, N four times, and the antipoints of N, A twice: but properly the point A is the only focus. And if there be a cusp K; then all the three foci A, B, C coincide with K; and the nine foci are K nine times; but in fact there is no proper focus.

140. A circular cubic cannot have two nodes unless it break up into a line and circle; and similarly a bicircular quartic cannot have two nodes (exclusive of course of the points I, J) unless it break up into two circles; the last-mentioned case will be considered in the sequel in reference to the problem of tactions.

Article No. 141. *As to the Analytical Theory for the Circular Cubic and the Bicircular Quartic respectively.*

141. It may be remarked in regard to the analytical theory about to be given, that although the investigation is very similar for the circular cubic and for the bicircular quartic, yet the former cannot be deduced from the latter case. In fact if for the bicircular quartic, using a form somewhat more general than that which is ultimately adopted, we suppose that for the two nodes respectively ($\xi = 0$, $z = 0$) and ($\eta = 0$, $z = 0$), then if $l\xi + mz = 0$, $l'\xi + m'z = 0$, $n\eta + pz = 0$, $n'\eta + p'z = 0$ are the tangents at the two nodes respectively, the equation will be

$$(l\xi + mz)(l'\xi + m'z)(n\eta + pz)(n'\eta + p'z) + ez^2\xi\eta + z^3(a\xi + b\eta) + cz^4 = 0,$$

and if (in order to make this equation divisible by z, and the curve so to break up into the line $z = 0$ and a cubic) we write $l = 0$ or $n = 0$, then the curve will indeed break up as required, but we shall have, not the general cubic through the two points ($\xi = 0$, $z = 0$), ($\eta = 0$, $z = 0$), but in each case a nodal cubic, viz., if $l = 0$ there will be a node at the point ($\eta = 0$, $z = 0$), and if $n = 0$ a node at the point ($\xi = 0$, $z = 0$).

Article Nos. 142 to 144. *Analytical Theory for the Circular Cubic.*

142. I consider then the two cases separately; and first the circular cubic. The equation may be taken to be

$$\xi\eta(p\xi + q\eta) + ez\xi\eta + z^2(a\xi + b\eta + cz) = 0,$$

or, what is the same thing,

$$\xi\eta\,(p\xi+q\eta+ez)+z^2\,(a\xi+b\eta+cz)=0,$$

viz. (ξ, η, z) being any coordinates whatever, this is the general equation of a cubic passing through the points $(\xi=0, z=0)$, $(\eta=0, z=0)$, and at these points touched by the lines $\xi=0$, $\eta=0$ respectively. And if $(\xi, \eta, z=1)$ be circular coordinates, then we have the general equation of a circular cubic having the lines $\xi=0$, $\eta=0$ for its asymptotes, or say the point $\xi=0$, $\eta=0$ for its centre; the equation of the remaining asymptote is evidently $p\xi+q\eta+ez=0$; to make the curve real we must have (p, q) and (a, b) conjugate imaginaries, e and c real.

143. Taking in any case the points I, J to be the points $\xi=0$, $z=0$ and $\eta=0$, $z=0$ respectively, for the equation of a tangent from I write $p\xi=\theta z$; then we have

$$\theta\eta\,(\theta z+q\eta+ez)+z\,(a\theta z+bp\eta+cpz)=0,$$

that is

$$z^2\,(a\theta+cp)+\eta z\,(\theta^2+e\theta+bp)+\eta^2\,.\,q\theta=0,$$

and the line will be a tangent if only

$$(\theta^2+e\theta+bp)^2-4q\theta\,(a\theta+cp)=0,$$

that is, the four tangents from I are the lines $p\xi=\theta z$, where θ is any root of this equation; similarly the four tangents from J are the lines $q\eta=\phi z$, where ϕ is any root of the equation

$$(\phi^2+e\phi+aq)^2-4p\phi\,(b\phi+cq)=0.$$

Writing the two equations under the forms

$$\left(\begin{array}{ll}6, & \\ 3e, & \\ e^2+2bp-4aq, & \\ 3ebp & -6cpq, \\ 6b^2p^2, & \end{array}\right)\!(\theta,\ 1)^4=0, \qquad \left(\begin{array}{ll}6, & \\ 3e, & \\ e^2+2aq-4bp, & \\ 3eaq & -6cpq, \\ 6a^2q^2, & \end{array}\right)\!(\phi,\ 1)^4=0,$$

the equations have the same invariants; viz., for the first equation the invariants are easily found to be

$$I=\ \ 3\,(e^2-4bp-4aq)^2+72\,(ce-2ab)\,pq,$$

$$J=-\ (e^2-4bp-4aq)^3-36\,(ce-2ab)\,pq\,(e^2-4bp-4aq)-216\,c^2p^2q^2,$$

and then by symmetry the other equation has the same invariants. The absolute invariant $I^3\div J^2$ has therefore the same value in the two equations; that is, the equations are linearly transformable the one into the other, which is the before-mentioned theorem that the two pencils are homographic.

144. The two equations will be satisfied by $\theta=\phi$, if only $bp=aq$; that is, if $p=\frac{a}{k}$, $q=\frac{b}{k}$; putting for convenience $\frac{e}{k}$ in place of e, the equation of the curve is then

$$\xi\eta\,(a\xi+b\eta+ez)+kz^2\,(a\xi+b\eta+cz)=0.$$

In this case the pencils of tangents are $a\xi = k\theta z$, $b\eta = k\theta z$, where θ is determined by a quartic equation, or taking the corresponding lines (which by their intersections determine the foci A, B, C, D) to be $(a\xi = k\theta_1 z$, $b\eta = k\theta_1 z)$, &c., these four points lie in the line $a\xi - b\eta = 0$, which is a line through the centre of the curve, or point $\xi = 0$, $\eta = 0$: the formulæ just obtained belong therefore to the symmetrical case of the circular cubic. Passing to rectangular coordinates, writing $z = 1$, and taking $y = 0$ for the equation of the axis, it is easy to see that the equation may be written

$$(x^2 + y^2)(x - a) + k(x - b) = 0;$$

or, changing the origin and constants,

$$xy^2 + (x - a)(x - b)(x - c) = 0.$$

Article Nos. 145 to 149. *Analytical Theory for the Bicircular Quartic.*

145. The equation for the bicircular quartic may be taken to be

$$k(\xi^2 - \alpha^2 z^2)(\eta^2 - \beta^2 z^2) + ez^2\xi\eta + z^3(a\xi + b\eta) + cz^4 = 0,$$

viz. (ξ, η, z) being any coordinates whatever, this is the equation of a quartic curve having a node at each of the points $(\xi = 0, z = 0)$ and $(\eta = 0, z = 0)$: the equations of the two tangents at the one node are $\xi - \alpha z = 0$, $\xi + \alpha z = 0$; and those of the two tangents at the other node are $\eta - \beta z = 0$, $\eta + \beta z = 0$; $\xi = 0$ is thus the harmonic of the line $z = 0$ in regard to the tangents at $(\xi = 0, z = 0)$, and $\eta = 0$ is the harmonic of the same line $z = 0$ in regard to the tangents at $(\eta = 0, z = 0)$. If $(\xi, \eta, z = 1)$ be circular coordinates, then we have the general equation of the bicircular quartic having the lines $\xi + \alpha z = 0$, $\xi - \alpha z = 0$ for one pair, and the lines $\eta - \beta z = 0$, $\eta + \beta z = 0$ for the other pair of parallel asymptotes; and therefore the point $\xi = 0$, $\eta = 0$ for centre, and the lines $\beta\xi - \alpha\eta = 0$, $\beta\xi + \alpha\eta = 0$ for nodal axes. In order that the curve may be real we must have (α, β), (a, b) conjugate imaginaries, k, e, c real. The points $(\xi = 0, z = 0)$ and $(\eta = 0, z = 0)$ are as before the points I, J. If $\alpha = 0$, the node at I becomes a cusp, and so if $\beta = 0$, the node at J becomes a cusp; the form thus includes the case of a bicuspidal or Cartesian curve.

146. To find the tangents from I, writing in the equation of the curve $\xi = \theta\alpha z$ we have

$$k\alpha^2(\theta^2 - 1)(\eta^2 - \beta^2 z^2) + e\alpha\theta\eta z + z(a\alpha\theta z + b\eta) + cz^2 = 0;$$

that is

$$\begin{aligned} &\eta^2 \,.\, k\alpha^2(\theta^2 - 1), \\ &+ \eta z \,.\, e\alpha\theta + b, \\ &+ z^2 \,.\, - k\alpha^2\beta^2(\theta^2 - 1) + a\alpha\theta + c = 0, \end{aligned}$$

and the condition of tangency is

$$4k(\theta^2 - 1)\{k\alpha^2\beta^2(\theta^2 - 1) - a\alpha\theta - c\} + \left(e\theta + \frac{b}{\alpha}\right)^2 = 0;$$

viz., the tangents from I are $\xi = \theta\alpha z$, where θ is any root of this equation. Similarly, if we have

$$4k(\phi^2 - 1)\{k\alpha^2\beta^2(\phi^2 - 1) - b\beta\phi - c\} + \left(e\phi + \frac{a}{\beta}\right)^2 = 0,$$

the tangents from J are $\eta = \phi\beta z$, where ϕ is any root of this equation.

147. The two equations may be written

$$\left(\begin{array}{l} 24k^2\alpha^2\beta^2, \\ -\ 6ka\alpha, \\ -\ 8k^2\alpha^2\beta^2 - \ 4kc + e^2, \\ \ \ 6a\alpha \qquad\qquad + 3e\dfrac{b}{\alpha}, \\ \ \ 24k^2\alpha^2\beta^2 + 24kc + 6\dfrac{b^2}{\alpha^2} \end{array}\right)(\theta,\ 1)^4 = 0, \qquad \left(\begin{array}{l} 24k^2\alpha^2\beta^2, \\ -\ 6kb\beta, \\ -\ 8k^2\alpha^2\beta^2 - \ 4kc + e^2, \\ \ \ 6kb\beta \qquad\qquad + 3e\dfrac{a}{\beta}, \\ \ \ 24k^2\alpha^2\beta^2 + 24kc + 6\dfrac{a^2}{\beta^2} \end{array}\right)(\phi,\ 1)^4 = 0,$$

which equations have the same invariants; in fact for the first equation the invariants are found to be as follows, viz., if for shortness $C = -8k^2\alpha^2\beta^2 - 4kc + e^2$, then

$$I = 576k^4\alpha^4\beta^4 + 576k^3c\alpha^2\beta^2 + 144k^2(a^2\alpha^2 + b^2\beta^2) + 72kab + 3C^2,$$

$$J = C\{576k^4\alpha^4\beta^4 + 576k^3c\alpha^2\beta^2 + 144k^2(a^2\alpha^2 + b^2\beta^2) + 36kea\beta - C^2\}$$
$$- 864k^3eab\alpha^2\beta^2 - 216k^2e^2(a^2\alpha^2 + b^2\beta^2) - 216k^2a^2b^2,$$

and then by symmetry the other equation has the same invariants. The absolute invariant $I^3 \div J^2$ has thus the same value in the two equations, that is, the equations are linearly transformable the one into the other, which is the before-mentioned theorem that the pencils are homographic.

148. The equations will be satisfied by $\theta = \phi$ if only $a\alpha = b\beta$, that is, if $a, b = m\beta, m\alpha$; or by $\theta = -\phi$ if only $a\alpha = -b\beta$, that is, if $a, b = m\beta, -m\alpha$: the equation of the curve is in these two cases respectively

$$k(\xi^2 - \alpha^2z^2)(\eta^2 - \beta^2z^2) + ez^2\xi\eta + mz^3(\beta\xi + \alpha\eta) + cz^4 = 0,$$

$$k(\xi^2 - \alpha^2z^2)(\eta^2 - \beta^2z^2) + ez^2\xi\eta + mz^3(\beta\xi - \alpha\eta) + cz^4 = 0.$$

If to fix the ideas we attend to the first case, then the equation in θ is

$$\left(\begin{array}{l} 24k^2\alpha^2\beta^2, \\ -\ 6km\alpha\beta, \\ -\ 8k^2\alpha^2\beta^2 - \ 4kc + e^2, \\ \ \ 6km\alpha\beta \qquad\qquad + 3me, \\ \ \ 24k^2\alpha^2\beta^2 + 24kc + 6m^2 \end{array}\right)(\theta,\ 1)^4 = 0;$$

and we may take as corresponding tangents through the two nodes respectively $\xi = \theta\alpha z$, $\eta = \theta\beta z$; the foci A, B, C, D, which are the intersections of the pairs of lines ($\xi = \theta_1\alpha z$,

$\eta = \theta_1 \beta z$), &c., lie, it is clear, in the line $\beta\xi - \alpha\eta = 0$, which is one of the nodal axes of the curve. Similarly, in the second case, if θ be determined by the foregoing equation, we may take as corresponding tangents through the two nodes respectively $\xi = \theta\alpha z$, $\eta = -\theta\beta z$; the foci (A, B, C, D), which are the intersections of the pairs of lines ($\xi = \theta_1\alpha z$, $\eta = -\theta_1\beta z$), &c., lie in the line $\beta\xi + \alpha\eta = 0$, which is the other of the nodal axes of the curve. In either case the foci A, B, C, D lie in a line, that is, we have the curve symmetrical; and, as we have just seen, the focal axis, or axis of symmetry, is one or other of the nodal axes.

149. In the case of the Cartesian, or when $\alpha = 0$, $\beta = 0$, viz., the equation $a\alpha = b\beta$ is satisfied identically, and this seems to show that the Cartesian is symmetrical; it is to be observed, however, that for $\alpha = 0$, $\beta = 0$ the foregoing formulæ fail, and it is proper to repeat the investigation for the special case in question. Writing $\alpha = 0$, $\beta = 0$, the equation of the curve is

$$k\xi^2\eta^2 + ez^2\xi\eta + z^3(a\xi + b\eta) + cz^4 = 0,$$

and then, taking $\xi = \theta bz$ for the equation of the tangent from I, we have

$$\begin{aligned}&\eta^2 \,.\, kb^2\theta^2\\ &+ \eta z \,.\, b(e\theta + 1)\\ &+ z^2 \,.\, ab\theta + c = 0,\end{aligned}$$

and the condition of tangency is

$$4k\theta^2(ab\theta + c) - (e\theta + 1)^2 = 0;$$

viz., we have here a cubic equation. Similarly, if we have $\eta = \theta az$ for the equation of a tangent from J, then

$$4k\phi^2(ab\phi + c) - (e\phi + 1)^2 = 0.$$

Hence θ being determined by the cubic equation as above, we may take $\phi = \theta$, and consequently the equations of the corresponding tangents will be $\xi = \theta bz$, $\eta = \theta az$, viz., the foci A, B, C will be given as the intersections of the pairs of lines ($\xi = \theta_1 bz$, $\eta = \theta_1 az$), &c. The foci lie therefore in the line $a\xi - b\eta = 0$; or the curve is symmetrical, the focal axis, or axis of symmetry, passing through the centre.

Article Nos. 150 to 158. *On the Property that the Points of Contact of the Tangents from a Pair of Concyclic Foci lie in a Circle.*

150. We have seen that the sixteen foci form four concyclic sets (A, B, C, D), (A_1, B_1, C_1, D_1), (A_2, B_2, C_2, D_2), (A_3, B_3, C_3, D_3), that is, A, B, C, D are in a circle. We may, if we please, say that any one focus is concyclic—viz., it lies in a circle with three other foci; but any two foci taken at random are not concyclic; it is only a pair such as (A, B) taken out of a set of four concyclic foci which are concyclic, viz., there exist two other foci lying with them in a circle. The number of such pairs is, it is clear $= 24$. Let A, B be any two concyclic foci, I say that the points of contact of the tangents AI, AJ, BI, BJ, lie in a circle.

151. Consider the case of the bicircular quartic, and take as before ($\xi=0,\ z=0$), and ($\eta=0,\ z=0$) for the coordinates of the points I, J respectively. Let the two tangents from the focus A be $\xi-\alpha z=0$, $\eta-\alpha' z=0$, say for shortness $p=0$, $p'=0$, then the equation of the curve is expressible in the form $pp'U=V^2$ [1], where $U=0$, $V=0$ are each of them a circle, viz., U and V are each of them a quadric function containing the terms z^2, $z\eta$, $z\xi$, and $\xi\eta$. Taking an indeterminate coefficient λ, the equation may be written

$$pp'(U+2\lambda V+\lambda^2 pp')=(V+\lambda pp')^2,$$

and then λ may be so determined that $U+2\lambda V+\lambda^2 pp'=0$, shall be a 0-circle, or pair of lines through I and J. It is easy to see that we have thus for λ a cubic equation, that is, there are three values of λ, for each of which the function $U+2\lambda V+\lambda^2 pp'$ assumes the form $(\xi-\beta z)(\eta-\beta' z)$, $=qq'$ suppose: taking any one of these, and changing the value of V so as that we may have V in place of $V+\lambda pp'$, the equation is $pp'qq'+V^2$, where $V=0$ is as before a circle, the equation shows that the points of contact of the tangents $p=0$, $p'=0$, $q=0$, $q'=0$ lie in this circle $V=0$. The circumstance that λ is determined by a cubic equation would suggest that the focus $q=0$, $q'=0$ is one of the three foci B, C, D concyclic with A; but this is the very thing which we wish to prove, and the investigation, though somewhat long, is an interesting one.

152. Starting from the form $pp'qq'=V^2$, then introducing as before an arbitrary coefficient λ, the equation may be written

$$pp'(qq'+2\lambda V+\lambda^2 pp')=(V+\lambda pp')^2,$$

and we may determine λ so that $qq'+2\lambda V+\lambda^2 pp'=0$ shall be a pair of lines. Writing $V=H\xi\eta-L\eta z-L'\xi z+Mz^2$, and substituting for pp' and qq' their values $(\xi-\alpha z)(\eta-\alpha' z)$ and $(\xi-\beta z)(\eta-\beta' z)$, the equation in question is

$$(1+2\lambda H+\lambda^2)\,\xi\eta-(\beta+2\lambda L+\lambda^2\alpha)\,\eta z-(\beta'+2\lambda L'+\lambda^2\alpha')\,\xi z+(\beta\beta'+2\lambda M+\lambda^2\alpha\alpha')\,z^2=0,$$

and the required condition is

$$(1+2\lambda H+\lambda^2)(\beta\beta'+2\lambda M+\lambda^2\alpha\alpha')=(\beta+2\lambda L+\lambda^2\alpha)(\beta'+2\lambda L'+\lambda^2\alpha');$$

or reducing, this is

$$\begin{aligned}&(2M+2H\beta\beta'-2L'\beta-2L\beta)\\ &+\lambda\left((\alpha-\beta)(\alpha'-\beta')+4HM-4LL'\right)\\ &+\lambda^2(2M+2H\alpha\alpha'-2L'\alpha-2L\alpha')=0,\end{aligned}$$

viz., λ is determined by a quadric equation. Calling its roots λ_1, and λ_2, the foregoing equation, substituting therein successively these values, becomes $(\xi-\gamma z)(\eta-\gamma' z)=0$, and $(\xi-\delta z)(\eta-\delta' z)=0$ respectively, say $rr'=0$ and $ss'=0$.

[1] This investigation is similar to that in Salmon's *Higher Plane Curves*, p. 196, in regard to the double tangents of a quartic curve.

153. We have to show that the four foci ($p=0$, $p'=0$), ($q=0$, $q'=0$), ($r=0$, $r'=0$), ($s=0$, $s'=0$) are a set of concyclic foci; that is, that the lines $p=0$, $q=0$, $r=0$, $s=0$ correspond homographically to the lines $p'=0$, $q'=0$, $r'=0$, $s'=0$; or, what is the same thing, that we have

$$\begin{vmatrix} 1, & \alpha, & \alpha', & \alpha\alpha' \\ 1, & \beta, & \beta', & \beta\beta' \\ 1, & \gamma, & \gamma', & \gamma\gamma' \\ 1, & \delta, & \delta', & \delta\delta' \end{vmatrix} = 0,$$

or, as it will be convenient to write this equation,

$$\frac{\alpha-\beta}{\alpha'-\beta'}\,\frac{\gamma-\delta}{\gamma'-\delta'}=\frac{\alpha-\delta}{\alpha'-\delta'}\,\frac{\beta-\gamma}{\beta'-\gamma'}.$$

154. We have

$$\gamma=\frac{\beta+2\lambda_1L+\lambda_1^2\alpha}{1+2H\lambda_2+\lambda_1^2},\quad \gamma'=\frac{\beta'+2\lambda_1L'+\lambda_1^2\alpha'}{1+2H\lambda_1+\lambda_1^2},$$

$$\delta=\frac{\beta+2\lambda_2L+\lambda_2^2\alpha}{1+2H\lambda_2+\lambda_2^2},\quad \delta'=\frac{\beta'+2\lambda_2L'+\lambda_2^2\alpha'}{1+2H\lambda_2+\lambda_2^2}.$$

The expressions of $\alpha-\delta$, &c., are severally fractions, the denominators of which disappear from the equation; the numerators are

$$\begin{aligned} \text{for } \alpha-\delta, &= \alpha(1+2\lambda_2H+\lambda_2^2)-(\beta+2\lambda_2L+\alpha\lambda_2^2), \\ &= \alpha-\beta+2\lambda_2(\alpha H-L); \\ \text{for } \beta-\gamma, &= \beta(1+2\lambda_1H+\lambda_1^2)-(\beta+2\lambda_1L+\alpha\lambda_1^2), \\ &= \lambda_1\{2(\beta H-L)(\alpha-\beta)\}; \\ \text{for } \gamma-\delta, &= (\beta+2L\lambda_1+\alpha\lambda_1^2)(1+2H\lambda_2+\lambda_2^2) \\ &\quad -(\beta+2L\lambda_2+\alpha\lambda_2^2)(1+2H\lambda_1+\lambda_1^2), \\ &= (\alpha'-\beta')\{2H^2\alpha\beta-2HL(\alpha+\beta)+2L^2+\tfrac{1}{2}(\alpha-\beta)^2\}; \end{aligned}$$

and it hence easily appears that the equation to be verified is

$$\frac{2H^2\alpha\beta-2HL(\alpha+\beta)+2L^2+\frac{1}{2}(\alpha-\beta)^2}{2H^2\alpha'\beta'-2HL'(\alpha'+\beta')+2L'^2+\frac{1}{2}(\alpha'-\beta')^2}=\frac{\alpha-\beta+2(\alpha H-L)\lambda_2}{\alpha'-\beta'+2(\alpha'H-L')\lambda_2}\cdot\frac{2(\beta H-L)-(\alpha-\beta)\lambda_1}{2(\beta'H-L')-(\alpha'-\beta')\lambda_2}.$$

155. This is

$$\frac{B-C}{B'-C'}=\frac{A+B\lambda_1+C\lambda_2+D\lambda_1\lambda_2}{A'+B'\lambda_1+C'\lambda_2+D'\lambda_1\lambda_2},$$

if for shortness

$$\begin{aligned} A &= 2(\alpha-\beta)(\beta H-L), & A' &= 2(\alpha'-\beta')(\beta'H-L'), \\ B &= -(\alpha-\beta)^2, & B' &= -(\alpha'-\beta')^2, \\ C &= 4(\alpha H-L)(\beta H-L), & C' &= 4(\alpha'H-L')(\beta'H-L'), \\ D &= -2(\alpha-\beta)(\alpha H-L), & D' &= -2(\alpha'-\beta')(\alpha'H-L'), \end{aligned}$$

and the equation then is

$$AB'-A'B+CA'-C'A-(\lambda_1+\lambda_2)(BC'-B'C)+\lambda_1\lambda_2\big(CD'-C'D-(BD'-B'D)\big).$$

156. Calculating $AB' - A'B$, $CA' - C'A$, $CD' - C'D$, $BD' - B'D$, these are at once seen to divide by $\{(\alpha\beta' - \alpha'\beta) H + L(\alpha' - \beta') - L'(\alpha - \beta)\}$; we have, moreover,

$$\begin{aligned} BC' - B'C &= -4(\alpha-\beta)^2(\alpha' H - L')(\beta' H - L') + 4(\alpha'-\beta')^2(\alpha H - L)(\beta H - L), \\ &= -\{(\alpha\alpha' - \beta\beta')H - L(\alpha'-\beta') - L'(\alpha-\beta)\}\{\alpha\beta' - \alpha'\beta)H + L(\alpha'-\beta') - L'(\alpha-\beta)\}, \end{aligned}$$

viz., this also contains the same factor; and omitting it, the equation is found to be

$$\begin{aligned} &\{(\alpha-\beta)(\alpha'-\beta') - 4(\beta H - L)(\beta' H - L')\} \\ -2&\{(\alpha\alpha' - \beta\beta')H - L(\alpha'-\beta') - L'(\alpha-\beta)\}(\lambda_1+\lambda_2) \\ +&\{-(\alpha-\beta)(\alpha'-\beta') + 4(\alpha H - L)(\alpha' H - L')\}\lambda_1\lambda_2 \quad = 0; \end{aligned}$$

viz., substituting for $\lambda_1 + \lambda_2$ and $\lambda_1\lambda_2$ their values, this is

$$\begin{aligned} &\{(\alpha-\beta)(\alpha'-\beta') - 4(\beta H - L)(\beta' H - L')\}(M + H\alpha\alpha' - L\alpha' - L'\alpha) \\ -&\{(\alpha\alpha' - \beta\beta')H - L(\alpha'-\beta')\}\{(\alpha-\beta)(\alpha'-\beta') + 4HM - 4LL'\} \\ +&\{-(\alpha-\beta)(\alpha'-\beta') + 4(\alpha H - L)(\alpha' H - L')\}\{M + H\beta\beta' - L\beta' - L'\beta\} = 0, \end{aligned}$$

which should be identically true. Multiplying by H, and writing in the form

$$\begin{aligned} &\{(\alpha-\beta)(\alpha'-\beta') - 4(\beta H - L)(\beta' H - L')\}(HM - LL' + (\alpha H - L)(\alpha' H - L')) \\ -&\{(\alpha H - L)(\alpha' H - L') - (\beta H - L)(\beta' H - L')\}((\alpha-\beta)(\alpha'-\beta') + 4(HM - LL')) \\ +&\{-(\alpha-\beta)(\alpha'-\beta') + 4(\alpha H - L)(\alpha' H - L')\}(HM - LL' + (\beta H - L)(\beta' H - L')) = 0, \end{aligned}$$

we at once see that this is so, and the theorem is thus proved, viz., that the equation being $pp'qq' = V^2$, the foci $(p=0,\ p'=0)$ and $(q=0,\ q'=0)$ are concyclic.

157. By what precedes, λ being a root of the foregoing quadric equation, we may write

$$qq' + 2\lambda V + \lambda^2 pp' = K^2 rr',$$

where the focus $r=0$, $r'=0$ is concyclic with the other two foci; but from the equation of the curve $V = \sqrt{pp'qq'}$, that is we have

$$qq' + 2\lambda\sqrt{pp'qq'} + \lambda^2 pp' = Krr',$$

or, what is the same thing,

$$\lambda\sqrt{pp'} + \sqrt{qq'} + K\sqrt{rr'} = 0,$$

viz., this is a form of the equation of the curve; substituting for p, p', q, q', r, r' their values, writing also

$$\begin{aligned} \mathfrak{A} &= (\xi - \alpha z)(\eta - \alpha' z), \\ \mathfrak{B} &= (\xi - \beta z)(\eta - \beta' z), \\ \mathfrak{C} &= (\xi - \gamma z)(\eta - \gamma' z), \end{aligned}$$

and changing the constants λ, K (viz. $\lambda : 1 : K = \sqrt{l} : \sqrt{m} : \sqrt{n}$) the equation is

$$\sqrt{l\mathfrak{A}} + \sqrt{m\mathfrak{B}} + \sqrt{n\mathfrak{C}} = 0,$$

viz., we have the theorem that for a bicircular quartic if ($\xi - \alpha z = 0$, $\eta - \alpha' z = 0$), ($\xi - \beta z = 0$, $\eta - \beta' z = 0$, ($\xi - \gamma z = 0$), $\eta - \gamma' z = 0$) be any three concyclic foci, then the equation is as just mentioned; that is, the curve is a trizomal curve, the zomals being the three given foci regarded as 0-circles. The same theorem holds in regard to the circular cubic, and a similar demonstration would apply to this case.

158. It may be noticed that we might, without proving as above that the two foci ($p = 0$, $p' = 0$), ($q = 0$, $q' = 0$) were concyclic, have passed at once from the form $pp'qq' = V^2$, to the form $\lambda\sqrt{pp'} + \sqrt{qq'} + K\sqrt{rr'} = 0$ (or $\sqrt{l\mathfrak{A}} = \sqrt{m\mathfrak{B}} = \sqrt{n\mathfrak{C}} = 0$), and then by the application of the theorem of the variable zomal (thereby establishing the existence of a fourth focus concyclic with the three) have shown that the original two foci were concyclic. But it seemed the more orderly course to effect the demonstration without the aid furnished by the reduction of the equation to the trizomal form.

PART IV. (Nos. 159 to 206). ON TRIZOMAL AND TETRAZOMAL CURVES WHERE THE ZOMALS ARE CIRCLES.

Article Nos. 159 to 165. *The Trizomal Curve—The Tangents at I, J, &c.*

159. I consider the trizomal

$$\sqrt{l\mathfrak{A}^\circ} + \sqrt{m\mathfrak{B}^\circ} + \sqrt{n\mathfrak{C}^\circ} = 0,$$

where A, B, C being the centres of three given circles, $\mathfrak{A}^\circ$, &c. denote as before, viz., in rectangular and in circular coordinates respectively, we have

$$\mathfrak{A}^\circ = (x - az)^2 + (y - a'z)^2 - a''^2z^2, \quad = (\xi - \alpha z)(\eta - \alpha' z) - a''^2z^2,$$
$$\mathfrak{B}^\circ = (x - bz)^2 + (y - b'z)^2 - b''^2z^2, \quad = (\xi - \beta z)(\eta - \beta' z) - b''^2z^2,$$
$$\mathfrak{C}^\circ = (x - cz)^2 + (y - c'z)^2 - c''^2z^2, \quad = (\xi - \gamma z)(\eta - \gamma' z) - c''^2z^2.$$

By what precedes, the curve is of the order $= 4$, touching each of the given circles twice, and having a double point, or node, at each of the points I, J; that is, it is a bicircular quartic: but if for any determinate values of the radicals $\sqrt{l}$, $\sqrt{m}$, $\sqrt{n}$, we have

$$\sqrt{l} + \sqrt{m} + \sqrt{n} = 0,$$

then there is a branch

$$\sqrt{l\mathfrak{A}^\circ} + \sqrt{m\mathfrak{B}^\circ} + \sqrt{n\mathfrak{C}^\circ} = 0,$$

containing ($z = 0$) the line infinity; and the order is here $= 3$: viz., the curve here passes through each of the points I, J and through another point at infinity (that is, there is an asymptote), and is thus a circular cubic.

160. I commence by investigating the equations of the nodal tangents at the points I, J respectively; using for this purpose the circular coordinates (ξ, η, $z = 1$), it is to be observed that, in the rationalised equation, for finding the tangents at ($\xi = 0$, $z = 0$) we have only to attend to the terms of the second order in (ξ, z), and

similarly for finding the tangents at ($\eta=0,\ z=0$) we have only to attend to the terms of the second order in (η, z). But it is easy to see that any term involving a'', b'', or c'' will be of the third order at least in (ξ, z), and similarly of the third order at least in (η, z); hence for finding the tangents we may reject the terms in question, or, what is the same thing, we may write a'', b'', c'' each $=0$, thus reducing the three circles to their respective centres. The equation thus becomes

$$\sqrt{l(\xi-\alpha z)(\eta-\alpha' z)}+\sqrt{m(\xi-\beta z)(\eta-\beta' z)}+\sqrt{n(\xi-\gamma z)(\eta-\gamma' z)}=0.$$

For finding the tangents at ($\xi=0,\ z=0$) we have in the rationalised equation to attend only to the terms of the second order in (ξ, z); and it is easy to see that any term involving α', β', γ' will be of the third order at least in (ξ, z), that is, we may reduce α', β', γ' each to zero; the irrational equation then becomes divisible by $\sqrt{\eta}$, and throwing out this factor, it is

$$\sqrt{l(\xi-\alpha z)}+\sqrt{m(\xi-\beta z)}+\sqrt{n(\xi-\gamma z)}=0,$$

viz., this equation which evidently belongs to a pair of lines passing through the point ($\xi=0,\ z=0$) gives the tangents at the point in question; and similarly the tangents at the point ($\eta=0,\ z=0$) are given by the equation

$$\sqrt{l(\eta-\alpha' z)}+\sqrt{m(\eta-\beta' z)}+\sqrt{n(\eta-\gamma' z)}=0.$$

161. To complete the solution, attending to the tangents at ($\xi=0,\ z=0$), and putting for shortness

$$\begin{aligned}\lambda &= \quad l-m-n,\\ \mu &= -l+m-n,\\ \nu &= -l-m+n,\\ \Delta &= \quad l^2+m^2+n^2-2mn-2nl-2lm,\end{aligned}$$

the rationalised equation is easily found to be

$$\begin{aligned}&\xi^2 . \Delta\\ &-2\xi z\,(l\lambda\alpha+m\mu\beta+n\nu\gamma)\\ &+\ z^2\,(l^2\alpha^2+m^2\beta^2+n^2\gamma^2-2mn\beta\gamma-2nl\gamma\alpha-2lm\alpha\beta)=0\,;\end{aligned}$$

and it is to be noticed that in the case of the circular cubic or when $\sqrt{l}+\sqrt{m}+\sqrt{n}=0$, then $\Delta=0$, so that the equation contains the factor z, and throwing this out, the equation gives a single line, which is in fact the tangent of the circular cubic.

162. Returning to the bicircular quartic, we may seek for the condition in order that the node may be a cusp: the required condition is obviously

$$\Delta\,(l^2\alpha^2+m^2\beta^2+n^2\gamma^2-2mn\beta\gamma-2nl\gamma\alpha-2lm\alpha\beta)-(l\lambda\alpha+m\mu\beta+n\nu\gamma)^2=0,$$

or observing that

$$\begin{aligned}\Delta-\lambda^2 &= -4mn, \ \&c.\\ \Delta+\mu\nu &= -2l\lambda, \ \&c.\end{aligned}$$

this is

$$l\alpha^2+m\beta^2+n\gamma^2+\lambda\beta\gamma+\mu\gamma\alpha+\nu\alpha\beta=0,$$

or substituting for λ, μ, ν, their values, it is

$$l(\alpha-\beta)(\alpha-\gamma)+m(\beta-\gamma)(\beta-\alpha)+n(\gamma-\alpha)(\gamma-\beta)=0,$$

or, as it is more simply written,

$$\frac{l}{\beta-\gamma}+\frac{m}{\gamma-\alpha}+\frac{n}{\alpha-\beta}=0.$$

163. If the node at $(\eta=0, z=0)$ be also a cusp, then we have in like manner

$$\frac{l}{\beta'-\gamma'}+\frac{m}{\gamma'-\alpha'}+\frac{n}{\alpha'-\beta'}=0.$$

Now observing that

$$(\gamma-\alpha)(\alpha'-\beta')-(\gamma'-\alpha')(\alpha-\beta),\ =\begin{vmatrix}\alpha, & \alpha', & 1\\ \beta, & \beta', & 1\\ \gamma, & \gamma', & 1\end{vmatrix}$$

$$=(\alpha-\beta)(\beta'-\gamma')-(\alpha'-\beta')(\beta-\gamma),$$
$$=(\beta-\gamma)(\gamma'-\alpha')-(\beta'-\gamma')(\gamma-\alpha),$$

$=\Omega$ suppose: the two equations give

$$l:m:n=\Omega(\beta-\gamma)(\beta'-\gamma'):\Omega(\gamma-\alpha)(\gamma'-\alpha'):\Omega(\alpha-\beta)(\alpha'-\beta');$$

or if Ω is not $=0$, then

$$l:m:n=\ (\beta-\gamma)(\beta'-\gamma'):\ (\gamma-\alpha)(\gamma'-\alpha'):\ (\alpha-\beta)(\alpha'-\beta').$$

164. If

$$\Omega=\begin{vmatrix}\alpha, & \alpha', & 1\\ \beta, & \beta', & 1\\ \gamma, & \gamma', & 1\end{vmatrix},=0,$$

or, what is the same thing, if

$$\begin{vmatrix}a, & a', & 1\\ b, & b', & 1\\ c, & c', & 1\end{vmatrix}=0,$$

the centres A, B, C are in a line; taking it as the axis of x, we have $\alpha=\alpha'=a$, $\beta=\beta'=b$, $\gamma=\gamma'=c$; and the conditions for the cusps at I, J respectively reduce themselves to the single condition

$$\frac{l}{b-c}+\frac{m}{c-a}+\frac{n}{a-b}=0,$$

so that this condition being satisfied, the curve

$$\sqrt{l\{(x-az)^2+y^2-a''^2z^2\}}+\sqrt{m\{(x-bz)^2+y^2-b''^2z^2\}}+\sqrt{n\{(x-cz)^2+y^2-c''^2z^2\}}=0$$

is a Cartesian; viz., given any three circles with their centres on a line, there are a singly infinite series of Cartesians, each touched by the three circles respectively;

the line of centres is the axis of the curve, but the centres A, B, C are not the foci, except in the case $a''=0$, $b''=0$, $c''=0$, where the circles vanish. The condition for l, m, n is satisfied if $l : m : n=(b-c)^2 : (c-a)^2 : (a-b)^2$; these values, writing $\sqrt{l} : \sqrt{m} : \sqrt{n} = b-c : c-a : a-b$, give not only $\sqrt{l}+\sqrt{m}+\sqrt{n}=0$, but also $a\sqrt{l}+b\sqrt{m}+c\sqrt{n}=0$; these are the conditions for a branch containing $(z^2=0)$ the line infinity twice; the equation

$$(b-c)\sqrt{(x-az)^2+y^2-a''^2z^2}+(c-a)\sqrt{(x-bz)^2+y^2-b''^2z^2}+(a-b)\sqrt{(x-cz)^2+y^2-c''^2z^2}=0,$$

is thus that of a conic, and if $a''=0$, $b''=0$, $c''=0$, then the curve reduces itself to $y^2=0$, the axis twice.

165. If Ω is not $=0$, then we have

$$l : m : n=(\beta-\gamma)(\beta'-\gamma') : (\gamma-\alpha)(\gamma'-\alpha') : (\alpha-\beta)(\alpha'-\beta'),$$

viz., l, m, n are as the squared distances $\overline{BC}^2$, $\overline{CA}^2$, $\overline{AB}^2$, say as $f^2 : g^2 : h^2$; or when the centres of the given circles A, B, C are not in a line, then f, g, h being the distances BC, CA, AB of these centres from each other, we have, touching each of the given circles twice, the *single* Cartesian

$$f\sqrt{\mathfrak{A}^\circ}+g\sqrt{\mathfrak{B}^\circ}+h\sqrt{\mathfrak{C}^\circ}=0,$$

which, in the particular case where the radii a'', b'', c'' are each $=0$, becomes

$$f\sqrt{\mathfrak{A}}+g\sqrt{\mathfrak{B}}+h\sqrt{\mathfrak{C}}=0,$$

viz., this is the circle through the points A, B, C, say the circle ABC, twice.

Article Nos. 166 to 169. *Investigation of the Foci of a Conic represented by an Equation in Areal Coordinates.*

166. I premise as follows: Let A, B, C be any given points, and in regard to the triangle ABC let the *areal* coordinates of a current point P be u, v, w; that is, writing PBC, &c., for the areas of these triangles, take the coordinates to be

$$u : v : w=PBC : PCA : PAB,$$

or, what is the same thing in the rectangular coordinates $(x, y, z=1)$, if

$$(a, a', 1), (b, b', 1), (c, c', 1),$$

be the coordinates of A, B, C respectively, take

$$u : v : w=\begin{vmatrix} x, & y, & z \\ b, & b', & 1 \\ c, & c', & 1 \end{vmatrix} : \begin{vmatrix} x, & y, & z \\ c, & c', & 1 \\ a, & a', & 1 \end{vmatrix} : \begin{vmatrix} x, & y, & z \\ a, & a', & 1 \\ b, & b', & 1 \end{vmatrix},$$

or in the circular coordinates (ξ, η, $z=1$), if (α, α', 1), (β, β', 1), (γ, γ', 1) be the coordinates of the three points respectively, then

$$u : v : w = \begin{vmatrix} \xi, & \eta, & z \\ \beta, & \beta', & 1 \\ \gamma, & \gamma', & 1 \end{vmatrix} : \begin{vmatrix} \xi, & \eta, & z \\ \gamma, & \gamma', & 1 \\ \alpha, & \alpha', & 1 \end{vmatrix} : \begin{vmatrix} \xi, & \eta, & z \\ \alpha, & \alpha', & 1 \\ \beta, & \beta', & 1 \end{vmatrix}.$$

167. For the point I we have $(\xi, \eta, z) = (0, 1, 0)$, and hence if its areal coordinates be (u_0, v_0, w_0), we have

$$u_0 : v_0 : w_0 = \beta - \gamma : \gamma - \alpha : \alpha - \beta,$$

and hence also, (u, v, w) referring to the current point P, we find

$$\begin{aligned} v_0 w - w_0 v = (\gamma - \alpha) &[(\alpha' - \beta')(\xi - \alpha z) - (\alpha - \beta)(\eta - \alpha' z)] \\ &- (\alpha - \beta)[(\gamma' - \alpha')(\xi - \alpha z) - (\gamma - \alpha)(\eta - \alpha' z)], \ = \Omega(\xi - \alpha z), \end{aligned}$$

if

$$\Omega = (\gamma - \alpha)(\alpha' - \beta') - (\alpha - \beta)(\gamma' - \alpha'), \ = \begin{vmatrix} \alpha, & \alpha', & 1 \\ \beta, & \beta', & 1 \\ \gamma, & \gamma', & 1 \end{vmatrix};$$

whence

$$v_0 w - w_0 v : w_0 u - w u_0 : u_0 v - u v_0 = \xi - \alpha z : \xi - \beta z : \xi - \gamma z,$$

and in precisely the same manner, if u_0', v_0', w_0' refer to the point J, then

$$u_0' : v_0' : w_0' = \beta' - \gamma' : \gamma' - \alpha' : \alpha' - \beta',$$

and

$$v_0' w - w_0' v : w_0' u - w u_0' : u_0' v - u v_0' = \eta - \alpha' z : \eta - \beta' z : \eta - \gamma' z.$$

168. Consider the conic

$$(a, b, c, f, g, h \between u, v, w)^2 = 0,$$

where u, v, w are any trilinear coordinates whatever; and take the inverse coefficients to be (A, B, C, F, G, H) ($A = bc - f^2$, &c.), then for any given point the coordinates of which are (u_0, v_0, w_0), the equation of the tangents from this point to the conic is, as is well known,

$$(A, B, C, F, G, H \between v_0 w - w_0 v, w_0 u - u_0 w, u_0 v - v_0 u)^2 = 0;$$

consequently for the conic

$$(a, b, c, f, g, h \between u, v, w)^2 = 0,$$

where (u, v, w) are areal coordinates referring, as above, to any three given points A, B, C, the equation of the pair of tangents from the point I to the conic is

$$(A, B, C, F, G, H \between \xi - \alpha z, \xi - \beta z, \xi - \gamma z)^2 = 0,$$

and that of the pair of tangents from J is

$$(A, B, C, F, G, H \between \eta - \alpha' z, \eta - \beta' z, \eta - \gamma' z)^2 = 0,$$

these two line-pairs intersecting, of course, in the foci of the conic.

169. In particular, if the conic is a conic passing through the points A, B, C, then taking its equation to be

$$lvw + mwu + nuv = 0,$$

the inverse coefficients are as $(l^2,\ m^2,\ n^2,\ -2mn,\ -2nl,\ -2lm)$, and we have for the equations of the two line-pairs

$$\sqrt{l(\xi - \alpha z)} + \sqrt{m(\xi - \beta z)} + \sqrt{n(\xi - \gamma z)} = 0,$$

$$\sqrt{l(\eta - \alpha' z)} + \sqrt{m(\eta - \beta' z)} + \sqrt{n(\eta - \gamma' z)} = 0.$$

Article No. 170. *The Theorem of the Variable Zomal.*

170. Consider the four circles

$$\mathfrak{A}^\circ = 0,\ \mathfrak{B}^\circ = 0,\ \mathfrak{C}^\circ = 0,\ \mathfrak{D}^\circ = 0\ \left(\mathfrak{A}^\circ = (x - az)^2 + (y - a'z)^2 - a''^2 z^2,\ \&c.\right),$$

which have a common orthotomic circle; so that as before

$$\mathrm{a}\mathfrak{A}^\circ + \mathrm{b}\mathfrak{B}^\circ + \mathrm{c}\mathfrak{C}^\circ + \mathrm{d}\mathfrak{D}^\circ = 0,$$

where

$$\mathrm{a} : \mathrm{b} : \mathrm{c} : \mathrm{d} = BCD : -CDA : DAB : -ABC.$$

I consider the first three circles as given, and the fourth circle as a variable circle cutting at right angles the orthotomic circle of the three given circles; this being so, attending only to the ratios a : b : c, we may write

$$\mathrm{a} : \mathrm{b} : \mathrm{c} \quad = DBC : \quad DCA : DAB,$$

that is, (a, b, c) are proportional to the areal coordinates of the centre of the variable circle in regard to the triangle ABC.

171. Suppose that the centre of the variable circle is situate on a given conic, then expressing the equation of this conic in areal coordinates in regard to the triangle ABC, we have between (a, b, c) the equation obtained by substituting these values for the coordinates in the equation of the conic; that is, the equation of the variable circle is

$$\mathrm{a}\mathfrak{A}^\circ + \mathrm{b}\mathfrak{B}^\circ + \mathrm{c}\mathfrak{C}^\circ \qquad = 0,$$

where (a, b, c) are connected by an equation

$$(a,\ b,\ c,\ f,\ g,\ h\between \mathrm{a},\ \mathrm{b},\ \mathrm{c})^2 = 0.$$

Hence $(A,\ B,\ C,\ F,\ G,\ H)$ being the inverse coefficients, the equation of the envelope of the variable circle is

$$(A,\ B,\ C,\ F,\ G,\ H\between \mathfrak{A}^\circ,\ \mathfrak{B}^\circ,\ \mathfrak{C}^\circ)^2 = 0,$$

and, in particular, if the conic be a conic passing through the points A, B, C, and such that its equation in the areal coordinates $(u,\ v,\ w)$ in regard to the triangle ABC is

$$lvw + mwu + nuv = 0,$$

then the equation of the envelope is

$$(l^2,\ m^2,\ n^2,\ -mn,\ -nl,\ -lm \between \mathfrak{A}^\circ,\quad \mathfrak{B}^\circ,\quad \mathfrak{C}^\circ)^2=0\,;$$

that is, it is

$$(1,\ \ 1,\ 1,\ -\ \ 1,\ -\ \ 1,\ -1 \between l\mathfrak{A}^\circ,\ m\mathfrak{B}^\circ,\ n\mathfrak{C}^\circ)^2=0,$$

or, what is the same thing, it is

$$\sqrt{l\mathfrak{A}^\circ}+\sqrt{m\mathfrak{B}^\circ}+\sqrt{n\mathfrak{C}^\circ}=0.$$

172. It has been seen that the equations of the nodal tangents at the points I, J respectively are respectively

$$\sqrt{l\,(\xi-\alpha z\,)}+\sqrt{m\,(\xi-\beta z\,)}+\sqrt{n\,(\xi-\gamma z\,)}=0,$$

$$\sqrt{l\,(\eta-\alpha' z)}+\sqrt{m\,(\eta-\beta' z)}+\sqrt{n\,(\eta-\gamma' z)}=0,$$

and that these are the equations of the tangents to the conic $lvw+mwu+nuv=0$ from the points I, J respectively. We have thus Casey's theorem for the generation of the bicircular quartic as follows:—The envelope of a variable circle which cuts at right angles the orthotomic circle of three given circles $\mathfrak{A}^\circ=0$, $\mathfrak{B}^\circ=0$, $\mathfrak{C}^\circ=0$, and has its centre on the conic $lvw+mwu+nuv=0$ which passes through the centres of the three given circles is the bicircular quartic, or trizomal

$$\sqrt{l\mathfrak{A}^\circ}+\sqrt{m\mathfrak{B}^\circ}+\sqrt{n\mathfrak{C}^\circ}=0,$$

which has its nodo-foci coincident with the foci of the conic.

173. To complete the analytical theory, it is proper to express the equation of the orthotomic circle by means of the areal coordinates $(u,\ v,\ w)$. Writing for shortness $a^2+a'^2-a''^2=a^{\backprime}$, &c., and therefore

$$\mathfrak{A}^\circ=x^2+y^2-2axz-2a'yz-a^{\backprime}z^2,\ \ \&\text{c.},$$

then if as before

$$u:v:w=\begin{vmatrix} x, & y, & z \\ b, & b', & 1 \\ c, & c', & 1 \end{vmatrix} : \begin{vmatrix} x, & y, & z \\ c, & c', & 1 \\ a, & a', & 1 \end{vmatrix} : \begin{vmatrix} x, & y, & z \\ a, & a', & 1 \\ b, & b', & 1 \end{vmatrix},$$

and therefore

$$x:y:z=au+bv+cw\ :\ a'u+b'v+c'w\ :\ u+v+w,$$

the equation of the orthotomic circle is

$$\begin{vmatrix} x-az, & y-a'z, & ax+a'y-a^{\backprime}z \\ x-bz, & y-b'z, & bx+b'y-b^{\backprime}z \\ x-cz, & y-c'z, & cx+c'y-c^{\backprime}z \end{vmatrix}=0,$$

viz., throwing out the factor z, this is

$$u\,(ax+a'y-a^{\backprime}z)+v\,(bx+b'y-b^{\backprime}z)+w\,(cx+c'y-c^{\backprime}z)=0,$$

or, what is the same thing, it is

$$(au+bv+cw)\,x+(a'u+b'v+c'w)\,y-(a^{\backprime}u+b^{\backprime}v+c^{\backprime}w)\,z \qquad =0,$$

viz., it is

$$(au+bv+cw)^2+(a'u+b'v+c'w)^2-(a^{\backprime}u+b^{\backprime}v+c^{\backprime}w)(u+v+w)=0,$$

that is, substituting for $a^{\backprime}$, $b^{\backprime}$, $c^{\backprime}$ their values, it is

$$\begin{aligned}
&a''^2u^2 \quad + b''^2v^2 \quad + c''^2w^2\\
&+\left(b''^2+c''^2-(b-c)^2-(b'-c')^2\right)vw\\
&+\left(c''^2+a''^2-(c-a)^2-(c'-a')^2\right)wu\\
&+\left(a''^2+b''^2-(a-b)^2-(a'-b')^2\right)uv=0,
\end{aligned}$$

and it may be observed that using for a moment α, β, γ to denote the angles at which the three circles taken in pairs respectively intersect, then we have $2b''c''\cos\alpha$ $=b''^2+c''^2-(b-c)^2-(b'-c')^2$, &c., and the equation of the orthotomic circle thus is

$$(1,\ 1,\ 1,\ \cos\alpha,\ \cos\beta,\ \cos\gamma\between a''u,\ b''v,\ c''w)^2=0.$$

174. We have in the foregoing enunciation of the theorem made use of the three given circles A, B, C, but it is clear that these are in fact *any* three circles in the series of the variable circle, and that the theorem may be otherwise stated thus:

The envelope of a variable circle which has its centre in a given conic, and cuts at right angles a given circle, is a bicircular quartic, such that its nodo-foci are the foci of the conic.

Article Nos. 175 to 177. *Properties depending on the relation between the Conic and Circle.*

175. I refer to the conic of the theorem simply as the conic, and to the fixed circle simply as the circle, or when any ambiguity might otherwise arise, then as the orthotomic circle. This being so, I consider the effect in regard to the trizomal curve, of the various special relations which may exist between the circle and the conic.

If the conic touch the circle, the curve has a node at the point of contact.

If the conic has with the circle a contact of the second order, the curve has a cusp at the point of contact.

If the centre of the circle lie on an axis of the conic, then the four intersections lie in pairs symmetrically in regard to this axis, or the curve has this axis as an axis of symmetry.

If the conic has double contact with the circle (this implies that the centre of the circle is situate on an axis of the conic) the curve has a node at each of the points of contact, viz., it breaks up into two circles intersecting in these two points.

The centres of the two circles respectively are the two foci of the conic, which foci lie on the axis in question. Observe that in the general case there are at each of the circular points at infinity two tangents, without any correspondence of the tangents of the one pair singly to those of the other pair, and there are thus four intersections, the four foci of the conic; in the present case, where the curve is a pair of circles, the two tangents to the same circle correspond to each other, and intersect in the two foci on the axis in question. The other two foci, or antipoints of these, are each of them the intersection of a tangent of the one circle by a tangent of the other circle.

If the conic has with the circle a contact of the third order (this implies that the circle is a circle of maximum or minimum curvature, at the extremity of an axis of the conic), then the curve has at this point a tacnode, viz., it breaks up into two circles touching each other and the conic at the point in question, and having their centres at the two foci situate on that axis of the conic respectively.

176. If the conic is a parabola, then the curve is a circular cubic having the four intersections of the parabola and circle for a set of concyclic foci, and having the focus of the parabola for *centre*. The like particular cases arise, viz.,

If the circle touch the parabola, the curve has a node at the point of contact.

If the circle has, with the parabola, a contact of the second order, the curve has a cusp at the point of contact.

If the centre of the circle is situate on the axis of the parabola, then the four intersections are situate in pairs symmetrically in regard to this axis, and the curve has this axis for an axis of symmetry.

If the circle has double contact with the parabola (which, of course, implies that the centre lies on the axis), then the curve has a node at each of the points of contact, viz., the curve breaks up into a line and circle intersecting at the two points of contact, and the circle has its centre at the focus of the parabola.

If the circle has with the parabola a contact of the third order (this implies that the circle is the circle of maximum curvature, touching the parabola at its vertex), then the curve has a tacnode, viz., it breaks up into a line and circle touching each other and the parabola at the vertex, that is, the line is the tangent to the parabola at its vertex, and the circle is the circle having the focus of the parabola for its centre, and passing through the vertex, or what is the same thing, having its radius $=\frac{1}{2}$ of the semi-latus rectum of the parabola.

177. If the conic be a circle, then the curve is a bicircular quartic such that its four nodo-foci coincide together at the centre of the circle; viz., the curve is a Cartesian having the centre of the conic for its cuspo-focus, that is, for the intersection of the cuspidal tangents of the Cartesian. The intersections of the conic with the other circle, or say with the orthotomic circle, are a pair of non-axial foci of the Cartesian; viz., the antipoints of these are two of the axial foci. The third axial focus is the centre of the orthotomic circle.

Article No. 178. *Case of Double Contact, Casey's Equation in the Problem of Tactions.*

178. In the case where the conic has double contact with the orthotomic circle, then (as we have seen) the envelope of the variable circle is a pair of circles, each touching the variable circle; or, if we start with three given circles and a conic through their centres, then the envelope is a pair of circles, each of them touching each of the three given circles; that is, we have a solution of the problem of tactions. Multiplying by 2, the equation found *ante*, No. 173, for the variable circle, and then for the moment representing it by $(\mathrm{a, b, c, f, g, h} \between u, v, w)^2 = 0$; then attributing any signs at pleasure to the radicals $\sqrt{\mathrm{a}}$, $\sqrt{\mathrm{b}}$, $\sqrt{\mathrm{c}}$, the equation of a conic through the centres of the given circles, and having double contact with the orthotomic circle, will be

$$(\mathrm{a, b, c, f, g, h} \between u, v, w)^2 - (u\sqrt{\mathrm{a}} + v\sqrt{\mathrm{b}} + w\sqrt{\mathrm{c}})^2 = 0,$$

viz., representing this equation as before by

$$lvw + mwu + nuv = 0,$$

we have

$$l : m : n = \mathrm{f} - \sqrt{\mathrm{bc}} : \mathrm{g} - \sqrt{\mathrm{ca}} : \mathrm{h} - \sqrt{\mathrm{ab}},$$

that is, substituting for a, b, c, f, g, h their values, and taking, for instance, a, b, c $= a''\sqrt{2}$, $b''\sqrt{2}$, $c''\sqrt{2}$, we find

$$\begin{aligned} l : m : n = \; & (b'' - c'')^2 - (b - c)^2 - (b' - c')^2 \\ : \; & (c'' - a'')^2 - (c - a)^2 - (c' - a')^2 \\ : \; & (a'' - b'')^2 - (a - b)^2 - (a' - b')^2, \end{aligned}$$

that is, l, m, n are as the squares of the tangential distances (direct) of the three circles taken in pairs, and this being so, the equation of a pair of circles touching each of the three given circles is $\sqrt{l\mathfrak{A}^\circ} + \sqrt{m\mathfrak{B}^\circ} + \sqrt{n\mathfrak{C}^\circ} = 0$. It is clear that, instead of taking the three direct tangential distances, we may take one direct tangential distance and two inverse tangential distances, viz., the tangential distances corresponding to any three centres of similitude which lie in a line; we have thus in all the equations of four pairs of circles, viz., of the eight circles which touch the three given circles. This is Casey's theorem in the problem of tactions.

Article No. 179. *The Intersections of the Conic and Orthotomic Circle are a set of four Concyclic Foci.*

179. The conic of centres intersects the orthotomic circle in four points, and for each of these the radius of the variable circle is $= 0$, that is, the points in question are a set of four concyclic foci (A, B, C. D) of the curve. Regarding the foci as given, the circle which contains them is of course the orthotomic circle; and there are a singly infinite series of curves, viz., these correspond to the singly infinite series of conics which can be drawn through the given foci. As for a given curve there are

four sets of concyclic foci, there are four different constructions for the curve, viz., the orthotomic circle may be any one of the four circles O, R, S, T, which contain the four sets of concyclic foci respectively; and the conic of centres is a conic through the corresponding set of four concyclic foci. We have thus four conics, but the foci of each of them coincide with the nodo-foci of the curve, that is, the conics are confocal; that such confocal conics exist has been shown, *ante*, Nos. 78 to 80.

Article Nos. 180 and 181. *Remark as to the Construction of the Symmetrical Curve.*

180. It is to be observed that in applying as above the theorem of the variable zomal to the construction of a symmetrical curve, the orthotomic circle made use of was one of the circles R, S, T, not the circle O, which is in this case the axis; in fact, we should then have the conic and the orthotomic circle each of them coinciding with the axis. And the variable circle, *quà* circle having its centre on the axis, cuts the axis at right angles whatever the radius may be; that is, the variable circle is no longer sufficiently determined by the theorem. The curve may nevertheless be constructed as the envelope of a variable circle having its centre on the axis; viz., writing $\mathfrak{A}^\circ = (x - az)^2 + y^2 - a''^2z^2$, &c., and starting with the form

$$\sqrt{l\mathfrak{A}^\circ} + \sqrt{m\mathfrak{B}^\circ} + \sqrt{n\mathfrak{C}^\circ} = 0,$$

then recurring to the demonstration of the theorem (*ante*, No. 47), the equation of the variable circle is $\mathrm{a}\mathfrak{A}^\circ + \mathrm{b}\mathfrak{B}^\circ + \mathrm{c}\mathfrak{C}^\circ = 0$, where a, b, c are any quantities satisfying $\frac{l}{\mathrm{a}} + \frac{m}{\mathrm{b}} + \frac{n}{\mathrm{c}} = 0$, or, what is the same thing, taking q an arbitrary parameter, and writing $\frac{l}{\mathrm{a}} = 1 + q$, $\frac{m}{\mathrm{b}} = 1 - q$, $\frac{n}{\mathrm{c}} = -2$, the equation of the variable circle is

$$\frac{1}{1+q}\, l\mathfrak{A}^\circ + \frac{1}{1-q}\, m\mathfrak{B}^\circ - \tfrac{1}{2} n\mathfrak{C}^\circ = 0.$$

Compare Nos. 118—123 for the like mode of construction of a conic; but it is proper to consider this in a somewhat different form.

181. Assume that the equation of the variable circle is

$$\mathfrak{D}^\circ = (x - dz)^2 + y^2 - d''^2z^2 = 0\,;$$

we have therefore identically

$$\mathrm{a}\mathfrak{A}^\circ + \mathrm{b}\mathfrak{B}^\circ + \mathrm{c}\mathfrak{C}^\circ + \mathrm{d}\mathfrak{D}^\circ = 0,$$

viz., this gives

$$\begin{aligned} \mathrm{a} \;\; + \mathrm{b} \;\; + \mathrm{c} \;\; &= -\mathrm{d}\,, \\ \mathrm{a}a + \mathrm{b}b + \mathrm{c}c &= -\mathrm{d}d, \\ \mathrm{a}\,(a^2 - a''^2) + \mathrm{b}\,(b^2 - b''^2) + \mathrm{c}\,(c^2 - c''^2) &= -\mathrm{d}\,(d^2 - d''^2), \end{aligned}$$

and from these equations we obtain a, b, c equal respectively to given multiples of d; substituting these values in the equation $\frac{l}{\mathrm{a}} + \frac{m}{\mathrm{b}} + \frac{n}{\mathrm{c}} = 0$, d divides out, and we have an equation involving the parameters of the given circles, and also d, d'', the parameters of the variable circle; viz., an equation determining d'', the radius of the variable circle, in terms of d, the coordinate of its centre. I consider in particular the case where the given circles are points; that is, where the given equation is

$$\sqrt{l\mathfrak{A}} + \sqrt{m\mathfrak{B}} + \sqrt{n\mathfrak{C}} = 0.$$

The equations here are

$$\begin{aligned} \mathrm{a} \;\; + \mathrm{b} \;\; + \mathrm{c} \;\; &= - \mathrm{d}, \\ \mathrm{a}a + \mathrm{b}b + \mathrm{c}c &= - \mathrm{d}d, \\ \mathrm{a}a^2 + \mathrm{b}b^2 + \mathrm{c}c^2 &= - \mathrm{d}\,(d^2 - d''^2), \end{aligned}$$

and from these we obtain

$$\begin{aligned} \mathrm{a}\,(a-b)\,(a-c) &= - \mathrm{d}\,\big((d-b)\,(d-c) - d''^2\big) \\ \mathrm{b}\,(b-c)\,(b-a) &= - \mathrm{d}\,\big((d-c)\,(d-a) - d''^2\big) \\ \mathrm{c}\,(c-a)\,(c-b) &= - \mathrm{d}\,\big((d-a)\,(d-b) - d''^2\big), \end{aligned}$$

so that the equation $\dfrac{l}{\mathrm{a}} + \dfrac{m}{\mathrm{b}} + \dfrac{n}{\mathrm{c}} = 0$ becomes

$$\frac{l\,(a-b)\,(a-c)}{(d-b)\,(d-c) - d''^2} + \frac{m\,(b-c)\,(b-a)}{(d-c)\,(d-a) - d''^2} + \frac{n\,(c-a)\,(c-b)}{(d-a)\,(d-b) - d''^2} = 0,$$

or, as this is more conveniently written,

$$\frac{l}{b-c}\,\frac{1}{(d-b)\,(d-c) - d''^2} + \frac{m}{c-a}\,\frac{1}{(d-c)\,(d-a) - d''^2} + \frac{n}{a-b}\,\frac{1}{(d-a)\,(d-b) - d''^2} = 0,$$

viz., considering d, d'' as the abscissa and ordinate of a point on a curve, and representing them by x, y respectively, the equation of this curve is

$$\frac{l}{b-c}\,\frac{1}{(x-b)\,(x-c) - y^2} + \frac{m}{c-a}\,\frac{1}{(x-c)\,(x-a) - y^2} + \frac{n}{c-a}\,\frac{1}{(x-a)\,(x-b) - y^2} = 0,$$

which is a certain quartic curve; and we have the original curve

$$\sqrt{l\mathfrak{A}} + \sqrt{m\mathfrak{B}} + \sqrt{n\mathfrak{C}} = 0,$$

as the envelope of a variable circle having for its diameter the double ordinate of this quartic curve.

Write for shortness $\dfrac{l}{b-c}$, $\dfrac{m}{c-a}$, $\dfrac{n}{a-b} = L,\ M,\ N$ respectively, then the equation of the quartic curve may be written

$$\Sigma L\;[(x-a)^2\,(x-b)\,(x-c) - y^2\,(x-a)\,(2x-b-c) + y^4] = 0,$$

viz., this is

$$\begin{aligned} \Sigma L\;[&x\,(x-a)\,(x-b)\,(x-c) \\ &- y^2\,\big(2x^2 - (a+b+c)\,x + (ab+ac+bc)\big) + y^4 \\ &- a\,(x-a)\,(x-b)\,(x-c) + y^2\,(ax+bc)] = 0, \end{aligned}$$

or what is the same thing, the equation is

$$\begin{aligned}(L+M+N)\,[x(x-a)(x-b)(x-c)-y^2\left(2x^2-(a+b+c)\,x+ab+ac+bc\right)+y^4]\\ -(La+Mb+Nc)(x-a)(x-b)(x-c)\\ +y^2\{(La+Mb+Nc)\,x+Lbc+Mca+Nab\}=0.\end{aligned}$$

In the particular case where $L+M+N=0$, that is, where

$$\frac{l}{b-c}+\frac{m}{c-a}+\frac{n}{a-b}=0,$$

the quartic curve becomes a cubic, viz., putting for shortness

$$-\delta=\frac{Lbc+Mca+Nab}{La+Mb+Nc},$$

the equation of the cubic is

$$y^2=\frac{(x-a)(x-b)(x-c)}{x-\delta},$$

viz., this is a cubic curve having three real asymptotes, and a diameter at right angles to one of the asymptotes, and at the inclinations $+45°$, $-45°$ to the other two asymptotes respectively—say that it is a "rectangular" cubic. The relation $\frac{l}{b-c}+\frac{m}{c-a}+\frac{n}{a-b}=0$ implies that the curve $\sqrt{l\mathfrak{A}}+\sqrt{m\mathfrak{B}}+\sqrt{n\mathfrak{C}}=0$ is a Cartesian, and we have thus the theorem that the envelope of a variable circle having for diameter the double ordinate of a rectangular cubic is a Cartesian.

I remark that using a particular origin, and writing the equation of the rectangular cubic in the form $y^2=x^2-2mx+\alpha+\frac{2A}{x}$, the equation of the variable circle is

$$(x-d)^2+y^2=d^2-2md+\alpha+\frac{2A}{d},$$

that is

$$x^2+y^2-\alpha-2d(x-m)-\frac{2A}{d}=0,$$

where d is the variable parameter. Forming the derived equation in regard to d, we have

$$x-m=\frac{A}{d^2},$$

and thence

$$x^2+y^2-\alpha=\frac{4A}{d},$$

$$(x^2+y^2-\alpha)^2=\frac{16A^2}{d^2}=16A\,(x-m),$$

that is, the equation of the envelope is $(x^2+y^2-\alpha)^2=16A\,(x-m)=0$, which is a known form of the equation of a Cartesian.

Article Nos. 182 and 183. *Focal Formulæ for the General Curve.*

182. Considering any three circles centres A, B, C, and taking $\mathfrak{A}^\circ$, &c., to denote as usual, let the equation of the curve be

$$\sqrt{l\mathfrak{A}^\circ} + \sqrt{m\mathfrak{B}^\circ} + \sqrt{n\mathfrak{C}^\circ} = 0\,;$$

then considering a fourth circle, centre D, a position of the variable circle, and having therefore the same orthotomic circle with the given circles, so that as before

$$\mathrm{a}\mathfrak{A}^\circ + \mathrm{b}\mathfrak{B}^\circ + \mathrm{c}\mathfrak{C}^\circ + \mathrm{d}\mathfrak{D}^\circ = 0,$$

the formulæ No. 47 (changing only U, V, W, T into $\mathfrak{A}^\circ$, $\mathfrak{B}^\circ$, $\mathfrak{C}^\circ$, $\mathfrak{D}^\circ$) are at once applicable to express the equation of the curve in terms of any three of the four circles A, B, C, D.

In particular, the circles may reduce themselves to the four points A, B, C, D, a set of concyclic foci, and here, the equation being originally given in the form

$$\sqrt{l\mathfrak{A}} + \sqrt{m\mathfrak{B}} + \sqrt{n\mathfrak{C}} = 0,$$

the same formulæ are applicable to express the equation in terms of any three of the four foci.

183. It is to be observed that in this case if the positions of the four foci are given by means of the circular coordinates $\left(\alpha, \frac{1}{\alpha}, 1\right)$, &c., which refer to the centre of the circle $ABCD$ as origin, and with the radius of this circle taken as unity, then the values of a, b, c, d (*ante*, No. 90), are given in the form adapted to the formulæ of No. 49, viz., we have

$$\mathrm{a} : \mathrm{b} : \mathrm{c} : \mathrm{d} = \alpha(\beta\gamma\delta) : -\beta(\gamma\delta\alpha) : \gamma(\delta\alpha\beta) : -\delta(\alpha\beta\gamma),$$

where $(\beta\gamma\delta) = (\beta-\gamma)(\gamma-\delta)(\delta-\beta)$, &c. The relation $\frac{l}{\mathrm{a}} + \frac{m}{\mathrm{b}} + \frac{n}{\mathrm{c}} = 0$, putting therein $l : m : n = \rho\alpha(\beta-\gamma)^2 : \sigma\beta(\gamma-\alpha)^2 : \tau\gamma(\alpha-\beta)^2$, (or, what is the same thing, taking the equation of the curve to be given in the form $(\beta-\gamma)\sqrt{\rho\alpha\mathfrak{A}} + (\gamma-\alpha)\sqrt{\sigma\beta\mathfrak{B}} + (\alpha-\beta)\sqrt{\tau\gamma\mathfrak{C}} = 0$), becomes

$$\rho(\beta-\gamma)(\alpha-\delta) + \sigma(\gamma-\alpha)(\beta-\delta) + \tau(\alpha-\beta)(\gamma-\delta) = 0,$$

viz., this equation, considering ρ, σ, τ, α, β, γ as given, determines the position of the fourth focus D, or when A, B, C, D are given, it is the relation which must exist between ρ, σ, τ; and the four forms of the equation are

$$\left(\begin{array}{cccc} . & \sqrt{\tau}(\delta-\gamma), & \sqrt{\sigma}(\beta-\delta), & \sqrt{\rho}(\gamma-\beta) \\ \sqrt{\tau}(\gamma-\delta), & . & \sqrt{\rho}(\delta-\alpha), & \sqrt{\sigma}(\alpha-\gamma) \\ \sqrt{\sigma}(\delta-\beta), & \sqrt{\rho}(\alpha-\delta), & . & \sqrt{\tau}(\beta-\alpha) \\ \sqrt{\rho}(\beta-\gamma), & \sqrt{\sigma}(\gamma-\alpha), & \sqrt{\tau}(\alpha-\beta), & . \end{array}\right)\left(\sqrt{\alpha\mathfrak{A}}, \sqrt{\beta\mathfrak{B}}, \sqrt{\gamma\mathfrak{C}}, \sqrt{\delta\mathfrak{D}}\right) = 0,$$

viz., the curve is represented by means of any one of these four equations involving each of them three out of the four given foci A, B, C, D.

Article Nos. 184 and 185. *Case of the Circular Cubic.*

184. In the case of a circular cubic, we must have

$$\rho(\beta-\gamma)(\alpha-\delta)+\sigma(\gamma-\alpha)(\beta-\delta)+\tau(\alpha-\beta)(\gamma-\delta)=0,$$

$$\sqrt{\alpha\rho}\,(\beta-\gamma)\qquad+\sqrt{\beta\sigma}\,(\gamma-\alpha)\qquad+\sqrt{\gamma\tau}\,(\alpha-\beta)\qquad=0,$$

which, when the foci A, B, C, D are given, determine the values of $\rho:\sigma:\tau$ in order that the curve may be a circular cubic. We see at once that there are two sets of values, and consequently two circular cubics having each of them the given points A, B, C, D for a set of concyclic foci. The two systems may be written

$$\sqrt{\rho}\,:\,\sqrt{\sigma}\,:\,\sqrt{\tau}=\sqrt{\alpha\delta}-\sqrt{\beta\gamma}\,:\,\sqrt{\beta\delta}-\sqrt{\gamma\alpha}\,:\,\sqrt{\gamma\delta}-\sqrt{\alpha\beta},$$

viz., it being understood that $\sqrt{\alpha\delta}$ means $\sqrt{\alpha}\,.\sqrt{\delta}$, &c., then, according as $\sqrt{\delta}$ has one or other of its two opposite values, we have one or other of the two systems of values of $\rho:\sigma:\tau$. To verify this, observe that writing the equation under the form

$$\sqrt{\alpha\rho}\,:\,\sqrt{\beta\sigma}\,:\,\sqrt{\gamma\tau}=\alpha\sqrt{\delta}-\sqrt{\alpha\beta\gamma}\,:\,\beta\sqrt{\delta}-\sqrt{\alpha\beta\gamma}\,:\,\gamma\sqrt{\delta}-\sqrt{\alpha\beta\gamma},$$

the second equation is verified; and that writing them under the form

$$\rho:\sigma:\tau=-(\beta+\gamma)(\alpha+\delta)+M\,:\,-(\gamma+\alpha)(\beta+\delta)+M\,:\,-(\alpha+\beta)(\gamma+\delta)+M,$$

where

$$M=\beta\gamma+\alpha\delta+\gamma\alpha+\beta\delta+\alpha\beta+\gamma\delta-2\sqrt{\alpha\beta\gamma\delta},$$

the second equation is also verified.

185. If we assume for a moment $\alpha=\cos a+i\sin a=e^{ia}$, &c., viz., if a, b, c, d be the inclinations to any fixed line of the radii through A, B, C, D respectively, then we have

$$\sqrt{\alpha\delta}\pm\sqrt{\beta\gamma}=e^{\frac{1}{4}(a+b+c+d)i}\,\{e^{\frac{1}{4}(a+d-b-c)i}\pm e^{-\frac{1}{4}(a+d-b-c)i}\},$$

$$\sqrt{\alpha}\,(\beta-\gamma)\;=e^{\frac{1}{2}(a+b+c)i}\quad\{e^{\frac{1}{2}(b-c)i}\qquad-e^{-\frac{1}{2}(b-c)i}\qquad\},$$

and thence

$$\sqrt{\alpha\rho}\,(\beta-\gamma)\,:\,\sqrt{\beta\sigma}\,(\gamma-\alpha)\,:\,\sqrt{\gamma\tau}\,(\alpha-\beta)=\cos\tfrac{1}{4}(a+d-b-c)\sin\tfrac{1}{2}(b-c)$$
$$:\cos\tfrac{1}{4}(b+d-c-a)\sin\tfrac{1}{2}(c-a)$$
$$:\cos\tfrac{1}{4}(c+d-a-b)\sin\tfrac{1}{2}(a-b);$$

or else

$$=\sin\tfrac{1}{4}(a+d-b-c)\sin\tfrac{1}{2}(b-c)$$
$$:\sin\tfrac{1}{4}(b+d-c-a)\sin\tfrac{1}{2}(c-a)$$
$$:\sin\tfrac{1}{4}(c+d-a-b)\sin\tfrac{1}{2}(a-b).$$

Putting in these formulæ,

$$\tfrac{1}{4}(a-b-c)=A,\quad\text{then we have}\quad B-C=\tfrac{1}{2}(b-c),$$
$$\tfrac{1}{4}(b-c-a)=B,\qquad\text{,,}\qquad C-A=\tfrac{1}{2}(c-a),$$
$$\tfrac{1}{4}(c-a-b)=C,\qquad\text{,,}\qquad A-B=\tfrac{1}{2}(a-b),$$

and for either set of values the verification of the relation

$$\sqrt{\alpha\rho}\,(\beta-\gamma)+\sqrt{\beta\sigma}\,(\gamma-\alpha)+\sqrt{\gamma\tau}\,(\alpha-\beta)=0,$$

will depend on the two identical equations

$$\sin A\sin(B-C)+\sin B\sin(C-A)+\sin C\sin(A-B)=0,$$
$$\cos A\sin(B-C)+\cos B\sin(C-A)+\cos C\sin(A-B)=0:$$

although the foregoing solution for the case of a circular cubic is the most elegant one, I will presently return to the question and give the solution in a different form.

Article No. 186. *Focal Formulæ for the Symmetrical Curve.*

186. In the symmetrical case, where the foci A, B, C, D are on a line, then if, as usual, a, b, c, d denote the distances from a fixed point, we have the expressions of $(\mathrm{a}, \mathrm{b}, \mathrm{c}, \mathrm{d})$ in a form adapted to the formulæ of No. 49, viz.,

$$\mathrm{a}:\mathrm{b}:\mathrm{c}:\mathrm{d}=(b-c)(c-d)(d-b):-(c-d)(d-a)(a-c):(d-a)(a-b)(b-d):-(a-b)(b-c)(c-a),$$

so that, assuming

$$l:m:n=\rho\,(b-c)^2:\sigma\,(c-a)^2:\tau\,(a-b)^2,$$

the equation

$$\frac{l}{\mathrm{a}}+\frac{m}{\mathrm{b}}+\frac{n}{\mathrm{c}}=0,$$

becomes

$$\rho\,(b-c)(a-d)+\sigma\,(c-a)(b-d)+\tau\,(a-b)(c-d)=0,$$

and the equation of the curve may be presented under any one of the four forms

$$\left(\begin{array}{cccc} . , & \sqrt{\tau}\,(d-c), & \sqrt{\sigma}\,(b-d), & \sqrt{\rho}\,(c-b) \\ \sqrt{\tau}\,(c-d), & . , & \sqrt{\rho}\,(d-a), & \sqrt{\sigma}\,(a-c) \\ \sqrt{\sigma}\,(d-b), & \sqrt{\rho}\,(a-d), & . , & \sqrt{\tau}\,(b-a) \\ \sqrt{\rho}\,(b-c), & \sqrt{\sigma}\,(c-a), & \sqrt{\tau}\,(a-b), & . \end{array}\right)(\sqrt{\mathfrak{A}}, \sqrt{\mathfrak{B}}, \sqrt{\mathfrak{C}}, \sqrt{\mathfrak{D}})=0.$$

Article No. 187. *Case of the Symmetrical Circular Cubic.*

187. For a circular cubic we must have

$$\rho\,(b-c)(a-d)+\sigma\,(c-a)(b-d)+\tau\,(a-b)(c-d)=0,$$
$$\sqrt{\rho}\,(b-c)\qquad+\sqrt{\sigma}\,(c-a)\qquad+\sqrt{\tau}\,(a-b)\qquad=0.$$

These equations give $\sqrt{\rho}:\sqrt{\sigma}:\sqrt{\tau}=1:1:1$ (values which obviously satisfy the two equations), or else

$$\sqrt{\rho}:\sqrt{\sigma}:\sqrt{\rho}=a+d-b-c:b+d-c-a:c+d-a-b.$$

In fact, these values obviously satisfy the second equation; and to see that they satisfy the first equation, we have only to write them under the form

$$\rho : \sigma : \tau = M - 4(b+c)(a+d) : M - 4(c+a)(b+d) : M - 4(a+b)(c+d),$$

where $M = (a+b+c+d)^2$. The first set gives for the curve

$$(b-c)\sqrt{\mathfrak{A}} + (c-a)\sqrt{\mathfrak{B}} + (a-b)\sqrt{\mathfrak{C}} = 0,$$

but this contains the line $z=0$ not once only, but twice; it in fact is $(y^2=0)$, the axis taken twice; the only proper cubic with the foci A, B, C, D *in lined* is therefore

$$(b-c)(a+d-b-c)\sqrt{\mathfrak{A}} + (c-a)(b+d-c-a)\sqrt{\mathfrak{B}} + (a-b)(c+d-a-b)\sqrt{\mathfrak{C}} = 0,$$

the equation of which is, of course, expressible in each of the other three forms.

Article Nos. 188 to 192. *Case of the General Circular Cubit.*

188. Returning to the general case of the circular cubic, the lines BC, AD meet in R, and if we denote by a_1, b_1, c_1, d_1, the distances from R of the four points respectively, so that $b_1c_1 = a_1d_1 = \text{rad.}^2 R$, then observing that a, b, c, d are proportional to the triangles BCD, CDA, DAB, ABC, with signs such that $\mathrm{a}+\mathrm{b}+\mathrm{c}+\mathrm{d}=0$, we find

$$\mathrm{a} : \mathrm{b} : \mathrm{c} : \mathrm{d} = -d_1(b_1-c_1) : c_1(a_1-d_1) : -b_1(a_1-d_1) : a_1(b_1-c_1);$$

and this being so, the equations $\frac{l}{\mathrm{a}}+\frac{m}{\mathrm{b}}+\frac{n}{\mathrm{c}}=0$, $\sqrt{l}+\sqrt{m}+\sqrt{n}=0$, give two systems of values of $\sqrt{l} : \sqrt{m} : \sqrt{n}$, viz., these are

$$\sqrt{l} : \sqrt{m} : \sqrt{n} = b_1-c_1 : c_1-a_1 : a_1-b_1,$$

and

$$= b_1-c_1 : c_1+a_1 : -a_1-b_1.$$

(To verify this, observe that for the first set we have

$$\frac{l}{\mathrm{a}}+\frac{m}{\mathrm{b}}+\frac{n}{\mathrm{c}} = \frac{(b_1-c_1)^2}{-d_1(b_1-c_1)} + \frac{(c_1-a_1)^2}{c_1(a_1-d_1)} + \frac{(a_1-b_1)^2}{-b_1(a_1-d_1)},$$

$$= \frac{b_1-c_1}{-d_1} + \frac{1}{a_1-d_1}\left(c_1+\frac{a_1^2}{c_1}-b_1-\frac{a_1^2}{b_1}\right),$$

$$= \frac{b_1-c_1}{-d_1} + \frac{b_1-c_1}{a_1-d_1}\left(\frac{a_1^2}{b_1c_1}-1\right),$$

$$= -\frac{b_1-c_1}{d_1} + \frac{b_1-c_1}{a_1-d_1}\left(\frac{a_1}{d_1}-1\right), = 0;$$

and the like as regards the second set.)

189. These values of $\sqrt{l} : \sqrt{m} : \sqrt{n}$ give the equations of the two circular cubics with the foci (A, B, C, D), the equation of each of them under a fourfold form, viz., we have

$$\left(\begin{array}{cccc} . , & d_1-c_1, & b_1-d_1, & c_1-b_1 \\ c_1-d_1, & . , & d_1-a_1, & a_1-c_1 \\ d_1-b_1, & a_1-d_1, & . , & b_1-a_1 \\ b_1-c_1, & c_1-a_1, & a_1-b_1, & . \end{array}\right)(\sqrt{\mathfrak{A}}, \sqrt{\mathfrak{B}}, \sqrt{\mathfrak{C}}, \sqrt{\mathfrak{D}})=0 \quad \text{(first curve)},$$

and

$$\left(\begin{array}{cccc} . , & -c_1-d_1, & d_1+b_1, & -b_1+c_1 \\ d_1+c_1, & . , & a_1-d_1, & c_1-a_1 \\ -b_1-d_1, & d_1-a_1, & . , & a_1+b_1 \\ b_1-c_1, & -c_1+a_1, & -a_1-b_1, & . \end{array}\right)(\sqrt{\mathfrak{A}}, \sqrt{\mathfrak{B}}, \sqrt{\mathfrak{C}}, \sqrt{\mathfrak{D}})=0 \quad \text{(second curve)}.$$

190. Similarly CA and BD meet in S, and if we denote by a_2, b_2, c_2, d_2 the distances from S of the four points respectively, so that $c_2a_2=b_2d_2=\text{rad.}^2S$ (observe that if as usual A, B, C, D are taken in order on the circle O, then A, C are on opposite sides of S, and similarly B, D are on opposite sides of S, so that taking a_2, b_2 positive c_2, d_2 will be negative), we have

$$\mathrm{a} : \mathrm{b} : \mathrm{c} : \mathrm{d} = c_2(b_2-d_2) : d_2(c_2-a_2) : -a_2(b_2-d_2) : -b_2(c_2-a_2),$$

and then the equations $\frac{l}{\mathrm{a}}+\frac{m}{\mathrm{b}}+\frac{n}{\mathrm{c}}=0$, $\sqrt{l}+\sqrt{m}+\sqrt{n}=0$, are satisfied by the two sets of values

$$\sqrt{l} : \sqrt{m} : \sqrt{n} = \quad b_2-c_2 : c_2-a_2 : a_2-b_2,$$

and

$$= -b_2-c_2 : c_2-a_2 : a_2+b_2,$$

and we have the equations of the same two cubic curves, each equation under a fourfold form, viz., these are

$$\left(\begin{array}{cccc} . , & -c_2+d_2, & -d_2+b_2, & -b_2+c_2 \\ c_2-d_2, & . , & d_2-a_2, & -c_2+a_2 \\ -b_2+d_2, & a_2-d_2, & . , & -a_2+b_2 \\ b_2-c_2, & c_2-a_2, & a_2-b_2, & . \end{array}\right)(\sqrt{\mathfrak{A}}, \sqrt{\mathfrak{B}}, \sqrt{\mathfrak{C}}, \sqrt{\mathfrak{D}})=0 \quad \text{(first curve)},$$

and

$$\left(\begin{array}{cccc} . , & c_2+d_2, & -d_2+b_2, & -b_2-c_2 \\ -d_2-c_2, & . , & a_2+d_2, & c_2-a_2 \\ -b_2+d_2, & -d_2-a_2, & . , & a_2+b_2 \\ b_2+c_2, & -c_2+a_2, & -a_2-b_2, & . \end{array}\right)(\sqrt{\mathfrak{A}}, \sqrt{\mathfrak{B}}, \sqrt{\mathfrak{C}}, \sqrt{\mathfrak{D}})=0 \quad \text{(second curve)}.$$

191. And again AB and CD meet in T, and denoting by a_3, b_3, c_3, d_3 the distances from T of the four points respectively, so that $a_3b_3=c_3d_3=\text{rad.}^2T$, we have

$$\mathrm{a} : \mathrm{b} : \mathrm{c} : \mathrm{d} = b_3(c_3-d_3) : -a_3(c_3-d_3) : -d_3(a_3-b_3) : c_3(a_3-b_3);$$

and the equations $\frac{l}{\mathrm{a}}+\frac{m}{\mathrm{b}}+\frac{n}{\mathrm{c}}=0$, $\sqrt{l}+\sqrt{m}+\sqrt{n}=0$, then give for $\sqrt{l}$, $\sqrt{m}$, $\sqrt{n}$ two sets of values, viz., these are

$$\sqrt{l} : \sqrt{m} : \sqrt{n} = b_3 - c_3 : \quad c_3 - a_3 : a_3 - b_3,$$

and

$$= b_3 + c_3 : -c_3 - a_3 : a_3 - b_3;$$

and we again obtain the equations of the two cubics, each equation under a fourfold form, viz., these are

$$\left(\begin{array}{cccc} . , & -c_3+d_3, & -d_3+b_3, & c_3-b_3 \\ -d_3+c_3, & . , & -a_3+d_3, & a_3-c_3 \\ -b_3+d_3, & -d_3+a_3, & . , & b_3-a_3 \\ b_3-c_3, & c_3-a_3, & a_3-b_3, & . \end{array}\right)(\sqrt{\mathfrak{A}}, \sqrt{\mathfrak{B}}, \sqrt{\mathfrak{C}}, \sqrt{\mathfrak{D}})=0,$$

and

$$\left(\begin{array}{cccc} . , & c_3-d_3, & d_3+b_3, & -c_3-b_3 \\ d_3-c_3, & . , & -a_3-d_3, & a_3+c_3 \\ -b_3-d_3, & d_3+a_3, & . , & b_3-a_3 \\ b_3+c_3, & -c_3-a_3, & a_3-b_3, & . \end{array}\right)(\sqrt{\mathfrak{A}}, \sqrt{\mathfrak{B}}, \sqrt{\mathfrak{C}}, \sqrt{\mathfrak{D}})=0.$$

192. The three systems have been obtained independently, but they may of course be derived each from any other of them: to show how this is, recollecting that we have

$$\begin{aligned} RA, RB, RC, RD &= a_1, b_1, \quad c_1, \quad d_1, \\ SA, SB, SC, SD &= a_2, b_2, -c_2, -d_2, \\ TA, TB, TC, TD &= a_3, b_3, \quad c_3, \quad d_3; \end{aligned}$$

then to compare

$$(a_1, b_1, c_1, d_1) \text{ and } (a_2, b_2, c_2, d_2);$$

the similar triangles

$$\begin{aligned} &SBC \text{ give } b_1 - c_1 : -c_2 : b_2, \\ &SAD \qquad = a_1 - d_1 : -d_2 : a_2, \end{aligned}$$

and the similar triangles

$$\begin{aligned} &RAC \text{ give } a_2 - c_2 : \quad c_1 : a_1, \\ &RBD \qquad = b_2 - d_2 : \quad d_1 : b_1; \end{aligned}$$

using these equations to determine the ratios of a_2, b_2, c_2, d_2 we have

$$\frac{a_2-c_2}{b_2-d_2}=\frac{c_1}{d_1}, \text{ or } d_1a_2 - d_1c_2 - c_1b_2 + c_1d_2 = 0;$$

that is

$$b_2\left\{-c_1+d_1\frac{a_1-d_1}{b_1-c_1}\right\}+c_2\left\{-d_1+c_1\frac{a_1-d_1}{b_1-c_1}\right\}=0;$$

and hence

$$b_2(-b_1c_1+c_1^2+a_1d_1-d_1^2)+c_2(-b_1d_1+c_1d_1+a_1c_1-c_1d_1)=0,$$

that is

$$b_2(c_1^2-d_1^2)+c_2(a_1c_1-b_1d_1)=0,$$

but

$$a_1c_1-b_1d_1=\frac{b_1}{d_1}(c_1^2-d_1^2),$$

or the equation gives $b_2+\frac{b_1}{d_1}c_2=0$, or say $b_2:c_2=b_1:-d_1$, and this with $\frac{b_1-c_1}{a_1-d_1}=\frac{c_2}{d_2}=\frac{b_2}{a_2}$, gives all the ratios, or we have

$$a_2:b_2:c_2:d_2=b_1(a_1-d_1):b_1(b_1-c_1):-d_1(a_1-d_1):-d_1(b_1-c_1).$$

We have then for example

$$b_2-c_2:c_2-a_2:a_2-b_2=b_1-c_1:c_1-a_1:a_1-b_1;\ \&c.,$$

showing the identity of the forms in (a_1, b_1, c_1, d_1) and (a_2, b_2, c_2, d_2).

Article No. 193. *Transformation to a New Set of Concyclic Foci.*

193. Consider the equation

$$\sqrt{l\mathfrak{A}}+\sqrt{m\mathfrak{B}}+\sqrt{n\mathfrak{C}}=0,$$

which refers to the foci A, B, C, and taking D the fourth concyclic focus, let (A_1, D_1) be the antipoints of (A, D) and (B_1, C_1) the antipoints of (B, C); so that (A_1, B_1, C_1, D_1) are another set of concyclic foci. We have $\mathfrak{B}_1 . \mathfrak{C}_1=\mathfrak{B} . \mathfrak{C}$, and it appears, *ante* No. 104, that we can find l_1, m_1, n_1, such that identically

$$-l\mathfrak{A}+m\mathfrak{B}+n\mathfrak{C}=-l_1\mathfrak{A}_1+m_1\mathfrak{B}_1+n_1\mathfrak{C}_1$$

and that $m_1n_1=mn$. The equation of the curve gives

$$-l\mathfrak{A}+m\mathfrak{B}+n\mathfrak{C}+2\sqrt{mn\mathfrak{B}\mathfrak{C}}=0,$$

we have therefore

$$-l_1\mathfrak{A}_1+m_1\mathfrak{B}_1+n_1\mathfrak{C}+2\sqrt{m_1n_1\mathfrak{B}_1\mathfrak{C}_1}=0,$$

that is,

$$\sqrt{l_1\mathfrak{A}_1}+\sqrt{m_1\mathfrak{B}_1}+\sqrt{n_1\mathfrak{C}_1}=0,$$

viz., this is the equation of the curve expressed in terms of the concyclic foci A_1, B_1, C_1.

Article No. 194. *The Tetrazomal Curve, Decomposable or Indecomposable.*

194. I consider the tetrazomal curve

$$\sqrt{l\mathfrak{A}^\circ}+\sqrt{m\mathfrak{B}^\circ}+\sqrt{n\mathfrak{C}^\circ}+\sqrt{p\mathfrak{D}^\circ}=0,$$

where the zomals are circles described about any given points A, B, C, D as centres.

There is not, in general, any identical equation $\mathrm{a}\mathfrak{A}^\circ + \mathrm{b}\mathfrak{B}^\circ + \mathrm{c}\mathfrak{C}^\circ + \mathrm{d}\mathfrak{D}^\circ = 0$, but when such relation exists, and when we have also $\frac{l}{\mathrm{a}} + \frac{m}{\mathrm{b}} + \frac{n}{\mathrm{c}} + \frac{p}{\mathrm{d}} = 0$, then the curve breaks up into two trizomals. When the conditions in question do not subsist, the curve is indecomposable. But there may exist between l, m, n, p relations in virtue of which a branch or branches ideally contain ($z^a = 0$) the line infinity a certain number of times, and which thus cause a depression in the order of the curve. The several cases are as follows:

Article No. 195. *Cases of the Indecomposable Curve.*

195. I. The general case; l, m, n, p not subjected to any condition. The curve is here of the order $= 8$; it has a quadruple point at each of the points I, J (and there is consequently no other point at infinity); it is touched four times by each of the circles A, B, C, D; and it has six nodes, viz., these are the intersections of the pairs of circles

$$\begin{aligned} &\sqrt{m\mathfrak{B}^\circ} + \sqrt{n\mathfrak{C}^\circ} = 0, \quad \sqrt{l\mathfrak{A}^\circ} + \sqrt{p\mathfrak{D}^\circ} = 0, \\ &\sqrt{n\mathfrak{C}^\circ} + \sqrt{l\mathfrak{A}^\circ} = 0, \quad \sqrt{m\mathfrak{B}^\circ} + \sqrt{p\mathfrak{D}^\circ} = 0, \\ &\sqrt{l\mathfrak{A}^\circ} + \sqrt{m\mathfrak{B}^\circ} = 0, \quad \sqrt{n\mathfrak{C}^\circ} + \sqrt{p\mathfrak{D}^\circ} = 0; \end{aligned}$$

the number of dps. is $6 + 2 . 6, = 18$, and there are no cusps, hence the class is $= 20$; and the deficiency is $= 3$.

II. We may have

$$\sqrt{l} + \sqrt{m} + \sqrt{n} + \sqrt{p} = 0;$$

there is in this case a single branch ideally containing ($z = 0$) the line infinity; the order is $= 7$. Each of the points I, J is a triple point, there is consequently one other point at infinity; viz., this is a real point, or the curve has a real asymptote. There are 6 nodes as before; dps. are $6 + 2 . 3, = 12$; class $= 18$, deficiency $= 3$.

III. We may have

$$\sqrt{l} + \sqrt{m} = 0, \quad \sqrt{n} + \sqrt{p} = 0;$$

there are then two branches each ideally containing ($z = 0$) the line infinity; the order is $= 6$. Each of the points I, J is a double point, and there are therefore two more points at infinity. These may be real or imaginary; viz., the curve may have (besides the asymptotes at I, J) two real or imaginary asymptotes. The circles $\sqrt{l\mathfrak{A}} + \sqrt{m\mathfrak{B}} = 0$, $\sqrt{n\mathfrak{C}} + \sqrt{p\mathfrak{D}} = 0$, each contain ($z = 0$) the line infinity, or they reduce themselves to two lines, so that in place of two nodes we have a single node at the intersection of these lines; number of nodes is $= 5$. Hence dps. are $5 + 2 . 1, = 7$. Class $= 16$, deficiency $= 3$.

IV. We may have

$$\sqrt{l} : \sqrt{m} : \sqrt{n} : \sqrt{p} = \mathrm{a} : \mathrm{b} : \mathrm{c} : \mathrm{d};$$

there is here a single branch containing ($z^2=0$) the line infinity twice; the order is $=6$. Each of the points I, J is a double point, and there are therefore two more points at infinity, that is (besides the asymptotes at I, J), there are two (real or imaginary) asymptotes. The number of nodes, as in the general case, is $=6$. Hence dps. are $6+2.1$, $=8$; class is $=14$; deficiency $=2$.

I notice the included particular case where the circles reduce themselves to their centres; viz., we have here the curve

$$\mathrm{a}\sqrt{\mathfrak{A}}+\mathrm{b}\sqrt{\mathfrak{B}}+\mathrm{c}\sqrt{\mathfrak{C}}+\mathrm{d}\sqrt{\mathfrak{D}}=0,$$

which (see *ante* No. 93) is in fact the curve which is the locus of the foci of the conics which pass through the four points A, B, C, D. It is at present assumed that the four points are not a circle; this case will be considered *post* No. 199. If we have BC, AD meeting in R; CA, BD in S, and AB, CD in T, then these points R, S, T are three of the six nodes. In fact, writing down the equations of the two circles

$$\mathrm{b}\sqrt{\mathfrak{B}}+\mathrm{c}\sqrt{\mathfrak{C}}=0, \quad \mathrm{a}\sqrt{\mathfrak{A}}+\mathrm{d}\sqrt{\mathfrak{D}}=0,$$

and observing that when the current point is taken at R, we have $\mathfrak{B} : \mathfrak{C}=\overline{RB}^2 : \overline{RC}^2$ $=(BAD)^2 : (CAD)^2=\mathrm{c}^2 : \mathrm{b}^2$, and similarly $\mathfrak{A} : \mathfrak{D}=\overline{RA}^2 : \overline{RD}^2=(ABC)^2 : (DBC)^2=\mathrm{d}^2 : \mathrm{a}^2$, we see that each of the two circles passes through the point R, or this point is a node. Similarly, the points S and T are each of them a node.

V. If

$$\sqrt{l}=\sqrt{m}=\sqrt{n}=\sqrt{p},$$

there are here three branches, each ideally containing ($z=0$) the line infinity; the order is thus $=5$. Each of the points I, J is an ordinary point on the curve; there are besides at infinity three points, all real, or one real and two imaginary; that is (besides the asymptotes at I, J) there are three asymptotes, all real, or one real and two imaginary. Each of the circles $\sqrt{\mathfrak{A}}+\sqrt{\mathfrak{B}}=0$, &c., contains the line infinity, and is thus reduced to a line; the number of nodes is therefore $=3$. Hence also, dps. $=3$; class $=14$; deficiency $=3$.

Article No. 196. *Cases of the Indecomposable Curve, the Centres being in a Line.*

196. There are some peculiarities in the case where the centres A, B, C, D are on a line; taking as usual (a, b, c, d) for the x-coordinates or distances of the four centres from a fixed point on the line, I enumerate the cases as follows:

I. No relation between l, m, n, p; corresponds to I. *supra*.

II. $\sqrt{l}+\sqrt{m}+\sqrt{n}+\sqrt{p}=0$; corresponds to II. *supra*.

III. $\sqrt{l}+\sqrt{m}=0$, $\sqrt{n}+\sqrt{p}=0$; corresponds to III. *supra*.

IV. $\sqrt{l}+\sqrt{m}+\sqrt{n}+\sqrt{p}=0$, $a\sqrt{l}+b\sqrt{m}+c\sqrt{n}+d\sqrt{p}=0$; corresponds to IV. *supra*, viz., there is a branch ideally containing $(z^2=0)$ the line infinity twice. But, observe that whereas in IV. *supra*, in order that this might be so, it was necessary to impose on l, m, n, p three conditions giving the definite systems of values $\sqrt{l} : \sqrt{m} : \sqrt{n} : \sqrt{p} = \mathrm{a} : \mathrm{b} : \mathrm{c} : \mathrm{d}$, in the present case only two conditions are imposed, so that a single arbitrary parameter is left.

V. $\sqrt{l}=\sqrt{m}=\sqrt{n}=\sqrt{p}$; corresponds to V. *supra*.

VI. $\sqrt{l}+\sqrt{m}=0$, $\sqrt{n}+\sqrt{p}=0$, $a\sqrt{l}+b\sqrt{m}+c\sqrt{n}+d\sqrt{p}=0$, or what is the same thing, $\sqrt{l} : \sqrt{m} : \sqrt{n} : \sqrt{p} = c-d : d-c : b-a : a-b$; the equation is thus $(c-d)(\sqrt{\mathfrak{A}^\circ}-\sqrt{\mathfrak{B}^\circ})-(a-b)(\sqrt{\mathfrak{A}^\circ}-\sqrt{\mathfrak{B}^\circ})=0$. There is here one branch ideally containing $(z^2=0)$ the line infinity twice, and another branch ideally containing $(z=0)$ the line infinity once; order is $=5$. Each of the points I, J is an ordinary point on the curve, the remaining points at infinity are a node $(\mathfrak{A}^\circ=\mathfrak{B}^\circ,\ \mathfrak{C}^\circ=\mathfrak{D}^\circ)$, as presently mentioned, counting as three points, viz., one branch has for its tangent the line infinity, and the other branch has for its tangent a line perpendicular to the axis; or what is the same thing, there is a hyperbolic branch having an asymptote perpendicular to the axis, and a parabolic branch ultimately perpendicular to the axis. The number of nodes is $=5$, viz., there is the node $\mathfrak{A}^\circ=\mathfrak{B}^\circ$, $\mathfrak{C}^\circ=\mathfrak{D}^\circ$ just referred to; and the two pairs of nodes $\big((c-d)\sqrt{\mathfrak{A}^\circ}-(a-b)\sqrt{\mathfrak{C}^\circ}=0,\ -(c-d)\sqrt{\mathfrak{B}^\circ}+(a-b)\sqrt{\mathfrak{D}^\circ}=0\big)$ and $\big((c-d)\sqrt{\mathfrak{A}^\circ}+(a-b)\sqrt{\mathfrak{D}^\circ}=0,\ (c-d)\sqrt{\mathfrak{B}^\circ}+(a-b)\sqrt{\mathfrak{C}^\circ}=0\big)$, each pair symmetrically situate in regard to the axis. Hence also dps. $=5$; class $=10$; deficiency $=1$.

And there is apparently a seventh case, which, however, I exclude from the present investigation, viz., this would be if we had

$$\left(\begin{array}{cccc} 1, & 1, & 1, & 1, \\ a, & b, & c, & d, \\ a^2, & b^2, & c^2, & d^2, \\ a''^2, & b''^2, & c''^2, & d''^2, \end{array}\right)(\sqrt{l},\ \sqrt{m},\ \sqrt{n},\ \sqrt{p})=0,$$

that is, a, b, c, d denoting as before, if we had

$$\sqrt{l} : \sqrt{m} : \sqrt{n} : \sqrt{p} = \mathrm{a} : \mathrm{b} : \mathrm{c} : \mathrm{d}, \quad \text{and} \quad \mathrm{a}a''^2+\mathrm{b}b''^2+\mathrm{c}c''^2+\mathrm{d}d''^2=0.$$

For observe that in this case we have

$$\mathrm{a}\mathfrak{A}^\circ+\mathrm{b}\mathfrak{B}^\circ+\mathrm{c}\mathfrak{C}^\circ+\mathrm{d}\mathfrak{D}^\circ=0, \quad \text{and} \quad \frac{l}{\mathrm{a}}+\frac{m}{\mathrm{b}}+\frac{n}{\mathrm{c}}+\frac{p}{\mathrm{d}}=0;$$

that is, the supposition in question belongs to the decomposable case.

Article No. 197. *The Decomposable Curve.*

197. We have next to consider the decomposable case, viz., when we have

$$\mathrm{a}\mathfrak{A}^\circ+\mathrm{b}\mathfrak{B}^\circ+\mathrm{c}\mathfrak{C}^\circ+\mathrm{d}\mathfrak{D}^\circ=0;$$

see *ante*, Nos. 87 *et seq.*—it there appears that (unless the centres A, B, C, D are in a line) the condition signifies that the four circles have a common orthotomic circle; and when we have also

$$\frac{l}{\mathrm{a}}+\frac{m}{\mathrm{b}}+\frac{n}{\mathrm{c}}+\frac{p}{\mathrm{d}}=0.$$

The formulæ for the decomposition are given *ante*, Nos. 42 *et seq.* Writing therein $\mathfrak{A}^\circ$, $\mathfrak{B}^\circ$, $\mathfrak{C}^\circ$, $\mathfrak{D}^\circ$ in place of U, V, W, T respectively, it thereby appears that the tetrazomal curve $\sqrt{l\mathfrak{A}^\circ}+\sqrt{m\mathfrak{B}^\circ}+\sqrt{n\mathfrak{C}^\circ}+\sqrt{p\mathfrak{D}^\circ}=0$, breaks up into the two trizomal curves

$$\sqrt{l_1\mathfrak{A}^\circ}+\sqrt{m_1\mathfrak{B}^\circ}+\sqrt{n_1\mathfrak{C}^\circ}=0,\quad \sqrt{l_2\mathfrak{A}^\circ}+\sqrt{m_2\mathfrak{B}^\circ}+\sqrt{n_2\mathfrak{C}^\circ}=0,$$

where

$$\sqrt{l_1}=\sqrt{l}+\frac{\mathrm{a}}{\mathrm{d}}\frac{p}{\sqrt{l}},\qquad \sqrt{l_2}=\sqrt{l}+\frac{\mathrm{a}}{\mathrm{d}}\frac{p}{\sqrt{l}},$$

$$\sqrt{m_1}=\sqrt{m}-\sqrt{\frac{\mathrm{a}}{\mathrm{bcd}}}\frac{p}{l}\,\mathrm{b}\sqrt{n},\quad \sqrt{m_2}=\sqrt{m}+\sqrt{\frac{\mathrm{a}}{\mathrm{bcd}}}\frac{p}{l}\,\mathrm{b}\sqrt{n},$$

$$\sqrt{n_1}=\sqrt{n}+\sqrt{\frac{\mathrm{a}}{\mathrm{bcd}}}\frac{p}{l}\,\mathrm{c}\sqrt{m},\quad \sqrt{n_2}=\sqrt{n}-\sqrt{\frac{\mathrm{a}}{\mathrm{bcd}}}\frac{p}{l}\,\mathrm{c}\sqrt{m},$$

and where we have

$$\frac{l_1}{\mathrm{a}}+\frac{m_1}{\mathrm{b}}+\frac{n_1}{\mathrm{c}}=0,\quad \frac{l_2}{\mathrm{a}}+\frac{m_2}{\mathrm{b}}+\frac{n_2}{\mathrm{c}}=0.$$

Article Nos. 198 to 203. *Cases of the Decomposable Curve, Centres not in a line.*

198. I assume, in the first instance, that the centres of the circles are not in a line; we have the following cases:

I. No further relation between l, m, n, p; the order of the tetrazomal is $=8$; the order of each of the trizomals is $=4$, that is each of them is a bicircular quartic.

II. $\sqrt{l}+\sqrt{m}+\sqrt{n}+\sqrt{p}=0$; the order of the tetrazomal is $=7$, that of one of the trizomals must be $=3$.

To verify this, observe that we have

$$\sqrt{l_1}+\sqrt{m_1}+\sqrt{n_1}=\sqrt{l}+\sqrt{m}+\sqrt{n}+\frac{\mathrm{a}p}{\mathrm{d}\sqrt{l}}+\frac{\sqrt{p}}{\sqrt{l}}\sqrt{\frac{\mathrm{a}}{\mathrm{bcd}}}(\mathrm{c}\sqrt{m}-\mathrm{b}\sqrt{n}),$$

or substituting for $\sqrt{l}+\sqrt{m}+\sqrt{n}$ the value $-\sqrt{p}$, this is

$$=\frac{\sqrt{p}}{\mathrm{d}\sqrt{l}}\left\{\mathrm{a}\sqrt{p}-\mathrm{d}\sqrt{l}+\sqrt{\frac{\mathrm{ad}}{\mathrm{bc}}}(\mathrm{c}\sqrt{m}-\mathrm{b}\sqrt{n})\right\},$$

and similarly for $\sqrt{l_2}+\sqrt{m_2}+\sqrt{n_2}$, the only change being in the sign of the radical $\sqrt{\frac{\mathrm{ad}}{\mathrm{bc}}}$. But from the two conditions satisfied by l, m, n, p it is easy to deduce

$$(\mathrm{a}\sqrt{p}-\mathrm{d}\sqrt{l})^2-\frac{\mathrm{ad}}{\mathrm{bc}}(\mathrm{c}\sqrt{m}-\mathrm{b}\sqrt{n})^2=0,$$

and hence one or other of the two functions

$$\sqrt{l_1}+\sqrt{m_1}+\sqrt{n_1},\ \sqrt{l_2}+\sqrt{m_2}+\sqrt{n_2} \text{ is } =0;$$

that is, one of the trizomal curves is a cubic.

III. $\sqrt{l}+\sqrt{p}=0$, $\sqrt{m}+\sqrt{n}=0$; order of the tetrazomal is $=6$; and hence order of each of the trizomals is $=3$. To verify this, observe that here

$$l\left(\frac{1}{a}+\frac{1}{d}\right)+m\left(\frac{1}{b}+\frac{1}{c}\right)=0,$$

which since $a+b+c+d=0$, gives $\frac{l}{m}=\frac{ad}{bc}$; so that, properly fixing the sign of the radical, we may write $\sqrt{l}+\sqrt{\frac{ad}{bc}}\sqrt{m}=0$. We have then

$$\sqrt{l_1}=\frac{a+d}{d}\sqrt{l},\quad \sqrt{m_1}+\sqrt{n_1}=\sqrt{\frac{a}{abc}}(b+c)\sqrt{m};$$

which last equation, using $\sqrt{\frac{ad}{bc}}$ to denote as above, but properly selecting the signification of $\pm$, may be written

$$\sqrt{m_1}+\sqrt{n_1}=\pm\frac{b+c}{d}\sqrt{\frac{ad}{bc}}\sqrt{m}.$$

Hence

$$\sqrt{l_1}\mp(\sqrt{m_1}+\sqrt{n_1})=\frac{1}{a}\left\{(a+d)\sqrt{l}+(b+c)\sqrt{\frac{ad}{bc}}\sqrt{m}\right\}$$

$$=\frac{a+d}{d}\left\{\sqrt{l}+\sqrt{\frac{ad}{bc}}\sqrt{m}\right\},\ =0,$$

viz., $\sqrt{l_1}\mp(\sqrt{m_1}+\sqrt{n_1})$ with a properly selected signification of the sign $\mp$ is $=0$; and similarly $\sqrt{l_2}\mp(\sqrt{m_2}+\sqrt{n_2})$ with a properly selected signification of the sign $\mp$ is $=0$; that is, each of the trizomals is a cubic.

199. IV. $\sqrt{l}:\sqrt{m}:\sqrt{n}:\sqrt{p}=a:b:c:d$ (values which, be it observed, satisfy of themselves the above assumed equation $\frac{l}{a}+\frac{m}{b}+\frac{n}{c}+\frac{p}{d}=0$); the order of the tetrazomal is $=6$; and the order of each of the trizomals is here again $=3$. We in fact have $\sqrt{l_1}=a+d$, $\sqrt{m_1}+\sqrt{n_1}=b+c$, and therefore $\sqrt{l_1}+\sqrt{m_1}+\sqrt{n_1}=0$; and similarly $\sqrt{l_2}+\sqrt{m_2}+\sqrt{n_2}=0$; that is, each of the trizomals is a cubic.

I attend, in particular, to the case where the four circles reduce themselves to the points A, B, C, D; these four points are then in a circle; and the curve under consideration is

$$a\sqrt{\mathfrak{A}}+b\sqrt{\mathfrak{B}}+c\sqrt{\mathfrak{C}}+d\sqrt{\mathfrak{D}}=0;$$

in the general case where the points A, B, C, D are not on a circle, this is, as has been seen, a sextic curve, the locus of the foci of the conics which pass through the four given points; in the case where the points are in a circle then the sextic breaks up into two cubics (viz., observing that the curve under consideration is $\sqrt{l\mathfrak{A}}+\sqrt{m\mathfrak{B}}+\sqrt{n\mathfrak{C}}+\sqrt{p\mathfrak{D}}=0$, where $\sqrt{l} : \sqrt{m} : \sqrt{n} : \sqrt{p} = \mathrm{a} : \mathrm{b} : \mathrm{c} : \mathrm{d}$, these values do of themselves satisfy the condition of decomposability $\frac{l}{\mathrm{a}}+\frac{m}{\mathrm{b}}+\frac{n}{\mathrm{c}}+\frac{p}{\mathrm{d}}=0$), that is, the locus of the foci of the conics which pass through four points on a circle is composed of two circular cubics, each of them having the four points for a set of concyclic foci. It is easy to see why the sextic, thus defined as a locus of foci, must break up into two cubics; in fact, as we have seen, the conics which pass through the four concyclic points A, B, C, D have their axes in two fixed directions; there is consequently a locus of the foci situate on the axes which are in one of the fixed directions, and a separate locus of the foci situate on the axes which lie in the other of the fixed directions; viz., each of these loci is a circular cubic.

200. Adopting the notation of No. 188, or writing

$$RA=a_1,\ RB=b_1,\ RC=c_1,\ RD=d_1,$$

(and therefore $b_1c_1=a_1d_1$) we have

$$\mathrm{a} : \mathrm{b} : \mathrm{c} : \mathrm{d} = -d_1(b_1-c_1) : c_1(a_1-d_1) : -b_1(a_1-d_1) : a_1(b_1-c_1).$$

Moreover

$$\sqrt{l_1} = \mathrm{a}+\mathrm{d}\quad,\quad \sqrt{l_2}=\mathrm{a}+\mathrm{d},$$

$$\sqrt{m_1}=\mathrm{b}+\sqrt{\frac{\mathrm{bcd}}{\mathrm{a}}},\quad \sqrt{m_2}=\mathrm{b}-\sqrt{\frac{\mathrm{bcd}}{\mathrm{a}}},$$

$$\sqrt{n_1}=\mathrm{c}-\sqrt{\frac{\mathrm{bcd}}{\mathrm{a}}},\quad \sqrt{n_2}=\mathrm{c}+\sqrt{\frac{\mathrm{bcd}}{\mathrm{a}}},$$

and we have

$$\frac{\mathrm{bcd}}{\mathrm{a}}=(a_1-d_1)^2\frac{\mathrm{a}_1\mathrm{b}_1\mathrm{c}_1}{d_1}=a_1^2(a_1-d_1)^2,\quad \sqrt{\frac{\mathrm{bcd}}{\mathrm{a}}}=-a_1(a_1-d_1)\ \text{suppose};$$

and thence

$$\sqrt{l_1}=(a_1-d_1)(b_1-c_1),\quad \sqrt{l_2}=(a_1-d_1)(\ b_1-c_1)$$

$$\sqrt{m_1}=(a_1-d_1)(c_1-a_1),\quad \sqrt{m_2}=(a_1-d_1)(\ c_1+a_1)$$

$$\sqrt{n_1}=(a_1-d_1)(a_1-b_1),\quad \sqrt{n_2}=(a_1-d_1)(-a_1-b_1),$$

that is

$$\sqrt{l_1} : \sqrt{m_1} : \sqrt{n_1} = b_1-c_1 : c_1-a_1 : \ a_1-b_1,$$

$$\sqrt{l_2} : \sqrt{m_2} : \sqrt{n_2} = b_1-c_1 : c_1+a_1 : -a_1-b_1,$$

agreeing with the formulæ No. 188.

The tetrazomal curve

$$-d_1(b_1-c_1)\sqrt{\mathfrak{A}}+c_1(a_1-d_1)\sqrt{\mathfrak{B}}-b_1(a_1-d_1)\sqrt{\mathfrak{C}}+a_1(b_1-c_1)\sqrt{\mathfrak{D}}=0$$

is thus decomposed into the two trizomals

$$(b_1 - c_1)\sqrt{\mathfrak{A}} + (c_1 - a_1)\sqrt{\mathfrak{B}} + (a_1 - b_1)\sqrt{\mathfrak{C}} = 0,$$

$$(b_1 - c_1)\sqrt{\mathfrak{A}} + (c_1 + a_1)\sqrt{\mathfrak{B}} - (a_1 + b_1)\sqrt{\mathfrak{C}} = 0.$$

201. Observe that the tetrazomal equation is a consequence of either of the trizomal equations; taking for instance the first trizomal equation, this gives the tetrazomal equation, and consequently any combination of the trizomal equation and the tetrazomal equation is satisfied if only the trizomal equation is satisfied. Multiply the trizomal equation by $-a_1 + d_1$ and add it to the tetrazomal equation; the resulting equation contains the factor a_1, and omitting this, it is

$$(b_1 - c_1)(-\sqrt{\mathfrak{A}} + \sqrt{\mathfrak{D}}) + (a_1 - d_1)(\sqrt{\mathfrak{B}} - \sqrt{\mathfrak{C}}) = 0,$$

where observe that $b_1 - c_1$ is the distance BC, and $a_1 - d_1$ the distance AD. But in like manner multiplying the second trizomal equation by $-a_1 + d_1$, and adding it to the original tetrazomal equation, the resulting equation, omitting the factor a_1, is

$$(b_1 - c_1)(-\sqrt{\mathfrak{A}} + \sqrt{\mathfrak{D}}) - (a_1 - d_1)(\sqrt{\mathfrak{B}} - \sqrt{\mathfrak{C}}) = 0;$$

viz., it is in fact the same tetrazomal equation as was obtained by means of the first trizomal equation.

The new tetrazomal equation, say

$$(b_1 - c_1)(-\sqrt{\mathfrak{A}} + \sqrt{\mathfrak{D}}) + (a_1 - d_1)(\sqrt{\mathfrak{B}} - \sqrt{\mathfrak{C}}) = 0,$$

is thus equivalent to the original tetrazomal equation; observe that it is an equation of the form $\sqrt{l\mathfrak{A}} + \sqrt{m\mathfrak{B}} + \sqrt{n\mathfrak{C}} + \sqrt{p\mathfrak{D}} = 0$, where

$$\sqrt{l} = -(b_1 - c_1), \quad \sqrt{m} = a_1 - d_1, \quad \sqrt{n} = (a_1 - d_1), \quad \sqrt{p} = b_1 - c_1,$$

and where consequently $\sqrt{l} + \sqrt{p} = 0$, $\sqrt{m} + \sqrt{n} = 0$, that is an equation of the form (198) III., decomposable, as it should be, into the equations of two circular cubics. Writing

$$\frac{-\sqrt{\mathfrak{A}} + \sqrt{\mathfrak{D}}}{a_1 - d_1} = \theta, \quad \frac{\sqrt{\mathfrak{B}} - \sqrt{\mathfrak{C}}}{b_1 - c_1} = \theta,$$

where θ is an arbitrary parameter, the curve is obtained as the locus of the intersections of two similar conics having respectively the foci (A, D) and the foci (B, C) (see Salmon, *Higher Plane Curves*, p. 174): whence we have the theorem, that if A, B, C, D are any four points on a circle, the two circular cubics which are the locus of the foci of the conics which pass through the four points A, B, C, D, are also the locus of the intersections of the similar conics, which have for their foci (A, D) and (B, C) respectively; and of the similar conics with the foci (B, D) and (C, A) respectively; and of the similar conics with the foci (C, D) and (A, B) respectively.

202. V. $\sqrt{l} = \sqrt{m} = \sqrt{n} = \sqrt{p}$. The order of the tetrazomal is $= 5$, whence those of the trizomals should be $= 3$ and $= 2$ respectively. To verify this observe that the

equation $\frac{l}{a}+\frac{m}{b}+\frac{n}{c}+\frac{p}{d}=0$ gives $\frac{1}{a}+\frac{1}{b}+\frac{1}{c}+\frac{1}{d}=0$, and combining with $a+b+c+d=0$, these are only satisfied by one of the systems $(a+b=0,\ c+d=0)$, $(a+c=0,\ b+d=0)$, $(a+d=0,\ b+c=0)$. Selecting to fix the ideas the first of these, or writing

$$(a,\ b,\ c,\ d)=(a,\ -a,\ c,\ -c),$$

so that we have identically

$$a(A^\circ-B^\circ)+c(C^\circ-D^\circ)=0,$$

an equation which signifies that the radical axis of the circles A, B is also the radical axis of the circles C, D; then, writing as we may do, $\sqrt{\frac{a}{bcd}\frac{p}{l}}\left(=\sqrt{\frac{1}{c^2}}\right)=\frac{1}{c}$, we have

$$\sqrt{l_1}\ =1-\frac{a}{c},\qquad \sqrt{l_2}\ =1-\frac{a}{c},$$

$$\sqrt{m_1}=1+\frac{a}{c},\qquad \sqrt{m_2}=1-\frac{a}{c},$$

$$\sqrt{n_1}\ =1+1,\ =2,\quad \sqrt{n_2}=1-1,\ =0.$$

Here $\sqrt{l_1}+\sqrt{m_1}-\sqrt{n_1}=0$, which gives one of the trizomals a cubic, viz., this is the trizomal

$$\left(1-\frac{a}{c}\right)\sqrt{\mathfrak{A}^\circ}+\left(1+\frac{a}{c}\right)\sqrt{\mathfrak{B}^\circ}+2\sqrt{\mathfrak{C}^\circ}=0.$$

The other trizomal reduces itself to the bizomal $\sqrt{\mathfrak{A}^\circ}+\sqrt{\mathfrak{B}^\circ}=0$, which regarded as a trizomal, or written under the form $(\sqrt{\mathfrak{A}^\circ}+\sqrt{\mathfrak{B}^\circ})^2=0$, is the line $\mathfrak{A}^\circ-\mathfrak{B}^\circ=0$ twice, viz., this is the radical axis of the circles A_1, B_1 twice; and the order is thus $=2$. By what precedes, the line in question is in fact the common radical axis of the circles A, B and of the circles C, D.

Article Nos. 203 to 205. *Cases of the Decomposable Curve, the Centres in a Line.*

203. We have yet to consider the decomposable case when the centres A, B, C, D are on a line; the equation $a\mathfrak{A}^\circ+b\mathfrak{B}^\circ+c\mathfrak{C}^\circ+d\mathfrak{D}^\circ=0$ here subsists universally, whatever be the radii a'', b'', c'', d''. We establish as before the relation $\frac{l}{a}+\frac{m}{b}+\frac{n}{c}+\frac{p}{d}=0$. The cases are as follows:

I. No further relation between l, m, n, p; order of tetrazomal $=8$, of trizomals 4 and 4.

II. $\sqrt{l}+\sqrt{m}+\sqrt{n}+\sqrt{p}=0$; order of tetrazomal $=7$; of trizomals $=4$ and 3; same as II. *supra*.

III. $\sqrt{l}+\sqrt{p}=0$, $\sqrt{m}+\sqrt{n}=0$; order of tetrazomal $=6$; of trizomals 3 and 3; same as III. *supra*.

204. IV. $\sqrt{l}+\sqrt{m}+\sqrt{n}+\sqrt{p}=0$, $\mathrm{a}\sqrt{l}+\mathrm{b}\sqrt{m}+\mathrm{c}\sqrt{n}+\mathrm{d}\sqrt{p}=0$; order of tetrazomal $=6$; this is a remarkable case, the orders of the trizomals are either 3, 3 or else 4, 2.

To explain how this is, it is to be noticed that in the absence of any special relation between the radii, the above conditions combined with $\frac{l}{\mathrm{a}}+\frac{m}{\mathrm{b}}+\frac{n}{\mathrm{c}}+\frac{p}{\mathrm{d}}=0$ give $\sqrt{l}:\sqrt{m}:\sqrt{n}:\sqrt{p}=\mathrm{a}:\mathrm{b}:\mathrm{c}:\mathrm{d}$[1]; when l, m, n, p have these values, the case is the same as IV. *supra*, and the orders of the trizomals are 3, 3. But if the radii of the circles satisfy the condition

$$\begin{vmatrix} 1, & 1, & 1, & 1 \\ a, & b, & c, & d \\ a^2, & b^2, & c^2, & d^2 \\ a''^2, & b''^2, & c''^2, & d''^2 \end{vmatrix}=0,$$

then the two conditions satisfy of themselves the remaining condition $\frac{l}{\mathrm{a}}+\frac{m}{\mathrm{b}}+\frac{n}{\mathrm{c}}+\frac{p}{\mathrm{d}}=0$, and the ratios $\sqrt{l}:\sqrt{m}:\sqrt{n}:\sqrt{p}$ instead of being determinate as above, depend on an arbitrary parameter.

We have

$$\sqrt{l_1}=\sqrt{l}+\frac{\mathrm{a}}{\mathrm{d}}\frac{p}{\sqrt{l}},\quad \sqrt{m_1}=\sqrt{m}-\sqrt{\frac{\mathrm{a}}{\mathrm{bcd}}\frac{p}{l}}\,\mathrm{b}\sqrt{n},\quad \sqrt{n_1}=\sqrt{n}+\sqrt{\frac{\mathrm{a}}{\mathrm{bcd}}\frac{p}{l}}\,\mathrm{c}\sqrt{m},$$

and between l, m, n, p only the relations

$$\sqrt{l}+\sqrt{m}+\sqrt{n}+\sqrt{p}=0,\quad a\sqrt{l}+b\sqrt{m}+c\sqrt{n}+d\sqrt{p}=0.$$

We find first

$$\begin{aligned}\sqrt{l_1}+\sqrt{m_1}+\sqrt{n_1}&=\sqrt{l}+\sqrt{m}+\sqrt{n}\\ &\quad+\frac{\sqrt{p}}{\sqrt{l}}\left\{\frac{\mathrm{a}}{\mathrm{d}}\sqrt{p}-\sqrt{\frac{\mathrm{a}}{\mathrm{bcd}}}(\mathrm{b}\sqrt{n}-\mathrm{c}\sqrt{m})\right\}\\ &=-\frac{\sqrt{p}}{\sqrt{l}}\left\{\frac{1}{\mathrm{d}}(\mathrm{d}\sqrt{l}-\mathrm{a}\sqrt{p})-\sqrt{\frac{\mathrm{a}}{\mathrm{bc}}}(\mathrm{b}\sqrt{n}-\mathrm{c}\sqrt{m})\right\},\end{aligned}$$

[1] Writing x^2, y^2, z^2, w^2 in place of $\sqrt{l}$, $\sqrt{m}$, $\sqrt{n}$, $\sqrt{p}$, we have to find x, y, z, w from the conditions

$$x+y+z+w=0,$$
$$ax+by+cz+dw=0,$$
$$\frac{x^2}{\mathrm{a}}+\frac{y^2}{\mathrm{b}}+\frac{z^2}{\mathrm{c}}+\frac{w^2}{\mathrm{d}}=0,$$

where the constants are connected by the relation

$$a\mathrm{a}+b\mathrm{b}+c\mathrm{c}+d\mathrm{d}=0.$$

It readily appears that the line represented by the first two equations *touches* the quadric surface in the point $x:y:z:w=\mathrm{a}:\mathrm{b}:\mathrm{c}:\mathrm{d}$, so that these are in general the only values of $\sqrt{l}:\sqrt{m}:\sqrt{n}:\sqrt{p}$. In the case next referred to in the text the line lies in the surface, and the values are not determined.

and then

$$(d-a)\sqrt{l} = (b-d)\sqrt{m} + (c-d)\sqrt{n},$$

$$(d-a)\sqrt{p} = (a-b)\sqrt{m} + (a-c)\sqrt{n},$$

whence

$$d\sqrt{l} - a\sqrt{p} = \frac{b-c}{d-a}(\mathrm{b}\sqrt{n} - \mathrm{c}\sqrt{m}),$$

and we have thus

$$\sqrt{l_1} + \sqrt{m_1} + \sqrt{n_1} = \frac{\sqrt{p}}{\mathrm{d}\sqrt{l}}\left(\frac{b-c}{d-a} - \sqrt{\frac{\mathrm{ad}}{\mathrm{bc}}}\right)(\mathrm{b}\sqrt{n} - \mathrm{c}\sqrt{m});$$

and similarly

$$\sqrt{l_2} + \sqrt{m_2} + \sqrt{n_2} = \frac{\sqrt{p}}{\mathrm{d}\sqrt{l}}\left(\frac{b-c}{d-a} + \sqrt{\frac{\mathrm{ad}}{\mathrm{bc}}}\right)(\mathrm{b}\sqrt{n} - \mathrm{c}\sqrt{m}):$$

(observe that in the case not under consideration $\mathrm{b}\sqrt{n} - \mathrm{c}\sqrt{m} = 0$, and therefore $\sqrt{l_1} + \sqrt{m_1} + \sqrt{n_1} = 0$, $\sqrt{l_2} + \sqrt{m_2} + \sqrt{n_2} = 0$).

In the present case we have

$\mathrm{a}:\mathrm{b}:\mathrm{c}:\mathrm{d} = (b-c)(c-d)(d-b) : -(c-d)(d-a)(a-c) : (d-a)(a-b)(b-d) : -(a-b)(b-c)(c-a),$

and thence

$$\frac{\mathrm{ad}}{\mathrm{bc}} = \frac{(b-c)^2}{(d-a)^2},$$

so that only one of the two sums $\sqrt{l_1} + \sqrt{m_1} + \sqrt{n_1}$, $\sqrt{l_2} + \sqrt{m_2} + \sqrt{n_2}$ is $=0$, viz., assuming

$$\sqrt{\frac{\mathrm{ad}}{\mathrm{bc}}} = \frac{b-c}{d-a},$$

we have $\sqrt{l_1} + \sqrt{m_1} + \sqrt{n_1} = 0$.

We have then also

$$a\sqrt{l_1} + b\sqrt{m_1} + c\sqrt{n_1} = a\sqrt{l} + b\sqrt{m} + c\sqrt{n}$$

$$+\frac{\sqrt{p}}{\sqrt{l}}\left\{\frac{a\mathrm{a}\sqrt{p}}{\mathrm{d}} - \sqrt{\frac{\mathrm{a}}{\mathrm{bcd}}}(b\mathrm{b}\sqrt{n} - c\mathrm{c}\sqrt{m})\right\}$$

$$= -\frac{\sqrt{p}}{\sqrt{l}}\left\{\frac{1}{\mathrm{d}}(d\mathrm{d}\sqrt{l} - a\mathrm{a}\sqrt{p}) - \sqrt{\frac{\mathrm{a}}{\mathrm{bcd}}}(b\mathrm{b}\sqrt{n} - c\mathrm{c}\sqrt{m})\right\};$$

but we find

$$d\mathrm{d}\sqrt{l} - a\mathrm{a}\sqrt{p} = \frac{b-c}{d-a}(b\mathrm{b}\sqrt{n} - c\mathrm{c}\sqrt{m}),$$

and thence

$$a\sqrt{l_1} + b\sqrt{m_1} + c\sqrt{n_1} = \frac{\sqrt{p}}{\mathrm{d}\sqrt{l}}\left(\frac{b-c}{d-a} - \sqrt{\frac{\mathrm{ad}}{\mathrm{bc}}}\right)(b\mathrm{b}\sqrt{n} - c\mathrm{c}\sqrt{m}),\ = 0,$$

in virtue of $\sqrt{\frac{\mathrm{ad}}{\mathrm{bc}}} = \frac{b-c}{d-a}$. Hence $\sqrt{l_1} : \sqrt{m_1} : \sqrt{n_1}, = b-c : c-a : a-b$, or the corresponding trizomal is a conic, but the other trizomal is a quartic.

205. V. $\sqrt{l}=\sqrt{m}=\sqrt{n}=\sqrt{p}$; order of tetrazomal is $=5$; orders of trizomals $=3,\ 2$; same as V. *supra.*

VI. $\sqrt{l}+\sqrt{p}=0,\ \sqrt{m}+\sqrt{n}=0,\ a\sqrt{l}+b\sqrt{m}+c\sqrt{n}+d\sqrt{p}=0$; order of tetrazomal $=5$; orders of trizomals are 3, 2.

We have here

$$\sqrt{l_1} = \frac{\mathrm{a}+\mathrm{d}}{\mathrm{d}}\sqrt{l},$$

$$\sqrt{m_1} = \sqrt{m}+\sqrt{\frac{\mathrm{a}}{\mathrm{bcd}}}\,\mathrm{b}\sqrt{m},$$

$$\sqrt{n_1} = \sqrt{m}+\sqrt{\frac{\mathrm{a}}{\mathrm{bcd}}}\,\mathrm{c}\sqrt{m},$$

or writing the values of $\sqrt{m_1}$, $\sqrt{n_1}$ in the form

$$\sqrt{m_1} = \sqrt{m}+\sqrt{\frac{\mathrm{ad}}{\mathrm{bc}}}\,\frac{\mathrm{b}}{\mathrm{d}}\sqrt{m},$$

$$\sqrt{n_1} = -\sqrt{m}+\sqrt{\frac{\mathrm{ad}}{\mathrm{bc}}}\,\frac{\mathrm{c}}{\mathrm{d}}\sqrt{m},$$

then observing that as before $l=\frac{\mathrm{ad}}{\mathrm{bc}}m$, if to fix the ideas we assume $\sqrt{l}=\sqrt{\frac{\mathrm{ad}}{\mathrm{bc}}}\sqrt{m}$, the equations are

$$\sqrt{l_1} = \frac{\mathrm{a}+\mathrm{d}}{\mathrm{d}}\sqrt{l} \text{ and similarly } \sqrt{l_2} = \frac{\mathrm{a}+\mathrm{d}}{\mathrm{d}}\sqrt{l}$$

$$\sqrt{m_1} = \sqrt{m}+\frac{\mathrm{b}}{\mathrm{d}}\sqrt{l}, \qquad \sqrt{m_2} = \sqrt{m}-\frac{\mathrm{b}}{\mathrm{d}}\sqrt{l},$$

$$\sqrt{n_1} = -\sqrt{m}+\frac{\mathrm{c}}{\mathrm{d}}\sqrt{l}, \qquad \sqrt{n_2} = \sqrt{m}-\frac{\mathrm{c}}{\mathrm{d}}\sqrt{l},$$

whence

$$\sqrt{l_1}+\sqrt{m_1}+\sqrt{n_1}=0, \quad \sqrt{l_2}-\sqrt{m_2}-\sqrt{n_2}=0.$$

We have moreover

$$a\sqrt{l_1}=\frac{a\mathrm{a}+d\mathrm{d}}{\mathrm{d}}\sqrt{l},$$

$$b\sqrt{m_1}+c\sqrt{n_1}=(b-c)\sqrt{m}+\frac{b\mathrm{b}+c\mathrm{c}}{\mathrm{d}}\sqrt{l},$$

and thence

$$a\sqrt{l_1}+b\sqrt{m_1}+c\sqrt{n_1}=(a-d)\sqrt{l}+(b-c)\sqrt{m}=0,$$

so that

$$\sqrt{l_1} : \sqrt{m_1} : \sqrt{n_1} = b-c : c-a : a-b;$$

the corresponding trizomal is thus a conic, and it has been seen that the other trizomal is a cubic.

VII. If we have $\begin{vmatrix} 1 & 1 & 1 & 1 \\ a & b & c & d \\ a^2 & b^2 & c^2 & d^2 \\ a''^2 & b''^2 & c''^2 & d''^2 \end{vmatrix} = 0$, and $\left(\begin{array}{cccc} 1 & 1 & 1 & 1 \\ a & b & c & d \\ a^2 & b^2 & c^2 & d^2 \\ a''^2 & b''^2 & c''^2 & d''^2 \end{array}\right)(\sqrt{l}, \sqrt{m}, \sqrt{n}, \sqrt{p}) = 0$,

the tetrazomal has a branch ideally containing ($z^3 = 0$) the line infinity 3 times; order is $= 5$; orders of the trizomals are 3, 2. We have here

$$\sqrt{l} : \sqrt{m} : \sqrt{n} : \sqrt{p} = \mathrm{a} : \mathrm{b} : \mathrm{c} : \mathrm{d},$$

and thence

$$\sqrt{l_1} = \mathrm{a} + \mathrm{d}\quad, \quad \sqrt{l_2} = \mathrm{a} + \mathrm{d}\quad,$$

$$\sqrt{m_1} = \mathrm{b} - \sqrt{\frac{\mathrm{bcd}}{\mathrm{a}}}, \quad \sqrt{m_2} = \mathrm{b} + \sqrt{\frac{\mathrm{bcd}}{\mathrm{a}}},$$

$$\sqrt{n_1} = \mathrm{c} + \sqrt{\frac{\mathrm{bcd}}{\mathrm{a}}}, \quad \sqrt{n_2} = \mathrm{c} - \sqrt{\frac{\mathrm{bcd}}{\mathrm{a}}},$$

which give

$$\sqrt{l_1} + \sqrt{m_1} + \sqrt{n_1} = 0, \quad \sqrt{l_2} + \sqrt{m_2} + \sqrt{n_2} = 0.$$

Moreover

$$\begin{aligned} a\sqrt{l_1} + b\sqrt{m_1} + c\sqrt{n_1} = &\ a(\mathrm{a}+\mathrm{d}) + b\mathrm{b} + c\mathrm{c} \\ &- (b-c)\sqrt{\frac{\mathrm{bcd}}{\mathrm{a}}} \\ = &\ (a-d)\,\mathrm{d} - (b-c)\sqrt{\frac{\mathrm{bcd}}{\mathrm{a}}} \\ = &\ \mathrm{d}\left\{(a-d) - (b-c)\sqrt{\frac{\mathrm{bc}}{\mathrm{ad}}}\right\}, \end{aligned}$$

and similarly

$$a\sqrt{l_2} + b\sqrt{m_2} + c\sqrt{n_2} = \mathrm{d}\left\{(a-d) + (b-c)\sqrt{\frac{\mathrm{bc}}{\mathrm{ad}}}\right\};$$

whence in virtue of

$$\frac{\mathrm{ad}}{\mathrm{bc}} = \frac{(b-c)^2}{(d-a)^2},$$

one of the two expressions is $= 0$; and the trizomals are thus a conic and a cubic.

Article No. 206. *The Decomposable Curve; Transformation to a different set of Concyclic Foci.*

206. Consider the decomposable case of

$$\sqrt{l\mathfrak{A}} + \sqrt{m\mathfrak{B}} + \sqrt{n\mathfrak{C}} + \sqrt{p\mathfrak{D}} = 0;$$

viz., the points A, B, C, D lie here in a circle, and we have $\frac{l}{\mathrm{a}} + \frac{m}{\mathrm{b}} + \frac{n}{\mathrm{c}} + \frac{p}{\mathrm{d}} = 0$. Taking (A_1, D_1) the antipoints of (A, D); (B_1, C_1) the antipoints of (B, C); then

$\mathfrak{A}_1\mathfrak{D}_1 = \mathfrak{A}\mathfrak{D}$, $\mathfrak{B}_1\mathfrak{C}_1 = \mathfrak{B}\mathfrak{C}$ (No. 65) and referring to the formulæ, *ante,* Nos. 100 *et seq.*, it appears that we can find l_1, m_1, n_1, p_1 such that identically

$$-l\mathfrak{A} + m\mathfrak{B} + n\mathfrak{C} - p\mathfrak{D} = -l_1\mathfrak{A}_1 + m_1\mathfrak{B}_1 + n_1\mathfrak{C}_1 - p_1\mathfrak{D}_1,$$

and moreover that $lp = l_1p_1$, $mn = m_1n_1$.

The equation of the curve gives

$$-l\mathfrak{A} + m\mathfrak{B} + n\mathfrak{C} - p\mathfrak{D} - 2\sqrt{lp\mathfrak{A}\mathfrak{D}} + 2\sqrt{mn\mathfrak{B}\mathfrak{C}} = 0,$$

which may consequently be written

$$-l_1\mathfrak{A}_1 + m_1\mathfrak{B}_1 + n_1\mathfrak{C}_1 - p_1\mathfrak{D}_1 - 2\sqrt{l_1p_1\mathfrak{A}_1\mathfrak{D}_1} + 2\sqrt{m_1n_1\mathfrak{B}_1\mathfrak{C}_1} = 0;$$

viz., this is

$$\sqrt{l_1\mathfrak{A}_1} + \sqrt{m_1\mathfrak{B}_1} + \sqrt{n_1\mathfrak{C}_1} + \sqrt{p_1\mathfrak{D}_1} = 0;$$

that is, the two trizomals expressed by the original tetrazomal equation involving the set of concyclic foci (A, B, C, D) are thus expressed by a new tetrazomal equation involving the different set of concyclic foci (A_1, B_1, C_1, D_1); and we might of course in like manner express the equation in terms of the other two sets of concyclic foci (A_2, B_2, C_2, D_2) and (A_3, B_3, C_3, D_3) respectively. It might have been anticipated that such a transformation existed, for we could as regards each of the component trizomals separately pass from the original set to a different set of concyclic foci, and the two trizomal equations thus obtained would, it might be presumed, be capable of composition into a single tetrazomal equation; but the direct transformation of the tetrazomal equation is not on this account less interesting.

Annex I. *On the Theory of the Jacobian.*

Consider any three curves $U=0$, $V=0$, $W=0$, of the same order r, then writing

$$J(U, V, W) = \frac{d(U, V, W)}{d(x, y, z)} = \begin{vmatrix} d_xU, & d_xV, & d_xW \\ d_yU, & d_yV, & d_yW \\ d_zU, & d_zV, & d_zW \end{vmatrix},$$

we have the Jacobian curve $J(U, V, W)=0$, of the order $3r-3$.

A fundamental property is that if the curves $U=0$, $V=0$, $W=0$ have any common point, this is a point on the Jacobian, and not only so, but it is a node, or double point, that is, for the point in question we have $J=0$, and we have also $d_xJ=0$, $d_yJ=0$, $d_zJ=0$.

It follows that for the three curves $l\Theta + L\Phi = 0$, $m\Theta + M\Phi = 0$, $n\Theta + N\Phi = 0$ ($\Theta = 0$ of the order $r-s'$, $\Phi = 0$ of the order $r-s$, $l=0$, $m=0$, $n=0$ each of the order s', $L=0$, $M=0$, $N=0$ each of the order s) which have in common the

$(r-s')(r-s)$ points of intersection of the curves $\Theta=0$, $\Phi=0$, each of these points is a node on the Jacobian, and hence that the Jacobian must be of the form

$$J(l\Theta+L\Phi,\ m\Theta+M\Phi,\ n\Theta+N\Phi)=A\Theta^2+2B\Theta\Phi+C\Phi^2=0,$$

where obviously the degrees of A, B, C must be $r+2s'-3$, $r+s+s'-3$, $r+2s-3$ respectively. In the particular case where $s'=0$, that is where l, m, n are constants, we have $A=0$; the Jacobian curve then contains as a factor $(\Phi=0)$, and throwing this out, the curve is $B\Theta+C\Phi=0$, viz., this is a curve of the order $2r+s-3$ passing through each of $r(r-s)$ points of intersection of the curves $\Theta=0$, $\Phi=0$.

In particular, if $r=2$, $s=1$, that is, if the curves are the conics $\Theta+L\Phi=0$, $\Theta+M\Phi=0$, $\Theta+N\Phi=0$, passing through the two points of intersection of the conic $\Theta=0$ by the line $\Phi=0$, then the Jacobian is a conic passing through these same two points, viz., its equation is of the form $\Theta+\Omega\Phi=0$. This intersects any one of the given conics, say $\Theta+L\Phi=0$ in the points $\Theta=0$, $\Phi=0$, and in two other points $\Theta+\Omega\Phi=0$, $\Omega-L=0$; at *each* of the last-mentioned points, the tangents to the two curves, and the lines drawn to the two points $\Theta=0$, $\Phi=0$, form a harmonic pencil.

Although this is, in fact, the known theorem that the Jacobian of three circles is their orthotomic circle, yet it is, I think, worth while to give a demonstration of the theorem as above stated in reference to the conics through two given points.

Taking $(z=0,\ x=0)$, $(z=0,\ y=0)$ for the two given points $\Theta=0$, $\Phi=0$, the general equation of a conic through the two points is a quadric equation containing terms in z^2, zx, zy, xy; taking any two such conics

$$cz^2+2fyz+2gzx+2hxy=0,$$
$$Cz^2+2Fyz+2Gzx+2Hxy=0,$$

these intersect in the two points $(x=0,\ z=0)$, $(y=0,\ z=0)$ and in two other points; let $(x,\ y,\ z)$ be the coordinates of either of the last-mentioned points, and take $(X,\ Y,\ Z)$ as current coordinates, the equations of the lines to the fixed points and of the two tangents are

$$Xz-Zx=0,\qquad Yz-Zy=0,$$
$$(hy+gz)(Xz-Zx)+(hx+fz)(Yz-Zy)=0,$$
$$(Hy+Gz)(Xz-Zx)+(Hx+Fz)(Yz-Zy)=0,$$

whence the condition for the harmonic relation is

$$(hy+gz)(Hx+Fz)+(hx+fz)(Hy+Gz)=0,$$

that is

$$(fG+gF)z^2+(hF+fH)yz+(gH+hG)zx+2hHxy=0,$$

but from the equations of the two conics multiplying by $\tfrac{1}{2}H$, $\tfrac{1}{2}h$ and adding, we have

$$\tfrac{1}{2}(cH+hC)z^2+(hF+fH)yz+(gH+hG)zx+2hHxy=0;$$

viz., the condition is thus reduced to

$$cH+hC-2(fG+gF)=0,$$

so that this condition being satisfied for one of the points in question, it will be satisfied for the other of them. Now for the three conics

$$\begin{aligned} cz^2 + 2f\ yz + 2g\ zx + 2h\ xy &= 0, \\ c'z^2 + 2f'\ yz + 2g'\ zx + 2h'xy &= 0, \\ c''z^2 + 2f''yz + 2g''zx + 2h''xy &= 0, \end{aligned}$$

forming the Jacobian, and throwing out the factor z, we may write the equation in the form

$$Cz^2 + 2Fyz + 2Gzx + 2Hxy = 0,$$

where the values are

$$\begin{aligned} C &= g\,(f'c'' - f''c') + g'\,(f''c - fc'') + g''\,(fc' - f'c), \\ H &= g\,(h'f'' - h''f') + g'\,(h''f - hf'') + g''\,(hf' - h'f), \\ 2F &= h\,(f'c'' - f''c') + h'\,(f''c - fc'') + h''\,(fc' - f'c\,), \\ 2G &= h\,(c'g'' - c''g') + h'\,(c''g - cg'') + h''\,(cg' - c'g\,); \end{aligned}$$

and we thence obtain

$$\begin{aligned} cH + hC &= -\,(fg' - f'g)\,(c''h - ch'') + (f''g - fg'')\,(ch' - c'h) \\ &= \ 2\,(fG + gF), \end{aligned}$$

viz., the condition is satisfied in regard to the Jacobian and the first of the three conics; and it is therefore also satisfied in regard to the Jacobian and the other two conics respectively.

I do not know any general theorem in regard to the Jacobian which gives the foregoing theorem of the orthotomic circle. It may be remarked that the use in the Memoir of the theorem of the orthotomic circle is not so great as would at first sight appear: it fixes the ideas to speak of the orthotomic circle of three given circles rather than of their Jacobian, but we are concerned with the orthotomic circle less as the circle which cuts at right angles the given circles than as a circle standing in a known relation to the given circles.

ANNEX II. *On* CASEY'S *Theorem for the Circle which touches three given Circles.*

The following two problems are identical:

1. To find a circle touching three given circles.

2. To find a cone-sphere (sphere the radius of which is $=0$) passing through three given points in space.

In fact, in the first problem if we use z to denote a given constant (which may be $=0$), then taking a, a' and $i\,(z-a'')$ for the coordinates of the centre and for the radius of one of the given circles; and similarly b, b', $i\,(z-b'')$; c, c', $i\,(z-c'')$ for the other two given circles; and S, S', $i(z-S'')$ for the required circle; the equations of the given circles will be

$$(x-a)^2+(y-a')^2+(z-a'')^2=0,$$
$$(x-b)^2+(y-b')^2+(z-b'')^2=0,$$
$$(x-c)^2+(y-c')^2+(z-c'')^2=0,$$

and that of the required circle will be

$$(x-S)^2+(y-S')^2+(z-S'')^2=0.$$

In order that this may touch the given circles, the distances of its centre from the centres of the given circles must be $i(S''-a'')$, $i(S''-b'')$, $i(S''-c'')$ respectively; the conditions of contact then are

$$(S-a)^2+(S'-a')^2+(S''-a'')^2=0,$$
$$(S-b)^2+(S'-b')^2+(S''-b'')^2=0,$$
$$(S-c)^2+(S'-c')^2+(S''-c'')^2=0,$$

or we have from these equations to determine S, S', S''. But taking (a, a', a''), (b, b', b''), (c, c', c'') for the coordinates of three given points in space, and (S, S', S'') for the coordinates of the centre of the cone-sphere through these points, we have the very same equations for the determination of (S, S', S''), and the identity of the two problems thus appears.

I will presently give the direct analytical solution of this system of equations. But to obtain a solution in the form required, I remark that the equation of the cone-sphere in question is nothing else than the relation that exists between the coordinates of any four points on a cone-sphere; to find this, consider any five points in space, 1, 2, 3, 4, 5; and let $\overline{12}$, &c. denote the distances between the points 1 and 2, &c.; then we have between the distances of the five points the relation

$$\begin{vmatrix} 0, & 1, & 1, & 1, & 1, & 1 \\ 1, & 0, & \overline{12}^2, & \overline{13}^2, & \overline{14}^2, & \overline{15}^2 \\ 1, & \overline{21}^2, & 0, & \overline{23}^2, & \overline{24}^2, & \overline{25}^2 \\ 1, & \overline{31}^2, & \overline{32}^2, & 0, & \overline{34}^2, & \overline{35}^2 \\ 1, & \overline{41}^2, & \overline{42}^2, & \overline{43}^2, & 0, & \overline{45}^2 \\ 1, & \overline{51}^2, & \overline{52}^2, & \overline{53}^2, & \overline{54}^2, & 0 \end{vmatrix} = 0;$$

whence taking 5 to be the centre of the cone-sphere through the points 1, 2, 3, 4, we have $\overline{15}=\overline{25}=\overline{35}=\overline{45}=0$; and the equation becomes

$$\begin{vmatrix} 0, & \overline{12}^2, & \overline{13}^2, & \overline{14}^2 \\ \overline{21}^2, & 0, & \overline{23}^2, & \overline{24}^2 \\ \overline{31}^2, & \overline{32}^2, & 0, & \overline{34}^2 \\ \overline{41}^2, & \overline{42}^2, & \overline{43}^2, & 0 \end{vmatrix} = 0,$$

which is the relation between the distances of any four points on a cone-sphere; this equation may be written under the irrational form

$$\overline{23}\,.\,\overline{14}+\overline{31}\,.\,\overline{24}+\overline{12}\,.\,\overline{34}=0.$$

Taking (a, a', a''), (b, b', b''), (c, c', c''), (x, y, z) for the coordinates of the four points respectively, we have

$$\begin{aligned}\overline{23}&=\sqrt{(b-c)^2+(b'-c')^2+(b''-c'')^2}, & \overline{14}&=\sqrt{(x-a)^2+(y-a')^2+(z-a'')^2},\\ \overline{31}&=\sqrt{(c-a)^2+(c'-a')^2+(c''-a'')^2}, & \overline{24}&=\sqrt{(x-b)^2+(y-b')^2+(z-b'')^2},\\ \overline{12}&=\sqrt{(a-b)^2+(a'-b')^2+(a''-b'')^2}, & \overline{34}&=\sqrt{(x-c)^2+(y-c')^2+(z-c'')^2},\end{aligned}$$

or the symbols having these significations, we have

$$\overline{23}\,.\,\overline{14}+\overline{31}\,.\,\overline{24}+\overline{12}\,.\,\overline{34}=0$$

for the equation of the cone-sphere through the three points; or rather (since the rational equation is of the order 4 in the coordinates (x, y, z)) this is the equation of the pair of cone-spheres through the three given points; and similarly it is in the first problem the equation of a pair of circles each touching the three given circles respectively.

In the first problem the radii of the given circles were $i(z-a'')$, $i(z-b'')$, $i(z-c'')$ respectively; denoting these radii by α, β, γ, or taking the equations of the given circles to be

$$\begin{aligned}(x-a)^2+(y-a')^2-\alpha^2&=0,\\ (x-b)^2+(y-b')^2-\beta^2&=0,\\ (x-c)^2+(y-c')^2-\gamma^2&=0,\end{aligned}$$

the symbols then are

$$\begin{aligned}\overline{23}&=\sqrt{(b-c)^2+(b'-c')^2-(\beta-\gamma)^2}, & \overline{14}&=\sqrt{(x-a)^2+(y-a')^2-\alpha^2},\\ \overline{31}&=\sqrt{(c-a)^2+(c'-a')^2-(\gamma-\alpha)^2}, & \overline{24}&=\sqrt{(x-b)^2+(y-b')^2-\beta^2},\\ \overline{12}&=\sqrt{(a-b)^2+(a'-b')^2-(\alpha-\beta)^2}, & \overline{34}&=\sqrt{(x-c)^2+(y-c')^2-\gamma^2},\end{aligned}$$

and the equation of the pair of circles is as before

$$\overline{23}\,.\,\overline{14}+\overline{31}\,.\,\overline{24}+\overline{12}\,.\,\overline{34}=0;$$

where it is to be noticed that $\overline{23}$, $\overline{31}$, $\overline{12}$ are the tangential distances of the circles 2 and 3, 3 and 1, 1 and 2 respectively; viz., if α, β, γ are the radii taken positively, then these are the direct tangential distances. By taking the radii positively or negatively at pleasure, we obtain in all four equations—the tangential distances being all direct as above, or else any one is direct, and the other two are inverse; we have thus the four pairs of tangent circles.

The cone-spheres which pass through a given circle are the two spheres which have their centres in the two antipoints of the given circle; and it is easy to see that the foregoing investigation gives the following (imaginary) construction of the tangent circles; viz., given any three circles A, B, C in the same plane, to draw the tangent circles. Taking the antipoints of the three circles, then selecting any three antipoints (one for each circle) so as to form a triad, we have in all four complementary pairs of triads. Through a triad, and through the complementary triad draw two circles, these are situate symmetrically on opposite sides of the plane; and combining each antipoint of the first circle with the symmetrically situated antipoint of the second circle, we have two pairs of points, the points of each pair being symmetrically situate in regard to the plane, and having therefore an anticircle in this plane; these two anticircles are a pair of tangent circles; and the four pairs of complementary triads give in this manner the four pairs of tangent circles.

I return to the equations

$$\begin{aligned}
(x-S)^2 &+ (y-S')^2 + (z-S'')^2 = 0,\\
(a-S)^2 &+ (a'-S')^2 + (a''-S'')^2 = 0,\\
(b-S)^2 &+ (b'-S')^2 + (b''-S'')^2 = 0,\\
(c-S)^2 &+ (c'-S')^2 + (c''-S'')^2 = 0;
\end{aligned}$$

by eliminating (S, S', S'') from these equations we shall obtain the equation of the pair of cone-spheres through the points (a, a', a''), (b, b', b''), (c, c', c''). Write $x-S$, $y-S'$, $z-S''=X$, Y, Z, then we have $X^2+Y^2+Z^2=0$, and, putting for shortness

$$\begin{aligned}
\mathfrak{A} &= (a-x)^2 + (a'-y)^2 + (a''-z)^2,\\
\mathfrak{B} &= (b-x)^2 + (b'-y)^2 + (b''-z)^2,\\
\mathfrak{C} &= (c-x)^2 + (c'-y)^2 + (c''-z)^2,
\end{aligned}$$

then, by means of the equation just obtained, the other three equations become

$$\begin{aligned}
\mathfrak{A} + 2\,[(a-x)\,X + (a'-y)\,Y + (a''-z)\,Z] &= 0,\\
\mathfrak{B} + 2\,[(b-x)\,X + (b'-y)\,Y + (b''-z)\,Z] &= 0,\\
\mathfrak{C} + 2\,[(c-x)\,X + (c'-y)\,Y + (c''-z)\,Z] &= 0.
\end{aligned}$$

These last equations give

$$\begin{aligned}
X : Y : Z = {} & \lambda\mathfrak{A} + \mu\mathfrak{B} + \nu\,\mathfrak{C}\\
& : \lambda'\mathfrak{A} + \mu'\mathfrak{B} + \nu'\,\mathfrak{C}\\
& : \lambda''\mathfrak{A} + \mu''\mathfrak{B} + \nu''\mathfrak{C},
\end{aligned}$$

where

$$\begin{aligned}
\lambda &= b'c'' - b''c' + (c'-b')\,z - (c''-b'')\,y,\\
\mu &= c'a'' - c''a' + (a'-c')\,z - (a''-c'')\,y,\\
\nu &= a'b'' - a''b' + (b'-a')\,z - (b''-a'')\,y,\\
\lambda' &= b''c - bc'' + (c''-b'')\,x - (c-b)\,z,\\
\mu' &= c''a - ca'' + (a''-c'')\,x - (a-c)\,z,\\
\nu' &= a''b - ab'' + (b''-a'')\,x - (b-a)\,z,\\
\lambda'' &= bc' - b'c + (c-b)\,y - (c'-b')\,x,\\
\mu'' &= ca' - c'a + (a-c)\,y - (a'-c')\,x,\\
\nu'' &= ab' - a'b + (b-a)\,y - (b'-a')\,x;
\end{aligned}$$

and the result of the elimination then is

$$(\lambda\mathfrak{A} + \mu\mathfrak{B} + \nu\mathfrak{C})^2 + (\lambda'\mathfrak{A} + \mu'\mathfrak{B} + \nu'\mathfrak{C})^2 + (\lambda''\mathfrak{A} + \mu''\mathfrak{B} + \nu''\mathfrak{C})^2 = 0.$$

But substituting for $\mathfrak{A}$, $\mathfrak{B}$, $\mathfrak{C}$ their values, and writing, for shortness,

$$\begin{aligned}
-i &= b'c'' - b''c' + c'a'' - c''a' + a'b'' - a''b,\\
-j &= b''c - bc'' + c''a - ca'' + a''b - a\,b'',\\
-k &= bc' - b'c + c\,a' - c'a + a\,b' - a'\,b,\\
\Delta &= a\,(b'c'' - b''c) + a'\,(b''c - bc'') + a''\,(bc' - b'c),\\
-p &= (b'c'' - b''c')\,(a^2 + a'^2 + a''^2) + (c'a'' - c''a')\,(b^2 + b'^2 + b''^2) + (a'b'' - a''b')\,(c^2 + c'^2 + c''^2),\\
-q &= (b''c - bc'')\,(a^2 + a'^2 + a''^2) + (c''a - ca'')\,(b^2 + b'^2 + b''^2) + (a''b - ab'')\,(c^2 + c'^2 + c''^2),\\
-r &= (bc' - b'c)\,(a^2 + a'^2 + a''^2) + (ca' - c'a)\,(b^2 + b'^2 + b''^2) + (ab' - ab')\,(c^2 + c'^2 + c''^2),\\
-l &= (c - b)\,(a^2 + a'^2 + a''^2) + (a - c)\,(b^2 + b'^2 + b''^2) + (b - a)\,(c^2 + c'^2 + c''^2),\\
-m &= (c' - b')\,(a^2 + a'^2 + a''^2) + (a' - c')\,(b^2 + b'^2 + b''^2) + (b' - a')\,(c^2 + c'^2 + c''^2),\\
-n &= (c'' - b'')\,(a^2 + a'^2 + a''^2) + (a'' - c'')\,(b^2 + b'^2 + b''^2) + (b'' - a'')\,(c^2 + c'^2 + c''^2),
\end{aligned}$$

we find

$$\begin{aligned}
&\lambda\mathfrak{A} + \mu\mathfrak{B} + \nu\mathfrak{C}\\
&\quad = -\ i\,(x^2 + y^2 + z^2)\\
&\qquad + 2i\,(x^2 + y^2 + z^2) - 2x\,(ix + jy + kz) - 2\Delta x + ny - mz - p,
\end{aligned}$$

with similar expressions for $\lambda'\mathfrak{A} + \mu'\mathfrak{B} + \nu'\mathfrak{C}$, $\lambda''\mathfrak{A} + \mu''\mathfrak{B} + \nu''\mathfrak{C}$, and the result is

$$\begin{aligned}
&\{i\,(x^2 + y^2 + z^2) - 2x\,(ix + jy + kz) - 2\Delta x + ny - mz - p\}^2\\
&+ \{j\,(x^2 + y^2 + z^2) - 2y\,(ix + jy + kz) - nx - 2\Delta y + lz - q\}^2\\
&+ \{k\,(x^2 + y^2 + z^2) - 2z\,(ix + jy + kz) + mx - ly - 2\Delta z - r\}^2 = 0,
\end{aligned}$$

viz., this is

$$\begin{aligned}
&(x^2 + y^2 + z^2)^2\,(i^2 + j^2 + k^2)\\
&+ (x^2 + y^2 + z^2)\,\{4\Delta\,(ix + jy + kz) + 2\,\big(i\,(ny - mz) + j\,(lz - nx) + k\,(mx - ly)\big)\\
&\qquad\qquad + 4\Delta^2 - 2\,(ip + jq + kr) + (l^2 + m^2 + n^2)\}\\
&- (lx + my + nz)^2 + 4\,(ix + jy + kz)\,(px + qy + rz)\\
&+ 4\Delta\,(px + qy + rz) - 2\,\big(p\,(ny - mz) + q\,(lz - nx) + r\,(mx - ly)\big)\\
&+ p^2 + q^2 + r^2 = 0,
\end{aligned}$$

viz., this is in the rational form the equation of the pair of cone-spheres. The function on the left-hand side must, it is clear, be save to a numerical factor the norm of

$$\begin{aligned}
&\sqrt{(b - c)^2 + (b' - c')^2 + (b'' - c'')^2}\,.\,\sqrt{(x - a)^2 + (y - a')^2 + (z - a'')^2}\\
&+ \sqrt{(c - a)^2 + (c' - a')^2 + (c'' - a'')^2}\,.\,\sqrt{(x - b)^2 + (y - b')^2 + (z - b'')^2}\\
&+ \sqrt{(a - b)^2 + (a' - b')^2 + (a'' - b'')^2}\,.\,\sqrt{(x - c)^2 + (y - c')^2 + (z - c'')^2},
\end{aligned}$$

the numerical factor of the expression in question is in fact $=-4$, that is, the norm is

$$=-4(x^2+y^2+z^2)^2(i^2+j^2+k^2)+\&c.;$$

so that attending only to the highest powers in (x, y, z) we ought to have

$$\text{Norm}\{\sqrt{(b-c)^2+(b'-c')^2+(b''-c'')^2}+\sqrt{(c-a)^2+(c'-a')^2+(c''-a'')^2}+\sqrt{(a-b)^2+(a'-b')^2+(a''-b'')^2}\}$$
$$=-4(i^2+j^2+k^2).$$

It is easy to see that the norm is in fact composed of the terms

$$2(b'-c')^2\{\ \ (b-c)^2-(c-a)^2-(a-b)^2\},$$
$$+\,2(c'-a')^2\{-(b-c)^2+(c-a)^2-(a-b)^2\},$$
$$+\,2(a'-b')^2\{-(b-c)^2-(c-a)^2+(a-b)^2\},$$

and of the similar terms (a, b, c), (a'', b'', c''), and in (a', b', c'), (a'', b'', c''); the above written terms are $=-4$ into

$$(b'-c')^2(a-b)(a-c)+(c'-a')^2(b-c)(b-a)+(a'-b')^2(c-a)(c-b),$$

which is

$$=a'^2(b-c)^2+b'^2(c-a)^2+c'^2(a-b)^2$$
$$+\,2b'c'(a-b)(c-a)+2c'a'(b-c)(a-b)+2a'b'(c-a)(b-c),$$
$$=\{a'(b-c)+b'(c-a)+c'(a-b)\}^2$$
$$=k^2;$$

and the value of the norm is thus $=-4(i^2+j^2+k^2)$, as it should be.

ANNEX III. *On the Norm of* $(b-c)\sqrt{\mathfrak{A}^\circ}+(c-a)\sqrt{\mathfrak{B}^\circ}+(a-b)\sqrt{\mathfrak{C}^\circ}$, *when the Centres are in a Line.*

The norm of $\sqrt{U}+\sqrt{V}+\sqrt{W}$ is

$$=(1,\ 1,\ 1,\ -1,\ -1,\ -1\between U,\ V,\ W)^2,$$

whence that of $\sqrt{U+U'}+\sqrt{V+V'}+\sqrt{W+W'}$ is

$$=\ \ (1,\ 1,\ 1,\ -1,\ -1,\ -1\between U,\ V,\ W)^2$$
$$+\ \ (1,\ 1,\ 1,\ -1,\ -1,\ -1\between U',\ V',\ W')^2$$
$$+\,2(1,\ 1,\ 1,\ -1,\ -1,\ -1\between U,\ V,\ W\between U',\ V',\ W'),$$

where the last term is $=2$ into

$$U'(U-V-W)+V'(-U+V-W)+W'(-U-V+W);$$

and the norm of $\sqrt{U+U'+U''}+\sqrt{V+V'+V''}+\sqrt{W+W'+W''}$ is obviously composed in a similar manner.

Now, applying the formula to obtain the norm of

$$(b-c)\sqrt{a^2+\theta+\alpha}+(c-a)\sqrt{b^2+\theta+\beta}+(a-b)\sqrt{c^2+\theta+\gamma},$$

the expression contains six terms, two of which are at once seen to vanish; and writing for shortness (,,) in place of $(1, 1, 1, -1, -1, -1)$ the remaining terms are

$$\begin{aligned}&(,,)\left((b-c)^2\alpha,\ (c-a)^2\beta,\ (a-b)^2\gamma\right)^2\\ +2\ &(,,)\left((b-c)^2\alpha,\ (c-a)^2\beta,\ (a-b)^2\gamma\between(b-c)^2a^2,\ (c-a)^2b^2,\ (a-b)^2c^2\right)\\ +2\theta&(,,)\left((b-c)^2\alpha,\ (c-a)^2\beta,\ (a-b)^2\gamma\between(b-c)^2\ \ ,\ (c-a)^2\ \ ,\ (a-b)^2\ \ \right)\\ +2\theta&(,,)\left((b-c)^2a^2,\ (c-a)^2b^2,\ (a-b)^2c^2\between(b-c)^2\ \ ,\ (c-a)^2\ \ ,\ (a-b)^2\ \ \right);\end{aligned}$$

the first of these terms requires no reduction; the second, omitting the factor 2, is

$$\begin{aligned}&(b-c)^2\alpha\,[\ \ (b-c)^2a^2-(c-a)^2b^2-(a-b)^2c^2]\\ +&(c-a)^2\beta\,[-(b-c)^2a^2+(c-a)^2b^2-(a-b)^2c^2]\\ +&(a-b)^2\gamma\,[-(b-c)^2a^2-(c-a)^2b^2+(a-b)^2c^2];\end{aligned}$$

which is

$$=2(a-b)(b-c)(c-a)\,[bc(b-c)\alpha+ca(c-a)\beta+ab(a-b)\gamma].$$

Similarly the third term, omitting the factor 2θ, is

$$\begin{aligned}&(b-c)^2\alpha\,[\ \ (b-c)^2-(c-a)^2-(a-b)^2]\\ +&(c-a)^2\beta\,[-(b-c)^2+(c-a)^2-(a-b)^2]\\ +&(a-b)^2\gamma\,[-(b-c)^2-(c-a)^2+(a-b)^2],\end{aligned}$$

which is

$$=2(a-b)(b-c)(c-a)\,[(b-c)\alpha+(c-a)\beta+(a-b)\gamma],$$

and for the last term, omitting the factor 2θ, this may be deduced therefrom by writing (a^2, b^2, c^2) in place of (α, β, γ), viz., it is

$$=-2(a-b)^2(b-c)^2(c-a)^2.$$

Hence, restoring the omitted factors, and collecting, we find

$$\begin{aligned}&\text{Norm }\{(b-c)\sqrt{a^2+\theta+\alpha}+(c-a)\sqrt{b^2+\theta+\beta}+(a-b)\sqrt{c^2+\theta+\gamma}\}\\ =&(b-c)^4\alpha^2+(c-a)^4\beta^2+(a-b)^4\gamma^2-2(c-a)^2(a-b)^2\beta\gamma-2(a-b)^2(b-c)^2\gamma\alpha-2(b-c)^2(c-a)^2\alpha\beta\\ &+4\theta(a-b)(b-c)(c-a)\,[\ \ (b-c)\alpha+\ \ (c-a)\beta+\ \ (a-b)\gamma]\\ &+4\ \ (a-b)(b-c)(c-a)\,[bc(b-c)\alpha+ca(c-a)\beta+ab(a-b)\gamma]\\ &-4\theta(a-b)^2(b-c)^2(c-a)^2.\end{aligned}$$

Hence, first writing $a-x$, $b-x$, $c-x$ in place of a, b, c; then y^2 for θ, and $(-a''^2, -b''^2, -c''^2)$ for (α, β, γ); and finally introducing z for homogeneity, we find

$$\begin{aligned}
&\text{Norm }\{(b-c)\sqrt{(x-az)^2+y^2-a''^2z^2}+(c-a)\sqrt{\ ,,}+(a-b)\sqrt{\ ,,}\}=z^2 \text{ into}\\
&\quad z^2\big((b-c)^4a''^4+(c-a)^4b''^4+(a-b)^4c''^4\\
&\qquad -2(c-a)^2(a-b)^2b''^2c''^2-2(a-b)^2(b-c)^2c''^2a''^2-2(b-c)^2(c-a)^2a''^2b''^2\big)\\
&\quad -4y^2\ (b-c)(c-a)(a-b)[\ (b-c)a''^2+(c-a)b''^2+(a-b)c''^2]\\
&\quad -4\ \ (b-c)(c-a)(a-b)\{\ (b-c)a''^2(z^2bc-zx(b+c)+x^2)\\
&\qquad\qquad +(c-a)b''^2(z^2ca-zx(c+a)+x^2)\\
&\qquad\qquad +(a-b)c''^2(z^2ab-zx(a+b)+x^2)\}\\
&\quad -4y\ \ (b-c)^2(c-a)^2(a-b)^2,
\end{aligned}$$

so that the equation $(b-c)\sqrt{\mathfrak{A}^\circ}+(c-a)\sqrt{\mathfrak{B}^\circ}+(a-b)\sqrt{\mathfrak{C}^\circ}=0$, in its rationalised form, contains $(z^2=0)$ the line infinity twice, and the curve is thus a conic. If $a''^2=b''^2=c''^2=k''^2$, then the expression of the norm is

$$=z^2 \text{ into } -4(a-b)^2(b-c)^2(c-a)^2(y^2-k''^2z^2),$$

viz., when the three circles have each of them the same radius k'', the curve is the pair of parallel lines $y^2-k''^2z^2=0$; and in particular when $k''=0$, or the circles reduce themselves each to a point, then the curve is $y^2=0$, the axis twice.

ANNEX IV. *On the Trizomal Curves $\sqrt{lU}+\sqrt{mV}+\sqrt{nW}=0$, which have a Cusp, or two Nodes.*

The trizomal curve $\sqrt{lU}+\sqrt{mV}+\sqrt{nW}=0$, has not in general any nodes or cusps: in the particular case where the zomal curves are circles, we have however seen how the ratios $l : m : n$ may be determined so that the curve shall acquire a node, two nodes, or a cusp; viz., regarding a, b, c as current areal coordinates, we have here a conic $\frac{l}{\mathrm{a}}+\frac{m}{\mathrm{b}}+\frac{n}{\mathrm{c}}=0$, the locus of the centres of the variable circle, and the solution depends on establishing a relation between this conic and the orthotomic circle or Jacobian of the three given circles. I have in my paper "Investigations in connection with Casey's Equation," *Quart. Math. Jour.* vol. VIII. (1867), pp. 334—342, [395] given, after Professor Cremona, a solution of the general question to find the number of the curves $\sqrt{lU}+\sqrt{mV}+\sqrt{nW}=0$, which have a cusp, or which have two nodes, and I will here reproduce the leading points of the investigation. I remark, that although one of the loci involved in it is the same as that occurring in the case of the three circles (viz., we have in each case the Jacobian of the given curves), the other two loci Σ and Δ, which present themselves, seem to have no relation to the conic of centres which is made use of in the particular case.

We have the curves $U=0$, $V=0$, $W=0$, each of the same order r; and considering a point the coordinates whereof are (l, m, n), we regard as corresponding to this point the curve $\sqrt{lU}+\sqrt{mV}+\sqrt{nW}=0$, say for shortness, the curve Ω, being as above a curve of the order $2r$, having r^2 contacts with each of the given curves $U=0$, $V=0$, $W=0$. As long as the point (l, m, n) is arbitrary, the curve Ω has not any node, and in order that this curve may have a node, it is necessary that the point (l, m, n) shall lie on a certain curve Δ; this being so, the node will, it is easy to see, lie on the curve J, the Jacobian of the three given curves; and the curves J and Δ will correspond to each other point to point, viz., taking for (l, m, n) any point whatever on the curve Δ, the curve Ω will have a node at some one point of J; and conversely, in order that the curve Ω may be a curve having a node at a given point of J, the point (l, m, n) must be at some one point of the curve Δ. The curve Δ has, however, nodes and cusps; each node of Δ corresponds to two points of J, viz., for (l, m, n) at a node of Δ, the curve Ω is a binodal curve having a node at each of the corresponding points of J; each cusp of Δ corresponds to two coincident points of J, viz. for (l, m, n) at a cusp of Δ, the curve Ω has a node at the corresponding point of J. The number of the binodal curves Ω is thus equal to the number of the nodes of Δ, and the number of the cuspidal curves Ω is equal to the number of the cusps of Δ; and the question is to find the Plückerian numbers of the curve Δ. This Professor Cremona accomplished in a very ingenious manner, by bringing the curve Δ into connexion with another curve Σ (viz., Σ is the locus of the nodes of those curves $lU+mV+nW=0$ which have a node), and the result arrived at is that for the curve Δ

Order $= 3(r-1)(3r-2)$,

Class $= 6(r-1)^2$,

Nodes $= \frac{3}{2}(r-1)(27r^3-63r^2+22r+16)$,

Cusps $= 3(r-1)(7r-8)$,

Double tangents $= \frac{3}{2}(r-1)(12r^3-36r^2+19r+16)$,

Inflexions $=12(r-1)(r-2)$;

so that, finally, the number of the cuspidal curves $\sqrt{lU}+\sqrt{mV}+\sqrt{nW}=0$, is found to be $=3(r-1)(7r-8)$, and the number of the binodal curves of the same form is found to be $=\frac{3}{2}(r-1)(27r^3-63r^2+22r+16)$. When the given curves are conics, or for $r=2$, these numbers are $=18$ and 36 respectively; but the formulæ are not applicable to the case where the conics have a point or points of intersection in common; nor, consequently, to the case of the three circles.

415.

CORRECTIONS AND ADDITIONS TO THE MEMOIR ON THE THEORY OF RECIPROCAL SURFACES (*Phil. Trans.* vol. CLIX. 1869, [411]).

[From the *Philosophical Transactions of the Royal Society of London*, vol. CLXII. (for the year 1872), pp. 83—87. Received July 22,—Read November 16, 1871.]

1. I AM indebted to Dr Zeuthen for the remark that although the "off-points" and "off-planes," as explained in the memoir, are real singularities, they are not the singularities to which the θ, θ' of the formulæ refer. The most convenient way of correcting this is to retain all the formulæ with θ, θ' as they stand, but to write ω, ω' for the number of "off-points" and "off-planes" respectively; viz. we thus have

ω, off-points,

θ, unexplained singular points,

and

ω', off-planes,

θ', unexplained singular planes,

the formulæ as they stand, taking account of the unexplained singularities θ and θ', but not taking any account at all of the off-points and off-planes ω, ω'. The extended formulæ in which these are taken into account are:

$$\begin{aligned}
a(n-2) &= \kappa - B + \rho + 2\sigma + 3\omega,\\
b(n-2) &= \rho + 2\beta + 3\gamma + 3t,\\
c(n-2) &= 2\sigma + 4\beta + \gamma + \theta + \omega,\\
a(n-2)(n-3) &= 2(\delta - C - 3\omega) + 3(ac - 3\sigma - \chi - 3\omega) + 2(ab - 2\rho - j),\\
b(n-2)(n-3) &= 4k + (ab - 2\rho - j) + 3(bc - 3\beta - 2\gamma - i),\\
c(n-2)(n-3) &= 6h + (ac - 3\sigma - \chi - 3\omega) + 2(bc - 3\beta - 2\gamma - i),
\end{aligned}$$

which replace Salmon's original formulæ (A) and (B).

2. In the formulæ

$$q = b^2 - b - 2k - 3\gamma - 6t,$$
$$r = c^2 - c - 2h - 3\beta,$$

it is assumed that the nodal curve has no actual multiple points other than the t triple points, and no stationary points other than the γ points which lie on the cuspidal curve; and similarly that the cuspidal curve has no actual multiple points, and no stationary points other than the β points which lie on the nodal curve; and this being so, q is the class of the nodal curve and r that of the cuspidal curve. But we may take the formulæ as *universally* true; viz. q may be considered as standing for $b^2 - b - 2k - 3\gamma - 6t$, and r as standing for $c^2 - c - 2h - 3\beta$; only then q and r are not in all cases the classes of the two curves respectively.

3. In the formulæ No. 6 *et seq.*, introducing the new singularity ω, we have as follows:

$$(a - b - c)(n - 2) = (\kappa - B - \theta + 2\omega) - 6\beta - 4\gamma - 3t,$$
$$(a - 2b - 3c)(n - 2)(n - 3) = 2(\delta - C - 3\omega) - 8k - 18h - 12(bc - 3\beta - 2\gamma - i);$$

and substituting these in $n' = a(a - 1) - 2b - 3c$, and writing for n' its value $= a(a - 1) - 2\delta - 3\kappa$, we have, as in the memoir,

$$\begin{aligned} n' = n(n-1)^2 - n(7b + 12c) + 4b^2 + 8b + 9c^2 + 15c \\ - 8k - 8h + 18\beta + 12\gamma + 12i - 9t \\ - 2C - 3B - 3\theta; \end{aligned}$$

viz. there is no term in ω.

Writing $(n - 2)(n - 3) = a + 2b + 3c + (-4n + 6)$ in the equations which contain $(n - 2)(n - 3)$, these become

$$\begin{aligned} a(-4n + 6) &= 2(\delta - C) - a^2 - 4\rho - 9\sigma - 2j - 3\chi - 15\omega, \\ b(-4n + 6) &= 4k - 2b^2 - 9\beta - 6\gamma - 3i - 2\rho - j, \\ c(-4n + 6) &= 6h - 3c^2 - 6\beta - 4\gamma - 2i - 3\sigma - \chi - 3\omega, \end{aligned}$$

(Salmon's equations (C)); and adding to each equation four times the corresponding equation with the factor $(n - 2)$, these become

$$\begin{aligned} a^2 - 2a &= 2(\delta - C) + 4(\kappa - B) - \sigma - 2j - 3\chi - 3\omega, \\ 2b^2 - 2b &= 4k - \beta + 6\gamma + 12t - 3i + 2\rho - j, \\ 3c^2 - 2c &= 6h + 10\beta + 4\theta - 2i + 5\sigma - \chi + \omega. \end{aligned}$$

Writing in the first of these $a^2 - 2a = n' + 2\delta + 3\kappa - a$, and reducing the other two by means of the values of q, r, the equations become

$$n' - a = -2C - 4B + \kappa - \sigma - 2j - 3\chi - 3\omega,$$
$$2q + \beta + 3i + j = 2\rho,$$
$$3r + c + 2i + \chi = 5\sigma + \beta + 4\theta + \omega.$$

The reciprocal of the first of these is

$$\sigma' = a - n + \kappa' - 2j' - 3\chi' - 2C' - 4B' - 3\omega';$$

viz. writing $a = n(n-1) - 2b - 3c$, and $\kappa = 3n(n-2) - 6b - 8c$, this is

$$\sigma' = 4n(n-2) - 8b - 11c - 2j' - 3\chi' - 2C' - 4B' - 3\omega';$$

and it thus appears that the order σ' of the spinode curve is reduced by 3 for each off-plane ω'.

4. As to the other two equations, writing for ρ, σ their values, these become

$$j + 6t + 3i + 5\beta + 6\gamma = b(2n - 4) - 2q,$$
$$2\chi + 3\omega + 4i + 18\beta + 5\gamma = c(5n - 12) - 6r + 3\theta,$$

equations which admit of a geometrical interpretation. In fact, when there is only a nodal curve, the first equation is

$$j + 6t = b(2n - 4) - 2q,$$

which we may verify when the nodal curve is a complete intersection, $P = 0$, $Q = 0$; for if the equation of the surface is $(A, B, C \between P, Q)^2 = 0$, where the degrees of A, B, C, P, Q are $n - 2f$, $n - f - g$, $n - 2g$, f, g respectively, then the pinch-points are given by the equations $P = 0$, $Q = 0$, $AC - B^2 = 0$, and the number j of pinch-points is thus

$$= fg(2n - 2f - 2g), \; = (2n-4)fg - 2fg(f + g - 2);$$

but for the curve $P = 0$, $Q = 0$ we have $t = 0$, and its order and class are $b = fg$, $q = fg(f + g - 2)$, or the formula is thus verified.

Similarly, when there is only a cuspidal curve, the second equation is

$$2\chi + 3\omega = c(5n - 12) - 6r + 3\theta,$$

which may be verified when the cuspidal curve is a complete intersection, $P = 0$, $Q = 0$; the equation of the surface is here $(A, B, C \between P, Q)^2 = 0$, where $AC - B^2 = MP + NQ$, and the points χ, ω are given as the intersections of the curve with the surface $(A, B, C \between N, -M)^2 = 0$.

Now $AC - B^2$ vanishing for $P = 0$, $Q = 0$ we must have $A = \Lambda\alpha^2 + A'$, $B = \Lambda\alpha\beta + B'$, $C = \Lambda\beta^2 + C'$, where A', B', C' vanish for $P = 0$, $Q = 0$; and thence $M = \Lambda M' + M''$, $N = \Lambda N' + N''$, where M'', N'' vanish for $P = 0$, $Q = 0$. The equation

$$(A, B, C \between N, -M)^2 = 0,$$

writing therein $P = 0$, $Q = 0$, thus becomes $\Lambda^3(N'\alpha - M'\beta)^2 = 0$; and its intersections with the curve $P = 0$, $Q = 0$ are the points $P = 0$, $Q = 0$, $\Lambda = 0$ each three times, and the points $P = 0$, $Q = 0$, $N'\alpha - M'\alpha = 0$ each twice; viz. they are the points $2\chi + 3\omega$.

But if the degree of Λ is $= \lambda$, then the degrees of N', M', α^2, $\alpha\beta$, β^2 are $2n - 3f - 2g - \lambda$, $2n - 2f - 3g - \lambda$, $n - 2f - \lambda$, $n - f - g - \lambda$, $n - 2g - \lambda$, whence the degree of $\Lambda^3(N'\alpha - M'\beta)$ is $= 5n - 6f - 6g$, and the number of points is $= fg(5n - 6f - 6g)$, viz. this is

$$= fg(5n - 12) - 6fg(f + g - 2),$$

or it is $= c(5n - 12) - 6r$; so that θ being $= 0$, the equation is verified.

5. It was also pointed out to me by Dr Zeuthen that in the value of $24t$ given in No. 10 the term involving χ should be -6χ instead of $+6\chi$, and that in consequence the coefficients of χ are erroneous in several others of the formulæ. Correcting these, and at the same time introducing the terms in ω, and writing down also the terms in θ as they stand, we have

$$\begin{aligned}
4i &= \ldots - 2\chi + 3\theta - 3\omega,\\
24t &= \ldots - 6\chi + 9\theta - 9\omega,\\
2\sigma &= \ldots - \theta - \omega,\\
8\rho &= \ldots + 6\chi - 9\theta + 9\omega,\\
8\kappa &= \ldots - 6\chi + 17\theta - 25\omega,\\
2\delta &= \ldots + 6\chi - 9\theta + 15\omega,\\
8n' &= \ldots - 30\chi + 21\theta - 45\omega,\\
c' &= \ldots - 12\chi + 10\theta - 20\omega.
\end{aligned}$$

The equations of No. 11, used afterwards, No. 53, should thus be

$$\begin{aligned}
4i + 6r &= (5n-12)\,c - 18\beta - 5\gamma - 2\chi + 3\theta - 3\omega,\\
-24t - 8q + 18r &= (-8n+16)\,b + (15n-36)\,c - 34\beta + 9\gamma + 4j - 6\chi + 9\theta - 9\omega;
\end{aligned}$$

and from these I deduce

$$44q + \tfrac{63}{2}r = (44n-88)\,b + (\tfrac{105}{4}n - 63)\,c - \tfrac{409}{2}\beta - \tfrac{633}{4}\gamma - 132\,t - 87i - 22j - \tfrac{21}{2}\chi + \tfrac{63}{4}\theta.$$

6. In No. 32 we have (without alteration) $\theta = 16$; but in the application (Nos. 40 and 41) to the surface $FP^2 + GR^2Q^3 = 0$ we have $\theta = 0$, and there are $\omega = fpq$ off-points, $F=0$, $P=0$, $Q=0$, and $\chi = gpq$ close-points, $G=0$, $P=0$, $Q=0$. The new equations involving ω are thus satisfied.

7. I have ascertained that the value of β' obtained, Nos. 51 to 64 of the memoir, is inconsistent with that obtained in the "Addition" by consideration of the deficiency, and that it is in fact incorrect. The reason is that, although, as stated No. 53, the values of two of the coefficients D, E may be assumed at pleasure, they cannot, in conjunction with a given system of values of A, B, C, be thus assumed at pleasure; viz. A, B, C being $=110$, 272, 44 respectively, the values of D, E are really determinate. I have no direct investigation, but by working back from the formula in the Addition I find that we must have $D = \frac{477}{4}$, $E = 315$; the values of the remaining coefficients then are

$$F = \tfrac{63}{2},\ G = -\tfrac{715}{2},\ H = -\tfrac{1005}{4},\ I = -198;$$

or the formula is

$$\begin{aligned}
\beta' = 2n\,(n-2)(11n-24)&\\
&-(110n-272)\,b + 44q\\
&-(\tfrac{477}{4}n - 315)\,c + \tfrac{63}{2}r\\
&+\tfrac{715}{2}\beta + \tfrac{1005}{4}\gamma + 198\,t\\
&-hC - gB - xi - \lambda j - \mu\chi - \nu\theta - f\omega\\
&-h'C' - g'B' - x'i' - \lambda'j' - \mu'\chi' - \nu'\theta' - f'\omega';
\end{aligned}$$

but I have not as yet any means of determining the coefficients f, f' of the terms in ω, ω'.

From the several cases of a cubic surface we obtain as in the memoir; but applying to the same surfaces the reciprocal equation for β, instead of the results of the memoir, we find

$$\begin{aligned} h' &= -4, \\ g'+16\nu &= -198, \\ g'+2\mu &= 45, \\ g+g' &= 18, \\ \lambda &= 5 \end{aligned}$$

(so that now $\lambda+\lambda'=-2$, as is also given by the cubic scroll). And combining the two sets of results, we have

$$\begin{aligned} h &= 24, \\ \lambda &= 5, \\ \mu &= \tfrac{27}{2}+\tfrac{1}{2}g, \\ \nu &= -\tfrac{27}{2}+\tfrac{1}{16}g, \\ h' &= -4, \\ g' &= 18-g, \\ \lambda' &= -7, \\ \mu' &= 6-\tfrac{1}{2}g, \\ \nu' &= \tfrac{9}{4}-\tfrac{1}{16}g; \end{aligned}$$

but the coefficients g, x, x', f, f' are still undetermined. To make the result agree with that of the Addition, I assume $x=-86$, $x'=-1$, $g=+28$; whence we have

$$\begin{aligned} \beta' = {} & 2n(n-2)(11n-24) \\ & -(110n-272)\,b+44q \\ & -(\tfrac{477}{4}n-315)\,c+\tfrac{63}{2}r \\ & +\tfrac{715}{2}\beta+\tfrac{1005}{4}\gamma+198t \\ & -24C-28B+86i-5j-\tfrac{55}{2}\chi+\tfrac{47}{4}\theta-f\omega \\ & +4C'+10B'+i'+7j'+8\chi'-\tfrac{1}{2}\theta'-f'\omega'; \end{aligned}$$

and if we substitute herein the foregoing value of $44q+\tfrac{63}{2}r$, we obtain

$$\begin{aligned} \beta' = {} & 2n(n-2)(11n-24) \\ & +(-66n+184)\,b \\ & +(-93n+252)\,c \\ & +153\beta+93\gamma+66t \\ & -24C-28B-i-27j-38\chi+\tfrac{55}{2}\theta-f\omega \\ & +4C'+10B'+i'+7j'+8\chi'-\tfrac{1}{2}\theta'-f'\omega', \end{aligned}$$

which, except as to the terms in ω, ω', the coefficients of which are not determined, agrees with the value given in the Addition.

Dr Zeuthen considers that in general $i'=i$; I presume this is so, but have not verified it.

416.

ON THE THEORY OF RECIPROCAL SURFACES.

[*Published as an* Addition by Prof. CAYLEY *in* Dr SALMON'S *Treatise on the Analytic Geometry of Three Dimensions*, 4th Ed. (8vo. Dublin, 1882), pp. 592—604.]

620. IN further developing the theory of reciprocal surfaces it has been found necessary to take account of other singularities, some of which are as yet only imperfectly understood. It will be convenient to give the following complete list of the quantities which present themselves:

n, order of the surface.

a, order of the tangent cone drawn from any point to the surface.

δ, number of nodal edges of the cone.

κ, number of its cuspidal edges.

ρ, class of nodal torse.

σ, class of cuspidal torse.

b, order of nodal curve.

k, number of its apparent double points.

f, number of its actual double points.

t, number of its triple points.

j, number of its pinch-points.

q, its class.

c, order of cuspidal curve.

h, number of its apparent double points.

θ, number of its points of an unexplained singularity.

χ, number of its close-points.

ω, number of its off-points.

r, its class.

β, number of intersections of nodal and cuspidal curves, stationary points on cuspidal curve.

γ, number of intersections, stationary points on nodal curve.

i, number of intersections, not stationary points on either curve.

C, number of cnicnodes of surface.

B, number of binodes.

And corresponding reciprocally to these:

n', class of surface.

a', class of section by arbitrary plane.

δ', number of double tangents of section.

κ', number of its inflexions.

ρ', order of node-couple curve.

σ', order of spinode curve.

b', class of node-couple torse.

k', number of its apparent double planes.

f', number of its actual double planes.

t', number of its triple planes.

j', number of its pinch-planes.

q', its order.

c', class of spinode torse.

h', number of its apparent double planes.

θ', number of its planes of a certain unexplained singularity.

χ', number of its close-planes.

ω', number of its off-planes.

r', its order.

β', number of common planes of node-couple and spinode torse, stationary planes of spinode torse.

γ', number of common planes, stationary planes of node-couple torse.

i', number of common planes, not stationary planes of either torse.

C', number of cnictropes of surface.

B', number of its bitropes.

In all 46 quantities.

621. In part explanation, observe that the definitions of ρ and σ agree with those given, Art. 609: the nodal torse is the torse enveloped by the tangent planes along the nodal curve; if the nodal curve meets the curve of contact a, then a tangent plane of the nodal torse passes through the arbitrary point, that is, ρ will be the number of these planes which pass through the arbitrary point, viz. the class of the torse. So also the cuspidal torse is the torse enveloped by the tangent planes along the cuspidal curve; and σ will be the number of these tangent planes which pass through the arbitrary point, viz. it will be the class of the torse. Again, as regards ρ' and σ': the node-couple torse is the envelope of the bitangent planes of the surface, and the node-couple curve is the locus of the points of contact of these planes; similarly, the spinode torse is the envelope of the *parabolic* planes of the surface, and the spinode curve is the locus of the points of contact of these planes; viz. it is the curve UH of intersection of the surface and its Hessian; the two curves are the reciprocals of the nodal and cuspidal torses respectively, and the definitions of ρ', σ' correspond to those of ρ and σ.

622. In regard to the nodal curve b, we consider k the number of its apparent double points (excluding actual double points); f the number of its actual double points (each of these is a point of contact of two sheets of the surface, and there is thus at the point a single tangent plane, viz. this is a plane f', and we thus have $f'=f$); t the number of its triple points; and j the number of its pinch-points—these last are not singular points of the nodal curve *per se*, but are singular in regard to the curve as nodal curve of the surface; viz. a pinch-point is a point at which the two tangent planes are coincident. The curve is considered as not having any stationary points other than the points γ, which lie also on the cuspidal curve; and the expression for the class consequently is $q = b^2 - b - 2k - 2f - 3\gamma - 6t$.

623. In regard to the cuspidal curve c we consider h the number of its apparent double points; and upon the curve, not singular points in regard to the curve *per se*, but only in regard to it as cuspidal curve of the surface, certain points in number θ, χ, ω respectively. The curve is considered as not having any actual double or other multiple points, and as not having any stationary points except the points β, which lie also on the nodal curve; and thus the expression for the class is $r = c^2 - c - 2h - 3\beta$.

624. The points γ are points where the cuspidal curve with the two sheets (or say rather half-sheets) belonging to it are intersected by another sheet of the surface; the curve of intersection with such other sheet belonging to the nodal curve of the surface has evidently a stationary (cuspidal) point at the point of intersection.

As to the points β, to facilitate the conception, imagine the cuspidal curve to be a semi-cubical parabola, and the nodal curve a right line (not in the plane of the curve) passing through the cusp; then intersecting the two curves by a series of parallel planes, any plane which is, say, above the cusp, meets the parabola in two real points and the line in one real point, and the section of the surface is a curve with two real cusps and a real node; as the plane approaches the cusp, these approach together, and, when the plane passes through the cusp, unite into a singular point in the nature of a triple point (=node + two cusps); and when the plane passes below the cusp, the two cusps of the section become imaginary, and the nodal line changes from crunodal to acnodal.

625. At a point i the nodal curve crosses the cuspidal curve, being on the side away from the two half-sheets of the surface acnodal, and on the side of the two half-sheets crunodal, viz. the two half-sheets intersect each other along this portion of the nodal curve. There is at the point a single tangent plane, which is a plane i'; and we thus have $i=i'$.

626. As already mentioned, a cnicnode C is a point where, instead of a tangent plane, we have a tangent quadricone; and at a binode B the quadricone degenerates into a pair of planes. A cnictrope C' is a plane touching the surface along a conic; in the case of a bitrope B', the conic degenerates into a flat conic or pair of points.

627. In the original formulæ for $a(n-2)$, $b(n-2)$, $c(n-2)$, we have to write $\kappa-B$ instead of κ, and the formulæ are further modified by reason of the singularities θ and ω. So in the original formulæ for $a(n-2)(n-3)$, $b(n-2)(n-3)$, $c(n-2)(n-3)$, we have instead of δ to write $\delta-C-3\omega$; and to substitute new expressions for $[ab]$, $[ac]$, $[bc]$, viz. these are

$$[ab]=ab-2\rho-j,$$
$$[ac]=ac-3\sigma-\chi-\omega,$$
$$[bc]=bc-3\beta-2\gamma-i.$$

The whole series of equations thus is

(1) $$a'=a.$$

(2) $$f'=f.$$

(3) $$i'=i.$$

(4) $$a=n(n-1)-2b-3c.$$

(5) $$\kappa'=3n(n-2)-6b-8c.$$

(6) $$\delta'=\tfrac{1}{2}n(n-2)(n^2-9)-(n^2-n-6)(2b+3c)+2b(b-1)+6bc+\tfrac{9}{2}c(c-1).$$

(7) $$a(n-2)=\kappa-B+\rho+2\sigma+3\omega.$$

(8) $$b(n-2)=\rho+2\beta+3\gamma+3t.$$

(9) $$c(n-2)=2\sigma+4\beta+\gamma+\theta+\omega.$$

(10) $$a(n-2)(n-3)=2(\delta-C-3\omega)+3(ac-3\sigma-\chi-3\omega)+2(ab-2\rho-j).$$

(11) $$b(n-2)(n-3)=4k+(ab-2\rho-j)+3(bc-3\beta-2\gamma-i).$$

(12) $$c(n-2)(n-3)=6h+(ac-3\sigma-\chi-3\omega)+2(bc-3\beta-2\gamma-i).$$

(13) $$q=b^2-b-2k-2f-3\gamma-6t.$$

(14) $$r=c^2-c-2h-3\beta.$$

Also, reciprocal to these

(15) $a' = n'(n'-1) - 2b' - 3c'$.

(16) $\kappa = 3n'(n'-2) - 6b' - 8c'$.

(17) $\delta = \tfrac{1}{2}n'(n'-2)(n'^2-9) - (n'^2-n'-6)(2b'+3c') + 2b'(b'-1) + 6b'c' + \tfrac{9}{2}c'(c'-1)$.

(18) $a'(n'-2) = \kappa' - B' + \rho' + 2\sigma' + 3\omega'$.

(19) $b'(n'-2) = \rho' + 2\beta' + 3\gamma' + 3t'$.

(20) $c'(n'-2) = 2\sigma' + 4\beta' + \gamma' + \theta' + \omega'$.

(21) $a'(n'-2)(n'-3) = 2(\delta' - C' - 3\omega') + 3(a'c' - 3\sigma' - \chi' - 3\omega') + 2(a'b' - 2\rho' - j')$.

(22) $b'(n'-2)(n'-3) = 4k' + (a'b' - 2\rho' - j') + 3(b'c' - 3\beta' - 2\gamma' - i')$.

(23) $c'(n'-2)(n'-3) = 6h' + (a'c' - 3\sigma' - \chi' - 3\omega') + 2(b'c' - 3\beta' - 2\gamma' - i')$.

(24) $q' = b'^2 - b' - 2k' - 2f' - 3\gamma' - 6t'$.

(25) $r' = c'^2 - c' - 2h' - 3\beta'$,

together with one other independent relation, in all 26 relations between the 46 quantities.

628. The new relation may be presented under several different forms, equivalent to each other in virtue of the foregoing 25 relations; these are

(26) $2(n-1)(n-2)(n-3) - 12(n-3)(b+c) + 6q + 6r + 24t + 42\beta + 30\gamma - \tfrac{3}{2}\theta = \Sigma$;

(27) $26n - 12c - 4C - 10B + \beta - 7j - 8\chi + \tfrac{1}{2}\theta - 4\omega = \Sigma$,

in each of which two equations Σ is used to denote the same function of the accented letters that the left-hand side is of the unaccented letters.

(28)
$$\begin{aligned}\beta' + \tfrac{1}{2}\theta' = {} & 2n(n-2)(11n-24) \\ & + (-66n+184)b \\ & + (-93n+252)c \\ & + 22(2\beta + 3\gamma + 3t) \\ & + 27(4\beta + \gamma + \theta) \\ & + \beta + \tfrac{1}{2}\theta \\ & - 24C - 28B - 27j - 38\chi - 73\omega \\ & + 4C' + 10B' + 7j' + 8\chi' - 4\omega'.\end{aligned}$$

Or, reciprocally,

(29)
$$\begin{aligned}\beta + \tfrac{1}{2}\theta = {} & 2n'(n'-2)(11n'-24) \\ & + (-66n'+184)b' \\ & + (-93n'+252)c' \\ & + 22(2\beta' + 3\gamma' + 3t') \\ & + 27(4\beta' + \gamma' + \theta') \\ & + \beta' + \tfrac{1}{2}\theta' \\ & - 24C' - 28B' - 27j' - 38\chi' - 73\omega' \\ & + 4C + 10B + 7j + 8\chi - 4\omega.\end{aligned}$$

The foregoing equation (26) in fact expresses that the surface and its reciprocal have the same deficiency; viz. the expression for the deficiency is

$$(30)\quad \text{Deficiency} = \tfrac{1}{6}(n-1)(n-2)(n-3) - (n-3)(b+c) + \tfrac{1}{2}(q+r) + 2t + \tfrac{7}{2}\beta + \tfrac{5}{2}\gamma + i - \tfrac{1}{8}\theta,$$
$$= \tfrac{1}{6}(n'-1)(n'-2)(n'-3) - \ \&c.$$

629. The equation (28) (due to Prof. Cayley) is the correct form of an expression for β', first obtained by him (with some errors in the numerical coefficients) from independent considerations, but which is best obtained by means of the equation (26); and (27) is a relation presenting itself in the investigation. In fact, considering a as standing for its value $n(n-1)-2b-3c$, we have from the first 25 equations

6	a	$=\Sigma$
$+2$	$3n-c-\kappa$	$=\Sigma$
-2	$a(n-2)-\kappa+B-\rho-2\sigma-3\omega$	$=\Sigma$
-4	$b(n-2)-\rho-2\beta-3\gamma-3t$	$=\Sigma$
-6	$c(n-2)-2\sigma-4\beta-\gamma-\theta-\omega$	$=\Sigma$
$+2$	$n+\kappa-\sigma-2C-4B-2j-3\chi-3\omega$	$=\Sigma$
-3	$2q-2\rho+\beta+j$	$=\Sigma$
-2	$3r+c-5\sigma-\beta-4\theta+\chi-\omega$	$=\Sigma$

and multiplying these equations by the numbers set opposite to them respectively, and adding, we find

$$-2n^3+12n^2+4n+b(12n-36)+c(12n-48)$$
$$-6q-6r-4C-10B-41\beta-30\gamma-24t-7j-8\chi+2\theta-4\omega=\Sigma,$$

and adding thereto (26) we have the equation (27); and from this (28), or by a like process, (29), is obtained without much difficulty. As to the 8 Σ-equations or symmetries, observe that the first, third, fourth, and fifth are in fact included among the original equations (for an expression which vanishes is in fact $=\Sigma$); we have from them moreover $3n-c=3a'-\kappa'$, and thence $3n-c-\kappa=3a'-\kappa-\kappa'$, which is $=\Sigma$, or we have thus the second equation; but the sixth, seventh, and eighth equations have yet to be obtained.

630. The equations (15), (16), (17) give

$$n' = a(a-1)-2\delta-3\kappa,$$
$$c' = 3a(a-2)-6\delta-8\kappa,$$
$$b' = \tfrac{1}{2}a(a-2)(a^2-9)-(a^2-a-6)(2\delta+3\kappa)+2\delta(\delta-1)+6\delta\kappa+\tfrac{9}{2}\kappa(\kappa-1);$$

from (7), (8), (9) we have

$$(a-\ b-\ c)(n-2) \quad = \quad \kappa-B \qquad -6\beta-\ 4\gamma-3t-\theta+2\omega,$$
$$(a-2b-3c)(n-2)(n-3)=2(\delta-C)-8k-18h-6bc+18\beta+12\gamma+6i-6\omega,$$

and substituting these values for κ and δ, and for a its value $=n(n-1)-2b-3c$ we obtain the values of n', c', b'; viz. the value of n' is

$$\begin{aligned} n' = n(n-1)^2 - n(7b+12c) + 4b^2 + 8b + 9c^2 + 15c \\ - 8k - 18h + 18\beta + 12\gamma + 12i - 9t \\ - 2C - 3B - 3\theta. \end{aligned}$$

Observe that the effect of a cnicnode C is to reduce the class by 2, and that of a binode B to reduce it by 3.

631. We have

$$(n-2)(n-3) = n^2 - n + (-4n+6) = a + 2b + 3c + (-4n+6),$$

and making this substitution in the equations (10), (11), (12), which contain $(n-2)(n-3)$, these become

$$\begin{aligned} a(-4n+6) &= 2(\delta - C) - a^2 - 4\rho - 9\sigma - 2j - 3\chi - 15\omega, \\ b(-4n+6) &= 4k - 2b^2 - 9\beta - 6\gamma - 3i - 2\rho - j, \\ c(-4n+6) &= 6h - 3c^2 - 6\beta - 4\gamma - 2i - 3\sigma - \chi - 3\omega, \end{aligned}$$

(the foregoing equations (C) Salmon p. 586); and adding to each equation four times the corresponding equation with the factor $(n-2)$, these become

$$\begin{aligned} a^2 - 2a &= 2(\delta - C) + 4(\kappa - B) - \sigma - 2j - 3\chi - 3\omega, \\ 2b^2 - 2b &= 4k - \beta + 6\gamma + 12t - 3i + 2\rho - j, \\ 3c^2 - 2c &= 6h + 10\beta + 4\theta - 2i + 5\sigma - \chi + \omega. \end{aligned}$$

Writing in the first of these $a^2 - 2a = n' + 2\delta + 3\kappa - a$, and reducing the other two by means of the values of q, r, the equations become

$$\begin{aligned} n' - a &= -2C - 4B + \kappa - \sigma - 2j - 3\chi - 3\omega, \\ 2q + \beta + 3i + j &= 2\rho, \\ 3r + c + 2i + \chi &= 5\sigma + \beta + 4\theta + \omega, \end{aligned}$$

which give at once the last three of the 8 Σ-equations.

The reciprocal of the first of these is

$$\sigma' = a - n + \kappa' - 2j' - 3\chi' - 2C' - 4B' - 3\omega',$$

or writing herein $a = n(n-1) - 2b - 3c$ and $\kappa' = 3n(n-2) - 6b - 8c$, this is

$$\sigma' = 4n(n-2) - 8b - 11c - 2j' - 3\chi' - 2C' - 4B' - 3\omega',$$

giving the order of the spinode curve; viz. for a surface of the order n without singularities this is $= 4n(n-2)$, the product of the orders of the surface and its Hessian.

632. Instead of obtaining the second and third equations as above, we may to the value of $b(-4n+6)$ add twice the value of $b(n-2)$; and to twice the value of $c(-4n+6)$ add three times the value of $c(n-2)$, thus obtaining equations free from ρ and σ respectively; these equations are

$$\begin{aligned} b(-2n+2) &= 4k - 2b^2 - 5\beta - 3i + 6t - j, \\ c(-5n+6) &= 12h - 6c^2 - 5\gamma - 4i - 2\chi + 3\theta - 3\omega, \end{aligned}$$

equations which, introducing therein the values of q and r, may also be written

$$\begin{aligned} b(2n-4) &= 2q + 5\beta + 6\gamma + 6t + 3i + j + 4f, \\ c(5n-12) + 3\theta &= 6r + 18\beta + 5\gamma + 4i + 2\chi + 3\omega. \end{aligned}$$

Considering as given, n the order of the surface; the nodal curve with its singularities b, k, f, t; the cuspidal curve and its singularities c, h; and the quantities β, γ, i which relate to the intersections of the nodal and cuspidal curves; the first of the two equations gives j, the number of pinch-points, being singularities of the nodal curve quoad the surface; and the second equation establishes a relation between θ, χ, ω, the numbers of singular points of the cuspidal curve quoad the surface.

In the case of a nodal curve only, if this be a complete intersection $P=0$, $Q=0$, the equation of the surface is $(A, B, C \between P, Q)^2 = 0$, and the first equation is

$$b(-2n+2) = 4k - 2b^2 + 6t - j;$$

or, assuming $t=0$, say $j = 2(n-1)b - 2b^2 + 4k$, which may be verified; and so in the case of a cuspidal curve only, when this is a complete intersection $P=0$, $Q=0$, the equation of the surface is $(A, B, C \between P, Q)^2 = 0$, where $AC - B^2 = MP + NQ$; and the second equation is

$$c(-5n+6) = 12h - 6c^2 - 2\chi + 3\theta - 3\omega,$$

or, say

$$2\chi + 3\omega = (5n-6)c - 6c^2 + 12h + 3\theta,$$

which may also be verified.

633. We may in the first instance out of the 46 quantities consider as given the 14 quantities

$$n; \quad b, k, f, t; c, h, \theta, \chi \quad ; \beta, \gamma, i; C, B,$$

then of the 26 relations, 17 determine the 17 quantities

$$\begin{array}{llll} a, \delta, \kappa, \rho, \sigma; j, q & ; r, \omega & ; & \\ n'; a', \delta', \kappa' & ; b', f' & ; c' & ; \quad i', \end{array}$$

and there remain the 9 equations

(18), (19), (20), (21), (22), (23), (24), (25), (28),

connecting the 15 quantities

$$\rho', \sigma'; k', t', j', q'; h', \theta', \chi', \omega', r'; \beta', \gamma'; C', B'.$$

Taking then further as given the 5 quantities j', χ', ω', C', B',

equations	(18) and (21)	give	ρ', σ',
equation	(19)	gives	$2\beta' + 3\gamma' + 3t'$,
„	(20)	„	$4\beta' + \gamma' + \theta'$,
„	(28)	„	$\beta' + \frac{1}{2}\theta'$,

so that taking also t' as given, these last three equations determine β', γ', θ'; and finally

equation	(22)	gives	k',
„	(23)	„	h',
„	(24)	„	q',
„	(25)	„	r',

viz. taking as given in all 20 quantities, the remaining 26 will be determined.

634. In the case of the general surface of the order n, without singularities, we have as follows:

$n = n,$

$a = n(n-1),$

$\delta = \frac{1}{2}n(n-1)(n-2)(n-3),$

$\kappa = n(n-1)(n-2),$

$n' = n(n-1)^2,$

$a' = n(n-1),$

$\delta' = \frac{1}{2}n(n-2)(n^2-9),$

$\kappa' = 3n(n-2),$

$b' = \frac{1}{2}n(n-1)(n-2)(n^3-n^2+n-12),$

$k' = \frac{1}{8}n(n-2)(n^{10}-6n^9+16n^8-54n^7+164n^6-288n^5+547n^4-1058n^3+1068n^2-1214n+1464),$

$t' = \frac{1}{6}n(n-2)(n^7-4n^6+7n^5-45n^4+114n^3-111n^2+548n-960),$

$q' = n(n-2)(n-3)(n^2+2n-4),$

$\rho' = n(n-2)(n^3-n^2+n-12),$

$c' = 4n(n-1)(n-2),$

$h' = \frac{1}{2}n(n-2)(16n^4-64n^3+80n^2-108n+156),$

$r' = 2n(n-2)(3n-4),$

$\sigma' = 4n(n-2),$

$\beta' = 2n(n-2)(11n-24),$

$\gamma' = 4n(n-2)(n-3)(n^3-3n+16),$

the remaining quantities vanishing.

635. The question of singularities has been considered under a more general point of view by Zeuthen, in the memoir "Recherche des singularités qui ont rapport à une droite multiple d'une surface," *Math. Annalen*, t. IV. pp. 1—20, 1871. He attributes to the surface:

A number of singular points, viz. points at any one of which the tangents form a cone of the order μ, and class ν, with $y+\eta$ double lines, of which y are tangents to branches of the nodal curve through the point, and $z+\zeta$ stationary lines, whereof z are tangents to branches of the cuspidal curve through the point, and with u double planes and v stationary planes; moreover, these points have only the properties which are the most general in the case of a surface regarded as a locus of points; and Σ denotes a sum extending to all such points. {The foregoing general definition includes the cnicnodes ($\mu=\nu=2$, $y=\eta=z=\zeta=u=v=0$), and [also, but not properly] the binodes ($\mu=2$, $\eta=1$, $\nu=y=$ &c. $=0$), [it includes also the off-points ($\mu=\nu=3$, $z=v=1$, $y=\eta=(=0)$)].}

And, further, a number of singular planes, viz. planes any one of which touches along a curve of the class μ' and order ν', with $y'+\eta'$ double tangents, of which y' are generating lines of the node-couple torse, $z'+\zeta'$ stationary tangents, of which z' are generating lines of the spinode torse, u' double points and v' cusps; it is, moreover, supposed that these planes have only the properties which are the most general in the case of a surface regarded as an envelope of its tangent planes; and Σ' denotes a sum extending to all such planes. {The definition includes the cnictropes ($\mu'=\nu'=2$, $y'=\eta'=z'=\zeta'=u'=v'=0$), and [also, but not properly] the bitropes ($\mu'=2$, $\eta'=1$, $\nu'=y'=$ &c. $=0$), [it includes also the off-planes ($\mu'=\nu'=3$, $z'=v'=1$, $y'=\eta'=\zeta'=0$)].}

636. This being so, and writing

$$x=\nu+2\eta+3\zeta,\quad x'=\nu'+2\eta'+3\zeta',$$

the equations (7), (8), (9), (10), (11), (12), contain in respect of the new singularities additional terms, viz. these are

$$\begin{aligned}
a(n-2)&=\ldots+\Sigma\,[x(\mu-2)-\eta-2\zeta],\\
b(n-2)&=\ldots+\Sigma\,[y(\mu-2)],\\
c(n-2)&=\ldots+\Sigma\,[z(\mu-2)],\\
a(n-2)(n-3)&=\ldots+\Sigma\,[x(-4\mu+7)+2\eta+4\zeta],\\
b(n-2)(n-3)&=\ldots+\Sigma\,[y(-4\mu+8)]-\Sigma'(4u'+3v'),\\
c(n-2)(n-3)&=\ldots+\Sigma\,[z(-4\mu+9)]-\Sigma'(2v'),
\end{aligned}$$

and there are of course the reciprocal terms in the reciprocal equations (18), (19), (20), (21), (22), (23). These formulæ are given without demonstration in the memoir just referred to: the principal object of the memoir, as shown by its title, is the consideration not of such singular points and planes, but of the multiple right lines of a surface; and in regard to these, the memoir should be consulted.

NOTES AND REFERENCES.

384. THE conclusion arrived at Nos. 27—30 that the transformed curve of the order $D+1$ depends upon $4D-6$ parameters is at variance with Riemann's theorem according to which the number of parameters is $3p-3$, (p Riemann $=D$ Cayley), $=3D-3$, and this last is the correct value. My erroneous conclusion is referred to in the preface to Clebsch and Gordan's *Theorie der Abel'schen Functionen* (Leipzig, 1866), "Unter den von Riemann behandelten Theilen der Theorie haben wir die Frage nach der Anzahl der Moduln einer Klasse von Abel'schen Functionen ausschliessen zu müssen geglaubt. Diese Frage ist durch die scharfsinnigen Betrachtungen des Herrn Cayley Gegenstand der Controverse geworden: sie ist überhaupt wohl zunächst nur durch tiefe algebraische Untersuchungen endgültig zu entscheiden, für deren Schwierigkeiten die gegenwärtig bekannten Methoden nicht mehr auszureichen scheinen." In the case D (or p) $=3$, my value is 10, Riemann's is 9: that the latter is correct was shown by a direct proof in the paper Brill, "Note bezüglich der Zahl der Moduln einer Klasse von algebraischen Gleichungen," *Math. Ann.*, t. I. (1869), pp. 401—406: the explanation of my error is given in the paper, Cayley, "Note on the Theory of Invariants," *Math. Ann.*, t. III. (1871), pp. 268—271.

400. The question here considered, viz., the expression of a binary sextic f in the form v^2-u^3, v and u a cubic and a quadric respectively, forms the basis of the very interesting investigations contained in the Memoir, Clebsch "Zur Theorie der binären Formen sechster Ordnung und zur Dreitheilung der hyperelliptischen Functionen," *Gött. Abh.*, t. XIV. (1869), pp. 1—59. Considering f as a given sextic it is remarked that the number of solutions, or what is the same thing the number of the functions u or v, although at first sight $=45$, is really $=40$; supposing that there is a given solution u, v, or that the sextic function is in the first instance given in the form v^2-u^3, then if any other solution is u', v', we have $v^2-u^3=v'^2-u'^3$, where v', u' are functions to be determined: there are in all 39 solutions, a set of 27 and a set of 12 solutions: viz. writing the equation in the form $(v+v')(v-v')=(u-u')(u-\epsilon u')(u-\epsilon^2u')$, ϵ an imaginary cube root of unity, then either the $v+v'$ and the $v-v'$ contain each of them as a factor one of the quadric functions $u-u'$, $u-\epsilon u'$, $u-\epsilon^2u'$ (which gives the set of 27 solutions) or else the $v+v'$ and the $v-v'$ are each of them the product of three linear factors of the quadric functions respectively (which gives the set of 12

solutions). It may be added that the 27 solutions form 9 groups of 3 each and that these 9 groups depend upon Hesse's equation of the order 9 for the determination of the inflexions of a cubic curve; and that the 12 solutions are determined by an equation of the order 12 which is the known resolvent of this order arising from Hesse's equation and is solved by means of a quartic equation with a quadrinvariant $=0$. As appears by the title of the memoir, the question is connected with that of the trisection of the hyperelliptic functions.

401, 403. On the subject of Pascal's theorem, see Veronese, "Nuove teoremi sull' hexagrammum mysticum," *R. Accad. dei Lincei* (1876—77), pp. 7—61; Miss Christine Ladd (Mrs Franklin), "The Pascal Hexagram," *Amer. Math. Jour.*, t. II. (1879), pp. 1—12, and Veronese, "Interprétations géométriques de la théorie des substitutions de n lettres, particulièrement pour $n=3$, 4, 5, en relation avec les groupes de l'Hexagramme Mystique," *Ann. di Matem.*, t. XI. 1882—83, pp. 93—236. See also Richmond, "A Symmetrical System of Equations of the Lines on a Cubic Surface which has a Conical Point," *Quart. Math. Jour.*, t. XXII. (1889), pp. 170—179, where the author discusses a perfectly symmetrical system of the lines on the cubic surface and deduces from them equations of the lines relating to a Pascal's hexagon: there are of course through the conical point 6 lines lying on a quadric cone and these by their intersections with the plane give the six points of the hexagon: the interest of the paper consists as well in the connexion established between the two theories as in the perfectly symmetrical form given to the equations.

406, 407. A correction was made by Halphen to the fundamental theorem of Chasles that the number of the conics $(X, 4Z)$ is $=\alpha\mu+\beta\nu$, he finds that a diminution is in some cases required, and thus that the general form is, Number of conics $(X, 4Z)=\alpha\mu+\beta\nu-\Gamma$: see Halphen's two Notes, *Comptes Rendus*, 4 Sep. and 13 Nov., 1876, t. LXXXIII. pp. 537 and 886, and his papers "Sur la théorie des caractéristiques pour les coniques," *Proc. Lond. Math. Soc.*, t. IX. (1877—1878), pp. 149—170, and "Sur les nombres des coniques qui dans un plan satisfont à cinq conditions projectives et indépendantes entre elles," *Proc. Lond. Math. Soc.*, t. X. (1878—79), pp. 76—87: also Zeuthen's paper "Sur la revision de la théorie des caractéristiques de M. Study," *Math. Ann.*, t. XXXVII. (1890), pp. 461—464, where the point is brought out very clearly and tersely.

The correction rests upon a more complete development of the notion of the line-pair-point, viz. this degenerate form of conic seems at first sight to depend upon three parameters only, the two parameters which determine the position of the coincident lines, and a third parameter which determines the position therein of the coincident points: but there is really a fourth parameter. {Compare herewith the point-pair, or indefinitely thin conic, which working with point-coordinates presents itself in the first instance as a coincident line-pair depending on two parameters only, but which really depends also on the two parameters which determine the position therein of the vertices.} As to the fourth parameter of the line-pair-point the most simple definition is a metrical one; taking the semiaxes of the degenerate conic to be a and b $(a=0,\ b=0)$ then we have two positive integers p and q prime to each other such that the ratio

$a^p : b^q$ is finite; and this being so the fractional or it may be integer number $p : q$ is the fourth parameter in question. But it is preferable to adopt Halphen's purely descriptive definition, viz. we consider a conic 1° in reference to three given points y, z, t on a given line, and take x, x' for the intersections of the conic with the line: we take $a=(y, z, t, x)-(y, z, t, x')$ for the difference of the corresponding anharmonic ratios of the three points with the points x, x' respectively; and 2° we consider the conic in reference to three given lines Y, Z, T through a given point and take X, X' for the tangents from the given point to the conic; we take $b=(Y, Z, T, X)-(Y, Z, T, X')$ for the difference of the corresponding anharmonic ratios of the three lines with the lines X, X' respectively $\left(\text{observe that these values are } a=\frac{x-x'}{z-x \,.\, z-x'} \div \frac{y-t}{z-y \,.\, z-t},\right.$ and $\left. b=\frac{X-X'}{Z-X \,.\, Z-X'} \div \frac{Y-T}{Z-Y \,.\, Z-T}\right)$. Here when the conic is a line-pair-point, $x=x'$ and $X=X'$, where $a=0$ and $b=0$, but we have as before the integers p and q such that $a^p : b^q$ is finite, and we have thus the fourth parameter $p : q$.

Halphen's correction is now as follows, starting from the formula number of conics $(X, 4Z)=\alpha\mu+\beta\nu$, we may have among the $\alpha\mu+\beta\nu$ conics line-pair-points any one of which if we disregard altogether the fourth parameter is a conic satisfying the five conditions, but which unless the fourth parameter thereof has its proper value is an improper solution of the problem and as such it has to be rejected: if the number of such solutions is $=\Gamma$, then there is this number to be subtracted, and the formula becomes, Number of conics $(X, 4Z)=\alpha\mu+\beta\nu-\Gamma$.

It may be asked in what way the fourth parameter comes into the question at all: as an illustration suppose that a, b denoting the semiaxes of a conic, or else the above mentioned descriptively defined quantities, then p, q, k denoting given quantities (p and q positive integers prime to each other) the condition X may be that the conic shall be such that $a^p \div b^q = k$; this implies $a^p : b^q$ finite, and hence clearly if the system of conics $(X, 4Z)$ contains line-pair-points, no such line-pair-point can be a proper solution unless this relation $a^p \div b^q = k$ is satisfied.

412. Zeuthen's Memoir of 1876 presently referred to contains applications to the theory of Cubic Surfaces, the numerical results given in the table p. 539 agree for the most part with those of the Memoir 412, see p. 363, but for the surfaces III, VI, IX and XII discrepancies occur in the values of r' and h' relating to the spinode develope. As to this observe that Zeuthen's h', or say $\bar{h}'$ includes actual as well as apparent double planes, and we have $r'=c'^2-c'-2\bar{h}'-3\beta'$, my h' relates to apparent double planes only, but as I assume that there are no actual double planes the formula is $r'=c'^2-c'-2h'-3\beta'$, and as the values of c' and β' agree we have in fact in each of the four cases $r'+2h'$ (Cayley) $=r'+2h'$ (Zeuthen). The values found are

		III	VI	IX	XII		III	VI	IX	XII
Cayley	n'	72	24	12	6	r'	42	24	32	9
Zeuthen	n'	84	30	24	7	r'	18	12	8	7

and assuming the correctness of Zeuthen's values it would seem to follow that the four forms of surface have

$$12, \quad 6, \quad 12, \quad 1$$

actual double planes respectively.

413. In the equation No. 36, $\Omega = AP + BQ + CR + .. = 0$, it is implicitly assumed that the number of terms P, Q, R,.. is finite, viz. the implied theorem is that any given k-fold relation whatever (k of course a finite number) there is always a *finite* number of functions P, Q, R,... such that every onefold relation included in the k-fold relation is of the form in question Ω, $= AP + BQ + CR + \dots$, $= 0$: this seems self-evident enough, but I never succeeded in finding a proof: a proof of the theorem has however been obtained by Hilbert, see his papers "Zur Theorie der algebraischen Gebilden (Erste Note)," *Gött. Nachr.* No. 16, (1888), pp. 450—457.

411, 415, 416. The first and second of these papers precede in date Zeuthen's Memoir of 1871 referred to in 416, but I ought in that paper to have referred also to his later Memoir, "Revision et extension des formules numériques de la théorie des surfaces réciproques," *Math. Ann.* t. x. (1876), pp. 446—546. I compare the notations as follows, viz. for the unaccented letters we have

Cayley.	Zeuthen.
$n, a, \delta, \kappa, \rho, \sigma$	$n, a, \delta, \kappa, \rho, \sigma$
b, q, k, t, γ	$b, q, \bar{k}, t, \gamma;\ s$
$c, r, h, \beta, \theta, \omega$	$c, r, \bar{h}, \beta\ ;\ m$
j, χ	j, χ
C, B	B, U, O
f, i	f, i, d, g, e
23 letters in all.	27 letters in all.

Here for Zeuthen's k, h, I have written $\bar{k}$, $\bar{h}$, viz. these numbers represent the Plückerian equivalents of the number of double points for the nodal and cuspidal curves respectively. Zeuthen considers also the general node, say $\mathfrak{C}\,(\mu,\ \nu,\ y+\eta,\ z+\zeta,\ u,\ v)$, see 416, this includes the cnicnode C and off-point ω, and accordingly he includes under it and takes no special notice of these singularities, but it does not properly include, and he takes special notice of, the binode B; it does not extend to the case where the tangent cone breaks up into cones each or any of them more than once repeated, and accordingly not to the case of a unode U where the tangent cone is a pair of coincident planes. He introduces this singularity, and also the singularity of the osculating point O which is understood rather more easily by means of the reciprocal singularity of the osculating plane O', this is a tangent plane meeting the surface in a curve having the point of contact for a triple point; and he disregards my unexplained singularity θ. The letters s, m do not denote singularities; s is the class of the envelope of the osculating planes of the nodal curve, m the class of the envelope of the osculating planes of the cuspidal curve. Finally d denotes the number of stationary points (cusps) of the nodal curve, exclusive of the points γ which lie on the cuspidal curve; and g and e denote, g the number of ordinary actual double points of the cuspidal curve, e the number of stationary points (cusps) of the same curve, exclusive of the points β which lie on the nodal curve.

Moreover with Zeuthen, the nodal curve has

$$3t + f + 3O' + \Sigma' \text{ double points}$$

($\bar{k} = k + 3t + f + 3O' + \Sigma'$, if k denotes, as with me, the number of apparent double points of the curve), and it has

$$\gamma + d + \Sigma' \text{ stationary points.}$$

The cuspidal curve has

$$g + 6\chi' + 12B' + U' + 4O' + \Sigma + \Sigma' \text{ double points}$$

($\bar{h} = h + g + 6\chi' + 12B' + U' + 4O' + \Sigma + \Sigma'$, if h denotes, as with me, the number of apparent double points of the curve), and it has

$$\beta + e + 2O' \text{ stationary points}$$

and the nodal and cuspidal curves intersect in

$$3\beta + 2\gamma + i + 12O' + \Sigma + \Sigma' \text{ points;}$$

where I have written Σ and Σ' to denote sums (different in the different equations) determined by Zeuthen, and depending on the singularities $\mathfrak{C}$ and $\mathfrak{C}'$ respectively.

For comparison of my formulæ with Zeuthen's it is thus proper in my formulæ to write $C = 0$, $\omega = 0$, $\theta = 0$ (but in the first instance I retain θ) and in his formulæ to write $U = 0$, $O = 0$, $d = 0$, $g = 0$, $e = 0$, $\Sigma = 0$, $\Sigma' = 0$. Doing this the last mentioned formulæ give as with me $3t + f$ double points and γ stationary points for the nodal curve, but they give for the cuspidal curve $6\chi' + 12\beta'$ (instead of 0) double points and β stationary points; and the two curves intersect (as with me) in $3\beta + 2\gamma + i$ points. There is a real discrepancy in the number $6\chi' + 12\beta'$ of double points on the cuspidal curve.

I compare his $(6+26+1=)$ 33 relations:

(1) $a=a'.\qquad d=d'.$
$f=f'.\qquad g=g'.$
$i=i'.\qquad h=h'.$

(6) $n(n-1)=a+2b+3c.$
(7) $a(a-1)=n+2\delta'+3\kappa'.$
(8) $c-\kappa'=3(n-a).$

(9) $b(b-1)=q+2\bar{k}+3\gamma+3d+\Sigma'.$
(10) $[3(b-q)=\gamma+d-s+\Sigma'$, determines $s]$.

(11) $c(c-1)=r+2\bar{h}+3\beta+6C'+3e.$
(12) $[3(c-r)=\beta+e-m+2C'+\Sigma'$, determines $m]$.

(13) $a(n-2)=\kappa-B+\rho+2\sigma \qquad +\Sigma.$
(14) $b(n-2)=\rho+2\beta+3\gamma+3t+9C'+\Sigma.$
(15) $c(n-2)=2\sigma+4\beta+\gamma+8\chi'+16B'+12C'+\Sigma.$

(16) $a(n-2)(n-3)=2(\delta-3U)+3(ac-3\sigma-\chi)+2(ab-2\rho-j).$
(17) $b(n-2)(n-3)=4(\bar{k}-3t-f)+(ab-2\rho-j)+3(bc-3\beta-2\gamma-i)+39C'+\Sigma+\Sigma'.$
(18) $c(n-2)(n-3)=6(\bar{h}-6\chi'-12B'-U'-4C'-g)+(ac-3\sigma-\chi)+2(bc-3\beta-2\gamma-i).$
$-30C'+\Sigma+\Sigma',$

with the like reciprocal equations (6) to (18);

(19) $\sigma+m-r-\beta-4j'-3\chi'-14U'+\Sigma'$
$=\sigma'+m'-r'-\beta'-4j-3\chi-14U+\Sigma.$

where $\bar{k}=$

$\bar{h}=$

and my (3 + 22 + 1 =) 26 relations as follows:

(1) $a = a'$.

(2) $f = f'$.

(3) $i = i'$.

(4) $a = n(n-1) - 2b - 3c$.

(5) $\kappa' = 3n(n-2) - 6b - 8c$.

(6) $\delta' = \frac{1}{2}n(n-2)(n^2-9) - \&c.$

(A) (13) $q = b^2 - b - 2k - 2f - 3\gamma - 6t$.

(B) (14) $r = c^2 - c - 2h - 3\beta$.

(C) (7) $a(n-2) = k - B + \rho + 2\sigma + 3\omega$.

(D) (8) $b(n-2) = \rho + 2\beta + 3\gamma + 3t$.

(E) (9) $c(n-2) = 2\sigma + 4\beta + \gamma + \theta + \omega$.

(F) (10) $a(n-2)(n-3) = 2(\delta - C - 3\omega) + 3(ac - 3\sigma - \chi - 3\omega) + 2(ab - 2\rho - j \quad)$.

(G) (11) $b(n-2)(n-3) = \quad 4k \quad + \quad (ab - 2\rho - j \quad) + 3(bc - 2\beta - 2\gamma - i)$.

(H) (12) $c(n-2)(n-3) = \quad 6h \quad + \quad (ac - 3\sigma - \chi - 3\omega) + 2(bc - 3\beta - 2\gamma - i)$,

with the like reciprocal equations (4) to (14);

(I) (26) $2(n-1)(n-2)(n-3) - 12(n-3)(b+c) + 6q + 6r + 24t + 42\beta + 30\gamma - \frac{3}{2}\theta$.

$= 2(n'-1)(n'-2)(n'-3) - 12(n'-3)(b'+c') + 6q' + 6r' + 24t' + 42\beta' + 30\gamma' - \frac{3}{2}\theta'$.

$k + 3f + 3t + 3O' + \Sigma$.

$h + g + 6\chi' + 12B' + U' + 4O' + \Sigma + \Sigma'$.

Substituting for $\bar{k}$, $\bar{h}$ their values we have instead of (A), (B), (C), (D) the equations

$$(A')\quad b^2-b \qquad = q+2k+2f+6t+6O'+3\gamma+3\delta+\Sigma+\Sigma'.$$

$$(B')\quad c^2-c \qquad = r+2h+2g+12\chi'+24B'+2U'+14O'+3\beta+3e+\Sigma+\Sigma'.$$

$$(G')\quad b(n-2)(n-3)=4k+27O'+(ab-2\rho-j)+3(bc-3\beta-2\gamma-i)+\Sigma+\Sigma'.$$

$$(H')\quad c(n-2)(n-3)=6h-30O'+(ac-3\sigma-\chi)+2(bc-3\beta-2\gamma-i)+\Sigma+\Sigma'.$$

Writing as before $C=0$, $\omega=0$; $U=0$, $O=0$, $d=0$, $g=0$, $e=0$, and neglecting the terms in Σ, Σ', the two equations (E) become

$$\text{Zeuthen}\qquad c(n-2)=2\sigma+4\beta+\gamma+8\chi'+16B',$$

$$\text{Cayley}\qquad c(n-2)=2\sigma+4\beta+\gamma+\theta,$$

which can be made to agree by writing $\theta=8\chi'+16B'$. But we have

$$\text{Zeuthen } (B')\qquad c^2-c \quad =r+2h+3\beta+12\chi'+24B',$$

$$\text{Cayley } (B)\qquad c^2-c \quad =r+2h+3\beta,$$

values which differ by the terms $12\chi'+24B'$, or if θ has the value just written down, the term $\frac{3}{2}\theta$.

I refrain from a comparison of the two equations (I.), and of the expressions for the deficiency given by these two equations respectively—but I notice here the expression for the deficiency obtained by Zeuthen in the last section (XIV.) of his Memoir, viz. this is

$$\begin{aligned}24(D+1)= \;& c'-12a+24n+\beta+3r-15c+2\sigma+6\chi+12\chi'+6g+9e\\ &+8B+24B'+18U+6U'+6O'\\ &+\Sigma(3\nu+3z+8\eta+13\zeta)+\Sigma'(6\zeta).\end{aligned}$$

The problem is a very difficult one, and it cannot be held that as yet a complete solution has been obtained. Take in plane geometry the question of reciprocal curves: here, using throughout point-coordinates, we start with a curve represented by the general equation $(x, y, z)^n=0$, such a curve has only isolated singularities, viz. the line-singularities of the inflexion and the double tangent, we know the expression in point-coordinates of any such singularity (inflexion or double tangent as the case may be), viz. we can at once write down the equation of a curve of the order n having a given stationary tangent and point of contact therewith, or a given double tangent and two points of contact therewith. Returning to the general curve $(x, y, z)^n=0$, we know that the reciprocal curve has other isolated singularities, viz. the point-singularities which correspond to these, the double point (or node) and the stationary point (or cusp), and we know the expression of any such singularity (node or cusp as the case may be), viz. we can at once write down the equation of a curve of the order n having at a given point a node with given tangents, or a cusp with given tangent. And then starting afresh with a curve of the order n having a node or a cusp we obtain the effect thereof as regards the line-singularities of the inflexion and the double tangent. We are thus led to consider as ordinary singularities in the theory the above-mentioned four singularities of the inflexion, the double tangent, the node and the cusp: and we know further that any other singularity whatever of a plane curve is compounded in a definite manner of a certain number of some or all of these singularities.

But in the theory of surfaces, starting in like manner with the general equation $(x, y, z, w)^n = 0$, such a surface has torse-singularities, the node-couple torse, and the spinode-torse; each of these is in general an indecomposable torse of a certain kind (but there is the new cause of complication that it may break into two or more separate torses), but we do not know the analytical expression of these singularities, nor consequently the analytical expression of the curve-singularities which correspond to them, the nodal curve and the cuspidal curve. Thus if we attempt to start with a surface $(x, y, z, w)^n = 0$ having a nodal curve, we can indeed write down the equation in its most general form, viz. if the nodal curve has for its complete expression the k equations $P = 0$, $Q = 0$, $R = 0$, &c. (viz. if the curve is such that every surface whatever through the curve is of the form $\Omega, = AP + BQ + CR + \ldots, = 0$) then the most general equation of the surface having this curve for a nodal curve is $(A, B, C, \ldots \between P, Q, R, \ldots)^2 = 0$, but this form is far too complicated to be worked with; and if for simplicity we take the nodal curve to be a complete intersection $P = 0$, $Q = 0$, and consequently the equation of the surface to be $(A, B, C \between P, Q)^2 = 0$, then it is by no means clear that we do not in this way introduce limitations extraneous to the general theory. The same difficulty applies of course, and with yet greater force, to the cuspidal curve; and even if we could deal separately with the cases of a surface having a given nodal curve, and a given cuspidal curve, this would in no wise solve the problem for the more general case of a surface having a given nodal curve and a given cuspidal curve. It is to be added that the general surface of the order n has no plane- or point-singularities, and thus that such singularities (which correspond most nearly to the singularities considered in the theory of reciprocal curves) present themselves in the theory of reciprocal surfaces as extraordinary singularities.

END OF VOL. VI.

CAMBRIDGE:
PRINTED BY C. J. CLAY, M.A. AND SONS,
AT THE UNIVERSITY PRESS.

CAMBRIDGE UNIVERSITY PRESS.

A Treatise on Elementary Dynamics. By S. L. LONEY, M.A., late Fellow of Sidney Sussex College. New and Enlarged Edition. Crown 8vo. 7*s*. 6*d*.

Solutions to the Examples in a Treatise on Elementary Dynamics. By the same Author. Crown 8vo. 7*s*. 6*d*.

Plane Trigonometry. By the same Author. Crown 8vo. 7*s*. 6*d*. Part I. up to and including the Solution of Triangles is published separately. 5*s*.

A History of the Study of Mathematics at Cambridge. By W. W. ROUSE BALL, M.A., Fellow and Lecturer on Mathematics in Trinity College, Cambridge. Crown 8vo. 6*s*.

The Electrical Researches of the Hon. H. Cavendish, F.R.S. Written between 1771 and 1781. Edited from the original MSS. in the possession of the Duke of Devonshire, K.G., by the late J. CLERK MAXWELL, F.R.S. Demy 8vo. 18*s*.

A Catalogue of the Portsmouth Collection of Books and Papers written by or belonging to Sir ISAAC NEWTON. Demy 8vo. 5*s*.

The Analytical Theory of Heat, by JOSEPH FOURIER. Translated, with Notes, by A. FREEMAN, M.A., formerly Fellow of St John's College, Cambridge. Demy 8vo. 12*s*.

A Short History of Greek Mathematics. By J. GOW, Litt.D., Fellow of Trinity College. Demy 8vo. 10*s*. 6*d*.

Diophantos of Alexandria; a Study in the History of Greek Algebra. By T. L. HEATH, M.A., Fellow of Trinity College, Cambridge. Demy 8vo. 7*s*. 6*d*.

A Treatise on the General Principles of Chemistry. By M. M. PATTISON MUIR, M.A., Fellow of Gonville and Caius College. Demy 8vo. *New Edition.* 15*s*.

Elementary Chemistry. By M. M. PATTISON MUIR, M.A., and CHARLES SLATER, M.A., M.B. Crown 8vo. 4*s*. 6*d*.

Practical Chemistry. A Course of Laboratory Work. By M. M. PATTISON MUIR, M.A., and D. J. CARNEGIE, M.A. Crown 8vo. 3*s*.

Notes on Qualitative Analysis. Concise and Explanatory. By H. J. H. FENTON, M.A., F.I.C., Demonstrator of Chemistry in the University of Cambridge. Cr. 4to. *New Edition.* 6*s*.

Lectures on the Physiology of Plants. By S. H. VINES, Sc.D., Professor of Botany in the University of Oxford. Demy 8vo. With Illustrations. 21*s*.

PITT PRESS MATHEMATICAL SERIES.

Arithmetic for Schools. By C. SMITH, M.A., Master of Sidney Sussex College, Cambridge. Extra Fcap. 8vo. 3*s*. 6*d*.

Elementary Algebra. By W. W. ROUSE BALL, M.A., Fellow and Mathematical Lecturer of Trinity College, Cambridge. Extra Fcap. 8vo. 4*s*. 6*d*.

Euclid's Elements of Geometry. Edited by H. M. TAYLOR, M.A., Fellow and formerly Tutor of Trinity College, Cambridge. Extra Fcap. 8vo. Books I.—IV. 3*s*. Books I.—VI. 4*s*. Books I. and II. 1*s*. 6*d*. Books III. and IV. 1*s*. 6*d*. Books V. and VI. 1*s*. 6*d*.

Solutions to the Exercises in Taylor's Euclid. Books I.—IV. By W. W. TAYLOR, M.A. 6*s*.

The Elements of Statics and Dynamics. By S. L. LONEY, M.A., late Fellow of Sidney Sussex College. Third Edition. Extra Fcap. 8vo. 7*s*. 6*d*.

Part I. ELEMENTS OF STATICS. Extra Fcap. 8vo. 4*s*. 6*d*.

Part II. ELEMENTS OF DYNAMICS. Extra Fcap. 8vo. 3*s*. 6*d*.

Solutions to the Examples in the Elements of Statics and Dynamics. By S. L. LONEY, M.A. 7*s*. 6*d*.

Mechanics and Hydrostatics for Beginners. By S. L. LONEY, M.A. 4*s*. 6*d*.

An Elementary Treatise on Plane Trigonometry. By E. W. HOBSON, Sc.D., Fellow of Christ's College, Cambridge, and University Lecturer in Mathematics, and C. M. JESSOP, M.A., Fellow of Clare College, and Assistant Master at Reading School. Extra Fcap. 8vo. 4*s*. 6*d*.

London: C. J. CLAY AND SONS,
CAMBRIDGE UNIVERSITY PRESS WAREHOUSE, AVE MARIA LANE.

www.ingramcontent.com/pod-product-compliance
Lightning Source LLC
LaVergne TN
LVHW011245110826
845149LV00001B/50

9781418185879